Advances in Environmental Remote Sensing

Sensors, Algorithms, and Applications

Taylor & Francis Series in
Remote Sensing Applications

Series Editor

Qihao Weng

Indiana State University
Terre Haute, Indiana, U.S.A.

Advances in Environmental Remote Sensing

Sensors, Algorithms, and Applications

Edited by

Qihao Weng

CRC Press
Taylor & Francis Group
Boca Raton London New York

CRC Press is an imprint of the
Taylor & Francis Group, an **informa** business

CRC Press
Taylor & Francis Group
6000 Broken Sound Parkway NW, Suite 300
Boca Raton, FL 33487-2742

© 2011 by Taylor and Francis Group, LLC
CRC Press is an imprint of Taylor & Francis Group, an Informa business

First issued in paperback 2017

No claim to original U.S. Government works

ISBN 13: 978-1-138-07291-6 (pbk)
ISBN 13: 978-1-4200-9175-5 (hbk)

Library of Congress Cataloging-in-Publication Data

Advances in environmental remote sensing : sensors, algorithms, and applications/ [edited by] Qihao Weng.
    p. cm. -- (Taylor & Francis series in remote sensing applications series)
    Includes bibliographical references and index.
    ISBN 978-1-4200-9175-5 (hardback)
    1. Environmental sciences--Remote sensing. I. Weng, Qihao. II. Title. III. Series.

GE45.R44A39 2011
628.028'4--dc22   2010028913

Visit the Taylor & Francis Web site at
http://www.taylorandfrancis.com

and the CRC Press Web site at
http://www.crcpress.com

Contents

Section I Sensors, Systems, and Platforms

Section II Algorithms and Techniques

Section III   Environmental Applications-Vegetation

Section IV   Environmental Applications: Air, Water, and Land

Acknowledgments

I extend my heartfelt thanks to all the contributors of this book for making this endeavor possible. Moreover, I offer my deepest appreciation to all the reviewers, who have taken precious time from their busy schedules to review the chapters submitted for this book. Finally, I am indebted to my family for their enduring love and support. It is my hope that the publication of this book will facilitate students to understand the state-of-the art knowledge of environmental remote sensing and to provide researchers with an update on the newest development in many sub-fields of this dynamic field. The reviewers of the chapters of this book are listed here in alphabetical order:

Thomas Blaschke
Hubo Cai
Toby Carlson
Paolo Gamba
Anatoly Gitelson
Chengquan Huang
Stefaan Lhermitte
Lin Li
Desheng Liu
Hua Liu
Dengsheng Lu
Janet Nichol
Ruiliang Pu

Dale Quattrochi
Yang Shao
Conghe Song
Junmei Tang
Xiaohua Tong
Guangxin Wang
George Xian
Xiangming Xiao
Jian-sheng Yang
Ping Yang
Zhengwei Yang
Fei Yuan
Yuyu Zhou

Editor

Dr. Qihao Weng is a professor of geography and the director of the Center for Urban and Environmental Change at Indiana State University. From 2008 to 2009, he visited the National Aeronautics and Space Administration (NASA) as a senior research fellow. He is also a guest/adjunct professor at Wuhan University and Beijing Normal University, and a guest research scientist at the Beijing Meteorological Bureau in China. His research focuses on remote sensing and GIS analysis of urban environmental systems, land-use and land-cover change, urbanization impacts, and human–environment interactions.

Dr. Weng is the author of more than 120 peer-reviewed journal articles and other publications and three books (*Urban Remote Sensing*, 2006, CRC Press; *Remote Sensing of Impervious Surfaces*, 2007, CRC Press; and *Remote Sensing and GIS Integration: Theories, Methods, and Applications*, 2009, McGraw-Hill Professional). He has been the recipient of some significant awards, including the Robert E. Altenhofen Memorial Scholarship Award (1998) from the American Society for Photogrammetry and Remote Sensing (ASPRS), the Best Student-Authored Paper Award from the International Geographic Information Foundation (1999), the Theodore Dreiser Distinguished Research Award from Indiana State University (2006), a NASA senior fellowship (2008), and the 2010 Erdas Award for Best Scientific Paper in Remote Sensing from ASPRS (first place). Dr. Weng has worked extensively with optical and thermal remote sensing data, with research support from the National Science Foundation (NSF), NASA, USGS, the U.S. Agency for International Development (USAID), the National Geographic Society, and the Indiana Department of Natural Resources. Professionally, Dr. Weng was a national director of ASPRS (2007–2010). He also serves as an associate editor of *ISPRS Journal of Photogrammetry and Remote Sensing*, and is the series editor for both the Taylor & Francis series in remote sensing applications, and the McGraw-Hill series in GIS&T.

Contributors

Thomas Blaschke
Z_GIS Centre for Geoinformatics and
   Department for Geography and Geology
University of Salzburg
Salzburg, Austria

Toby N. Carlson
Department of Meteorology
Penn State University
University Park, Pennsylvania

Venkateswarlu Dheeravath
World Food Program
United Nations Joint Logistic Center
Juba, South Sudan, Sudan

Giorgio Franceschetti
Eureka Aerospace
Pasadena, California

Feng Gao
NASA Goddard Space Flight Center
Greenbelt, Maryland

Anatoly A. Gitelson
CALMIT, School of Natural Resources
University of Nebraska
Lincoln, Nebraska

Edward P. Glenn
Department of Soil, Water, and
   Environmental Science
University of Arizona
Tucson, Arizona

Peng Gong
Department of Environmental Science,
   Policy, and Management
University of California
Berkeley, California

Joshua M. Gray
Department of Geography
University of North Carolina
Chapel Hill, North Carolina

Muralikrishna Gumma
International Rice Research Center
Manila, Philippines

Daniela Gurlin
CALMIT, School of Natural Resources
University of Nebraska
Lincoln, Nebraska

Munir A. Hanjra
International Centre of Water for Food
   Security
Charles Stuart University
Wagga Wagga, NSW, Australia

Scott Hetrick
Anthropological Center for Training and
   Research on Global Environmental
   Change
Indiana University
Bloomington, Indiana

Collin Homer
USGS Earth Resources Observation and
   Science Center
Sioux Falls, South Dakota

Chengquan Huang
Department of Geography
University of Maryland
College Park, Maryland

Alfredo R. Huete
Department of Plant Functional Biology
   and Climate Change Cluster
University of Technology
Sydney, NSW, Australia

Kasper Johansen
Centre for Spatial Environmental Research,
School of Geography, Planning and
   Environmental Management
University of Queensland
Brisbane, Australia

Joshua Kalfas
Department of Botany and Microbiology
Center for Spatial Analysis
University of Oklahoma
Norman, Oklahoma

Guiying Li
Anthropological Center for Training and
    Research on Global Environmental
    Change
Indiana University
Bloomington, Indiana

Dengsheng Lu
Anthropological Center for Training and
    Research on Global Environmental
    Change
Indiana University
Bloomington, Indiana

Ross S. Lunetta
National Exposure Research Laboratory
    (NERL)
U.S. Environmental Protection Agency
Research Triangle Park, North Carolina

Emilio Moran
Anthropological Center for Training and
    Research on Global Environmental
    Change
Indiana University
Bloomington, Indiana

Wesley J. Moses
Naval Research Laboratory
Washington, DC

Janet Nichol
Department of Land Surveying and
    Geo-Informatics
Hong Kong Polytechnic University
Hunghom, Kowloon, Hong Kong

George P. Petropoulos
Regional Analysis Division
Foundation for Research and Technology
Hellas Institute of Applied and
    Computational Mathematics
Heraklion, Crete, Greece

Sorin C. Popescu
Department of Ecosystem Science and
    Management
Texas A&M University
College Station, Texas

Ruiliang Pu
Department of Geography
University of South Florida
Tampa, Florida

Rudolf Richter
DLR–German Aerospace Center
DFD–Remote Sensing Data Center
Wessling, Germany

Yang Shao
National Research Council
U.S. Environmental Protection
    Agency
Research Triangle Park, North Carolina

Conghe Song
Department of Geography
University of North Carolina
Chapel Hill, North Carolina

Gregory N. Taff
Department of Earth Sciences
University of Memphis
Memphis, Tennessee

James Z. Tatoian
Eureka Aerospace
Pasadena, California

Prasad S. Thenkabail
Southwest Geographic Science
    Center
U.S. Geological Survey
Flagstaff, Arizona

Dirk Tiede
Z_GIS Centre for Geoinformatics and
    Department for Geography and
    Geology
University of Salzburg
Salzburg, Austria

Thierry Toutin
Natural Resources Canada
Canada Centre for Remote Sensing
Ottawa, Ontario, Canada

Qihao Weng
Department of Geography
Center for Urban and Environmental
    Change
Indiana State University
Terre Haute, Indiana

Man Sing Wong
Department of Land Surveying and
    Geo-Informatics
Hong Kong Polytechnic University
Hunghom, Kowloon, Hong Kong

George Xian
ARTS/USGS Earth Resources Observation
    and Science Center
Sioux Falls, South Dakota

Xiangming Xiao
Department of Botany and Microbiology,
    College of Arts and Sciences
Center for Spatial Analysis, College of
    Atmospheric and Geographic Science
University of Oklahoma
Norman, Oklahoma

Yosef Z. Yacobi
Israel Oceanographic & Limnological
    Research
Yigal Allon Kinneret Limnological
    Laboratory
Migdal, Israel

Huimin Yan
Institute of Geographic Science and
    Natural Resources Research
Chinese Academy of Sciences
Beijing, China

Limin Yang
USGS Earth Resources Observation and
    Science Center
Sioux Falls, South Dakota

Stephen R. Yool
Department of Geography and
    Development
University of Arizona
Tucson, Arizona

Qingyuan Zhang
Goddard Space Flight Center
NASA
Greenbelt, Maryland

Introduction to Recent Advances in Remote Sensing of the Environment

Qihao Weng

Aims and Scope

The main purpose of compiling such a book is to provide an authoritative supplementary text for upper-division undergraduate and graduate students, who may have chosen a textbook from a variety of choices in the market. This book collects two types of articles: (1) comprehensive review articles from leading authorities to examine the developments in concepts, methods, techniques, and applications in a subfield of environmental remote sensing, and (2) focused review articles regarding the latest developments in a hot topic with one to two concise case studies. Because of the nature of articles collected, this book can also serve as a good reference book for researchers, scientists, engineers, and policy-makers who wish to keep up with new developments in environmental remote sensing.

Synopsis of the Book

This book is divided into four sections. Section I deals with various sensors, systems, or sensing using different regions of wavelengths. Section II exemplifies recent advances in algorithms and techniques, specifically in image preprocessing and thematic information extraction. Section III focuses on remote sensing of vegetation and related features of the Earth's surface. Finally, Section IV examines developments in the remote sensing of air, water, and other terrestrial features.

The chapters in Section I provide a comprehensive overview of some important sensors and remote sensing systems, with the exception of Chapter 5. By reviewing key concepts and methods and illustrating practical uses of particular sensors/sensing systems, these chapters provide insights into the most recent developments and trends in remote sensing and further identify the major existing problems of these trends. These remote sensing systems utilize visible, reflected infrared, thermal infrared, and microwave spectra, and include both passive and active sensors. In Chapter 1, Song and his colleagues evaluate one of the longest remote sensing programs in the world, that is, the U.S. Landsat program, and discuss its applications in vegetation studies. With a mission of long-term monitoring of vegetation and terrestrial features, Landsat has built up a glorious history. The remote sensing literature is filled with a large number of articles in vegetation classification and change detection. However, remote sensing of vegetation remains a great challenge, especially the sensing of biophysical parameters such as leaf area index (LAI), biomass, and forest successional stages (Song, Gray, and Gao 2010). A remarkable strength of the Landsat program is its time-series data, especially when considering the addition of the upcoming Landsat Data Continuity Mission (LDCM); however, these data are not a panacea for vegetation studies. Song, Gray, and Gao (2010) suggest that the synergistic use of data from other remote sensors may provide complimentary vegetation information to Landsat data, such as high spatial resolution (<10 m) satellite images that provide textural information, radar sensors that provide information on the dielectric properties of the surface and are capable of penetrating clouds, light detection and ranging (LiDAR, which provides geometric information), and coarse spatial but high temporal resolution sensors (e.g., Moderate Resolution Imaging Spectroradiometer [MODIS]). Chapter 1 provides an excellent example for the integrated use of Landsat and MODIS data by introducing the spatial and temporal adaptive reflectance fusion model (Chapter 1, Section 1.3.5; Gao et al. 2006).

In Chapter 2, Shao and his colleagues provide a comprehensive review of selected data products, algorithms, and applications of MODIS. MODIS has its roots in earlier sensors such as the Advanced Very High Resolution Radiometer (AVHRR) and coastal zone color scanner (CZCS), but provides substantial improvements over these earlier sensing systems (Lillesand, Kiefer, and Chipman 2008). MODIS provides a wide range of data products applicable to land, ocean, and atmosphere. Chapter 2 focuses on the examination of land products and applications, in particular, application studies at the global and regional levels. For each data product, the contributors document most recent advances, but also point out the product's limitations in data quality and validation.

LiDAR has been increasingly used in many geospatial applications due to its high data resolution, low consumption of time and cost, compared to many traditional remote sensing technologies. Unlike other remotely sensed data, LiDAR data focus solely on geometry rather than on radiometry. Many researchers have used LiDAR in conjunction with optical remote sensing and geographic information system (GIS) data in urban, environment, and resource studies (Weng 2009). Chapter 3 offers a detailed introduction of the basic concept of LiDAR, and types of sensors and platforms. Based on the works of the author this chapter further provides a review of LiDAR remote sensing applications in estimating forest biophysical parameters and surface and canopy fuels, and for characterizing wildlife habitats.

Synthetic aperture radar (SAR) has been a key sensing system for various environmental applications, and the Earth and planetary exploration. In Chapter 4, Franceschetti and Tatoian introduce to the reader two new concepts of SAR imaging: (1) impulse SAR and (2) polychromatic SAR. The theoretical foundations of the two systems are presented with some preliminary experimental data for validating the theory. The authors further discuss the distinct advantages of these systems over conventional microwave imaging sensors and their potential applications, and speculate on future research directions.

Hyperspectral remote sensing, as a cutting-edge technology, has been widely applied in vegetation and ecological studies. Chapter 5 provides an overview of spectral characteristics for a set of plant biophysical and biochemical parameters. A wide range of techniques are reviewed, including such spectral analysis techniques as spectral derivative analysis, spectral matching, spectral index analysis, spectral absorption features and spectral position variables, hyperspectral transformation, spectral unmixing analysis, and hyperspectral classifications. Further, two general analytical approaches are discussed: (1) empirical/statistical methods and (2) physically based modeling. The chapter concludes with the authors' perspectives on the future directions of hyperspectral remote sensing of vegetation biophysical parameters.

Thermal infrared (TIR) remote sensing techniques have been applied in urban climate and environmental studies. Chapter 6 examines the current practices, problems, and prospects of this particular field of study, especially the applications of remotely sensed TIR data in urban studies. It is suggested that the majority of previous researches have focused on land-surface temperature (LST) patterns and their relationships with urban-surface biophysical characteristics, especially with vegetation indices and land-use/land-cover types. Less attention has been paid to the derivation of urban heat island (UHI) parameters from LST data and to the use of remote sensing techniques to estimate surface energy fluxes. Major recent advances, future research directions, and the impacts of planned TIR sensors with LDCM and HyspIRI missions are outlined in the chapter.

Section II presents new developments in algorithms and techniques, specifically in image preprocessing, thematic information extraction, and digital change detection. Chapter 7 conducts a concise review of atmospheric correction algorithms for the optical remote sensing of land. This review focuses on physical models of atmospheric correction that describe the radiative transfer in the Earth's atmosphere, instead of empirical methods. The author presents sequentially the correction algorithms for hyperspectral, thermal, and multispectral sensors, then discusses the combined method for performing topographic and atmospheric corrections, and ends with examples of correcting non-standard atmospheric conditions, including haze, cirrus, and cloud shadow. The chapter concludes with the author's perspective on major challenges and future research needs in atmospheric and topographic correction. In addition, the chapter includes a brief survey and a comparison of capacity among commercially available atmospheric correction software/modules, which will be very useful for students.

Geometric correction is more important now than ever due mainly to the growing need for off-nadir and high-resolution imaging, fully digital processing and interpretation of remote sensing images, and image fusion and remote sensing–GIS data integration in practical applications (Toutin 2010). Three-dimensional (3D) geometric processing and correction of Earth observation (EO) satellite data is a key issue in multisource, multiformat data integration, management, and analysis for many EO and geomatic applications (Toutin 2010). Chapter 8 first reviews the source of geometric distortions (with relation to platform, sensor, other measuring instruments, Earth, and atmosphere), and then compares different mathematical models for correcting geometric distortions (e.g., 2D/3D polynomial, 3D rational functions, and physical and deterministic models). Subsequently, the methods and algorithms in each processing step of the geometric correction are examined in detail, supplemented with plentiful literature. This type of examination allows the tracking of error propagation from the input data to the final output product.

Image classification is a fundamental protocol in digital image processing and provides crucial information for subsequent environmental and socioeconomic applications. Generating a satisfactory classification image from remote sensing data is not a straightforward task. Many factors contribute to this difficulty, including the characteristics of a study area, availability of suitable remote sensing data, ancillary and ground reference data, proper use of variables and classification algorithms, and the analyst's experience (Lu and Weng 2007). Chapter 9 provides a brief overview of the major steps in image classification, and examines the techniques for improving classification performance, including the use of spatial information, multitemporal and ancillary data, and image fusion. A case study is further presented that explores the role of vegetation indices and textural images in improving vegetation classification performance in a moist tropical region of the Brazilian Amazon with Landsat Thematic Mapper (TM) imagery.

Object-based image analysis (OBIA; or GEOBIA for geospatial OBIA) is becoming a new paradigm among the mapping sciences (Blaschke 2010). With the improvement of OBIA software capacity and the increased availability of high spatial resolution satellite images and LiDAR data, vegetation-mapping capabilities are expected to grow rapidly in the near future in terms of both the accuracy and the amount of biophysical vegetation parameters that can be retrieved (Blaschke, Johansen, and Tiede 2010). Chapter 10 reviews the development of OBIA and the current status of its application in vegetation mapping. Two case studies are provided to illustrate this mapping capacity. The first case uses LiDAR data to map riparian zone extent and to estimate plant project cover (PPC) within the riparian zone in central Queensland, Australia. Whereas PPC was calculated at the pixel level, OBIA was used for mapping the riparian zone extent and validating the PPC results. The second case study aims at extracting individual tree crowns from a digital surface model (DSM) by using OBIA and grid computing techniques in the federal state of Upper Austria, Austria. Finally, the contributors share their insights on the existing problems and development trends of OBIA with respect to automation, the concept of scale, transferability of rules, and the impacts of improved remote sensing capacities.

Digital change detection requires the careful design of each step, including the statement of research problems and objectives, data collection, preprocessing, selection of suitable detection algorithms, and evaluation of the results (Lu et al. 2010). Errors or uncertainties may emerge from any of these steps, but it is important to understand the relationship among these steps and to identify the weakest link in the image-processing chain (Lu et al. 2010). In Chapter 11, Lu and his colleagues update earlier research (Lu et al. 2004) by re-examining the essential steps in change detection and by providing a case study for detecting urban land-use/land-cover in a complex urban–rural frontier in Mato Grosso state, Brazil, based on the comparison of extracted impervious surface data from multi-temporal Landsat TM images. They conclude that the selection of a change detection procedure, whether a per-pixel, a subpixel, or an object-oriented method, must conform to the research objectives, remote sensing data used, and geographical size of the study area.

The remaining sections of the book focus on various environmental applications of remote sensing technology. Section III centers on the remote sensing of vegetation, but each chapter has a very different approach or perspective. Chapter 12 reviews many of the advancements made in the remote sensing of ecosystem structure, processes, and function, and also notes that there exist important trade-offs and compromises in characterizing ecosystems from space related to spatial, spectral, and temporal resolutions of the imaging sensors. Huete and Glenn (2010) suggest that an enormous mismatch exists between leaf-level and species-level ecological variables and satellite spatial resolutions, and this mismatch makes it difficult to validate satellite-derived products. They further assert that high temporal resolution hyperspectral remote sensing satellite measurements provide powerful monitoring tools for the characterization of landscape phenology and ecosystem processes, especially when these remote sensing measurements are used in conjunction with calibrated, time-series-based *in situ* data sets from surface sensor networks.

In the western United States, wildfire is a major threat to both humans and the natural environment. Dr. Steve Yool and his colleagues at the University of Arizona have been taking great efforts to study the dynamic relationships among fire, climate, and people from an interdisciplinary perspective, which has been termed "pyrogeography" (Yool 2009). In Chapter 13, Yool introduces a remote sensing method to estimate and to map a fuel moisture stress index by standardizing normalized difference vegetation index (NDVI) with the Z transform. This index can be employed as a spatial and temporal fine-scale metric to determine fire season (Yool 2010). Based on a case study conducted in southeastern Arizona, the author demonstrate that the onset and length of the fire season depend on elevation and other microclimatic factors. Fire-season summary maps derived from the fuel moisture stress index may potentially provide lead time to plan for future fire seasons (Yool 2010).

Knowledge of forest disturbance and regrowth has obvious scientific significance in the context of global environmental change. Forest change analysis by using time-series analysis of Landsat images is a logical approach, given the long history of Landsat data records (see Chapter 1 for details). Chapter 14 introduces an approach for reconstructing forest disturbance history using Landsat data records. Major steps include the development of Landsat time-series stacks (Huang et al. 2009), and performing change analysis using vegetation-change tracker algorithm (Huang et al. 2010). This approach has been used to produce disturbance products for many areas in the United States (Huang 2010). The author thus further presents two examples of application of this approach to the states of Mississippi and Alabama and the seven national forests in the eastern United States. The application of this approach for an area outside the United States is possible if the area has a long-term satellite data record of quality and temporally frequent acquisitions, and an inventory of Landsat holdings at international ground-receiving stations (Huang 2010).

Satellite-based modeling of the gross primary production (GPP) of terrestrial ecosystems requires high-quality satellite data, extensive field measurements, and effective radiative transfer models. Current satellite-based GPP models are largely founded on the concept of light-use efficiency (Xiao et al. 2010). Such production efficiency models (PEMs) may be grouped into two categories based on how they calculate the absorption of light for photosynthesis: (1) those models using the fraction of photosynthetically active radiation absorbed by vegetation canopy, and (2) those using the fraction of photosynthetically active radiation absorbed by chlorophyll (Xiao et al. 2010). Chapter 15 provides a review of satellite-based PEMs and highlights the major differences between these two approaches. The authors conclude that further research efforts are needed in the validation of satellite-based production efficiency models (PEMs) and the error reduction of GPP estimates from net ecosystem exchange (NEE) data using a consistent method.

In Chapter 16, Thenkabail and colleagues discuss the maps and statistics of global croplands and the associated water use determined by remote sensing and nonremote-sensing approaches. Sources of uncertainty in the areas and limitations of existing cropland maps are further examined. Thenkabail et al. (2010) conclude that among four major cropland area maps and statistics at the global level, one study employed a mainly multisensor remote sensing approach, whereas the others used a combination of national statistics and geospatial techniques. However, the uncertainties in these major maps and statistics, as well as the geographic locations of croplands, are quite high. They suggest that it is necessary to utilize higher spatial and temporal resolution satellite images to generate global cropland maps with greater geographic precision, crop types, and cropping intensities.

Section V presents examples of applications of remote sensing technology for studies of air, water, and land. This section starts with atmospheric remote sensing, which has great significance in the estimation of aerosol and microphysical properties of the atmosphere in order to understand aerosol climatic issues at scales ranging from local and regional to global. Aerosol monitoring at the local scale is more challenging due to relatively weak atmospheric signals, coarse spatial resolution images, and the spectral confusion between urban bright surfaces and aerosols. Chapter 17 reviews MODIS algorithms for aerosol retrieval at both global and local scales, and illustrates them with a research involving the retrieval of aerosol optical thickness (AOT) over Hong Kong and the Pearl River Delta region, China, by using 500-m MODIS data. The feasibility of using 500-m AOT for mapping urban anthropogenic emissions, monitoring changes in regional aerosols, and pinpointing biomass-burning locations is also demonstrated. Wong and Nichol (2010) suggest that due to the high temporal resolution of MODIS imagery, aerosol retrieval can be accomplished on a routine basis for the purpose of air quality monitoring over megacities.

The quality of inland, estuarine, and coastal waters is of high ecological and economical importance (Gitelson et al. 2010). Chapter 18 demonstrates the development, evaluation, and validation of algorithms for the remote estimation of chlorophyll-a (Chl-a) concentration in turbid, productive, inland, estuarine, and coastal waters, a pigment universally found in all phytoplankton species and routinely used as a substitute for biomass in all types of aquatic environments. The rationale behind the bio-optical algorithms is presented and the suitability of the developed algorithms for accurate estimation of Chl-a concentration is examined. Gitelson et al. (2010) assert that their algorithms, which are developed by a semi-analytical method and calibrated in a restricted geographic area, can be applied to diverse aquatic ecosystems without the need for further parameterization.

Chapter 19 is concerned with the interaction between the Earth's land surface and the atmosphere. Here, Petropoulos and Carlson provide a concise review of the development of remote sensing-based methods currently used in the estimation of surface energy fluxes, that is, the one-layer model, two-layer model, and the "triangle" method (Gillies and Carlson 1995; Gillies et al. 1997), by examining the main characteristics and by comparing their strengths and limitations. Next, remote sensing methods for estimation of soil-water content are assessed, which use visible, TIR, and microwave data, or their combinations. The remaining half of this chapter provides a detailed account of the triangle method, its theoretical background, implementation, and validation; and the soil–vegetation–atmosphere transfer (SVAT) model, which is essential for the implementation of the protocol.

Urban environmental problems have become unprecedentedly significant in the twenty-first century. The National Research Council Decadal Survey suggests that urban environment should be defined as a "new science" to be focused on the U.S. satellite missions of the near future (National Research Council 2007). As such, remote sensing of urban and suburban areas has recently become a new scientific frontier (Weng and Quattrochi 2006). Chapter 20 reviews remote sensing approaches to measure the biophysical features of the urban environment, and examines the most important concepts and recent research progresses. This chapter ends with the author's prospects on future developments and emerging trends in urban remote sensing, particularly, in the aspect of algorithms.

The U.S. Geological Survey (USGS) National Land-Cover Database (NLCD) has been developed over the past two decades. NLCD products provide timely, accurate, and spatially explicit national land cover at 30-m resolution, and have proven effective for addressing issues such as ecosystem health, biodiversity, climate change, and land management policy. Chapter 21 summarizes major scientific and technical issues in the development of NLCD 1992, NLCD 2001, and NLCD 2006 products. Experiences and lessons learned from the development of NLCD in terms of project design, technical approaches, and project implementation are documented. Further, future improvements are discussed for the development of next-generation NLCD products, that is, the NLCD 2011.

MATLAB® is a registered trademark of The MathWorks, Inc. For product information, please contact:

The MathWorks, Inc.
3 Apple Hill Drive
Natick, MA 01760-2098 USA
Tel: 508 647 7000
Fax: 508-647-7001
E-mail: info@mathworks.com
Web: www.mathworks.com

References

Blaschke, T. 2010. Object based image analysis for remote sensing. *ISPRS Int J Photogramm Remote Sens* 65(1):2–16.

Blaschke, T., K. Johansen, and D. Tiede. 2011. Object based image analysis for vegetation mapping and monitoring. In *Advances in Environmental Remote Sensing: Sensors, Algorithms, and Applications*, ed. Q. Weng, 245–276, chap. 10. Boca Raton, FL: CRC Press.

Gao, F., J. Masek, M. Schwaller, and H. Forrest. 2006. On the blending of the Landsat and MODIS surface reflectance: Predict daily Landsat surface reflectance. *IEEE Trans Geosci Remote Sens* 44(8):2207–18.

Gillies, R. R., and T. N. Carlson. 1995. Thermal remote sensing of surface soil water content with partial vegetation cover for incorporation into climate models. *J Appl Meteorol* 34:745–56.

Gillies, R. R., T. N. Carlson, J. Cui, W. P. Kustas, and K. S. Humes. 1997. Verification of the "triangle" method for obtaining surface soil water content and energy fluxes from remote measurements of the Normalized Difference Vegetation Index (NDVI) and surface radiant temperature. *Int J Remote Sens* 18:3145–66.

Gitelson, A., D. Gurlin, W. Moses, and Y. Yacobi. 2011. Remote estimation of chlorophyll-a concentration in inland, estuarine, and coastal waters. In *Advances in Environmental Remote Sensing: Sensors, Algorithms, and Applications*, ed. Q. Weng, 449–478, chap. 18. Boca Raton, FL: CRC Press.

Huang, C. 2011. Forest change analysis using time series Landsat observations. In *Advances in Environmental Remote Sensing: Sensors, Algorithms, and Applications*, ed. Q. Weng, 348–374, chap. 14. Boca Raton, FL: CRC Press.

Huang, C., S. N. Goward, J. G. Masek, F. Gao, E. F. Vermote, N. Thomas, K. Schleeweis et al. 2009. Development of time series stacks of Landsat images for reconstructing forest disturbance history. *Int J Digital Earth* 2(3):195–218.

Huang, C., S. N. Goward, J. G. Masek, N. Thomas, Z. Zhu, and J. E. Vogelmann. 2010. An automated approach for reconstructing recent forest disturbance history using dense Landsat time series stacks. *Remote Sens Environ* 114(1):183–98.

Huete, A. R., and E. P. Glenn. 2011. Remote sensing of ecosystem structure and function. In *Advances in Environmental Remote Sensing: Sensors, Algorithms, and Applications*, ed. Q. Weng, 295–324, chap. 12. Boca Raton, FL: CRC Press.

Lillesand, T. M., R. W. Kiefer, and J. W. Chipman. 2008. *Remote Sensing and Image Interpretation.* 6th ed., 473. Hoboken, NJ: John Wiley & Sons.

Lu, D., P. Mausel, E. Brondizios, and E. Moran. 2004. Change detection techniques. *Int J Remote Sens* 25:2365–407.

Lu, D., E. Moran, S. Hetrick, and G. Li. 2011. Land use and cover change detection. In *Advances in Environmental Remote Sensing: Sensors, Algorithms, and Applications*, ed. Q. Weng, 277–292, chap. 11. Boca Raton, FL: CRC Press.

Lu, D., and Q. Weng. 2007. A survey of image classification methods and techniques for improving classification performance. *Int J Remote Sens* 28(5):823–70.

National Research Council. 2007. *Earth Science and Applications from Space: National Imperatives for the Next Decade and Beyond.* Washington, DC: The National Academy Press.

Song, C., J. M. Gray, and F. Gao. 2011. Landsat imagery for vegetation studies. In *Advances in Environmental Remote Sensing: Sensors, Algorithms, and Applications*, ed. Q. Weng, 3–30, chap. 1. Boca Raton, FL: CRC Press.

Thenkabail, P. S., M. A. Hanjra, V. Dheeravath, and M. Gumma. 2011. Global croplands and their water use: Remote sensing and non-remote sensing perspectives. In *Advances in Environmental Remote Sensing: Sensors, Algorithms, and Applications*, ed. Q. Weng, 391–428, chap. 16. Boca Raton, FL: CRC Press.

Toutin, T. 2011. 3D geometric correction of earth observation satellite data. In *Advances in Environmental Remote Sensing: Sensors, Algorithms, and Applications*, ed. Q. Weng, 177–222, chap. 8. Boca Raton, FL: CRC Press.

Weng, Q. 2009. *Remote Sensing and GIS Integration: Theories, Methods, and Applications*. 183–208. New York: McGraw-Hill.

Weng, Q., and D. A. Quattrochi. 2006. An introduction to urban remote sensing. In *Urban Remote Sensing*, ed. Q. Weng, and D. Quattrochi, 1–4. Boca Raton, FL: CRC Press.

Wong, M. S., and J. E. Nichol. 2011. Remote sensing of aerosols from space: A review of aerosol retrieval using MODIS. In *Advances in Environmental Remote Sensing: Sensors, Algorithms, and Applications*, ed. Q. Weng, 431–448, chap. 17. Boca Raton, FL: CRC Press.

Xiao, X., H. Yan, J. Kalfas, and Q. Zhang. 2011. Satellite-based modeling of gross primary production of terrestrial ecosystems. In *Advances in Environmental Remote Sensing: Sensors, Algorithms, and Applications*, ed. Q. Weng, 375–390, chap. 15. Boca Raton, FL: CRC Press.

Yool, S. R. 2009. Pygeography. In *Encyclopedia of Geography*, ed. B. Warf. Thousand Oaks, CA: Sage Reference.

Yool, S. R. 2011. Remote sensing of live fuel moisture. In *Advances in Environmental Remote Sensing: Sensors, Algorithms, and Applications*, ed. Q. Weng, 325–346, chap. 13. Boca Raton, FL: CRC Press.

Section I

Sensors, Systems, and Platforms

1

Remote Sensing of Vegetation with Landsat Imagery

Conghe Song, Joshua M. Gray, and Feng Gao

CONTENTS

1.1 Introduction

The U.S. Landsat program is one of the most successful remote-sensing programs in the world. The launch of the Landsat series of satellites marked the beginning of a new era in remote sensing (Williams, Goward, and Arvidson 2006). Due to the critical role played by vegetation in the terrestrial ecosystem and the emphasis of Landsat sensors on vegetation reflectance characteristics, Landsat data greatly enhanced our understanding of the dynamics of vegetation and its functions in the terrestrial ecosystem (Cohen and Goward 2004). The first Landsat satellite, initially called the Earth Resource Technology Satellite, was launched in 1972. To date, seven Landsat satellites have been launched (Table 1.1). Except Landsat 6, all other satellites in the series were successfully put in orbit. Table 1.2 shows the history of sensors deployed on the Landsat satellites. The first three Landsat satellites had similar onboard sensors, including return beam vidicon (RBV) and multispectral scanners (MSSs). Starting with Landsat 4, thematic mapper (TM) sensors were deployed and RBV was removed. The TM sensors have 30 × 30 m spatial resolution for reflective

TABLE 1.1

Brief History of Landsat Satellites

Satellite	Launch Date	Decommission Date	Orbit Height (km)	Temporal Resolution (days)
Landsat 1	July 23, 1972	January 6, 1978	900	18
Landsat 2	January 22, 1975	February 25, 1982	900	18
Landsat 3	March 5, 1978	March 31, 1983	900	18
Landsat 4	July 16, 1982	–	705	16
Landsat 5	March 2, 1984	–	705	16
Landsat 6	October 5, 1993	Failure	705	16
Landsat 7	April 15, 1999	–	705	16
LCDM	December 2012	–	705	16

TABLE 1.2

Sensors Used or to Be Used in Landsat Series Satellites

Sensor	Satellite	Band Width (μm)	Spatial Resolution (m)
RBV	Landsat 1, 2	0.475–0.575	80
		0.580–0.680	80
		0.690–0.830	80
	Landsat 3	0.505–9.750	30
MSS	Landsat 1–5	0.50–0.60	79 (1–3)/82 (4–5)
		0.60–0.70	79/82
		0.70–0.80	79/82
		0.80–0.11	79/82
	Landsat 3	10.4–12.6	240
TM	Landsat 4, 5	0.45–0.52	30
		0.52–0.60	30
		0.63–0.69	30
		0.76–0.90	30
		1.55–1.75	30
		10.4–12.5	120
		2.08–2.35	30
ETM	Landsat 6	Same as TM	Same as TM
		0.50–0.90	15
ETM+	Landsat 7	Same as TM	30 (60 m thermal)
		0.50–0.90	15
LCDM	LCDM	0.433–0.453	30
		0.450–0.515	30
		0.525–0.600	30
		0.630–0.680	30
		0.845–0.885	30
		1.560–1.660	30
		2.100–2.300	30
		0.5–0.680	15
		1.360–1.390	30

bands, and 120 × 120 m for the thermal band on the ground. This intermediate spatial resolution imagery provides land-surface information detailed enough for most scientific and application needs; the spatial resolution also allows the sensor to cover ground areas large enough for regional planning and management with a single scene (185 × 175 km).

The longest-serving satellite to date among the Landsat series is Landsat 5. It was launched in 1984, and remains in operation (as of December 1, 2009), with the exception of a few temporary technical glitches. The TM sensors were upgraded to Enhanced Thematic Mapper (ETM) sensors for the ill-fated Landsat 6, and the ETM sensor was further improved to ETM+ onboard Landsat 7. The ETM+ sensor maintained the same multispectral bands as TM at the same spatial resolution with the addition of a panchromatic band (15 × 15 m spatial resolution). This band offers the opportunity to sharpen the other bands. With the advance of technology, the thermal band on Landsat 7 was refined to 60 × 60 m from its earlier 120 × 120 m spatial resolution. Unfortunately, the scan-line corrector on Landsat 7 permanently malfunctioned since May 2003, causing a loss of approximately 25% of the data, most of which was located between scan lines toward the scene edges. Although some gap-filling remedy operations can recover most of the data lost, the gap-filled data cannot be guaranteed to have a quality equivalent to that of the original data. Fortunately, the Landsat Data Continuity Mission (LDCM), the follow-up Landsat satellite, is currently scheduled to launch in late 2012 (http://ldcm.nasa.gov). The LDCM sensors added two more reflective bands for coastal and cirrus clouds needs, but dropped the thermal band (Table 1.2). The Landsat image collection, spanning nearly four decades, is the longest continuous data record of land-surface conditions. Landsat data has contributed significantly to the understanding of the Earth's environment (Williams, Goward, and Arvidson 2006). A complete review of the applications of Landsat images cannot be achieved within a single book chapter. This chapter primarily focuses on the use of Landsat images in extracting biophysical information of vegetation, with an emphasis on forests, which are the biggest challenges faced by remote-sensing scientists.

1.2 Spectral Information of Vegetation in Landsat Thematic Mapper/Enhanced Thematic Mapper+ Bands

The spectral information of vegetation in Landsat TM/ETM+ imagery is primarily determined by the designation of spectral bands as seen in Table 1.2. The first three bands of TM/ETM+ sensors are in the visible spectrum. In the first three bands, reflected energy from vegetation is determined by the concentration of leaf pigments. Leaves strongly absorb solar radiation in the visible spectrum, particularly the red spectrum, for photosynthesis. The fourth band is in the near-infrared (NIR) region of the solar spectrum, to which healthy green leaves are highly reflective. The contrast in leaf reflectance between the red and NIR spectra is the physical basis for numerous vegetation indices using optical remote sensing. The two mid-infrared bands relate to the moisture content in healthy vegetation.

Vegetation indices produced by the combination of reflectance in red and NIR bands are perhaps the most commonly used data in vegetation mapping using Landsat data. The two mid-infrared bands are also very useful for vegetation monitoring. Horler and Ahern (1986) found that the two mid-infrared bands are very sensitive to vegetation density, especially in the early stages of clear-cut regeneration. Fiorella and Ripple (1993a) found that the TM ratio 4:5 is highly correlated with the age of young Douglas fir stands in the western Cascade mountains of Oregon. Kimes et al. (1996) were able to map the ages of young forest stands using TM 3, 4, 5 along with elevation, slope, and aspect in the H. J. Andrews Experimental Forest. Jakubauskas (1996) also found that the mid-infrared bands of Landsat TM images were useful in differentiating early successional stages of lodgepole pine stands in Yellowstone National Park.

The spectral information from Landsat TM/ETM+ reflective bands are not independent of each other, but are highly correlated. Two statistical approaches are often used to reduce information redundancy in the imagery. One commonly used approach is the principal component analysis (Richards 1984; Fung and Ledrew 1987; Seto et al. 2002), in which the image information from all six bands is compressed into the first few principal components. Because the principal components are orthogonal to each other, there is no information redundancy among the components. For Landsat imagery, more than 95% of the variation can be compressed into the first three components. Thus, principal component analysis can significantly reduce data volume with little information loss. However, the principal component transformation of remotely sensed data is image dependent, that is, the transformation coefficients vary from image to image and are sometimes difficult to interpret. A similar approach, the tasseled cap transformation, is often applied to compress information from the six reflective bands into three meaningful indices: brightness, greenness, and wetness (Crist and Cicone 1984). The tasseled cap transformation concept was originally developed by Kauth and Thomas (1976) for Landsat MSS data. The advantages of tasseled cap transformation over principal component analysis include (1) the resulting components are meaningful; and (2) the transformation coefficients are preset, that is, not dependent on images.

The tasseled cap indices, brightness, greenness, and wetness were extensively used in extracting vegetation information. Fiorella and Ripple (1993b) found that although all three indices can be used to separate old-growth forests from mature forests, wetness was more significant than brightness and greenness. Cohen, Spies, and Fiorella (1995) reached a similar conclusion that the tasseled cap wetness can be used to distinguish forest age classes for closed-canopy conifer forests in the western Cascade mountains of Oregon. The tasseled cap transformation was further developed by Collins and Woodcock (1996) to become the multitemporal tasseled cap transformation. Using this approach, they were able to detect tree mortality in the Lake Tahoe region.

1.3 Applications

1.3.1 Vegetation Cover

Vegetation-cover information in remote sensing usually involves one of two scales. On the regional scale, land surface is classified as either vegetated or nonvegetated, and the fraction of the vegetated area over the total area is referred to as vegetation cover. This regional vegetation cover can be obtained in a relatively straightforward manner through conventional classification of remotely sensed data, in which each pixel of the remotely sensed data is labeled as a land-cover type. A tally of all the vegetated pixels among the total pixels provides the vegetation cover. On the pixel scale, vegetation cover usually refers to the fraction of a single pixel occupied by green vegetation. Conventional classification labels a pixel as one and only one land-cover type; thus, it cannot provide subpixel information. The spatial resolution of Landsat imagery often leads to multiple components of land-cover types in a single pixel. It is particularly common in complex landscapes, such as the urban environment, challenging the conventional classification approach in estimating vegetation content. Subpixel vegetation cover is needed in order to accurately measure the vegetation cover of these areas. Obtaining subpixel vegetation-cover information requires the use of an analytical approach called *spectral mixture analysis* (SMA).

SMA makes the following assumptions: (1) the landscape is composed of a few fundamental components, referred to as *endmembers*, each of which is spectrally distinct from the others; (2) the endmember spectral signatures do not change within the area of interest; and (3) the composite remotely sensed signal for a mixed pixel is linearly related to the fractions of endmember presence (Sabol, Adams, and Smith 1992). The key step in SMA is appropriate endmember selection, including the number of endmembers and their corresponding spectral signatures (Tompkins et al. 1997; Elmore et al. 2000; Theseira et al. 2003). Although Landsat TM/ETM+ imagery has six reflective bands, the number of endmembers used for SMA is often only three or four due to the limitations in the dimensionality of Landsat imagery. Smith et al. (1990) used three endmembers, vegetation, soil, and shade, to map vegetation cover in a desert environment with Landsat imagery. Ridd (1995) developed a three-endmember model, vegetation-impervious-soil (VIS), to map urban structure for Salt Lake City, Utah. The VIS model was later applied to Bangkok, Thailand (Madhavan et al. 2001) and Brisbane, Australia (Phinn et al. 2002). Small (2001) modified the VIS model to a vegetation low albedo and high albedo (VLH) model for New York City after analyzing a time series of Landsat TM imagery. Wu and Murray (2003) added a soil endmember to the VLH model and it became a four-endmember model to describe the urban structure for Columbus, Ohio.

The endmember signatures can be obtained from "pure" pixels in the image over which the mixture analysis is performed. Endmembers whose spectral signatures are obtained in this manner are called *image endmembers*. The advantage of image endmembers is that the endmember spectral signatures are at the same relative measurement scale as the image to be analyzed. The challenge is to identify the pure pixels that can be treated as endmembers. An alternative approach is to obtain the endmember signature from a spectral signature reference library that was developed from spectroradiometer measurements on the ground. Endmembers whose spectral signatures are obtained from a reference spectral library are called *reference endmembers*. Although the reference endmember signatures can be very accurate, care must be taken when using them for SMA as the signature data and the image data are measured by two instruments under very different conditions. The assumption that the endmember spectral signatures do not change within the area of interest is an oversimplification of the real world. There are significant endmember signature variations. For example, the vegetation endmember can be grass, coniferous, and broadleaf trees, each of which has a very different spectral signature from the others. To accommodate the variations of endmember signatures, Roberts et al. (1998) developed the multiple endmember SMA (MESMA), in which the spectral signatures of endmembers were dynamically selected from a spectral library containing hundreds of reference endmembers. Song (2005) developed a Bayesian SMA (BSMA) to account for the effect of endmember signature variation. In BSMA, an endmember spectral signature is no longer a single or enumerable spectral signature, but a probability distribution function. The BSMA is an effective approach that accounts for endmember spectral signature variation and helps reduce error in extracting subpixel vegetation fraction from Landsat imagery (Song 2005).

1.3.2 Leaf Area Index

1.3.2.1 Measuring Leaf Area Index on the Ground

Leaves are the interface for energy and gaseous exchanges between the terrestrial ecosystem and the atmosphere. The amount of leaves in a given area is measured by leaf area index (LAI), which is generally defined as the one-sided total leaf area divided by the ground area over which the leaves are distributed (Monteith and Unsworth 1973). This definition is applicable to broadleaf trees. For coniferous trees, a projected leaf area is used (Myneni, Nemani, and Running 1997). LAI is considered to be the most important land-surface biophysical parameter in understanding terrestrial ecosystem functions (Running and Hunt 1993). Therefore, the continuous estimation of LAI over a large geographic area via remotely sensed data is of high interest to scientists. In fact, it is the only viable option for estimating LAI continuously over the Earth's land surface.

Estimating LAI from remotely sensed data is highly challenging due to a number of factors. It is very difficult to obtain accurate LAI on the ground for model development and validation using remotely sensed data, particularly for forested areas. Two approaches can be used to obtain LAI on the ground, as reviewed multiple times (Breda 2003; Weiss et al. 2004; Jonckheere et al. 2004): direct and indirect approaches. The direct approach involves direct measurements of leaf area. The most destructive direct approach is complete harvesting of all vegetation within a delimited area. This approach is applicable for herbs and crops, but impractical in forests. For forests, a destructive sampling approach is often used, in which a standard tree is identified for each species and size class. The standard tree is then harvested so that its total leaf area can be accurately measured and an allometric relationship between total individual leaf area and the tree-stem diameter at breast height (DBH) can be developed. The allometric relationship is then applied to estimate the total leaf area for all individual trees within a sampling plot; then LAI can be calculated. This is perhaps the most accurate measure of LAI, but it is also very labor intensive. Few studies can afford this kind of sampling. Moreover, the allometric relationships developed at one place do not transfer well to other places. The least destructive, but time-consuming, direct approach to measure LAI is the litter-trap approach, in which multiple litter traps of preset size are deployed in the forest stands. Leaves that fall into the traps are periodically harvested and their areas measured. For a deciduous forest, the maximum LAI can be estimated at the end of the growing season. However, for a coniferous forest, one needs multiple years of data to estimate the peak LAI. This approach is time-consuming and requires that constant attention be paid to the litter traps (McCarthy et al. 2007). An intermediate destructive approach takes into consideration sapwood cross-sectional areas. Pipe theory (Shinozaki et al. 1964) provides the theoretical basis for this approach. Marshall and Waring (1986) found that using sapwood cross-sectional areas to estimate LAI was more accurate than using DBH.

Indirect approaches using optical instruments are more efficient in measuring LAI. Jonckheere et al. (2004) reviewed the theory and performance of optical instruments used in estimating LAI, including LAI-2000 (Licor, Inc., Lincoln, NE), TRAC (3rd Wave Engineering, Ontario, Canada), DEMON (CSIRO, Canberra, Australia), Ceptometer (Decagon Devices, Inc., Pullman, WA), and a digital hemispherical camera. The theoretical basis for the optical measurements of LAI is Beer's law. Assuming random leaf distribution within the canopy space, Beer's law predicts canopy gap fraction as

$$P(\theta) = \exp(-G(\theta)\Omega L/\cos\theta) \qquad (1.1)$$

where θ is the solar zenith angle, $P(\theta)$ is the canopy gap fraction in the direction of θ, and $G(\theta)$ is the leaf projection factor of unit LAI in the direction of θ. The clumping index is Ω, and the LAI is L in Equation 1.1. Most of the optical instruments measure $P(\theta)$ for the canopy. Given $P(\theta)$ and certain assumption for $G(\theta)$, we can obtain ΩL, that is, the effective LAI (L_e), but not L. The gap fractions measured by optical instruments are the combined effects of leaf and woody components. To obtain LAI, one needs to correct the measured effective foliage area index for woody areas and the leaf clumping effect in Beer's law as follows:

$$L = (1-\alpha)L_e/\Omega \tag{1.2}$$

where α is the woody to total area ratio, which depends on the vegetation type. Gower, Kucharik, and Norman (1999) provided α values for some common tree species, varying from 0.03 to 0.22. Chen and Cihlar (1996) used the LAI-2000 device to measure L_e and the TRAC device to measure Ω to estimate L. For conifer species, there is an additional level of clumping, at the needle-to-shoot scale. The needle-to-shoot area ratio, γ, is needed to correct for the clumping index Ω. Gower, Kucharik, and Norman (1999) provided γ values for a few common needleleaf trees, ranging from 1.20 to 2.08. Kucharik, Norman, and Gower (1999) designed an imaging device to estimate γ. Therefore, for conifer forests, LAI can be derived from effective LAI measured with the optical instruments as

$$L = (1-\alpha)L_e\gamma/\Omega \tag{1.3}$$

1.3.2.2 Mapping Leaf Area Index with Landsat Imagery

Landsat TM/ETM+ imagery has a unique advantage over many other satellite images in mapping LAI because its spatial resolution is fine enough to identify individual stands. In the meantime, the image covers a sufficiently large area to meet most application needs. Because there are numerous other factors influencing remotely sensed signals received at Landsat TM/ETM+ sensors, including LAI, leaf angle distribution, leaf clumping, sun and viewing angles, and background conditions, LAI cannot be inverted analytically from remotely sensed signals (Gobron, Pinty, and Verstraete 1997; Eklundh, Harrie, and Kuusk 2001). Most studies that map LAI using Landsat imagery have been based on empirical models. The mapping of LAI using Landsat imagery based on empirical models generally takes place in three steps: (1) measuring LAI for sampling plots on the ground, (2) developing an empirical model between LAI for the sampling plots and some spectral measurements for the same locations, and (3) applying the empirical model spatially within the area of interest. The most commonly used spectral measurements include the normalized difference vegetation index (NDVI) and the simple ratio (SR) vegetation index. For Landsat TM imagery, NDVI is calculated from the surface reflectance values of the red (TM3) and NIR (TM4) bands as

$$NDVI = \frac{\rho_{TM4} - \rho_{TM3}}{\rho_{TM4} + \rho_{TM3}} \tag{1.4}$$

where ρ_{TM3} and ρ_{TM4} are surface reflectances for TM3 and TM4, respectively. The SR vegetation index is

$$SR = \frac{\rho_{TM4}}{\rho_{TM3}} \tag{1.5}$$

One of the earliest studies that used Landsat TM–type data was by Peterson, Westman, and Stephenson (1986); they used Airborne Thematic Mapper simulator data to study the potential of Landsat TM imagery for mapping LAI. Their study was based on 18 conifer stands with LAI values ranging from 0.6 to 16.1. These stands were distributed across west central Oregon along an environmental gradient with a wide range of moisture and temperature. Using atmospherically corrected surface reflectance, both linear ($R^2 = 0.83$) and log-linear ($R^2 = 0.91$) regression explained the variation in LAI well. However, Peterson, Westman, and Stephenson (1986) cautioned the use of the empirical relationships they developed for a particular vegetation zone within the region. A study by Spanner (1994) found that the empirical relationship between LAI and spectral vegetation indices strongly depends on canopy cover and understory condition. To reduce the canopy cover effect, Nemani et al. (1993) used the mid-infrared band to correct NDVI, resulting in an improved relationship between NDVI and LAI.

Chen and Cihlar (1996) evaluated the potential of both NDVI and SR vegetation index in mapping LAI using Landsat TM imagery. They found that NDVI and SR vegetation index are better correlated to effective LAI than LAI. Due to the influence of understory vegetation, midsummer Landsat TM imagery is not as good as late-spring imagery in extracting LAI. Turner et al. (1999) compared spectral vegetation indices with different radiometric correction levels across three temperate zones, and found that NDVI based on surface reflectance best correlates with LAI. However, the NDVI–LAI relationship reaches an asymptote when the LAI value reaches 3–5. They also found that the sensitivity of spectral vegetation indices to LAI differs between coniferous and deciduous forests. Thus, it is desirable to stratify land-cover classes in order to achieve local accuracy using spectral vegetation indices to estimate LAI. The study by Fassnacht et al. (1997) reports similar conclusions.

Both NDVI and SR vegetation index make use of information in only two of the six bands from Landsat TM/ETM+ imagery. Nemani et al. (1993) used an additional band, the mid-infrared band, to reduce canopy openness effect in NDVI, leading to an improved empirical model. Brown et al. (2000) applied the same mid-infrared band to the SR vegetation index. Because the mid-infrared correction leads to a lower SR, Brown et al. (2000) called the corrected SR the reduced SR (RSR). Chen et al. (2002) suggested that RSR can unify coniferous and deciduous vegetation cover types in mapping LAI. Although RSR was not initially developed based on Landsat TM imagery, Chen et al. (2002) used the RSR approach to develop a fine-resolution LAI surface based on Landsat TM imagery and scaled up the algorithm with coarse spatial resolution imagery to produce an LAI surface covering Canada. In order to make full use of the spectral information available in all bands and to account for uncertainty in reflectance measurements, Cohen et al. (2003) proposed a reduced major axis (RMA) regression approach to link LAI with spectral information through canonical transformation. The RMA approach can significantly improve the relationship between LAI and spectral information from Landsat imagery.

Because of the empirical nature of the approaches used to map LAI with Landsat imagery, the fitness of the model varies significantly from study to study, as shown in Table 1.3. These empirical models generally do not transfer well to places outside the area in which they were developed. Therefore, for any new applications, one still has to develop his or her own empirical models, and he or she should not expect the same good performance of certain empirical models to reappear. There is still a significant amount of trial-and-error efforts needed before an appropriate empirical LAI model can be developed. In the future, mapping of LAI should not be limited to Landsat data only. The recent abundance of high spatial resolution imagery offers new opportunities for mapping LAI (Colombo et al. 2003;

TABLE 1.3

Regression Models in the Literature Using Landsat TM/ETM+ Images to Map LAI

Vegetation Index	Ground LAI	Model	R^2	Source
SR	Allometry	SR = 1.23 + 0.614 LAI	0.82	Running et al. 1986
SR	Allometry	SR = 1.92 SR$^{0.583}$	0.91	Peterson et al. 1986
NDVI	Allometry	LAI = −1431 + 32.25 NDVI	0.86	Curran et al. 1992
NDVI$_c$	Allometry	NDVI$_c$ = 0.70 exp (0.70 LAI)	0.64	Nemani et al. 1993
SR	Ceptometer	SR = 3.1196 + 4.5857 log (LAI)	0.97	Spanner et al. 1994
SR	LAI-2000/TRAC	SR = 2.781 + 0.843 LAI	0.53	Chen and Cihlar 1996
NDVI	Allometry	NDVI = 0.607 + 0.0377 LAI	0.72	Fassnacht et al. 1997
NDVI	Allometry	NDVI = 0.5724 + 0.0989 LAI − 0.0114 LAI^2 + 0.0004 LAI^3	0.74	Turner et al. 1999
RSR	LAI-2000/TRAC	RSR = α + β LAI[a]	0.55	Brown et al. 2000
RSR	LAI-2000/TRAC	RSR = 1.0743 LAI + 1.2843	0.63	Chen et al. 2002
CI	Allometry	LAI = 4.19 − 1.68 CI	0.72	Berterretche et al. 2005
NDVI	LAI-2000	LAI = α exp (βNDVI)[a]	0.77	Soudani et al. 2006

Note: The spectral indices in the table include simple ratio (SR), reduced simple ratio (RSR), normalized difference vegetation index (NDVI), corrected normalized difference vegetation index (NDVI$_c$), and canonical index (CI).

[a] Model parameters were not provided in the paper.

Soudani et al. 2006; Song and Dickinson 2008). In addition, remotely sensed data from lidar sensors can provide valuable information for mapping LAI (Riano et al. 2004; Roberts et al. 2005; Morsdorf et al. 2006), although lidar remote sensing does not cover the area in a wall-to-wall fashion as optical remote sensing does. The synergistic use of information from multiple sensors, each of which provides complementary information, should be adopted in the future for accurate mapping of LAI.

1.3.3 Biomass

Biomass refers to the total dry weight of all parts that make up a live plant, including those above (e.g., leaves, branches, and stems) and below (e.g., fine and coarse roots) ground. It is the accumulation of the annual net primary production over the plant life after litter fall and mortality. The information of forest biomass is of great scientific and economic value, particularly over large areas. Obtaining biomass for individual plants requires destructive sampling of aboveground components and excavation of belowground components. Destructive sampling is relatively easy to perform for herbaceous plants, but it is extremely laborious and time-consuming to perform for forests (Whittaker et al. 1974). Moreover, destructive sampling cannot be used to obtain biomass over large areas, particularly for forests. A common approach to estimate areal-based biomass for forest ecosystems is to develop an allometric relationship between the easily measured stem diameter at breast height (DBH), and the individual biomass sampled on a species-specific basis, and then apply this allometry to each individual within a sampling plot to estimate the areal-based biomass. Tremendous efforts have been devoted to developing species-specific allometric relationships for biomass in the past (Grier and Logan 1977; Gholz et al. 1979; Ter-Mikaelian and Korzukhin 1997; Smith, Heath and Jenkins 2003; Jenkins et al. 2003). However, the application of such species-specific biomass allometry to sampling plots cannot provide spatially explicit distribution of biomass over large areas. Remotely sensed data provide the potential to scale up biomass from sampling plots to spatially explicit biomass over a region. Three types of remotely sensed data are investigated in the literature for their potential use in mapping biomass: optical (Sader et al. 1989; Foody et al. 1996), radar (Dobson et al. 1995), and lidar (Lefsky et al. 1999). The mapping of biomass using remotely sensed data from radar and lidar sensors is beyond the scope of this chapter.

Optical remotely sensed signals over a vegetated area are primarily energy reflected by the leaves; that is, biomass does not have a direct remote-sensing signal. However, LAI usually reaches asymptote soon after canopy closure, whereas biomass can continue to increase for many years (Song, Woodcock, and Li 2002). Figure 1.1 shows the results of coupled GORT-ZELIG modeling from the Geometric Optical Radiative Transfer (GORT) model with the ZELIG forest succession model for a typical stand in the H. J. Andrews Experimental Forest. Forest biomass increases almost linearly in the first 100 years. However, the remotely sensed signals are only sensitive to biomass change when biomass is below 100 Mg/ha. Moreover, the relationships of NDVI and tasseled cap greenness with biomass are influenced by background conditions. Tasseled cap wetness is resistant to background noise, but all indices suffer from signal saturation problems. It is interesting to note that the threshold for signal saturation from GORT-ZELIG simulation is very similar to the threshold value for saturation from empirical studies (Steininger 2000).

The most common approach used for mapping biomass with Landsat TM/ETM+ imagery is to develop an empirical model that directly relates remotely sensed signals (e.g., surface reflectance or vegetation indices) to biomass derived on the ground, and then apply this empirical model spatially to the area of interest (Foody 2003; Zheng et al. 2004). Numerous successful applications of this approach have been reported (Anderson, Hanson, and Haas 1993; Roy and Ravan 1996; Fazakas, Nilsson, and

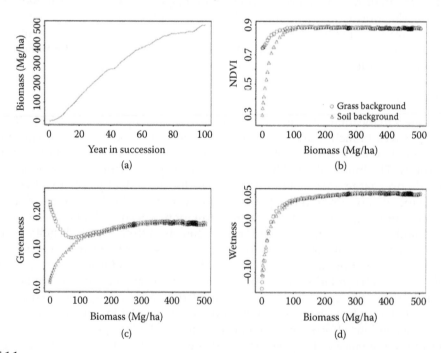

FIGURE 1.1
The GORT-ZELIG model results for biomass and its relationship with spectral indices: (a) temporal trajectory of biomass for a typical stand in H. J. Andrews Experimental Forest; (b) normalized difference vegetation index (NDVI) versus biomass; (c) tasseled cap greenness versus biomass; and (d) tasseled cap wetness versus biomass.

Olsson 1999; Steininger 2000; Tomppo et al. 2002). However, these successful applications were performed in areas with low biomass. When the biomass is high, the remotely sensed signals no longer respond to biomass increase (Sader et al. 1989; Trotter, Dymond, and Goulding 1997). Lu (2005) reviewed the potential of using Landsat TM imagery for mapping aboveground biomass in the Brazilian Amazon, and found that the spectral signals are suitable for aboveground biomass for forests with simple structure. He also indicated that spatial information is useful in mapping aboveground biomass, although other studies found that spatial information from Landsat TM imagery provides little help in extracting canopy structure because the spatial resolution is too coarse compared to the size of trees (Cohen, Spies, and Bradshaw 1990; Song and Woodcock 2002).

Overall, the mapping of biomass remains a major challenge in remote sensing. Both optical and radar remote sensing suffer from a signal saturation problem (Sader et al. 1989; Dobson et al. 1995). An alternative is to use remotely sensed data from lidar sensors. Lidar data provide canopy height information, from which canopy biomass can be derived using allometry. Use of lidar remote sensing overcomes the signal saturation problem. However, the height–biomass allometry is species specific. Lidar can only provide canopy height, but not species-specific information. Moreover, lidar data does not provide wall-to-wall coverage except for small footprint lidar for a small area. Synergistic use of multiple sensors is needed in the future for mapping biomass accurately with remotely sensed data.

1.3.4 Monitoring Forest Successional Stages with Landsat Imagery

1.3.4.1 Forest Succession

Forest ecosystems are the most complex terrestrial ecosystems on Earth, providing key ecological goods and services for many other plants and animals, as well as for humans (Dixon et al. 1994; Dobson, Bradshaw, and Baker 1997; Noble and Dirzo 1997; Myers et al. 2000). Forests are constantly undergoing changes, even without human disturbance. This process is called *forest succession* (Clements 1916). Forest succession is a complex ecological process that involves multidimensional changes, including, but not limited to, the growth and mortality of individual trees as well as the establishment of new individuals. Depending on the initial condition, forest succession can be classified into primary succession and secondary succession. Primary succession begins in an area that has not been previously occupied by a vegetation community, whereas secondary succession occurs in an area from which a community was removed (Odum 1953). The ecological goods and services provided by the forest ecosystem are highly dependent on forest successional stages (Song and Woodcock 2003a; Pregitzer and Euskirchen 2004; Lamberson et al. 1992). Therefore, it is not only important to know the location and size of forest areas, but it is also crucial to know its successional stages in order to accurately understand their current ecological functions or to predict their future ecological roles. Remote sensing offers the potential to monitor forest successional stages over large areas.

1.3.4.2 Empirical Approaches

Two kinds of change occur in forest ecosystems: the gradual change of forest succession, and the sudden change of deforestation due to anthropogenic (e.g., timber harvesting) or natural (e.g., fire) disturbances. It is usually quite straightforward to map deforestation with Landsat TM/ETM+ imagery as a result of dramatic change in surface reflectance before and after the disturbance (Skole and Tucker 1993; Cohen et al. 1998; Woodcock et al.

2001). A common empirical approach used to map forest successional stages is supervised image classification. This approach first breaks the continuous successional sere into a discrete set of successional stages. Then, a training set for each successional stage is identified in the image, and a classifier is trained with the training set to classify the entire image. Hall et al. (1991) studied the pattern of forest succession in Superior National Forest with two Landsat MSS images (dated July 3, 1973 and June 18, 1983) after correcting the atmospheric, seasonal, and sensor differences for the two images. Two sets of reference data were used. One set was developed through ground observations, and the other was based on aerial photography and high-resolution airborne digital imagery. These data were plotted in the Cartesian space of MSS bands 1 and 4, and the spectral space for each successional stage was delineated and applied to the rest of the image. Jakubauskas (1996) classified the lodgepole pine forests into six successional stages with a Landsat TM image based on 69 ground control sites. Helmer, Brown, and Cohen (2000) were able to differentiate secondary and old-growth forests through supervised classification with multidate Landsat images for montane tropical forests. Fiorella and Ripple (1993b) used unsupervised classification to sort a Landsat TM image into 99 spectral clusters, and then regrouped these clusters into five successional stages. Cohen, Spies, and Fiorella (1995) were able to separate the closed-canopy conifer forests into two or three age classes with regression analysis. Kimes et al. (1996) were able to map forest stand ages for young stands (age <50 years) by combining Landsat TM data with ancillary data for a neural network classifier. For recently regenerated secondary forests, it is possible to extract the forest age based on the time when deforestation occurred (Foody et al. 1996; Lucas et al. 2002; Kennedy, Cohen, and Schroeder 2007; Huang et al. 2009). However, this approach works only for relatively young secondary forests. These successful empirical applications do not provide much guidance for new applications elsewhere. More sophisticated approaches for monitoring forest succession should be built on physical-based algorithms (Hall, Shimabukuro, and Huemmrich 1995).

1.3.4.3 Physical-Based Approaches

1.3.4.3.1 Li–Strahler Model

Remotely sensed signals are essentially reflected energy within the sensor instantaneous field of view recorded at the given sun–sensor geometry within a particular wavelength range. For a forested scene, the structure and composition of the canopy as well as the background condition determine how much energy is received at the satellite sensor. Numerous models have been developed to understand the relationship between scene structure and the energy it reflects (Suits 1972; Verhoef 1984; Li and Strahler 1985). Most of these models are forward models, that is, the model can predict the energy reflected given the scene structure and sun–sensor geometry. Among such models, the Li–Strahler model (Li and Strahler 1985) can be inverted for mean crown size and canopy cover over a stand, thus providing information for forest succession. The Li–Strahler model assumes the reflected spectral energy for a pixel is the area-weighted average of the first scattering of four scene components: sunlit crown (C), shaded crown (T), sunlit background (G), and shaded background (Z), that is,

$$S = K_c C + K_z Z + K_g G + K_t T \qquad (1.6)$$

where S is the ensemble reflected spectral energy from a pixel, and the Ks are the areal fractions of the corresponding scene components. Li and Strahler (1985) provided mathematical models describing the scene-component fractions based on optical theory given the sun–sensor and tree crown geometry. Thus, the model is also called the *geometric–optical* model. Li and Strahler (1985) showed that the average tree crown radius for a forest stand can be inverted from the remotely sensed images as follows:

$$R^2 = \frac{V(m) - \omega M^2}{(1+\omega)M} \tag{1.7}$$

where R is the expected value of tree crown horizontal radius, and $\omega = (1+C_r^2)^4 - 1$ with C_r being the coefficient of variation of the crown radius. The parameter m is called the "tree-ness" factor, which is defined as the ratio of the sum of squared crown radii of all trees in a pixel to the area of the pixel (A), that is, $m = \left(\sum_{i=1}^{n} r_i^2\right)/A = nR^2/A$, where n is the number of trees in the pixel. $V(m)$ and M are the interpixel variance and the mean value of m within a forest stand, respectively. The treeness factor (m) for a given pixel can be derived from remotely sensed data as follows:

$$m = \frac{\|GS\|}{\Gamma \|GX\|} \tag{1.8}$$

where $\|GS\|$ is the Euclidean distance between G (sunlit background reflectance) and S (ensemble pixel reflectance) in the spectral space, and X is the gravity center of the triangle CTZ. Similarly, $\|GX\|$ is the Euclidean distance between G and X; Γ is a scalar of geometry factor. The Li–Strahler model assumes the pixel size is significantly larger than the tree crown size, yet there is significant variation in tree counts among the pixels covering a forest stand. Thus, the forest stand is significantly larger than the pixel size. The spatial resolution of Landsat TM/ETM+ data meets the aforementioned requirements well. Franklin and Strahler (1988) and Wu and Strahler (1994) achieved some success in estimating tree crown size with the Li–Strahler model. However, in more comprehensive studies, Woodcock et al. (1994, 1997) showed that although tree cover can be mapped effectively with the Li–Strahler model, separation of crown cover into tree crowns based on the inversion of the Li–Strahler model was poor.

1.3.4.3.2 GORT-ZELIG Model

The Li–Strahler model assumes that tree crowns are three-dimensional opaque objects randomly distributed in the scene. Multiple scattering of photons within the forest canopy and between the background and the canopy was significantly simplified. Li, Strahler, and Woodcock (1995) further improved the model to account for the multiple scattering of photons by integrating the geometric–optical model with a traditional turbid medium radiative transfer model (GORT). They also modified the crown shape from the previously considered cone to the more flexible ellipsoid. The ellipsoid is a more realistic abstraction for most tree crowns (Peddle, Hall, and LeDrew 1999). Ni et al. (1999) further simplified the original GORT model to become an analytical model. The analytical GORT is relatively simple to apply in modeling the bidirectional reflectance distribution function (BRDF) for a forest scene, and also integrates the strength of both geometric–optical and radiative transfer models.

Song, Woodcock, and Li (2002) coupled the GORT model with a gap-type forest successional model, ZELIG (Urban 1990), which was in turn developed based on the JABOWA (Botkin, Janak, and Wallis 1972) and the FORENA (Shugart and West 1977) models. The ZELIG model provides canopy structure to GORT, which provides canopy reflectance under a given sun–sensor geometry. Song, Woodcock, and Li (2002) simulated a Douglas fir/western hemlock stand for the first 50 years of succession and produced the canopy reflectance for the six reflectance bands of Landsat TM sensors under two contrasting

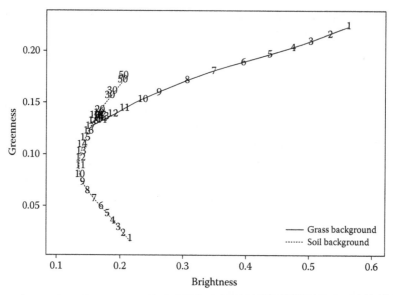

FIGURE 1.2

The modeled temporal trajectory of forest succession with GORT-ZELIG in the tasseled cap brightness/greenness space for a typical stand in the H. J. Andrews Experimental Forest with two contrasting background conditions. The numbers on the lines indicate years in succession. (Reprinted from *Remote Sensing of Environment*, 82, Song, C., Woodcock, C. E., and Li, X, The spectral/temporal manifestation of forest succession in optical imagery: The potential of multitemporal imagery, 285–302. Copyright (2002), with permission from Elsevier.)

background conditions. Figure 1.2 shows the spectral–temporal trajectories associated with forest succession in the tasseled cap brightness/greenness space. The spectral–temporal trajectory of forest succession is highly nonlinear, indicating that the monitoring of forest succession requires multiple images in time to determine the forest's successional stage. Background conditions strongly influence the canopy reflectance before canopy closure. For a bright grass background, the establishment of trees leads to a rapid decrease in brightness due to the shadows cast. However, for a dark soil background, the establishment of new trees causes a rapid increase in greenness but a minimal change in brightness. The spectral trajectories from the two contrasting backgrounds converge when the canopy closes, minimizing the influence of background conditions.

To validate the nonlinearity of forest succession, spectral–temporal trajectories were constructed from multiple Landsat images for several stands with similar ages but different growth conditions in the H. J. Andrews Experimental Forest. Figures 1.3a–c show that the observed spectral–temporal trajectories for a few well-regenerated young stands, constructed from a series of multitemporal Landsat images, do possess the modeled nonlinearity. However, the one stand (Figure 1. 3d) that was not well regenerated did not show the modeled spectral–temporal trajectory. Biophysical modeling, such as GORT-ZELIG, provides a theoretical basis for understanding the manifestation of forest succession in optical imagery through time.

A complete forest succession sere can span several centuries, whereas Landsat TM imagery dates only as far back as 1984. There are no satellite images that provide coverage for a complete forest succession sere. A similar strategy that was used in traditional forest succession studies can be used in monitoring forest succession with satellite imagery, that is, the "substitute space for time" strategy. This strategy reconstructs a complete forest succession sere with forests at different successional stages at the same time, but in different

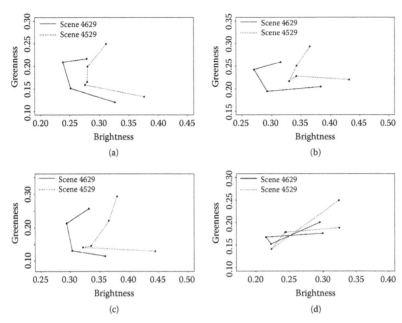

FIGURE 1.3
Observed successional trajectories for four stands identified on the ground in the H. J. Andrews Experimental Forest: (a), (b), and (c) are three successfully regenerated stands and (d) is a poorly regenerated stand. The successional trajectories were constructed from two overlapping Landsat thematic mapper scenes (4629: path = 46, row = 29; 4529: path = 45, row = 29). (Reprinted from *Remote Sensing of Environment*, 82, Song, C., Woodcock, C. E., and Li, X, The spectral/temporal manifestation of forest succession in optical imagery: The potential of multitemporal imagery, 285–302. Copyright (2002), with permission from Elsevier.)

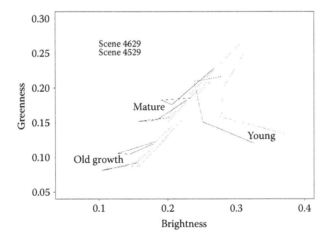

FIGURE 1.4
Spectral–temporal trajectories of forest succession from young to old-growth forests reconstructed using the substitute space for time strategy in the H. J. Andrews Experimental Forest. The successional trajectories were constructed from two overlapping Landsat thematic mapper scenes (4629: path = 46, row = 29; 4529: path = 45, row = 29). (Reprinted from *Remote Sensing of Environment*, 82, Song, C., Woodcock, C. E., and Li, X, The spectral/temporal manifestation of forest succession in optical imagery: The potential of multitemporal imagery, 285–302. Copyright (2002), with permission from Elsevier.)

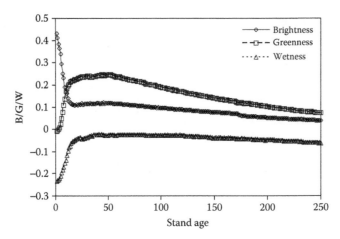

FIGURE 1.5
Modeled temporal trajectories with GORT-ZELIG for tasseled cap brightness, greenness, and wetness associated with forest succession from young to old-growth stages for a typical stand in the H. J. Andrews Experimental Forest. (Reprinted from *Remote Sensing of Environment*, 106, Song, C., Schroeder, T. A., and Cohen, W. B., Predicting temperate conifer forest successional stage distrubtions with multitemporal Landsat Thematic Mapper imagery, 228–237. Copyright (2007), with permission from Elsevier.)

places. Figure 1.4 shows the spectral–temporal trajectories for a complete forest succession sere reconstructed from a multitemporal Landsat TM image series for several stands. The spectral–temporal trajectories for a complete forest succession sere are more complicated than the modeled trajectories for young stands.

Song, Schroeder, and Cohen (2007) further improved the GORT-ZELIG simulation by introducing a two-layer canopy structure, an understory and an overstory, so that the simulation can continue to the old-growth successional stage. They also introduced leaf spectral signature changes from mature to old-growth forests. Figure 1.5 shows the non-linear spectral–temporal trajectory for a typical stand on a flat surface in the H. J. Andrews Experimental Forest. The tasseled cap brightness index decreases rapidly in the first 10–15 years and then slowly with stand age. The tasseled cap greenness and wetness indices increase relatively rapidly with stand age in the first 10–15 years and then decrease gradually with stand age.

Song, Schroeder, and Cohen (2007) used more than 1000 stands with known age classes from the U.S. Forest Service forest inventory and analysis (FIA) data in western Oregon and multiple Landsat images to validate the modeled successional trajectory (Figure 1.6). Because of the long time involved, the substitute space for time strategy was used to construct a successional trajectory for a complete forest succession sere. Each age class in the FIA plots represents a span of 10 years. Therefore, the initial rapid change in the brightness, greenness, and wetness indices as modeled in Figure 1.5 cannot be seen. However, the gradual decrease in brightness and greenness is clear from the mean values of all stands at the same age class despite tremendous variations in the spectral signature. The decrease in tasseled cap wetness is not seen when all the stands are put together. The decreasing trend became clear after the stands were separated into coastal ranges and western Cascades (Figure 1.7). Song, Schroeder, and Cohen (2007) also did some regression analysis to predict the age class of stands. They found that using spectral information from multiple Landsat images improved the prediction of stand age based on the adjusted R^2 in the analysis.

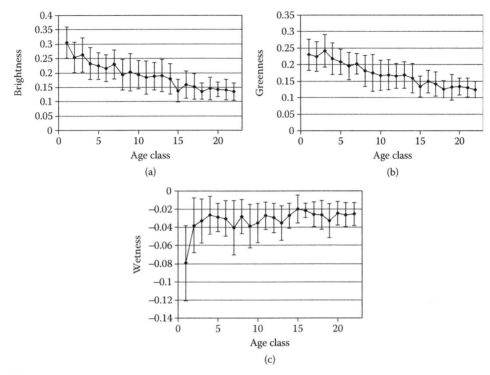

FIGURE 1.6
Observed mean successional trajectory reconstructed from a single Landsat thematic mapper imagery in western Oregon based on U.S. Forest Service forest inventory and analysis plot data. The vertical lines indicate standard deviation: (a) brightness, (b) greenness, and (c) wetness. (Reprinted from *Remote Sensing of Environment*, 106, Song, C., Schroeder, T. A., and Cohen, W. B., Predicting temperate conifer forest successional stage distributions with multitemporal Landsat Thematic Mapper imagery, 228–237. Copyright (2007), with permission from Elsevier.)

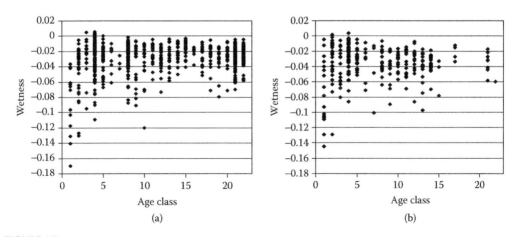

FIGURE 1.7
The observed temporal trajectory for wetness for the same forest inventory and analysis plots in Figure 1.6c, after separating the plots into geographic regions: (a) coastal ranges of Oregon and (b) western Cascades of Oregon. (Reprinted from *Remote Sensing of Environment*, 106, Song, C., Schroeder, T. A., and Cohen, W. B. Predicting temperate conifer forest successional stage distributions with multitemporal Landsat Thematic Mapper imagery, 228–237. Copyright (2007), with permission from Elsevier.)

1.3.4.4 Factors of Uncertainty

Several factors contribute to the noise in Landsat remotely sensed data for monitoring forest succession, including sensor degradation, atmospheric effects, phenology, topography, and sun–sensor geometry (Song and Woodcock 2003b). Landsat 5 sensor degradation is well known (Thome et al. 1997; Teillet et al. 2001; Chander, Markham, and Helder 2009). In the past, the data user had to sort through the literature to determine the sensor gain for a particular image. In this Internet era, the time-dependent sensor gains of Landsat 5 can be obtained online, and images are also provided. Landsat 7 ETM+ sensors were found to be stable (Teillet et al. 2001). Among the numerous uncertain factors, when and how to correct for atmospheric effects on Landsat images are the most confusing issues faced by data users, particularly relatively new data users. Song et al. (2001) evaluated the commonly used correction approaches for classification and change detection. They found that the more complicated approach for atmospheric correction did not necessarily lead to higher classification and change detection accuracies. They further evaluated such approaches for monitoring forest succession (Song and Woodcock 2003b). The effect of atmospheric correction depends on the spectral information used. For example, the tasseled cap wetness index is not sensitive to different algorithms, whereas the tasseled cap greenness index and NDVI are quite sensitive to the algorithm used.

Forests often occur in mountainous areas on Earth. Although trees always grow upright regardless of the slope of a surface, topography changes the sun–object–sensor geometry, thereby influencing the proportions of shaded and sunlit objects seen by the sensor (Schaaf, Li, and Strahler 1994). Moreover, remotely sensed images collected by Landsat sensors over different years from the same place are often affected by seasonal variations, which give rise to noise from multiple confounding factors. First, due to phenology, the amount of leaves that reflects solar radiation to the sensor varies with the season. Therefore, the same forest can have very different spectral signals in different seasons (Song and Woodcock 2003b). Second, the position of the sun can change significantly in different seasons, causing changes in the proportions of sunlit and shaded objects viewable by the sensors. Variations in local topography can further complicate the problem. The sun–object–sensor geometry effect can be modeled by biophysical models, such as the GORT-ZELIG model (Song, Woodcock and Li 2002); but the phenological effect is difficult to incorporate in these models to account for changes in canopy reflectance. Thus, modeling forest succession using multitemporal images is best done with images collected close to the anniversary date.

1.3.5 Landsat and Moderate Resolution Imaging Spectroradiometer Data Fusion

Landsat TM/ETM+ data provide enough spatial details for monitoring land-cover and land-use change. However, the 16-day revisit cycle has limited their use for studying global biophysical processes, which evolve rapidly during the growing season. In cloudy areas of the Earth, the problem is much worse; researchers are fortunate to get two to three clear images per year, and often they get none at all. In the meantime, Moderate Resolution Imaging Spectroradiometer (MODIS) sensors aboard the Terra and Aqua platforms provide daily global observations that are valuable in capturing rapid surface changes. However, spatial resolutions of 250 × 1000 m may not be good enough for heterogeneous areas. To better utilize Landsat and MODIS data, one solution is to combine the spatial resolution of Landsat with the temporal frequency of MODIS.

The Terra platform crosses the equator at about 10:30 A.M., local time, roughly 30 minutes later than Landsat 7. Their orbital parameters are identical; thus, the viewing (near-nadir) and solar geometries for the Terra platform are close to those of the corresponding Landsat acquisition. The MODIS observations include 250-m spatial resolution for red (band 1) and NIR (band 2) wavebands and 500-m spatial resolution for the other five MODIS land bands (bands 3–7). The MODIS land bands have corresponding spectral means to the Landsat ETM+ sensor except their bandwidths are narrower than ETM+. Comparisons between MODIS and Landsat surface reflectance data reveal that they are very consistent (Masek et al. 2006).

Traditional image fusion methods such as the intensity–hue–saturation (IHS) transformation, principal component substitution (PCS), and wavelet decomposition focus on producing new multispectral images that combine high-resolution panchromatic data with multispectral observations acquired simultaneously at coarser resolutions. They are useful for generating pan-sharpened images. However, they are not effective in synthesizing spatial resolution and temporal coverage when input data sources are acquired from different dates that may be affected by larger geolocation errors, larger coarse-to-fine resolution ratio, and dynamic land-surface changes.

In order to combine Landsat and MODIS data, a spatial and temporal adaptive reflectance fusion model (STARFM) was developed (Gao et al. 2006). It provides valuable information for applications that require high resolution in both time and space (Hilker et al. 2009a). This model uses a weighting function to fuse MODIS and Landsat data by introducing additional information from spectrally similar neighboring pixels. The changes of reflectance from coarse-resolution homogeneous pixels are applied to the fine-resolution image. Simulations and predictions based on actual Landsat and MODIS images show that STARFM can predict reflectance well if coarse-resolution homogeneous pixels exist in the image (Gao et al. 2006). This approach makes several reasonable assumptions. First, atmospherically corrected surface reflectance values are assumed to be comparable from time to time and location to location. Second, similar adjacent land-cover areas are assumed to have similar spectral patterns and temporal change patterns over a limited area. Third, the surface reflectance of a homogeneous land-cover type is assumed to be identical for both coarse and fine spatial resolutions.

Figure 1.8 shows the STARFM-predicted surface reflectance (e) on September 17, 2001 at Landsat spatial resolution from MODIS images of the same day (b) and ETM+/MODIS image pairs on August 12, 2001 (a and d) and September 29, 2002 (c and f). The model-predicted image captures rapid seasonal changes from MODIS data while retaining the Landsat spatial details. Clear land and water boundaries can be predicted from the additional spatial information from neighboring pixels. Linear objects such as roads are obvious in the predicted images.

The STARFM method does not explicitly handle the directional dependence of reflectance in MODIS products. It uses either MODIS surface reflectance (Vermote, El Saleous, and Justice 2002) from nadir view or MODIS nadir BRDF-adjusted reflectance (Schaaf et al. 2002) as inputs. Roy et al. (2008) considered a semiphysical fusion approach that uses the MODIS BRDF/albedo product and Landsat ETM+ data to predict ETM+ reflectance. This method assumes that the MODIS modulation term c is representative of the reflectance variation at Landsat ETM+ scale, which may not hold when reflectance change occurs in a spatially heterogeneous manner at scales larger than the 30-m Landsat pixels and smaller than the 500-m MODIS pixels (Roy et al. 2008).

The STARFM algorithm relies on the spectrally similar pixels from Landsat image for prediction. It cannot predict disturbance events if the changes caused by disturbances are

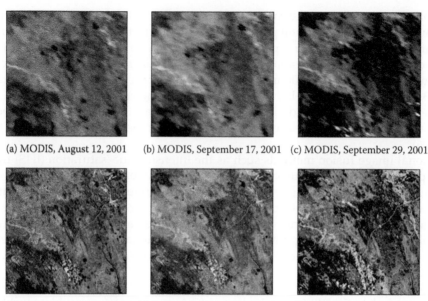

(a) MODIS, August 12, 2001   (b) MODIS, September 17, 2001   (c) MODIS, September 29, 2001

(d) ETM+, August 12, 2001   (e) Prediction, September 17, 2001   (f) ETM+, September 29, 2001

FIGURE 1.8
(See color insert following page 426.) Predicted Landsat surface reflectance (e) using STARFM from daily Moderate Resolution Imaging Spectroradiometer (MODIS) reflectance imagery (b) and Landsat/MODIS image pairs (a and d, c and f). (Reprinted from Gao, F., Masek, J., Schwaller, M., and Forrest, H., On the blending of the Landsat and MODIS surface reflectance: Predicting daily Landsat surface reflectance, *IEEE Trans Geosci Remote Sens* 44(8):2207–18. © (2006) IEEE.)

transient and not recorded in the base Landsat images. Hilker et al. (2009b) proposed a new fusion algorithm based on the STARFM algorithm called *spatial temporal adaptive algorithm for mapping reflectance change* (STAARCH). The STAARCH algorithm uses the MODIS-derived change sequence to identify the dates of the disturbance events, with which the STAARCH algorithm can choose the optimal Landsat base data and thus improve the accuracy of the synthetic Landsat images for each available date of MODIS imagery.

The STARFM algorithm can predict fine-resolution data well if homogeneous coarse-resolution pixels exist in the image (Gao et al. 2006). It is less ideal if the prediction area is complex and most coarse-resolution pixels are mixed. To solve this problem, enhanced STARFM (ESTARFM) was developed by considering conversion coefficients in the model based on the pixel unmixing theory (Zhu et al. 2010) so that homogeneous pixels and heterogeneous pixels have different conversion coefficients in the prediction. The ESTARFM algorithm also has the potential to be applied to different data sources/sensors that may not be consistent due to the differences in sensor characteristics or data processing.

1.4 Conclusions

The role of Landsat imagery in monitoring vegetation is irreplaceable. The spatial resolution of Landsat TM/ETM+ imagery is fine enough to provide the spatial details of vegetation, and coarse enough to allow a single Landsat scene to cover 185 × 175 km, meeting most regional application needs. The design of the spectral bands best captures the reflectance characteristics of vegetation. In addition, nearly four decade's worth of image archive has been continuously recorded, which is the longest data record among all remote-sensing programs, and the temporal information has proven invaluable for monitoring vegetation conditions. Landsat data have been successfully used for monitoring vegetation area changes by land-use/land-cover classification and change detection, SMA, and extracting biophysical parameters, such as LAI, biomass, and forest successional stages. Two types of approaches were used in the literature: empirical approaches and biophysical models. Although classification and change detection with Landsat imagery for areal changes in vegetation have been well established in the literature, the extraction of biophysical parameters, particularly LAI, biomass, and forest successional stages, remains a challenging task, primarily because of signal saturation. The best use of Landsat data in the future requires synergistic use of data from different sensors, including optical sensors at higher spatial resolution that provide texture information or coarse spatial resolutions that provide temporal information, and lidar/radar sensors, which provide complimentary vegetation information unavailable from Landsat. Unquestionably, LCDM will greatly enhance the value of the Landsat data series for scientific investigations, which will be unrivaled by any other sensor. The Landsat series data will continue to play a pivotal role in enhancing our understanding of vegetation spatial patterns as well as the ecological functions of the vegetation in the future.

References

Anderson, G. L., J. D. Hanson, and R. H. Haas. 1993. Evaluating Landsat thematic mapper derived vegetation indexes for estimating above-ground biomass on semiarid rangelands. *Remote Sens Environ* 45(2):165–75.

Berterretche, M., A. T. Hudak, W. B. Cohen, T. K. Maiersperger, S. T. Gower, and J. Dungan. 2005. Comparison of regression and geostatistical methods for mapping Leaf Area Index (LAI) with Landsat ETM+ data over a boreal forest. *Remote Sens Environ* 96(1):49–61.

Botkin, D. B., J. F. Janak, and J. R. Wallis. 1972. Rationale, limitations, and assumptions of a northeastern forest growth simulator. *IBM J Res Dev* 16(2):101–16.

Breda, N. J. J. 2003. Ground-based measurements of leaf area index: A review of methods, instruments and current controversies. *J Exp Bot* 54(392):2403–17.

Brown, L., J. M. Chen, S. G. Leblanc, and J. Cihlar. 2000. A shortwave infrared modification to the simple ratio for LAI retrieval in boreal forests: An image and model analysis. *Remote Sens Environ* 71(1):16–25.

Chander, G., B. L. Markham, and D. L. Helder. 2009. Summary of current radiometric calibration coefficients for Landsat MSS, TM, ETM+, and EO-1 ALI sensors. *Remote Sens Environ* 113(5):893–903.

Chen, J. M., and J. Cihlar. 1996. Retrieving leaf area index of boreal conifer forests using Landsat TM images. *Remote Sens Environ* 55(2):153–62.

Chen, J. M., G. Pavlic, L. Brown, J. Cihlar, S. G. Leblanc, H. P. White, R. J. Hall et al. 2002. Derivation and validation of Canada-wide coarse-resolution leaf area index maps using high-resolution satellite imagery and ground measurements. *Remote Sens Environ* 80(1):165–84.

Clements, F. E. 1916. *Plant Succession: An Analysis of the Development of Vegetation.* Washington, DC: Carnegie Institute Publication 242.

Cohen, W. B., and S. N. Goward. 2004. Landsat's role in ecological applications of remote sensing. *Bioscience* 54(6):535–45.

Cohen, W. B., M. Fiorella, J. Gray, E. Helmer, and K. Anderson. 1998. An efficient and accurate method for mapping forest clearcuts in the Pacific Northwest using Landsat imagery. *Photogramm Eng Remote Sensing* 64(4):293–300.

Cohen, W. B., T. K. Maiersperger, S. T. Gower, and D. P. Turner. 2003. An improved strategy for regression of biophysical variables and Landsat ETM+ data. *Remote Sens Environ* 84(4):561–71.

Cohen, W. B., T. A. Spies, and G. A. Bradshaw. 1990. Semivariograms of digital imagery for analysis of conifer canopy structure. *Remote Sens Environ* 34(3):167–78.

Cohen, W. B., T. A. Spies, and M. Fiorella. 1995. Estimating the age and structure of forests in a multi-ownership landscape of western Oregon, USA. *Int J Remote Sens* 16(4):721–46.

Collins, J. B., and C. E. Woodcock. 1996. An assessment of several linear change detection techniques for mapping forest mortality using multitemporal Landsat TM data. *Remote Sens Environ* 56(1):66–77.

Colombo, R., D. Bellingeri, D. Faolini, and C. M. Marino. 2003. Retrieval of leaf area index in different vegetation types using high resolution satellite data. *Remote Sens Environ* 86(1):120–31.

Crist, E. P., and R. C. Cicone. 1984. A physically-based transformation of Thematic Mapper data: The TM Tasseled Cap. *IEEE Trans Geosci Remote Sens* GE-22(3):256–63.

Curran, P. J., J. L. Dungan, and H. L. Gholz. 1992. Seasonal LAI in slash pine estimated with Landsat TM. *Remote Sens Environ* 39(1):3–13.

Dixon, R. K., S. Brown, R. A. Houghton, A. M. Solomon, M. C. Trexler, and J. Wisniewski. 1994. Carbon pools and flux of global forest ecosystems. *Science* 263(5144):185–90.

Dobson, A. P., A. D. Bradshaw, and A. J. M. Baker. 1997. Hopes for the future: Restoration ecology and conservation biology. *Science* 277(5325):515–22.

Dobson, M. C., F. T. Ulaby, L. E. Pierce, T. L. Sharik, K. M. Bergen, J. Kellndorfer, J. R. Kendra et al. 1995. Estimation of forest biophysical characteristics in northern Michigan with SIR-C/X-SAR. *IEEE Trans Geosci Remote Sens* 33(4):877–95.

Eklundh, L., L. Harrie, and A. Kuusk. 2001. Investigating relationships between Landsat ETM plus sensor data and leaf area index in a boreal conifer forest. *Remote Sens Environ* 78(3):239–51.

Elmore, A. J., J. F. Mustard, S. J. Manning, and D. B. Lobell. 2000. Quantifying vegetation change in semiarid environments: Precision and accuracy of spectral mixture analysis and the normalized difference vegetation index. *Remote Sens Environ* 73(1):87–102.

Fassnacht, K. S., S. T. Gower, M. D. MacKenzie, E. V. Mordheim, and T. M. Lillesand. 1997. Estimating the leaf area index of north central Wisconsin forests using the Landsat Thematic Mapper. *Remote Sens Environ* 61(2):229–45.

Fazakas, Z., M. Nilsson, and H. Olsson. 1999. Regional forest biomass and wood volume estimation using satellite data and ancillary data. *Agric For Meteorol* 98–9:417–25.

Fiorella, M., and W. J. Ripple. 1993a. Analysis of conifer forest regeneration using Landsat Thematic Mapper data. *Photogr Eng Remote Sens* 59(9):1383–8.

Fiorella, M., and W. J. Ripple. 1993b. Determining successional stage of temperate coniferous forests with Landsat satellite data. *Photogr Eng Remote Sens* 59(2):239–46.

Foody, G. M. 2003. Remote sensing of tropical forest environments: Towards the monitoring of environmental resources for sustainable development. *Int J Remote Sens* 24(20):4035–46.

Foody, G. M., G. Palubinskas, R. M. Lucas, P. J. Curran, and M. Honzak. 1996. Identifying terrestrial carbon sinks: Classification of successional stages in regenerating tropical forest from Landsat TM data. *Remote Sens Environ* 55(3):205–16.

Franklin, J., and A. H. Strahler. 1988. Invertible canopy reflectance modeling of vegetation structure in semi-arid woodland. *IEEE Trans Geosci Remote Sens* 26(6):809–25.

Fung, T., and E. Ledrew. 1987. Application of principal components-analysis to change detection. *Photogramm Eng Remote Sensing* 53(12):1649–58.

Gao, F., J. Masek, M. Schwaller, and H. Forrest. 2006. On the blending of the Landsat and MODIS surface reflectance: Predicting daily Landsat surface reflectance. *IEEE Trans Geosci Remote Sens* 44(8):2207–18.

Gholz, H. L., C. C. Grier, A. G. Campbell, and A. T. Brown. 1979. *Equations for Estimating Biomass and Leaf Area of Plants in the Pacific Northwest.* Forest Research Laboratory, School of Forestry, Oregon State University. Corvallis, Oregon.

Gobron, N., B. Pinty, and M. M. Verstraete. 1997. Theoretical limits to the estimation of the leaf area index on the basis of visible and near-infrared remote sensing data. *IEEE Trans Geosci Remote Sens* 35(6):1438–45.

Gower, S. T., J. K. Kucharik, and J. M. Norman. 1999. Direct and indirect estimation of leaf area index, f(APAR), and net primary production of terrestrial ecosystems. *Remote Sens Environ* 70(1):29–51.

Grier, C. C., and R. S. Logan. 1977. Old-growth Pseudotsuga-Menziesii of a western Oregon watershed: Biomass distribution and production budgets. *Ecol Monogr* 47(4):373–400.

Hall, F. G., D. B. Botkin, D. E. Strebel, K. D. Woods, and S. J. Goetz. 1991. Large-scale patterns of forest succession as determined by remote sensing. *Ecology* 72(2):628–40.

Hall, F. G., Y. Shimabukuro, and K. F. Huemmrich. 1995. Remote sensing of forest biophysical structure using mixture decomposition and geometric reflectance models. *Ecol Appl* 5(4):993–1013.

Helmer, E. H., S. Brown, and W. B. Cohen. 2000. Mapping montane tropical forest successional stage and land use with multi-date Landsat imagery. *Int J Remote Sens* 21(11):2163–83.

Hilker, T., M. A. Wulder, N. C. Coops, J. Linke, G. McDermid, J. G. Masek, F. Gao, and J. C. White. 2009b. A new data fusion model for high spatial- and temporal-resolution mapping of forest disturbance based on Landsat and MODIS. *Remote Sens Environ* 113(8):1613–27.

Hilker, T., M. A. Wulder, N. C. Coops, N. Seitz, J. C. White, F. Gao, J. G. Masek, and G. Stenhouse. 2009a. Generation of dense time series synthetic Landsat data through data blending with MODIS using a spatial and temporal adaptive reflectance fusion model. *Remote Sens Environ* 113(9):1988–99.

Horler, D. N. H., and F. J. Ahern. 1986. Forestry information content of Thematic Mapper data. *Int J Remote Sens* 7(3):405–28.

Huang, C. Q., S. N. Goward, J. G. Masek, F. Gao, E. F. Vermote, N. Thomas, K. Schleeweis et al. 2009. Development of time series stacks of Landsat images for reconstructing forest disturbance history. *Int J Digit Earth* 2(3):195–218.

Jakubauskas, M. E. 1996. Thematic mapper characterization of lodgepole pine seral stages in Yellowstone National Park, USA. *Remote Sens Environ* 56(2):118–32.

Jenkins, J. C., D. C. Chojnacky, L. S. Heath, and R. A. Birdsey. 2003. National-scale biomass estimators for United States tree species. *For Sci* 49(1):12–35.

Jonckheere, I., S. Fleck, K. Nackaerts, B. Muys, P. Coppin, M. Weiss, and F. Baret. 2004. Review of methods for in situ leaf area index determination Part I: Theories, sensors and hemispherical photography. *Agric For Meteorol* 121(1–2):19–35.

Kauth, R. J., and G. S. Thomas. 1976. The Tasseled Cap: A graphic description of the spectral-temporal development of agricultural crops as seen by Landsat. *Proceedings of the Symposium on Machine Processing of Remotely Sensed Data,* pp. 4B41–51. West Lafayette, IN: Purdue University.

Kennedy, R. E., W. B. Cohen, and T. A. Schroeder. 2007. Trajectory-based change detection for automated characterization of forest disturbance dynamics. *Remote Sens Environ* 110(3):370–86.

Kimes, D. S., B. N. Holben, J. E. Nickeson, and W. A. McKee. 1996. Extracting forest age in a Pacific Northwest forest from Thematic Mapper and topographic data. *Remote Sens Environ* 56(2):133–40.

Kucharik, C. J., J. M. Norman, and S. T. Gower. 1999. Characterization of radiation regimes in nonrandom forest canopies: Theory, measurements, and a simplified modeling approach. *Tree Physiol* 19(11):695–706.

Lamberson, R. H., R. Mckelvey, B. R. Noon, and C. Voss. 1992. A dynamic analysis of northern spotted owl viability in a fragmented forest landscape. *Conserv Biol* 6(4):505–12.

Lefsky, M. A., W. B. Cohen, S. A. Acker, G. G. Parker, T. A. Spies, and D. Harding. 1999. Lidar remote sensing of the canopy structure and biophysical properties of Douglas fir western hemlock forests. *Remote Sens Environ* 70(3):339–61.

Li, X., and A. H. Strahler. 1985. Geometric-optical modeling of a conifer forest canopy. *IEEE Trans Geosci Remote Sens* GE-23:705–21.

Li, X., A. H. Strahler, and C. E. Woodcock. 1995. A hybrid geometric optical-radiative transfer approach for modeling albedo and directional reflectance of discontinuous canopies. *IEEE Trans Geosci Remote Sens* 33(2):466–80.

Lu, D. 2005. Aboveground biomass estimation using Landsat TM data in the Brazilian Amazon. *Int J Remote Sens* 26(12):2509–25.

Lucas, R. M., M. Honzak, I. D. Amaral, P. J. Curran, and G. M. Foody. 2002. Forest regeneration on abandoned clearances in central Amazonia. *Int J Remote Sens* 23(5):965–88.

Madhavan, B. B., S. Kubo, N. Kurisaki, and T. V. L. N. Sivakumar. 2001. Appraising the anatomy and spatial growth of the Bangkok metropolitan area using a vegetation-impervious-soil model through remote sensing. *Int J Remote Sens* 22(5):789–806.

Marshall, J. D., and R. H. Waring. 1986. Comparison of methods of estimating leaf-area index in old-growth Douglas-fir. *Ecology* 67(4):975–9.

Masek, J. G., E. F. Vermote, N. E. Saleous, R. Wolfe, F. G. Hall, F. Huemmrich, F. Gao, J. Kutler, and T. K. Lim. 2006. A Landsat surface reflectance data set for North America, 1990–2000. *IEEE Geosci Remote Sens Lett* 3(1):69–72.

McCarthy, H. R., R. Oren, A. C. Finzi, D. S. Ellsworth, H. S. Kim, K. H. Johnsen, and B. Millar. 2007. Temporal dynamics and spatial variability in the enhancement of canopy leaf area under elevated atmospheric CO_2. *Glob Chang Biol* 13(12):2479–97.

Monteith, J. L., and M. H. Unsworth. 1973. *Principles of Environmental Physics*. 2nd ed. London: Edward Arnold.

Morsdorf, F., B. Kotz, E. Meier, K. I. Itten, and B. Allgower. 2006. Estimation of LAI and fractional cover from small footprint airborne laser scanning data based on gap fraction. *Remote Sens Environ* 104(1):50–61.

Myers, N., R. A. Mittermeier, C. G. Mittermeier, G. A. B. da Fonseca, and J. Kent. 2000. Biodiversity hotspots for conservation priorities. *Nature* 403(6772):853–8.

Myneni, R. B., R. R. Nemani, and S. W. Running. 1997. Estimation of global leaf area index and absorbed par using radiative transfer models. *IEEE Trans Geosci Remote Sens* 35(6):1380–93.

Nemani, R., L. Pierce, S. Running, and L. Band. 1993. Forest ecosystem processes at the watershed scale: Sensitivity to remotely-sensed Leaf Area Index estimates. *Int J Remote Sens* 14(13):2519–34.

Ni, W., X. Li, C. E. Woodcock, M. R. Caetano, and A. H. Strahler. 1999. An analytical hybrid GORT model for bidirectional reflectance over discontinuous plant canopies. *IEEE Trans Geosci Remote Sens* 37(2):987–99.

Noble, I. R., and R. Dirzo. 1997. Forests as human-dominated ecosystems. *Science* 277(5325):522–5.

Odum, E. P. 1953. *Fundamentals of Ecology*. Philadelphia, PA: W. B. Saunders Company.

Peddle, D. R., F. G. Hall, and E. F. LeDrew. 1999. Spectral mixture analysis and geometric-optical reflectance modeling of boreal forest biophysical structure. *Remote Sens Environ* 67(3):288–97.

Peterson, D. L., W. E. Westman, and N. J. Stephenson. 1986. Analysis of forest structure using thematic mapper simulator data. *IEEE Trans Geosci Remote Sens* 24(1):113–21.

Phinn, S., M. Stanford, P. Scarth, A. T. Murray, and P. T. Shyy. 2002. Monitoring the composition of urban environments based on the vegetation-impervious surface-soil (VIS) model by subpixel analysis techniques. *Int J Remote Sens* 23(20):4131–53.

Pregitzer, K. S., and E. S. Euskirchen. 2004. Carbon cycling and storage in world forests: Biome patterns related to forest age. *Glob Chang Biol* 10(12):2052–77.

Riano, D., F. Valladares, S. Condes, and E. Chuvieco. 2004. Estimating of leaf area index and covered ground from airborne laser scanner (lidar) in two contrasting forests. *Agric For Meteorol* 124(3–4):269–75.

Richards, J. A. 1984. Thematic mapping from multitemporal image data using the principal components transformation. *Remote Sens Environ* 16(1):35–46.

Ridd, M. K. 1995. Exploring a V-I-S (vegetation-impervious surface-soil) model for urban ecosystem analysis through remote sensing: Comparative anatomy for cities. *Int J Remote Sens* 16(12):2165–85.

Roberts, S. D., T. J. Dean, D. L. Evans, J. W. McCombs, R. L. Harrington, and P. A. Glass. 2005. Estimating individual tree leaf area in loblolly pine plantations using LiDAR-derived measurements of height and crown dimensions. *For Ecol Manage* 213(1–3):54–70.

Roberts, D. A., M. Gardner, R. Church, S. Ustin, G. Scheer, and R. O. Green. 1998. Mapping Chaparral in the Santa Monica Mountains using multiple endmember spectral mixture models. *Remote Sens Environ* 65(3):267–79.

Roy, D. P., J. Ju, P. Lewis, C. Schaaf, F. Gao, M. Hansen, and E. Lindquist. 2008. Multitemporal MODIS-Landsat data fusion for relative radiometric normalization, gap filling, and prediction of Landsat data. *Remote Sens Environ* 112(6):3112–30.

Roy, P. S., and S. A. Ravan. 1996. Biomass estimation using satellite remote sensing data: An investigation on possible approaches for natural forest. *J Biosci* 21(4):535–61.

Running, S. W., and E. R. Hunt Jr. 1993. Generalization of forest ecosystem process model for other biomes, BIOME-BGC, and an application for global-scale models. In *Scaling Physiological Processes: Leaf to Globe*, ed. J. R. Ehleringer and C. B. Field. San Diego, CA: Academic Press.

Running, S. W., D. L. Peterson, M. A. Spanner, and K. B. Teuber. 1986. Remote sensing of coniferous forest leaf area. *Ecology* 67(1):273–6.

Sabol, D. E., J. B. Adams Jr., and M. O. Smith. 1992. Quantitative subpixel spectral detection of targets in multispectral images. *J Geophys Res* 97(E2):2659–72.

Sader, S. A., R. B. Waide, W. T. Lawrence, and A. T. Joyce. 1989. Tropical forest biomass and successional age class relationships to a vegetation index derived from Landsat TM data. *Remote Sens Environ* 28:143–56.

Schaaf, C. B., F. Gao, A. H. Strahler, W. Lucht, X. Li, T. Tsang, N. C. Strugnell et al. 2002. First operational BRDF, albedo and nadir reflectance products from MODIS. *Remote Sens Environ* 83(1–2):135–48.

Schaaf, C. B., X. Li, and A. H. Strahler. 1994. Topographic effects on bidirectional and hemispherical reflectance calculated with a geometric-optical canopy model. *IEEE Trans Geosci Remote Sens* 32(6):1186–93.

Seto, K. C., C. E. Woodcock, C. Song, X. Huang, J. Lu, and R. K. Kaufmann. 2002. Measuring land-use change with Landsat TM: Evidence from Pearl River Delta. *Int J Remote Sens* 23(10):1985–2004.

Shinozaki, K., K. Yoda, K. Hozumi, and T. Kira. 1964. A quantitative theory of plant form–the pipe model theory. I. Basis analysis. *Jpn J Ecol* 14:97–105.

Shugart, H. H., and D. C. West. 1977. Development of an Appalachian deciduous forest succession model and its application to assessment of the impact of the chestnut blight. *J Environ Manage* 5(2):161–79.

Skole, D., and C. Tucker. 1993. Tropical deforestation and habitat fragmentation in the Amazon: Satellite data from 1978 to 1988. *Science* 260(5116):1905–10.

Small, C. 2001. Estimation of urban vegetation abundance by spectral mixture analysis. *Int J Remote Sens* 22(7):1305–34.

Smith, J. E., L. S. Heath, and J. C. Jenkins. 2003. Forest volume-to-biomass models and estimates of mass for live and standing dead trees of US forests. General Technical Report NE-298, USDA Forest Service, Northeastern Research Station. Newtown Square, PA.

Smith, M. O., S. L. Ustin, J. B. Adams, and A. R. Gillespie. 1990. Vegetation in deserts: I. A regional measure of abundance from multispectral images. *Remote Sens Environ* 31(1):1–26.

Song, C. 2005. Spectral mixture analysis for subpixel vegetation fractions in the urban environment: How to incorporate endmember variability? *Remote Sens Environ* 95(2):248–63.

Song, C., and M. B. Dickinson. 2008. Extracting forest canopy structure from spatial information of high resolution optical imagery: Tree crown size and leaf area index. *Int J Remote Sens* 29(19):5605–22.

Song, C., and C. E. Woodcock. 2002. The spatial manifestation of forest succession in optical imagery: The potential of multiresolution imagery. *Remote Sens Environ* 82(2–3):271–84.

Song, C., and C. E. Woodcock. 2003a. A regional forest ecosystem carbon budget model: Impacts of forest age structure and landuse history. *Ecol Modell* 164(1):33–47.

Song, C., and C. E. Woodcock. 2003b. Monitoring forest succession with multitemporal Landsat images: Factors of uncertainty. *IEEE Trans Geosci Remote Sens* 41(11):2557–67.

Song, C., C. E. Woodcock, and X. Li. 2002. The spectral/temporal manifestation of forest succession in optical imagery: The potential of multitemporal imagery. *Remote Sens Environ* 82(2–3):285–302.

Song, C., C. E. Woodcock, K. C. Seto, M. Pax-Lenney, and S. A. Macomber. 2001. Classification and change detection using Landsat TM data: When and how to correction for atmospheric effects? *Remote Sens Environ* 75:230–44.

Song, C., T. A. Schroeder, and W. B. Cohen. 2007. Predicting temperate conifer forest successional stage distributions with multitemporal Landsat Thematic Mapper imagery. *Remote Sens Environ* 106(2):228–37.

Soudani, K., C. Francois, G. L. LeMaire, V. LeDantec, and E. Dufrene. 2006. Comparative analysis of IKONOS, SPOT, and ETM+ data for leaf area index estimation in temperate coniferous and deciduous forest stands. *Remote Sens Environ* 102(1–2):161–75.

Spanner, M., L. Johnson, J. Miller, R. McCreight, J. Freemantle, J. Runyon, and P. Gong. 1994. Remote sensing of seasonal leaf area index across the Oregon transect. *Ecol Appl* 4(2):258–71.

Steininger, M. K. 2000. Satellite estimation of tropical secondary forest above-ground biomass: Data from Brazil and Bolivia. *Int J Remote Sens* 21(6–7):1139–57.

Suits, G. H. 1972. The calculation of the directional reflectance of a vegetative canopy. *Remote Sens Environ* 2:117–25.

Teillet, P. M., J. L. Barker, B. L. Markham, R. R. Irish, G. Fedosejevs, and J. C. Storey. 2001. Radiometric cross-calibration of the Landsat-7 ETM+ and Landsat-5 TM sensors based on tandem data sets. *Remote Sens Environ* 78(1–2):39–54.

Ter-Mikaelian, M. T., and M. D. Korzukhin. 1997. Biomass equations for sixty-five North America tree species. *For Ecol Manage* 97(1):1–24.

Theseira, M. A., G. Thomas, J. C. Taylor, F. Gemmell, and J. Varjo. 2003. Sensitivity of mixture modeling to end-member selection. *Int J Remote Sens* 24(7):1559–75.

Thome, K., B. Markham, J. Barker, P. Slater, and S. Biggar. 1997. Radiometric calibration of Landsat. *Photogramm Eng Remote Sensing* 63(7):853–8.

Tompkins, S., J. F. Mustard, C. M. Pieters, and D. W. Forsyth. 1997. Optimization of endmembers for spectral mixture analysis. *Remote Sens Environ* 59(3):472–89.

Tomppo, E., M. Nilsson, M. Rosengren, P. Aalto, and P. Kennedy. 2002. Simultaneous use of Landsat-TM and IRS-1C WiFS data in estimating large area tree stem volume and aboveground biomass. *Remote Sens Environ* 82(1):156–71.

Trotter, C. M., J. R. Dymond, and C. J. Goulding. 1997. Estimation of timber volume in a coniferous plantation forest using Landsat TM. *Int J Remote Sens* 18(10):2209–23.

Turner, D. P., W. B. Cohen, R. E. Kennedy, K. S. Fassnacht, and J. M. Briggs. 1999. Relationships between leaf area index and Landsat TM spectral vegetation indices across three temperate zone sites. *Remote Sens Environ* 70(1):52–68.

Urban, D. L. 1990. *A Versatile Model to Simulate Forest Pattern: A User's Guide to ZELIG Version 1.0.* Charlottesville, VA: University of Virginia.

Verhoef, W. 1984. Light scattering by leaf layers with application to canopy reflectance modeling: The SAIL model. *Remote Sens Environ* 16(2):125–41.

Vermote, E. F., N. El Saleous, and C. Justice. 2002. Atmospheric correction of MODIS data in the visible to middle infrared: First results. *Remote Sens Environ* 83(1–2):97–111.

Weiss, M., F. Baret, G. J. Smith, I. Jonckheere, and P. Coppin. 2004. Review of methods for in situ leaf area index (LAI) determination. Part II. Estimation of LAI, errors and sampling. *Agric For Meteorol* 121(1–2):37–53.

Whittaker, R. H., F. H. Bormann, G. E. Likens, and T. G. Siccama. 1974. Hubbard Brook Ecosystem Study: Forest biomass and production. *Ecol Monogr* 44(2):233–54.

Williams, D. L., S. Goward, and T. Arvidson. 2006. Landsat: Yesterday, today, and tomorrow. *Photogramm Eng Remote Sensing* 72(10):1171–8.

Woodcock, C. E., J. B. Collins, S. Gopal, V. D. Jakabhazy, X. Li, S. A. Macomber, S. Ryherd et al. 1994. Mapping forest vegetation using Landsat TM imagery and a canopy reflectance model. *Remote Sens Environ* 50(3):240–54.

Woodcock, C. E., J. B. Collins, V. D. Jakabhazy, X. Li, S. A. Macomber, and Y. Wu. 1997. Inversion of the Li-Strahler canopy reflectance model for mapping forest structure. *IEEE Trans Geosci Remote Sens* 35(2):405–14.

Woodcock, C. E., S. A. Macomber, M. Pax-Lenney, and W. B. Cohen. 2001. Monitoring large areas for forest change using Landsat: Generalization across space, time and Landsat sensors. *Remote Sens Environ* 78(1–2):194–203.

Wu, C., and A. T. Murray. 2003. Estimating impervious surface distribution by spectral mixture analysis. *Remote Sens Environ* 84(4):493–505.

Wu, Y., and A. H. Strahler. 1994. Remote estimation of crown size, stand density and foliage biomass on the Oregon transect. *Ecol Appl* 4(2):299–312.

Zheng, D., J. Rademacher, J. Chen, T. Crow, M. Bresee, J. Le Moine, and S. Ryu. 2004. Estimating aboveground biomass using Landsat 7 ETM+ data across a managed landscape in northern Wisconsin, USA. *Remote Sens Environ* 93(3):402–11.

Zhu, X., J. Chen, F. Gao, and J. Masek. 2010. An enhanced spatial and temporal adaptive reflectance fusion model for complex heterogeneous regions. *Remote Sens Environ* 114(11):2610–2623.

2

Review of Selected Moderate-Resolution Imaging Spectroradiometer Algorithms, Data Products, and Applications

Yang Shao, Gregory N. Taff, and Ross S. Lunetta

CONTENTS

2.1 Introduction

The Moderate-Resolution Imaging Spectroradiometer (MODIS) is one of the key instruments designed as part of the National Aeronautics and Space Administration (NASA)'s Earth-Observing System (EOS) to provide long-term global observation of the Earth's land, ocean, and atmospheric properties (Asrar and Dokken 1993). The instrument was developed based on experiences with the Advanced Very High Resolution Radiometer (AVHRR) and the Landsat Thematic Mapper (TM). MODIS was designed not only for providing continuous global observations but also as a new-generation sensor with an increased combination of spectral, spatial, radiometric, and temporal resolutions. In addition to emphasizing the advances in the sensor instrument, the MODIS mission also emphasizes the development of operational data-processing algorithms to generate global remote-sensing spectral data sets and a variety of value-added products spanning both the optical and biophysical domains. The motivation is to provide MODIS standard products to the general scientific community to support both theoretical and practical applications. Two MODIS instruments were initially scheduled for launch on the EOS-AM and EOS-PM platforms in June 1998 and December 2000, respectively (Running et al. 1994). The actual launch dates were December 18, 1999 (EOS-Terra) and May 4, 2002 (EOS-Aqua). Terra MODIS data have been available since February 2000. Subsequently, numerous scientific papers have been published on MODIS data, algorithms, validation, and applications.

This chapter provides a review of selected MODIS data products and algorithms. We review a large number of MODIS algorithm theoretical basis documents (ATBDs) developed by individual MODIS science teams and scientific papers published over the last 10–15 years. Our main interest is to review MODIS algorithms in order to increase understanding of the standard data products, document advances and limitations, and identify data quality and validation issues. The general organization of this chapter is as follows: First, we briefly describe the MODIS sensor characteristics. Then, we review selected MODIS data products and algorithms for land, atmosphere, and ocean disciplines. Our focus is on the MODIS land product because of its relatively wider use among the three. Finally, we review a wide range of applications and research activities that emphasize the broad range of MODIS products.

2.1.1 Moderate-Resolution Imaging Spectroradiometer Sensor Characteristics

Both EOS-Terra and EOS-Aqua are polar-orbiting sun-synchronous platforms. The orbit height of EOS platforms is 705 km at the equator. Terra's equatorial crossing time (descending) is 10:30 A.M. local time, approximately 30 minutes later than the Landsat 7 satellite. Aqua crosses (ascends) the equator at approximately 1:30 P.M. Each MODIS instrument has a two-sided scan mirror that operates perpendicular to the spacecraft track. The mirror scanning extends 55° at either sides of the nadir, providing a nominal swath of 2330 km. The wide swath allows nearly global coverage to be obtained by each instrument every 1–2 days.

In addition to high temporal resolution, the MODIS sensor has high spectral, spatial, and radiometric resolutions compared to previous sensor systems, such as the AVHRR. A total of 36 spectral bands are carefully positioned across the 0.412–14.235 µm spectral region. Among the 36 spectral bands, the first two bands are located in the red (0.648 µm) and near-infrared (NIR; 0.858 µm) regions with a spatial resolution of 250 m. There are five additional bands (bands 3–7: 0.470 µm, 0.555 µm, 1.240 µm, 1.640 µm, and 2.13 µm) with a spatial resolution of 500 m located in the visible to shortwave infrared (SWIR) spectral regions. The remaining 29 spectral bands (bands 8–36) have 1000-m spatial resolution, and are located in the middle and long-wave thermal infrared (TIR) regions. The MODIS instrument also has a 12-bit radiometric resolution and an advanced onboard calibration subsystem that ensures high calibration accuracy (Guenther et al. 1998; Justice et al. 1998). The sensor characteristics are considered to be substantially improved over other similar observation systems (Townshend and Justice 2002). Unlike the AVHRR (mainly designed for monitoring the atmosphere), the MODIS sensor, is well suited for a wide range of research applications intended to improve the understanding of land, ocean, and atmosphere processes, domain interactions, and the impacts of human activity on the global environment. Table 2.1 shows

TABLE 2.1

MODIS Technical Specifications Including Primary Use, Band Numbers, Bandwidths, Spectral Radiance, Spatial Resolutions, and SNR

Primary Use	Band	Bandwidth (μm)	Spectral Radiance (W/m²·μm·sr)	SNR	Spatial Resolution at Nadir (m)
Land/cloud boundaries	1	0.620–0.670	21.8	128	250
	2	0.841–0.876	24.7	201	
Land/cloud properties	3	0.459–0.479	35.3	243	500
	4	0.545–0.565	29.0	228	
	5	1.230–1.250	5.4	74	
	6	1.628–1.652	7.3	275	
	7	2.105–2.155	1.0	110	
Ocean color/phytoplankton/ biogeochemistry	8	0.405–0.420	44.9	880	1000
	9	0.438–0.448	41.9	838	
	10	0.483–0.493	32.1	802	
	11	0.526–0.536	27.9	754	
	12	0.546–0.556	21.0	750	
	13	0.662–0.672	9.5	910	
	14	0.673–0.683	8.7	1087	
	15	0.743–0.753	10.2	586	
	16	0.862–0.877	6.2	516	
Atmospheric water vapor	17	0.890–0.920	10.0	167	1000
	18	0.931–0.941	3.6	57	
	19	0.915–0.965	15.0	250	

Primary Use	Band	Bandwidth (μm)	Spectral Radiance (W/m²·μm·sr)	Required NEΔT(K)[a]	Spatial Resolution at Nadir (m)
Surface/cloud temperature	20	3.660–3.840	0.45	0.05	1000
	21	3.929–3.989	2.38	2	
	22	3.929–3.989	0.67	0.07	
	23	4.020–4.080	0.79	0.07	
Atmospheric temperature	24	4.433–4.598	0.17	0.25	1000
	25	4.482–4.549	0.59	0.25	
Cirrus clouds	26	1.360–1.390	6.00	150[b]	1000 m
Water vapor	27	6.535–6.895	1.16	0.25	1000
	28	7.175–7.475	2.18	0.25	
	29	8.400–8.700	9.58	0.05	
Ozone	30	9.580–9.880	3.69	0.25	1000
Surface/cloud temperature	31	10.780–11.280	9.55	0.05	1000
	32	11.770–12.270	8.94	0.05	
Cloud top altitude	33	13.185–13.485	4.52	0.25	1000
	34	13.485–13.785	3.76	0.25	
	35	13.785–14.085	3.11	0.25	
	36	14.085–14.385	2.08	0.35	

[a] NEΔT(K) = noise-equivalent temperature difference.
[b] SNR.

MODIS technical specifications including primary use, band numbers, bandwidths, spectral radiance, spatial resolutions, and signal-to-noise ratio (SNR).

2.1.2 Data Products and Algorithms

The MODIS instrument calibration, algorithm development, and standard data products are provided by the MODIS science team. The science team consists of over 70 American and international scientists, divided into four discipline groups for calibration, land, atmosphere, and ocean. Each discipline group has clearly defined scientific responsibilities, and close interactions between the groups are maintained throughout algorithm development, data processing, evaluation, and product distribution.

MODIS data products are broadly categorized into five levels from level 0 to level 4. The MODIS level-0 data set is the initial data set automatically converted from the instrumental raw format. The level-0 data are subsequently split into granules, and an Earth location algorithm is employed to add geodetic position information to each MODIS granule. This creates the MODIS level-1A product that contains geodetic information, such as latitude, longitude, height, satellite zenith/azimuth angle, and solar zenith/azimuth angle (Nishihama et al. 1997). Level-1A data are further processed to generate level-1B product (calibrated radiance for all bands and surface reflectance values for selected bands). Additional information such as data quality flags and error estimates are also provided. The MODIS level-1B data are still considered to be instrument data. The data are used primarily as input to derive higher-order MODIS geophysical products (levels 2–4). For example, MODIS level 2G is a gridded product that stores level-2 data in an Earth-based uniform grid system. Level-3 data provide an estimation of optical or biophysical variables for each grid element for predefined spatial and temporal resolutions (e.g., daily, eight-day, and monthly). Algorithms for level-3 products often include spatial resampling, averaging, and temporal composition. Finally, level-4 data are generated through a variety of algorithms, models, and statistical methods. Generally, additional ancillary data are required to generate level-4 data (e.g., MODIS net primary production [NPP] product).

MODIS data products are also labeled by collection version. Each collection version indicates a complete set of MODIS files corresponding to a specific data updating or reprocessing stage. At the time of preparation of this chapter, the MODIS science team had completed the processing of collection-5 data. The MODIS team anticipates that another round of data processing will be conducted in 2010, subject to the availability of new MODIS algorithms. The distribution of MODIS land, atmosphere, and ocean data is primarily supported by three data centers: the Goddard Space Flight Center in Greenbelt, MD (level 2, level 2G, ocean color, sea-surface temperature); the U.S. Geological Survey EROS Data Center in Sioux Falls, SD (land products); and the National Snow and Ice Data Center (NSIDC) in Boulder, CO (snow and sea ice). The MODIS level-1 and atmosphere products are distributed through the Level-1 and Atmosphere Archive and Distribution System Web site.

2.2 Moderate-Resolution Imaging Spectroradiometer Land Products

MODIS land products are developed by the MODIS land discipline group (MODLAND). Standard land products include both remote sensing surface variables (i.e., radiance, surface reflectance) and a wide range of derived variables such as vegetation indexes (VIs), leaf area index (LAI), fraction of photosynthetically active radiation (fPAR), bidirectional reflectance distribution function (BRDF), land-surface temperature (LST), NPP, fire and burn scar, land cover and land-cover change, and snow and sea ice cover (Justice et al. 1998; Running et al. 1994). Detailed descriptions of MODIS land products are provided by Justice et al. (1998) and ATBDs developed by the MODIS science team. Selected MODIS land products, algorithms, and validation issues will be reviewed in this chapter.

2.2.1 Surface Reflectance

The core of the MODIS surface reflectance algorithm is atmospheric correction. Atmospheric gases, aerosols, and clouds have direct impacts on solar radiation though absorption and scattering. The atmospheric effects may modify pixel brightness and change wavelength dependence on radiance (Herman and Browning 1975; Kaufman 1989). The objective of atmospheric correction is to remove atmospheric effects, and thus extract surface reflectance values as if they were measured at ground level. The successful retrieval of surface reflectance values is important for improving remote-sensing data quality and subsequent data analysis and applications (Gordon, Brown, and Evans 1988; Liang et al. 2002; Tanre, Holben, and Kaufman 1992).

One of the principal challenges for an operational atmospheric correction algorithm is the large variations of aerosols and water vapor in space and time. Often, the optical characteristics of aerosols are very difficult to model because of large variations in aerosol loadings, particle sizes, and distributions. Due to the lack of available data on aerosol characteristics, previous operational atmospheric correction algorithms often assume standard atmosphere with zero or constant aerosol loading to simplify the problem. The main advantage of the MODIS atmospheric correction algorithm is that it derives atmospheric characteristics from the MODIS data itself. The MODIS-derived data on aerosol optical thickness and water vapor content are coupled with MODIS spectral information and other ancillary data (i.e., a digital elevation model) in a radiative transfer model to derive surface reflectance values. The direct implementation of the radiative transfer model at a per-pixel level is impossible for daily global MODIS data, because of the high computational cost involved. Therefore, a lookup table (LUT) approach is used to simplify the radiative transfer computation. A number of atmospheric effect quantities, such as path radiance, atmospheric reflectance for isotropic light, and diffuse transmittance, are precalculated for different aerosol loadings and sun-view geometries using the second simulation of a satellite signal in the solar spectrum (6S) code (Vermote et al. 1997). Surface reflectance values are then estimated using a second-degree equation. The detailed mathematical equations and algorithms are described by Vermote and Vermeulen (1999).

It should be noted that the MODIS atmospheric correction algorithm also considers adjacent effects, BRDF, and atmosphere coupling effects. The adjacent effects occur when the reflectance of a target pixel is mixed with those from surrounding pixels (Tanre, Herman, and Deschamps 1981). These effects should not be ignored for heterogeneous ground surfaces, especially for fine-resolution pixels (i.e., 250 m). The MODIS atmospheric correction algorithm employs an inverting approach to correct the adjacent effects under linear combination assumptions (Tanre, Herman, and Deschamps 1981). The coupling of BRDF with atmospheric correction is implemented using a priori estimates of surface BRDF. The MODIS algorithm uses the BRDF from a previous 16-day period (Strahler et al. 1996), which increases accuracy compared to the commonly used Lambertian assumption.

MODIS surface reflectance values are derived for MODIS bands 1–7 using the atmospheric correction algorithms. The major advantage of this approach is that MODIS-derived atmospheric optical properties are used to achieve automated and operational correction at the global level (Kaufman and Tanre 1996). The quality of MODIS surface reflectance is highly dependent on a number of MODIS-derived input data products (i.e., atmospheric properties) and on radiative transfer models that incorporate various theoretical assumptions. The validation of MODIS surface reflectance products has been conducted by intensive field campaigns and continuous validation at various validation sites. Liang et al. (2002) suggested that the direct comparison of MODIS surface reflectance values and ground point measurements is unrealistic due to scale mismatch. They proposed deriving surface reflectance values using higher-resolution remote sensing data (e.g., Landsat data) along with field calibration data, and then upscaling (i.e., degrading) the high-resolution surface reflectance values to the MODIS spatial resolution. In their validation work, MODIS surface reflectance values appear to have reasonable accuracy (±5%) when compared to the degraded Landsat-derived surface reflectance values. Note that this validation effort was conducted mostly for vegetated areas on relatively clear days. Additional continuous validation is needed for different land-cover conditions and aerosol loadings. It is important to incorporate additional validation results to further improve the quality of the MODIS surface reflectance data product, because the product serves as an important input to many higher-level MODIS algorithms that produce MODIS land products such as VIs, land-cover classification, change detection, fire products, and others.

2.2.2 Vegetation Indexes

It has been widely shown that VIs provide valuable measurements of vegetation activity and conditions (Tucker 1979; Tucker, Townshend, and Goff 1985). The normalized difference VI (NDVI) is probably the most commonly used VI, because it is highly correlated with many other biophysical parameters related to vegetation canopy properties, processes, and functions (Curran 1980; Tucker et al. 1981; Asrar et al. 1984; Goward, Tucker, and Dye 1985). Mathematically, NDVI is a simple ratio of two linear combinations of spectral reflectance values of NIR and red bands,

$$NDVI = \frac{\rho_{NIR} - \rho_{red}}{\rho_{NIR} + \rho_{red}} \qquad (2.1)$$

where ρ_{NIR} and ρ_{red} denote surface reflectance values at the NIR and red wavelength intervals, respectively. NDVI data is one of the standard MODIS VI products (Justice et al. 1998; Huete et al. 2002). This data set is also referred to as *continuity* data, which extend the AVHRR's long-term NDVI records.

In addition to the NDVI product, MODIS VI products also include a newly developed enhanced VI (EVI) (Huete et al. 2002),

$$EVI = G \times \frac{\rho_{NIR} - \rho_{red}}{\rho_{NIR} + C_1 \times \rho_{red} - C_2 \times \rho_{blue} + L} \qquad (2.2)$$

where G is the gain factor, C_1 and C_2 are aerosol resistance coefficients, and L is the canopy background adjustment. The numeric values for these coefficients are 2.5, 6.0, 7.5, and 1.0, respectively (Huete, Justice, and Liu 1994; Liu and Huete 1995). Compared to NDVI, EVI provides improved sensitivity of vegetation signals in high biomass or dense forest regions (Huete et al. 2002). The EVI is also better correlated with tree canopy structure characteristics such as LAI (Gao et al. 2000). The finest spatial resolution of the MODIS VI product is 250 m. It should be noted that there is no 250-m blue band for the MODIS instrument; thus, the 500-m blue-band surface reflectance values are used as replacements to generate 250-m EVI products. Also, water, clouds/shadows, and pixels with heavy aerosol loadings are masked out for the VI products, since VI values are not robust for these cover types.

The MODIS standard VI products are provided at 250 m, 500 m, 1.0 km, and 0.05° (5600 m) resolutions through 16-day data composites. The MODIS VI data composite algorithm was developed based on the experiences gained from the AVHRR-NDVI composite algorithm. The motivation was to generate cloud-free and consistent NDVI products at the global scale. The AVHRR-NDVI composite algorithm selects the maximum NDVI value for a pixel within each 14-day time interval. This is commonly referred to as the *maximum value compositing* (MVC) algorithm. One main drawback of this algorithm is that it favors pixels with large view angles. Such pixels often have higher NDVI values than the nadir-view pixels, but they may not be cloud free (Goward et al. 1991). The MODIS science team developed two new approaches to solve this problem: the constrained-view angle–MVC (CV–MVC) approach and the BRDF-composite (BRDF-C) approach. The CV–MVC compares the two highest NDVI/EVI values and selects the one with the smaller view angle for compositing, which typically improves the spatial consistency for VI time-series data. The BRDF-C algorithm is considered to be more complicated. It requires a minimum of five valid VI values for each pixel to mathematically interpolate nadir-view reflectance values and VIs (Walthall et al. 1985). This largely limits its applicability in regions with frequent cloud cover; thus, it can be considered a region-dependent algorithm. Currently, CV–MVC is used as the primary compositing algorithm for MODIS VI products with MVC as a backup algorithm. The BRDF-C algorithm is not used due to its regional dependency.

The results of the validation of MODIS VIs have been reported by a number of researchers (Huete et al. 2002; Gao et al. 2003; Brown et al. 2006). Gao et al. (2003) compared MODIS VIs with those from high spatial resolution images through a scaling-up approach. It was found that both MODIS 1-day VI and 16-day composited VI matched well with the values derived from higher spatial resolution data sets. Huete et al. (2002) conducted validation work in four field campaigns across the United States and at sites in North America and South America. They compared MODIS NDVI and EVI with regard to temporal (seasonal) vegetation profiles, dynamic range and saturation, and their relationships with biophysical variables such as LAI, biomass, canopy cover, and fraction of absorbed photosynthetically active radiation (APAR). The MODIS NDVI and EVI temporal profiles matched during the vegetation growing season in selected biomes. One noticeable difference between MODIS NDVI and EVI was the dynamic range. Whereas MODIS NDVI appears to be saturated (e.g., >0.9) in high biomass regions, EVI shows more sensitivities in such regions without suffering data saturation. The latter is also more advantageous in that it differentiates forest types such as broadleaf and needleleaf forests, whereas MODIS NDVI shows very similar signals for these forest types. These differences can have direct impacts on VI-based land-cover mapping applications. A comparison between MODIS-NDVI and AVHRR-NDVI also showed interesting results (Huete et al. 2002). These two time-series products have very similar signals for arid and semiarid regions in dry seasons; however, MODIS-NDVI products have much higher values in wet seasons. Brown et al. (2006) further suggested that the differences between these two NDVI products are land cover–dependent and they cannot simply be interchanged for analyses. These studies suggest the challenge of data "continuity" between AVHRR-NDVI and MODIS-NDVI data records. The contributing factors include differences in sensor band characteristics and the atmospheric correction and compositing algorithms used. Further research is needed to link AVHRR-NDVI and MODIS-NDVI in a more consistent manner for monitoring global vegetation conditions and changes.

2.2.3 Land Cover and Change Detection Products

Timely and accurate global land-cover information is important for a wide range of studies, including those on global climate change, carbon and hydrologic balance, terrestrial ecosystems, and human impacts on the natural earth system (Townshend and Justice 2002). Operational global land-cover mapping, however, is extremely challenging due to limitations in training data, a high computational cost, and intrinsic spectral confusion between land-cover classes. Historically, global land-cover maps have been compiled by a number of research institutions and organizations (Friedl et al. 2002). The first remote sensing–based global map was produced by DeFries and Townshend (1994) using time-series AVHRR-NDVI monthly composite data at a 1.0-degree spatial resolution. AVHRR-based global maps at finer spatial resolutions (e.g., 1–8 km) have been subsequently developed using a variety of classification algorithms (Loveland et al. 2000). The main concern regarding the AVHRR-derived land-cover data products is related to AVHRR sensor characteristics, which were not configured for land-cover mapping. The MODIS science team has high expectations for MODIS-derived land-cover map products, mainly due to the improved sensor characteristics (spatial, spectral, and radiometric resolutions), advances in computer algorithms, such as those on atmospheric correction and image classification, and improved quality and quantity of training data sites. Land-cover mapping and land-cover change was identified as the most important task of the MODIS land science team (Asrar and Dokken 1993; Running et al. 1994).

MODIS land-cover classification follows the International Geosphere-Biosphere Programme classification scheme. A total of 17 land-cover classes are defined including 11 natural vegetation classes, 3 nonvegetation classes, and 3 human-altered classes (Friedl et al. 2002). The training data points are designed to ensure global representation through the system for terrestrial ecosystem parameterization (Muchoney et al. 1999). This global site database includes more than 1373 sites. Training data points are developed mainly through visual interpretation of high-resolution remote sensing imagery. Additional ancillary data were also used to augment training data points. Note that the global site database is dynamic and needs to be updated continually to meet the requirements of operational global land-cover mapping. The inputs for MODIS land-cover classification include the 16-day composite of MODIS surface reflectance values (bands 1–7) and the EVI. Two image-classification algorithms were considered for land-cover classification by the MODIS science team. A supervised decision-tree algorithm (Quinlan 1993) was selected over a neural network (Carpenter et al. 1992) algorithm, based on global operational considerations. An advanced boosting algorithm (Freund 1995) was integrated with the decision-tree algorithm. This provided more robust estimates of per-pixel probabilities of class membership. Currently, standard MODIS land-cover products are provided at 500-m and 0.05-degree spatial resolutions on annual intervals.

The validation of MODIS land-cover data products is an ongoing process. Initial results from Friedl et al. (2002) suggest improved classification performance over AVHRR-derived products. This can be attributed to increased MODIS sensor characteristics, advances in atmospheric correction, and improved classification algorithms. The accuracy of MODIS land-cover products, however, does appear to have high regional differences. The quality of MODIS land-cover products at high latitudes is particularly questionable due to the deterioration of MODIS inputs at such latitudes (e.g., low solar zenith angles). Considerable classification confusion may occur between agriculture and natural vegetation. In a recent study, Giri, Zhu, and Reed (2005) compared the MODIS global land-cover data and the Global Land Cover 2000 (GLC-2000) data. These two global land-cover data sets are derived using very different input data and classification algorithms. Although a general agreement was found at the class aggregated level, there were substantial differences for individual classes. Moreover, the agreements were highly variable across different biomes. This calls for further studies in the development of land-cover classification schemes and classification algorithms.

The MODIS land-cover change algorithm does not use a postclassification comparison approach. The main reason is that the classification errors associated with two individual image classifications can be accumulated during postclassification comparisons, which may seriously impact change detection performances (Singh 1989). Instead, the MODIS land-cover change algorithm relies on the analysis of multitemporal image stacks or time-trajectories to assess the land-cover dynamics caused by processes such as deforestation, agricultural expansion, and urbanization. Change-vector analysis (Lambin and Strahler 1994) is the primary change detection technique used in the MODIS land-cover change algorithm. The input data for the change-vector analysis include a variety of MODIS-derived spectral–spatial variables such as VIs, surface temperature, and spatial structure indexes. To detect the annual land-cover change between consecutive years, these variables are compiled for each individual year by monthly (32-day) composites. The land-cover states of the two consecutive years can be treated as two points located in a multitemporal feature space. A change vector can thus be generated by linking these two points in the multitemporal feature space. The direction and magnitude of the change vector are assessed to identify potential land-cover changes (Lambin and Strahler 1994). The main advantages of using change-vector analysis are that it can overcome the error accumulation problem and identify subtle land-cover changes. Currently, the MODIS land-cover change product is provided at 1.0-km spatial resolution. In addition to the annual land-cover change product, Zhan et al. (2002) developed the vegetative cover conversion product as a global alarm product of land-cover change caused by anthropogenic activities and extreme natural events. The spatial resolution of the land-cover change alarm product is 250 m. The MODIS level-1B data was used as input for decision trees to detect wildfire, flood, and deforestation activities. Furthermore, the MODIS research team at the University of Maryland is actively developing enhanced land-cover and land-cover change products. Such products include the global 250-m land-cover change indicator product, the global 500-m vegetation continuous fields (VCF) product, and the global 1.0-km land-cover product. The validation of MODIS land-cover change products is an ongoing process. A review of the recent literature suggests that very few studies have been performed for the validation of MODIS land-cover change products at the local, regional, and global levels.

2.2.4 Fire Products

MODIS fire products consist of both fire detection and burn scar products. The theoretical background of the fire detection algorithm is provided by Kaufman et al. (1992). The MODIS fire detection algorithm also benefits from the rich experiences gained from the AVHRR and visible and infrared (IR) scanner (Giglio, Kendall, and Tucker 2000). The main objective of the algorithm is to automatically detect locations where active burning is occurring. The primary inputs for the fire detection algorithm are MODIS spectral signals at 4 and 11 μm. The MODIS channel at 4 μm is considered to be the most sensitive channel for both fire flaming and fire smoldering, whereas the channel around 11 μm (TIR) detects strong emissions from fires (Dozier 1981). The MODIS fire detection algorithm consists of multiple processing steps to identify fire pixels. The initial step removes obvious nonfire pixels through a preliminary classification; potential fire pixels are then identified through the thresholding of brightness temperatures (T_4 and T_{11}) derived from the MODIS channels at 4 and 11 μm. The threshold values of T_4 are specified as 310 K and 305 K for daytime and nighttime pixels, respectively. In addition, the difference between T_4 and T_{11} needs to be larger than 10 K for a pixel to be labeled as a potential fire pixel. The MODIS spectral values at bands 1 (0.648 μm), 2 (0.858 μm), and 7 (2.13 μm) are also incorporated in the decision rules to reduce false alarms (e.g., sun glint) and confusion caused by clouds (Nath, Rao, and Rao 1993).

Within the potential fire pixels, the MODIS fire algorithm further considers two approaches to identify unambiguous fire pixels. The first approach relies on high threshold values of brightness temperatures to identify actual fire pixels. The second approach examines the contextual information of neighboring pixels (from 3×3 to 21×21) to identify active fire pixels. At least eight valid neighboring pixels are required for the background contextual analysis using 4 and 11 μm brightness temperature values. The brightness temperature values for focal pixels are compared with the background contextual statistics to make decisions. The final fire products are labeled using the following categories: missing data, cloud, water, nonfire, fire, or unknown (Giglio et al. 2003). The fire radiative power is also computed for each fire pixel using the empirical relationship developed by Kaufman et al. (1998). A range of standard MODIS fire products are provided at various processing levels (level 2, level 2G, and level 3) with different spatial (1.0 km and 0.5°) and temporal resolutions (daily, eight-day, and monthly composite).

The MODIS burn scar algorithm was developed by Roy et al. (2002). Burn scar products identify the spatial extent of the recent burn area, in contrast to the identification of active fire in the MODIS fire algorithm. The identification of burn scars at the global scale is an extremely challenging task since the spectral signals of burn areas are very similar to those of other land-cover types such as flooding area and shadows from clouds and surface relief. The current MODIS burn scar algorithm can be considered a change detection approach through a statistical and temporal modeling of bidirectional reflectance variables. For each pixel, the bidirectional reflectance values within a predefined temporal window (i.e., 16-day) are used in a statistical model to predict a subsequent reflectance value. This predicted value is then compared to the actual observed surface reflectance value to identify the chance of change. Threshold values are specified to identify pixels with large decreases of surface reflectance values. The primary inputs to the MODIS burn scar algorithm are MODIS bands 2 (841–876 nm) and 5 (1230–1250 nm), which are the most sensitive to burning and postfire reflectance change. Additionally, simple band relationships among MODIS bands 2, 5, and 7 are used in the MODIS burn scar algorithm to reduce false alarms such as cloud shadow or soil moisture changes.

The validation of MODIS file product has been conducted by several researchers using Advanced Spaceborne Thermal Emission and Reflection Radiometer (ASTER)-derived fire products as references (Morisette et al. 2005; Csiszar, Morisette, and Giglio 2006). These studies concluded that approximately 50% of the large fire clusters (45–60 ASTER pixels) were correctly identified. Ellicott et al. (2009) validated the MODIS-derived fire products (during 2001–2007) and found a slight underestimation in fire extent. They further analyzed the spatial distribution and found that Africa and South America account for about 70% of global fires annually, suggesting high rates of biomass burning in those regions. For the validation of burn scar products, Chang and Song (2009) compared the standard MODIS burn scar products with burned areas derived in the SPOT-based L3JRC product for the years 2000–2007. The spatial and temporal patterns of these two products were found to be consistent, especially during the fire season. The research also suggested that MODIS burn scar products performed better than L3JRC products when compared with selected ground-based measurements in Canada, China, Russia, and the United States. One noticeable problem with the MODIS burn scar product is the underestimation of burn area in boreal forests.

2.2.5 Snow and Sea Ice Cover

Data regarding spatial extents and dynamics of global snow cover are important for studies pertaining to hydrologic and biogeochemical cycling, surface albedo, global energy balances, and climate change (Robinson, Dewey, and Heim 1993). Although large-scale hemispheric snow maps are routinely developed by the National Environmental Satellite Data and Information Service and the Interactive Multisensor Snow and Ice Mapping System (IMS), their spatial resolutions are generally coarse (e.g., IMS product at a resolution of 25 km). The MODIS snow-cover algorithm, or Snowmap, was developed as an automated computer algorithm that can be used to identify snow cover at higher spatial resolutions (e.g., 500 m) globally (Hall, Riggs, and Salomonson 2001; Hall et al. 2002).

Snow cover has distinct spectral signals that can be clearly differentiated from most other natural land-cover types. The primary confusion is with clouds, but previous research suggests that snow and cloud cover have different spectral responses at visible and SWIR channels. Specifically, snow cover has a strong reflectance in the visible range, but a low reflectance in the SWIR spectral region. On the other hand, clouds typically have strong reflectance values in both spectral regions (Dozier 1989). A ratio-based normalized difference snow index (NDSI) has been developed for snow mapping with Landsat data (Dozier 1989). The NDSI is also one of the primary algorithms used for MODIS Snowmap products.

$$\text{NDSI} = \frac{\text{band } 4 - \text{band } 6}{\text{band } 4 + \text{band } 6} \tag{2.3}$$

The MODIS Snowmap algorithm uses sensor reflectance values in bands 4 and 6 to compute the NDSI (Equation 2.3). A pixel is labeled as snow if the NDSI value is larger than the threshold value of 0.4. Additional decision rules in the Snowmap include the thresholding of MODIS bands 2 (>0.11) and 4 (>0.10). Generally, the NDSI value decreases as the purity of snow pixels is reduced. In order to identify a partial snow pixel (e.g., >50%) in a forested region, Snowmap incorporates MODIS-NDVI to map the snow pixel. For instance, pixels might be labeled as snow in cases of NDVI = 0.1 approximately and NDSI <0.4 (Hall et al. 2002). Currently, standard MODIS snow products are provided at daily and temporal compositing of 8-day and monthly intervals. The temporal compositing algorithm simply selects a maximum value within a specified temporal interval. A similar decision-rules technique used in Snowmap has also been employed for the MODIS sea ice product through the sea ice mapping algorithm (Icemap).

Validation of the Snowmap product has been difficult due to limited reference data and scale mismatch between various remote sensing–derived snow map products. Klein and Barnett (2003) conducted a snow map validation study for the upper Rio Grande River basin of Colorado and New Mexico using snow-cover map products developed by the National Operational Hydrologic Remote Sensing Center (NOHRSC). A high overall agreement of 86% was reported. The MODIS and NOHRSC snow-cover maps were also compared with in situ snow measurements. The MODIS Snowmap product performed best, with a 94% agreement. Noticeable MODIS Snowmap errors occurred in locations where snow depths were less than 4 cm. Ault et al. (2006) compared MODIS snow products with data from a number of observation stations that included amateur observations across the Laurentian Great Lakes region. The MODIS snow-cover map matched very well with observational data sets. Major errors identified in the MODIS snow-cover map occurred in forested areas. Hall and Riggs (2007) reported that the accuracy of selected MODIS snow product images (500 m) was approximately 93%. Confusion between snow and cloud was a major problem. Although the Snowmap algorithm successfully differentiates a majority of snow and cloud pixels at a 500-m spatial resolution, there were large uncertainties at the partial or subpixel level. Additional uncertainties were attributed to thin snow cover. The snow-cover composite data is believed to be less accurate due to error accumulation from the daily snow product (Hall and Riggs 2007).

2.2.6 Leaf Area Index and Fraction of Photosynthetically Active Radiation

The term LAI denotes one-side green leaf area per unit ground area. It is a plant-canopy attribute that is often used in process-based ecosystem, hydrology, and global climate models (Sellers et al. 1997). The term fPAR denotes the fraction of photosynthetically active radiation absorbed by plant canopies. A large amount of research has been conducted to study the relationships among plant-canopy reflectance, spectral VIs, LAI, and fPAR (Asrar et al. 1984; Asrar, Myneni, and Kanematsu 1989). One common approach estimates LAI and fPAR by developing empirical models based on remote sensing surface reflectance or VIs such as NDVI (Asrar, Myneni, and Kanematsu 1989).

The MODIS LAI/fPAR algorithm relies on a three-dimensional (3D) radiative transfer model and an LUT approach to estimate LAI/fPAR. A global biome map is developed to allocate land-cover types to six broad biomes, including grasses and cereal crops, shrubs, broadleaf crops, savannas, broadleaf forests, and needleleaf forests (Myneni et al. 2002). This simplifies a number of assumptions and input parameters for the radiative transfer model. The 3D radiative transfer model generates several spectral and angular signatures that can be compared to the MODIS directional surface reflectance values through a LUT. The MODIS LAI/fPAR algorithm then derives location-specific results by incorporating the law of energy conservation (Knyazikhin et al. 1998). Further details about the MODIS LAI/fPAR algorithm and its theoretical background can be found in work by Knyazikhin et al. (1999). Standard MODIS LAI/fPAR products include 1-km spatial resolution data for both the daily and eight-day maximum value composite data set.

Privette et al. (2002) conducted initial validation work for MODIS LAI products using field-sampled data in southern Africa and found that the accuracy of these products is within an acceptable level. The MODIS LAI products successfully depicted the structural and phenological variability in semiarid woodlands and savannas. Wang et al. (2004) conducted LAI validation work in a needle-leaf forest site near Ruokolahti, Finland. Field-based LAI measures were first linked to high-resolution Landsat images and then aggregated to match the MODIS spatial resolution. The MODIS LAI products showed a higher variation than expected. The values were also overestimated compared to the field-based LAI measures. The authors suggest that the understory vegetation might cause such uncertainties. Iiames (2008) assessed MODIS LAI products for the evergreen needleleaf biome in the southeastern United States. The major challenges were attributed to the uncertainties in the creation of the high-resolution LAI reference map, land-cover classification, and the influences from vegetative understory. Yang et al. (2006) further addressed the sources of

MODIS LAI uncertainties, including the inputs of land-cover maps, surface reflectance, and LUTs used in the MODIS LAI algorithm. Kanniah et al. (2009) assessed the accuracy of LAI and fPAR for a north Australian savanna site and found that the MODIS products captured the seasonal variation in LAI and fPAR well, especially the most recent collection-5 data. However, Xiao et al. (2009) raised concerns related to the spatial–temporal discontinuity of MODIS LAI products for many locations. They proposed a new algorithm for estimating LAI from time-series MODIS reflectance data to increase temporal continuity and improve accuracy.

2.2.7 Net Primary Productivity

In addition to developing standard products linked to plant-canopy structure and bio-optical properties, the MODIS science team also emphasizes the development of algorithms and standard products for plant productivity and processes. One of the standard MODIS products that provides a key measure of vegetation productivity is NPP. It denotes the rate of net carbon gain by vegetation over a specified time period and can also be represented as the difference between gross primary production (GPP) and plant respiration. It is commonly measured at monthly, annual, or longer temporal intervals. The estimation of NPP requires the integration of ecological principles, remote sensing data, and other ancillary surface data sets. Potter et al. (1993) found that NPP can be estimated as a product of APAR and an efficiency of radiation use. The theoretical basis of the relationship between APAR and NPP is provided by Monteith (1972, 1977).

Theoretically, NPP values can be estimated based on an empirical relationship between APAR and NPP that has been demonstrated in numerous studies (Asrar et al. 1984; Goward, Tucker, and Dye 1985). However, the relationship between the two variables is also dependent on vegetation type and numerous other control factors, such as concentration of photosynthetic enzymes, canopy structure, and soil-water availability (Russell, Jarvis, and Monteith 1989; Running et al. 1999). This represents a considerable challenge to the development of an operational MODIS NPP algorithm using the APAR-based approach. The current MODIS NPP algorithm relies on an alternative approach that computes the difference between GPP and plant respiration. The basis for this approach is that APAR is actually more closely related to GPP than to NPP (Running et al. 1999). A detailed algorithm flowchart can be found in the work of Running et al. (1999). The primary algorithm can be broken down into two subroutines: The first estimates the daily GPP using standard MODIS fPAR products and ancillary surface meteorological measures as inputs. Different radiation conversion efficiency parameters are also provided as inputs using a LUT (stratified by biome types). The second subroutine estimates daily plant respiration. MODIS LAI is used as one of the inputs to estimate leaf mass, which is further used as an input in estimating plant respiration. The results from the two subroutines (estimated GPP and plant respiration) are used to derive daily NPP. The daily NPP product is provided at a spatial resolution of 1.0 km. In addition to the daily NPP, the MODIS algorithm also provides annual NPP. The annual NPP is estimated by integrating daily NPP and subtracting a number of respiration parameters for live woody tissue, leaves, and fine roots (Running et al. 1999).

Turner et al. (2006) evaluated MODIS NPP and GPP products across multiple biomes. The GPP at eddy covariance flux towers and plot-level measurements of NPP were scaled up to 25 km^2 and compared with the MODIS products. The authors report large variations in results over different biome types and land uses. The MODIS products overestimated NPP and GPP at low-productivity sites and underestimated those values at high-productivity sites. One of the main error sources was attributed to the inputs (e.g., fPAR estimates) to the MODIS NPP algorithm (Turner et al. 2006).

2.3 Moderate-Resolution Imaging Spectroradiometer Atmosphere and Ocean Products

2.3.1 Aerosols

MODIS atmosphere and ocean products are developed by the MODIS atmosphere discipline group and ocean discipline group, respectively.

Aerosols, especially human-made aerosols, may lead to large reductions in the amount of solar irradiance reaching the Earth's surface and increases in the solar heating of the atmosphere (Ramanathan et al. 2001). Aerosol loadings and distributions are often poorly characterized, because they are highly variable in space and time. Remote sensing–based characterization is generally performed by estimating aerosol optical depth or thickness. To account for the very different surface reflectance properties associated with oceans and the land surface, the MODIS products incorporate two independent algorithms to retrieve aerosol optical depth (Kaufman et al. 1997).

The aerosol algorithm over ocean integrates a radiative transfer model and LUT to produce aerosol optical depth estimates. The radiative transfer model has been run under a range of predefined aerosol conditions that describe particle modes (whether fine or coarse particles), total loadings, sensor–sun geometry angles, wind speed, and other parameters computed from ancillary data (Ahmad and Fraser 1982). The theoretical background is provided by Wang and Gordon (1994), who use fine or coarse particle modes to model multiple scattering process of radiance. The radiative transfer model produces an LUT that can link spectral reflectance values to aerosol spectral properties or optical depth estimates. The observed MODIS surface reflectance values are simply compared to the values in the LUT to find the best fit using a least-squares algorithm.

Aerosols over the land surface are more concentrated compared to those over the ocean surface, because the majority of aerosol sources are located on land (Kaufman et al. 1997). The estimation of aerosol optical depth over land surface is considered to be more challenging due to the highly variable reflective properties associated with different land-cover types. The radiance components from the land surface cannot be easily separated from those of aerosols (note that the ocean surface is generally darker and water-leaving radiance can often be assumed to be zero). This is one of the major reasons that aerosol optical depth was not routinely estimated at the global level before the use of MODIS data (Kaufman et al. 1997).

The MODIS aerosol algorithm over land relies on the accurate identification of dark surface pixels. VI-based dark pixel detection was found to be unreliable for global applications, because VIs themselves are affected by the presence of aerosols (Holben et al. 1986). For the MODIS aerosol algorithm over land, two MODIS spectral bands at 2.1 and 3.8 μm are used to detect dark pixels (Kaufman et al. 1997). The spectral band at 2.1 μm is preferred, especially when the reflectance value for this band is lower than 0.05. The wavelengths of these two spectral bands are considerably longer than those of typical aerosol particles; thus, the surface reflectance retrieved for these spectral bands can be considered free from aerosol impacts. Under aerosol-free conditions, there are stable relationships between surface reflectance in the visible bands (0.47 and 0.66 μm) and that in the SWIR bands (2.1 and 3.8 μm). Thus, the surface reflectance values in visible bands can be estimated from those derived for the SWIR channels (Kaufman et al. 1997). The difference between the estimated and the MODIS-derived surface reflectance values in visible bands can be attributed to the presence of aerosols. This is the fundamental assumption of the MODIS aerosol algorithm for land surfaces.

Validations of aerosol optical depth estimates have been conducted by a number of researchers. Remer et al. (2002) compared 8000 MODIS-derived optical depth values and aerosol robotic network (AERONET) measurements. MODIS estimates were reported to be within the acceptable uncertainty levels over ocean and land surfaces. Chu et al. (2002) compared the MODIS-derived aerosol optical depths and measurements from 30 AERONET sites. They found that the levels of consistency were higher for continental inland regions than for coastal regions. The partial water surface may have contaminated the aerosol optical depth estimation in the coastal regions. The authors also suggest that the lack of AERONET sites in East Asia, India, and Australia makes global validation of MODIS aerosol optical depths particularly challenging. Aloysius et al. (2009) compared MODIS-derived aerosol optical depths and National Centers for Environmental Prediction reanalysis data over the southeast Arabian Sea. They reported high correlations ($R^2 = 0.96$) between the two data sets. At the local level, Li et al. (2005) suggested that the standard MODIS 10-km aerosol optical depth estimates are insufficient to characterize the local aerosol variation over urban areas. They modified the MODIS aerosol algorithm and derived aerosol optical depth at 1.0-km spatial resolution over Hong Kong. High accuracies were reported compared to field measures. This suggests that there is considerable potential for using MODIS data in the estimation of aerosol optical depth at a higher spatial resolution over local areas.

2.3.2 Clouds

Clouds play major roles in the Earth's radiation budget and climate change research (Ramanathan 1987). The MODIS atmosphere science team has developed a variety of algorithms to generate MODIS cloud products, including a cloud mask and cloud physical and optical properties. The review provided here focuses on the MODIS cloud detection, or cloud mask algorithm. The MODIS cloud mask algorithm employs an automated and threshold-based approach to identify clouds. The algorithm is based on previous cloud detection research and experiences from the International Satellite Cloud Climatology Project (ISCCP; Rossow and Garder 1993) and the AVHRR processing scheme over cloud, land, and ocean (APOLLO; Gesell 1989) cloud detection algorithm (Ackerman et al. 2006). These algorithms primarily use multiple radiance thresholds testing to label pixels as cloudy or clear. The ISCCP algorithm also integrates spatial and temporal information in its decision rules.

The primary inputs to the MODIS cloud detection algorithm include 19 MODIS visible and IR radiance values. Additional ancillary data sets include sun-sensor geometry angles, ecosystem classifications, land and water distributions, elevation above mean sea level, daily snow and ice maps from NSIDC, and the daily sea ice concentration product from the National Oceanic and Atmospheric Administration (NOAA). The ancillary data provide a basis to segment the Earth's surface into a range of surface conditions over time, including daytime land, daytime water, nighttime land, nighttime water, daytime desert, and daytime and nighttime snow or ice surfaces (Ackerman et al. 2006). The MODIS cloud detection algorithm employs different threshold testing for different surface conditions over time. For a specific surface condition at a given time, each 1.0-km pixel is put through a variety of radiance and temperature-based threshold tests, which can be classified into the following five groups: simple IR threshold tests, brightness temperature differences, solar reflectance tests,

NIR thin cirrus, and IR thin cirrus testing. One advantage of the MODIS cloud detection algorithm is the inclusion of a confidence level for each threshold test, rather than providing simple categorical labels such as cloudy or clear. The confidence level is computed based on the distance of the pixel from the threshold value, and a continuous value is derived for each test (high confidence of clear pixel = 1; high confidence of cloudy pixel = 0). For each threshold testing group, a minimum confidence value is determined. The final confidence level is then integrated from the results of the five groups. As a result, the MODIS algorithm provides multiple levels of "confidence" for the cloud mask product (i.e., cloudy, probably clear, confidently clear, and uncertain). This allows users to develop their own decision rules while processing or using the standard MODIS cloud mask product.

Berendes et al. (2004) compared MODIS-derived daytime cloud products with observations from ground-based instrumentation located in northern Alaska. They report agreement within ±20% between the two data sets. In their study, the MODIS cloud mask appeared to be more accurate than ground-based instruments in the detection of thin cirrus clouds. However, other researchers suggest that the detection of cirrus cloud cover still remains a major challenge to MODIS cloud masking. Dessler and Yang (2003) analyzed MODIS cloud mask products for two 3-day periods from December 2000 and June 2001. They report that approximately one-third of the pixels flagged as cloud free by the MODIS cloud mask contained detectable thin cirrus clouds. Further research is needed to improve the detection of thin cirrus clouds by the MODIS cloud algorithm.

2.3.3 Ocean

Numerous standard MODIS ocean data products are provided by the MODIS science team, including *normalized water-leaving radiance,* pigment concentration, chlorophyll fluorescence, chlorophyll-a pigment concentration, photosynthetically available radiation, suspended solids concentration, organic matter concentration, ocean water attenuation coefficient, ocean primary productivity, sea-surface temperature, phycoerythrin concentration, and ocean aerosol properties.

Many MODIS ocean algorithms were developed from experiences with the coastal zone color scanner (Gordon and Voss 1999). A common perception is that water color (spectral measures) can be used to derive important biophysical parameters related to phytoplankton pigment concentration, primary productivity, and sea-surface temperature. One main challenge of ocean-color characterization is that the retrieval of the relevant signal from the total radiance is difficult, because the water-leaving radiance is quite small (<10%) compared to the total radiance received at the sensor. In other words, at-sensor radiance is dominated by atmospheric effects over the ocean surface. It is, therefore, necessary to conduct an atmospheric correction for the MODIS ocean-color products. A detailed atmospheric correction algorithm is provided by Gordon and Voss (1999). The output of the algorithm is called normalized water-leaving radiance, which approximates water-leaving radiance (sun at zenith) free of atmospheric impacts for most oceanic conditions. The normalized water-leaving radiance is further used as an input to generate almost all other MODIS ocean products. For instance, the current MODIS pigment concentration and bio-optical properties are largely dependent on empirical or semiempirical relationships derived between spectral and biophysical measures obtained from the same field observations. Therefore, the normalized water-leaving radiance over a large ocean area can be compared to those spectral measures obtained at field observations to generate estimates of pigment concentration or other biophysical properties.

2.3.4 Other Algorithms

It must be noted that the MODIS science team has developed a large number of algorithms over the period of MODIS instrument design, prelaunch, and postlaunch phases. Some of these algorithms are continually updated, which leads to several MODIS data reprocessing procedures. Because this chapter only reviews some selected MODIS algorithms and products, it is by no means a complete description of all MODIS algorithms and products. There is a range of MODIS standard products that are not discussed in this chapter, particularly in the atmosphere and ocean disciplines. The ATBDs developed by the MODIS science team are probably the best resource for readers interested in a more in-depth review of MODIS algorithms, their theoretical backgrounds, and the available data products.

2.4 Moderate-Resolution Imaging Spectroradiometer Applications

Since the launch of MODIS-Terra, hundreds of scientific papers have been published on the application of MODIS data at global, regional, and local levels. The remote sensing literature has covered research on the following topics: global climate models (Oleson et al. 2003; Tian et al. 2004), land cover and change detection (Lunetta et al. 2006; Zhang et al. 2008; Gill et al. 2009), forest disturbance and vegetation dynamics (Evrendilek and Gulbeyaz 2008; Hansen et al. 2008; Hilker et al. 2009; Maeda et al. 2009), vegetation and crop phenology monitoring (Zhang et al. 2003; Sakamoto et al. 2005), terrestrial ecosystem carbon exchange (Garbulsky et al. 2008; Xiao et al. 2009), ecohydrologic analysis (Hwang et al. 2008), crop mapping and crop yield estimation (Doraiswamy et al. 2004; Sakamoto et al. 2009), human health issues (Hu 2009), air quality assessment (Gupta and Christopher 2008), water quality monitoring and assessment (Hu et al. 2004), and species and habitat distribution (Vina et al. 2008).

At the global level, various MODIS data products have been used as primary inputs to climate models, and as reference data to validate the climate models (Oleson et al. 2003; Tian et al. 2004). For example, Tian et al. (2004) compared the land surface albedo from the community land model (CLM; Bonan et al. 2002) with MODIS albedo products (Gao et al. 2005) under two land-surface scenarios. The first land scenario used older standard parameters in the CLM for a "control run." The second scenario used a range of newly derived MODIS land parameters such as VCF, LAI, land cover, and plant functional type as the model inputs. Improved CLM results are reported when the MODIS-derived products were used as land-surface parameters. Lawrence and Chase (2007) developed new CLM land-surface parameters based on MODIS land products and found that the new model had substantial improvements in surface albedo estimation, which further improved the simulation of precipitation and near-surface air temperature. Although the MODIS data algorithms and products show much promise for climate modeling, Dickinson (2008) suggests that one of the major challenges faced by the current remote sensing data are spatial and temporal discontinuities. For example, general land cover on the Earth's surface should be quite stable over time, except for some small random changes caused by human or natural disturbances. However, spatial and temporal discontinuities often occur in remote sensing–derived land-surface parameters as a result of system limitations or systematic errors. Future research should address these problems, mainly through algorithm improvements.

The use of MODIS data for applications in forest disturbance, vegetation dynamics, urban development, agricultural expansion, and crop mapping and management generally relies on image-classification and change detection techniques. Instead of using the standard MODIS global data, researchers often need to develop their own classification and change detection algorithms for local and regional applications. There are three motivations for researchers to develop their own products using MODIS spectral information: First, the information desired at local and regional levels is generally more detailed than those given in the MODIS global data sets. Second, the spatial resolution of standard MODIS global data might be too coarse for local applications. Third, the accuracy of the MODIS global data sets varies across regions. It is often possible to improve accuracy using an increased number of training data points, ancillary data, and algorithms that fit better with local conditions.

The desire to obtain more detailed land-use and cover-type information can be illustrated by a number of research projects that focus on crop mapping using MODIS data. The standard MODIS land-cover product does not include specific crop types in its mapping scheme. Recent studies suggest that MODIS data has sufficient spatial and temporal resolution to identify major crop types such as corn, soybean, and wheat in intensive agricultural regions in the United States (Wardlow, Egbert, and Kastens 2007; Wardlow and Egbert 2008; Shao et al. 2010). These studies often rely on the use of MODIS time-series NDVI or a phenology-based analysis for land-cover and crop identification. Xiao et al. (2005) found that the MODIS-NDVI profiles were also useful in characterizing rice distributions, mainly due to the unique NDVI profiles associated with rice transplanting, growing, and fallow periods. The results from these studies suggest that the unique combination of spatial, spectral, and temporal resolutions associated with MODIS data results in more detailed land-use and cover-type classification at regional and local scales.

The 500-m or 1.0-km spatial resolution land-cover products may be too coarse for many regional- or local-scale applications. This is particularly evident for areas with complex or heterogeneous land-cover patterns at finer spatial scales (Lobell and Asner 2004; Knight et al. 2006). Many researchers have employed spectral mixture analysis to unmix MODIS pixels, and thus derive proportional land cover at the subpixel level. Chang et al. (2007) estimated proportional corn and soybean cover within MODIS 500-m data. Knight et al. (2006) examined the potential of subpixel land-cover estimation using multitemporal MODIS-NDVI 250-m data. The subpixel land-cover mapping problem was also addressed by the MODIS science team. It is actually designed as a part of the MODIS enhanced land-cover and land-cover change products. Hansen et al. (2002) employed a regression tree algorithm to derive subpixel tree-cover products at 500-m spatial resolution. His subpixel classification approach relied on training pixels that contain tree-cover proportions derived from high-resolution satellite images. The regression tree was trained to model the relationship between MODIS signals and tree proportions at the subpixel level. The assessment of the subpixel tree-cover estimation accuracy was extremely challenging due to the lack of reference data sets, especially at regional or global scales. The trend toward subpixel analysis is not limited to land-cover classification; researchers are also actively working on subpixel cloud detection and subpixel snow-cover mapping (Salomonson and Appel 2004). The relationship between sensor spatial resolution and ground surface features continues to be a challenging topic for the remote-sensing research community.

MODIS-based change detection has been employed by many researchers to study deforestation, urbanization, and agricultural expansion (Lunetta et al. 2006; Zhang et al. 2008; Gill et al. 2009). Most of the change detection algorithms have been developed for the 250-m MODIS data, because many human-introduced land-cover changes occur at fine spatial scales. Lunetta et al. (2006) developed an automated land-cover change alarm product in the Albemarle-Pamlico Estuary System region of the United States. The approach relied on detecting pixels that have experienced significant changes in the annually-integrated NDVI values. A large drop in annually-integrated NDVI may suggest possible land-cover changes such as urban development or vegetation clear-cutting. Jin and Sader (2005) also used the MODIS 250-m VIs to detect forest harvest disturbance in northern Maine. They found that although the MODIS single-day and 16-day composite NDVI data showed no significant difference in overall detection accuracies, the single-day NDVI actually performed better when disturbed patch sizes were smaller.

Zhang et al. (2003) examined vegetation phenology using a time-series MODIS VI. They used a series of piecewise logistic functions to detect the transition dates of vegetation activity on an intraannual basis. Sakamoto et al. (2005) analyzed time-series data of EVI. Subsequent to data smoothing, the points of maximum, minimal, and inflection were identified to examine phenological stages of paddy rice, which were then used to evaluate crop productivity and management. Soudani et al. (2008) examined vegetation phenological dates for deciduous forest stands using 250-m daily MODIS-NDVI data. Key phenological dates (e.g., onset of green-up) matched well with in situ observations. The level of temporal uncertainty in MODIS-NDVI data is approximately 8 days. This MODIS-derived vegetation phenology can be particularly useful for research in vegetation–climate interactions and modeling (Pettorelli et al. 2005).

One potential source of uncertainty in time-series studies is the error of misregistration. Although the MODIS science team has substantially increased the registration accuracy over several reprocessing procedures, the 75–100 m misregistration error is still a substantial challenge for performing time-series analysis at the 250-m spatial resolution (Tan et al. 2006). The impacts of misregistration in time-series composite data can be even larger due to a potential "multiplier effect" and the selection of pixels under different sun-sensor geometry angles. Therefore, it is important for users to understand these potential error sources. Additional research is needed to further our understanding of the cumulative impacts associated with MODIS data quality, sun-sensor geometry information, and misregistration errors.

2.5 Summary

The MODIS instrumental characteristics represent a new generation of sensor systems for global observation. Global coverage of MODIS data are obtained every 1–2 days. The spectral, spatial, and radiometric resolutions are also substantially improved in MODIS compared to previous global sensor systems such as the AVHRR. In addition to the spectral products commonly provided for all remote sensing platforms, the MODIS science team devoted tremendous efforts in developing a wide range of MODIS-derived scientific data sets that are readily available for the scientific community. The MODIS data represent not only a "continuous" remote sensing data record that extends previous sensor systems, but also a substantial improvement by integrating the most advanced remote-sensing theory, algorithm development, data processing, validation, and distribution.

A majority of the current MODIS algorithms are operational at the global level. Data quality has been improving over several data reprocessing cycles. Validation of MODIS standard products is an ongoing effort undertaken by both the MODIS science team and independent researchers. Most validation efforts suggest a high level of data quality for the MODIS products. This can be attributed to the improvement of spectral, spatial, temporal, and radiometric resolutions, as well as improvements made in algorithm development by the MODIS science team. The success of MODIS is also evident from the exponential growth of applications that use MODIS data products at global, regional, and local levels. Future development of MODIS data and algorithms may integrate more feedback from continuous data quality validation and applications. These include many potential topics such as subpixel analysis, scaling problems, biophysical applications, in situ data integration (of cloud, ice, water, and land data), and optical and climate modeling.

Acknowledgments

The U.S. Environmental Protection Agency (EPA) partially funded and contributed to the development of this chapter. Although this work was reviewed by the EPA and has been approved for publication, it may not necessarily reflect official agency policy. Mention of any trade names or commercial products does not constitute endorsement or recommendation for use.

References

Ackerman, S. et al. 2006. Discriminating clear-sky from cloud with MODIS algorithm theoretical basis document (MOD35). Available online at http://modis-atmos.gsfc.nasa.gov/_docs/atbd_mod06.pdf, 125p.

Ahmad, Z., and R. S. Fraser. 1982. An iterative radiative-transfer code for ocean-atmosphere systems. *J Atmos Sci* 39:656–65.

Aloysius, M. et al. 2009. Validation of MODIS derived aerosol optical depth and an investigation on aerosol transport over the South East Arabian Sea during ARMEX-II. *Ann Geophys* 27:2285–96.

Asrar, G. et al. 1984. Estimating absorbed photosynthetic radiation and leaf-area index from spectral reflectance in wheat. *Agron J* 76:300–6.

Asrar, G., and D. J. Dokken. 1993. *EOS Reference Handbook.* Greenbelt, MD: NASA.

Asrar, G., R. B. Myneni, and E. T. Kanematsu. 1989. Estimation of plant canopy attributes from spectral reflectance measurements. In *Theory and Applications of Optical Remote Sensing*, ed. G. Asrar. New York: Wiley.

Ault, T. W. et al. 2006. Validation of the MODIS snow product and cloud mask using student and NWS cooperative station observations in the Lower Great Lakes Region. *Remote Sens Environ* 105:341–53.

Berendes, T. A. et al. 2004. Cloud cover comparisons of the MODIS daytime cloud mask with surface instruments at the North Slope of Alaska ARM site. *IEEE Trans Geosci Remote Sens* 42:2584–93.

Bonan, G. B. et al. 2002. The land surface climatology of the community land model coupled to the NCAR community climate model. *J Clim* 15:3123–49.

Brown, M. E. et al. 2006. Evaluation of the consistency of long-term NDVI time series derived from AVHRR, SPOT-Vegetation, SeaWiFS, MODIS, and Landsat ETM+ sensors. *IEEE Trans Geosci Remote Sens* 44:1787–93.

Carpenter, G. A. et al. 1992. Fuzzy ART: A neural network architecture for incremental supervised learning of analog multidimensional maps. *IEEE Trans Neural Netw* 3:698–713.

Chang, D., and Y. Song. 2009. Comparison of L3JRC and MODIS global burned area products from 2000 to 2007. *J Geophys Res Atmos*, 114:D16106, doi:10.1029/2008JD011361.

Chang, J. et al. 2007. Corn and soybean mapping in the United States using MODN time-series data sets. *Agron J* 99:1654–64.

Chu, D. A. et al. 2002. Validation of MODIS aerosol optical depth retrieval over land. *Geophys Res Lett* 29:8007, doi:10.1029/2001GL013205.

Csiszar, I. A., J. T. Morisette, and L. Giglio. 2006. Validation of active fire detection from moderate-resolution satellite sensors: The MODIS example in northern Eurasia. *IEEE Trans Geosci Remote Sens* 44:1757–64.

Curran, P. J. 1980. Multispectral remote sensing of vegetation amount. *Prog Phys Geog* 4:175–84.

Defries, R. S., and J. R. G. Townshend. 1994. NDVI-derived land-cover classifications at a global-scale. *Int J Remote Sens* 15:3567–86.

Dessler, A. E., and P. Yang. 2003. The distribution of tropical thin cirrus clouds inferred from terra MODIS data. *J Clim* 16:1241–7.

Dickinson, R. E. 2008. Applications of terrestrial remote sensing to climate modeling. In *Advances in land remote sensing: system, modelling, inversion and application*, ed. S. Liang, 498. Springer Press.

Doraiswamy, P. C. et al. 2004. Crop condition and yield simulations using Landsat and MODIS. *Remote Sens Environ* 92:548–59.

Dozier, J. 1981. A method for satellite identification of surface-temperature fields of subpixel resolution. *Remote Sens Environ* 11:221–9.

Dozier, J. 1989. Spectral signature of Alpine snow cover from the LANDSAT Thematic Mapper. *Remote Sens Environ* 28:9–22.

Ellicott, E. et al. 2009. Estimating biomass consumed from fire using MODIS FRE. *Geophys Res Lett* 36:1–5, doi:10.1029/2009GL038581.

Evrendilek, F., and O. Gulbeyaz. 2008. Deriving vegetation dynamics of natural terrestrial ecosystems from MODIS NDVI/EVI data over Turkey. *Sensors* 8:5270–302.

Freund, Y. 1995. Boosting a weak learning algorithm by majority. *Inf Comput* 121:256–85.

Friedl, M. A. et al. 2002. Global land cover mapping from MODIS: Algorithms and early results. *Remote Sens Environ* 83:287–302.

Gao, X. et al. 2000. Optical-biophysical relationships of vegetation spectra without background contamination. *Remote Sens Environ* 74:609–20.

Gao, F. et al. 2003. Detecting vegetation structure using a kernel-based BRDF model. *Remote Sens Environ* 86:198–205.

Gao, F. et al. 2005. MODIS bidirectional reflectance distribution function and albedo Climate Modeling Grid products and the variability of albedo for major global vegetation types. *J Geophys Res Atmos* 110:D01104, doi:10.1029/2004JD005190.

Garbulsky, M. F. et al. 2008. Remote estimation of carbon dioxide uptake by a Mediterranean forest. *Glob Chang Biol* 14:2860–7.

Gesell, G. 1989. An algorithm for snow and ice detection using AVHRR data-an extension to the APOLLO software package. *Int J Remote Sens* 10:897–905.

Giglio, L., J. D. Kendall, and C. J. Tucker. 2000. Remote sensing of fires with the TRMM VIRS. *Int J Remote Sens* 21:203–7.

Giglio, L. et al. 2003. An enhanced contextual fire detection algorithm for MODIS. *Remote Sens Environ* 87:273–82.

Gill, T. K. et al. 2009. Estimating tree-cover change in Australia: Challenges of using the MODIS vegetation index product. *Int J Remote Sens* 30:1547–65.

Giri, C., Z. L. Zhu, and B. Reed. 2005. A comparative analysis of the Global Land Cover 2000 and MODIS land cover data sets. *Remote Sens Environ* 94:123–32.

Gordon, H. R., J. W. Brown, and R. H. Evans. 1988. Exact Rayleigh scattering calculations for use with the NIMBUS-7 Coastal Zone Color Scanner. *Appl Opt* 27:862–71.

Gordon, H. R., and K. J. Voss. 1999. MODIS normalized water-leaving radiance, Version 4, MODIS Algorithm Theoretical Basis Document (ATBD). Technical Report OD 18, NAS5-31363, University of Miami, Coral Gables, FL,93p.

Goward, S. N., C. J. Tucker, and D. G. Dye. 1985. North-American vegetation patterns observed with the NOAA-7 advanced very high-resolution radiometer. *Vegetatio* 64:3–14.

Goward, S. N. et al. 1991. Normalized difference vegetation index measurements from the advanced very high-resolution radiometer. *Remote Sens Environ* 35:257–77.

Guenther, B. et al. 1998. Prelaunch algorithm and data format for the Level 1 calibration products for the EOS-AM1 Moderate-Resolution Imaging Spectroradiometer (MODIS). *IEEE Trans Geosci Remote Sens* 36:1142–51.

Gupta, P., and S. A. Christopher. 2008. An evaluation of Terra-MODIS sampling for monthly and annual particulate matter air quality assessment over the Southeastern United States. *Atmos Environ* 42:6465–71.

Hall, D. K., and G. A. Riggs. 2007. Accuracy assessment of the MODIS snow products. *Hydrol Processes* 21:1534–47.

Hall, D. K., G. A. Riggs, and V. V. Salomonson. 2001. Algorithm Theoretical Basis Document (ATBD) for the MODIS snow and sea ice-mapping algorithms. Available at http://modis.gsfc.nasa.gov/data/atbd/atbd_mod10.pdf, 45p.

Hall, D. K. et al. 2002. MODIS snow-cover products. *Remote Sens Environ* 83:181–94.

Hansen, M. C. et al. 2002. Towards an operational MODIS continuous field of percent tree cover algorithm: Examples using AVHRR and MODIS data. *Remote Sens Environ* 83:303–19.

Hansen, M. C. et al. 2008. Humid tropical forest clearing from 2000 to 2005 quantified by using multi-temporal and multiresolution remotely sensed data. *Proc Natl Acad Sci U S A* 105:9439–44.

Herman, B. M., and S. R. Browning. 1975. Effect of aerosols on earth-atmosphere albedo. *J Atmos Sci* 32:1430–45.

Hilker, T. et al. 2009. A new data fusion model for high spatial- and temporal-resolution mapping of forest disturbance based on Landsat and MODIS. *Remote Sens Environ* 113:1613–27.

Holben, B. N. 1986. Characteristics of maximum-value composite images from temporal AVHRR data. *Int J Remote Sens* 7:1417–34.

Hu, Z. Y. 2009. Spatial analysis of MODIS aerosol optical depth, PM2.5, and chronic coronary heart disease. *Int J Health Geogr* 8:27, doi:10.1186/1476-072X-8-27.

Hu, C. M. et al. 2004. Assessment of estuarine water-quality indicators using MODIS medium-resolution bands: Initial results from Tampa Bay, FL. *Remote Sens Environ* 93:423–41.

Huete, A., C. Justice, and H. Liu. 1994. Development of vegetation and soil indexes for MODIS-EOS. *Remote Sens Environ* 49:224–34.

Huete, A. et al. 2002. Overview of the radiometric and biophysical performance of the MODIS vegetation indices. *Remote Sens Environ* 83:195–213.

Hwang, T. et al. 2008. Evaluating drought effect on MODIS Gross Primary Production (GPP) with an eco-hydrological model in the mountainous forest, East Asia. *Glob Chang Biol* 14:1037–56.

Iiames J. S. et al. 2008. Validation of an integrated estimation of loblolly pine (*Pinus taeda* L.) leaf area index using two indirect optical methods in the southeastern United States. *Southern Journal of Applied Forestry* 32:101–110.

Jin, S. M., and S. A. Sader. 2005. MODIS time-series imagery for forest disturbance detection and quantification of patch size effects. *Remote Sens Environ* 99:462–70.

Justice, C. O. et al. 1998. The Moderate-Resolution Imaging Spectroradiometer (MODIS): Land remote sensing for global change research. *IEEE Trans Geosci Remote Sens* 36:1228–49.

Kanniah, K. D. et al. 2009. Evaluation of Collections 4 and 5 of the MODIS Gross Primary Productivity product and algorithm improvement at a tropical savanna site in northern Australia. *Remote Sens Environ* 113:1808–22.

Kaufman, Y. J. 1989. The atmospheric effect on remote sensing and its correction. In *Theory and Applications of Optical Remote Sensing*, ed. G. Asrar, 336–428. New York: John Wiley & Sons.

Kaufman, Y. J., and D. Tanre. 1996. Strategy for direct and indirect methods for correcting the aerosol effect on remote sensing: From AVHRR to EOS-MODIS. *Remote Sens Environ* 55:65–79.

Kaufman, Y. J. et al. 1992. Biomass burning airborne and spaceborne experiment in the Amazonas (BASE-A). *J Geophys Res Atmos* 97:14581–99.

Kaufman, Y. J. et al. 1997. The MODIS 2.1-mu m channel-correlation with visible reflectance for use in remote sensing of aerosol. *IEEE Trans Geosci Remote Sens* 35:1286–98.

Kaufman, Y. J. et al. 1998. Potential global fire monitoring from EOS-MODIS. *J Geophys Res Atmos* 103:32215–38.

Klein, A. G., and A. C. Barnett. 2003. Validation of daily MODIS snow cover maps of the Upper Rio Grande River Basin for the 2000–2001 snow year. *Remote Sens Environ* 86:162–76.

Knight, J. F. et al. 2006. Regional scale land cover characterization using MODIS-NDVI 250-m multi-temporal imagery: A phenology-based approach. *Giscience Remote Sens* 43:1–23.

Knyazikhin, Y. et al. 1998. Synergistic algorithm for estimating vegetation canopy leaf area index and fraction of absorbed photosynthetically active radiation from MODIS and MISR data. *J Geophys Res Atmos* 103:32257–75.

Knyazikhin, Y. et al. 1999. MODIS Leaf Area Index (LAI) and Fraction of Photosynthetically Active Radiation Absorbed by Vegetation (FPAR) Product (MOD15) algorithm theoretical basis document. Available at http://modis.gsfc.nasa.gov/data/atbd/atbd_mod15.pdf, 130p.

Lambin, E. F., and A. H. Strahler. 1994. Change-vector analysis: A tool to detect and categorize land-cover change processes using high temporalresolution satellite data. *Remote Sens Environ* 48:231–44.

Lawrence, P. J., and T. N. Chase. 2007. Representing a new MODIS consistent land surface in the Community Land Model (CLM 3.0). *J Geophys Res Biogeosci* 112:G01023, doi:10.1029/2006JG000168.

Li, C. C. et al. 2005. Retrieval, validation, and application of the 1-km aerosol optical depth from MODIS measurements over Hong Kong. *IEEE Trans Geosci Remote Sens* 43:2650–8.

Liang, S. L. et al. 2002. Validating MODIS land surface reflectance and albedo products: Methods and preliminary results. *Remote Sens Environ* 83:149–62.

Liu, H. Q., and A. Huete. 1995. A feedback based modification of the NDVI to minimize canopy background and atmospheric noise. *IEEE Trans Geosci Remote Sens* 33:457–65.

Lobell, D. B., and G. P. Asner. 2004. Cropland distributions from temporal unmixing of MODIS data. *Remote Sens Environ* 93:412–22.

Loveland, T. R. et al. 2000. Development of a global land cover characteristics database and IGBP DISCover from 1 km AVHRR data. *Int J Remote Sens* 21:1303–30.

Lunetta, R. S. et al. 2006. Land-cover change detection using multi-temporal MODIS NDVI data. *Remote Sens Environ* 105:142–54.

Maeda, E. E. et al. 2009. Predicting forest fire in the Brazilian Amazon using MODIS imagery and artificial neural networks. *Int J Appl Earth Obs Geoinf* 11:265–72.

Monteith, J. L. 1972. Solar radiation and productivity in tropical ecosystems. *J Appl Ecol* 9:747–66.

Monteith, J. L. 1977. Climate and efficiency of crop production in Britain. *Philos Trans R Soc Lond B Biol Sci* 281:277–94.

Morisette, J. T. et al. 2005. Validation of the MODIS active fire product over Southern Africa with ASTER data. *Int J Remote Sens* 26:4239–64.

Muchoney, D. et al. 1999. The IGBP DISCover confidence sites and the system for terrestrial ecosystem parameterization: Tools for validating global land-cover data. *Photogramm Eng Remote Sensing* 65:1061–7.

Myneni, R. B. et al. 2002. Global products of vegetation leaf area and fraction absorbed PAR from year one of MODIS data. *Remote Sens Environ* 83:214–31.

Nath, A. N., M. V. Rao, and K. H. Rao. 1993. Observed high-temperatures in the sunglint area over the North Indian-Ocean. *Int J Remote Sens* 14:849–53.

Nishihama, M. et al. 1997. *MODIS Level 1A Earth Location: Algorithm Theoretical Basis Document version 3.0, SDST-092.* MODIS Science Data Support Team. Maryland, USA: NASA Goddard Spaceflight Centre, 147p.

Oleson, K. W. et al. 2003. Assessment of global climate model land surface albedo using MODIS data. *Geophys Res Lett* 30:1443, doi:10.1029/2002GL016749.

Pettorelli, N. et al. 2005. Using the satellite-derived NDVI to assess ecological responses to environmental change. *Trends Ecol Evol* 20:503–10.

Potter, C. S. et al. 1993. Terrestrial ecosystem production-a process model based on global satellite and surface data. *Global Biogeochem Cycles* 7:811–41.

Privette, J. L. et al. 2002. Early spatial and temporal validation of MODIS LAI product in the Southern Africa Kalahari. *Remote Sens Environ* 83:232–43.

Quinlan, J. R. 1993. *C4.5: Programs for Machine Learning*. San Mateo, CA: Morgan Kaufmann.

Ramanathan, V. 1987. The role of Earth radiation budget studies in climate and general-circulation research. *J Geophys Res Atmos* 92:4075–95.

Ramanathan, V. et al. 2001. Atmosphere-aerosols, climate, and the hydrological cycle. *Science* 294:2119–24.

Remer, L. A. et al. 2002. Validation of MODIS aerosol retrieval over ocean. *Geophys Res Lett* 29:8008, doi:10.1029/2001GL013204.

Robinson, D. A., K. F. Dewey, and R. R. Heim. 1993. Global snow cover monitoring-an update. *Bull Am Meteorol Soc* 74:1689–96.

Rossow, W. B., and L. C. Garder. 1993. Cloud detection using satellite measurements of infrared and visible radiances for ISCCP. *J Clim* 6:2341–69.

Roy, D. P., P. E. Lewis, and C. O. Justice. 2002. Burned area mapping using multi-temporal moderate spatial resolution data-a bi-directional reflectance model-based expectation approach. *Remote Sens Environ* 83:263–86.

Running, S. W. et al. 1994. Terrestrial remote-sensing science and algorithms planned for EOS MODIS. *Int J Remote Sens* 15:3587–620.

Running, S. W. et al. 1999. MODIS daily photosynthesis (PSN) and annual net primary production (NPP) product (MOD17): Algorithm theoretical basis documents. Available at http://www.ntsg.umt.edu/modis/ATBD/ATBD_MOD17_v21.pdf, 59p.

Russell, G., P. G. Jarvis, and J. L. Monteith. 1989. Absorption of radiation by canopies and stand growth. In *Plant Canopies: Their Growth, Form and Function*, ed. G. Russell, B. Marshall and P. G. Jarvis, 21–39. Cambridge, UK: Cambridge University Press.

Sakamoto, T. et al. 2005. A crop phenology detection method using time-series MODIS data. *Remote Sens Environ* 96:366–74.

Sakamoto, T. et al. 2009. Agro-ecological interpretation of rice cropping systems in flood-prone areas using MODIS imagery. *Photogramm Eng Remote Sensing* 75:413–24.

Salomonson, V. V., and I. Appel. 2004. Estimating fractional snow cover from MODIS using the normalized difference snow index. *Remote Sens Environ* 89:351–60.

Sellers, P. J. et al. 1997. Modeling the exchanges of energy, water, and carbon between continents and the atmosphere. *Science* 275:502–9.

Shao, Y. et al. 2010. Mapping cropland and major crop types across the Great Lakes Basin using MODIS-NDVI data. *Photogramm Eng Remote Sensing* 75:73–84.

Singh, A. 1989. Digital change detection techniques using remotely-sensed data. *Int J Remote Sens* 10:989–1003.

Soudani, K. et al. 2008. Evaluation of the onset of green-up in temperate deciduous broadleaf forests derived from Moderate-Resolution Imaging Spectroradiometer (MODIS) data. *Remote Sens Environ* 112:2643–55.

Strahler, A. H. et al. 1996. MODIS BRDF/albedo product: Algorithm theoretical basis documentation. Version 4.0. NASA/EOS ATBD, 94p.

Tan, B. et al. 2006. The impact of gridding artifacts on the local spatial properties of MODIS data: Implications for validation, compositing, and band-to-band registration across resolutions. *Remote Sens Environ* 105:98–114.

Tanre, D., M. Herman, and P. Y. Deschamps. 1981. Influence of the background contribution upon space measurements of ground reflectance. *Applied Optics* 20(20):3676–84.

Tanre, D., B. N. Holben, and Y. J. Kaufman. 1992. Atmospheric correction algorithm for NOAA-AVHRR products - theory and application. *IEEE Trans Geosci Remote Sens* 30:231–48.

Tian, Y. et al. 2004. Land boundary conditions from MODIS data and consequences for the albedo of a climate model. *Geophys Res Lett* 31:L05504, doi:10.1029/2003GL019104.

Townshend, J. R. G., and C. O. Justice. 2002. Towards operational monitoring of terrestrial systems by moderate-resolution remote sensing. *Remote Sens Environ* 83:351–9.

Tucker, C. J. 1979. Red and photographic infrared linear combinations for monitoring vegetation. *Remote Sens Environ* 8:127–50.

Tucker, C. J. et al. 1981. Remote-sensing of total dry-matter accumulation in winter-wheat. *Remote Sens Environ* 11:171–89.

Tucker, C. J., J. R. G. Townshend, and T. E. Goff. 1985. African land-cover classification using satellite data. *Science* 227:369–75.

Turner, D. P. et al. 2006. Evaluation of MODIS NPP and GPP products across multiple biomes. *Remote Sens Environ* 102:282–92.

Vermote, E. F., and A. Vermeulen. 1999. MODIS Algorithm Technical Background Document, atmospheric correction algorithm: Spectral reflectances (MOD09). NASA contract NAS5-96062 University of Maryland, USA.

Vermote, E. F. et al. 1997. Second simulation of the satellite signal in the solar spectrum, 6S: An overview. *IEEE Trans Geosci Remote Sens* 35:675–86.

Vina, A. et al. 2008. Evaluating MODIS data for mapping wildlife habitat distribution. *Remote Sens Environ* 112:2160–9.

Walthall, C. L. et al. 1985. Simple equation to approximate the bidirectional reflectance from vegetative canopies and bare soil surfaces. *Appl Opt* 24:383–7.

Wang, M. H., and H. R. Gordon. 1994. Radiance reflected from the ocean-atmosphere system-synthesis from individual components of the aerosol-size distribution. *Appl Opt* 33:7088–95.

Wang, Y. J. et al. 2004. Evaluation of the MODIS LAI algorithm at a coniferous forest site in Finland. *Remote Sens Environ* 91:114–27.

Wardlow, B. D., and S. L. Egbert. 2008. Large-area crop mapping using time-series MODIS 250-m NDVI data: An assessment for the U.S. Central Great Plains. *Remote Sens Environ* 112:1096–116.

Wardlow, B. D., S. L. Egbert, and J. H. Kastens. 2007. Analysis of time-series MODIS 250-m vegetation index data for crop classification in the U.S. Central Great Plains. *Remote Sens Environ* 108:290–310.

Xiao, X. M. et al. 2005. Mapping paddy rice agriculture in southern China using multi-temporal MODIS images. *Remote Sens Environ* 95:480–92.

Xiao, Z. Q. et al. 2009. A temporally integrated inversion method for estimating leaf area index from MODIS data. *IEEE Trans Geosci Remote Sens* 47:2536–45.

Yang, W. Z. et al. 2006. MODIS leaf area index products: From validation to algorithm improvement. *IEEE Trans Geosci Remote Sens* 44:1885–98.

Zhan, X. et al. 2002. Detection of land cover changes using MODIS 250-m data. *Remote Sens Environ* 83:336–50.

Zhang, X. Y. et al. 2003. Monitoring vegetation phenology using MODIS. *Remote Sens Environ* 84:471–5.

Zhang, X. et al. 2008. Land cover classification of the North China Plain using MODIS_EVI time series. *ISPRS J Photogramm Remote Sens* 63:476–84.

3

Lidar Remote Sensing

Sorin C. Popescu

CONTENTS

3.1 Foundations of Laser Theory

Laser ranging systems are commonly referred to as "lidar" systems. Lidar is an acronym describing light detection and ranging systems, which are sometimes also referred to as "ladar," either from laser detection and ranging or from laser radar. A universally accepted terminology does not exist, but most commonly we refer to these systems as lidar systems, although the spelling of lidar may differ—lidar, LiDAR, or LIDAR.

Lasers have been one of the greatest scientific developments of the twentieth century. After five decades of achievements in this field, lasers are still a symbol of high technology. The word laser is an acronym that summarizes the nature of laser light—light amplification by the stimulated emission of radiation. Therefore, a laser is a special type of light source with certain characteristics related to its wavelength, output power, duration of emission, beam divergence, coherence, and the systems and materials that generate it.

Albert Einstein developed the foundation of stimulated emission of radiation and published his findings in 1916 and 1917. In essence, Einstein demonstrated that atoms can absorb and emit radiation spontaneously and that atoms in certain excited states can be induced to emit radiation. For about 40 years after Einstein's theoretical work on stimulated emission was published, the concept was used only in theoretical discussions and had little relevance in experimental work. The first successful production of stimulated emission was achieved by Charles H. Townes between 1951 and 1953, who was then at Columbia University; he built a device called a "maser"—microwave amplification by the stimulated emission of radiation. This device produced a coherent beam of microwaves. Later, in 1964, Townes shared the Nobel Prize in physics with two other maser pioneers, Nikolai Basov and Aleksander Prokhorov. A collaborator of Townes, Arthur Schawlow, also received the Nobel Prize in physics in 1981 for research done on lasers. However, the winner of the laser invention race, who is accredited with the development of the first ruby laser in 1960, is Theodore H. Maiman, with what was at that time the Hughes Aircraft Corporation research laboratory (Maiman 1960). The ruby laser is a good example of what we expect a laser to be. The wavelength of the ruby laser is toward the end of the red region of the electromagnetic spectrum, at 694 nm, and it emits coherent waves in short pulses in a concentrated beam of light.

There are many types of lasers. Depending on how they operate, laser sources can emit light in a pulsed mode or as steady beams; the latter are also known as "continuous-wave" (CW) lasers. Laser pulses are characterized by pulse duration and repetition rates. The pulse duration can range from milliseconds to femtoseconds, that is, from 10^{-3} to 10^{-15} seconds (Hecht 1992, p. 13). Because the human eye's response is much slower than the laser pulse frequency, some lasers that may look continuous to the eye are actually pulsed lasers.

Lasers can also be differentiated based on power output, which spans a wide range from milliwatts—thousandths of a watt—to kilowatts. Nevertheless, lasers cannot adjust their power output on demand, but they may be able to adjust it over a limited range. Power output is a characteristic of the materials that produce lasers. Moreover, each type of laser-producing material emits laser light with a characteristic wavelength or a range of wavelengths. Table 3.1 presents a list of the most common laser types based on materials

TABLE 3.1

Most Common Laser Types

Laser Material Type	Laser Light Wavelength (nm)
Organic dye dissolved in solvent	300–1000 (tunable laser)
Rare gas ions (e.g., argon ion)	450–530
Helium neon	543 (red), 632.8 (green), 1150 (near-infrared)
Semiconductor	670–680 and 750–900
Nd: YAG	1064 (near-infrared)
Hydrogen fluoride	2600–3000

producing laser light and their wavelengths, in nanometers, although the actual list is much longer with lasers ranging from the ultraviolet region to the microwave region.

3.1.1 How Is Laser Light Generated?

A laser is a light source with unique properties. As the expansion of the acronym suggests, a laser amplifies light signals that stimulate emission of radiation. The stimulated emission occurs in an amplifying medium contained in an optical resonator or cavity, which holds the amplified light and redirects it through the medium for repeated amplifications. A set of two mirrors feed the light back into the amplifier medium. One cavity mirror reflects essentially all of the light back into the amplifying medium, whereas the other mirror transmits a constant fraction of the light, for example, 10%, which becomes the laser beam, and reflects the rest back into the medium.

3.2 Laser Light Properties

Laser light has important properties that differentiate it from white or ordinary light, most notably coherence, wavelength and spectral purity, directionality, beam divergence, power modulation, and polarization. Probably the best known property of laser light is coherence.

3.2.1 Coherence of Laser Light

Figure 3.1 illustrates the concept of coherence, when light waves are in phase with one another, which means their peaks are lined up at the same point in time. To have coherent waves, light waves must start with the same phase at the same position, and they also need to have the same wavelength, that is, to be spectrally pure. Perfect coherence is difficult to achieve, and not all types of laser light are equally coherent. Coherence can be characterized as spatial or temporal. Laser light waves may encounter different optical path conditions, which make them drift out of phase. As such, temporal coherence is defined by how long the laser light waves remain in phase as they travel. Spatial coherence measures the area over which light waves are coherent, and it is the essential prerequisite that gives a strong directionality to laser beams. Probably the most fundamental difference between laser light and radiation from other light sources, such as ordinary light, is that laser light has the potential to generate beams with very high temporal and spatial coherence.

Coherent light waves in phase with each other

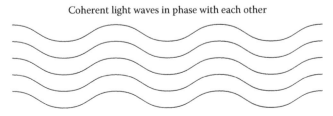

FIGURE 3.1
Coherent electromagnetic waves.

3.2.2 Laser Wavelength and Spectral Purity

Laser light is commonly considered monochromatic, meaning that all photons have nearly the same wavelength. Although lasers normally emit a range of wavelengths, the bandwidth of even the broadest-band laser is much narrower than that of ordinary light.

3.2.3 Laser Beam Divergence

Laser light can form tightly focused beams that travel long distances without spreading out like ordinary light. The most common definition of beam divergence is based on the spreading angle measured in milliradians (mrad). The divergence of most CW lasers is around 1 mrad, whereas for pulsed lasers it may be slightly larger. For reference purposes, a full circle, or 360°, equals 2π radians, 1 radian equals approximately 57.30°, and 1 mrad corresponds to 0.057°. Laser beam divergence is usually reported for the far field, at large distances from the laser, and the divergence angle is normally measured from the center of the beam to the edge (Hecht 1992). Most commonly, beam divergence is considered the angle between the beam sides (Baltsavias 1999). No matter how divergence is measured, calculating the size of the beam or the laser footprint diameter is a trigonometric problem (see Equation 3.1). Where do we consider the edge of the beam? Laser beam propagation can be approximated by assuming that the beam is a Gaussian-type beam, which means that the intensity profile follows a Gaussian function, with the transverse irradiance profile shown in Figure 3.2. This profile shows that the beam intensity gradually drops off toward the sides of the beam, and the beam edge is considered where intensity has fallen to $1/e^2$ or 13.5% of its peak, or maximum axial value (Hecht 1992).

Figure 3.3a shows an exaggerated divergence of a laser beam in a simplified representation that ignores the near range of the laser beam where the light rays remain parallel, sometimes called the "Rayleigh range."

$$D = 2H \tan(\theta/2) \tag{3.1}$$

where

D is the beam diameter (diameter of illuminated area or footprint), H is the distance from the laser to the illuminated spot (flying height for airborne laser scanning), and θ is the divergence angle.

For a small divergence angle and large distances occurring in airborne laser applications, the angle in radians offers a good enough approximation of its tangent function; therefore, a commonly used formula is

$$D = H\theta \tag{3.2}$$

For example, for a beam with a divergence of 1 mrad and a distance of 1000 m, the footprint diameter becomes 1 m. With airborne laser scanning, the illuminated footprint size and shape is also affected by the scanning angle and the slope of the terrain. With airborne laser scanning, Equation 3.2 can be used for calculating footprint diameter of laser beams at nadir, but for laser beams at a certain scan angle on a flat terrain, Equation 3.3 provides a more appropriate calculation using trigonometry with triangles ABC and ABD shown in Figure 3.3b:

$$D = H\left[\tan\left(\theta_{scan} + \frac{\theta}{2}\right) - \tan\left(\theta_{scan} - \frac{\theta}{2}\right)\right] \tag{3.3}$$

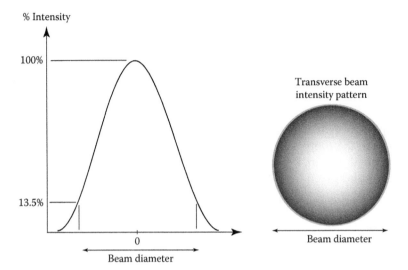

FIGURE 3.2
Transverse beam intensity pattern and beam diameter.

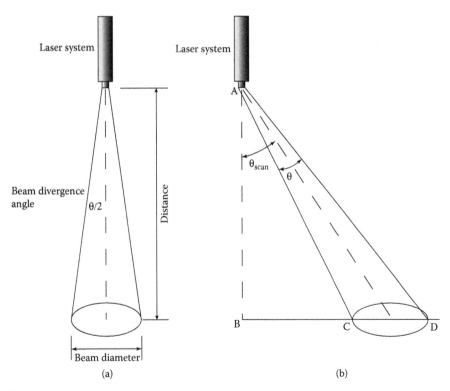

FIGURE 3.3
Beam divergence at (a) nadir and (b) a certain scan angle.

A formula that is easier to use can be derived from Equation 3.3 by considering the laser path distance equal to $H/\cos(\theta_{scan})$. Projecting the footprint on the flat terrain gives the following formula:

$$D = \frac{H}{\cos^2\theta_{scan}}\theta \qquad (3.4)$$

For the example given for Equation 3.4, for a beam with a divergence of 1 mrad and a flying height of 1000 m, the footprint diameter at a scanning angle of 20° from nadir becomes 1.13 m. For inclined terrain, the footprint size calculation becomes more complicated, with details and formulas given by Baltsavias (1999).

3.3 Laser Ranging

Laser range finding uses the same principles as radar distance measurements, with the major difference being the use of shorter wavelengths of the electromagnetic spectrum. The basic principle of laser ranging is the measurement of the time it takes for a laser signal to travel from the transmitter to the reflecting surface of a target and back to the receiver, although two major physical effects are used: For pulsed lasers, the traveling time of light pulse is measured and converted to a distance estimate, whereas for CW lasers, ranging is obtained by measuring the phase difference between the transmitted and the received signals. These range-finding techniques belong to time-of-flight (TOF) methods.

Soon after lasers were invented, precise distance measurements were obtained through laser range finding. In the late 1960s, the National Aeronautics and Space Administration (NASA) used lasers to measure the distance from the Earth to reflectors installed on the Moon by Apollo missions. Armed forces use lasers to measure distances to targets on the battlefield, whereas a plethora of handheld laser range finders are used in hunting, golf, archery, and other sports. Terrestrial field surveyors and engineers also use range finders, more recently coupled with theodolites in total stations. With respect to laser ranging for remote-sensing purposes, laser sensors are installed on air- or spaceborne platforms, which most commonly employ pulsed laser systems with scanning technology. Ground-based laser sensors are installed on tripods and are capable of scanning targets on the ground from various angles.

Most laser ranging applications use pulsed lasers, usually solid-state lasers with high power outputs. A common laser type is the neodymium-doped yttrium aluminum garnet (Nd: YAG) laser, which emits light with a wavelength of 1064 nm, in the infrared portion of the electromagnetic spectrum, with pulse widths around 10 nanoseconds and several megawatts of power (Wehr and Lohr 1999).

For range measurements with pulsed lasers, the laser system measures the traveling time between the emitted pulse and the received echo, and the distance between the ranging unit and the target surface is calculated by

$$D = c\frac{t}{2} \qquad (3.5)$$

where c is the speed of light and t is the pulse travel time; t is divided by two since the pulse travels twice the distance to the target, that is, from transmitter to target and from target to receiver.

Travel time is measured by a time counter relative to the leading edge of the pulse (Figure 3.4). The leading edge is not well defined, but generally it is considered as a fraction of the signal peak to avoid issues caused by various pulse amplitudes (Baltsavias 1999).

For accurate range measurements, the laser pulse should be short. Equation 3.6 relates range resolution (ΔD) and time resolution (Δt):

$$\Delta D = c \frac{\Delta t}{2} \tag{3.6}$$

Equation 3.6 shows that the range resolution is determined by the resolution of the time interval measurement. As such, for a 10-nanosecond pulse, the range resolution is 3×10^5 km/s $\times 10^{-8}$ s/2 = 1.5 m. Equations 3.5 and 3.6 show that range measurement accuracy does not depend on the distance. The term "resolution" should not be confused with range measurement accuracy. Range resolution refers to the smallest change in the distance that can be resolved with the TOF laser. Range accuracy refers to the largest total error in measuring distances and is usually in the order of centimeters for airborne laser range finders, although it differs in the vertical and horizontal axes.

For CW lasers, ranging is obtained by modulating the laser intensity with a sinusoidal signal. The traveling time is proportional to the phase difference between the transmitted and the received signals, and the distance information is extracted from the received signal by comparing its modulation phase with that of the emitted signal. Due to laser complexity in achieving a similar ranging performance to pulsed lasers, CW lasers are rarely used.

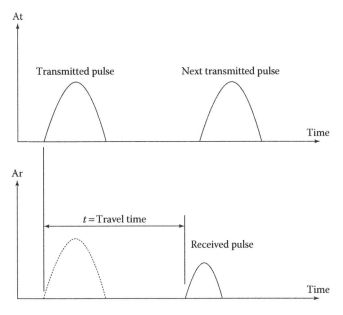

FIGURE 3.4
Variation of beam power level with time.

3.4 Laser Ranging Power Balance

Airborne and satellite laser range measurements are influenced by atmospheric conditions, laser power, and the reflectivity of the target. The power of the laser echo received at the sensor is directly proportional to target reflectivity. Table 3.2 shows the typical reflectivity of various materials for a laser wavelength of 900 nm. Range and reflectivity are directly related, or more specifically, the range is proportional to the square root of reflectivity.

$$R \propto \sqrt{\rho} \qquad (3.7)$$

Equation 3.7 can be used to determine a correction factor for maximum laser range depending on the reflectivity of the target, as shown in Figure 3.5. As the figure shows, targets with a reflectivity of 40% restrict maximum range to about 70% of the maximum range for a target with 80% reflectivity. When flying an airborne laser scanning system over forests with mixed species, coniferous and deciduous, it is important to be aware of the maximum range limitations for the two species types. For coniferous trees with a typical reflectivity of about 30%, the maximum range is approximately 60% of that for deciduous trees, which have a typical reflectivity of about 60%.

The reflectivity of a target also affects the minimum size of a detectable object. For example, if we ignore the influence of other factors, such as atmospheric conditions, target shape, or terrain slope, a laser system that is capable of measuring the distance to a target with a reflectivity of 30% should be capable of detecting a target with a reflectivity of 60% that is half the size of the less-reflective target.

TABLE 3.2

Reflectivity Values for Various Diffuse Reflecting Materials and Surfaces, Natural and Human-Made, for a Laser Wavelength of 900 nm

Material	Reflectivity
White paper	Up to 100%
Dimension lumber (pine, clean, dry)	94%
Snow	80%–90%
White masonry	85%
Limestone, clay	Up to 75%
Deciduous trees	Typically 60%
Coniferous trees	Typically 30%
Carbonate sand (dry)	57%
Carbonate sand (wet)	41%
Beach sands, bare areas in dessert	Typically 50%
Rough wood pallet (clean)	25%
Concrete, smooth	24%
Asphalt with pebbles	17%
Lava	8%

Source: Adapted from Riegl, U. S. A. (n.d.). http://www
.rieglusa.com (retrieved January 16, 2008).

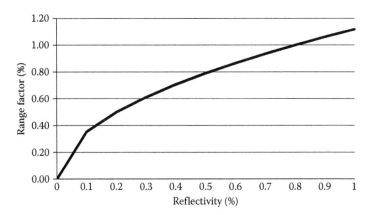

FIGURE 3.5
Correction factor for maximum laser range based on target reflectivity (normalized for 80% reflectivity).

3.5 Enabling Technologies

In the late 1960s, NASA used lasers to measure the distance from the Earth to reflectors installed on the Moon by Apollo missions. About 3 decades later, since the mid-1990s, laser ranging using airborne and terrestrial scanners became accepted as a proven technology with multiple applications for the surveying and mapping communities. There are no major differences between the optical and mechanical principles of airborne, spaceborne, and terrestrial lidar scanning systems, other than those in the mounting platforms and the complexity of additional technologies for determining sensor position and orientation. These additional or enabling technologies, along with advances in laser sensor technology, have defined the developmental stages of scientific and commercial laser scanning systems.

3.5.1 Global Positioning System Unit

Global positioning system (GPS) units have become essential components of navigation systems and surveying tools. This system is a key component of "direct georeferencing," which consists of the direct recording of the position and orientation parameters of a remote sensing instrument used for registering the acquired data to a geographic coordinates system. Mapping applications of direct georeferencing include aerial photogrammetry and airborne lidar.

3.5.2 Inertial Measurement Unit

The inertial measurement unit (IMU) is sometimes referred to as a part of the inertial navigation system (INS), which integrates other components in addition to the IMU, such as a navigation processor to handle navigational computations, a GPS, an electronic compass, or a barometric system. The IMUs detect motion with respect to a hypothetical stationary reference system and normally contain three gyroscopes and three accelerometers, all orthogonal, measuring angular velocities and linear accelerations, respectively. By processing the signals from these devices, normally recorded at a frequency of 50–1000 Hz, it is possible to track the position and orientation of the device, the current rate of acceleration, and changes in rotational attributes, including pitch, roll, and yaw.

In an airborne lidar application, the IMU is used to measure the orientation of the laser beam at the exact time of a range measurement. Most commonly, the IMU is mounted rigidly to the lidar sensor housing in order to provide orientation parameters with respect to the laser reference point, which can be the laser scanning mirror or the fiber optic bundle, depending on the sensor scanning principle. A calibration process known as "boresight calibration" corrects for the mounting misalignment between the IMU and the lidar reference frame. Typically, the boresight calibration of a sensor requires airborne calibration using a reference surface test to correct for pitch, roll, and heading offsets.

The GPS-aided INS provides direct measurement of the position and orientation parameters and is also referred to as a direct georeferencing system. When used with a remote sensor, such as a lidar sensor or a digital camera, direct georeferencing provides all the information needed to register the acquired data in geographic coordinates. Since the mid-1990s, direct georeferencing has become an alternative to aerial triangulation by either totally replacing it or complementing it. Aerial triangulation is used to solve for aerial photography camera exterior orientation parameters, which convey the information necessary to tie image measurements to ground coordinates for planimetric and topographic map compilation, orthophoto production, and digital terrain model editing. Direct georeferencing systems are integral components of airborne remote sensing systems, including lidar, interferometric synthetic aperture radar, and digital cameras.

3.6 Components of a Lidar System

A lidar system may include different components depending on the mounting platform. These can be air- or spaceborne components or ground-based components. Ground-based lidar systems, also referred to as "terrestrial" lidar or laser scanning systems, can be mounted on mobile and fixed, but portable, platforms. Air- or spaceborne instruments can fly on rotary or fixed-wing platforms and satellites, respectively. The basic components of an airborne lidar system are shown in Figure 3.6. For airborne systems, the three main components include (1) a laser ranging unit, (2) an orientation unit, most commonly referred to as the IMU and (3) the GPS unit. Computer hardware and software integrate data streams coming from all components and provide data storage and a variety of post-acquisition registration, processing, and export functions. Terrestrial lidar systems vary in complexity and may include the same components as an airborne system, especially for mobile units mounted on vehicles or boats, or may have a simpler construction for fixed units mounted on a tripod. The latter types may include only the laser ranging unit, a computer, data storage components, and an optional digital camera.

The IMU describes the orientation or the attitude of the unit in terms of roll, pitch, and yaw (Figure 3.7) and serves the characterization of flight dynamics and the derivation of accurate ground coordinates for each laser shot. The GPS unit consists of an onboard differential GPS receiver, which is commonly assisted by one or more ground stations for improving the accuracy of laser footprint coordinates after post-processing.

By knowing the location of the sensor platform and the sensor in three-dimensional (3D) coordinates (GPS-provided data), the trajectory of the laser beam provided by the orientation of the sensor (IMU data), the angle of the laser pulse relative to the sensor, recorded by the laser scanning device, and the range to targets on the Earth's surface as measured by the laser ranging unit, we can compute accurate 3D coordinates for each laser footprint

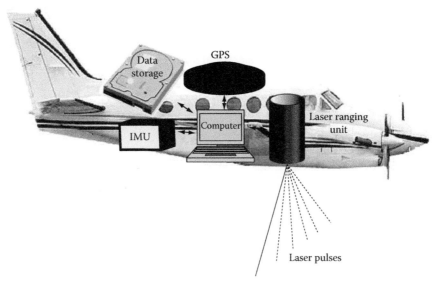

FIGURE 3.6
Basic components of an airborne lidar system. GPS = global positioning system; IMU = inertial measurement unit.

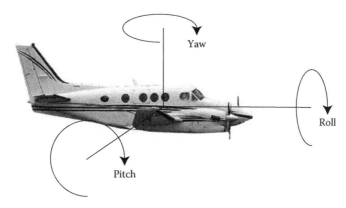

FIGURE 3.7
Orientation unit (inertial measurement unit) detects changes in roll, pitch, and yaw.

on the ground. All these data sources that allow the calculation of the 3D coordinates in a post-processing mode are integrated by computer hardware and software and linked together using a time stamp.

3.6.1 Laser Ranging Unit

The principles of laser ranging are described in Section 3.3. The pulse ranging measurement principle is employed by most airborne, satellite, and terrestrial systems, and they commonly include a laser transmitter and a receiver, each with their associated optics. The laser ranging unit may be coupled with an optical and mechanical scanning unit that deflects the laser beams across their flight path to collect a swath of ranging data.

The transmitter part of the laser ranging unit expands the laser beam to reduce the area density of the laser-pulse-transmitted energy and controls the divergence of the laser beam (Fujii and Fukuchi 2005). The receiver part works as a photodiode by converting the backscattered laser light intensity into electrical impulses. The received laser power of the backscattered echo is only a small fraction of the transmitted power.

3.7 Types of Laser Sensors

Different lidar sensors may have similar components, but the recorded data may be of distinct formats. This section categorizes lidar sensors based on their ability to record discrete returns or waveform data and presents the three different platforms used for acquiring lidar data, terrestrial, airborne, and satellite-based.

3.7.1 Discrete-Return Lidar Sensors

The design of the receiver part of the laser ranging unit is particularly important as it may determine the type of lidar data the receiver records—discrete return measurements or the full waveform. In the first case, a laser pulse may provide multiple returns depending on the type of surface it intercepts. When the laser beam hits porous objects, such as the forest canopy, it may intercept foliage or tree branches over part of the laser footprint, which may backscatter enough energy to trigger the recording of the travel time by the laser receiver (Figure 3.8). After hitting the top of the canopy, part of the laser beam may continue its travel through openings in the canopy until it again hits another layer of foliage or branches, or possibly the ground, which may generate secondary returns of the same pulse. Depending on the complexity of the forest canopy and the settings of the laser receiver, a laser pulse may generate up to four or five discrete returns, sometimes with less dependence on the limitations imposed by the receiver.

Ideally, a laser pulse hitting the forest canopy would provide a return from the top of the canopy—the first return in Figure 3.8—and it would still be able to penetrate to the ground and record a last return from the forest floor—the third return in Figure 3.8. Such measurements allow us to accurately characterize vegetation height and the terrain elevation under the canopy. Some of the laser pulses intercepting the canopy may provide only one return when foliage, branches, or tree trunks block the entire footprint or when these pulses hit the bare ground without intercepting tall layers of vegetation. Similarly, when the laser footprint covers completely nonporous objects, such as roofs, sides of buildings, or other human-made structures, the laser pulse will provide only one return. Therefore, discrete-returns lidar data include first returns, intermediate returns, and last returns.

Most discrete-returns lidar sensors use constant fraction discriminators (CFDs) to minimize the "range walk" or systematic variation in range with signal level. Backscattered laser signals have varying amplitudes depending on the initial pulse energy, size of the intercepted object, and target reflectance characteristics. In order to handle such variations, most laser receivers use a constant amplitude ratio to identify a laser return and record its travel time and power, denoted as amplitude or intensity. The CFD is used to define the leading edge of the pulse (Figure 3.9), which, as explained in Section 3.3, is not well defined but generally considered to be a fraction of the signal peak to avoid issues caused by various pulse amplitudes (Baltsavias 1999).

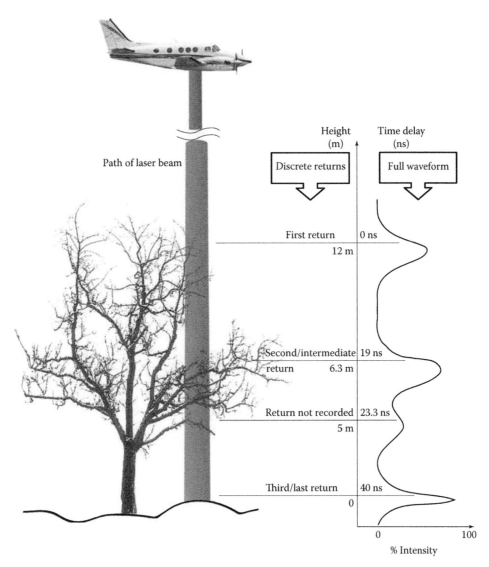

FIGURE 3.8
Laser beam interaction with vegetation and variation of the backscattered laser signal.

The CFD-based receivers need to reset their detectors to prepare for the next pulse or the next echo returned from the same pulse; therefore, there is a time separation between returns recorded for the same echo. Although the reset time, sometimes referred to as "nominal dead time," varies with sensors and manufacturers, it is most commonly around 8–10 nanoseconds. This reset time translates to a range separation of 1.2–1.5 m between the recorded returns of the same pulse, when considering to- and from-target travel times.

The reset time and the minimum range separation between multiple returns have implications for detecting ground covered by vegetation. When the ground is covered by tall grasses or shrubs with heights less than 1.2–1.5 m, the laser beam may provide a return from the top of the vegetation cover and may penetrate to the ground and generate a secondary ground return. This ground return may not be detected due to the fact

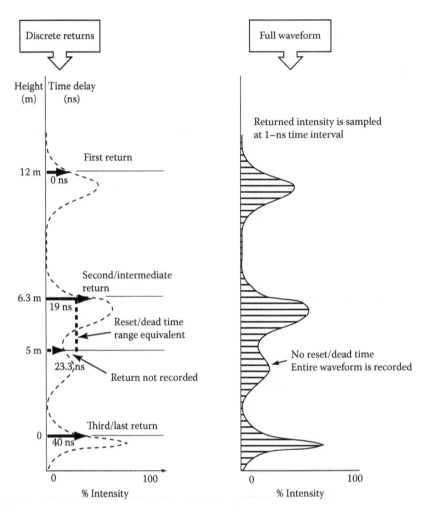

FIGURE 3.9
Conceptual differences between discrete-return and waveform lidar systems.

that no returns are recorded by the receiver within the reset time. This situation does not mean that characterizing ground elevation is always biased toward higher elevation values when there is a low layer of vegetation. A significant number of laser pulses will penetrate to the ground, and laser point classification algorithms will identify lower pulses that most likely hit the ground and use them to generate digital elevation models.

Some airborne lidar sensors manufactured during the mid-1990s (e.g., Optech ALTM 1020, Optech, Inc., Vaughan, Ontario, Canada) could be toggled to record either the first or the last return, and two flights over the same area were necessary to get the bare ground terrain model and the top of the canopy surface, when flown over forest vegetation. Surveys in the U.S. Pacific Northwest carried out using the Optech ALTM 1020 scanning system indicated a minimum 2030% penetration of coniferous canopies (Flood and Gutelius 1997). In the same region, with conifer-dominated stands and dense overstory, Means (2000) observed a very low penetration to the ground of only 1–5%, for a small-footprint lidar. Kraus and Pfeifer (1998) estimated a penetration rate of less than 25% for their lidar study in the Vienna Woods (Wienerwald), in Austria, using an Optech ALTM 1020 lidar system.

A study by Popescu, Wynne, and Nelson (2002) conducted in Virginia over forests of varying age classes including deciduous, coniferous, and mixed stands estimated the penetration rate for the last return laser hits, or first return when there was only one return, to be approximately 4%. The laser point density on the ground, for one flight line, was 0.47 points per square meter for the first return, and the last return when there was only one return; 0.20 points per square meter for the second return, less than half compared to the first return point density; 0.02 points per square meter for the third return; and 0.0001 points per square meter for the fourth return. None of the pulses were able to produce a fifth return for the given vegetation conditions, although the sensor, an AeroScan system that later became Leica's ALS40 sensor (Leica Geosystems, Inc., Heerbrugg, Switzerland), was configured to receive up to five returns.

Since the early 1990s, discrete-returns lidar sensors have experienced major technological advances, reflected mainly in an increased pulse frequency, the recording of multiple returns for each pulse, the recording of intensity information, and the positional accuracy. The latest generation of airborne laser scanners has added waveform-recording capability and the ability to handle multiple pulses in the air. Systems that are able to track echoes from multiple pulses in the air have the potential to significantly increase the productivity of airborne lidar data acquisition systems as these systems, do not depend on receiving the target reflection before starting the next range measurement cycle. More pulses providing range measurements will enable lidar data users to fly a notably wider swath while maintaining the same point densities as conventional systems, or acquire significantly increased point densities for the same swath widths, leading to appreciably reduced flight costs in the end. Due to such innovative technological achievements developed by commercial laser systems manufacturers and the increased number of service providers, airborne lidar is used routinely for topographic mapping, vegetation assessment and forest inventory, 3D urban modeling, wireless communications planning, corridor mapping of power lines and oil pipes, and transportation planning, to mention just a few of the applications. Ground-based laser scanners have been used mainly for surveying and industrial 3D mapping.

Despite the advances in scanning lidar technology, a number of research groups are using airborne lidar profiling systems to extract elevation profiles along flight lines, mainly due to the lower cost of the sensor and the reduced data volume acquired during flight time. Such a system has been developed at NASA for forest research, called a "portable airborne laser system" (PALS), by Nelson, Short, and Valenti (2003). This system is in fact based on off-the-shelf components, including a Riegl laser range finder, a Garmin GPS receiver, and a video camera.

3.7.2 Waveform Lidar Systems

Lidar sensors able to record the entire backscatter amplitude of the laser pulse are referred to as waveform lidar systems. Such sensors have been used from both airborne and satellite platforms.

3.7.2.1 Airborne Lidar Systems

Whereas discrete-returns lidar systems record, for each laser pulse, the time of travel and the intensity of every return, waveform lidar systems record the time-varying intensity of the returned energy from each laser pulse and therefore provide information on the height distribution of the returned intensity (Figure 3.9). The shape of the returned intensity offers a direct description of the vertical distribution of surfaces illuminated by the laser pulse (Harding et al. 2001; Dubayah et al. 2000) and is useful in characterizing complex targets, such as forest canopy. The returned energy is digitized at equal time intervals, such as for every nanosecond.

Terrestrial applications using waveform lidar have been documented from the early 1980s. Aldred and Bonner (1985) were the first to describe an application of waveform lidar to study forest canopies in Canada. For their study, they used a laser system originally developed for characterizing bathymetric water depth and measured forest biophysical properties, such as tree height, canopy cover, and type of tree species—hardwoods, conifers, or mixed. They also studied the effect of different laser beam footprint sizes on stand-height estimates and found that the footprint diameter was not critical when estimating stand height. An interesting investigation in their study looked at different methods of estimating tree height by analyzing waveform start and end points and concluded that the leading edge threshold, followed by the peak-to-peak and trailing edge values, provided the best forest height estimates. Another bathymetric lidar system with dual frequency, 532 and 1064 nm, and waveform recording was used by Nilsson (1996) to measure tree heights and timber volume. The pulse length was 7 nanoseconds, the sampling interval was 2.5 nanoseconds, and the digitized waveform was a combination of the two, green and infrared, returns.

It has been proven that waveform lidar systems are most successful for vegetation analysis. NASA has developed experimental sensors that record the complete waveform from medium- and large-footprint lasers with a ground beam diameter between 5 m and tens of meters as predecessors of spaceborne lidar systems. Two NASA airborne research lidar systems have been used to characterize vegetation: the scanning lidar imager of canopies by echo recovery (SLICER; Blair et al. 1994; Harding et al. 1994) and the laser vegetation imaging sensor (LVIS; Blair and Hofton 1999). The LVIS sensor emits laser pulses with a duration of 10 nanoseconds at full width half maximum (FWHM) and digitizes the detected return energy at 500 megasamples per second or every 0.5 nanoseconds. The LVIS beam diameter depends on the flying height, with a typical size of 10–25 m for its footprint. This sensor has a scan angle of about 12° and can cover 2-km swaths from an altitude of 10 km.

3.7.2.2 *Commercial Waveform-Recording Small-Footprint Lidar*

A discrete-returns lidar system has limitations with respect to the number of echoes it can record from a single pulse. A waveform-recording lidar overcomes this constraint by recording the entire laser pulse energy as a function of time. This approach to recording the laser backscatter amplitude with high frequency affords a better characterization of the vertical distribution of reflecting surfaces within the laser footprint, which for most of the commercial airborne sensors is smaller than 1 m in diameter.

As explained in Section 3.7.1, discrete-returns systems are affected by the reset time between separate returns of the same pulse, which translates to a range separation of 1.2–1.5 m between consecutive echoes, when considering to- and from-target travel times. This has implications for detecting ground covered by vegetation, such as tall grasses or shrubs, when the laser beam may provide a return from the top of the vegetation cover and penetrate to the ground and generate a secondary ground return.

By using adequate modeling of the recorded waveform, it has been shown that full-waveform analysis enables the extraction of additional information compared to discrete-return systems, such as the range to the ground peak underneath tall grasses, shrubs, or forest vegetation (Gutierrez, Neuenschwander, and Crawford 2005). Most often, waveform analysis extracts range, elevation variation, and reflectance properties from the pulse width and amplitude. In their study conducted in 2005, Gutierrez, Neuenschwander, and Crawford compared the elevations derived from a conventional discrete-return system with waveform data collected using the same sensor, an Optech ALTM instrument, and found that elevation data agree well between the two datasets. They also concluded that the waveform data provided increased information about the vertical distribution of reflecting surfaces. A common approach to extracting information from waveform data is to model the waveform as a series of Gaussian distribution functions, as demonstrated for LVIS by Hofton, Minster, and Blair (2000), or Persson et al. (2005).

Some of the commercial airborne lidar systems are able to collect waveforms for small-footprint laser beams, such as the Riegl LMS-Q680 (Riegl USA, Orlando, FL), TopEye Mark II (Blom, Sweden), or Optech's ALTM 3100 (Vaughn, Ontario, Canada). Some sensor manufacturers, such as Optech and Leica, provide the option of waveform digitizer modules that can be integrated with their discrete-return systems to allow full-waveform digitization.

3.7.2.3 Spaceborne Lidar Systems

The lidar waveform-recording technology developed for the NASA airborne systems made use of prototypes of methods and techniques later used by spaceborne altimeter systems such as the shuttle laser altimeter (SLA; Garvin et al. 1998), which in 1996–1997 provided the first global-scale laser altimeter dataset. In 1997, the Mars orbiter laser altimeter (MOLA), an instrument used aboard the *Mars Global Surveyor* spacecraft, acquired its first pass across the surface of Mars. The altimeter obtained measurements of topographic profiles, surface reflectivity, and backscattered laser pulse width, with surface spot sizes of 70–300 m (Smith et al. 1998). The next space-based system was the geoscience laser altimeter system (GLAS) carried on the ice, cloud and land elevation satellite (ICESat), which was launched on January 13, 2003 from the Vandenberg Air Force Base in California.

An overview of the ICESat mission is provided by Schutz et al. (2005). The ICESat laser measurements were designed with the primary objective of monitoring ice sheets mass balance. Measurements are currently distributed in 15 science data products, which have interdisciplinary applications, including the characterization of land topography and vegetation canopy heights. The system operates by sending laser pulses with a frequency of 40 Hz and pulse duration of approximately 5 nanoseconds. The returning laser echo is sampled every nanosecond, and the digitized pulses are referred to as laser waveforms. The ICESat platform orbits at an altitude of approximately 600 km, and from that height above the ground, the laser footprints have approximately a 64-m circular diameter. More precisely, the footprints are elliptical, with their size and ellipticity varying during the course of the mission. Along one orbital transect, the footprints are spaced at about 172-m intervals (Schutz et al. 2005). The GLAS surface elevations are reported with respect to the TOPEX/Poseidon reference ellipsoid. Among all GLAS standard products, the level-1 altimetry products, GLA01, contain waveforms digitized in 544 bins with a bin size of 1 nanosecond or equivalently 15 cm; however, beginning with the data acquisition phase L3A (October 2004), the bin size of BIN 1-151 has been changed to 4 nanoseconds (60 cm) to reduce the risk of waveform truncation. The level-2 global land-surface products, GLA14, provide an alternate fitting that locates up to six Gaussian components (mode, amplitude, and sigma) to characterize the shape of the total waveform. The ICESat spacecraft allows for off-nadir pointing of the laser, by up to ±5°, in order to target areas of interest or to compensate for orbit drift (Schutz et al. 2005).

3.7.3 Flash Lidar

An alternative to scanning the target of interest with pulses of laser light is offered by the flash lidar technology. Rather than using one receiver to detect echoes from each laser pulse, flash lidar uses a focal plane array as a detector to acquire a frame of 3D data from a laser pulse that floods the scene. The concept of the flash lidar detector is similar to that of the focal plane array of a two-dimensional optical digital camera. Each pixel in the array configuration can independently measure the travel time for each laser pulse.

The sensor captures an entire frame of range data from a single pulse of laser light with a certain frequency, such as 60 frames per second. Just like scanning lidar sensors, flash lidar sensors are capable of capturing both range and intensity data.

At the time of this writing, little information is available about the flash lidar technology. Few companies pursue the development of this technology, such as Ball Aerospace & Technologies Corp., Boulder, CO, and Advanced Scientific Concepts, Inc., Santa Barbara, CA.

3.7.4 Ground-Based Lidar

Over the last 30 years, lasers have been incorporated into surveying instruments such as simple range finders or more complex total stations. Such uses of lasers have led to the development of ground-based or terrestrial lidar scanners, which are capable of scanning the landscape surrounding their location. Most of the time, such sensors are set up on a tripod or on vehicles, and they are respectively called "static" and "dynamic" or "mobile" systems. Static systems do not require the integration of supporting technologies or units, such as GPS and IMUs, whereas mobile systems need direct georeferencing through the use of GPS and IMUs.

Ground-based lidar systems have developed considerably over the last decade, and the use of such sensors has resulted in the proliferation of a large number of applications, from surveying, architecture, accident scene reconstruction, monitoring of buildings and bridges, measurement of complex industrial facilities, monitoring of quarries and open mines, and recording of building and monument facades, to geological structures and vegetation analysis.

Most of the ground-based lidar systems utilize the TOF principle for range measurements, although a few employ the phase measuring technique. Depending on the coverage they are capable of illuminating with lidar points, ground-based lidar systems can be differentiated as panoramic, hybrid, or camera-type scanners (Steiger 2003). Panoramic scanners cover the surrounding landscape in a systematic pattern with 360° coverage in the horizontal plane and more than 270° in the vertical plane, practically missing only the area below the instrument's tripod in covering a full spherical field of view. Although the hybrid scanners are capable of scanning a 360° field of view in the horizontal plane, they may have limited scanning angles toward the zenith, since most such scanners are used for topographic applications and are not required to scan objects overhead. The camera-type scanners normally have a limited field of view in both horizontal and vertical planes. Panoramic scanners are the most versatile for indoor or outdoor applications and can be set to cover a limited viewing angle, if so desired.

With respect to the range over which ground-based systems can be used, depending on the manufacturer and the intended application, such systems can record ranges from 100 m to 1 km. The most common terrestrial sensors are manufactured by the same companies that build airborne lidar sensors, such as Leica, Optech, and Riegl, although there are other systems as well, such as Trimble, Topcon, or research systems like Echidna.

With the development of dynamic terrestrial laser scanners, mobile mapping literally takes on new dimensions. The mobile lidar technology has in fact many similarities with airborne lidar, mainly in requiring continuous georeferencing of the moving vehicle that carries one or more sensors. As such, the mobile lidar technology integrates GPS and IMU components. Such systems are mainly used in the urban environment for reproducing facades of buildings from the ground level, which can be integrated with airborne datasets for producing accurate and complete 3D urban models.

3.8 Lidar Data Format

Until recently, discrete-return lidar data were provided in text or binary format, which was usually proprietary, most commonly with geographic coordinates and intensity recordings for multiple returns and pulses making up the point cloud. The drawback of this approach was the lack of portability and consistency among software tools used for processing the datasets from different providers or different sensors. The first version of a standard lidar file format, the LAS 1.0 (Graham 2005), was released in 2002 with the intention of allowing different lidar hardware and software tools to output data in a common format. The initial LAS specification was a relatively compact binary encoding of point location and point attribute data. The third revision of the LAS format specification was released in July 2009, and it is owned by the American Society for Photogrammetry & Remote Sensing (ASPRS). The LAS 1.3 specification includes a noteworthy improvement over previous specifications, that is, the possibility of encoding lidar waveform data. In addition, the LAS 1.3 includes important information regarding the sensor used to collect lidar data, processing software, number of lidar points, point coordinates, intensity, classification, and other relevant data. Table 3.3 shows the standard lidar point classes in LAS 1.3.

TABLE 3.3

Standard Lidar Point Classes in the ASPRS LAS 1.3 Data Format

Classification Value (Bits 0:4)	Meaning
0	Created, never classified
1	Unclassified 1
2	Ground
3	Low vegetation
4	Medium vegetation
5	High vegetation
6	Building
7	Low point (noise)
8	Model key point (mass point)
9	Water
10	Reserved for ASPRS definition
11	Reserved for ASPRS definition
12	Overlap points 2
13–31	Reserved for ASPRS definition

3.9 Examples of Environmental Applications of Lidar Remote Sensing

Environmental applications of lidar remote sensing cover a wide spectrum, such as applications in environmental engineering, mapping geologic faults under the forest canopy, monitoring coastal changes, assessing landslide hazards, quantifying the growth and retreat of ice sheets, and estimating vegetation structural attributes, biophysical parameters, and habitat characterization. Developments in lidar remote-sensing applications for environmental studies are occurring rapidly, and they are driven by intensive research and increasing availability of lidar data from commercial and governmental sources. Two general application trends can be observed: (1) characterizing the topographic features and (2) assessing the 3D structure of vegetation canopies. Topography mapping with lidar remote sensing is potentially the fastest-growing area of environmental applications. Most environmental studies need topographic information, and lidar has proven its ability to acquire highly accurate and detailed elevations, which have a strong influence on the structure, spatial extent, composition, and function of ecological systems. Most often, topographic applications use discrete-returns lidar data provided by commercial remote sensing companies. When deriving topographic information, a substantial number of lidar points in the point cloud, mainly representing vegetation hits, are discarded in the step known as "vegetation removal." On the contrary, for most ecological applications that use discrete-returns lidar data, the lidar returns from the canopy are of the highest interest. In addition to the discrete-return airborne systems, waveform lidar data have been used for characterizing vegetation structure over large areas. Since topographic lidar applications have been described in great detail in other texts, such as the works of Maune (2007) or Shan and Toth (2009), Sections 3.9.1 through 3.9.3 provide examples of lidar remote sensing applications for environmental studies of vegetation and habitat characteristics.

3.9.1 Lidar for Estimating Forest Biophysical Parameters

The use of remote sensing in mapping the spatial distribution of canopy characteristics allows an accurate and efficient estimation of tree dimensions and canopy properties at local, regional, and even global scales. In particular, lidar remote sensing has the capability to acquire direct 3D measurements of the forest structure that are useful for estimating a variety of forest biophysical parameters, such as tree height; crown dimensions, canopy closure, leaf area index, tree density, forest volume, and forest biomass, and in mapping fire risk by assessing surface and canopy fuels.

During the late 1980s, a number of lidar studies for estimating tree height, forest biomass, and carbon date were conducted, for example, studies by Maclean and Krabill (1986), Nelson, Swift, and Krabill (1988), and Nelson, Krabill, and Tonelli (1988). These first studies used profiling lidar systems and developed models to predict stem volume and dry biomass based on forest canopy height and closure as measured by airborne lidar. Since then, numerous researchers have used a variety of lidar systems and sampling techniques to quantify tree dimensions, standing timber volume, aboveground biomass, and carbon date, mainly with scanning systems.

Previous lidar studies, whether using waveform or discrete-return lidar data, attempted to derive measurements, such as tree height and crown dimensions, at stand level (Næsset and Bjerknes 2001; Hall et al. 2005), plot level (Holmgren, Nilsson, and Olsson 2003; Hyyppä et al. 2001; Lim and Treitz 2004; Popescu, Wynne, and Scrivani 2004), or individual tree level (Persson, Holmgren, and Söerman 2002; Coops et al. 2004; Yu et al. 2004; Holmgren and Persson 2004; Roberts et al. 2005; Chen et al. 2006; Koch, Heyder, and Weinacker 2006; Popescu 2007) and then use allometric relationships or statistical analysis to estimate other characteristics, such as biomass, volume, crown bulk density, and canopy fuel parameters. Figure 3.10 shows a lidar point cloud with a point density of 8 points per square meter collected by a discrete-return sensor over coniferous forests in the western United States. Figure 3.11 displays the ground-based lidar data acquired from a tripod system in Mesquite forests in central Texas.

Forest canopy structure was estimated using data from scanning lasers that provided lidar data with full-waveform digitization (Harding et al. 1994, 2001; Lefsky et al. 1997; Means et al. 1999). Small-footprint, discrete-returns systems were used to estimate canopy characteristics, with many studies focusing on tree height (Næsset 1997; Magnussen and Boudewyn 1998; Magnussen, Eggermont, and LaRiccia 1999; Næsset and Økland 2002; Popescu, Wynne, and Nelson 2002; McCombs, Roberts, and Evans 2003; Maltamo et al. 2004; Popescu and Wynne 2004) or crown dimensions, such as the study conducted by Popescu, Wynne, and Nelson (2003). Figure 3.12 shows a portion of a canopy height model of mixed forest conditions in the southern United States. The canopy model has been processed automatically with methods described by Popescu and Wynne (2004) in identifying individual trees, and their heights and crown dimensions have been measured.

After more than two decades of research in vegetation assessment with lidar, the following four aspects could be concluded: First, with waveform lidar systems having large footprints, robust regressions can be developed to predict volume and biomass over large area extents. The R^2 values for plot-level models range from 0.8 to 0.9 (Lefsky et al. 1999, 2002;

FIGURE 3.10
(See color insert following page 426.) Lidar point cloud over coniferous forests in the western United States.

FIGURE 3.11
(See color insert following page 426.) Ground-based lidar data collected over Mesquite trees in central Texas.

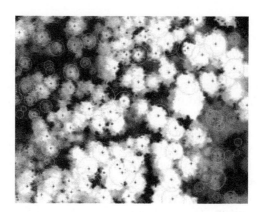

FIGURE 3.12
(See color insert following page 426.) Automatically measuring individual trees on a lidar-derived canopy height model. Circles represent computer-measured crown diameters, whereas each cross sign indicates identified individual trees.

Drake et al. 2002). With discrete-returns systems, that is, scanning lidar systems with small footprints, usually of submeter range, the strength of the prediction models for volume and biomass are more variable, with R^2 values ranging from 0.4 to 0.9 (Nilsson 1996; Næsset 1997, 2002; Nelson, Short, and Valenti 2003; Popescu, Wynne, and Nelson 2003; Popescu, Wynne, and Scrivani 2004; Zhao, Popescu, and Nelson 2009; Popescu and Zhao, 2009). Second, conifer attributes can be estimated with higher accuracy than hardwood parameters. The evidence for this statement is found scattered throughout the literature and may be attributed to the more complex canopy structure of deciduous stands and individual tree growth form, which make height–volume or biomass relationships noisier for hardwoods (Lefsky et al. 1999; Popescu, Wynne, and Scrivani 2004; Næsset 2004). Third, despite intense research efforts and few operational uses, there is a lack of lidar processing tools and, thus, investigators are spending considerable efforts on developing software. Fourth, airborne lidar data can be used to inventory biomass and carbon at scales from local to regional and global. With scanning lidar, biomass and carbon can be accurately estimated at local scales; for examples, see the studies by Popescu et al. (2003, 2004). Using profiling lidar data, as in the studies of Nelson, Short, and Valenti (2003), biomass and carbon can be estimated over large areas, whereas satellite lidar (e.g., ICESat/GLAS) can be used for global estimates of canopy properties (Ranson et al. 2004).

3.9.2 Lidar Applications for Estimating Surface and Canopy Fuels

Few lidar studies focus on assessing canopy structure and characteristics, such as fuel weight, canopy and crown base height, and crown bulk density (Pyysalo and Hyyppä 2002; Holmgren and Persson 2004; Riaño et al. 2003, 2004; Andersen, McGaughey, and Reutebuch 2005; Mutlu et al. 2008; Mutlu, Popescu, and Zhao 2008; Popescu and Zhao 2008). Among these studies, there seems to be a unanimous acceptance that airborne lidar overestimates crown base height for individual trees or plot-level canopy base height, which is an intuitive finding given the fact that airborne lidar portrays crowns from above, and lower branches have a reduced probability of being intercepted by laser pulses that might be blocked by higher branches (Holmgren and Persson 2004; Andersen, McGaughey, and Reutebuch 2005).

3.9.3 Lidar Remote Sensing for Characterizing Wildlife Habitat

Ecologists have long recognized the importance of vegetation structure for characterizing wildlife habitat, but field methods for gathering such information are time consuming and challenging. Vertical forest structure is related to biodiversity and habitat. "In general, the more vertically diverse a forest is the more diverse will be its biota ..." (Brokaw and Lent 1999). Remote sensing techniques provide an attractive alternative (e.g., Turner et al. 2003), especially when 3D data are acquired directly with sensors such as lidar.

Hinsley et al. (2002) and Hill et al. (2003) employed an airborne laser system to assess bird habitat. They used an airborne laser scanning system to map forest structure across a 157-hectare deciduous woodland in the eastern United Kingdom. The researchers related laser-based forest canopy heights to chick mass (i.e., nestling weight), a surrogate for breeding success, which, in turn, is a function of "territory quality." They found that for one species, chick mass increased with increasing forest canopy height, and for a second species, chick mass decreased. Hill et al. (2003) concludes that airborne laser scanning data can be used to predict habitat quality and to map species distributions as a function of habitat structure.

Nelson, Keller, and Ratnaswamy (2005) mapped and estimated the areal extent of Delmarva fox squirrel (DFS) habitat using an airborne profiling lidar flown over Delaware. The study results indicated that (1) systematic airborne lidar data can be used to screen extensive areas to locate potential DFS habitat; (2) 78% of sites meeting certain minimum length, height, and canopy closure criteria will support DFS populations, according to a habitat suitability model; (3) airborne lidar can be used to calculate county and state acreage estimates of potential habitat; and (4) the linear transect data can be used to calculate selected patch statistics.

Hyde et al. (2005) used a large-footprint (12.5 m) scanning lidar to map California spotted owl habitat across a 60,000-hectare study area in the Sierra Nevada, California. They looked at forest canopy height, canopy cover, and biomass in the mountainous forests. Their ultimate objective was to produce maps for the U.S. Forest Service for wildlife habitat and forest resource management and to conclude that lidar provides "important metrics that have been exceptionally difficult to measure over large areas."

Recent studies, such as the ones conducted by Clawges et al. (2008) or Vierling et al. (2008), show the potential of using airborne lidar in studying animal–habitat relationships and in quantifying the vegetation structural attributes important for wildlife species. Clawges et al. used lidar to assess avian species diversity, density, and occurrence in a pine aspen forest in South Dakota. They concluded that lidar data can provide an alternative to field surveys for some vegetation structure indices, such as total vegetation volume, shrub density index, and foliage height diversity. They calculated different foliage height diversity indices using various foliage height categories and found that habitat assessment may be enhanced by using lidar data in combination with spectral data.

3.10 Lidar Systems for Atmospheric Studies

Although this chapter focuses on lidar remote sensing for environmental applications, laser remote sensing technologies are also used efficiently for providing four-dimensional—space and time—measurements of the atmosphere and its constituents. Range-resolved measurements of the atmosphere have been carried out from the ground, air, and space.

In fact, Middleton and Spilhaus (1953), who are credited with coining the lidar acronym, did so in the context of meteorological instruments, but without expressly mentioning what it could be the acronym of. Fiocco and Smullin described atmospheric measurements with a ruby laser in 1963. Currently, lidar systems used in atmospheric studies observe spatial and temporal distribution of atmospheric gases, atmospheric pressure, temperature, turbulence, and wind (Weitkamp, 2005). Physical processes observed with these lidars include laser backscattering by aerosols and clouds (Mie scattering), laser backscattering by molecules (Rayleigh scattering), absorption by atoms and molecules (differential absorption lidar [DIAL]), Raman scattering, fluorescence, and Doppler shift by aerosols and clouds (Doppler lidar). These lidar systems are not discussed further in this book. Similarly, this book does not discuss short-distance laser remote sensing technologies used in industrial, security, and medical applications. Interested readers can find a relatively rich lidar literature for atmospheric studies both in book and scientific articles formats, for example, studies by Fujii and Fukuchi (2005).

3.11 Conclusions

Lidar data availability is increasing along with the spectrum of lidar applications in environmental remote sensing at a multitude of scales and the user's need for up-to-date information on sensors, processing algorithms, and applications. As such, the goal of this chapter is to provide the fundamentals of lidar remote sensing technology and some examples of environmental applications of this technology for characterizing the 3D structure of vegetation canopies.

The present widespread use of lidar remote sensing offers an optimistic vision of the future for environmental applications and research investigations. Intrinsic lidar data structure allows the integration of data acquired by different platforms, terrestrial, airborne, and spaceborne, as complementary or validation tools for applications at multiple scales from local to regional and global. In addition, the fusion of lidar and optical or radar data aims at reducing the limitations of each technology and utilizing their synergistic characteristics for complex environmental assessment. In the context of global climate and environmental changes, lidar proves to be an important technology that makes possible the analysis of the 3D structure of vegetation canopies and facilitates operational applications and scientific discovery. There is no doubt that lidar will continue to be one of the most important geospatial data acquisition technologies subject to continuous developments of all its components: acquisitions systems and hardware, data formats, processing algorithms and software, operational principles, quality, accuracy, and standards.

References

Aldred, A., and M. Bonner. 1985. *Application of Airborne Lasers to Forest Surveys*, p. 62. Canadian Forestry Service, Petawawa National Forestry Centre, Information Report PI-X-51. Ottawa, Canada.

Andersen, H. E., R. J. McGaughey, and S. E. Reutebuch. 2005. Estimating forest canopy fuel parameters using LiDAR data. *Remote Sens Environ* 94:441–9.

Baltsavias, E. 1999. Airborne laser scanning: Basic relations and formulas. *ISPRS J Photogramm Remote Sens* 54:199–214.

Blair, J. B., D. B. Coyle, J. L. Bufton, and D. J. Harding. 1994. Optimization of an airborne laser altimeter for remote sensing of vegetation and tree canopies. In *Proceedings of IGARSS'94*, vol. II, 939–41. Pasadena, CA: IEEE Geoscience and Remote Sensing Society.

Blair, J. B., and M. A. Hofton. 1999. Modeling laser altimeter return waveforms over complex vegetation using high-resolution elevation data. *Geophys Res Lett* 26(16):2509–12.

Brokaw, N., and R. Lent. 1999. Vertical structure. In *Maintaining Biodiversity in Forest Ecosystems*, ed. M. L. Hunter Jr., 373–99. New York: Cambridge University Press.

Chen, Q., D. Baldocchi, P. Gong, and M. Kelly. 2006. Isolating individual trees in a savanna woodland using small footprint LiDAR data. *Photogramm Eng Remote Sens* 72(8):923–32.

Clawges, R., K. T. Vierling, L. A. Vierling, and E. Rowell. 2008. Use of airborne lidar for assessment of avian habitat and estimation of select vegetation indices in the Black Hills National Forest, South Dakota, USA. *Remote Sens Environ* 112:2065–73.

Coops, N. C., M. A. Wulder, D. S. Culvenor, and B. St-Onge. 2004. Comparison of forest attributes extracted from fine spatial resolution multispectral and lidar data. *Can J Remote Sens* 30(6):855–66.

Drake, J. B., R. Dubayah, R. Knox, D. B. Clark, and J. B. Blair. 2002. Sensitivity of large-footprint lidar to canopy structure and biomass in a neotropical rainforest. *Remote Sens Environ* 81:378–92.

Dubayah, R., R. Knox, M. Hofton, J. B. Blair, and J. Drake. 2000. Land surface characterization using lidar remote sensing. In *Spatial Information for Land Use Management*, ed. M. J. Hill and R. J. Aspinall, 25–38. Singapore: International Publishers Direct.

Flood, M., and Gutelius, B. 1997. Commercial implication of topographic terrain mapping using scanning airborne laser radar. ISPRS Journal of *Photogrammetry & Remote Sensing* 63(4): 327–66.

Fujii, T., and Fukuchi, T., eds. 2005. *Laser Remote Sensing*. Boca Raton, FL: Taylor & Francis/CRC Press.

Graham, L. 2005. The LAS 1.1 data format. *Photogramm Eng Remote Sensing* 71(7):777–80.

Gutierrez, R., A. Neuenschwander, and M. Crawford. 2005. Development of laser waveform digitization for airborne lidar topographic mapping instrumentation. *Proc IEEE Int Geosci Remote Sens Symp* 2:1154–7.

Hall, S. A., I. C. Burke, D. O. Box, M. R. Kaufmann, and J. M. Stoker. 2005. Estimating stand structure using discrete-return LiDAR: An example from low density, fire prone ponderosa pine forests. *For Ecol Manage* 208:189–209.

Harding, D. J., J. B. Blair, J. B. Garvin, and W. T. Laurence. 1994. Laser altimetry waveform measurement of vegetation canopy structure. *Proceedings of the International Geoscience and Remote Sensing Symposium—IGARSS '94*, pp. 1251–3. Noordwijk, the Netherlands: ESA Scientific and Technical Publication.

Harding, D. J., M. A. Lefsky, G. G. Parker, and J. B. Blair. 2001. Lidar altimeter measurements of canopy structure: Methods and validation for closed-canopy, broadleaf forests. *Remote Sens Environ* 76:283–97.

Hecht, J. 1992. *Understanding Lasers: An Entry Level Guide*. New York: IEEE Press.

Hill, R. A., S. A. Hinsley, P. E. Bellamy, and H. Balzter. 2003. Ecological applications of airborne laser scanner data: Woodland bird habitat modeling. In *Proceedings of ScandLaser Scientific Workshop on Airborne Laser Scanning of Forests*, ed. J. Hyyppä, E. Næsset, H. Olsson, T. Granqvist Pahlén and H. Reese, 78–87. Sweden: Umeå.

Hinsley, S. A., R. A. Hill, D. L. A. Gaveau, and P. E. Bellamy. 2002. Quantifying woodland structure and habitat quality for birds using airborne laser scanning. *Funct Ecol* 16:851–7.

Hofton, M. A., J. -B. Minster, and J. B. Blair. 2000. Decomposition of laser altimeter waveforms. *IEEE Trans Geosci Remote Sens* 38:1989–96.

Holmgren, J., M. Nilsson, and H. Olsson. 2003. Estimation of tree height and stem volume on plots using airborne laser scanning. *For Sci* 49(3):419–28.

Holmgren, J., and Å. Persson. 2004. Identifying species of individual trees using airborne laser scanner. *Remote Sens Environ* 90:415–23.

Hyde, P., R. Dubayah, B. Peterson, J. B. Blair, M. Hoften, C. Hunsaker, R. Knox, and W. Walker. 2005. Mapping forest structure for wildlife habitat analysis using waveform lidar: Validation of montane ecosystems. *Remote Sens Environ* 96:427–37.

Hyyppä, J., O. Kelle, M. Lehikoinen, and M. Inkinen. 2001. A segmentation-based method to retrieve stem volume estimates from 3-D tree height models produced by laser scanners. *IEEE Trans Geosci Remote Sens* 39(5):969–75.

Koch, B., U. Heyder, and H. Weinacker. 2006. Detection of individual tree crowns in airborne LiDAR data. *Photogramm Eng Remote Sensing* 72(4):357–64.

Kraus, K., and N. Pfeifer, 1998. Determination of Terrain Models in Wooded Areas with Airborne Laser Scanner Data. ISPRS-Journal of *Photogrammetry and Remote Sensing* 53:193–203.

Lefsky, M. A., W. B. Cohen, S. A. Acker, T. A. Spies, G. G. Parker, and D. Harding. 1997. LiDAR remote sensing of forest canopy structure and related biophysical parameters at the H. J. Andrews experimental forest, Oregon, USA. In *Natural Resources Management Using Remote Sensing and GIS*, ed. J. D. Greer, 79–91. Washington, DC: ASPRS.

Lefsky, M. A., W. B. Cohen, D. J. Harding, G. G. Parker, S. A. Acker, and S. T. Gower. 2002. LIDAR remote sensing of above-ground biomass in three biomes. *Glob Ecol Biogeogr* 11:393–9.

Lefsky, M. A., D. J. Harding, W. B. Cohen, G. G. Parker, and H. H. Shugart. 1999. Surface LIDAR remote sensing of basal area biomass in deciduous forests of eastern Maryland, USA. *Remote Sens Environ* 67:83–98.

Lim, K. S., and P. M. Treitz. 2004. Estimation of aboveground forest biomass from airborne discrete return laser scanner data using canopy-based quantile estimators. *Scand J For Res* 19(6): 558–70.

Maclean, G. A., and W. B. Krabill. 1986. Gross merchantable timber volume estimation using an airborne LIDAR system. *Can J Remote Sens* 12:7–18.

Magnussen, S., and P. Boudewyn. 1998. Derivations of stand heights from airborne laser scanner data with canopy-based quantile estimators. *Can J For Res* 28:1016–31.

Magnussen, S., P. Eggermont, and V. N. LaRiccia. 1999. Recovering tree heights from airborne laser scanner data. *For Sci* 45(3):407–22.

Maiman, T. 1960. Stimulated optical emission in ruby. *Nature* 197:493–4.

Maltamo, M., K. Eerikäinen, J. Pitkänen, J. Hyyppä, and M. Vehmas. 2004. Estimation of timber volume and stem density based on scanning laser altimetry and expected tree size distribution functions. *Remote Sensing of Environment* 90(3):319–30.

Maune, D. F. 2007. *Digital Elevation Model Technologies and Applications: The Dem Users Manual*. ASPRS Publications. Washington, DC: U.S.A.

McCombs, J. W., S. D. Roberts, and D. L. Evans. 2003. Influence of fusing LiDAR and multispectral imagery on remotely sensed estimates of stand density and mean tree height in a managed loblolly pine plantation. *For Sci* 49(3):457–66.

Means, J. E., S. A. Acker, D. J. Harding, J. B. Blair, M. A. Lefsky, W. B. Cohen, M. E. Harmon, and W. A. McKee. 1999. Use of large-footprint scanning airborne LiDAR to estimate forest stand characteristics in the Western Cascades of Oregon. *Remote Sens Environ* 67:298–308.

Means, J. E. 2000. Comparison of large-footprint and small-footprint lidar systems: design, capabilities, and uses. In *Proceedings: Second International Conference on Geospatial Information in Agriculture and Foretry*, Lake Buena Vista, Florida, 10–12 January 2000:I-185–92.

Middleton, W. E. K., and A. F. Spilhaus, 1953. *Meteorological Instruments*. Toronto, University of Toronto Press, 254–63.

Mutlu, M., S. C. Popescu, C. Stripling, and T. Spencer. 2008. Assessing surface fuel models using LiDAR and multispectral data fusion. *Remote Sens Environ* 112(1):274–85.

Mutlu, M., S. C. Popescu, and K. Zhao. 2008. Sensitivity analysis of fire behavior modeling with lidar-derived surface fuel maps. *For Ecol Manage* 256:289–94.

Næsset, E. 1997. Estimating timber volume of forest stands using airborne laser scanner data. *Remote Sens Environ* 61(2):246–53.

Næsset, E. 2002. Predicting forest stand characteristics with airborne scanning laser using a practical two-stage procedure and field data. *Remote Sens Environ* 80:88–99.

Næsset, E. 2004. Practical large-scale forest stand inventory using small-footprint airborne scanning laser. *Scand J For Res* 19:164–79.

Næsset, E., and K. -O. Bjerknes. 2001. Estimating tree heights and number of stems in young forest stands using airborne laser scanner data. *Remote Sens Environ* 78:328–40.

Næsset, E., and T. Økland. 2002. Estimating tree height and tree crown properties using airborne scanning laser in a boreal nature reserve. *Remote Sens Environ* 79:105–15.

Nelson, R., C. Keller, and M. Ratnaswamy. 2005. Locating and estimating the extent of Delmarva fox squirrel habitat using an airborne LiDAR profiler. *Remote Sens Environ* 96:292–301.

Nelson, R. F., W. Krabill, and J. Tonelli. 1988. Estimating forest biomass and volume using airborne laser data. *Remote Sens Environ* 24:247–67.

Nelson, R. F., E. A. Short, and M. A. Valenti. 2003. A multiple resource inventory of Delaware using airborne laser data. *BioScience* 53(10):981–92.

Nelson, R. F., R. Swift, and W. Krabill. 1988. Using airborne lasers to estimate forest canopy and stand characteristics. *J For* 86:31–8.

Nilsson, M. 1996. Estimation of tree heights and stand volume using an airborne LIDAR system. *Remote Sens Environ* 56:1–7.

Persson, Å., J. Holmgren, and U. Söerman. 2002. Detecting and measuring individual trees using an airborne laser scanner. *Photogramm Eng Remote Sensing* 68(9):925–32.

Persson, Å., U. Söderman, J. Töpel, and S. Ahlberg. 2005. Visualization and analysis of full-waveform airborne laser scanner data. *Int Arch Photogramm Remote Sens Spat Inf Sci* ISPRS WG III/3, III/4, V/3, pp. 103–108.

Popescu, S. C. 2007. Estimating biomass of individual pine trees using airborne lidar. *Biomass Bioenergy* 31(9):646–55.

Popescu, S. C., and R. H. Wynne. 2004. Seeing the trees in the forest: Using lidar and multispectral data fusion with local filtering and variable windowsize for estimating tree height. *Photogrammetric Engineering and Remote Sensing*, 70:589–604.

Popescu, S. C., R. H. Wynne, and R. H. Nelson. 2002. Estimating plot-level tree heights with LIDAR: Local filtering with a canopy-height based variable window size. *Comput Electron Agric* 37(1–3):71–95.

Popescu, S. C., R. H. Wynne, and R. H. Nelson. 2003. Measuring individual tree crown diameter with LIDAR and assessing its influence on estimating forest volume and biomass. *Can J Remote Sens* 29(5):564–77.

Popescu, S. C., R. H. Wynne, and J. A. Scrivani. 2004. Fusion of small-footprint lidar and multispectral data to estimate plot-level volume and biomass in deciduous and pine forests in Virginia, USA. *For Sci* 50(4):551–65.

Popescu, S. C., and K. Zhao. 2008. A voxel-based lidar method for assessing crown base height. *Remote Sens Environ* 112(3):767–81.

Pyysalo, U., and H. Hyyppä. 2002. Geometric shape of the tree extracted from laser scanning data. In *International Society for Photogrammetry and Remote Sensing—ISPRS Commission III Symposium (PCV'02)*, p. 4. Austria: Graz.

Ranson, K. J., G. Sun, K. Kovacs, and V. I. Kharuk. 2004. Landcover attributes from ICESat GLAS data in Central Siberia. In *Proceedings, Geoscience and Remote Sensing Symposium*, 753–6. IGARSS '04. Anchorage, Alaska.

Riaño, D., E. Meier, B. Algöwer, E. Chuvieco, and S. L. Ustin. 2003. Modeling airborne laser scanning data for the spatial generation of critical forest parameters in fire behavior modeling. *Remote Sens Environ* 86:177–86.

Riaño, D., Valladares, F., Condes, S., and Chuvieco, E. 2004. Estimation of leaf area index and covered ground from airborne laser scanner (Lidar) in two contrasting forests. *Agricultural and Forest Meteorology*, 124:269–275.

Riegl, U. S. A. 2008. http://www.rieglusa.com (retrieved January 16, 2008).

Roberts, S. D., T. J. Dean, D. L. Evans, J. W. McCombs, R. L. Harrington, and P. A. Glass. 2005. Estimating individual tree leaf area in loblolly pine plantations using LiDAR-derived measurements of height and crown dimensions. *For Ecol Manage* 213:54–70.

Schutz, B. E., H. J. Zwally, C. A. Shuman, D. Hancock, and J. P. DiMarzio. 2005. Overview of the ICESat mission. *Geophysical Research Letters* 32, L21S01.

Shan, J., and C. K. Toth. 2009. *Topographic laser ranging and scanning: Principles and processing.* Boca Raton, FL: Taylor & Francis/CRC Press.

Smith, D. E. Topography of the northern hemisphere of Mars from the Mars Orbiter Laser Altimeter (MOLA). *Science* 279:1686–1962.

Steiger, R. 2003. *Terrestrial Laser Scanning—Technology, Systems, and Applications*, p. 10. Marrakech, Morocco: Second FIG Regional Conference.

Turner, W., S. Spector, N. Gardeiner, M. Fladeland, E. Sterling, and M. Steininger. 2003. Remote sensing for biodiversity science and conservation. *Trends Ecol Evol* 18(6):306–14.

Vierling, K. T., L. A. Vierling, S. Martinuzzi, W. Gould, and R. Clawges. 2008. Lidar: Shedding new light on habitat modeling. *Front Ecol Environ* 6:90–8.

Wehr, A., and U. Lohr. 1999. Airborne laser scanning—an introduction and overview. *ISPRS J Photogramm Remote Sens* 54:68–82.

Weitkamp, C., 2005. In *Laser Remote Sensing*, ed. T. Fujii and T. Fukuchi. Boca Raton, FL: Taylor and Francis/CRC Press.

Yu, X., J. Hyyppä, H. Kaartinen, and M. Maltamo. 2004. Automatic detection of harvested trees and determination of forest growth using airborne laser scanning. *Remote Sens Environ* 90:451–62.

Zhao, K., and S. C. Popescu. 2009. Lidar-based mapping of leaf area index and its use for validating GLOBCARBON satellite LAI product in a temperate forest of the southern USA. *Remote Sens Environ* 113(8):1628–45.

Zhao, K., S. C. Popescu, and R. F. Nelson. 2009. Lidar remote sensing of forest biomass: A scale-invariant estimation approach using airborne lasers. *Remote Sens Environ* 113(1):182–96.

4

Impulse Synthetic Aperture Radar

Giorgio Franceschetti and James Z. Tatoian

CONTENTS

4.1 Scenario

Synthetic aperture radar (SAR) is one of the key sensors currently positioned on satellites for Earth and planetary exploration. It is also widely used on airplanes for imaging of Earth's surface without the time and space constraints imposed by the satellites' prescribed orbits. Additional applications, such as subsurface imaging after earthquakes and through-the-wall detection and identification of criminals and/or terrorist activities, are now emerging in the homeland security area (Amin 2010).

A SAR system radiates chirped pulses from different equispaced positions along the azimuth, which is usually a straight line; a linear array is thus synthesized. After raw data processing, the (microwave) image of the illuminated area is obtained, where the range and azimuth resolutions are $\Delta r = [2\lambda/(\Delta f / f)]$ and $\Delta x = L/2$, respectively. In the above expressions, λ and f are the wavelength and frequency of the pulse carrier, respectively, Δf is the chirp bandwidth, and L is the effective length of the radiating antenna. The azimuth resolution Δx requires the length of the synthesized array to be equal to $\lambda r/L$, where r is the distance between the sensor and the ground (assumed to be flat), and the spacing between the synthesized array elements does not exceed $L/2$. The conventional way to derive these results is by using the Doppler shift, in which a reference is made to the movement of the antenna platform with constant velocity along the azimuth. Although this procedure may lead to correct results (Curlander and McDonough 1991), it is essentially inappropriate from a physical viewpoint. Actually, the platform movement does not play any significant role in the SAR imaging (see image parameters Δr and Δx: only the array parameters are relevant and not the platform velocity). Accordingly, the proper way to trace the SAR system performance is to make a reference to the array, with the synthesizing procedure being just a technical detail; the subsequent processing procedure corresponds to a near-to-far field transformation (Franceschetti and Lanari 1999) by means of a proper beam-forming technique. This viewpoint allows for rational and sound extensions of the SAR performance in interferometric or tomographic three-dimensional (Fornaro and Serafino 2006) and forward-looking applications currently being developed. The latter correspond to end-fire phased array synthesis (instead of broadside, as in the conventional SAR), and the former correspond to two-dimensional phased array synthesis.

The above-mentioned extensions are simply generalizations of the conventional SAR systems, easily understandable if the phased array model is adopted in place of the Doppler shift. This chapter presents a completely new type of SAR system that radiates short carrierless pulses, namely the *Impulse Synthetic Aperture Radar* (ImpSAR). Again, the SAR system model is fully adopted, with the only difference being the use of synthezised *timed arrays* (Franceschetti, Tatoian, and Gibbs 2005), and not phased arrays. This is done for two reasons: (1) the term "phase" has no meaning in the time domain, and (2) the use of a frequency approach is convenient and helpful for narrowband signals, but can be misleading if the waveform bandwidth becomes large, that is, for ultra wideband (UWB) signals. The latter approach is unreasonably complicated, not transparent from a physical viewpoint, and also requires essentially inane classifications of the signals with reference to their relative bandwidth. An extension of ImpSAR to 3D imaging is also presented, where two-dimensional timed arrays are employed, and no phase unwrapping procedures take place, because the concept of phase in this context is irrelevant.

Higher range resolution of ImpSAR requires shorter radiated pulses, implying lower average radiated power, while radiated pulse amplitude is dictated by the currently available hardware technology. This limitation may be mitigated by employing larger antenna arrays in many applications where sensing from large standoff distances is required, including homeland security scenarios involving airborne and ground-based SAR systems.

Finally, the raw ImpSAR data may provide additional valuable information. Because each pulse instantaneously spans a very large bandwidth, it follows that the latter can be sliced among several subbands. Each subband can be independently processed, generating a number of conventional narrowband SAR images equal to the number of subbands. The name of the algorithm, which requires only software assets (the hardware being that of the ImpSAR), is *polychromatic SAR*, as it produces microwave images at different microwave frequency bands (different colors). This is particularly attractive in the low-frequency range, where the use of a conventional SAR usually presents weight- and size-related problems, especially for the airborne systems.

In this chapter, the full theory of ImpSAR and polychromatic SAR is presented, together with some preliminary experimental data that validate the theory, suggesting a promising future for these innovative systems.

4.2 Timed Array Analysis

Let us consider an array, composed of 2N+1 elements, as depicted in Figure 4.1. Each element is excited by the same signal $f(t)$. In the far field, the radiated field is proportional to

$$F(t, \vartheta) = \sum_{n=-N}^{N} f(t - n\Delta t) \qquad (4.1)$$

where $\Delta t = a \sin \vartheta / c$, and c is the speed of light in vacuum.

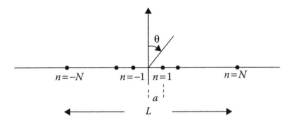

FIGURE 4.1
Array composed of 2*N*+1 identical elements. (From Franceschetti, G., J. Tatoian, and G. Gibbs, Timed arrays in a nutshell, *IEEE Trans Antennas Propagat* 53(12):4073–82), © (2005) IEEE.)

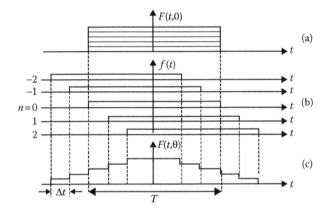

FIGURE 4.2
Timed array excited by a rectangular pulse and the resulting far-field radiated signals: (a) Received and superposed pulses along the broadside direction; (b) received pulses along the direction θ; (c) superposed received pulses along the direction θ. (From Franceschetti, G., J. Tatoian, and G. Gibbs, Timed arrays in a nutshell, *IEEE Trans Antennas Propagat* 53(12):4073–82), © (2005) IEEE.)

A definition of the *radiation diagram* similar to that of sinusoidal field is possible by substituting its radiated power density with energy density as follows (Franceschetti, Tatoian, and Gibbs 2005):

$$g(\vartheta) = \frac{\int F^2(t,\vartheta)\,\mathrm{d}t}{\int F^2(t,0)\,\mathrm{d}t} \qquad (4.2)$$

As an illustrative example, the case $N = 2$ is depicted in Figure 4.2, where the feeding signal $f(t)$ is the rectangular pulse, $f(t) = \mathrm{rect}[\,t/T\,]$, T being the pulsewidth.

Furthermore, in the far field, that is, in the properly defined Fraunhofer region of the array in time domain, $r > (2L^2/2cT)$, we have $2N\Delta t = (2Na/c)\sin\theta \cong (L/c)\sin\theta \le T$, $L = (2N+1)a \cong 2Na$ being the array length. These equalities are valid for large arrays, which is the case hereafter.

Computation of the radiation diagram, Equation 4.2, is now in order. From Equation 4.1, it follows that

$$\int F^2(t,0)\,\mathrm{d}t = (2N+1)^2 T \qquad (4.3)$$

and

$$
\int F^2(t,\theta)\,dt = (T-2N\Delta t)(2N+1)^2 + 2\Delta t \sum_{n=1}^{2N} n^2
$$

$$
= (T-2N\Delta t)(2N+1)^2 + 2\Delta t \frac{2N(2N+1)(4N+1)}{6}
$$

$$
= (2N+1)^2 T - 2N\Delta t \left[(2N+1)^2 - \frac{(2N+1)(4N+1)}{3} \right]
$$

$$
= (2N+1)^2 T - [(2N+1)-1]\Delta t (2N+1) \frac{[(2N+1)+1]}{3} \qquad (4.4)
$$

$$
= (2N+1)^2 T - \frac{1}{3}(2N+1)^3 \Delta t + \frac{1}{3}\Delta t (2N+1)
$$

$$
= (2N+1)^2 T - \frac{1}{3}\Delta t (2N+1)^3 \left[1 - \frac{1}{(2N+1)^2} \right]
$$

$$
\cong (2N+1)^2 T - \frac{1}{3}(2N+1)^3 \Delta t
$$

where the last equality is valid for large arrays. Computation of Equation 4.4 makes reference to the diagrams depicted in Figure 4.2. The integral is broken down into two parts: the first is the time interval where the pulses are synchronized; the second part is relative to the remaining time-interval, and is obtained by the evaluation of a finite summation (Franceschetti, Tatoian, and Gibbs 2005).

Dividing Equation 4.4 by Equation 4.3 leads to the formal expression of the radiation diagram; see Equation 4.2, hence

$$
g(\vartheta) = 1 - \frac{1}{3}\frac{(2N+1)\Delta t}{T} = 1 - \frac{1}{3}\frac{(2N+1)a\sin\theta}{cT} = 1 - \frac{1}{3}\frac{L\sin\theta}{cT} \qquad (4.5)
$$

Letting $g(\Theta) = 1/2$ and solving for the (conventional) 3-dB beamwidth 2Θ, we obtain

$$
\sin\Theta \cong \Theta = \frac{3cT}{2L}, \qquad 2\Theta = \frac{3cT}{L} \qquad (4.6)
$$

The definition in Equation 4.2 for the radiation diagram emphasizes the power content of the radiated beam. The alternative definition

$$
\xi(\vartheta) = 1 - \frac{\int [F(t,0)-F(t,\vartheta)]^2 \, dt}{\int F^2(t,0)\,dt} = 1 - \frac{1}{6}\frac{L\sin\vartheta}{cT} \qquad (4.7)
$$

refers to the shape of the pulse, ξ being the *similarity factor* in a quadratic norm (Franceschetti, Tatoian, and Gibbs 2005). Assuming as convenient value for the similarity factor $\xi(\Theta) = 0.917$, and solving for the beamwidth 2Θ, we obtain

$$
\sin\Theta \cong \Theta = \frac{cT}{2L}, \qquad 2\Theta = \frac{cT}{L} \qquad (4.8)
$$

Examination of the results in Equations 4.6 and 4.8 suggests that a suitable definition of timed-array 3-dB beamwidth is

$$\Theta = \frac{2cT}{L} \tag{4.9}$$

which is essentially the same as the expression used in the narrowband case where wavelength λ is substituted by $2cT$—twice the spatial extension of the pulse. In passim, this correspondence $2cT \leftrightarrow \lambda$ turns out to be valid for most (for instance, see the previously quoted Fraunhofer region definition), if not all parameters describing the performance of pulsed antennas and arrays (Franceschetti, Tatoian, and Gibbs 2005), and was a conjecture advocated over 30 years ago (Franceschetti and Papas 1974).

Consider the synthesized timed array as depicted in Figure 4.3. The 3-dB beamwidth of the array element is given by $2cT/l$, l being its effective length. The array length for the best attainable resolution is equal to the illuminated swath dimension, hence $L = (2cT/l)r$, r being the distance of the array from the ground. Accordingly, the azimuth resolution of the timed array is given by

$$\Delta x = \frac{2cT}{L} r = \frac{2cT}{(2cT/l)2r} r = \frac{l}{2} \tag{4.10a}$$

where the additional factor 2 in the denominator of the intermediate expression accounts for the round-trip propagation: the time delay Δt between the pulses radiated by nearby elements of the array doubles, virtually reducing the beamwidth of the synthetic array, as shown in Equations 4.5 and 4.6. The final result is identical to that of a conventional SAR. The range resolution is obviously given by

$$\Delta r = \frac{cT}{2} \tag{4.10b}$$

which is the standard expression for a radiated pulse.

All the above derivations, leading to Equations 4.10a and b, are made under the assumption that the imaged point, P, in Figure 4.3 is in the far field, which is not always the case. Accordingly, some processing of the raw data is necessary in order to transform the received near-field data to the far-field data; this can be implemented by a

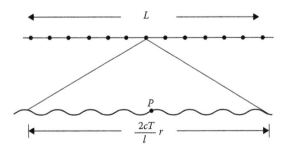

FIGURE 4.3
The synthetic timed array. (From Franceschetti, G., J. Tatoian, and G. Gibbs, Timed arrays in a nutshell, *IEEE Trans Antennas Propagat* 53(12):4073–82, © (2005) IEEE.)

simple *shift-and-add* procedure. If one denotes $f_n(t)$, $-N \le n \le N$, to be the pulse radiated by the array element n, then the received pulse is

$$f_n(t') = f_n\left(t - 2\frac{\sqrt{r^2 + (na)^2}}{c}\right)$$

(4.11)

whereas in the far field it is

$$f_n\left(t - 2\frac{r + |n|a\sin\theta}{c}\right) = f_n\left(t - 2\frac{r + \{(na)^2/r\}}{c}\right) = f_n\left(t - 2\frac{\{r^2 + (na)^2\}/r}{c}\right)$$

(4.12)

the r value being determined by the arrival time of the pulse radiated and received by the array element $n = 0$. Letting

$$\rho_n = \frac{r^2 + (na)^2}{r} - \sqrt{r^2 + (na)^2}$$

(4.13)

and substituting $2\rho_n/c$ into Equation 4.11 lead to the conclusion that the azimuthally compressed image of P is given by

$$g(P) = \sum_{n=-N}^{N} f_n\left(t' - 2\frac{\rho_n}{c}\right)$$

(4.14)

which justifies naming the procedure *shift-and-add*. Note that

$$\rho_n = \frac{r^2 + (na)^2}{r} - \sqrt{r^2 + (na)^2} \cong \frac{(na)^2}{2r}$$

(4.15)

if $(na/r)^2 \ll 1$. This may somehow simplify the procedure, but not significantly.

An example of an experimentally obtained ImpSAR image of an M16 rifle, along with its optical image, is shown in Figure 4.4. In the experiment, the width of the radiated pulse is 100 picoseconds, the length of the timed array is 4.5 m, and the distance between the target and the antenna is 6.0 m.

FIGURE 4.4
Optical (top) and microwave (bottom) images of the M16 rifle.

These parameters imply that range and azimuth resolutions are 1.3 and 8.0 cm, respectively, as shown in Equation 4.10. Note that the image not only resolves fine details of the target, but also has intrinsic peculiarities not present in conventional microwave images. This may be due to the wide bandwidth of the incident signal, so that the illuminated target cannot be modeled simply as a collection of point scatterers. Its resonant response should also be taken into account. This is an open problem worth exploring along two lines: improving the quality of the processed image and extracting value-added information from it (Franceschetti, Tatoian, and Gibbs 2009).

4.3 Radiated Pulse

In order to introduce the transmitted pulse, let us assume that the signal applied to the terminals of the transmitting antenna is

$$u(t) = \left[\exp\left(-\frac{t}{T}\right) - 1 + \frac{t}{T} \right] \exp\left(-\frac{t}{C}\right) U(t) \tag{4.16}$$

with C and T being the design parameters discussed in this section. The radiated signal is proportional to the derivative of the input signal (Franceschetti 1997), namely

$$g(t) = \frac{du}{dt} = \frac{1}{T} \left\{ \frac{C+T}{C} \left[1 - \exp\left(-\frac{t}{T}\right) \right] \exp\left(-\frac{t}{C}\right) U(t) - \frac{t}{C} \exp\left(-\frac{t}{C}\right) U(t) \right\} \tag{4.17}$$

$$= g_1(t) - g_2(t)$$

The radiated signal can be viewed as the superposition of two pulses, as detailed in the following equation. A prepulse

$$g_1(t) = \frac{1}{T} \frac{C+T}{C} \left[1 - \exp\left(-\frac{t}{T}\right) \right] \exp\left(-\frac{t}{C}\right) U(t) \tag{4.18a}$$

is the first term in Equation 4.17, which is dominant for $t < C$.
  A postpulse

$$g_2(t) = \frac{1}{T} \frac{t}{C} \exp\left(-\frac{t}{C}\right) U(t) \tag{4.18b}$$

is the second term in Equation 4.17, which is dominant for $t > C$.
  Both pulses start at $t = 0$, and decay to 0 as $t \to \infty$. For the following analysis, it is convenient to introduce the parameter $\xi = T/C$ and normalize the time to C, that is, $t \to t/C$. The graph of $g(t)$ for $\xi = 0.1$, ignoring the scaling factor $1/T$, is depicted in Figure 4.5.
  Figure 4.5 shows that the pulse $g(t)$ is composed of a narrow positive pulse followed by a wide negative pulse. The total integrated area of the signal is zero, as it should be (Franceschetti 1997), so that the first pulse is tall in contrast to the second one. The width t_0 of the first pulse is obtained by solving the relation $g_1(t_0) = g_2(t_0)$, leading to the equation

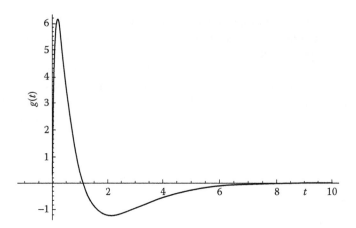

FIGURE 4.5
Graph of the function $g(t)$ for $\xi = 0.1$ and $T = 0.1$ in arbitrary time units.

$$\exp\left(-\frac{t_0}{\xi}\right) = 1 - \frac{1}{1+\xi}t_0 \tag{4.19}$$

Equation 4.19 represents the intersection of an exponential and a linear function whose slopes at the origin are $-1/\xi$ and $-1/(1 + \xi)$, respectively. In view of the expected small value of ξ, the slope of the exponential is close to $90°$, whereas that of the straight line is close to $45°$. It follows that the intersection of the two curves takes place where the exponential is close to zero, and thus, the point of the intersection is approximated by

$$t_0 \cong 1 + \xi \tag{4.20}$$

Equation 4.20 also provides the overall zero-to-zero (normalized) pulsewidth, as shown in Figure 4.5. In nonnormalized units, this pulsewidth is

$$\Delta T = (1 + \xi)C \tag{4.21}$$

Again, the design of a short pulse favors small values for the parameter ξ. This assumption leads to simpler and more understandable expressions for the derived relations.

The maximum value of the first pulse is expected to be close to the maximum of the function $g_1(t)$, because $g_2(t)$ is small in the time interval $t < 1$. This maximum value can be computed to be

$$g_1(t_{1M}) = \frac{1}{T}\left(\frac{\xi}{1+\xi}\right)^{\xi}, \quad t_{1M} = \xi \ln\left(\frac{1+\xi}{\xi}\right) \tag{4.22}$$

For subsequent analysis, it is convenient to note that for $\xi \to 0$

$$\xi \ln\left(\frac{1+\xi}{\xi}\right) \to \xi(\xi - \ln \xi) \to -\ln \xi^{\xi} \to 0 \tag{4.23a}$$

whereas for $\xi \to \infty$

$$\xi \ln\left(\frac{1+\xi}{\xi}\right) = \xi \ln\left(1 + \frac{1}{\xi}\right) \rightarrow \xi \frac{1}{\xi} = 1 \tag{4.23b}$$

It follows that the function $\xi \ln [(1 + \xi)/\xi]$ is a steadily increasing function of ξ, starting from 0 at $\xi = 0$ and approaching 1 for $\xi \rightarrow \infty$. Similarly, the second term in Equation 4.22,

$$\left(\frac{\xi}{1+\xi}\right)^{\xi} = \exp\left\{\ln\left(\frac{\xi}{1+\xi}\right)^{\xi}\right\} = \exp\left\{\xi \ln\left(\frac{\xi}{1+\xi}\right)\right\} \tag{4.24}$$

is a steadily decreasing function of ξ, starting from 1 at $\xi = 0$ and approaching $1/e = 0.368$ for $\xi \rightarrow \infty$. The above results, again, favor the choice of small values for the parameter ξ. Thus, for $\xi = 0.1$ we get $t_{1M} = 0.240$, which is consistent with the graph in Figure 4.5. Letting $T = 0.1$, $g_1(t_{1M}) = 7.87$, whereas an examination of Figure 4.5 suggests a smaller value $g(t_M) = 6.20$ due to the negative contribution of the term $g_2(t)$.

Evaluation of the spectrum of the radiated pulse follows. In the normalized Laplace domain $p \rightarrow pL$, the computation is straightforward, leading to

$$G(p) = \frac{1}{\xi}\left\{\frac{(1+\xi)}{p+1} - \frac{(1+\xi)}{p+\frac{(1+\xi)}{\xi}} - \frac{1}{(p+1)^2}\right\} \tag{4.25}$$

$G(0) = 0$, as it should be, because radiated fields cannot contain DC frequency components.

We can show that the equivalent compact expression for the spectrum of the radiated pulse is

$$G(p) = \left(\frac{1}{\xi}\right)^2 \frac{p}{[p+1]^2\left[p+\frac{(1+\xi)}{\xi}\right]} \tag{4.26}$$

In the Fourier domain, $p = i\omega$, $\omega C \rightarrow \omega$ being the normalized angular frequency, and the squared modulus of the spectrum is given by

$$|G(\omega)|^2 = \left(\frac{1}{\xi}\right)^4 \frac{\omega^2}{[\omega^2+1]^2\left[\omega^2+\left[\frac{(1+\xi)}{\xi}\right]^2\right]} \tag{4.27}$$

Equation 4.27 is symmetric with respect to the frequency axis. Considering the positive branch of the spectrum and equating its derivative, with respect to ω^2, to zero gives

$$2\omega^4 + \left[\frac{(1+\xi)}{\xi}\right]^2 \omega^2 - \left[\frac{(1+\xi)}{\xi}\right]^2 = 0 \tag{4.28}$$

whose solution

$$\omega_M^2 = \frac{1}{4}\left(\frac{1+\xi}{\xi}\right)^2\left\{\sqrt{1+8\left(\frac{\xi}{1+\xi}\right)^2}-1\right\} \cong \frac{1}{4}\left(\frac{1+\xi}{\xi}\right)^2 4\left(\frac{\xi}{1+\xi}\right)^2 = 1 \tag{4.29}$$

determines the maximum value of the spectrum

$$|G_{MAX}|^2 = |G(\omega_M)|^2 \cong |G(1)|^2$$

$$= \left(\frac{1}{\xi}\right)^4 \frac{1}{[1+1]^2\left[1+\left[\frac{(1+\xi)}{\xi}\right]^2\right]} = \left(\frac{1}{2\xi}\right)^2 \frac{1}{1+2\xi} \cong \left(\frac{1}{2\xi}\right)^2 \tag{4.30}$$

A graph of the power spectrum versus the normalized angular frequency is depicted in Figure 4.6 for the case $\xi = 0.1$. In Equation 4.30 the final two results report its limiting expressions for small values of the parameter ξ.

A possible estimate of the bandwidth of the power spectrum $|G(\omega)|^2$ is obtained by enforcing the condition $|G(\omega)|^2 < \dfrac{|G(\omega_M)|^2}{2}$. Referring to Equations 4.27 and 4.30, the angular-frequency bounds of the bandwidth are obtained by solving the following equation:

$$\left(\frac{1}{\xi}\right)^4 \frac{\omega^2}{[\omega^2+1]^2\left[\omega^2+\left[\frac{(1+\xi)}{\xi}\right]^2\right]} = \frac{1}{8}\left(\frac{1}{\xi}\right)^2 \tag{4.31}$$

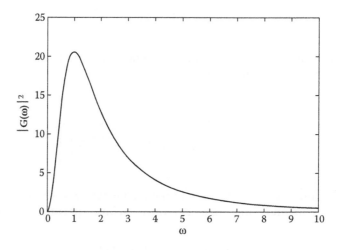

FIGURE 4.6
Graph of the power spectrum of the signal $g(t)$ depicted in Figure 4.5 associated with $\xi = 0.1$. The spectrum is symmetric about y axis; therefore, only positive normalized frequencies are depicted.

The expected solutions of the equation are on the order of unity due to the small assumed value of ξ. Equation 4.31 can be simplified

$$\left(\frac{1}{\xi}\right)^4 \frac{\omega^2}{[\omega^2+1]^2\left[\omega^2+\left[\frac{(1+\xi)}{\xi}\right]^2\right]} \cong \left(\frac{1}{\xi}\right)^4 \frac{\omega^2}{[\omega^2+1]^2\left[\frac{(1+\xi)}{\xi}\right]^2} \cong \left(\frac{1}{\xi}\right)^2 \frac{\omega^2}{[\omega^2+1]^2}$$

$$= \frac{1}{8}\left(\frac{1}{\xi}\right)^2 \tag{4.32}$$

and we immediately obtain the following values for the normalized upper and lower frequency bounds of the bandwidth and the bandwidth itself:

$$\omega_U = 2.41; \quad \omega_L = 0.41; \quad \Delta\omega = \omega_U - \omega_L = 2 \tag{4.33}$$

4.4 Polychromatic Synthetic Aperture Radar

In order to implement the polychromatic SAR imaging system, the bandwidth of the scattered pulse must be broken down into subbands, which are processed independently. The resulting spectrum of the "chopped" signal is obtained by applying a suitable filter function, $H(\omega - \Omega)$, to the spectrum of the radiated pulse $G(\omega)$ defined in Equation 4.25. Here, Ω is the value of the normalized angular frequency ω at the center of the chosen subband, subject to the constraints

$$\Omega + \gamma \leq 2.41 \tag{4.34a}$$

and

$$\Omega - \gamma \geq 0.41 \tag{4.34b}$$

where 2γ is the normalized filter bandwidth. These constraints assure that the subdivided bandwidth falls inside the radiated pulse bandwidth $\Delta\omega$, as shown in Equation 4.33.

The selected bell-shaped filter function

$$H(\omega) = \frac{\gamma^2}{\omega^2 + \gamma^2} \tag{4.35}$$

exhibits a maximum at $\omega = 0$, where it attains the value $H(0) = 1$, and two symmetric inflection points at $\omega = \pm\gamma/\sqrt{3}$, where the function and its square attain the values

$$H\left(\frac{\gamma}{\sqrt{3}}\right) = \frac{3}{4} = 0.75 \tag{4.36}$$

and

$$H^2\left(\frac{\gamma}{\sqrt{3}}\right) = \frac{9}{16} = 0.56 \tag{4.37}$$

Equation 4.37 shows that 2γ is essentially coincident with the 3-dB bandwidth of the filter.

For the SAR processing design, it is convenient to determine the shape and compute the parameters of the "chopped" signal. Letting $G_\Omega(\omega) = G(\omega) H(\omega - \Omega)$ be the bandwidth of the "chopped" signal, this signal is computed as the (inverse) Fourier Transform of its spectrum $G_\Omega(\omega)$, namely

$$g_\Omega(t) = \frac{1}{2\pi}\int_{-\infty}^{\infty} G_\Omega(\omega)\exp(i\omega t)\,d\omega = \frac{1}{2\pi}\int_{-\infty}^{\infty} G(\omega)H(\omega - \Omega)\exp(i\omega t)\,d\omega \tag{4.38}$$

where positive and negative values of Ω are used in the positive and negative ranges of ω, respectively. The signal represented by Equation 4.38 is the same when radiated only by the bandwidth $G_\Omega(\omega)$, and not by $G(\omega)$, and coincides with the signal scattered by a point target located inside the illuminated area, except for a scaling factor and time delay.

To proceed further, let us compute the inverse Fourier transform of $H(\omega - \Omega)$, as follows:

$$
\begin{aligned}
h_\Omega(t) &= \frac{1}{2\pi}\int_{-\infty}^{+\infty} H(\omega - \Omega)\cdot\exp(i\omega t)\,d\omega \\[4pt]
&= \frac{1}{2\pi}\int_{-\infty}^{+\infty}\frac{\gamma^2}{(\omega - \Omega)^2 + \gamma^2}U(\omega)\cdot\exp(i\omega t)\,d\omega + \frac{1}{2\pi}\int_{-\infty}^{+\infty}\frac{\gamma^2}{(\omega + \Omega)^2 + \gamma^2}U(-\omega)\cdot\exp(i\omega t)\,d\omega \\[4pt]
&= \frac{1}{2\pi}\int_{-\infty}^{+\infty}\frac{\gamma^2 U(\omega)}{(\omega - \Omega + i\gamma)(\omega - \Omega - i\gamma)}\cdot\exp(i\omega t)\,d\omega + \frac{1}{2\pi}\int_{-\infty}^{+\infty}\frac{\gamma^2 U(-\omega)}{(\omega + \Omega + i\gamma)(\omega + \Omega - i\gamma)}\cdot\exp(i\omega t)\,d\omega \\[4pt]
&= \frac{\gamma}{2}\exp(i\Omega t)\exp(-\gamma|t|) + \frac{\gamma}{2}\exp(-i\Omega t)\exp(-\gamma|t|) = \gamma\exp(-\gamma|t|)\cos(\Omega t)
\end{aligned} \tag{4.39}
$$

where the Fourier integrals have been computed by closing the integration contour in the upper and lower halves of the complex plane $\omega + i\omega'$ with half circles of infinite radius for $t > 0$ and $t < 0$, respectively.

Equation 4.39 represents the *impulse response* of the filter function centered at $\omega = \Omega$, whose expression in the phasor domain is

$$\hat{h}_\Omega(t) = \gamma\exp(-\gamma|t|)\exp(i\Omega\,t) \tag{4.40}$$

In order to be consistent with the usual procedure used in conventional SAR processing that utilizes I- and Q-channels, we move to the phasor domain for the continuation of our analysis.

Examination of Equation 4.38 shows that

$$\hat{g}_\Omega(t) = g(t)\otimes\hat{h}_\Omega(t) \tag{4.41}$$

where the symbol $\otimes$ is the convolution operator, and the hat symbol indicates the phasor quantities. Though this convolution is rather elaborate, it can be computed and examined (Eureka Aerospace 2008). Only a simplified analysis is reported hereafter in order to present and point out the basic features and qualifying parameters of polychromatic SAR.

The design value of γ for the filter function is dictated by the usual choices for its relative bandwidth, $2\gamma/\Omega$. In a conventional SAR system, this relative bandwidth is usually between 1% and 10% in airborne and spaceborne applications. Raw polychromatic SAR data is processed, yielding a number of microwave images, which are coincident (or at least similar) to those obtainable with conventional SAR systems. For the large value of $\Omega = 2$, this normalized bandwidth is at most 0.2, which is much smaller than the signal bandwidth of 2 given in Equation 4.33. The conclusion is that

$$G_\Omega(\omega) = G(\omega)H(\omega - \Omega) \cong G(\Omega)H(\omega - \Omega) \tag{4.42}$$

so that

$$\hat{g}_\Omega(t) \propto \hat{h}_\Omega(t) = \gamma \exp(-\gamma|t|)\exp(i\Omega t) \tag{4.43}$$

The signal represented by Equation 4.43 is proportional to the signal that would be transmitted if only the subbands around Ω were used. The return signal scattered by a point target at range r is proportional to $\exp(-\gamma|t - t'|)\exp[i\Omega(t - t')]$, where $t' = 2r/c$. An estimate of the attainable range resolution is obtained by compressing the raw data, implemented by removing the $\exp(i\Omega t)$ term via heterodyning and evaluating the convolution

$$\hat{s}(t) = \exp(-\gamma|t - t'|)\exp(-i\Omega t') \otimes \exp(-\gamma|t|) \tag{4.44}$$

For $t{-}t' = \eta \geq 0$, we get

$$
\int_{-\infty}^{+\infty} \exp(-\gamma|\eta - \tau|)\exp(-\gamma|\tau|)d\tau = \left\{ \begin{array}{l} \left[\int_{-\infty}^{0} \exp(-\gamma[\eta - \tau])\exp(\gamma\tau)d\tau + \int_{0}^{\eta} \exp[-\gamma(\eta - \tau)]\exp(-\gamma\tau)d\tau \right. \\ \left. + \int_{\eta}^{+\infty} \exp[\gamma(\eta - \tau)]\exp(-\gamma\tau)d\tau \right] \end{array} \right\}
$$
$$
= \frac{\exp(-\gamma\eta)}{2\gamma} + \eta\exp(-\gamma\xi) + \frac{(-\gamma\eta)}{2\gamma} = \frac{1 + \gamma\eta}{\gamma}\exp(-\gamma\eta) \tag{4.45}
$$

and we get the same result by substituting $\eta \rightarrow |\eta|$ when $\eta \leq 0$. We conclude that the processed signal is given, except for a multiplicative constant, by

$$\hat{s}(t) = \frac{1 + \gamma|t - t'|}{\gamma}\exp(-\gamma|t - t'|)\exp(-i\Omega t') \tag{4.46}$$

The modulus of the signal attains its maximum value, $|\hat{s}_M| = 1/\gamma$, at $t - t' = 0$, steadily decreases for $|t - t'| > 0$, and exhibits two inflection points at $t - t' = \pm 1/\gamma$, where its value is $2|\hat{s}_M|\exp(-1) = 0.736|\hat{s}_M| \cong 0.707|\hat{s}_M|$. The latter result provides an estimate of the effective

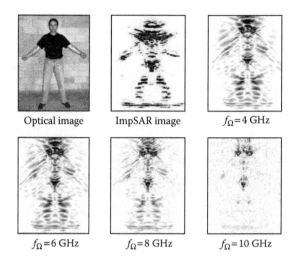

FIGURE 4.7
A comparison between impulse synthetic aperture radar (ImpSAR) and polychromatic synthetic aperture radar (SAR) images. The figure on the top left is the optical image of the target, and the second one is the ImpSAR image. The other figures are polychromatic SAR images, with the "chopped" bandwidth set to 3 GHz and centered at the frequencies indicated.

pulsewidth, $2/\gamma$, centered at $t = t'$, of the compressed signal, leading to the evaluation of the attainable range resolution.

Referring to nonnormalized quantities, we get

$$\Delta r = \frac{2c}{\gamma} = \frac{4c/\Omega}{(2\gamma)/\Omega} = 4\frac{\lambda_\Omega/2\pi}{(2\gamma)/\Omega} = \frac{2}{\pi}\frac{\lambda_\Omega/2}{(2\gamma)/\Omega} = 1.27\frac{\lambda_\Omega/2}{(2\gamma)/\Omega} \tag{4.47}$$

where $\lambda_\Omega = 2\pi c/\Omega$ is the wavelength of the center (carrier) angular frequency of the "chopped" bandwidth signal. The result given by Equation 4.47 mirrors that of the range resolution attainable by a conventional chirped SAR system, namely $(\lambda/2)/(\Delta f / f)$, where $f \to \Omega/2\pi$ and $\Delta f \to 2\gamma$ are the carrier frequency and chirp bandwidth, respectively. An example of polychromatic SAR imaging is depicted in Figure 4.7.

The difference between the ImpSAR image, which uses the entire 12-GHz bandwidth of the radiated signal, and polychromatic SAR images, each limited to a 3-GHz bandwidth centered at the frequencies indicated, is clearly pronounced. Different responses of the large target to different frequencies are also observed—a result that is open to further analysis.

4.5 Conclusions

In this chapter, two novel concepts of SAR imaging, namely impulse SAR and polychromatic SAR, were discussed at length. The theoretical foundation of the two systems has been presented and validated by experimental results. These two sensors exhibit promising features and have a wide range of potential applications where they have distinct advantages over conventional microwave imaging systems.

Impulse SAR has a number of attractive features. Its high range resolution is easily achievable with very short carrierless pulses; the absence of phase infers the absence of the grating lobes that are convenient for stereometric applications; signal processing is very fast, because it is directly implemented in time domain using the shift-and-add procedure. Finally, compared to conventional SAR systems, ImpSAR hardware is simple, and the system design and integration are straightforward. Moreover, its wide bandwidth allows an easy extension to polychromatic SAR.

Polychromatic SAR has the useful capability of generating multiple images simultaneously, which is of particular importance to the target detection and identification process. It is implemented purely in software (as it runs on existing ImpSAR hardware), and its utility can be easily extended to low-frequency ranges. This feature is particularly attractive as it extends SAR utility to ground-penetrating applications, including detection and identification of buried mines, unexploded ordnance, improvised explosive devices, pipes, and underground structures.

There is no doubt that, for the time being, the use of these sensors is limited to ground and airborne operations. However, this issue is only due to the limits of attainable radiating power using available solid-state pulsers and is expected to be solved with the increasing demand for impulse imaging technology.

Additional theoretical analysis is required to improve these systems, in particular, a deeper examination of the scattering of large bodies by very narrow pulses, when the pulse and the target are on two different spatial scales. This difference has not been explored on purpose, as it is believed that the problem should be modeled and solved directly in the time domain, without passing through the frequency domain, which is an ill-suited approach to the presented problem. This theoretical exploration is in progress.

Acknowledgments

The authors would like to thank George Gibbs of the U.S. Marine Corps System Command (MARCORSYSCOM) in Quantico, Virginia, and Martin Kruger and Andre des Rosiers of the Office of Naval Research (ONR) in Arlington, Virginia for their unstinting support, encouragement, and guidance throughout the entire ImpSAR development effort.

References

Amin, M., ed. 2010. *Through the Wall Radar Imaging*. Boca Raton, FL: CRC Press.

Curlander, J. C., and R. N. McDonough. 1991. *Synthetic Aperture Radar: Systems and Signal Processing (Wiley Series in Remote Sensing)*. New York: John Wiley & Sons.

Eureka Aerospace. 2008. Polychromatic SAR: A new concept in imaging radar. In *Technical Report to MARCORSYSCOM, contract M67854-07-C-1122*, February 29, 2008.

Fornaro, G., and F. Serafino. 2006. Imaging of single and double scatterers in urban areas via SAR tomography. *IEEE Trans Geosci Remote Sens* 44:3497–505.

Franceschetti, G. 1997. *Electromagnetics: Theory, Techniques, and Engineering Paradigms*. New York: Plenum Press.

Franceschetti, G., and R. Lanari. 1999. *SAR Processing Techniques*. Boca Raton, FL: CRC Press.

Franceschetti, G., and C. H. Papas. 1974. Pulsed antennas. *IEEE Trans Antennas Propag* 22:651–61.

Franceschetti, G., J. Tatoian, and G. Gibbs. 2005. Timed arrays in a nutshell. *IEEE Trans Antennas Propag* 53(12):4073–82.

Franceschetti, G., J. Tatoian, and G. Gibbs. 2009. Looking into transient scattering. In *Proceedings of PIERS 2009*, Xi'an, China. Cambridge, MA: The Electromagnetics Academy.

5

Hyperspectral Remote Sensing of Vegetation Bioparameters

Ruiliang Pu and Peng Gong

CONTENTS

5.1 Introduction

Imaging spectroscopy, as a new remote-sensing technique (i.e., "hyperspectral remote sensing"), is of growing interest to Earth remote sensing. Hyperspectral remote sensing refers to a special type of imaging technology that collects image data in many narrow contiguous spectral bands (<10-nm bandwidth) throughout the visible and solar-reflected infrared portions of the spectrum (Goetz et al. 1985). Since many Earth surface materials show diagnostic absorption features that are from 20- to 40-nm spectral resolution (Hunt 1980), spectral imaging systems, which acquire spectral data in contiguous narrow bands at <10-nm resolution, can produce data with sufficient resolution for direct identification of those materials with diagnostic spectral features. However, traditional remote sensing systems, which usually are called "multispectral remote sensing" systems and acquire data in a few discrete wide bands (usually >50-nm bandwidth), cannot resolve these spectral features (Goetz et al. 1985; Vane and Goetz 1988). Therefore, the value of hyperspectral remote sensing lies in its ability to acquire a complete reflectance spectrum for each pixel in an image, and it is developed for improving identification of materials and quantitative determination of physical and chemical properties of targets of interest, such as minerals, water, vegetation, soils, and human-made materials.

Imaging spectroscopy was developed for mineral mapping in the early 1980s (Goetz et al. 1985). The first imaging spectrometer, named the Airborne Imaging Spectrometer (AIS), was developed by the Jet Propulsion Laboratory (JPL) with a total of 128 spectral bands covering the spectral range between 0.9 and 2.4 μm in late 1982. The data made it possible to identify the minerals kaolinite and limestone unambiguously, which proved that direct mineral identification from orbit was possible (Goetz 1995). Funded by the National Aeronautics and Space Administration (NASA) and proposed by the JPL, the second generation of imaging spectrometers, represented by the Airborne Visible/Infrared Imaging Spectrometer (AVIRIS), came into being in 1987. The AVIRIS was the first imaging spectrometer to cover the solar reflected spectrum from 0.4 to 2.5 μm with a swath of 614 pixels. This spectrometer collects upwelling radiance through 224 contiguous spectral bands at approximately 10-nm bandwidth across the spectrum (Green et al. 1998). The AVIRIS has acquired and provided a large number of hyperspectral images for scientific research and applications every year since 1987 (Vane and Goetz 1993; Green et al. 1998). In parallel, following the AIS system, the fluorescence line imager (FLI; Hollinger et al. 1988), Advanced Solid-State Array Spectrometer (ASAS; Huegel 1988), Compact Airborne Spectrographic Imager (CASI; http://www.itres.com/Home), hyperspectral digital image collection experiment (HYDICE; Basedow et al. 1993), and Airborne Hyperspectral Scanners (HyMap; http://www.intspec.com) also provided a large number of hyperspectral images to researchers and practitioners. In addition to airborne hyperspectral systems, NASA and the European Space Agency (ESA) started developing the first generation of spaceborne hyperspectral sensor systems in 2000. Earth Observing-1 (EO-1; http://eo1.gsfc.nasa.gov/technology/) was launched on November 21, 2000. The three primary EO-1 instruments are the Advanced Land Imager (ALI), Hyperion, and a linear etalon imaging spectrometer array (LEISA) atmospheric corrector (AC). Among the three sensors, Hyperion and LAC are both hyperspectral sensors. The Hyperion instrument provides a new class of Earth observation data for improving Earth surface characterization. It has a high-resolution hyperspectral imager capable of resolving 220 spectral bands (from 0.4 to 2.5 μm) with a 30-m spatial resolution (Ungar et al. 2003). The Compact High-Resolution Imaging Spectrometer (CHRIS) is a new imaging spectrometer used aboard the ESA's PROBA satellite launched on October 22, 2001 (http://earth.esa.int/missions/thirdpartymission/proba.html).

All the aforementioned hyperspectral sensor systems have provided a large amount of valuable hyperspectral image data for various research and applications. The initial motivation for the development of imaging spectrometry was mineral identification, although early experiments were also conducted in botanical remote sensing (Goetz et al. 1985). However, since 1988, imaging spectrometry has been successfully applied to a wide range of disciplines including geology, ecology and vegetation, atmospheric science, hydrology, and oceanography.

Ecology and the study of terrestrial vegetation are important application fields for hyperspectral remote sensing (Green et al. 1998). A number of forest ecosystem variables, including leaf area index (LAI), absorbed fraction of photosynthetically active radiation (fPAR), canopy temperature, and community type are correlated with remotely sensed data or their derivatives (Johnson, Hlavka, and Peterson 1994). However, sensors in common use, such as the Landsat Multispectral Scanner (MSS) and the Thematic Mapper (TM), which integrate radiance data over wide bands of the electromagnetic spectrum, have limited value in studying the dominant canopy reflectance features such as the red spectral absorption band, near-infrared (NIR) reflectance band, and mid-infrared water absorption band (Wessman, Aber, and Peterson 1989). Moreover, the extraction of red edge and other optical parameters (e.g., Miller, Hare, and Wu 1990; Miller et al. 1991; Pu, Foschi, and Gong 2004) that are related to plant stress or senescence is impossible with broadband sensors.

Many minerals found on the Earth's surface have unique and diagnostic spectral reflectance signatures. Plants, on the other hand, are composed of the same few compounds and therefore should have similar spectral signatures (Vane and Goetz 1993). Indeed, major features of "peaks and valleys" along the spectral reflectance curve of a plant are due to the presence of pigments (e.g., chlorophyll [Chl]), water, and other chemical constituents. Therefore, characterization of diagnostic absorption features in plant spectra with hyperspectral data as done in geological mapping and mineral identification can also be done for extraction of the biochemical and biophysical parameters of plants (e.g., Wessman, Aber, and Peterson 1989; Johnson, Hlavka, and Peterson 1994; Curran, Windham, and Gholz 1995; Jacquemoud et al. 1996; Gong et al. 2003; Pu and Gong 2004; Cheng et al. 2006; Asner and Martin 2008).

Hyperspectral sensors aboard different types of platforms have made it possible to acquire higher spectral resolution data that contain more information on the subtle spectral features of plant canopies. The use of narrow (1–10 nm) instead of broad (50–200 nm) spectral bands could offer new potentials for remote sensing applied to vegetation (Guyot, Baret, and Jacquemond 1992). Hyperspectral data have been proven to be more useful in estimating biochemical content and concentration at both the leaf and canopy levels (e.g., Peterson et al. 1988; Johnson, Hlavka, and Peterson 1994; Darvishzadeh, Skidmore et al. 2008; Asner and Martin 2008) and some other ecosystem components such as LAI, plant species composition, and biomass (e.g., Gong, Pu, and Miller 1995; Gong, Pu, and Yu 1997; Martin et al. 1998; le Maire et al. 2008) than traditional remotely sensed data. Therefore, besides classification and identification of vegetation types, in terrestrial ecosystem study, hyperspectral remote sensing can be applied to the estimation of biochemical and biophysical parameters and to the evaluation of ecosystem functions.

In this chapter, we focus on a review of hyperspectral remote sensing techniques for extraction and assessment of plant biophysical and biochemical parameters. The objectives of this chapter are

- Provide an overview of the spectral characteristics of typical biophysical and biochemical parameters.
- Review information extraction and assessment techniques and methods specifically developed for analyzing imaging spectrometer data.

5.2 Spectral Characteristics of Typical Bioparameters

The spectral reflectance properties and characteristics of a list of typical plant bio parameters, including the biophysical and biochemical parameters (Table 5.1), have been the subject of systematic plant spectral reflectance studies. Typical biophysical parameters for their spectral analysis consist of vegetation canopy LAI, specific leaf area (SLA), crown closure

TABLE 5.1

Typical Plant Biophysical and Biochemical Parameters

Biophysical Parameter	Definition and Description	Spectral Response and Characteristics
LAI	The total one-sided area of all leaves in the canopy per unit area of ground.	The absorption spectral features caused by pigments in the visible region and by water content and other biochemicals in the SWIR region are useful for extracting and mapping LAI and CC.
SLA	Projected leaf area per unit leaf dry mass (cm^2/g).	Not directly related to water absorption bands, but SLA is a leaf structural property linked to the entire constellation of foliar chemicals and photosynthetic processes.
CC	Percentage of land area covered by the vertical projection of plants (tree crowns).	Same as that for LAI.
Species	Various plant species and species composition.	Spectral differences due to differences and variation in phenology/ physiology, internal leaf structure, biochemicals, and ecosystem type.
Biomass	The total of absolute amount of vegetation present (often considered in terms of the aboveground biomass) per unit area of ground.	Spectral responses to LAI, stand/ community structure, species and species composition, and image textural information.
NPP	The net flux of carbon between the atmosphere and terrestrial vegetation can be expressed on an annual basis in terms of net biomass accumulation, or NPP (Goetz and Prince 1996).	Spectra reflect vegetation condition and changes in LAI or canopy light absorption through time in visible and NIR regions.
fPAR	Effective absorbed fPAR in the visible region.	In the visible spectral region 400–700 nm, most absorbed by plant pigments, such as Chl-a and -b, Cars, and Anths; and leaf water and N contents for photosynthesis.
Chls (Chl-a, Chl-b)	Green pigments Chl-a and Chl-b for plant photosynthesis processing, found in green photosynthetic organisms, (mg/m^2 or $nmol/cm^2$).	Chl-a absorption features are near 430 and 660 nm, and Chl-b absorption features are near 450 and 650 nm in vivo (Lichtenthaler 1987; Blackburn 2006). But it is known that in situ Chl-a absorbs at both 450 and 670 nm.
Cars	Any of a class of yellow to red pigments, including carotenes and xanthophylls (mg/m^2).	Cars absorption feature in the blue region is near 445 nm in vivo (Lichtenthaler 1987). But it is known that in situ Cars absorb at 500 nm and even at a little bit longer wavelength.
Anths	Any of various water-soluble pigments that impart to flowers and other plant parts colors ranging from violet and blue to most shades of red (mg/m^2).	Anths absorption feature in the green region is at 530 nm in vivo, but in situ Anths absorb around 550 nm (Gitelson et al. 2001, 2009; Blackburn 2006).
N	Plant nutrient element (%).	The central wavelengths of N absorption features are near 1.51, 2.06, 2.18, 2.30, and 2.35 μm.

TABLE 5.1 (*Continued*)

Biophysical Parameter	Definition and Description	Spectral Response and Characteristics
P	Plant nutrient element (%).	No direct and significant absorption features across 0.40–2.50 µm, but it does indirectly affect the spectral characteristics of other biochemical compounds.
K	Plant nutrient element (%).	Foliar K concentration has only a slight effect on sclerenhyma cell walls, and thus on NIR reflectance.
W	Leaf or canopy water content or concentration (%).	The central wavelengths of those absorption features are near 0.97, 1.20, 1.40, and 1.94 µm.
Lignin	A complex polymer, the chief noncarbohydrate constituent of wood, which binds to cellulose fibers and hardens and strengthens the cell walls of plants (%).	The central wavelengths of lignin absorption features are near 1.12, 1.42, 1.69, and 1.94 µm.
Cellulose	A complex carbohydrate, which is composed of glucose units, and forms the main constituent of the cell wall in most plants (%).	The central wavelengths of cellulose absorption features are near 1.20, 1.49, 1.78, 1.82, 2.27, 2.34, and 2.35 µm.
Protein	Any of a group of complex organic macromolecules that contain carbon, hydrogen, oxygen, N, and usually sulfur, and are composed of one or more chains of amino acids (%).	The central wavelengths of protein absorption features are near 0.91, 1.02, 1.51, 1.98, 2.06, 2.18, 2.24, and 2.30 µm.

(CC), vegetation species and composition, biomass, effective absorbed fPAR, and net primary productivity (NPP), which reflect photosynthesis rate. Typical biochemical parameters are major pigments (Chls, carotenoids [Cars], and anthocyanins [Anths]), nutrients (nitrogen [N], phosphorus [P], and potassium [K]), leaf or canopy water content (W), and other biochemicals (e.g., lignin, cellulose, and protein). Analysis results are useful for determining the physicochemical properties of plants derived from spectral data and helpful for extracting bioparameters in order to assess vegetation and ecosystem conditions. Some analysis results of spectral characteristics for the list of typical biophysical and biochemical parameters from hyperspectral data are summarized in Sections 5.2.1 through 5.2.7.

5.2.1 Leaf Area Index, Specific Leaf Area, and Crown Closure

The LAI, SLA, and CC are important structural parameters for quantifying the energy and mass exchange characteristics of terrestrial ecosystems such as photosynthesis, respiration, transpiration, the carbon and nutrient cycle, and rainfall interception. The LAI parameter quantifies the amount of live green leaf material present in the canopy per unit ground area, whereas SLA describes the amount of leaf dry mass present in the plant canopy. The CC parameter can only quantify the percentage of area covered by the vertical projection of live green leaf material present in the canopy. The physiological and structural characteristics of plant leaves determine their typically low visible-light reflectance, except in green light. The high NIR reflectance of vegetation allows optical remote sensing to capture detailed information about the live, photosynthetically active forest canopy structure, and thus help understand the mass exchange between the atmosphere and the plant ecosystem (Zheng and Moskal 2009). As LAI and CC increase, many absorption features become significant due to changes in their amplitude, width, or location. The absorption features, including those caused by pigments in the visible region and by water content and other biochemicals in the shortwave infrared (SWIR) region (Curran 1989; Elvidge 1990), are useful in extracting and mapping LAI and CC. Different from LAI and CC, the spectral properties of SLA are not directly related to water absorption bands in the full range of a vegetation spectrum. However, SLA has a leaf structural property linked to the entire constellation of foliar chemicals and photosynthetic processes (Wright et al. 2004; Niinemets and Sack 2006). It is related to the NIR spectral reflectance that is dominated by the amount of leaf water content and leaf thickness (Jacquemoud and Baret 1990). Thus, at the leaf level, SLA is highly correlated with leaf spectral reflectance (Asner and Martin 2008).

Optical remote sensing, especially hyperspectral remote sensing, is aimed at retrieving the spectral characteristics of leaves, quantified by LAI, SLA, and CC, which are determined by the internal biochemical structure and pigments content of leaves. Currently, many spectral analysis techniques and methods (see reviews for individual methods and techniques in Section 5.3) are available for extracting and assessing the biophysical parameters LAI, SLA, and CC from various hyperspectral sensors, especially imaging spectrometers, such as spectral derivatives (e.g., Gong, Pu, and Miller 1992; Gong, Pu, and Miller 1995), spectral position variables (e.g., Miller, Hare, and Wu 1990; Pu, Gong et al. 2003), spectral indices (e.g., Gong et al. 2003; Delalieux et al. 2008), and physically based models (e.g., Schlerf and Atzberger 2006; Asner and Martin 2008; Darvishzadeh, Roshanak et al. 2008).

5.2.2 Species and Composition

Foliage spectral variability among individual species, or even within a single crown, is attributed not only to differences in internal leaf structure and biochemicals (e.g., water, Chl content, epiphyll cover, and herbivory; Clark, Roberts, and Clark 2005) but also to difference and variation in the phenology/physiology of plant species. In addition, the relative importance of these biochemical and structural properties among individual species is also dependent on measured wavelength, pixel size, and ecosystem type (Asner 1998). Few studies have been systematically carried out to determine the best wavelengths suitable for species recognition in the field. This obviously depends on species-specific biochemical characteristics that are related to foliar chemistry (Martin et al. 1998). Martin and Aber (1997) used AVIRIS data to estimate the N and lignin content in forest canopy foliage. Although either of the two by itself is insufficient to identify species, combined information can differentiate between species. For example, red pine and hemlock were reported to have very similar N concentration, but very different levels of lignin (Martin et al. 1998). Pu (2009) used 30 selected spectral variables evaluated by analysis of variance (ANOVA) from in situ hyperspectral data to identify 11 broadleaf species in an urban environment. Among the 30 selected spectral variables, most of the spectral variables are directly related to leaf chemistry. For example, some selected spectral variables are related to water absorption bands around 0.97, 1.20, and 1.75 μm, and the others are related to spectral absorption features of Chls, red-edge optical parameters, simple ratio (SR), vegetation index (VI), and reflectance at 680 nm, and other biochemicals such as lignin (near 1.20 and 1.42 μm), cellulose (near 1.20 and 1.49 μm), and N (near 1.51 and 2.18 μm; Curran 1989). In identifying invasive species in Hawaiian forests from native and other introduced species by remote sensing, Asner et al. (2008) confirmed the viewpoint that the observed differences in canopy spectral signatures are linked to relative differences in measured leaf pigments (Chls and Cars), nutrients (N and P), and structural (SLA) properties, as well as to canopy LAI.

5.2.3 Biomass

Leaf canopy biomass is calculated as the product of the leaf dry mass per area (LMA; unit: g/m^2, or the inverse of SLA) and LAI. Therefore, based on the spectral responses to LAI and LMA, both biophysical parameters can be estimated from hyperspectral data; thus, the leaf mass of the entire canopy is estimated (le Maire et al. 2008). Many VIs, such as the normalized difference VI (NDVI) and the SR constructed with NIR and red bands have been developed and directly applied to estimate leaf or canopy biomass. It has been recommended that VIs remove variability caused by canopy geometry, soil background, sun view angles, and atmospheric conditions when measuring biophysical properties (Elvidge and Chen 1995; Blackburn and Steele 1999). Broadband VIs use, in principle, average spectral information over a wide range, resulting in the loss of critical spectral information available in specific narrow (hyperspectral) bands (Hansena and Schjoerring 2003). Since many narrow bands are available for constructing VIs, selection of the correct wavelengths and bandwidths is important. When some VIs derived from hyperspectral data are used to estimate some biophysical parameters, narrow bands (10 nm) perform better than broadband (e.g., TM bands) using standard red/NIR and green/NIR NDVIs ($NDVI_{green}$; e.g., Gong et al. 2003; Hansena and Schjoerring 2003). For example, $NDVI_{SWIR}$ constructed with reflectances at wavelengths 1540 and 2160 nm is the best index for leaf mass estimation (le Maire et al. 2008); many hyperspectral bands in the SWIR region and some in the NIR region have the greatest potential to form spectral indices for LAI estimation (e.g., most effective band wavelengths centered around 820, 1040, 1200, 1250, 1650, 2100, and 2260 nm with bandwidths ranging from 10 to 300 nm; Gong et al. 2003).

5.2.4 Pigments: Chlorophylls, Carotenoids, and Anthocyanins

The Chls (Chl-a and Chl-b) are Earth's most important organic molecules, as they are the most important pigments necessary for photosynthesis. The second major group of plant pigments, composed of carotene and xanthophylls, is Cars, whereas Anths are water-soluble flavonoids, which form the third major group of pigments in leaves, but there is no unified explanation for their presence and function (Blackburn 2007b). Published spectral absorption wavelengths of isolated pigments show that Chl-a absorption features are around 430 and 660 nm and Chl-b absorption features are around 450 and 650 nm in vivo (Lichtenthaler 1987; Blackburn 2007b). But it is known that in situ Chl-a absorbs at both 450 and 670 nm. Cars absorption feature in the blue region is at 445 nm in vivo and β-carotene at 470 nm (Lichtenthaler 1987; Blackburn 2007b) in vivo. But it is also known that *in situ* Cars absorb at 500 nm and even at wavelengths that are a little bit longer. The absorption feature of Anths in the green region is at 530 nm in vivo, but *in situ* Anths absorb around 550 nm (Gitelson, Merzlyak, and Chivkunova 2001; Gitelson, Chivkunova, and Merzlyak 2009; Blackburn 2007b; Ustin et al. 2009).

Based on the spectral properties of the pigments, some researchers have used red edge (e.g., Curran, Windham, and Gholz 1995; Cho, Skidmore, and Atzberger 2008) optical parameters to estimate plant leaf and canopy Chls content and concentration. However, most of them have developed and used various VIs, constructed in either ratios or normalized difference ratios of two narrow bands in the visible and NIR regions, to estimate the major plant pigments Chls, Cars, and Anths at leaf or canopy levels (e.g., Gitelson and Merzlyak 1994; Blackburn 1998; Gitelson, Merzlyak, and Chivkunova 2001; Gitelson et al. 2002; Gitelson, Keydan, and Merzlyak 2006; Richardson, Duigan, and Berlyn 2002; Rama Rao et al. 2008). In addition, many researchers also employ physically based models at leaf or canopy levels to retrieve the pigments (e.g., Asner and Martin 2008; Feret et al. 2008) and use data transform approaches like wavelet analysis to retrieve Chl concentration from leaf reflectance spectra (Blackburn and Ferwerda 2008). (For a more detailed description and review of concrete analysis methods and techniques, see Section 5.3.)

5.2.5 Nutrients: Nitrogen, Phosphorous, and Potassium

The foliage and canopy N is related to a variety of ecological and biochemical processes (Martin et al. 2008). It is the most important nutrient element needed by plants for growth. The second and third most limiting nutrient constituents, P and K, are essential in all phases of plant growth; they are used in cell division, fat formation, energy transfer, seed germination, and flowering and fruiting (Milton, Eiswerth, and Ager 1991; Jokela et al. 1997). Among the three basic nutrient elements, N has significant absorption features that have been found in the visible, NIR, and SWIR regions. According to Curran (1989), N absorption features in their isolated form are located around 1.51, 2.06, 2.18, 2.30, and 2.35 µm. Since many biochemical compounds comprise N, such as Chls and protein, their spectral properties are also characterized by N concentration in plant leaves. It seems that P has no direct and significant absorption features across the visible, NIR, and SWIR regions, but it does indirectly affect the spectral characteristics of other biochemical compounds. The documented spectral changes include a higher reflectance in the green and yellow portions of the electromagnetic spectrum in P-deficient plants and a difference in the position of the long-wavelength edge (the red edge) of Chl absorption band centered around 0.68 µm (Milton, Eiswerth, and Ager 1991). Foliar K concentration has only a slight effect on needle morphology, thereby affecting NIR reflectance. This is because the sclerenchyma cell walls are thicker, with a high K concentration, which leads to higher NIR reflectance of leaves (Jokela et al. 1997).

To estimate nutrient concentrations from hyperspectral data, including in situ spectral measurements and imaging data, many analysis techniques and methods (see reviews for such individual methods and techniques in Section 5.3) have been developed. They include spectral derivatives (Milton, Eiswerth, and Ager 1991; Gong, Pu, and Heald 2002), spectral indices (Gong, Pu, and Heald 2002; Serrano, Peñuelas, and Ustin 2002; Hatfield et al. 2008; Rama Rao et al. 2008), spectral position variables (Gong, Pu, and Heald 2002; Cho and Skidmore 2006), continuum-removal method (Huber et al. 2008), statistical regression (LaCapra et al. 1996; Martin and Aber 1997; Martin et al. 2008), and inversion of physically based models (Asner and Martin 2008; Cho, Skidmore, and Atzberger 2008).

5.2.6 Leaf or Canopy Water Content

The evaluation of water status in vegetation is an important component of hyperspectral remote sensing (Goetz et al. 1985; Curran, Kupiec, and Smith 1997). Previous work on assessing the plant water status mainly depended on water spectral absorption features in the 0.40–2.50 µm region. According to Curran (1989), the central wavelengths of the absorption features are around 0.97, 1.20, 1.40, and 1.94 µm. In addition, the reflectance of dry vegetation shows an absorption feature centered at 1.78 µm by other chemicals (cellulose, sugar, and starch; Curran 1989) rather than by water, because pure water does not cause such an absorption feature (Palmer and Williams 1974). In general, the reflectance spectra of green and yellow leaves in those absorption bands are quickly saturated and solely dominated (Elvidge 1990) by changes in the leaf water content.

To extract these spectral absorption features, one of the most important techniques is to make use of VIs (Peñuelas et al. 1993; Peñuelas, Filella, and Sweeano 1996; Pu, Ge et al. 2003; Cheng et al. 2006; Colombo et al. 2008). Other analysis techniques (see reviews of these individual methods and techniques in Section 5.3) include spectral derivatives (Pu, Ge et al. 2003; Pu, Foschi, and Gong 2004), spectral position variables (Pu, Foschi, and Gong 2004), continuum-removal method (Pu, Ge et al. 2003; Huber et al. 2008), statistical regression (Curran, Kupiec, and Smith 1997; Colombo et al. 2008), and inversion of physically based models (Ustin et al. 1998; Clevers, Kooistra, and Schaepman 2008; Colombo et al. 2008).

5.2.7 Other Biochemicals: Lignin, Cellulose, and Protein

The spectral absorption features of other biochemicals are mostly located in the SWIR region (1.00–2.50 μm). According to Curran (1989), the central wavelengths of lignin absorption features are around 1.12, 1.42, 1.69, and 1.94 μm; the central wavelengths of cellulose absorption features are around 1.20, 1.49, 1.78, 1.82, 2.27, 2.34, and 2.35 μm; and the central wavelengths of protein absorption features are around 0.91, 1.02, 1.51, 1.98, 2.06, 2.18, 2.24, and 2.30 μm. So far, most techniques (see reviews for individual methods and techniques in Sections 5.3.1 through 5.3.9) for estimating the concentrations of lignin, cellulose, and protein from hyperspectral data use derivative spectra (Peterson et al. 1988; Wessman, Aber, and Peterson 1989; Curran, Kupiec, and Smith 1997), logarithm spectra (Card, Peterson, and Matson 1988; Peterson et al. 1988; Zagolski et al. 1996), spectral indices (Gastellu-etchegorry et al. 1995; Serrano, Peñuelas, and Ustin 2002), and/or statistical regression (Gastellu-etchegorry et al. 1995; LaCapra et al. 1996; Curran, Kupiec, and Smith 1997; Martin and Aber 1997).

5.3 Analysis Techniques and Methods

There are many analysis techniques and methods that currently are available to be used for extracting and assessing bioparameters from various hyperspectral data. A total of nine types or categories of the techniques and methods are reviewed in following Sections 5.3.1 through 5.3.9.

5.3.1 Derivative Analysis

In situ data or imaging hyperspectral data obtained in the field are rarely from a single object. They are contaminated by illumination variations caused by terrain relief, cloud, and viewing geometry. The spectral reflectance of a target of interest could also be affected by radiometric contributions from background materials like soil spectra. Derivative analysis has been considered a desirable tool in removing or compressing the effect of illumination variations (Demetriades-Shah, Steven, and Clark 1990; Tsai and Philpot 1998). It has also proven effective in reducing background effects when the spectral pattern of background materials has a lower frequency of variation (Gong, Pu, and Miller 1992; Li et al. 1993). For derivative analysis of hyperspectral data, a finite approximation (Tsai and Philpot 1998) can be applied to calculate the first- and second-order derivative spectra as follows:

$$\rho'(\lambda_i) \approx [\rho(\lambda_{i+1}) - \rho(\lambda_{i-1})] / \Delta\lambda \tag{5.1}$$

and

$$\rho''(\lambda_i) \approx [\rho'(\lambda_{i+1}) - \rho'(\lambda_{i-1})] / \Delta\lambda$$
$$\approx [\rho(\lambda_{i+1}) - 2\rho(\lambda_i) + \rho(\lambda_{i-1})] / \Delta\lambda^2 \qquad (5.2)$$

where $\rho'(\lambda_i)$ and $\rho''(\lambda_i)$ are the first and second derivatives, respectively, $\rho(\lambda_i)$ is reflectance at a wavelength (band) i, and $\Delta\lambda$ is the wavelength interval between λ_{i+1} and λ_{i-1} equal to twice the bandwidth in this case.

Derivative spectra have been successfully employed in hyperspectral data analysis for biophysical and biochemical parameter extraction (e.g., Gong, Pu, and Heald 2002; Pu, Ge et al. 2003; Huang et al. 2004; Galvão, Formaggio, and Tisot 2005; Laba et al. 2005; Cho and Skidmore 2006; Asner et al. 2008; Lucas and Carter 2008). It is believed that the accuracy of derivative analysis is sensitive to the signal-to-noise ratio of hyperspectral data and higher-order spectral derivative processing is susceptible to noise (Cloutis 1996). Lower-order derivatives (e.g., the first-order derivative) are less sensitive to noise and hence more effective in operational remote sensing. For example, Gong, Pu, and Yu (1997, 2001) report that the first derivative of tree spectra could considerably improve the accuracy of recognizing six conifer species commonly found in northern California.

5.3.2 Spectral Matching

Researchers van der Meer and Bakker (1997) developed a cross-correlogram spectral matching (CCSM) technique, taking into consideration the correlation coefficient between a target spectrum and a reference spectrum, the skewness of the spectra, and criterion of correlation significance. A cross-correlogram (i.e., CCSM) is constructed by calculating the cross-correlation at different match positions between a test (target) spectrum and a reference (a laboratory or pixel spectrum known to characterize a target of interest) spectrum, and is suitable for processing hyperspectral data. Further, van der Meer (2006) compared spectral angle mapper (SAM) with the vector CCSM between a known reference and an unknown target spectrum and the spectral information divergence (SID; Chang 2000) in differentiating the minerals alunite, kaolinite, montmorillonite, and quartz using both synthetic and real (i.e., AVIRIS) hyperspectral data of a (artificial or real) hydrothermal alteration system. The SID measures the discrepancy in probability distributions between two pixel vectors. His results suggest that SID and CCSM outperform SAM, and that SID is more effective in mapping the four minerals.

Given two spectral signature curves, $\rho_r = (\rho_{r1}, \rho_{r2}, ..., \rho_{rL})^T$ and $\rho_t = (\rho_{t1}, \rho_{t2}, ..., \rho_{tL})^T$, these measures are defined as follows:

$$\text{Cross-correlation } r_m = \frac{n \sum \rho_r \rho_t - \sum \rho_r \sum \rho_t}{\sqrt{\left[n \sum \rho_r^2 - (\sum \rho_r)^2 \right] \left[n \sum \rho_t^2 - (\sum \rho_t)^2 \right]}} \qquad (5.3)$$

where the cross-correlation r_m, at each match position m, is equivalent to the linear correlation coefficient and is defined as the ratio of covariance to the product of the sum of standard deviations; n is the effective number of bands when calculating the CCSM; and L is total number of bands ($n < L$).

SID is given by

$$\text{SID}_{(\rho_r, \rho_t)} = D(\rho_r \| \rho_t) + D(\rho_t \| \rho_r) \qquad (5.4)$$

where

$$D(\rho_t \| \rho_r) = \sum_{l=1}^{L} q_l D_l(\rho_t \| \rho_r) = \sum_{l=1}^{L} q_l [I_l(\rho_r) - I_l(\rho_t)] \tag{5.5}$$

and

$$D(\rho_r \| \rho_t) = \sum_{l=1}^{L} p_l D_l(\rho_r \| \rho_t) = \sum_{l=1}^{L} p_l [I_l(\rho_t) - I_l(\rho_r)] \tag{5.6}$$

Equations 5.5 and 5.6 are derived from the probability vectors $p = (p_1, p_2, ..., p_L)^T$ and $q = (q_1, q_2, ..., q_L)^T$ for the spectral signatures of vectors ρ_r and ρ_t, where $p_k = \rho_{rk} / \sum_{l=1}^{L} \rho_{rl}$, $q_k = \rho_{tk} / \sum_{l=1}^{L} \rho_{tl}$, $I_l(\rho_t) = -\log q_l$, and similarly $I_l(\rho_r) = -\log p_l$. Measures $I_l(\rho_t)$ and $I_l(\rho_r)$ are referred to as the "self-information" of ρ_t for band l. Note that Equations 5.5 and 5.6 represent the relative entropy of ρ_t with respect to ρ_r (indicated by the $\|$ symbol).

In the study of the spectroscopic determination of two health levels of the coast live oak leaves, Pu, Kelly et al. (2008) used the CCSM algorithm to discriminate between healthy and infected leaves by matching unknown leaf spectra with known infected leaf spectra in association with water stress. Wang et al. (2009) also classified land-cover types with the CCRM spectral matching technique. In spectral matching, it should be noted that the accuracy of spectral matching techniques (e.g., CCSM) is directly affected by geometry of sensors' observations and target size. This effect can be minimized by performing spectral normalization before conducting spectral matching (Pieters 1983). In general, such matching techniques are more useful for change detection of scene components than for identification of the unknown scene components (Yasuoka et al. 1990).

5.3.3 Spectral Index Analysis

When multispectral data is used to construct various spectral VIs, the advantage of VIs is their ease of use. When using hyperspectral data to conduct spectral VI analysis, hyperspectral remote sensing has the added advantage of increased chance and flexibility to choose spectral bands. With multispectral data, one may have only the choice to use the red and NIR bands. However, with hyperspectral data, one can choose many such red and NIR narrowband combinations (Gong et al. 2003). Accordingly, spectral VIs applied to hyperspectral data are called "narrowband VIs" (Zarco-Tejada et al. 2001; Eitel et al. 2006; He, Guo, and Wilmshurst 2006). Table 5.2 lists a set of 66 VIs that are developed for hyperspectral data. These VIs frequently appear in the literature on extracting and evaluating plant biophysical and biochemical parameters from hyperspectral data. The 66 VIs are grouped into five categories so that readers can conveniently locate a VI (or a group of VIs), based on the characteristics and functions of the VIs: (1) multiple bioparameters, (2) pigments (Chls, Cars, and Anths), (3) foliar chemistry, (4) water, and (5) stress. Within individual categories, the VIs are arranged in alphabetical order. A brief review of these VIs is given in this section.

Specifically, the use of VIs for extracting and assessing vegetation LAI, SLA, and CC includes the use of enhanced VI (EVI), two-band enhanced VI (EVI2), greenness index (GI), LAI determining index (LAIDI), modified Chl absorption ratio index 1 (MCARI1), modified Chl absorption ratio index 2 (MCARI2), modified SR (MSR), modified triangular VI 1 (MTVI1), modified triangular VI 2 (MTVI2), normalized difference infrared index

TABLE 5.2

Summary of 66 Spectral Indices Extracted from Hyperspectral Data, Collected from the Literature

Spectral Index	Characteristics and Functions	Definition	Reference
Multiple Bioparameters			
ATSAVI	Less affected by soil background and better for estimating homogeneous canopy.	$a(R_{800} - aR_{670} - b)/[(aR_{800} + R_{670} - ab + X(1 + a^2)]$, where $X = 0.08$, $a = 1.22$, and $b = 0.03$	Baret and Guyot 1991
EVI	Estimate vegetation LAI, biomass, and water content, and improve sensitivity in high-biomass regions.	$2.5(R_{NIR} - R_{red})/(R_{NIR} + 6R_{red} - 7.5R_{blue} + 1)$	Huete et al. 2002
EVI2	Similar to EVI, but without blue band and good for atmospherically corrected data.	$2.5(R_{NIR} - R_{red})/(R_{NIR} + 2.4R_{red} + 1)$	Jiang et al. 2008
GI	Estimate biochemical constituents and LAI at leaf and canopy levels.	R_{554}/R_{677}	Zarco-Tejada, Berjon et al. 2005
LAIDI	Sensitive to LAI variation at canopy level with a saturation point >8.	R_{1250}/R_{1050}	Delalieux et al. 2008
Improved Soil Adjusted Vegetation Index (MSAVI)	A more sensitive indicator of vegetation amount than SAVI at canopy level.	$0.5[2R_{800} + 1 - ((2R_{800} + 1)^2 - 8(R_{800} - R_{670}))^{1/2}]$	Qi et al. 1994
MSR	More linearly related to vegetation parameters than RDVI.	$(R_{800}/R_{670} - 1)/(R_{800}/R_{670} + 1)^{1/2}$	Chen 1996; Haboudane et al. 2004
MTVI1	More suitable for LAI estimation than TVI.	$1.2[1.2(R_{800} - R_{550}) - 2.5(R_{670} - R_{550})]$	Haboudane et al. 2004
MTVI2	Preserves sensitivity to LAI and resistance to Chl influence.	$\{1.5[1.2(R_{800} - R_{550}) - 2.5(R_{670} - R_{550})]\}/\{(2R_{800} + 1)^2 - [6R_{800} - 5(R_{670})^{1/2}] - 0.5\}^{1/2}$	Haboudane et al. 2004
NDVI	Responds to change in the amount of green biomass and more efficiently in vegetation with low to moderate density.	$(R_{NIR} - R_R)/(R_{NIR} + R_R)$	Rouse et al. 1973
Optimized soil-adjusted VI (OSAVI)	Similar to MSAVI, but more applicable in agricultural applications, whereas MSAVI is recommended for more general purposes.	$1.16(R_{800} - R_{670})/(R_{800} + R_{670} + 0.16)$	Rondeaux et al. 1996
PSND	Estimate LAI and Cars at leaf or canopy level.	$(R_{800} - R_{470})/(R_{800} + R_{470})$	Blackburn 1998

PVI_{hyp}	More efficiently quantify the low amount of vegetation by minimizing soil background influence on vegetation spectrum.	$(R_{1148} - aR_{807} - b)/(1 + a^2)^{1/2}$, where $a = 1.17$ and $b = 3.37$	Schlerf et al. 2005
RDVI	Suitable for low to high LAI values.	$(R_{800} - R_{670})/(R_{800} + R_{670})^{1/2}$	Reujean and Breon 1995; Haboudane et al. 2004
sLAIDI	Sensitive to LAI variation at canopy level with a saturation point >8.	$S(R_{1050} - R_{1250})/(R_{1050} + R_{1250})$, where S = 5	Delalieux et al. 2008
SPVI	Estimate LAI and canopy Chls.	$0.4[3.7(R_{800} - R_{670}) - 1.2\,|R_{530} - R_{670}|\,]$	Vincini et al. 2006
TCARI	Similar to OSAVI, but very sensitive to Chls content variations and very resistant to the variations of LAI and solar zenith angle.	$3[(R_{700} - R_{670}) - 0.2(R_{700} - R_{550})(R_{700}/R_{670})]$	Haboudane et al. 2002
SR	Same as NDVI.	R_{NIR}/R_R	Jordan, 1969
Visible atmospherically resistant index for green ref. ($VARI_{green}$)	Estimate green vegetation fraction (VF) with minimally sensitive to atmospheric effects with an error of <10%; better than NDVI for moderate to high VF values of VF.	$(R_{green} - R_{red})/(R_{green} + R_{red})$	Gitelson, Kaufman et al. 2002
VARI for red edge ref. ($VARI_{red-edge}$)	Same as $VARI_{green}$.	$(R_{red-edge} - R_{red})/(R_{red-edge} + R_{red})$	Gitelson, Kaufman et al. 2002
WDRVI	Estimate LAI, vegetation cover, biomass; better than NDVI.	$(0.1R_{NIR} - R_{red})/(0.1R_{NIR} + R_{red})$	Gitelson 2004
Pigments (Chls, Cars, and Anths)			
ARI	Estimate Anths content from reflectance changes in the green region at leaf level.	$ARI = (R_{550})^{-1} - (R_{700})^{-1}$	Gitelson, Merzlyak, and Chivkunova 2001
BGI	Estimate Chls and Cars content at leaf and canopy levels.	R_{450}/R_{550}	Zarco-Tejada, Berjon et al. 2005
BRI	Estimate Chls and Cars content at leaf and canopy levels.	R_{450}/R_{690}	Zarco-Tejada, Berjon et al. 2005
CARI	Quantify Chls concentration at leaf level.	$(|(a670 + R_{670} + b)|/(a^2 + 1)^{1/2}) \times (R_{700}/R_{670})$ $a = (R_{700} - R_{550})/150,\ b = R_{550} - (a \times 550)$	Kim et al. 1994
Chl_{green}	Estimate Chls content in anthocyanin-free leaves if NIR is set.	$(R_{760-800}/R_{540-560})^{-1}$	Gitelson, Keydan, and Merzlyak 2006

(Continued)

TABLE 5.2 (*Continued*)

Summary of 66 Spectral Indices Extracted from Hyperspectral Data, Collected from the Literature

Spectral Index	Characteristics and Functions	Definition	Reference
$Chl_{red-edge}$	Estimate Chls content in anthocyanin-free leaves if NIR is set.	$(R_{760-800}/R_{690-720})^{-1}$	Gitelson, Keydan, and Merzlyak 2006
CRI	Sufficient to estimate total Cars content in plant leaves.	$CRI_{550} = (R_{510})^{-1} - (R_{550})^{-1}$, $CRI_{700} = (R_{510})^{-1} - (R_{700})^{-1}$	Gitelson et al. 2002
DD	Estimate total Cars content in plant leaves.	$(R_{750} - R_{720}) - (R_{700} - R_{670})$	le Maire, Francois, and Dufrene 2004
DmSR	Quantify Chls content at leaf level.	$(DR_{720} - DR_{500})/(DR_{720} + DR_{500})$, where DR_λ is first derivative of ref. at wavelength λ	le Maire, Francois, and Dufrene 2004
EPI	Correlate best with Chl-a, Chls, and total Car contents.	$a \times R_{672}/(R_{550} \times R_{708})^\beta$	Datt 1998
LCI	Estimate Chl content in higher plants, sensitive to variation in reflectance caused by Chl absorption.	$(R_{850} - R_{710})/(R_{850} + R_{680})$	Datt 1999
mARI	Estimate anthocyanin content from reflectance changes in the green region at leaf level.	$mARI = ((R_{530-570})^{-1} - (R_{690-710})^{-1}) \times R_{NIR}$	Gitelson, Keydan, and Merzlyak 2006
MCARI	Respond to Chl variation and estimate Chl absorption.	$[(R_{701} - R_{671}) - 0.2(R_{701} - R_{549})]/(R_{701}/R_{671})$	Daughtry et al. 2000
MCARI1	Less sensitive to Chl effects; more responsive to green LAI variation.	$1.2[2.5(R_{800} - R_{670}) - 1.3(R_{800} - R_{550})]$	Haboudane et al. 2004
MCARI2	Preserves sensitivity to LAI and resistance to Chl influence.	$\{1.5[2.5(R_{800} - R_{670}) - 1.3(R_{800} - R_{550})]\}/\{(2R_{800} + 1)^2 - [6R_{800} - 5(R_{670})^{1/2}] - 0.5\}^{1/2}$	Haboudane et al. 2004
mCRI	Estimate Car pigment contents in foliage.	$mCRI_G = ((R_{510-520})^{-1} - (R_{560-570})^{-1}) \times R_{NIR}$, $mCRI_{RE} = ((R_{510-520})^{-1} - (R_{690-700})^{-1}) \times R_{NIR}$	Gitelson, Keydan, and Merzlyak 2006
mND_{680}	Quantify Chl content and sensitive to low content at leaf level.	$(R_{800} - R_{680})/(R_{800} + R_{680} - 2R_{445})$	Sims and Gamon 2002

mND_{705}	$(R_{750} - R_{705})/(R_{750} + R_{705} - 2R_{445})$	Quantify Chl content and sensitive to low content at leaf level; mND_{705} performance better than mND_{680}.	Sims and Gamon 2002
mSR_{705}	$(R_{750} - R_{445})/(R_{705} - R_{445})$	Quantify Chl content and sensitive to low content at leaf level.	Sims and Gamon 2002
NPCI	$(R_{680} - R_{430})/(R_{680} + R_{430})$	Assess Cars/Chl ratio at leaf level.	Peñuelas et al. 1994
NPQI	$(R_{415} - R_{435})/(R_{415} + R_{435})$	Detect variation of Chl concentration and Cars/Chl ratio at a leaf level.	Barnes et al. 1992 Peñuelas, Filella et al. 1995
PBI	R_{810}/R_{560}	Retrieve leaf total Chl and N concentrations from satellite hyperspectral data.	Rama Rao et al. 2008
PRI	$(R_{531} - R_{570})/(R_{531} + R_{570})$	Estimate Car pigment contents in foliage.	Gamon et al. 1992
PSSR	R_{800}/R_{500}	Estimate Car pigment contents in foliage.	Blackburn 1998
RARS	R_{760}/R_{500}	Estimate Car pigment contents in foliage.	Chappelle et al. 1992
RGR	R_{Red}/R_{Green}	Estimate anthocyanin content with a green and a red band.	Gamon and Surfus 1999; Sims and Gomon 2002
SIPI	$(R_{800} - R_{445})/(R_{800} - R_{680})$	Estimate Car pigment content change in foliage; related to a ratio, Cars:Chls.	Peñuelas, Baret et al. 1995
TVI	$0.5[120(R_{750} - R_{550}) - 200(R_{670} - R_{550})]$	Characterize the radiant energy absorbed by leaf pigments (Chls); note that the increase of Chls concentration also results in the decrease of the green reflectance.	Broge and Leblanc 2000; Haboudane et al. 2004
Foliar Chemistry			
CAI	$0.5(R_{2020} + R_{2220}) - R_{2100}$	Cellulose and lignin absorption features, discriminates plant litter from soils.	Nagler et al. 2000
NDLI	$[\log(1/R_{1754}) - \log(1/R_{1680})] / [\log(1/R_{1754}) + \log(1/R_{1680})]$	Quantify variation of canopy lignin concentration in native shrub vegetation.	Serrano et al. 2002
NDNI	$[\log(1/R_{1510}) - \log(1/R_{1680})] / [\log(1/R_{1510}) + \log(1/R_{1680})]$	Quantify variation of canopy N concentration in native shrub vegetation.	Serrano et al. 2002

(Continued)

TABLE 5.2 (*Continued*)

Summary of 66 Spectral Indices Extracted from Hyperspectral Data, Collected from the Literature

Spectral Index	Characteristics and Functions	Definition	Reference
Water			
DSWI	Detect water-stressed crops at a canopy level.	$(R_{802} + R_{547})/(R_{1657} + R_{682})$	Galvão et al. 2005
LWVI-1	Estimate leaf water content, an NDWI variant.	$(R_{1094} - R_{893})/(R_{1094} + R_{893})$	Galvão et al. 2005
LWVI-2	Estimate leaf water content, an NDWI variant.	$(R_{1094} - R_{1205})/(R_{1094} + R_{1205})$	Galvão et al. 2005
MSI	Detect variation of leaf water content.	R_{1600}/R_{819}	Hunt and rock 1989; Ceccato et al. 2001
NDII	Detect variation of leaf water content.	$(R_{819} - R_{1600})/(R_{819} + R_{1600})$	Hardinsky et al. 1983
NDWI	Improve the accuracy in retrieving the vegetation water content at both leaf and canopy levels.	$(R_{860} - R_{1240})/(R_{860} + R_{1240})$	Datt et al. 2003; Gao 1996
RATIO$_{1200}$	Estimate relative water content <60% at leaf level.	$2 \times R_{1180-1220}/(R_{1090-1110} + R_{1265-1285})$	Pu, Ge et al. 2003
RATIO$_{975}$	Estimate relative water content <60% at leaf level.	$2 \times R_{960-990}/(R_{920-940} + R_{1090-1110})$	Pu, Ge et al. 2003
RVI$_{hyp}$	Quantify LAI and water content at canopy level.	R_{1088}/R_{1148}	Schlerf et al. 2005
SIWSI	Estimate leaf or canopy water stress, especially in the semiarid environment.	$(R_{860} - R_{1640})/(R_{860} + R_{1640})$	Fensholt and Sandholt 2003
SRWI	Detect vegetation water content at leaf or canopy level.	R_{860}/R_{1240}	Zarco-Tejada et al. 2003
WI	Quantify relative water content at leaf level.	R_{900}/R_{970}	Peñuelas et al. 1997
Stress			
Plant senescence reflectance index (PSRI)	Sensitive to theCar/Chl ratio and used as a quantitative measure of leaf senescence and fruit ripening.	$(R_{680} - R_{500})/R_{750}$	Merzlyak et al. 1999
RVSI	Assess vegetation community stress at canopy level.	$[(R_{712} + R_{752})/2] - R_{732}$	Merton and Huntington 1999

(NDII), normalized difference VI (NDVI), pigment-specific normalized difference (PSND), hyperspectral perpendicular VI (PVI_{hyp}), renormalized difference VI (RDVI), hyperspectral ratio VI (RVI_{hyp}), standard of LAIDI (sLAIDI), spectral polygon VI (SPVI), SR, and wide dynamic range VI (WDRVI). For example, Gong et al. (2003) and Weihs et al. (2008) used PVI_{hyp}, SR, NDVI, RDVI, and RVI_{hyp}, constructed from hyperspectral image data Hyperion and HyMap, to estimate forest LAI. He, Guo, and Wilmshurst (2006) and Darvishzadeh, Skidmore et al. (2008) estimated LAI of grassland ecosystems with VIs: RDVI, MCARI2, and NDVI. With LAIDI and sLAIDI VIs, Delalieux et al. (2008) determined LAI in orchards. And Li et al. (2008) used MTVI2 to map LAI over an agricultural area from CASI hyperspectral image data.

Some VIs, including adjusted transformed soil-adjusted VI (ATSAVI), leaf water VI 1(LWVI-1), leaf water VI 2 (LWVI-2), NDVI, SR, triangular VI (TVI), and modified SR (mSR_{705}), can be used for identifying and mapping plant species and composition. For example, Galvão, Formaggio, and Tisot (2005) developed and used VIs, LWVI-1, LWVI-2, and NDVI to discriminate five sugarcane varieties in southern Brazil with EO-1 Hyperion data. Hestir et al. (2008) used mSR_{705} VI to map invasive species with airborne hyperspectral data (HyMap). Further, Lucas and Carter (2008) assessed vascular plant species richness on Horn Island, Mississippi, with various SR VIs constructed from HyMap hyperspectral image data. For estimating biomass from hyperspectral data, some VIs, such as EVI, modified normalized difference (mND_{705}), mSR_{705}, NDVI, SR, and WDRVI, are very useful. For example, Hansena and Schjoerring (2003) and le Maire et al. (2008) used various narrowband NDVIs and SRs to estimate wheat crop and broadleaf forest biomass, respectively.

With hyperspectral data, many VIs were developed for estimating plant pigments, especially for Chls (Chl-a and Chl-b). They are blue green pigment index (BGI), blue red pigment index (BRI), Chl absorption ratio index (CARI), Chl index using green reflectance (Chl_{green}), Chl index using red edge reflectance ($Chl_{red-edge}$), modified SR of derivatives (DmSR), leaf Chl index (LCI), modified Chl absorption in reflectance index (MCARI), mND_{705}, mSR_{705}, normalized total pigment to Chl index (NPCI), normalized phaeophytinization index (NPQI), plant biochemical index (PBI), photochemical/physiological reflectance index (PRI), PSND, red edge vegetation stress index (RVSI), structural independent pigment index (SIPI), TVI, NDVI, and SR. The VIs specifically developed for estimating Cars contents at leaf level include Car reflectance index (CRI), double difference (DD), eucalyptus pigment indexes (EPIs), modified Car reflectance index (mCRI), PRI, pigment-specific SR (PSSR), ratio analysis of reflectance spectra (RARS), and SIPI. A few VIs were designed for estimating Anths contents in foliage. They are anthocyanin reflectance index (ARI), modified ARI (mARI), and red-green ratio (RGR). These VIs were developed from various hyperspectral data and have been applied for estimating plant pigments by researchers (e.g., Gitelson and Merzlyak 1994; Blackburn 1998; Gamon and Surfus 1999; Gitelson, Buschmann, and Lichtenthaler 1999; Gitelson, Merzlyak, and Chivkunova 2001; Richardson, Duigan, and Berlyn 2002; Rama Rao et al. 2008). For instance, Blackburn (1998) used various narrowband SR, PSND, and SIPI VIs to quantify Chls and Cars of *Pteridium aquilinum* grass at leaf and canopy scales. Gitelson et al. (2001, 2006) developed mCRI, ARI, and mARI VIs with in situ spectral measurements taken from tree leaves to estimate Chls, Cars, and Anths contents. Rama Rao et al. (2008) developed a new VI, named PBI, for improved estimation of plant biochemicals from spaceborne hyperspectral data. The VI PBI is an SR of reflectances at 810 and 560 nm. It has the potential to retrieve leaf total Chls and N concentrations of various crops and at different geographical locations. Hatfield et al. (2008) used PSND and PRI to determine the pigments of agricultural crops. A study by le Maire et al. (2008) estimated leaf Chls content of broadleaf forest with NDVI and SR VIs derived from in situ and Hyperion hyperspectral data. Chappelle, Kim, and McMurtrey (1992) recommended the use of R_{760}/R_{500} as a quantitative measure of Cars. Peñuelas, Baret, and Filella (1995) proposed the use of SIPI for estimating Cars. For Anths estimation, Gamon and Surfus (1999) used a ratio of red-green reflectances $R_{600-700}/R_{500-600}$, and Gitelson, Merzlyak, and Chivkunova (2001), Gitelson, Keydan, and Merzlyak (2006) used an ARI and an mARI to estimate Anths content at the plant leaf level. However, Sims and Gamon (2002, page 352) concluded, "estimation of Cars and Anths contents remains more difficult than estimation of Chls content."

Many narrowband VIs were designed for estimating water content at the leaf and canopy levels. These VIs include disease water stress index (DSWI), LWVI-1, LWVI-2, moisture stress index (MSI), NDII, normalized difference water index (NDWI), PVI_{hyp}, 3-band ratio at 1200 nm ($RATIO_{1200}$), 3-band ratio at 975 nm ($RATIO_{975}$), RVI_{hyp}, RVSI, SWIR water stress index (SIWSI), SR water index (SRWI), and water index (WI). For example, Peñuelas et al. (1993, 1996) studied the reflectances of gerbera, pepper, bean plants, and wheat in the 950–970 nm region as an indicator of water status. Their results showed that the ratio of the reflectance at 970 nm, one of the water absorption bands, to the reflectance at 900 nm as the reference wavelength (R_{970}/R_{900} or WI) closely tracked changes in relative water content (RWC), leaf water potential, stomatal conductance, and cell wall elasticity. Cheng et al. (2006) and Clevers, Kooistra, and Schaepman (2008) used NDWI, WI, and SIWSI to estimate vegetation water content for different canopy scenarios with hyperspectral AVIRIS data. Colombo et al. (2008) estimated leaf and canopy water content in a poplar plantation using SRWI, NDII, and MSI derived from airborne hyperspectral image data. Pu, Ge et al. (2003) determined water status in coastal live oak leaves with $RATIO_{1200}$ and $RATIO_{975}$ indices derived from hyperspectral measurements.

A few VIs are designed for estimating nutrient constituents and concentrations of other biochemicals, such as lignin and cellulose. They are cellulose absorption index (CAI), normalized difference N index (NDNI), normalized difference lignin index (NDLI), NDVI, PBI, and SR. For example, Serrano, Peñuelas, and Ustin (2002) proposed NDNI and NDLI to assess N and lignin concentrations in chaparral vegetation using AVIRIS hyperspectral image data. Gong, Pu, and Heald (2002) and Hansena and Schjoerring (2003) used narrowband NDVI and SR indices to assess nutrient constituent concentrations (N, P, and K) in a conifer species and N status in wheat crops from hyperspectral data. Further, Rama Rao et al. (2008) estimated leaf N concentration of cotton and rice crops with PBI derived from Hyperion hyperspectral data.

5.3.4 Analysis of Absorption Features and Spectral Position Variables

Analysis of spectral absorption features is one step further toward the recognition of some essential properties of a target of interest. Quantitative characterization of absorption features allows for abundance estimation from hyperspectral data. Spectral absorption features are caused by a combination of factors inside and outside the matter surface, including electronic processes, molecular vibrations, abundance of chemical constituents, granular size and physical structure, and surface roughness relative to electromagnetic wavelength. Figure 5.1 shows the major absorption and reflectance features for vegetation.

In order to analyze the absorption features of a spectral reflectance curve, one needs to normalize the spectral curve so that only the spectral values inside the absorption features will be less than 1(100%). This can be done using a continuum-removal technique proposed by Clark and Roush (1984). As shown in Figure 5.2, a continuum is defined for each spectral curve by finding the high points (local maxima) along the curve and fitting straight

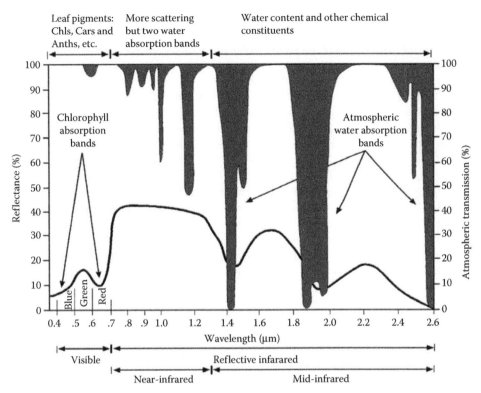

FIGURE 5.1
Major absorption features of vegetation. Leaves from different species may have different strengths of absorption. (Modified from Jensen, J. R. *Remote Sensing of the Environment: An Earth Resource Perspective*, 2nd ed., Prentice Hall, Upper Saddle River, NJ, 2007.)

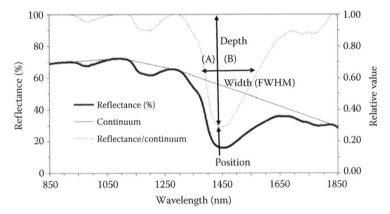

FIGURE 5.2
A part of the coastal live oak leaf spectrum adjusted by a continuum-removal technique (Data from Clark, R. N., and T. L. Roush. 1984. *J Geophys Res* 89:6329–40.) and the definitions of three absorption features. Depth measures the deepest absorption. Width measures full width half maximum (FWHM) absorption. Position marks the wavelength at the deepest absorption and areas (A or B) on each side of the deepest absorption are used for calculating asymmetry. (Modified from Pu, R., Ge, S., Kelly, N. M, Gong, P., *Int J Remote Sens*, 24, 9, 2003b.)

line segments between these points. This can be done either manually or automatically. The normalized curve is obtained by dividing the original spectral value at each band location with the value on the straight line segments at the corresponding wavelength location. Quantitative measures can be determined from each absorption peak after normalization of the raw spectral reflectance curve. An asymmetric term can also be defined by subtracting area A from area B (Figure 5.2; Kruse, Lefkoff, and Dietz 1993). The quantitative measures shown in Figure 5.2 can be used to determine the abundances of certain compounds in a pixel. For example, Pu, Ge et al. (2003) explored the effectiveness of these absorption parameters in correlation with the leaf water content of oak trees at various stages of disease infection. Galvão, Formaggio, and Tisot (2005) successfully used some absorption features extracted with this technique and other spectral indices from EO-1 Hyperion data to discriminate the five sugarcane varieties in southeastern Brazil. Huber et al. (2008) also estimated foliar biochemistry (the concentrations of N and carbon, and the content of water) from hyperspectral HyMap data in mixed forest canopy using such a continuum-removal technique.

Some absorption features or spectral position variables can also be modeled. For example, the red edge of vegetation between 670 and 780 nm has been widely modeled by a number of researchers. Based on the spectral properties of the pigments, some researchers have used red edge optical parameters (e.g., Curran, Windham, and Gholz 1995; Belanger, Miller, and Boyer 1995; Cho, Skidmore, and Atzberger 2008) to estimate plant leaf and canopy Chls content and concentration. Guyot, Baret, and Jacquemond (1992) proposed a four-point interpolation method to find the wavelength position of the inflection point on the red edge position and the red well position. Other methods include polynomial fitting (Pu, Gong et al. 2003), Lagrangian interpolation (Dawson and Curran 1998), inverted Gaussian model fitting (Miller, Hare, and Wu 1990), and linear extrapolation techniques (Cho and Skidmore 2006). The red edge optical parameters can be used for estimating Chls concentrations (Belanger, Miller, and Boyer 1995; Curran, Windham, and Gholz 1995), nutrient constituent concentrations (Gong, Pu, and Heald 2002; Cho, Skidmore, and Atzberger 2008), leaf relative water content (Pu, Ge et al. 2003; Pu, Foschi, and Gong 2004), and forest LAI (Pu, Gong et al. 2003). In addition, Pu, Foschi, and Gong (2004) proposed to extract 20 spectral variables (10 maximum-first derivatives plus 10 corresponding wavelength-position variables) from 10 slopes defined across a reflectance curve from 0.4 to 2.5 μm for estimating oak leaf relative water content. All these efforts can help extract absorption feature measures and other spectral features from original hyperspectral data for estimating vegetation parameters.

5.3.5 Hyperspectral Transformation

The principal component (PC) analysis (PCA) technique has been applied to reduce the data dimension and feature extraction from hyperspectral data for assessing leaf or canopy biophysical and biochemical parameters (e.g., Gong, Pu, and Heald 2002; Pu and Gong 2004). With a covariance (or correlation) matrix calculated from vegetated pixels only, it is commonly believed that the eigenvalues and corresponding eigenvectors computed from the covariance (or correlation) matrix are able to enhance vegetation variation information in the first several PCs. Because the PCA does not always produce images that show steadily decreasing image quality with increasing component number, Green et al. (1988) developed one transform method called "maximum noise fraction" (MNF) transform to maximize the signal-to-noise ratio when choosing PCs with increasing component number. Then, several MNFs to maximize the signal-to-noise ratio are selected for further analysis of hyperspectral data, such as for determining endmember spectra for spectral mixture analysis (Pu, Gong et al. 2008; Walsh et al. 2008) and hyperspectral mosaic (Hestir et al. 2008).

"Canonical discriminant analysis" (CDA) also is a dimension-reduction technique equivalent to canonical correlation analysis that can be used to determine the relationship between the quantitative variables and a set of dummy variables coded from the class variable in a low-dimensional discriminant space (Khattree and Naik 2000; Zhao and Maclean 2000). Given a classification variable and several quantitative variables, CDA derives canonical variables, linear combinations of the quantitative variables that summarize between-class variation in much the same way that PCA summarizes most variation in the first several PCs. In other words, CDA involves human effort and knowledge derived from training samples, whereas PCA performs a relatively automatic data transformation and tries to concentrate the majority of data variance in the first several PCs. However, unlike PCA, CDA is only occasionally analyzed and tested as a data transformation technique by researchers in the remote-sensing community for dimensional reduction and feature extraction (e.g., Zhao and Maclean 2000; van Aardt and Wynne 2001, 2007).

The wavelet transform (WT) is a relatively new signal-processing tool that provides a systematic means for analyzing signals at various scales or resolutions and shifts. In the past two decades, WT has been successfully applied to image processing, data compression, pattern recognition (Mallat 1998), image texture feature analysis (Fukuda and Hirosawa 1999), and feature extraction (Simhadri et al. 1998; Pittner and Kamarthi 1999). Wavelets have proven to be quite powerful in these remote-sensing application areas. This is attributed to the facts that the WT can decompose a spectral signal into a series of shifted and scaled versions of the mother wavelet function, and that the local energy variation (represented as peaks and valleys) of a spectral signal in different bands at each scale can be detected automatically and provide some useful information for further analysis of hyperspectral data (Pu and Gong 2004). With continuous WT (CWT), one can analyze both single-dimensional and multidimensional signals, such as hyperspectral image cubes, across a continuum of scales. With discrete WT (DWT), signals are analyzed over a discrete set of scales, typically dyadic ($2j$, j = 1, 2, 3, ...), and the transforms can be realized using a variety of fast algorithms and customized hardware (Bruce, Morgan, and Larsen 2001). The WT can decompose signals over dilated (scaled) and translated (shifted) wavelets (Mallat 1989; Rioul and Vetterli 1991). There are many different types of mother wavelets and wavelet bases to be selected for use. In practice, researchers need to test most of the wavelet families to find the most useful wavelet family in a particular project. After a set of DWT coefficients for each level or scale of a pixel-based spectrum is calculated, the energy feature of the wavelet decomposition coefficients is computed at each scale for both approximation and details and is used to form an energy feature vector (Pittner and Kamarthi 1999; Bruce, Morgan, and Larsen 2001; Li et al. 2001; Pu and Gong 2004). This can become a feature extraction through a dimension reduction. With hyperspectral data of vegetation and the WT technique, several studies already demonstrate the benefits of wavelet analysis. For example, Pu and Gong (2004) used the mother wavelet function db3 in MATLAB® (Misiti et al. 1996) to transform Hyperion data (167 available bands in their analysis) for extracting features through a dimension reduction for mapping forest LAI and CC. By using the wavelet analysis method, Blackburn (2007a) and Blackburn and Ferwerda (2008) retrieved plant pigments (Chls and Cars) concentration from leaf and canopy spectra, although further work is needed to refine this approach. Hsu and Tseng (2000) and Henry et al. (2004) used the wavelet analysis method (multiscale transform) to extract useful spectral features from hyperspectral data (AVIRIS and in situ spectral measurements) for plant/crop-type classification. They concluded that using spectral features extracted with the wavelet analysis method from hyperspectral data resulted in higher classification accuracy than using features with other methods (e.g., PCA and multiple VIs).

5.3.6 Spectral Unmixing Analysis

Unlike laboratory and in situ spectral reflectances, which are usually measured from "pure materials," a large portion of remotely sensed data is spectrally mixed. In order to identify various pure materials and to determine their spatial proportions from the remotely sensed data, the spectral mixing process has to be properly modeled. Once the spectral mixing process is modeled, the model can be inverted to derive the spatial proportions and spectral properties of pure materials. There are two types of spectral mixing: (1) linear spectral mixing and (2) nonlinear spectral mixing. Both linear and nonlinear spectral mixing models are simple tools used to describe spectral mixing processes. A real spectral mixing process could be complicated and can be more explicitly dealt with using radiative transfer (RT) models (e.g., Li and Strahler 1985, 1992); also their solutions are often difficult to obtain (Liang and Strahler 1993; Gong, Wang, and Liang 1999). Linear spectral mixing model (LSM) and its inversion have been widely used since the late 1980s. An LSM has been extensively applied to extract the abundance of various components within mixed pixels. The nonlinear spectral mixture model can be found detailed in the works of Sasaki et al. (1984) and Zhang et al. (1998). In addition, an artificial neural network (ANN) algorithm has been tested to unmix mixed pixels into fractional abundances of endmembers in some studies (Foody 1996; Wang and Zhang 1998; Flanagan and Civco 2001; Pu, Gong et al. 2008).

In the spectral mixture analysis, a typical LSM at pixel (i, j) can be expressed as follows:

$$\mathbf{R}_{ij} = \mathbf{M}\mathbf{F}_{ij} + \varepsilon_{ij} \tag{5.7}$$

where $\mathbf{R}_{ij}$ is a K-dimension reflectance (or digital number) vector, $\mathbf{F}_{ij}$ is an L-dimension fraction vector, $\mathbf{M}$ is a $K \times L$ endmember spectral matrix, and ε_{ij} is a K-dimension error vector representing residual error. The goal of spectral unmixing is to solve for $\mathbf{F}_{ij}$ with $\mathbf{R}_{ij}$ and $\mathbf{M}$ known. When the number of endmembers in pixel (i, j) are appropriately accounted for, $\mathbf{F}_{ij}$ should satisfy the following conditions:

$$\sum_{l=1}^{L} \mathbf{F}_l = 1, \text{ and } \mathbf{F}_l \geq 0 \tag{5.8}$$

It is well known that the inversion of Equation 5.7 (i.e., spectral unmixing) can be achieved with a least-squares solution (LSS) when $K > L$ (e.g., Adams, Smith, and Gillespie 1989; Sohn and McCoy 1997; Maselli 1998; Pu, Gong et al. 2008).

A feed-forward ANN algorithm is a nonlinear solution to the LSM, used for unmixing mixed pixels. The network training mechanism is an error-propagation algorithm (Rumelhart, Hinton, and Williams 1986; Pao 1989). In a layered structure, the input to each node is the sum of the weighted outputs of the nodes in the prior layer, except for the nodes in the input layer, which are connected to the feature values. The nodes in the last layer output a vector that corresponds to similarities in each class, or fractions of endmembers within a mixed pixel. One layer between the input and output layers is usually sufficient for most learning purposes. The learning procedure is controlled by a learning rate and a momentum coefficient, which need to be specified empirically based on the results of a limited number of tests. Network training is done by repeatedly presenting training samples (pixels) with the known fractions of endmembers. Training is terminated when the network output meets a minimum error criterion or optimal test accuracy is achieved. The trained network can then be used to estimate the fraction of each endmember in a mixed pixel.

This simple mixing model (LSM) has an advantage in that it is relatively simple and provides a physically meaningful measure of abundance in mixed pixels. However, there are a number of limitations to the simple mixing concept: The endmembers used in LSM are the same for each pixel, regardless of whether the materials represented by the endmembers are present in the pixel; it fails to account for the fact that the spectral contrast between those materials is variable; the LSM cannot account for subtle spectral differences among materials efficiently; and the maximum number of components that an LSM can map is limited by the number of bands in the image data (Li and Mustard 2003). Therefore, Roberts et al. (1998) introduced multiple endmember spectral mixture analysis (MESMA), a technique for identifying materials in a hyperspectral image using endmembers from a spectral library. The MESMA technique overcomes the limitations of the simple mixing model. Using the MESMA, the number of endmembers and their types are allowed to vary for each pixel in the image. The general MESMA procedure starts with a series of two-endmember candidate models, evaluates each model based on selection criteria and then, if required, constructs candidate models that incorporate more endmembers (Roberts et al. 1998).

The key to successful spectral mixture analysis is the selection of appropriate endmembers (Gong, Miller, and Spanner 1994; Tompkins et al. 1997). Determination of endmembers involves identifying the number of endmembers and extracting their corresponding spectral signatures. The pixel purity index (PPI), according to Boardman (1993), can be combined with the use and interpretation of scatter plots of MNF (Green et al. 1988) to characterize the relative abundance of endmembers across a scene to help determine endmember spectra. The ability to detect different surface materials in the endmember analysis of remotely sensed data is a function of spectral contrast among endmembers, noise, and spectral resolution (Shipman and Adams 1987; Sabol, Adams, and Smith 1990). Sufficient spectral information from hyperspectral data ensures the successful unmixing of mixed pixels. To select endmembers during the processing of MESMA, three selection criteria are fraction, root mean square error (RMSE), and the residuals of contiguous bands (Roberts et al. 1998). The minimum RMSE model is assigned to each pixel, and it can be used to map materials and fractions within the image (Painter et al. 1998) with the MESMA approach.

A number of researchers have applied LSM to hyperspectral data to estimate the abundance of general vegetation cover or specific vegetation species (Asner and Heidebrecht 2003; Miao et al. 2006; Judd et al. 2007; Hestir et al. 2008; Walsh et al. 2008; Pignatti et al. 2009). A neural network (NN)–based nonlinear solution also was applied to hyperspectral data to estimate the abundance of specific vegetation species (Pu, Gong et al. 2008; Walsh et al. 2008). Several researchers have applied the MESMA approach in a variety of environments for vegetation mapping. For example, Roberts et al. (1998, 2003) used MESMA and AVIRIS hyperspectral image data to map vegetation species and land-cover types in southern California chaparral. Using AVIRIS image data and the MESMA approach, Li, Ustin, and Lay (2005) and Rosso, Ustin, and Hastings (2005) mapped coastal salt marsh vegetation in China and the marshland vegetation of San Francisco Bay, California, respectively. In addition, Fitzgerald et al. (2005) successfully mapped multiple shadow fractions in a cotton canopy with MESMA approach and hyperspectral imagery.

5.3.7 Hyperspectral Image Classification

Traditional multispectral classifiers can be used, but they may have a less than expected effect as they face difficulties caused by the high dimensionality of hyperspectral data and the high correlation of adjacent bands with a limited number of training samples. In order to overcome these problems, a feature extraction preprocessing before classification is necessary. Feature extraction schemes such as PCA (or its noise-adjusted version, MNF), Fisher's linear discriminant analysis (LDA), or CDA) have been applied in transforming and reducing the data dimension by maximizing the ordered variance of the whole data set or the ratio of between-class variance and within-class variance of the training samples. Jia and Richards (1999) proposed a segmented PC transformation (i.e., segmented PCA or segPCA) to reduce the computation cost by selecting subsets of the covariance matrix in a lower segmented dimension. Penalized discriminant analysis (PDA) was suggested to deal with the high correlation among the bands more efficiently by penalizing the high within-class variance and to improve the performance of LDA (Yu et al. 1999). Jia and Richards (1994, 2002) first segmented the whole spectral space into several subspaces using a spectral correlation matrix and then used the maximum likelihood classifier, called "simplified maximum likelihood classification," to classify an image scene. Jimenez and Landgrebe (1998) segmented and transformed the whole spectrum into several subspectra, estimated training statistics at the subspaces, and iteratively updated an orthogonal projection matrix until a minimum Bhattacharyya distance (BD) was obtained among the classes.

Fisher's LDA and CDA search for successive linear combinations of data to maximize the ratio of between-class variance and within-class variance of training samples in an expectation of spreading the means or the cluster centers of different classes as much as possible while keeping the within-class variation at a similar level for all classes (Yu et al. 1999; Xu and Gong 2007; Pu and Liu 2010). It is based on an assumption of reliable estimation of training statistics. Segmented LDA (segLDA) first divides the whole spectrum into subblocks, with each block containing a set of continuous highly correlated spectral bands. Denote the dimension of the kth subblock as I_k, and $I_1 + \cdots + I_k + \cdots + I_K = I$. For each subblock of spectral bands, estimate the between-class covariance matrix and the within-class covariance matrix in a subspace that has a dimension equal to the number of bands in the subblock. Then, apply LDA to each subblock to generate new component images (features) with a number of $min(C - 1, I_k)$, where C is the number of classes and k is the kth subblock. This projection is supposed to spread the means of the classes as much as possible. With the newly projected images for each subblock, we could either select the first few feature images from each subblock to generate a combined pool of new features that can be subsequently used for classification, or select more feature images less than $min(C - 1, I_k)$ from the kth subblock for $k = 1, \ldots, K$ to form a new subspace. The LDA approach can be applied multiple times to reduce the data dimension in the search for an optimal set of orthogonal subspaces for use in final classification. The PDA introduces a penalty matrix Ω to the within-class covariance matrix to penalize and limit the effect that a band with high within-class variation may have in the case of LDA, while reserving the low within-class variation band. The function of the penalty matrix was geometrically interpreted by Yu et al. (1999). The matrix unequally smooths within-class variation for all the classes in the hyperspectral space. The realization of segmented PDA (segPDA) and segmented CDA (segCDA) is similar to that of segLDA in the sense that segmentation is done before applying PDA, except that PDA adds a penalty term to the estimation of the within-class covariance matrix. Similar to segPCA, segLDA, segCDA, and segPDA all save significant computation time.

Xu and Gong (2007) compared several feature extraction algorithms used for band reduction of Hyperion data. These include PCA, segPCA, LDA, segLDA, PDA, and seg-PDA. Feature reductions were all followed by classification of Hyperion images using a minimum distance (MD) classifier. With segPDA, segLDA, PDA, and LDA, similar accuracies were achieved, whereas the segPDA and segLDA newly proposed by Xu and Gong (2007) greatly improved computation efficiency. They also outperformed segPCA and PCA in classification accuracy due to the use of specific intra- and interclass covariance information. Similar to the conclusion drawn by Xu and Gong (2007), Pu and Liu (2010) also concluded that segCDA outperformed segPCA and segmented stepwise discriminant analysis (SDA) when 13 tree species were discriminated using in situ hyperspectral data and segCDA. Based on the study by Pu and Liu (2010), CDA or segCDA (under the condition of limited training samples) should be applied broadly in mapping forest-cover types, species identification, and other land use/land-cover classification practices with multi/hyperspectral remote sensing data, because it is superior to PCA and SDA for selection of features that are used for image classification.

Support vector machines (SVMs) as a new type of classifiers have been successfully applied to the classification of hyperspectral remote-sensing data. Traditionally, classifiers first model the density of various classes and then find a separating surface for classification. However, the estimation of density for various classes with hyperspectral data suffers from the Hughes phenomenon (Hughes 1968): For a limited number of training samples, the classification rate decreases as the dimension increases. The SVM approach does not suffer from this limitation because it directly seeks a separating surface through an optimization procedure that finds so-called support vectors that form the boundaries of the classes. This is an interesting property of hyperspectral image processing because usually there is only a set of limited training samples available to define the separating surface for classification. Further, the properties of SVMs make them well suited to hyperspectral image classification since they can handle data efficiently in high dimensionality, deal with noisy samples in a robust way, and make use of only those most characteristic samples as support vectors in the construction of classification models. Melgani and Bruzzone (2004) provided a detailed introduction of SVMs for the classification of hyperspectral imagery. SVMs are considered to be kernel-based classifiers that are based on mapping data from the original input feature space to a kernel feature space of higher dimensionality and then solving a linear problem in that space (Burges 1998). Camps-Valls and Bruzzone (2005) introduced several other kernel-based classifiers, including kernel Fisher discriminant analysis, regularized radial basis function NN, and a regularized boosting algorithm. They compared them with the SVMs and reported comparable accuracies in classifying the same agricultural AVIRIS scene as used by Melgani and Bruzzone (2004).

The SVM approach can significantly improve classification accuracy with hyperspectral data. For example, Melgani and Bruzzone (2004) tested four SVM strategies for multiclass discrimination including the "one against all," "one against one," "binary hierarchical tree balanced branches," and "binary hierarchical tree one against all" algorithms. They applied these algorithms to an AVIRIS image acquired over an agricultural area with nine classes and compared their performances with radial basis function NNs and K-nearest neighbor (K-NN) algorithms. They reported overall accuracies greater than 90% with an accuracy improvement of 7–12% over the NN and K-NN algorithms. Pal and Mather (2004) used a multiclass SVM for land-cover classification of Digital Airborne Imaging Spectrometer (DAIS) hyperspectral image data. Results showed that SVM outperforms maximum likelihood, univariate decision tree, and back propagation NN classifiers. For classification purposes with hyperspectral HyMap data, Camps-Valls et al. (2004) used SVMs for a six-class crop classification and analyzed their performance in terms of efficiency and robustness as compared to extensively used NNs and fuzzy methods. They concluded that SVMs yield better outcomes than NNs and fuzzy methods in terms of classification accuracy, simplicity, and robustness.

5.3.8 Empirical/Statistical Analysis Methods

Most researchers have employed statistical analysis methods to correlate biophysical or biochemical parameters with spectral reflectance, VIs, or derivative spectra in the visible, NIR, and SWIR wavelengths of hyperspectral data at leaf, canopy, or plant community level (Peterson et al. 1988; Wessman et al. 1988; Bolstad and Lillesand 1991; Smith et al. 1991; Gong, Pu, and Miller 1992, 1995; Gong, Pu, and Yu 1997; Franklin and McDermind 1993; Banninger, Johnson, and Peterson 1994; Johnson, Hlavka, and Peterson 1994; Matson et al. 1994; Pinel et al. 1994; Yoder and Waring 1994; Gastellu–Etchegorry et al. 1995; Gamon et al. 1995; Yoder and Pettigrew-Crosby 1995; Grossman et al. 1996; Zagolski et al. 1996; LaCapra et al. 1996; Gitelson and Merzlyak 1997; Martin and Aber 1997; Chen, Elvidge, and Groeneveld 1998; Blackburn 1998; Fourty and Baret 1998; Martin et al. 1998; Datt 1998; Serrano, Peñuelas, and Ustin 2002; Galvão, Formaggio, and Tisot 2005; Colombo et al. 2008; Darvishzadeh, Skidmore et al. 2008; Hestir et al. 2008; Huber et al. 2008). Johnson, Hlavka, and Peterson (1994) determined predictive relationships for biochemical concentrations using regressions between the chemical composition of forest canopy and the AVIRIS reflectance. Using data from AVIRIS and a CASI, Matson et al. (1994) demonstrated that canopy biochemicals carried information about forest ecosystem processes and suggested that some of this chemical information might be estimated remotely using hyperspectral data collected by airborne sensors. They found that the first differences were in the range of 1525–1564 nm, which figured prominently in all N equations. After correlating VIs of R_{NIR}/R_{700} and R_{NIR}/R_{550} with Chl content, Gitelson and Merzlyak (1996, 1997) demonstrated that the indices for Chl assessment were important for two deciduous species, maple and chestnut. In spectral feature analysis associated with N, P, and K deficiencies in *Eucalyptus saligna* seedling leaves, Ponzoni and Goncalves (1999) proved that spectral reflectance can be better estimated using a combination of nutrient constituents (N, P, and K) as independent variables with the results from simple and multiple regression. Martin et al. (1998) determined forest species composition using high spectral resolution remote-sensing data with an approach that combined forest species–specific chemical characteristics and previously derived relationships between hyperspectral data (AVIRIS) and foliar chemistry. They classified 11 forest-cover types, including pure and mixed stands of deciduous and conifer species, with an overall accuracy of 75%. With EO-1 Hyperion hyperspectral image data, Galvão, Formaggio, and Tisot (2005) successfully discriminated five sugarcane varieties in southeastern Brazil using a multiple discriminant analysis method that produced a classification accuracy of 87.5%. With multiple linear regression models, continuum-removal technique, and normalized HyMap spectra, Huber et al. (2008) estimated foliar concentrations of N and carbon, and content of water in a mixed forest canopy.

Partial least-squares regression (PLSR) is a technique that reduces the large number of measured collinear spectral variables to a few noncorrelated latent variables or PCs. The PCs represent the relevant structural information present in the measured reflectance spectra and are used to predict the dependent variables (i.e., biophysical and biochemical parameters; Darvishzadeh, Skidmore et al. 2008). The PLSR approach is different from PC regression (PCR) in the methods used in extracting factors (also called "components," "latent vectors," or "latent variables"). In short, PCR produces the weight (coefficient) matrix reflecting the covariance structure between the predictor variables, whereas PLSR produces the weight (coefficient) matrix reflecting the covariance structure between the predictor and response variables. In other words, PCR extracts factors to explain as much predictor sample variation as possible, whereas PLSR balances the two objectives of explaining both response variation and predictor variation as much as possible. Recently, there has been increasing interest in applying the PLSR approach to calibrate relationships between spectral variables, often derived from hyperspectral data and a set of bioparameters (Hansena and Schjoerring 2003; Asner and Martin 2008; Darvishzadeh, Skidmore et al. 2008; Martin et al. 2008; Weng, Gong, and Zhu 2008; Prieto-Blanco et al. 2009). For example, using spectral measurements taken from leaves and bioparameter data (Chl-a, Chl-b, Cars, Anths, water, N, P, and SLA) collected from 162 Australian tropical forest species, along with PLSR approach and canopy RT modeling, Asner and Martin (2008) concluded that a suite of leaf properties among tropical forest species can be estimated using full-range leaf spectra of fresh foliage collected in the field. Hansena and Schjoerring (2003) used two-band combinations in the normalized difference VIs constructed from in situ spectral measurements taken from wheat crop canopy and PLSR approach to estimate canopy green biomass and N status. They concluded that PLSR analysis may be a useful exploratory and predictive tool when applied to hyperspectral reflectance data analysis. The optimal number of PCs was determined by the guidelines described by Esbensen (2000). The basic PLSR algorithm will not be introduced here, but further information on the PLSR model can be found in the work of Ehsani et al. (1999).

Although univariate and multiple regression analysis methods are relatively simple and their modeling results frequently have higher estimation accuracy, empirical or statistical relationships are often site, species, and sensor specific, and thus cannot be directly applied to other study areas since the plant canopy structure and sensors' viewing geometry may vary among different sites and species. Therefore, during the last two decades, physically based modeling approaches have attracted the attention of many researchers, who have retrieved biophysical and biochemical parameters by inversing various physically based models from simulated spectra or real imaging data.

5.3.9 Physically Based Modeling

The theoretical basis of physically based models consists of developing a leaf or canopy scattering and absorption model that involves biochemistry and biophysics. These models, including RT and geometric–optical (GO) models, consider the underlying physics and complexity of the leaf internal structure and therefore are robust and have the potential to replace statistically based approaches (Zhang et al. 2008a, 2008b). In the context of the remote sensing of bioparameters, such models have been used in the forward mode to calculate leaf or canopy reflectance and transmittance and in the inversion mode to estimate leaf or canopy chemical and physical properties. For example, many researchers employ physically based models at leaf or canopy level to retrieve biochemical parameters, including leaf pigments from either simulated spectra or hyperspectral image data (Asner and Martin 2008; Feret et al. 2008; Zhang et al. 2008a, 2008b).

A number of RT models have been developed at leaf and canopy levels. They mostly simulate leaf reflectance and transmittance spectra between 0.4 and 2.50 μm. Among models focusing on leaf optical properties, the most important RT models may include the Propriétés Spectrales (PROSPECT) model (Jacquemoud and Baret 1990; Jacquemoud et al. 1996; Fourty et al. 1996; Demarez et al. 1999; le Maire, Francois, and Dufrene 2004), the leaf incorporating biochemistry exhibiting reflectance and transmittance yields (LIBERTY) model (Dawson, Curran, and Plummer 1998; Coops and Stone 2005), and the leaf experimental absorptivity feasibility model (LEAFMOD; Ganapol et al. 1998). Among those focusing on canopy optical properties, the most popular RT models are the scattering by arbitrary inclined leaves (SAIL; Verhoef 1984; Asner 1998) model and its improved versions that have been adapted to account for some heterogeneity within the vegetation canopy, for example, GeoSAIL (Verhoef and Bach 2003), 2M-SAIL (Weiss et al. 2001; Le Maire et al. 2008), and 4SAIL2 (Verhoef and Bach 2007). The other important canopy reflectance models include fast canopy reflectance (FCR; Kuusk 1994), the new advanced discrete model (NADIM; Jacquemoud et al. 2000; Ceccato et al. 2002), the Markov chain canopy reflectance model (MCRM; Kuusk 1995) adapted for row crops (Cheng et al. 2006), and the four models used for simulating discontinuous forest canopies, including discrete anisotropic RT (DART; Demarez and Gastellu-Etchegorry 2000), spreading of photons for radiation interception (SPRINT; Zarco-Tejada, Miller, Harron et al. 2004), forest light interaction model (FLIM; Zarco-Tejada, Miller, Morales et al. 2004), and three-dimensional forest light interaction (FLIGHT; Koetz et al. 2004). In addition, during the last two decades, researchers have developed some leaf-canopy-coupled models, including PROSAIL (Baret et al. 1992; Broge and Leblance 2000), LEAFMOD +CANMOD (Ganapol et al. 1999), LIBERTY+FLIGHT (Dawson et al. 1999), and LIBERTY+SAIL (Dash and Curran 2004). Among the RT models, based on the literature searched and analyzed by Jacquemoud et al. (2009), the most popular and important RT models on leaf, canopy, and leaf-canopy-coupled optical properties are PROSPECT, SAIL, and PROSAIL, as well as their modified versions.

The PROSPECT models, including the latest versions PROSPECT-4 and -5 (Feret et al. 2008), can provide specific absorption and scattering coefficients of leaf components. The model is widely used and well validated (Fourty et al. 1996). The SAIL model is a four-stream RT model developed by Verhoef (1984). It was later modified by Kuusk (1991) to take the hot spot feature into account. Linking the two models into PROSAIL allowed description of both the spectral and directional variation of canopy reflectance as a function of leaf biochemistry (mainly Chls, water, and dry matter contents) and canopy architecture (primarily LAI, LAD, and relative leaf size; Jacquemoud et al. 2009). The coupled leaf-canopy and other RT models are used to understand the way in which leaf reflectance properties are influenced by the larger number of controlling factors at canopy scale (Demarez and Gastellu-Etchegorry 2000). Coupled models have enabled the development and refinement of spectral indices that are insensitive to factors such as canopy structure, illumination geometry, and soil/litter reflectance (Broge and Leblanc 2000; Daughtry et al. 2000). Such approaches have also been used in defining predictive relationships that have been applied to hyperspectral imagery to generate maps of Chl (Haboudane et al. 2002; Zarco-Tejada, Miller, Morales et al. 2004, Zarco-Tejada, Berjon et al. 2005).

The GO models belong to one type of RT models developed to capture the variation of remote sensing signals on the Earth's surface with illumination and observation angles. Since GO models emphasize the effect of canopy architecture, they are very effective in capturing the angular distribution pattern of the reflected radiance, and are thus used widely in remote-sensing applications (Chen and Leblanc 2001) as aforementioned RT models. There are a lot of different types of GO models. For example, a model developed by Li and Strahler (1985) described the vegetation canopy using opaque geometric shapes (cones or cylinders), which cast shadows on the ground. Consequently, crown transparency is assumed to be zero. These GO models are mainly used to describe (sparse) forests or shrublands, where shadowing plays an important role.

Physically based models must be inverted to retrieve vegetation characteristics from the observed reflectance data. So far, different inversion techniques for physically based models mainly include iterative optimization methods (Goel and Thompson 1984; Liang and Strahler 1993; Jacquemoud et al. 1995; Jacquemoud et al. 2000; Meroni, Colombo, and Panigada 2004), lookup table (LUT) approaches (Knyazikhin et al. 1998; Weiss et al. 2000; Combal, Baret, and Weiss 2002; Combal et al. 2003; Gastellu-Etchegorry, Gascon, and Esteve 2003), and ANNs (Gong, Wang, and Liang 1999; Weiss and Baret 1999; Walthall et al. 2004; Schlerf and Atzberger 2006). In the iterative optimization approach, a stable and optimum inversion is not guaranteed. Moreover, the traditional iterative method is time-consuming and often requires a simplification of the models when processing large datasets. This may result in a decrease of the inversion accuracy and makes the retrieval of biophysical and biochemical variables unfeasible for large geographic areas (Houborg, Soegaard, and Boegh 2007). Methods employing LUTs can partially overcome this drawback. They operate using a database of simulated canopy reflectance variables in structural and radiometric properties. However, LUT creation can be complicated and requires an extensive set of reliable field measurements. The ANN technique, proposed in the forward and inverse modeling of RT models for retrieving bioparameters, is expected to reduce such complexity of inversion. For proper training (ANN) and representation (LUT), the techniques basically rely on a large database of simulated canopy reflectance spectra to achieve a high degree of accuracy. This increases the computational time for identifying the most appropriate LUT entry and the time required for training the ANN (Kimes et al. 2000; Liang 2004).

5.4 Summary and Future Directions

Hyperspectral remote sensing, or imaging spectroscopy, is a cutting-edge technology that can be utilized in ecological studies for extracting and assessing vegetation characterization. In this chapter, the spectral characteristics, properties, and/or responses of a set of plant biophysical and biochemical parameters were reviewed. These bioparameters mainly include typical biophysical parameters (LAI, SLA, CC, species/composition, biomass, NPP, and fPAR) and biochemical parameters (plant pigments such as Chl-a and Chl-b, Cars, and Anths, plant nutrients such as N, P, and K, leaf or canopy water content, and other chemicals such as lignin and cellulose; and protein concentration). To extract and assess typical bioparameters from various hyperspectral data, including laboratory and in situ hyperspectral measurements, spectra synthesized and/or simulated from physically based models, and airborne and spaceborne hyperspectral image data, relatively speaking, a wide range of analysis techniques and approaches that have already been developed and demonstrated are extensively reviewed in this chapter. The spectral analysis techniques cover spectral derivative analysis, spectral matching, spectral index analysis, spectral absorption features and spectral position variables, hyperspectral transformation, spectral unmixing analysis, and hyperspectral classifications; and the two general categories of analysis methods include empirical/statistical methods and physically based models. Advantages and disadvantages, or merits and drawbacks, for some specific analysis techniques and approaches were also discussed here. Data from imaging spectroscopy have repeatedly been shown to produce accurate estimates of many biochemical parameters and physical characteristics related to key ecological processes. Imaging spectroscopy is the only technology available to measure many important environmental properties over large regions, particularly canopy water content, dry plant residues, and soil biochemical properties (Ustin et al. 2004).

In the future, the richness of information available in the continuous spectral coverage afforded by both airborne and spaceborne imaging spectrometers will make it possible to address questions regarding vegetation bioparameters more correctly and accurately. Since hyperspectral data can provide richer and more delicate spectral information than multi-spectral data, spectral unmixing and automatic target detection remain important information extraction tasks in hyperspectral data analysis, and the use of PCA, mathematical programming, and factor analysis need to be further assessed in solving the linear mixing problem. Inversion of physically based RT models with hyperspectral data assisted by analysis of multiangular data will be useful in solving nonlinear spectral mixing problems because the angular data can be used to retrieve the structural information of vegetation.

When using various spectral VIs to estimate different bioparameters, the use of optimized VIs should be considered because there are many potential narrow bands ready to be used for developing various VIs from hyperspectral data. Experience has proven that with some optimized VIs for estimating some bioparameters, the estimation accuracy can be significantly increased (e.g., Gong et al. 2003). When attempting to identify a robust, generic solution, there is currently only limited evidence available with which one can rank the performance of the range of existing hyperspectral analysis approaches in quantifying plant bioparameters. Therefore, it is necessary to conduct intercomparison of hyperspectral approaches (Blackburn 2007b) across a large number of bioparameters using a large number of different analysis techniques. A sensitivity study is needed to determine the set of variables that can be retrieved with a reasonable accuracy for available imaging spectroscopy systems. Finally, although many analysis techniques have been developed and are available in some applications for estimating biochemicals from hyperspectral data at the leaf scale, in order to exploit the opportunities offered by imaging spectrometry for synoptic, consistent, and spatially continuous information, it is important to develop suitable methods that can also derive estimates of foliar biochemical concentrations from canopy-scale reflectance spectra. For this case, several strategies are available for the analysis of canopy spectra (Zarco-Tejada et al. 2001). This is a scaling issue, a problem encountered frequently in ecological studies.

Acknowledgments

The comments and suggestions of two anonymous reviewers were greatly valuable in improving the chapter. The authors sincerely appreciate their efforts.

References

Adams, J. B., M. O. Smith, and A. R. Gillespie. 1989. Simple models for complex natural surfaces: A strategy for the hyperspectral era of remote sensing. In *Proceedings of the 1989 International Geoscience and Remote Sensing Symposium*, 16–21. *IEEE*: Vancouver, BC, Canada.

Asner, G. P. 1998. Biophysical and biochemical sources of variability in canopy reflectance. *Rem Sens Environ* 64:134–53.

Asner, G. P., and K. B. Heidebrecht. 2003. Imaging spectroscopy for desertification studies: Comparing AVIRIS and EO-1 Hyperion in Argentina drylands. *IEEE Trans Geosci Rem Sens* 41:1283–96.

Asner, G. P., M. O. Jones, R. E. Martin, D. E. Knapp, and R. F. Hughes. 2008. Remote sensing of native and invasive species in Hawaiian forests. *Rem Sens Environ* 112:1912–26.

Asner, G. P., and R. E. Martin. 2008. Spectral and chemical analysis of tropical forests: Scaling from leaf to canopy levels. *Rem Sens Environ* 112:3958–70.

Banninger, C., L. Johnson, and D. Peterson. 1994. Determination of biochemical changes in conifer canopies with airborne visible/infrared imaging spectrometer (AVIRIS) data. SPIE 2318:2–9.

Baret, F., and G. Guyot. 1991. Potentials and limits of vegetation indices for LAI and APAR assessment. *Rem Sens Environ* 35:161–73.

Baret, F., S. Jacquemoud, G. Guyot, and C. Leprieur. 1992. Modeled analysis of the biophysical nature of spectral shifts and comparison with information content of broad bands. *Rem Sens Environ* 41:133–42.

Barnes, J. D., L. Balaguer, E. Manrique, S. Elvira, and A. W. Davison. 1992. A reappraisal of the use of DMSO for the extraction and determination of chlorophylls a and b in lichens and higher plants. *Environ Exp Bot* 32:85–100.

Basedow, R., P. Silverglate, W. Rappoport, R. Rockwell, D. Rosenberg, and K. Shu. 1993. The HYDICE instrument design and its application to planetary instruments. In *Lunar and Planetary Inst., Workshop on Advanced Technologies for Planetary Instruments*, ed. J. Appleby, Part 1, 1. Houston, TX: Lunar and Planetary Institute. (SEE N93-28764 11-91).

Belanger, M. J., J. R. Miller, and M. G. Boyer. 1995. Comparative relationships between some red edge parameters and seasonal leaf chlorophyll concentrations. *Can J Rem Sens* 21(1):16–21.

Blackburn, G. A. 1998. Quantifying chlorophylls and carotenoids at leaf and canopy scales: An evaluation of some hyperspectral approaches. *Rem Sens Environ* 66:273–85.

Blackburn, G. A. 2007a. Hyperspectral remote sensing of plant pigments. *J Exp Bot* 58(4):855–67.

Blackburn, G. A. 2007b. Wavelet decomposition of hyperspectral data: A novel approach to quantifying pigment concentrations in vegetation. *Int J Rem Sens* 28:2831–55.

Blackburn, G. A., and J. G. Ferwerda. 2008. Retrieval of chlorophyll concentration from leaf reflectance spectra using wavelet analysis. *Rem Sens Environ* 112:1614–32.

Blackburn, G. A., and C. M. Steele. 1999. Towards the remote sensing of Matorral vegetation physiology: Relationships between spectral reflectance, pigment, and biophysical characteristics of semiarid bushland canopies. *Rem Sens Environ* 70:278–92.

Boardman, J. W. 1993. Automated spectral unmixing of AVIRIS data using convex geometry concepts. In *Summaries 4th Jet Propulsion Laboratory*, ed. R. O. Green, 1:11–4. Pasadena, CA: Airborne Geoscience Workshop.

Bolstad, P. V., and T. M. Lillesand. 1991. Rapid maximum likelihood classification. *Photogramm Eng Rem Sensing* 57(1):67–74.

Broge, N. H., and E. Leblanc. 2000. Comparing prediction power and stability of broadband and hyperspectral vegetation indices for estimation of green leaf area index and canopy chlorophyll density. *Rem Sens Environ* 76:156–72.

Bruce, L. M., C. Morgan, and S. Larsen. 2001. Automated detection of subpixel hyperspectral targets with continuous and discrete wavelet. *IEEE Trans Geosci Rem Sens* 39:2217–26.

Burges, C. J. C. 1998. A tutorial on support vector machines for pattern recognition. *Data Min Knowl Discov* 2(2):121–67.

Camps-Valls, G., and L. Bruzzone. 2005. Kernel-based methods for hyperspectral image classification *IEEE Trans Geosci Rem Sens* 43(6):1351–62.

Camps-Valls, G., L. Gómez-Chova, J. Calpe-Maravilla, J. D. Martín-Guerrero, E. Soria-Olivas, L. Alonso-Chordá, and J. Moreno. 2004. Robust support vector method for hyperspectral data classification and knowledge discovery. *IEEE Trans Geosci Rem Sens* 42(7):1530–42.

Card, D. H., D. L. Peterson, and P. A. Matson. 1988. Prediction of leaf chemistry by use of visible and near infrared reflectance spectroscopy. *Rem Sens Environ* 26:123–47.

Ceccato, P., N. Gobron, S. Flasse, B. Pinty, and S. Tarantola. 2002. Designing a spectral index to estimate vegetation water content from remote sensing data: Part 1. Theoretical approach. *Rem Sens Environ* 82:188–97.

Ceccato, P., S. Flasse, S. Tarantola, S. Jacquemoud, and J. M. Gregoire. 2001. Detecting vegetation leaf water content using reflectance in the optical domain. *Remote Sensing of Environment* 77:22–33.

Chang, C. T. 2000. An information-theoretic approach to spectralvariability, similarity, and discrimination for hyperspectral image analysis. *IEEE Trans Geosci Rem Sens* 46:1927–32.

Chappelle, E. W., M. S. Kim, and J. E. McMurtrey III. 1992. Ratio analysis of reflectance spectra (RARS): An algorithm for the remote estimation of the concentrations of chlorophyll a, chlorophyll b, and carotenoids in soybean leaves. *Rem Sens Environ* 39:239–47.

Chen, J. M. 1996. Evaluation of vegetation indices and a modified simple ratio for boreal applications. *Can J Rem Sens* 22:229–42.

Chen, J. M., and S. G. Leblanc. 2001. Multiple-scattering scheme useful for geometric optical modeling. *IEEE Trans Geosci Rem Sens* 39:1061–71.

Chen, Z., C. D. Elvidge, and D. P. Groeneveld. 1998. Monitoring seasonal dynamics of arid land vegetation using AVIRIS data. *Rem Sens Environ* 65:255–66.

Cheng, Y. B., P. J. Zarco-Tejada, D. Riaño, C. A. Rueda, and S. L. Ustin. 2006. Estimating vegetation water content with hyperspectral data for different canopy scenarios: Relationships between AVIRIS and MODIS indexes. *Rem Sens Environ* 105:354–66.

Cho, M. A., and A. K. Skidmore. 2006. A new technique for extracting the red edge position from hyperspectral data: The linear extrapolation method. *Rem Sens Environ* 101:181–93.

Cho, M. A., A. K. Skidmore, and C. Atzberger. 2008. Towards red-edge positions less sensitive to canopy biophysical parameters for leaf chlorophyll estimation using properties optique spectrales des feuilles (PROSPECT) and scattering by arbitrarily inclined leaves (SAILH) simulated data. *Int J Rem Sens* 29(8):2241–55.

Clark, M. L., D. A. Roberts, and D. B. Clark. 2005. Hyperspectral discrimination of tropical rain forest tree species at leaf to crown scales. *Rem Sens Environ* 96:375–98.

Clark, R. N., and T. L. Roush. 1984. Reflectance spectroscopy: Quantitative analysis techniques for remote sensing applications. *J Geophys Res* 89:6329–40.

Clevers, J. G. P. W., L. Kooistra, and M. E. Schaepman. 2008. Using spectral information from the NIR water absorption features for the retrieval of canopy water content. *Int J Appl Earth Obs Geoinf* 10:388–97.

Cloutis, E. A. 1996. Hyperspectral geological remote sensing: Evaluation of analytical techniques. *Int J Rem Sens* 17(12):2215–42.

Colombo, R., M. Meroni, A. Marchesi, L. Busetto, M. Rossini, C. Giardino, and C. Panigada. 2008. Estimation of leaf and canopy water content in poplar plantations by means of hyperspectral indices and inverse modeling. *Rem Sens Environ* 112:1820–34.

Combal, B., F. Baret, and M. Weiss. 2002. Improving canopy variables estimation from remote sensing data by exploiting ancillary information: Case study on sugar beet canopies. *Agronomie* 22(2):205–15.

Combal, B., F. Baret, M. Weiss, A. Trubuil, D. Mace, A. Pragnere, R. Myneni, Y. Knyazikhin, and L. Wang. 2003. Retrieval of canopy biophysical variables from bidirectional reflectance: Using prior information to solve the ill-posed inverse problem. *Rem Sens Environ* 84(1):1–15.

Coops, N. C., and C. Stone. 2005. A comparison of field-based and modelled reflectance spectra from damaged *Pinus radiata* foliage. *Aust J Bot* 53:417–29.

Curran, P. J. 1989. Remote sensing of foliar chemistry. *Rem Sens Environ* 30:271–8.

Curran, P. J., J. A. Kupiec, and G. M. Smith. 1997. Remote sensing the biochemical composition of a slash pine canopy. *IEEE Trans Geosci Rem Sens* 35(2):415–20.

Curran, P. J., W. R. Windham, and H. L. Gholz. 1995. Exploring the relationship between reflectance red edge and chlorophyll content in slash pine leaves. *Tree Physiol* 15:203–6.

Darvishzadeh, R., A. Skidmore, M. Schlerf, and C. Atzberger. 2008. Inversion of a radiative transfer model for estimating vegetation LAI and chlorophyll in a heterogeneous grassland. *Rem Sens Environ* 112:2592–604.

Darvishzadeh, R., A. Skidmore, M. Schlerf, C. Atzberger, F. Corsi, and M. Cho. 2008. LAI and chlorophyll estimation for a heterogeneous grassland using hyperspectral measurements. *ISPRS J Photogramm Rem Sens* 63:409–26.

Dash, J., and P. J. Curran. 2004. The MERIS terrestrial chlorophyll index. *Int J Rem Sens* 25:5403–13.

Datt, B. 1998. Remote sensing of chlorophyll a, chlorophyll b, chlorophyll a+b, and total carotenoid content in Eucalyptus leaves. *Rem Sens Environ* 66:111–21.

Datt, B. 1999. A new reflectance index for remote sensing of chlorophyll content in higher plants: Tests using Eucalyptus leaves. *J Plant Physiol* 154:30–6.

Datt, B., T. R. McVicar, T. G. Van Niel, D. L. B. Jupp, and J. S. Pearlman. 2003. Preprocessing EO-1 Hyperion hyperspectral data to support the application of agricultural indexes. *IEEE Trans Geosci Rem Sens* 41:1246–59.

Daughtry, C. S. T., C. L. Walthall, M. S. Kim, E. B. Colstoun, and J. E. McMurtrey. 2000. Estimating corn leaf chlorophyll concentration from leaf and canopy reflectance. *Rem Sens Environ* 74:229–39.

Dawson, T. P., and P. J. Curran. 1998. A new technique for interpolating the reflectance red edge position. *Int J Rem Sens* 19:2133–9.

Dawson, T. P., P. J. Curran, P. R. J. North, and S. E. Plummer. 1999. The propagation of foliar biochemical absorption features in forest canopy reflectance: A theoretical analysis. *Rem Sens Environ* 67:147–59.

Dawson, T. P., P. J. Curran, and S. E. Plummer. 1998. LIBERTY-modeling the effects of leaf biochemical concentration on reflectance spectra. *Rem Sens Environ* 65:50–60.

Delalieux, S., B. Somers, S. Hereijgers, W. W. Verstraeten, W. Keulemans, and P. Coppin. 2008. A near-infrared narrow-waveband ratio to determine Leaf Area Index in orchards. *Rem Sens Environ* 112:3762–72.

Demarez, V., and J. P. Gastellu-Etchegorry. 2000. A modeling approach for studying forest chlorophyll content. *Rem Sens Environ* 71:226–38.

Demarez, V., J. P. Gastellu-Etchegorry, E. Mougin, G. Marty, C. Proisy, E. Dufrene, and V. Le Dantec. 1999. Seasonal variation of leaf chlorophyll content of a temperate forest inversion of the PROSPECT model. *Int J Rem Sens* 20(5):879–94.

Demetriades-Shah, T. H., M. D. Steven, and J. A. Clark. 1990. High-resolution derivative spectra in remote-sensing. *Rem Sens Environ* 33(1):55–64.

Ehsani, M. R., S. K. Upadhyaya, D. Slaughter, S. Shafii, and M. Pelletier. 1999. A NIR technique for rapid determination of soil mineral nitrogen. *Precis Agric* 1:217–34.

Eitel, J. U. H., P. E. Gessler, A. M. S. Smith, and R. Robberecht. 2006. Suitability of existing and novel spectral indices to remotely detect water stress in Populus spp. *For Ecol Manage* 229(1–3): 170–82.

Elvidge, C. D. 1990. Visible and near infrared reflectance characteristics of dry plant materials. *Int J Rem Sens* 11:1775–95.

Elvidge, C. D., and Z. Chen. 1995. Comparison of broad-band and narrow-band red and near-infrared vegetation indices. *Rem Sens Environ* 54:38–48.

Esbensen, K. H. 2000. *Multivariate Data Analysis—In Practice*, 598. Corvallis: CAMO.

Fensholt, R., and I. Sandholt. 2003. Derivation of a shortwave infrared water stress index from MODIS near-and shortwave infrared data in a semiarid environment. *Rem Sens Environ* 87(1):111–21.

Feret, J. B., C. François, G. P. Asner, A. A. Gitelson, R. E. Martin, L. P. R. Bidel, S. L. Ustin, G. le Maire, and S. Jacquemoud. 2008. PROSPECT-4 and 5: Advances in the leaf optical properties model separating photosynthetic pigments. *Rem Sens Environ* 112:3030–43.

Fitzgerald, G. J., P. J. Pinter Jr., D. J. Hunsaker, and T. R. Clarke. 2005. Multiple shadow fractions in spectral mixture analysis of a cotton canopy. *Rem Sens Environ* 97:526–39.

Flanagan, M., and D. L. Civco. 2001. Subpixel impervious surface mapping. In *Proceedings of American Society for Photogrammetry and Remote Sensing Annual Convention*, American Society for Photogrammetry and Remote Sensing, 23–7. St. Louis, MO.

Foody, G. M. 1996. Relating the land-cover composition of mixed pixels to artificial neural network classification. *Photogramm Eng Rem Sens* 62:491–9.

Fourty, T., and F. Baret. 1998. On spectral estimates of fresh leaf biochemistry. *Int J Rem Sens* 19(7):1283–97.

Fourty, T., F. Baret, S. Jacquemoud, G. Schmuck, and J. Verdebout. 1996. Leaf optical properties with explicit description of its biochemical composition: Direct and inverse problems. *Rem Sens Environ* 56:104–17.

Franklin, S. E., and G. J. McDermid. 1993. Empirical relations between digital SPOT HRV and CASI spectral response and lodgepole pine (*Pine contorta*) forest stand parameters. *Int J Rem Sens* 14(12):2331–48.

Fukuda, S., and H. Hirosawa. 1999. A wavelet-based texture feature set applied to classification of multifrequency polarimetric SAR images. *IEEE Trans Geosci Rem Sens* 37:2282–6.

Galvão, L. S., A. R. Formaggio, and D. A. Tisot. 2005. Discrimination of sugarcane varieties in Southeastern Brazil with EO-1 Hyperion data. *Rem Sens Environ* 94:523–34.

Gamon, J. A., C. B. Field, M. L. Goulden, K. L. Griffin, A. E. Hartley, G. Joel, J. Penuelas, and R. Valentini. 1995. Relationships between NDVI, canopy structure, and photosynthesis in three California vegetation types. *Ecol Appl* 5(1):28–41.

Gamon, J. A., J. Peñuelas, and C. B. Field. 1992. A narrow waveband spectral index that tracks diurnal changes in photosynthetic efficiency. *Rem Sens Environ* 41:35–44.

Gamon, J. A., and J. S. Surfus. 1999. Assessing leaf pigment content and activity with a reflectometer. *New Phytol* 143:105–17.

Ganapol, B. D., L. F. Johnson, P. D. Hammer, C. A. Hlavka, and D. L. Peterson. 1998. LEAFMOD: A new within-leaf radiative transfer model. *Rem Sens Environ* 63:182–93.

Ganapol, B. D., L. F. Johnson, C. A. Hlavka, D. L. Peterson, and B. Bond. 1999. LCM2: A coupled leaf/canopy radiative transfer model. *Rem Sens Environ* 70:153–66.

Gao, B. C. 1996. NDWI: A normalized difference water index for remote sensing of vegetation liquid water from space. *Rem Sens Environ* 58:257–66.

Gastellu-Etchegorry, J. P., F. Gascon, and P. Esteve. 2003. An interpolation procedure for generalizing a look-up table inversion method. *Rem Sens Environ* 87(1):55–71.

Gastellu-Etchegorry, J. P., F. Zagolski, E. Mougin, G. Marty, and G. Giordano. 1995. An assessment of canopy chemistry with AVIRIS case study in the Landes Forest, South-West France. *Int J Rem Sens* 16(3):487–501.

Gitelson, A. A. 2004. Wide dynamic range vegetation index for remote quantification of crop biophysical characteristics. *J Plant Physiol* 161:165–73.

Gitelson, A. A., C. Buschmann, and H. K. Lichtenthaler. 1999. The chlorophyll fluorescence ratio F735/F700 as an accurate measure of the chlorophyll content in plants. *Rem Sens Environ* 69:296–302.

Gitelson, A. A., O. B. Chivkunova, and M. N. Merzlyak. 2009. Nondestructive estimation of anthocyanins and chlorophylls in anthocyanic leaves. *Am J Bot* 96(10):1861–8.

Gitelson, A. A., Y. J. Kaufman, R. Stark, and D. Rundquist. 2002. Novel algorithms for remote estimation of vegetation fraction. *Rem Sens Environ* 80:76–87.

Gitelson, A. A., G. P. Keydan, and M. M. Merzlyak. 2006. Three-band model for non-invasive estimation of chlorophyll, carotenoids and anthocyanin contents in higher plant leaves. *Geophys Res Lett* 33:L11402. doi:10.1029/2006GL026457.

Gitelson, A. A., and M. N. Merzlyak. 1994. Quantitative estimation of chlorophyll-a using reflectance spectra: Experiments with autumn chestnut and maple leaves. *J Photochem Photobiol* 22:247–52.

Gitelson, A. A., and M. N. Merzlyak. 1996. Signature analysis of leaf reflectance spectra: Algorithm development for remote sensing of chlorophyll. *J Plant Phys* 148:494–500.

Gitelson, A. A., and M. N. Merzlyak. 1997. Remote estimation of chlorophyll content in higher plant leaves. *Int J Rem Sens* 18:2691–7.

Gitelson, A. A., M. N. Merzlyak, and O. B. Chivkunova. 2001. Optical properties and nondestructive estimation of anthocyanin content in plant leaves. *Photochem Photobiol* 74:38–45.

Gitelson, A. A., Y. Zur, O. B. Chivkunova, and M. N. Merzlyak. 2002. Assessing carotenoid content in plant leaves with reflectance spectroscopy. *Photochem Photobiol* 75:272–81.

Goel, N. S., and R. L. Thompson. 1984. Inversion of vegetation canopy reflectance models for estimating agronomic variables: Part V. Estimation of leaf area index and average leaf inclination angle using measured canopy reflectance. *Rem Sens Environ* 15:69–85.

Goetz, A. F. H. 1995. Imaging spectrometry for remote sensing: Vision to reality in 15 years. *Proc SPIE Imaging Spectrom* 2480:2–13.

Goetz, S. J., and S. D. Princ. 1996. Remote sensing of net primary production in boreal forest stands. *Agric For Meteorol* 78:149–79.

Goetz, A. F. H., G. Vane, J. E. Solomon, and B. N. Rock. 1985. Imaging spectrometry for earth remote sensing. *Science* 228(4704):1147–53.

Gong, P., J. R. Miller, and M. Spanner. 1994. Forest canopy closure from classification and spectral unmixing of scene components multisensor evaluation of an open canopy. *IEEE Trans Geosci Rem Sens* 32(5):1067–80.

Gong, P., R. Pu, G. S. Biging, and M. Larrieu. 2003. Estimation of forest leaf area index using vegetation indices derived from Hyperion hyperspectral data. *IEEE Trans Geosci Rem Sens* 41(6):1355–62.

Gong, P., R. Pu, and R. C. Heald. 2002. Analysis of in situ hyperspectral data for nutrient estimation of giant sequoia. *Int J Rem Sens* 23(9):1827–50.

Gong, P., R. Pu, and J. R. Miller. 1992. Correlating leaf area index of ponderosa pine with hyperspectral CASI data. *Can J Rem Sens* 18:275–82.

Gong, P., R. Pu, and J. R. Miller. 1995. Coniferous forest leaf area index estimation along the Oregon transect using compact airborne spectrographic imager data. *Photogramm Eng Rem Sensing* 61(9):1107–17.

Gong, P., R. Pu, and B. Yu. 1997. Conifer species recognition: An exploratory analysis of in situ hyperspectral data. *Rem Sens Environ* 62:189–200.

Gong, P., R. Pu, and B. Yu. 2001. Conifer species recognition: Effect of data transformation. *Int J Rem Sens* 22(17):3471–81.

Gong, P., D. Wang, and S. Liang. 1999. Inverting a canopy reflectance model using an artificial neural network. *Int J Rem Sens* 20(1):111–22.

Green, A. A., M. Berman, P. Switzer, and M. D. Craig. 1988. A transformation for ordering multispectral data in terms of image quality with implications for noise removal. *IEEE Trans Geosci Rem Sens* 26:65–74.

Green, R. O., M. L. Eastwood, C. M. Sarture, T. G. Chrien, M. Aronsson, B. J. Chippendale, J. A. Faust, B. E. Pavri, C. J. Chovit, M. Solis et al. 1998. Imaging spectroscopy and the airborne visible/infrared imaging spectrometer (AVIRIS). *Rem Sens Environ* 65:227–48.

Grossman, Y. L., S. L. Ustin, S. Jacquemoud, E. W. Sanderson, G. Schmuck, and J. Verdebout. 1996. Critique of stepwise multiple linear regression for the extraction of leaf biochemistry information from leaf reflectance data. *Rem Sens Environ* 56:182–93.

Guyot, G., F. Baret, and S. Jacquemond. 1992. *Imaging Spectroscopy: Fundamentals and Prospective Application*, ed. F. Toselli and J. Bodechtel, 145–65. Dordrecht, The Netherlands: Kluwer Academic Publishers.

Haboudane, D., J. R. Miller, E. Pattery, P. J. Zarco-Tejad, and I. B. Strachan. 2004. Hyperspectral vegetation indices and novel algorithms for predicting green LAI of crop canopies: Modeling and validation in the context of precision agriculture. *Rem Sens Environ* 90:337–52.

Haboudane, D., J. R. Miller, N. Tremblay, P. J. Zarco-Tejada, and L. Dextraze. 2002. Integrated narrow-band vegetation indices for prediction of crop chlorophyll content for application to precision agriculture. *Rem Sens Environ* 81(2–3):416–26.

Hansena, P. M., and J. K. Schjoerring. 2003. Reflectance measurement of canopy biomass and nitrogen status in wheat crops using normalized difference vegetation indices and partial least squares regression. *Rem Sens Environ* 86:542–53.

Hardinsky, M. A., V. Lemas, and R. M. Smart. 1983. The influence of soil salinity, growth form, and leaf moisture on the spectral reflectance of *Spartina alternifolia* canopies. *Photogramm Eng Rem Sens* 49:77–83.

Hatfield, J. L., A. A. Gitelson, J. S. Schepers, and C. L. Walthall. 2008. Application of spectral remote sensing for agronomic decisions. *Agron J* 100:S-117–31.

He, Y., X. Guo, and J. Wilmshurst. 2006. Studying mixed grassland ecosystems I: Suitable hyperspectral vegetation indices. *Can J Rem Sens* 32(2):98–107.

Henry, W. B., D. R. Shaw, K. R. Reddy, L. M. Bruce, and H. D. Tamhankar. 2004. Remote sensing to detect herbicide drift on crops. *Weed Technol* 18:358–68.

Hestir, E. L., S. Khanna, M. E. Andrew, M. J. Santos, J. H. Viers, J. A. Greenberg, S. S. Rajapakse, and S. L. Ustin. 2008. Identification of invasive vegetation using hyperspectral remote sensing in the California Delta ecosystem. *Rem Sens Environ* 112:4034–47.

Hollinger, A. B., L. H. Gray, J. F. R. Gower, and H. R. Edel. 1988. The fluorescence line imager: An imaging spectrometer for ocean and remote sensing. *Proc SPIE Imaging Spectrosc II* 834:2–11.

Houborg, R., H. Soegaard, and E. Boegh. 2007. Combining vegetation index and model inversion methods for the extraction of key vegetation biophysical parameters using Terra and Aqua MODIS reflectance data. *Rem Sens Environ* 106(1):39–58.

Hsu, P. H., and Y. H. Tseng. 2000. Multiscale analysis of hyperspectral data using wavelets for spectral feature extraction. In *21st Asian Conference on Remote Sensing*, Taipei, Taiwan. Online proceedings: http://www.gisdevelopment.net/aars/acrs/2000, accessed on September 15, 2010.

Huang, Z., B. J. Turner, S. J. Dury, I. R. Wallis, and W. J. Foley. 2004. Estimating foliage nitrogen concentration from HYMAP data using continuum removal analysis. *Rem Sens Environ* 93(1–2):18–29.

Huber, S., M. Kneubühler, A. Psomas, K. Itten, and N. E. Zimmermann. 2008. Estimating foliar biochemistry from hyperspectral data in mixed forest canopy. *For Ecol Manage* 256:491–501.

Huegel, F. G. 1988. Advanced solid state array spectroradiometer: Sensor and calibration improvements. *Proc SPIE Imaging Spectrosc II* 834:12–21.

Huete, A., K. Didan, T. Miura, E. P. Rodriguez, X. Gao, and L. G. Ferreira. 2002. Overview of the radiometric and biophysical performance of the MODIS vegetation indices. *Rem Sens Environ* 83:195–213.

Hughes, G. F. 1968. On the mean accuracy of statistical pattern recognizers. *IEEE Trans Inf Theory* 14(1):55–63.

Hunt, G. R. 1980. Electromagnetic radiation: The communication link in remote sensing. In *Remote Sensing in Geology*, ed. B. Siegal and A. Gillespia, 702. New York: John Wiley & Sons.

Hunt, E. R., and B. N. Rock. 1989. Detection of changes in leaf water content using near and middle-infrared reflectances. *Remote Sensing of Environment* 30:43–54.

Jacquemoud, S., C. Bacour, H. Poilve, and J. -P. Frangi. 2000. Comparison of four radiative transfer models to simulate plant canopies reflectance: Direct and inverse mode. *Rem Sens Environ* 74(3):471–81.

Jacquemoud, S., and F. Baret. 1990. PROSPECT: A model of leaf optical properties spectra. *Rem Sens Environ* 34:75–91.

Jacquemoud, S., F. Baret, B. Andrieu, F. M. Danson, and K. Jaggard. 1995. Extraction of vegetation biophysical parameters by inversion of the PROSPECT + SAIL models on sugar beet canopy reflectance data: Application to TM and AVIRIS sensors. *Rem Sens Environ* 52(3):163–72.

Jacquemoud, S., S. L. Ustin, J. Verdebout, G. Schmuck, G. Andreoli, and B. Hosgood. 1996. Estimating leaf biochemistry using the PROSPECT leaf optical properties model. *Rem Sens Environ* 56:194–202.

Jacquemoud, S., W. Verhoef, F. Baret, C. Bacour, P. J. Zarco-Tejada, G. P. Asner, C. François, and S. L. Ustin. 2009. PROSPECT+SAIL models: A review of use for vegetation characterization. *Rem Sens Environ* 113:S56–66.

Jensen, J. R. 2007. *Remote Sensing of the Environment: An Earth Resource Perspective*, 2nd ed, 592. Upper Saddle River, NJ: Prentice Hall.

Jia, X., and J. A. Richards. 1994. Efficient maximum likelihood classification for imaging spectrometer data sets. *IEEE Trans Geosci Rem Sens* 32:274–81.

Jia, X., and J. A. Richards. 1999. Segmented principal components transformation for efficient hyperspectral remote-sensing image display and classification. *IEEE Trans Geosci Rem Sens* 37:538–42.

Jia, X., and J. A. Richards. 2002. Cluster-space representation for hyperspectral data classification. *IEEE Trans Geosci Remote Sens* 40(3):593–8.

Jiang, Z., A. R. Huete, K. Didan, and T. Miura. 2008. Development of a two-band enhanced vegetation index without a blue band. *Rem Sens Environ* 112:3833–45.

Jimenez, L. O., and D. A. Landgrebe. 1998. Supervised classification in high-dimensional space: Geometrical, statistical, and asymptotical properties of multivariate data. *IEEE Trans Syst Man Cybern C Appl Rev* 28:39–54.

Johnson, L. F., C. A. Hlavka, and D. L. Peterson. 1994. Multivariate analysis of AVIRIS data for canopy biochemical estimation along the Oregon transect. *Rem Sens Environ* 47:216–30.

Jokela, A., T. Sarjala, S. Kaunisto, and S. Huttunen. 1997. Effects of foliar potassium concentration on morphology, ultrastructure and polyamine concentrations of Scots pine needles. *Tree Physiol* 17:677–85.

Jordan, C. F. 1969. Derivation of leaf area index from quality of light on the forest floor. *Ecology* 50:663–6.

Judd, C., S. Steinberg, F. Shaughnessy, and G. Crawford. 2007. Mapping salt marsh vegetation using aerial hyperspectral imagery and linear unmixing in Humboldt Bay, California. *Wetlands* 27(4):1144–52.

Khattree, R., and D. N. Naik. 2000. *Multivariate Data Reduction and Discrimination with SAS Software*, 558. Cary, NC: SAS Institute Inc.

Kim, M. S., C. S. T. Daughtry, E. W. Chappelle, and J. E. McMurtrey. 1994. The use of high spectral resolution bands for estimating absorbed photosynthetically active radiation (APAR). In *Proceedings of the 6th International Symposium on Physical Measurements and Signatures in Remote Sensing*, 299–306. France: Val d'Isere.

Kimes, D. S., Y. Knyazikhin, J. L. Privette, A. A. Abuelgasim, and F. Gao. 2000. Inversion methods for physically based models. *Rem Sens Rev* 18(2–4):381–439.

Knyazikhin, Y., J. V. Martonchik, D. Diner, R. B. Myneni, M. M. Verstraete, B. Pinty, and N. Gobron. 1998. Estimation of vegetation canopy leaf area index and fraction of absorbed photosynthetically active radiation from atmosphere-corrected MISR data. *J Geophys Res* 103(D24):32239–56.

Koetz, B., M. Schaepman, F. Morsdorf, P. Bowyer, K. Itten, and B. Allgöwer. 2004. Radiative transfer modeling within heterogeneous canopy for estimation of forest fire fuel properties. *Rem Sens Environ* 92:332–44.

Kruse, F. A., A. B. Lefkoff, and J. B. Dietz. 1993. Expert system-based mineral mapping in northern Death Valley, California/Nevada, using the airborne visible/infrared imaging spectrometer (AVIRIS). *Rem Sens Environ* 44:309–36.

Kuusk, A. 1991. The angular distribution of reflectance and vegetation indices in barley and clover canopies. *Rem Sens Environ* 37:143–51.

Kuusk, A. 1994. A multispectral canopy reflectance model. *Rem Sens Environ* 50:75–82.

Kuusk, A. 1995. A Markov chain model of canopy reflectance. *Agric For Meteorol* 76:221–36.

Laba, M., F. Tsai, D. Ogurcak, S. Smith, and M. E. Richmond. 2005. Field determination of optimal dates for the discrimination of invasive wetland plant species using derivative spectral analysis. *Photogramm Eng Rem Sens* 71(5):603–11.

LaCapra, V. C., J. M. Melack, M. Gastil, and D. Valeriano. 1996. Remote sensing of foliar chemistry of inundated rice with imaging spectrometry. *Rem Sens Environ* 55:50–8.

le Maire, G., C. Francois, and E. Dufrene. 2004. Towards universal broad leaf chlorophyll indices using PROSPECT simulated database and hyperspectral reflectance measurements. *Rem Sens Environ* 89:1–28.

le Maire, G., C. François, K. Soudani, D. Berveiller, J. -Y. Pontailler, N. Bréda, H. Genet, H. Davi, and E. Dufrêne. 2008. Calibration and validation of hyperspectral indices for the estimation of broadleaved forest leaf chlorophyll content, leaf mass per area, leaf area index and leaf canopy biomass. *Rem Sens Environ* 112:3846–64.

Li, J., L. M. Bruce, J. Byrd, and J. Barnett. 2001. Automated detection of *Pueraria montana* (Kudzu) through Haar analysis of hyperspectral reflectance data. *IEEE Int Geosci Rem Sens Symp*, Sydney, Australia, pp. 2247–9, July 9–13, 2001.

Li, Y., T. H. Demetriadesshah, E. T. Kanemasu, J. K. Shultis, and M. B. Kirkham. 1993. Use of 2nd derivatives of canopy reflectance for monitoring prairie vegetation over different soil backgrounds. *Rem Sens Environ* 44(1):81–7.

Li, Q., B. Hu, and E. Pattey. 2008. A scale-wise model inversion method to retrieve canopy biophysical parameters from hyperspectral remote sensing data. *Can J Rem Sens* 34(3):311–9.

Li, L., and J. F. Mustard. 2003. Highland contamination in lunar mare soils: Improved mapping with multiple end-member spectral mixture analysis (MESMA). *J Geophys Res* 108(E6):5053, doi:10.1029/2002JE001917.

Li, X., and A. H. Strahler. 1985. Geometric-optical modeling of a conifer forest canopy. *IEEE Trans Geosci Remote Sens* GE23:705–21.

Li, X., and A. H. Strahler. 1992. Geometric-optical bidirectional reflectance modeling of the discrete-crown canopy: Effect of crown shape and mutual shadowing. *IEEE Trans Geosci Rem Sens* GE30:276.

Li, L., S. L. Ustin, and M. Lay. 2005. Application of multiple endmember spectral mixture analysis (MESMA) to AVIRIS imagery for coastal salt marsh mapping: A case study in China Camp, CA, USA. *Int J Rem Sens* 26:5193–207.

Liang, S. 2004. *Quantitative Remote Sensing of Land Surfaces*. Wiley Praxis Series in Remote Sensing, Hoboken, NJ: John Wiley & Sons.

Liang, S., and A. H. Strahler. 1993. Calculation of the angular radiance distribution for a coupled atmosphere and leaf canopy. *IEEE Trans Geosci Rem Sens* GE31:1081.

Lichtenthaler, H. K. 1987. Chlorophyll and carotenoids: Pigments of photosynthetic biomembranes. *Methods Enzymol* 148:331–82.

Lucas, K. L., and G. A. Carter. 2008. The use of hyperspectral remote sensing to assess vascular plant species richness on Horn Island, Mississippi. *Rem Sens Environ* 112:3908–15.

Mallat, S. G. 1989. A theory for multiresolution signal decomposition: The wavelet representation. *IEEE Trans Pattern Anal Mach Intell* 11:674–93.

Mallat, S. G. 1998. *A Wavelet Tour of Signal Processing*. San Diego: Academic Press.

Martin, M. E., and J. D. Aber. 1997. Estimation of canopy lignin and nitrogen concentration and ecosystem processes by high spectral resolution remote sensing. *Ecol Appl* 7:431–43.

Martin, M. E., S. D. Newman, J. D. Aber, and J. C. Congalton. 1998. Determining forest species composition using high spectral resolution remote sensing data. *Rem Sens Environ* 65:249–54.

Martin, M. E., L. C. Plourde, S. V. Ollinger, M. -L. Smith, and B. E. McNeil. 2008. A generalizable method for remote sensing of canopy nitrogen across a wide range of forest ecosystems. *Rem Sens Environ* 112:3511–9.

Maselli, F. 1998. Multiclass spectral decomposition of remotely sensed scenes by selective pixel unmixing. *IEEE Trans Geosci Rem Sens* 36:1809–20.

Matson, P. A., L. F. Johnson, J. R. Miller, C. R. Billow, and R. Pu. 1994. Seasonal changes in canopy chemistry across the Oregon transect: Patterns and spectral measurement with remote sensing. *Ecol Appl* 4:280–98.

Melgani, F., and L. Bruzzone. 2004. Classification of hyperspectral remote sensing images with support vector machines. *IEEE Trans Geosci Rem Sens* 42(8):1778–90.

Meroni, M., R. Colombo, and C. Panigada. 2004. Inversion of a radiative transfer model with hyperspectral observations for LAI mapping in poplar plantations. *Rem Sens Environ* 92(2):195–206.

Merton, R., and J. Huntington. 1999. Early simulation of the ARIES-1 satellite sensor for multi-temporal vegetation research derived from AVIRIS. In *Summaries of the Eight JPL Airborne Earth Science Workshop*, Ed: R. O. Green, 299–307. Pasadena, CA: JPL Publication.

Merzlyak, M. N., A. A. Gitelson, O. B. Chivkunova, and V. Y. Rakitin. 1999. Nondestructive optical detection of pigment changes during leaf senescence and fruit ripening. *Physiol Plant* 106:135–41.

Miao, X., P. Gong, S. Swope, R. Pu, R. Carruthers, G. L. Anderson, J. S. Heaton, and C. R. Tracy. 2006. Estimation of yellow starthistle abundance through CASI-2 hyperspectral imagery using linear spectral mixture models. *Rem Sens Environ* 101:329–41.

Miller, J. R., E. W. Hare, and J. Wu. 1990. Quantitative characterization of the vegetation red edge reflectance. 1. An inverted-Gaussian reflectance model. *Int J Rem Sens* 11:1775–95.

Miller, J. R., J. Wu, M. G. Boyer, M. Belanger, and E. W. Hare. 1991. Season patterns in leaf reflectance red edge characteristics. *Int J Remote Sens* 12(7):1509–23.

Milton, N. M., B. A. Eiswerth, and C. M. Ager. 1991. Effect of phosphorus deficiency on spectral reflectance and morphology of soybean plants. *Rem Sens Environ* 36:121–7.

Misiti, M., Y. Misiti, G. Oppenheim, and J. M. Poggi. 1996. *Wavelet Toolbox User's Guide*. Natick, MA: The Math Works Inc.

Nagler, P. L., C. S. T. Daughtry, and S. N. Goward. 2000. Plant litter and soil reflectance. *Rem Sens Environ* 71:207–15.

Niinemets, U., and L. Sack. 2006. Structural determinants of leaf light harvesting capacity and photosynthetic potentials. *Prog Bot* 67:385–419.

Painter, T. H., D. A. Roberts, R. O. Green, and J. Dozier. 1998. The effect of grain size on spectral mixture analysis of snow-covered area from AVIRIS data. *Rem Sens Environ* 65:320–32.

Pal, M., and P. M. Mather. 2004. Assessment of the effectiveness of support vector machines for hyperspectral data. *Future Gene Comput Syst* 20:1215–25.

Palmer, K. F., and D. Williams. 1974. Optical properties of water in the near infrared. *J Opt Soc Am* 64:1107–10.

Pao, Y. 1989. *Adaptive Pattern Recognition and Neural Networks*. New York: Addison and Wesley.

Peñuelas, J., F. Baret, and I. Filella. 1995. Semi-empirical indices to assess carotenoids/chlorophyll a ratio from leaf spectral reflectance. *Photosynthetica* 31:221–30.

Peñuelas, J., I. Filella, C. Biel, L. Sweeano, and R. Save. 1993. The reflectance at the 950–70 nm region as an indicator of plant water status. *Int J Rem Sens* 14:1887–905.

Peñuelas, J., I. Filella, P. Lloret, F. Muñoz, and M. Vilajeliu. 1995. Reflectance assessment of mite effects on apple trees. *Int J Remote Sens* 16:2727–33.

Peñuelas, J., I. Filella, and L. Sweeano. 1996. Cell wall elasticity and water index (R970 nm/R900 nm) in wheat under different nitrogen availabilities. *Int J Rem Sens* 17:373–82.

Peñuelas, J., J. A. Gamon, A. L. Fredeen, J. Merino, and C. B. Field. 1994. Reflectance indices associated with physiological changes in nitrogen- and water-limited sunflower leaves. *Rem Sens Environ* 48:135–46.

Peñuelas, J., J. Piñol, R. Ogaya, and I. Filella. 1997. Estimation of plant water concentration by the reflectance water index WI (R900/R970). *Int J Rem Sens* 18:2869–75.

Peterson, D. L., J. D. Aber, P. A. Matson, D. H. Card, N. Swanberg, C. Wessman, and M. A. Spanner. 1988. Remote sensing of forest canopy and leaf biochemical contents. *Rem Sens Environ* 24:85–108.

Pieters, C. M. 1983. Strength of mineral absorption features in the transmitted component of near-infrared reflected light: First results from RELAB. *J Geophys Res* 88:9534–44.

Pignatti, S., R. M. Cavalli, V. Cuomo, L. Fusilli, S. Pascucci, M. Poscolieri, and F. Santini. 2009. Evaluating Hyperion capability for land-cover mapping in a fragmented ecosystem: Pollino National Park, Italy. *Rem Sens Environ* 113:622–34.

Pinel, V., F. Zagolski, J. P. Gastellu-Etchgorry, G. Giordano, J. Romier, G. Marty, E. Mougin, and R. Joffre. 1994. An assessment of forest chemistry with ISM. *SPIE* 2318:40–51.

Pittner, S., and S. V. Kamarthi. 1999. Feature extraction from wavelet coefficients for pattern recognition tasks. *IEEE Trans Pattern Anal Mach Intell* 21:83–8.

Ponzoni, F. J., J. L. de, and M. Goncalves. 1999. Spectral features associated with nitrogen, phosphorus, and potassium deficiencies in *Eucalyptus saligna* seedling leaves. *Int J Rem Sens* 20:2249–64.

Prieto-Blanco, A., P. R. J. North, M. J. Barnsley, and N. Fox. 2009. Satellite-driven modelling of Net Primary Productivity (NPP): Theoretical analysis. *Rem Sens Environ* 113(1):137–47.

Pu, R. 2009. Broadleaf species recognition with in situ hyperspectral data. *Int J Rem Sens* 30(11):2759–79.

Pu, R., L. Foschi, and P. Gong. 2004. Spectral feature analysis for assessment of water status and health level of coast live oak (*Quercus agrifolia*) leaves. *Int J Rem Sens* 25(20):4267–86.

Pu, R., S. Ge, N. M. Kelly, and P. Gong. 2003. Spectral absorption features as indicators of water status in Quercus Agrifolia leaves. *Int J Rem Sens* 24(9):1799–810.

Pu, R., and P. Gong. 2004. Wavelet transform applied to EO-1 hyperspectral data for forest LAI and crown closure mapping. *Rem Sens Environ* 91:212–24.

Pu, R., P. Gong, G. S. Biging, and M. R. Larrieu. 2003. Extraction of red edge optical parameters from Hyperion data for estimation of forest leaf area index. *IEEE Trans Geosci Rem Sens* 41(4):916–21.

Pu, R., P. Gong, R. Michishita, and T. Sasagawa. 2008. Spectral mixture analysis for mapping abundance of urban surface components from the Terra/ASTER data. *Rem Sens Environ* 112:939–54.

Pu, R., P. Gong, Y. Tian, X. Miao, and R. Carruthers. 2008. Invasive species change detection using artificial neural networks and CASI hyperspectral imagery. *Environ Monit Assess* 140:15–32.

Pu, R., N. M. Kelly, Q. Chen, and P. Gong. 2008. Spectroscopic determination of health levels of Coast Live Oak (*Quercus agrifolia*) leaves. *Geocarto Int* 23(1):3–20.

Pu, R., and D. Liu. In press. Segmented canonical discriminant analysis of in situ hyperspectral data for identifying thirteen urban tree species. *Int J Rem Sens*.

Qi, J., A. Chehbouni, A. R. Huete, Y. H. Kerr, and S. Sorooshian. 1994. A modified soil adjusted vegetation index. *Rem Sens Environ* 48:119–26.

Rama Rao, N., P. K. Garg, S. K. Ghosh, and V. K. Dadhwal. 2008. Estimation of leaf total chlorophyll and nitrogen concentrations using hyperspectral satellite imagery. *J Agr Sci* 146:65–75.

Reujean, J., and F. Breon. 1995. Estimating PAR absorbed by vegetation from bidirectional reflectance measurements. *Rem Sens Environ* 51:375–84.

Richardson, A. D., S. P. Duigan, and G. P. Berlyn. 2002. An evaluation of noninvasive methods to estimate foliar chlorophyll content. *New Phytol* 153:185–94.

Rioul, O., and M. Vetterli. 1991. Wavelet and signal processing. *IEEE Signal Process Mag* 8:14–38.

Roberts, D. A., P. E. Dennison, M. Gardner, Y. Hetzel, S. L. Ustin, and C. Lee. 2003. Evaluation of the potential of Hyperion for fire danger assessment by comparison to the Airborne Visible/ Infrared Imaging Spectrometer. *IEEE Trans Geosci Rem Sens* 41(6):1297–310.

Roberts, D. A., M. Gardner, R. Church, S. Ustin, G. Scheer, and R. O. Green. 1998. Mapping Chaparral in the Santa Monica Mountains using multiple endmember spectral mixture models. *Rem Sens Environ* 65:267–79.

Rondeaux, G., M. Steven, and F. Baret. 1996. Optimization of soil-adjusted vegetation indices. *Rem Sens Environ* 55:95–107.

Rosso, P. H., S. L. Ustin, and A. Hastings. 2005. Mapping marshland vegetation of San Francisco Bay, California, using hyperspectral data. *Int J Rem Sens* 26:5169–91.

Rouse, J. W., R. H. Haas, J. A. Schell, and D. W. Deering. 1973. Monitoring vegetation systems in the Great Plains with ERTS. *Proc 3rd ERTS Symp* 1:48–62.

Rumelhart, D. E., G. E. Hinton, and R. J. Williams. 1986. Learning internal representations by error propagation. In *Parallel Distributed Processing-Explorations in the Microstructure of Cognition*, vol. 1, 318–62. Cambridge, MA: MIT Press.

Sabol Jr., D. E., J. B. Adams, and M. O. Smith. 1990. Predicting the spectral detectability of surface materials using spectral mixture analysis. In *Proceedings of IGARSS'90*, ed. Ronnie Mills, Piscataway, NJ: IEEE. 967–70. College Park, Maryland.

Sasaki, K., S. Kawata, and S. Minami. 1984. Estimation of component spectral curves from unknown mixture spectra. *Appl Opt* 23:1955–9.

Schlerf, M., and C. Atzberger. 2006. Inversion of a forest reflectance model to estimate structural canopy variables from hyperspectral remote sensing data. *Rem Sens Environ* 100:281–94.

Schlerf, M., C. Atzberger, and J. Hill. 2005. Remote sensing of forest biophysical variables using HyMap imaging spectrometer data. *Rem Sens Environ* 95:177–94.

Serrano, L., J. Peñuelas, and S. L. Ustin. 2002. Remote sensing of nitrogen and lignin in Mediterranean vegetation from AVIRIS data: Decomposing biochemical from structural signals. *Rem Sens Environ* 81:355–64.

Shipman, H., and J. B. Adams. 1987. Detectability of minerals on desert alluvial fans using reflectance spectra. *J Geophys Res* 92:10391–402.

Simhadri, K. K., S. S. Iyengar, R. J. Holyer, M. Lybanon, and J. M. Zachary. 1998. Wavelet-based feature extraction from oceanographic images. *IEEE Trans Geosci Rem Sens* 36:767–78.

Sims, D. A., and J. A. Gamon. 2002. Relationships between leaf pigment content and spectral reflectance across a wide range of species, leaf structures and developmental stages. *Rem Sens Environ* 81:337–54.

Smith, N. J., G. A. Boratad, D. A. Hill, and R. C. Kerr. 1991. Using high-resolution airborne spectral data to estimate forest leaf area and stand structure. *Can J For Res* 21:1127–32.

Sohn, Y., and R. W. McCoy. 1997. Mapping desert shrub rangeland using spectral unmixing and modeling spectral mixtures with TM data. *Photogramm Eng Rem Sens* 63:707–16.

Tompkins, S., J. F. Mustard, C. M. Pieters, and D. W. Forsytth. 1997. Optimization of endmembers for spectral mixture analysis. *Rem Sens Environ* 59:472–89.

Tsai, F., and W. Philpot. 1998. Derivative analysis of hyperspectral data. *Rem Sens Environ* 66(1):41–51.

Ungar, S., J. Pearlman, J. Mendenhall, and D. Reuter. 2003. Overview of the Earth Observing One (EO-1) mission. *IEEE Trans Geosci Rem Sens* 41:1149–59.

Ustin, S. L., A. A. Gitelson, S. Jacquemoud, M. Schaepman, G. P. Asner, J. A. Gamon, and P. Zarco-Tejada. 2009. Retrieval of foliar information about plant pigment systems from high resolution spectroscopy. *Rem Sens Environ* 113:S67–77.

Ustin, S. L., D. A. Roberts, J. A. Gamon, G. P. Asner, and R. O. Green. 2004. Using imaging spectroscopy to study ecosystem processes and properties. *Bioscience* 54(6):523–34.

Ustin, S. L., D. A. Roberts, S. Jacquemoud, J. Pinzón, M. Gardner, G. Scheer, C. M. Castaneda, and A. Palacios-Orueta. 1998. Estimating canopy water content of chaparral shrubs using optical methods. *Rem Sens Environ* 65(3):280–91.

van Aardt, J. A. N., and R. H. Wynne. 2001. Spectral separability among six southern tree species. *Photogramm Eng Rem Sens* 67(12):1367–75.

van Aardt, J. A. N., and R. H. Wynne. 2007. Examining pine spectral separability using hyperspectral data from an airborne sensor: An extension of field-based results. *Int J Rem Sens* 28(2):431–6.

van der Meer, F. 2006. The effectiveness of spectral similarity measures for the analysis of hyperspectral imagery. *Int J Appl Earth Obs Geoinf* 8(1):3–17.

van der Meer, F., and W. Bakker. 1997. CCSM: Cross correlogram spectral matching. *Int J Rem Sens* 18(5):1197–201.

Vane, G., and A. F. H. Goetz. 1988. Terrestrial imaging spectroscopy. *Rem Sens Environ* 24:1–29.

Vane, G., and A. F. H. Goetz. 1993. Terrestrial imaging spectrometry: Current status, future trends. *Rem Sens Environ* 44:117–26.

Verhoef, W. 1984. Light scattering by leaf layers with application to canopy reflectance modeling: The SAIL model. *Rem Sens Environ* 16:125–41.

Verhoef, W., and H. Bach. 2003. Simulation of hyperspectral and directional radiance images using coupled biophysical and atmospheric radiative transfer models. *Rem Sens Environ* 87:23–41.

Verhoef, W., and H. Bach. 2007. Coupled soil-leaf-canopy and atmosphere radiative transfer modeling to simulate hyperspectral multi-angular surface reflectance and TOA radiance data. *Rem Sens Environ* 109:166–82.

Vincini, M., E. Frazzi, and P. D'Alessio. 2006. Angular dependence of maize and sugar beet VIs from directional CHRIS/PROBA data. In *4th ESA CHRIS PROBA Workshop*, ed. B. Hoersh and M. Cutter, 19–21. Frascati, Italy: ESRIN.

Walsh, S. J., A. L. McCleary, C. F. Mena, Y. Shao, J. P. Tuttle, A. González, and R. Atkinson. 2008. QuickBird and Hyperion data analysis of an invasive plant species in the Galapagos Islands of Ecuador: Implications for control and land use management. *Rem Sens Environ* 112:1927–41.

Walthall, C., W. Dulaney, M. Anderson, J. Norman, H. Fang, and S. Liang. 2004. A comparison of empirical and neural network approaches for estimating corn and soybean leaf area index from Landsat ETM+ imagery. *Rem Sens Environ* 92(4):465–74.

Wang, L., J. Chen, P. Gong, H. Shimazaki, and M. Tamura. 2009. Land-cover change detection with a cross-correlogram spectral matching algorithm. *Int J Rem Sens* 30(12):3259–73.

Wang, X., and Y. Zhang. 1998. The study on decomposing AVHRR mixed pixels by means of neural network model. *J Remote Sens Chin* 2:51–6.

Weihs, P., F. Suppan, K. Richter, R. Petritsch, H. Hasenauer, and W. Schneider. 2008. Validation of forward and inverse modes of a homogeneous canopy reflectance model. *Int J Rem Sens* 29(5):1317–38.

Weiss, M., and F. Baret. 1999. Evaluation of canopy biophysical variable retrieval performances from the accumulation of large swath satellite data. *Rem Sens Environ* 70(3):293–306.

Weiss, M., F. Baret, R. B. Myneni, A. Pragnere, and Y. Knyazikhin. 2000. Investigation of a model inversion technique to estimate canopy biophysical variables from spectral and directional reflectance data. *Agronomie* 20(1):3–22.

Weiss, M., D. Troufleau, F. Baret, H. Chauki, L. Prévot, A. Olioso, N. Bruguier, and N. Brisson. 2001. Coupling canopy functioning and radiative transfer models for remote sensing data assimilation. *Agric For Meteorol* 108:113–28.

Weng, Y. L., P. Gong, and Z. L. Zhu. 2008. Soil salt content estimation on the Yellow River Delta with satellite hyperspectral data. *Can J Rem Sens* 34(3):259–70.

Wessman, C. A., J. D. Aber, and D. L. Peterson. 1989. An evaluation of imaging spectrometry for estimating forest canopy chemistry. *Int J Rem Sens* 10:1293–316.

Wessman, C. A., J. D. Aber, D. L. Peterson, and J. M. Melillo. 1988. Remote sensing of canopy chemistry and nitrogen cycling in temperate forest ecosystems. *Nature* 335:154–6.

Wright, I. J., P. B. Reich, M. Westoby, D. D. Ackerly, Z. Baruch, F. Bongers, J. Cavender-Bares et al. 2004. The worldwide leaf economics spectrum. *Nature* 428:821–7.

Xu, B., and P. Gong. 2007. Land use/cover classification with multispectral and hyperspectral EO-1 data. *Photogramm Eng Rem Sens* 73(8):955–65.

Yasuoka, Y., T. Yokota, T. Miyazaki, and Y. Iikura. 1990. Detection of vegetation change from remotely sensed images using spectral signature similarity. In *Proceedings of the International Geoscience and Remote Sensing Symposium (IGRSS'90)*, 1609–12. College Park, MD, Piscataway, NJ: IEEE.

Yoder, B. J., and R. E. Pettigrew-Crosby. 1995. Predicting nitrogen and chlorophyll content and concentration from reflectance spectra (400–2500 nm) at leaf and canopy scales. *Rem Sens Environ* 53:199–211.

Yoder, B. J., and R. H. Waring. 1994. The normalized difference vegetation index of small douglas-fir canopies with varying chlorophyll concentrations. *Rem Sens Environ* 49:81–91.

Yu, B., M. Ostland, P. Gong, and R. Pu. 1999. Penalized linear discriminant analysis for conifer species recognition. *IEEE Trans Geosci Rem Sens* 37(5):2569–77.

Zagolski, F., V. Pinel, J. Romier, D. Alcayde, J. Fontanari, J. Gasstellu-Etchegorry, G. Giodano, G. Marty, E. Mougin, and R. Joffre. 1996. Forest canopy chemistry with high spectral resolution remote sensing. *Int J Rem Sens* 17(6):1107–28.

Zarco-Tejada, P. J., A. Berjón, R. López-Lozano, J. R. Miller, P. Martín, V. Cachorro, M. R. González, and A. Frutos. 2005. Assessing vineyard condition with hyperspectral indices: Leaf and canopy reflectance simulation in a row-structured discontinuous canopy. *Rem Sens Environ* 99:271–87.

Zarco-Tejada, P. J., J. R. Miller, J. Harron, B. Hu, T. L. Noland, N. Goel, G. H. Mohammed, and P. H. Sampson. 2004. Needle chlorophyll content estimation through model inversion using hyperspectral data from Boreal conifer forest canopies. *Rem Sens Environ* 89:1989–99.

Zarco-Tejada, P. J., J. R. Miller, A. Morales, A. Berjón, and J. Agüera. 2004. Hyperspectral indices and model simulation for chlorophyll estimation in open-canopy tree crops. *Rem Sens Environ* 90:463–76.

Zarco-Tejada, P. J., J. R. Miller, T. L. Noland, G. H. Mohammed, and P. H. Sampson. 2001. Scaling-up and model inversion methods with narrowband optical indices for chlorophyll content estimation in closed forest canopies with hyperspectral data. *IEEE Trans Geosci Rem Sens* 39(7):1491–507.

Zarco-Tejada, P. J., C. A. Rueda, and S. L. Ustin. 2003. Water content estimation in vegetation with MODIS reflectance data and model inversion methods. *Rem Sens Environ* 85:109–24.

Zhang, Y., J. M. Chen, J. R. Miller, and T. L. Noland. 2008a. Leaf chlorophyll content retrieved from airborne hyperspectral remote sensing imagery. *Rem Sens Environ* 112:3234–47.

Zhang, Y., J. M. Chen, J. R. Miller, and T. L. Noland. 2008b. Retrieving chlorophyll content in conifer needles from hyperspectral measurements. *Can J Rem Sens* 34(3):296–310.

Zhang, L., D. Li, Q. Tong, and L. Zheng. 1998. Study of the spectral mixture model of soil and vegetation in Poyang Lake area, China. *Int J Rem Sens* 19:2077–84.

Zhao, G., and A. L. Maclean. 2000. A comparison of canonical discriminant dnalysis and principal component analysis for spectral transformation. *Photogramm Eng Rem Sens* 66(7):841–7.

Zheng, G., and L. M. Moskal. 2009. Retrieving leaf area index (LAI) using remote sensing: Theories, methods and sensors. *Sensors* 9:2719–45.

6

Thermal Remote Sensing of Urban Areas: Theoretical Backgrounds and Case Studies

Qihao Weng

CONTENTS

6.1 Introduction

Remote-sensing thermal infrared (TIR) data have been widely used in urban climate and environmental studies (Weng 2009). A series of satellite and airborne sensors have been developed to collect TIR data from the Earth's surface, such as the Heat Capacity Mapping Mission (HCMM), Landsat Thematic Mapper (TM)/Enhanced TM (ETM+), Advanced Very High Resolution Radiometer (AVHRR), Advanced Spaceborne Thermal Emission and Reflection Radiometer (ASTER), TIR Multispectral Scanner (TIMS), and Moderate Resolution Imaging Spectroradiometer (MODIS). In addition to land-surface temperature (LST) measurements, these TIR sensors may be utilized to obtain emissivity data from different surfaces with varied resolutions and accuracies. Both LST and emissivity data are used in urban studies mainly for analyzing LST patterns and their relationship with surface characteristics, assessing the urban heat island (UHI) phenomenon, and relating LSTs with surface energy fluxes for characterizing landscape properties, patterns, and processes (Quattrochi and Luvall 1999).

By examining the recent literature and providing case studies, this chapter reviews methods and applications of thermal remote sensing applied to urban areas. The emphasis is on the summarization of major advances and problems in the LST–vegetation relationship, UHI modeling with remotely sensed TIR data, and the estimation of urban surface heat fluxes. The last part of the chapter offers the author's viewpoint on the prospects of TIR remote sensing systems.

6.2 Relationship between Land-Surface Temperature and Vegetation Abundance

The LST is an important parameter in urban thermal environment and dynamics studies. This parameter modulates the air temperature of the lower layer of the urban atmosphere, and is a primary factor in determining surface radiation and energy exchange, internal climate of buildings, and human comfort in cities (Voogt and Oke 1998). The physical properties of various types of urban surfaces, their color, the sky view factor, street geometry, traffic loads, and anthropogenic activities are important factors that determine LSTs in urban environments (Chudnovsky, Ben-Dor, and Saaroni 2004). The LST of urban surfaces corresponds closely to the distribution of land use and land-cover (LULC) characteristics (Lo, Quattrochi, and Luvall 1997; Weng 2001, 2003; Weng, Lu, and Schubring 2004). To study urban LSTs, some sophisticated numerical and physical models have been developed, including energy balance models (Oke et al. 1999; Tong et al. 2005), laboratory models (Cendese and Monti 2003), three-dimensional (3D) simulations (Saitoh, Shimada, and Hoshi 1996), Gaussian models (Streutker 2002), and other numerical simulations. Among these models and simulations, statistical analysis plays an important role in linking LST to surface characteristics, especially at larger geographic scales (Bottyán and Unger 2003). Previous studies have linked LST to biophysical and meteorological factors, such as built-up area and height (Bottyán and Unger 2003), urban and street geometry (Eliasson 1996), LULC (Dousset and Gourmelon 2003), and vegetation (Weng, Lu, and Schubring 2004), as well as population distribution (Fan and Sailor 2005; Weng, Lu, and Liang 2006; Xiao et al. 2008) and the intensity of human activities (Elvidge et al. 1997). However, it is the relationship between LST and various vegetation indices that has been the most extensively documented in the literature.

The LST–vegetation index relationship has been used by Carlson, Gillies, and Perry (1994) to retrieve surface biophysical parameters, by Kustas et al. (2003) to extract subpixel thermal variations, and by Lambin and Ehrlich (1996) and Sobrino and Raissouni (2000) to analyze land-cover dynamics. Many studies observe a negative relationship between LST and vegetation indices. This finding has pushed research in two major directions: (1) statistical analysis of LST–vegetation abundance relationship and (2) the thermal-vegetation index (TVX) approach. The latter by definition is a multispectral method of combining LST and a vegetation index in a scatter plot to observe their associations (Quattrochi and Ridd 1994).

6.2.1 Statistical Analysis of the Land-Surface Temperature: Vegetation Abundance Relationship

To understand the statistical relationship between LST and vegetation cover, different vegetation indices have been employed in search of a representative index. Goward, Xue, and Czajkowski (2002) showed that different spectral vegetation indices, such as normalized difference vegetation index (NDVI) and simple ratio, were related to leaf area index (LAI) and green biomass. For a long time now, NDVI has been used to quantify vegetation patterns and dynamics within cities, and has been incorporated with LST to measure the impacts of urbanization (Weng and Lu 2008). The relationship between NDVI and fractional vegetation cover (Fr) is not singular. Small (2001) suggested that NDVI did not provide areal estimates of the amount of vegetation. The NDVI measurements are a function of the visible and near-infrared reflectance from the plant canopy, reflectance of the same spectra from the soil, and atmospheric reflectance, and they are subject to the influence of errors related to observational and other errors (Yang, Yang, and Merchnat 1997). Plant species, leaf area, soil background, and shadow can all contribute to NDVI variability (Jasinski 1990). The relationship between NDVI and other measures of vegetation abundance (e.g., LAI values greater than 3) is well known to be nonlinear (Asrar et al. 1984). This nonlinearity and the platform dependency of NDVI suggest that this index may not be a good indicator for quantitative analyses of vegetation (Small 2001), and the relationship between NDVI and LST needs further calibration. More quantitative, physically based measures of vegetation abundance are called for, especially in applications that require biophysical measures (Small 2001). The importance of spatial resolution for detecting landscape patterns and changes should also be emphasized (Frohn 1998), and the relationship between NDVI variability and pixel size should be further investigated (Jasinski 1990).

More recent investigations are directed at finding a surrogate to NDVI. Weng, Lu, and Schubring (2004) derived the vegetation fraction at different scales (pixel aggregation levels), made a comparison between NDVI and vegetation fraction in terms of their effectiveness as an indicator of urban thermal patterns, and found a stronger negative correlation between vegetation fraction and LST than between NDVI and LST. Yuan and Bauer (2007) made a similar correlation analysis between impervious surface area (ISA) and NDVI, suggested that ISA showed higher stability and lower seasonal variability, and recommended it as a complementary measure to NDVI. Xian and Crane (2006) supported the aforementioned observations by suggesting that the combined use of ISA, NDVI, and LST can explain temporal thermal dynamics across cities.

6.2.2 Thermal-Vegetation Index Approach to the Land-Surface Temperature–Vegetation Relationship

The combination of LST and NDVI by a scatter plot results in a triangular shape (Carlson, Gillies, and Perry 1994; Gillies and Carlson 1995; Gillies et al. 1997). Several methods have been developed to interpret the LST–NDVI space, including the "triangle" method using a "soil–vegetation–atmosphere transfer" (SVAT) model (Carlson, Gillies, and Perry 1994; Gillies and Carlson 1995; Gillies et al. 1997), in situ measurement method (Friedl and Davis 1994), and remote sensing–based method (Betts et al. 1996). However, difficulties still exist in interpreting LST for sparse canopies because the measurements combine the temperature of the soil and that of vegetation, and the combinations are often nonlinear (Sandholt, Rasmussen, and Andersen 2002). Different versions of the TVX approach have been developed over the past decades. Price (1990) found that radiant surface temperature showed more variations in sparsely vegetated areas than in densely vegetated areas. This behavior results in the atypical triangular shape or, as observed by Moran et al. (1994), in a trapezoidal shape for large heterogeneous regions under conditions of strong sunlight (Gillies et al. 1997). In Chapter 19, Carlson and Petropoulos provide a comprehensive review of the triangle method for estimating surface evapotranspiration and soil moisture. The slope of the LST–NDVI curve has been related to soil moisture conditions (Carlson, Gillies, and Perry 1994; Gillies and Carlson 1995; Gillies et al. 1997; Goetz 1997; Goward, Xue, and Czajkowski 2002), the evapotranspiration of the surface (Boegh et al. 1998), and other applications in shaping the TVX concept. Ridd (1995) and Carlson, Gillies, and Perry (1994) interpreted different sections of the triangle and related them to different LULC types. Lambin and Ehrlich (1996) presented a comprehensive interpretation of the TVX space. Carlson and Arthur (2000) gave a physical meaning to the TVX space. Further, Goward, Xue, and Czajkowski (2002) provided a detailed analysis of the underlying biophysics of the observed TVX relationship, and suggested that the relationship was the result of modulation of radiant surface temperature by vegetation cover. The TVX approach was the subject of studies focusing on the development of new applications, and the patterns and dynamics of different vegetation types at all scales from local to global. Researchers used the TVX concept to develop new indices and estimated parameters. Moran et al. (1994) used the TVX trapezoid to develop a new index called a "water-deficit index" (WDI) to estimate evapotranspiration in the absence of meteorological data using the difference between surface and air temperatures. Lambin and Ehrlich (1996) proposed radiant surface temperature—NDVI ratios in the TVX space—and showed its usefulness in land-cover mapping. Owen, Carlson, and Gillies (1998) used the same space and suggested a land-cover index (LCI) for assessing UHI. Carlson and Arthur (2000) extended the TVX approach to calculate ISA and surface runoff. Jiang and Islam (2001), by linear decomposition of TVX scatter plot, estimated the "α" parameter of the Priestly–Taylor equation in the absence of ground meteorological data. Sandholt, Rasmussen, and Andersen (2002) proposed a "temperature-vegetation dryness index" (TVDI) based on the relationship between surface temperature and NDVI, and showed the effectiveness of TVDI by explaining larger spatial variations better than hydrologic models. Nishida et al. (2003) estimated evapotranspiration fraction (EF) using a new TVX algorithm to provide global time-series coverage of EF from MODIS data. Chen et al. (2006) investigated the relationship between temperature and various newly developed indices, and found that NDVI presented a limited range.

Apart from the introduction of new indices, much research has been carried out in the extraction of new TVX metrics. Several studies have focused on the slope of the LST–NDVI fit line (Nemani and Running 1989; Smith and Choudhury 1991). Variations in slope and intercept of the TVX space have been interpreted in relation to surface parameters. Nemani and Running (1989) related the slope of the TVX correlation to the stomatal resistance and evapotranspiration in a deciduous forest. Sandholt, Rasmussen, and Andersen (2002) linked TVX correlation slope to the evapotranspiration rate and used this relationship to estimate air temperature. The TVX concept has further been used to anaylze pixel trajectories. The idea emerged over the past decade that land-surface parameters associated with individual pixels can be visualized as vectors tracing out paths in a multiparameter space (Lambin and Ehrlich 1994). Several studies verified that urbanization is the major cause of the observed migration of pixels within the TVX space (Owen, Carlson, and Gillies 1998; Carlson and Sanchez-Azofiefa 1999). Owen, Carlson, and Gillies (1998) found that the initial location of the migrating pixels in the TVX triangle determined the magnitude and direction of the path. Carlson and Sanchez-Azofeifa (1999) used the TVX method to assess how surface climate was affected by rapid urbanization and deforestation in San Jose, Costa Rica. They found that urbanization was more effective in causing changes in surface climate than deforestation, and that different development styles followed different paths in the space. Carlson and Arthur (2000) compared average trajectories of different development styles, and showed that in the advanced stages of development, the paths come closer and indistinguishable from one another.

Finally, the TVX approach has been used in the so-called triangle inversion method to derive surface parameters. Carlson, Gillies, and Perry (1994) used an SVAT model to show the feasibility of extracting surface parameters such as soil moisture content and Fr from the analysis of the TVX space without ground data. This inversion method was used to impose physical limits on a solution of the SVAT model parameterized for a test site to remotely sense variables used in the model to derive surface biophysical variables. Gillies et al. (1997) verified that the borders of the triangle constrained the solutions for determining surface energy fluxes. Goward, Xue, and Czajkowski (2002) used the TVX approach as a means for assessing soil moisture conditions from satellite data. Owen, Carlson, and Gillies (1998) used this method to assess the impacts of urbanization on surface parameters. Some authors, however, have drawn attention to the problems presented by the TVX space. Goward, Xue, and Czajkowski (2002) showed that plant stomatal function confused the interpretation of the TVX space given by experimental studies to use TVX slope to assess soil moisture conditions. Nishida et al. (2003) discussed four main difficulties of the TVX method used for evapotranspiration (ET) estimation: (1) the method's dependency on meteorological data, (2) computational difficulties encountered in the inversion of numerical models on a global scale, (3) problems involved in accurate estimation in dense vegetation, and (4) estimation difficulties faced in complex landscapes. While trying to establish guidelines in order to overcome the aforementioned problems by a new model, they suggested their model was effective for urbanization monitoring since EF is able to capture variations in surface energy partitioning (Nishida et al. 2003).

6.2.3 Case Study: TVX Space and Its Temporal Trajectory Analysis in Tabriz, Iran, Using Landsat TM/ETM+ Images

Amiri et al. (2009) examined the spatial and temporal dynamics of LST in relation to LULC change in the TVX space by using Landsat TIR and reflective data. A methodology was developed to detect and monitor urban expansion and to trace the changes in biophysical parameters such as NDVI and LST resulting from changes in LULC. The Tabriz metropolitan area ($38°05'$, $46°17'$) in Iran was selected as the study area. Multitemporal images acquired by Landsat 4 TM, Landsat 5 TM, and Landsat 7 ETM+ sensors on June 30, 1989, August 18, 1998, and August 2, 2001, respectively, were processed to extract LULC classes and LST. The relationship between the temporal dynamics of LST and LULC was then examined. The TVX space was constructed in order to study the temporal variability of thermal data and vegetation cover.

Figure 6.1a shows the Fr/T^* scatter plot (TVX space) with sample LULC classes based on the Landsat TM image of August 18, 1998. To create the plot, the cloud-contaminated pixels were first excluded. The NDVI values were rescaled between bare soil ($NDVI_0$) and dense vegetation ($NDVI_S$), following a method suggested by Owen, Carlson, and Gillies (1998). The Fr was then calculated as the square of the rescaled value N^*. Areas with high and low temperatures (T_{max} and T_0) were selected from the bare and wet soils, respectively, and their data were used to calculate the normalized temperature values of T^* (Gillies et al. 1997). The resulting Fr/T^* scatter plot showed a typical triangular pattern, with a clear "warm edge" defined by the right side of the pixel envelope.

The temporal trajectory of pixels in the TVX space made it possible to observe most changes due to urbanization as the pixels migrated from the low-temperature dense

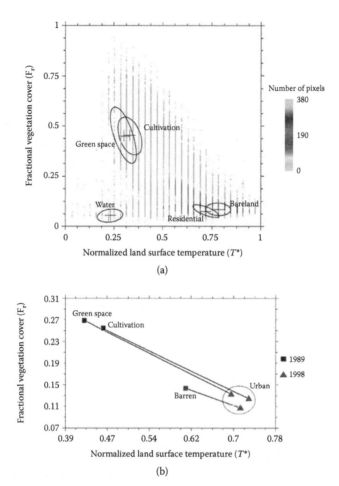

FIGURE 6.1
(See color insert following page 426.) Fractional vegetation cover (Fr)/T^* scatter plot (thermal-vegetation index [TVX] space) with sample land use and land-cover (LULC) classes from a Landsat thematic mapper (TM) image of the city of Tabriz and change trajectory in the TVX space for a specific period: (a) The scatter plot with sample LULC classes from a Landsat TM image of Tabriz (38°05′, 46°17′) in northwestern Iran, which was acquired on August 18, 1998; (b) change trajectory in the TVX space for a long (1989–1998) period (June 30, 1989–August 18, 1998). The vectors show the magnitude of change associated with LULC change from green space, cultivation, and barren pixels to urbanized pixels. (From Amiri, R., Q. Weng, A. Alimohammadi, and S. K. Alavipanah, *Remote Sens Environ*, 113, 12, 2009. With permission.)

vegetation condition to the high-temperature sparse vegetation condition in the TVX space (Figure 6.1b). Our result further showed that in the late stages of urbanization, affected pixels tend to converge and entirely lose their initial characteristics in the TVX space. The uncertainty analysis revealed that trajectory analysis in the TVX space involved a class-dependant noise component. This uncertainty emphasized the need for multiple LULC control points in the TVX space. In addition, this case study suggests that the use of multitemporal satellite data together with the examination of changes in the TVX space is effective and useful in urban LULC change monitoring and analysis of urban surface temperature conditions as long as the uncertainty issue is addressed.

6.3 Use of Remotely Sensed Data to Characterize and Model Urban Heat Islands

6.3.1 Background

Remotely sensed TIR data is a unique source of information in defining surface heat islands, which are related to canopy layer heat islands. In situ data (in particular, permanent meteorological station data) offers a high temporal resolution and long-term coverage but lacks spatial details. Observations and measurements by moving vehicles overcome this limitation to some extent, but do not provide a synchronized view over a city. Only remotely sensed TIR data can provide a continuous and simultaneous view of the whole city, which is of prime importance in the detailed investigation of urban surface climate. Rao (1972) was the first to assess the possibility of detecting the thermal footprint of urban areas. Since then, a wide range of TIR sensors have been developed and employed to study LST and UHI; they offer several improvements over their ancestors. However, in many of the previous studies, there is confusion between LST patterns and UHIs. A "satellite-derived" heat island is largely an artifact of the low-spatial-resolution imagery used, and the term "surface temperature patterns" is more meaningful than surface heat island (Nichol 1996). It remains a valid scientific issue how satellite-derived LSTs can be utilized to derive UHI parameters, and to model and simulate the UHI over space and time.

Previous studies of urban thermal landscapes and UHIs have been conducted using National Oceanic and Atmospheric Administration (NOAA) AVHRR data (Kidder and Wu 1987; Balling and Brazell 1988; Roth, Oke, and Emery 1989; Gallo et al. 1993; Gallo and Owen 1998; Streutker 2002). However, for all these studies, the 1.1-km spatial resolution AVHRR data were found suitable only for large-area urban temperature mapping and not for establishing accurate and meaningful relationships between image-derived values and those measured on the ground. The 120-m resolution Landsat TM (and later ETM+ data of 60-m resolution) TIR data have also been extensively utilized to derive LSTs and to study UHIs. Carnahan and Larson (1990) used the TM TIR data to observe mesoscale temperature differences between urban and rural areas in Indianapolis, Indiana, whereas Kim (1992) studied similar phenomena in Washington, DC. Nichol (1994) utilized TM TIR data to monitor microclimate for housing estates in Singapore, and further calculated LSTs of building walls based on a 3D geographic information system (GIS) model (Nichol 1998). Weng (2001, 2003) examined LST patterns and their relationships with land cover in Guangzhou and in the urban clusters in the Pearl River Delta of China. Weng, Lu, and Schubring (2004) utilized a Landsat ETM+ image to examine the LST–vegetation abundance relationship in Indianapolis. More recently, Lu and Weng (2006) applied spectral mixture analysis (SMA) to ASTER images in order to derive hot-object and cold-object fractions from the TIR bands of the sensor and biophysical variables from the nonthermal bands. Statistical analyses were then conducted to examine the relationship between LST and the derived fraction variables across the resolution from 15 to 90 m.

The most recent advances include development and utilization of quantitative surface descriptors for assessing the interplay between urban material fabric and urban thermal behavior (Weng, Lu, and Schubring 2004; Weng, Lu, and Liang 2006; Lu and Weng 2006; Weng and Lu 2008). Moreover, the landscape ecology approach was employed to assess this interplay across various spatial resolutions and to identify the operational scale at which both LST and LULC processes interacted to generate the urban thermal landscape patterns (Weng, Liu, and Lu 2007; Liu and Weng 2008). Because an ASTER sensor collects both daytime and nighttime TIR images, analysis of LST spatial patterns has also (Xiao et al. 2008) been conducted for a diurnal contrast (Nichol 2005).

A key issue in the application of TIR remote-sensing data in urban climate studies is the use of LST measurements at the microscale to characterize and quantify UHIs observed at the mesoscale. Streutker (2002, 2003) used AVHRR data to quantify the UHI of Houston, Texas, as a continuously varying two-dimensional (2D) Gaussian surface superimposed on a planer rural background, and derived the UHI parameters of magnitude (i.e., intensity), spatial extent, orientation, and central location. Rajasekar and Weng (2009) applied a nonparametric model by applying fast Fourier transformation (FFT) to MODIS imagery for characterization of the UHI over space, so that UHI magnitude and other parameters may be derived. Despite these advances, estimation of UHI parameters from multitemporal and multilocation TIR imagery still remains a promising research direction and will continue to be so in the years to come, given the increased interest of the urban climate community in using remote-sensing data.

6.3.2 Case Study: Characterizing an Urban Heat Island in Beijing, China, Using Advanced Spaceborne Thermal Emission and Reflection Radiometer Images

This section briefly introduces a method for characterizing UHIs using remotely sensed LST data and explains its application in Beijing, China. Higdon (2002) explained the process convolution for a one-dimensional (1D) process and made suggestions for its extension to two or three dimensions. In this case study, the process convolution model was extended to model the UHI of Beijing as a 2D Gaussian process using ASTER LST data. The procedure is detailed next.

Let $y_{(1,1)}, \ldots, y_{(i,j)}$ (where q is a 2D matrix of $(1,1), \ldots, (i,j)$) be data recorded over the 2D spatial locations $s_{(1,1)}, \ldots, s_{(i,j)}$ in S. In this research, the spatial method represents data as the sum of an overall mean μ, a spatial process $z = (z_{(1,1)}, \ldots, z_{(i,j)})^T$, and Gaussian white noise $\varepsilon = (\varepsilon_{(1,1)}, \ldots, \varepsilon_{(i,j)})^T$ with variance σ^2_ε:

$$y = s + z + \varepsilon \tag{6.1}$$

Here, the elements of z are the restriction of the spatial process $z(s)$ to the 2D data locations $s_{(1,1)}, \ldots, s_{(i,j)}$; $z(s)$ is defined to be a mean zero Gaussian process. But rather than specifying $z(s)$ through its covariance function, it is determined by the latent process $x(s)$ and the smoothing kernel $k(s)$. The latent process $x(s)$ is restricted to be nonzero at the 2D spatial sites $\omega_{(1,1)}, \ldots, \omega_{(a,b)}$, also in S, and $x = (x_{(1,1)}, \ldots, x_{(a,b)})^T$ where $x\omega_p = x(\omega_p); p = (1,1), \ldots, (a,b)$. Each x_p is then modeled as an independent draw from an $N(0,\sigma^2_\omega)$ distribution. The resulting continuous Gaussian process is then

$$z(S) = \sum_{p=(1,1)}^{(a,b)} X_j k(s - \omega_p) \tag{6.2}$$

where $k(s - \omega_p)$ is a kernel centered at ω_p. This gives the following linear model:

$$y = \mu l_{(i,j)} + Kx + \varepsilon \tag{6.3}$$

where $l_{(i,j)}$ is the (i,j)th vector of ls; the elements of K are given by

$$K_{pq} = k(s_p - \omega_q)x_q \tag{6.4}$$

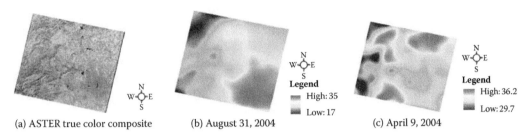

(a) ASTER true color composite　　　(b) August 31, 2004　　　(c) April 9, 2004

FIGURE 6.2
(See color insert following page 426.) The results of kernel convolution for two advanced spaceborne thermal emission and reflection radiometer (ASTER) images of Beijing: (a) A true color composite of Beijing using ASTER acquired on August 31, 2004; (b) and (c) the results of convoluted images (with a smoothing parameter of 0.6) showing thermal landscape pattern of Beijing on August 31, 2004 and April 9, 2004, respectively. The temperature is given in degrees Celsius.

$$x \sim N(0, \sigma^2{}_x l_{(a,b)}) \text{ and} \tag{6.5}$$

$$\varepsilon \sim N(0, \sigma^2{}_x l_{(i,j)}) \tag{6.6}$$

The results of convolution modeling can then be analyzed for patterns over space and time. The results of the kernel convolution can be compared to examine how UHI magnitude, center, and spatial extent change over space and time. In order to determine UHI magnitude, the mean temperature value within each image is considered as a background temperature and all values less than the mean are brought to the same level.

Figure 6.2 shows the results of kernel convolution (with a smoothing parameter of 0.6) for two ASTER images of Beijing acquired on August 31, 2004 and April 9, 2004, respectively. The August image clearly displays a UHI (Figure 6.2a) with a magnitude of 7°C. The built-up area had a higher temperature than the surrounding rural areas, and there was a temperature gradient from the urban areas in the southeastern corner to the mountainous area in the northwest. In contrast, the April image shows an urban heat sink in central Beijing (Figure 6.2b). According to our computation, the intensity of the heat sink was about 3°C. Higher temperatures corresponded to the three suburban agricultural/residential areas in the north, northwest, and south. The lowest temperature was detected in the western mountainous area.

6.4 Estimation of Urban Heat Fluxes Using Remote Sensing Data

Knowledge of urban surface energy balance is fundamental to the understanding of UHIs and urban thermal behavior (Oke 1982, 1988). Three items of information are needed in order to estimate land surface energy fluxes: (1) energy driving forces (i.e., incident solar energy, albedo, and resulting net radiation), (2) soil moisture availability and the vegetation–soil interaction, and (3) capacity of the atmosphere to absorb the flux, which depends on surface air temperature, vapor pressure gradients, and surface winds (Schmugge, Hook, and Coll 1998). Previous studies have focused on the methods for estimating variables related to the first two items from satellite remote sensing data, but little has been done to estimate the surface atmospheric parameters (Schmugge, Hook, and Coll 1998). These parameters are measured in the traditional way in the network of meteorological stations or by in situ field measurements.

Remote sensing TIR data can be applied to relate LSTs with surface energy fluxes for characterizing landscape properties, patterns, and processes (Quattrochi and Luvall 1999). Remotely sensed thermal imagery has the advantage of providing a time-synchronized dense grid of temperature data over a whole city, whereas optical sensing data have been used to monitor discrete land-cover types and to estimate biophysical variables (Steininger 1996). Together, remote sensing data can be used to estimate surface parameters related to the soil–vegetation system and surface soil moisture, radiation forcing components, and indicators of the surface's response to them (i.e., LST; Schmugge, Hook, and Coll 1998). If the advantage of time-sequential observations of satellite sensors (some sensors can even scan a specific geographic location twice a day—one at daytime and one at nighttime) is considered, remote-sensing data have great potential for studying the urban surface energy budget, as well as the spatial pattern and temporal dynamics of urban thermal landscapes. One of the earliest studies that combined surface energy modeling and remote sensing approaches was conducted by Carlson et al. (1981). They used satellite temperature measurements in conjunction with a 1D boundary layer model to analyze the spatial patterns of turbulent heat fluxes, thermal inertia, and ground moisture availability in Los Angeles, CA, and St. Louis, MO. This method was later applied in Atlanta by using AVHRR data, in which the net urban effect was determined as the difference between urban and rural simulations (Hafner and Kidder 1999). Because analyses of surface energy flux are extensively conducted over vegetated and agricultural areas, successful methods have been applied to urban areas (Zhang, Aono, and Monji 1998; Chrysoulakis 2003). Zhang, Aono, and Monji (1998) used Landsat TM data, in combination with routine meteorological data and field measurements, to estimate the urban surface energy fluxes in Osaka, Japan, and to analyze their spatial variability in both summer and winter. Chrysoulakis (2003) used ASTER imagery, together with in situ spatial data, to determine the spatial distribution of all-wave surface net radiation balance in Athens, Greece. Kato and Yamaguchi (2005) combined ASTER and Landsat ETM+ data with ground meteorological data to investigate the spatial patterns of surface energy fluxes in Nagoya, Japan, over four distinct seasons. Furthermore, this study separated anthropogenic heat discharge and natural heat radiation from sensible heat flux.

The seasonal and spatial variability of surface heat fluxes is crucial to the understanding of UHI phenomenon and dynamics, which has not been thoroughly addressed by previous studies. In a recent study, based on the two-source energy balance (TSEB) algorithm, we developed a method to estimate urban heat fluxes by the combined use of multispectral ASTER images and routine meteorological data, and applied it to the city of Indianapolis, for understanding the seasonal changes in the heat fluxes. The ASTER images of the four seasons were acquired and processed with atmospheric, radiometric, and geometric corrections before using them for the analysis. The ASTER data pertaining to surface kinetic temperature, surface spectral emissivity, and surface reflectance (VNIR and SWIR) was used. All the images were resampled to a resolution of 90 m. The nonvegetation and vegetation areas were separated for estimating heat fluxes based on computed NDVI values. The needed meteorological data was obtained from the Indiana State Climate Office, including data regarding shortwave radiation, air temperature, relative humidity, air pressure, and wind speed. Shortwave radiation data was obtained from the National Solar Radiation Database.

Figure 6.3 shows the estimated net radiation, sensible heat flux, latent heat flux, and ground heat flux on October 13, 2006, recorded in Indianapolis. The mean values and standard deviations of the surface heat fluxes by LULC type are displayed in Table 6.1. This study found that the estimated surface heat fluxes showed a strong seasonality, with the highest net radiation recorded in summer, followed by spring, fall, and winter. Sensible heat flux tended to change largely with surface temperature, whereas latent heat was largely modulated by the change in vegetation abundance and vigor over a year and the accompanying moisture condition. The fluctuation in all heat fluxes tended to be high in the summer months and low in the winter months. Sensible and latent heat fluxes showed

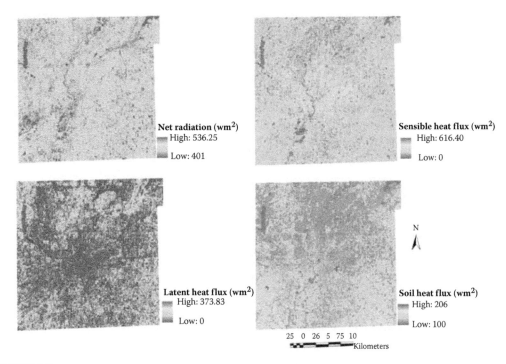

FIGURE 6.3
(See color insert following page 426.) Net radiation, sensible heat flux, latent heat flux, and soil heat flux on October 13, 2006 in Indianapolis estimated by the combined use of advanced spaceborne thermal emission and reflection radiometer image and ground meteorological data.

TABLE 6.1

Statistics of Surface Heat Fluxes by LULC Type in Indianapolis on October 13, 2006 (Unit: W/m^2)

Heat Fluxes	Urban and Built-Up Land	Agricultural Land	Forest Land	Grassland	Water	Bare Ground	Overall
Net radiation	377.87 (40.97)	394.98 (33.30)	426.37 (20.15)	378.76 (25.99)	484.40 (28.61)	363.61 (56.88)	396.47 (40.73)
Soil heat flux	151.15 (16.39)	118.49 (9.99)	63.96 (3.02)	113.63 (7.80)	169.54 (10.01)	109.08 (17.07)	113.09 (37.30)
Sensible heat flux	293.34 (41.95)	183.35 (31.91)	269.26 (95.34)	243.82 (73.24)	77.49 (11.86)	91.10 (14.51)	242.62 (100.31)
Latent heat flux	0.94 (8.78)	65.99 (39.88)	63.73 (34.70)	39.67 (37.31)	231.50 (52.90)	150.20 (49.28)	39.53 (52.15)

a stronger spatial variability than net radiation and ground heat flux. By computing heat fluxes by LULC type, we further investigated the geographic pattern and spatial variability of urban surface energy balance. The variations in net radiation among the LULC types were found to be attributable mainly to surface albedo and temperature, whereas the within-class variability in turbulent heat fluxes were more associated with changes in vegetation, water bodies, and other surface factors.

6.5 Future Prospects of Thermal Infrared Sensors

There has been significant progress in the studies focusing on LST–vegetation relationship, UHI modeling with remotely sensed TIR data, and estimation of urban surface heat fluxes. However, urban climate and environmental studies will be difficult, if not impossible, without TIR sensors having a global imaging capacity. At present, there are few sensors that have such TIR capabilities. The TM sensor aboard Landsat 5 has been acquiring images of the Earth nearly continuously from July 1982 to the present, with a TIR band of 120-m resolution, and is thus long overdue. On April 2, 2007, updates to the radiometric calibration of Landsat 5 TM data processed and distributed by the U.S. Geological Survey (USGS) Earth Resources Observation System (EROS) created an improved Landsat 5 TM data product that is now more comparable radiometrically to Landsat 7 ETM+ and provides the basis for continued long-term studies of the Earth's land surfaces. Another TIR sensor that has global imaging capacity is with Landsat 7 ETM+. On May 31, 2003, the ETM+ scan-line corrector (SLC) failed permanently. Although it is still capable of acquiring useful image data with the SLC turned off, particularly within the central part of any given scene, the National Aeronautics and Space Administration (NASA) has teamed up with USGS to focus on the Landsat Data Continuity Mission (LDCM), which is most likely not to have a TIR imager. In addition, Terra's ASTER TIR bands of 90-m resolution have been increasingly used in urban climate and environmental studies in recent years. The ASTER is an on-demand instrument, which means that data are acquired only over requested locations. The Terra satellite, launched in December 1999 as part of NASA's Earth Observing System, has a life expectancy of 6 years, and is now also overdue. The scientific and user community is looking forward to a Landsat ETM–like TIR sensor. The draft requirements for the LDCM thermal imager indicate that two thermal bands (10.3–11.3 μm and 11.5–12.5 μm) of 90 m or better spatial resolution are preferred (for details, readers are referred to the LDCM Web site, http://ldcm.nasa.gov/procurement/TIRimagereqs051006.pdf). The National Research Council Decadal Survey indicates the need for such a TIR sensor. The Hyperspectral Infrared Imager (HyspIRI) is defined as a mission with tier-2 priority to be launched in the next 8–10 years. Because of its hyperspectral visible and shortwave infrared bandwidths and its multispectral TIR capabilities, HyspIRI will be well suited for deriving land-cover and other biophysical attributes for urban climate and environmental studies (for more information, the readers are referred to the HyspIRI Web site, http://hyspiri.jpl.nasa.gov/). Its TIR imager is expected to provide seven bands between 7.5 and 12 μm and one band at 4 μm, all with 60-m spatial resolution. This TIR sensor is intended for the imaging of global land and shallow water (less than 50 m) with a 5-day revisit at the equator (1 day and 1 night imaging). These improved capabilities would allow for a more accurate estimation of

LST and emissivity, and for deriving unprecedented information on biophysical characteristics and even socioeconomic information such as population, quality of life indicators, and human settlements. Such information cannot be obtained from the current generation of satellites devices in orbit, such as MODIS, Landsat, or ASTER. Two major areas of application identified by the HyspIRI science team are urbanization and human health through the combined use of visible to shortwave infrared (VSWIR) and TIR data. Until then, we may have to bear with Landsat and ASTER for medium-resolution TIR data, and MODIS and AVHRR for coarse-resolution data. It is from this perspective that international collaborations on Earth resources satellites become very important.

Acknowledgments

This research is supported by the National Science Foundation (BCS-0521734) for a project entitled "Role of Urban Canopy Composition and Structure in Determining Heat Islands: A Synthesis of Remote Sensing and Landscape Ecology Approach." The author would also like to acknowledge the following individuals for their contributions in various capacities to this chapter: Dale Quattrochi, Xuefei Hu, Rongbo Xiao, Reza Amiri, and Umamaheshwaren Rajasekar.

References

Amiri, R., Q. Weng, A. Alimohammadi, and S. K. Alavipanah. 2009. The spatial-temporal dynamics of land surface temperatures in relation to fractional vegetation cover and land use/cover in the Tabriz urban area, Iran. *Remote Sens Environ* 113(12):2606–17.

Asrar, G., M. Fuchs, E. T. Kanemasu, and J. L. Hatfield. 1984. Estimating absorbed photosynthetic radiation and leaf area index from spectral reflectance in wheat. *Agron J* 76(22):300–6.

Balling, R. C., and S. W. Brazell. 1988. High resolution surface temperature patterns in a complex urban terrain. *Photogramm Eng Remote Sens* 54(9):1289–93.

Betts, A., J. Ball, A. Beljaars, M. Miller, and P. Viterbo. 1996. The land surface-atmosphere interaction: A review based on observational and global modeling perspectives. *J Geophys Res* 101(D3):7209–25.

Boegh, E., H. Soegaard, N. Hanan, P. Kabat, and L. Lesch. 1998. A remote sensing study of the NDVI-T$_s$ relationship and the transpiration from sparse vegetation in the Sahel based on high resolution satellite data. *Remote Sens Environ* 69(3):224–40.

Bottyán, Z., and J. Unger. 2003. A multiple linear statistical model for estimating the mean maximum urban heat island. *Theor Appl Climatol* 75(3–4):233–43.

Carlson, T. N., and S. T. Arthur. 2000. The impact of land use–land cover changes due to urbanization on surface microclimate and hydrology: A satellite perspective. *Global Planet Change* 25(1–2):49–65.

Carlson, T. N., J. K. Dodd, S. G. Benjamin, and J. N. Cooper. 1981. Satellite estimation of the surface energy balance, moisture availability and thermal inertia. *J Appl Meteorol* 20(1):67–87.

Carlson, T. N., R. R. Gillies, and E. M. Perry. 1994. A method to make use of thermal infrared temperature and NDVI measurements to infer surface water content and fractional vegetation cover. *Remote Sens Rev* 9(1–2):161–73.

Carlson, T. N., and G. A. Sanchez-Azofeifa. 1999. Satellite remote sensing of land use changes in and around San José, Costa Rica. *Remote Sens Environ* 70(3):247–56.

Carnahan, W. H., and R. C. Larson. 1990. An analysis of an urban heat sink. *Remote Sens Environ* 33(1):65–71.

Cendese, A., and P. Monti. 2003. Interaction between an inland urban heat island and a sea-breeze flow: A laboratory study. *J Appl Meteorol* 42(11):1569–83.

Chen, X. -L., H. -M. Zhao, P. -X. Li, and Z. -Y. Yin. 2006. Remote sensing image-based analysis of the relationship between urban heat island and land use/cover changes. *Remote Sens Environ* 104(2):133–46.

Chrysoulakis, N. 2003. Estimation of the all-wave urban surface radiation balance by use of ASTER multispectral imagery and in situ spatial data. *J Geophys Res D Atmos* 108(18):ACL 8-1–10.

Chudnovsky, A., E. Ben-Dor, and H. Saaroni. 2004. Diurnal thermal behavior of selected urban objects using remote sensing measurements. *Energy Build* 36(11):1063–74.

Dousset, B., and F. Gourmelon. 2003. Satellite multi-sensor data analysis of urban surface temperature and land cover. *ISPRS J Photogramm Remote Sens* 58(1–2):43–54.

Eliasson, I. 1996. Urban nocturnal temperatures, street geometry and land use. *Atmos Environ* 30(3):379–92.

Elvidge, C. D., K. E. Baugh, E. A. Kihn, H. W. Kroehl, E. R. Davis, and C. W. Davis. 1997. Relation between satellite-observed visible-near infrared emissions, population, economic activity and electric power consumption. *Int J Remote Sens* 18(6):1373–9.

Fan, H., and D. J. Sailor. 2005. Modeling the impacts of anthropogenic heating on the urban climate of Philadelphia: A comparison of implementations in two PBL schemes. *Atmos Environ* 39(1):73–84.

Friedl, M. A., and F. W. Davis. 1994. Sources of variation in radiometric surface temperature over a tallgrass prairie. *Remote Sens Environ* 48(1):1–17.

Frohn, R. C. 1998. *Remote Sensing for Landscape Ecology*. Boca Raton, FL: Lewis Publishers.

Gallo, K. P., A. L. McNab, T. R. Karl, J. F. Brown, J. J. Hood, and J. D. Tarpley. 1993. The use of NOAA AVHRR data for assessment of the urban heat island effect. *J Appl Meteorol* 32(5):899–908.

Gallo, K. P., and T. W. Owen. 1998. Assessment of urban heat island: A multi-sensor perspective for the Dallas-Ft. Worth, USA region. *Geocarto Int* 13(4):35–41.

Gillies, R. R., and T. N. Carlson. 1995. Thermal remote sensing of surface soil water content with partial vegetation cover for incorporation into climate models. *J Appl Meteorol* 34(4):745–56.

Gillies, R. R., T. N. Carlson, J. Cui, W. P. Kustas, and K. S. Humes. 1997. A verification of the "triangle" method for obtaining surface soil water content and energy fluxes from remote measurements of the Normalized Difference Vegetation index (NDVI) and surface radiant temperature. *Int J Remote Sens* 18(5):3145–66.

Goetz, S. J. 1997. Multisensor analysis of NDVI, surface temperature and biophysical variables at a mixed grassland site. *Int J Remote Sens* 18(1):71–94.

Goward, S. N., Y. Xue, and K. P. Czajkowski. 2002. Evaluating land surface moisture conditions from the remotely sensed temperature/vegetation index measurements: An exploration with the simplified biosphere model. *Remote Sens Environ* 79(2–3):225–42.

Hafner, J., and S. Q. Kidder. 1999. Urban heat island modeling in conjunction with satellite-derived surface/soil parameters. *J Appl Meteorol* 38(4):448–65.

Higdon, D. 2002. Space and space-time modeling using process convolution. In: Anderson, C. W., Barnett, V., Chatwin, P. C., El-Shaarawi, A. H. (Eds.), *Quantitative Methods for Current Environmental Issues*. Springer-Verlag, London, pp. 37–56.

Jasinski, M. F. 1990. Sensitivity of the normalized difference vegetation index to subpixel canopy cover, soil albedo, and pixel scale. *Remote Sens Environ* 32(2–3):169–87.

Jiang, L., and S. Islam. 2001. Estimation of surface evaporation map over southern Great Plains using remote sensing data. *Water Resour Res* 37(2):329–40.

Kato, S., and Y. Yamaguchi. 2005. Analysis of urban heat-island effect using ASTER and ETM+ data: Separation of anthropogenic heat discharge and natural heat radiation from sensible heat flux. *Remote Sens Environ* 99(1–2):44–54.

Kidder, S. Q., and H. T. Wu. 1987. A multispectral study of the St. Louis area under snow covered conditions using NOAA-7 AVHRR data. *Remote Sens Environ* 22(2):159–72.

Kim, H. H. 1992. Urban heat island. *Int J Remote Sens* 13(12):2319–36.

Kustas, W. P., J. M. Norman, M. C. Anderson, and A. N. French. 2003. Estimating subpixel surface temperatures and energy fluxes from the vegetation index-radiometric temperature relationship. *Remote Sens Environ* 85(4):429–40.

Lambin, F. F., and D. Ehrlich. 1996. The surface temperature-vegetation index space for land use and land cover change analysis. *Int J Remote Sens* 17(3):463–87.

Liu, H., and Q. Weng. 2008. Seasonal variations in the relationship between landscape pattern and land surface temperature in Indianapolis, USA. *Environ Monit Assess* 144(1–3):199–219.

Lo, C. P., D. A. Quattrochi, and J. C. Luvall. 1997. Application of high-resolution thermal infrared remote sensing and GIS to assess the urban heat island effect. *Int J Remote Sens* 18(2):287–304.

Lu, D., and Q. Weng. 2006. Spectral mixture analysis of ASTER imagery for examining the relationship between thermal features and biophysical descriptors in Indianapolis, Indiana. *Remote Sens Environ* 104(2):157–67.

Moran, M. S., T. R. Clarke, Y. Inoue, and A. Vidal. 1994. Estimating crop water deficit using the relation between surface-air temperature and spectral vegetation index. *Remote Sens Environ* 49(3):246–63.

Nemani, R. R., and S. W. Running. 1989. Estimation of regional surface resistance to evapotranspiration from NDVI and thermal-IR AVHRR data. *J Appl Meteorol* 28(4):276–84.

Nichol, J. E. 1994. A GIS-based approach to microclimate monitoring in Singapore's high-rise housing estates. *Photogramm Eng Remote Sens* 60(10):1225–32.

Nichol, J. E. 1996. High-resolution surface temperature patterns related to urban morphology in a tropical city: A satellite-based study. *J Appl Meteorol* 35(1):135–46.

Nichol, J. E. 1998. Visualization of urban surface temperature derived from satellite images. *Int J Remote Sens* 19(9):1639–49.

Nichol, J. E. 2005. Remote sensing of urban heat islands by day and night. *Photogramm Eng Remote Sens* 71(5):613–23.

Nishida, K., R. R. Nemani, J. M. Glassy, and S. W. Running. 2003. Development of an evapotranspiration index from Aqua/MODIS for monitoring surface moisture status. *IEEE Trans Geosci Remote Sens* 41(2):493–501.

Oke, T. R. 1982. The energetic basis of the urban heat island. *Q J R Meteorolog Soc* 108(455):1–24.

Oke, T. R. 1988. The urban energy balance. *Prog Phys Geog* 12(4):471–508.

Oke, T. R., R. A. Spronken-Smith, E. Jauregui, and C. S. B. Grimmond. 1999. The energy balance of central Mexico City during the dry season. *Atmos Environ* 33(24–25):3919–30.

Owen, T. W., T. N. Carlson, and R. R. Gillies. 1998. An assessment of satellite remotely-sensed land cover parameters in quantitatively describing the climatic effect of urbanization. *Int J Remote Sens* 19(9):1663–81.

Price, J. C. 1990. Using spatial context in satellite data to infer regional scale evapotranspiration. *IEEE Trans Geosci Remote Sens* 28(5):940–8.

Quattrochi, D. A., and N. S. Goel. 1995. Spatial and temporal scaling of thermal remote sensing data. *Remote Sens Rev* 12(3–4):255–86.

Quattrochi, D. A., and J. C. Luvall. 1999. Thermal infrared remote sensing for analysis of landscape ecological processes: Methods and applications. *Landsc Ecol* 14(6):577–98.

Quattrochi, D. A., and M. K. Ridd. 1994. Measurement and analysis of thermal energy responses from discrete urban surface using remote sensing data. *Int J Remote Sens* 15(10):1999–2022.

Rajasekar, U., and Q. Weng. 2009. Urban heat island monitoring and analysis by data mining of MODIS imageries. *ISPRS J Photogramm Remote Sens* 64(1):86–96.

Rao, P. K. 1972. Remote sensing of urban heat islands from an environmental satellite. *Bull Am Meteorol Soc* 53:647–8.

Ridd, M. K. 1995. Exploring a V-I-S (Vegetation-Impervious Surface-Soil) model for urban ecosystem analysis through remote sensing: comparative anatomy for cities. *Int J Remote Sens* 16(12):2165–85.

Roth, M., T. R. Oke, and W. J. Emery. 1989. Satellite derived urban heat islands from three coastal cities and the utilisation of such data in urban climatology. *Int J Remote Sens* 10:1699–720.

Saitoh, T. S., T. Shimada, and H. Hoshi. 1996. Modeling and simulation of the Tokyo urban heat island. *Atmos Environ* 30:3431–42.

Sandholt, I., K. Rasmussen, and J. Andersen. 2002. A simple interpretation of the surface temperature/vegetation index space for assessment of surface moisture status. *Remote Sens Environ* 79:213–24.

Schmugge, T., S. J. Hook, and C. Coll. 1998. Recovering surface temperature and emissivity from thermal infrared multispectral data. *Remote Sens Environ* 65:121–31.

Small, C. 2001. Estimation of urban vegetation abundance by spectral mixture analysis. *Int J Remote Sens* 22:1305–34.

Sobrino, J. A., and N. Raissouni. 2000. Toward remote sensing methods for land cover dynamic monitoring: Application to Morocco. *Int J Remote Sens* 21:353–66.

Steininger, M. K. 1996. Tropical secondary forest regrowth in the Amazon: Age, area and change estimation with Thematic Mapper data. *Int J Remote Sens* 17(1):9–27.

Streutker, D. R. 2002. A remote sensing study of the urban heat island of Houston, Texas. *Int J Remote Sens* 23:2595–608.

Streutker, D. R. 2003. Satellite-measured growth of the urban heat island of Houston, Texas. *Remote Sens Environ* 85:282–9.

Tong, H., A. Walton, J. Sang, and J. C. L. Chan. 2005. Numerical simulation of the urban boundary layer over the complex terrain of Hong Kong. *Atmos Environ* 39:3549–63.

Voogt, J. A., and T. R. Oke. 1998. Effects of urban surface geometry on remotely sensed surface temperature. *Int J Remote Sens* 19:895–920.

Weng, Q. 2001. A remote sensing-GIS evaluation of urban expansion and its impact on surface temperature in the Zhujiang Delta, China. *Int J Remote Sens* 22:1999–2014.

Weng, Q. 2003. Fractal analysis of satellite-detected urban heat island effect. *Photogramm Eng Remote Sens* 69:555–66.

Weng, Q. 2009. Thermal infrared remote sensing for urban climate and environmental studies: Methods, applications, and trends. *ISPRS J Photogramm Remote Sens* 64(4):335–44.

Weng, Q., H. Liu, and D. Lu. 2007. Assessing the effects of land use and land cover patterns on thermal conditions using landscape metrics in city of Indianapolis, United States. *Urban Ecosyst* 10:203–19.

Weng, Q., and D. Lu. 2008. A sub-pixel analysis of urbanization effect on land surface temperature and its interplay with impervious surface and vegetation coverage in Indianapolis, United States. *Int J Appl Earth Obs Geoinf* 10:68–83.

Weng, Q., D. Lu, and B. Liang. 2006. Urban surface biophysical descriptors and land surface temperature variations. *Photogramm Eng Remote Sensing* 72(11):1275–86.

Weng, Q., D. Lu, and J. Schubring. 2004. Estimation of land surface temperature-vegetation abundance relationship for urban heat island studies. *Remote Sens Environ* 89:467–83.

Xian, G., and M. Crane. 2006. An analysis of urban thermal characteristics and associated land cover in Tampa Bay and Las Vegas using Landsat satellite data. *Remote Sens Environ* 104(2):147–56.

Xiao, R., Q. Weng, Z. Ouyang, W. Li, E. W. Schienke, and W. Zhang. 2008. Land surface temperature variation and major factors in Beijing, China. *Photogramm Eng Remote Sensing* 74(4):451–61.

Yang, W., L. Yang, and J. W. Merchnat. 1997. An analysis of AVHRR/NDVI-ecoclimatological relations in Nebraska, USA. *Int J Remote Sens* 18:2161–80.

Yuan, F., and M. E. Bauer. 2007. Comparison of impervious surface area and normalized difference vegetation index as indicators of surface urban heat island effects in Landsat imagery. *Remote Sens Environ* 106(3):375–86.

Zhang, X., Y. Aono, and N. Monji. 1998. Spatial variability of urban surface het fluxes estimated from Landsat TM data under summer and winter conditions. *J Agric Meteorol* 54(1):1–11.

Section II

Algorithms and Techniques

7

Atmospheric Correction Methods for Optical Remote Sensing Imagery of Land

Rudolf Richter

CONTENTS

The optical part of the electromagnetic spectrum covers wavelengths from 100 nm to 1 mm. However, only a small part of the optical spectrum can be used for remote sensing from airborne and spaceborne platforms, because of the characteristics of the scattering, absorption, and emission of radiation by the terrestrial atmosphere. Figure 7.1 presents a typical atmospheric transmittance curve in those spectral regions that can be exploited with remote sensing techniques. Basically, there exist three large spectral intervals: 0.4–2.5 µm, 3–5 µm (mid-infrared or MIR), and 8–14 µm (thermal infrared or TIR). For technical reasons, the first region is often split into the visible to near-infrared (VIS to NIR or VNIR; 0.4–1.0 µm; no detector cooling required) and short-wave infrared (SWIR; 1.0–2.5 µm; detector cooling required) regions. The main absorbing gases in the atmosphere are water vapor, ozone, carbon dioxide, and oxygen; the most variable gas in space and time is water vapor.

The 0.4–3.0 µm region is often referred to as a "reflective" or "solar" region. The reflected solar radiation dominates in this region compared to ambient emitted radiation, whereas the emitted TIR radiation dominates in the 8–14 µm domain (Figure 7.2). The reflected solar radiation is plotted for three surface reflectance (ρ) levels, and the emitted radiation for a 300-K blackbody. The atmospheric influence is neglected in this figure.

In the MIR interval, reflected solar and emitted thermal radiations have the same order of magnitude, and both contributions have to be considered during atmospheric correction (AC). As the majority of existing high spatial resolution instruments does not possess MIR channels, we will not discuss this case but refer to the works of Hook et al. (2001) and Mushkin, Balick, and Gillespie (2005).

Table 7.1 contains an overview of some typical multispectral and hyperspectral instruments covering the reflective region, the TIR region with one channel, and the TIR region with more than 10 channels. Currently, all high spatial resolution (footprint <100 m)

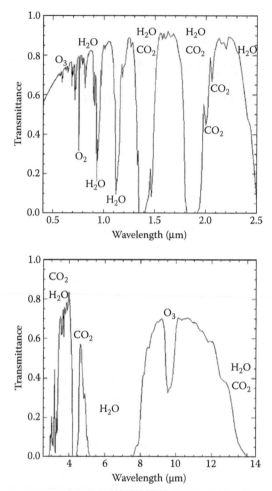

FIGURE 7.1
Atmospheric transmittance in the 0.4–2.5 μm and 3–14 μm regions.

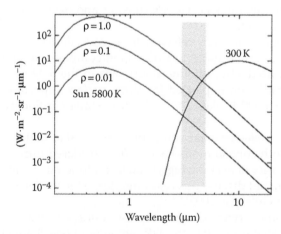

FIGURE 7.2
Solar or reflective region, mid-infrared (3–5 μm) region, and thermal infrared (8–14 μm) region.

TABLE 7.1

Prototypes of Multispectral and Hyperspectral Sensors

Platform	Multispectral		Hyperspectral	
	VNIR/SWIR	**VNIR/TIR**	**VNIR/SWIR**	**VNIR/TIR**
Satellite	SPOT, Ikonos	Landsat TM, ETM+, ASTER	Hyperion, CHRIS	–
Aircraft	ADS80	Daedalus ATM	AVIRIS, APEX	MASTER

SPOT: System Pour l'Observation de la Terre, http://en.wikipedia.org/wiki/SPOT_(satellite)
Ikonos: http://en.wikipedia.org/wiki/ikonos
ADS80: http://www.leica-geosystems.com
Landsat: http://landsat.gsfc.nasa.gov
ASTER: http://asterweb.jpl.nasa.gov
Daedalus ATM: http://www.nasa.gov/centers/dryden/research/Airsci/ER-2/tms.html
Hyperion: http://eo1.usgs.gov
CHRIS: http://earth.esa.int/proba
AVIRIS: http://aviris.jpl.nasa.gov
APEX: http://www.apex-esa.org/modules/APEX
MASTER: MODIS ASTER Simulator, http://masterweb.jpl.nasa.gov
Web sites accessed 16/09/2010.

hyperspectral instruments in orbit are restricted to the VNIR/SWIR region as they lack thermal channels. Airborne instruments covering the solar region and possessing at least a few thermal channels are still rare, and this is even more true for hyperspectral systems.

AC methods can be grouped into empirical approaches and physical models describing radiative transfer (RT) in the Earth's atmosphere. Here, we will only discuss RT-based approaches. As AC algorithms necessarily depend on the spectral regions covered by an instrument and also on the available number of channels, we will present the retrieval algorithms beginning with hyperspectral systems and terminating with a few channel multispectral instruments.

7.1 Atmospheric Correction for Hyperspectral Instruments (Solar Region)

Hyperspectral instruments are characterized by a large number (100 or more) of contiguous channels with a narrow spectral bandwidth (typically, 3–20 nm). A review article on this subject was published recently (Gao et al. 2009). The RT equation for a homogeneous surface under clear sky conditions can be formulated as follows:

$$L = L_\mathrm{p} + \frac{\tau E_\mathrm{g} \rho / \pi}{1 - \rho s} \qquad (7.1)$$

where L, L_p, τ, E_g, ρ, and s are at-sensor radiance, path radiance, ground-to-sensor transmittance, total solar flux on the ground, surface reflectance, and spherical albedo of the atmosphere, respectively. For brevity, the dependence on wavelength, solar and viewing geometry, and atmospheric parameters has been omitted. The at-sensor radiance is

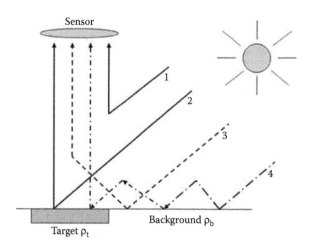

FIGURE 7.3
Schematic sketch of the adjacency effect. Radiation component 1 represents path radiance, component 2 represents direct plus diffuse flux on the target, component 3 represents volume scattering of adjacency, and component 4 represents atmospheric backscattering of adjacency effect.

measured by the sensor, whereas the relevant atmospheric parameters (aerosol type, aerosol optical thickness, water vapor column) can be retrieved from the imagery, enabling the calculation of L_p, τ, E_g, and s. With this knowledge, Equation 7.1 can be solved for ρ. In practice, scenes always consist of heterogeneous fields of different reflectance. Then, atmospheric cross talk occurs among such fields (Dave 1980; Richter et al. 2006), and radiation from the neighboring terrain spills over from the background to the considered target (see the schematic sketch in Figure 7.3).

In this case, Equation 7.1 has to be solved iteratively. Frequently used RT codes are DISORT (Stamnes et al. 1988), 6S (Vermote et al. 1997), MODTRAN (Berk et al. 1998), and libRadtran (Mayer and Kyling 2005). These codes calculate the at-sensor radiance for specified sun and observer geometries and atmospheric parameters. AC uses inverse modelling to retrieve the surface reflectance. Whereas 6S is restricted to the solar spectral region, the other three codes also cover the thermal domain.

As the recorded image data is digitized and rescaled to fit into an 8 or 16 bit/pixel encoding, a linear equation with an offset (c_0) and gain (c_1) has to be applied per channel to convert the scaled digital number (DN) to the corresponding at-sensor radiance, as follows:

$$L = c_0 + c_1 \mathrm{DN} \tag{7.2}$$

The metafile of an image usually contains these radiometric calibration coefficients. Care has to be taken as different units are frequently used, for example, $\mathrm{W} \cdot \mathrm{m}^{-2} \cdot \mathrm{sr}^{-1} \cdot \mu\mathrm{m}^{-1}$, $\mathrm{mW} \cdot \mathrm{cm}^{-2} \cdot \mathrm{sr}^{-1} \cdot \mu\mathrm{m}^{-1}$, and others.

For aerosol retrieval over land, different approaches exist depending on the spectral coverage of the instrument; whereas Kaufman et al. (1997) require SWIR bands, Guanter, Alonso, and Moreno (2005) suggest VNIR bands. The former method first masks dark land pixels in a 2.1- or 1.6-μm channel and then uses the following empirical correlation of SWIR surface reflectance values with reflectances in the blue/red region:

$$\rho(\text{red}) = 0.5\rho(2.1 \ \mu\text{m}) \quad \text{and} \quad \rho(\text{blue}) = 0.5\rho(\text{red}) \tag{7.3}$$

Several water vapor retrieval techniques have been published, using atmospheric window channels as a reference and channels in the absorption regions to measure the water vapor column (Gao and Goetz 1990; Carrere and Conel 1993; Schläpfer et al. 1998).

Another important issue is the removal of the effect of thin cirrus clouds. These can usually not be detected with channels in the VIS spectrum, but a narrow channel in the 1.38-μm region (for example, those available from Moderate Resolution Imaging Spectroradiometer [MODIS] and Airborne Visible/Infrared Imaging Spectrometer [AVIRIS] instruments) can be employed for cirrus detection and removal (Gao et al. 2002).

A typical artifact of pushbroom spectrometers is the spectral smile, an optical aberration that causes the spectrometer entrance slit, which represents the across-track swath, to be projected as a curve on the rectilinear detector array (Mouroulis, Green, and Chrien 2000). Therefore, for accurate water vapor and surface reflectance retrieval, one has to perform AC on a per-column basis as the channel center wavelength varies with the across-track position.

7.2 Atmospheric Correction for the Thermal Region

In the thermal spectrum, atmospheric transmittance is mainly influenced by water vapor, ozone (around 9.6 μm), and carbon dioxide (14 μm; see Figure 7.1). The aerosol influence still exists, but it is strongly reduced compared to the solar domain because of the much larger wavelength. Therefore, an accurate estimate of the water vapor column is required for retrieving surface properties, that is, spectral emissivity and surface temperature. Figure 7.4 shows a sketch of the radiation components in the thermal region: path radiance $(L_1 = L_p)$, emitted surface radiance (L_2), and reflected radiance (L_3). Thermal path radiance occurs due to emitted radiation from the atmosphere. The atmosphere also generates a hemispherical downwelling thermal flux F on the ground. As the surface emissivity ε is smaller than 1, the radiation reflected from the ground is $(1 - \varepsilon)F/\pi$, assuming an opaque surface, that is, $\rho = 1 - \varepsilon$.

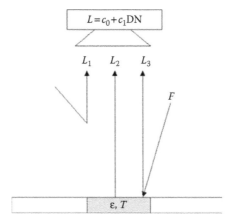

FIGURE 7.4
Radiation components in the thermal region.

Therefore, the at-sensor radiance can be written as

$$L = L_p + \tau \varepsilon B(T) + \tau(1-\varepsilon)F/\pi \qquad (7.4)$$

where B is Planck's blackbody function. The problem with this equation is that it contains two unknowns: the emissivity ε and temperature T. So for n thermal channels, we always have $n + 1$ unknowns, namely n emissivities and one temperature, which is an underdetermined set of equations. For most natural surfaces, the emissivity in the 10–12-μm region ranges between 0.95 and 0.99. If only a single thermal band is available (Landsat Thematic Mapper [TM], Enhanced TM plus [ETM+]), the emissivity is usually fixed at a constant value, say $\varepsilon = 0.97$, and Equation 7.4 is solved for the surface leaving radiance $B(T)$. The temperature T is then calculated by inverting Planck's function with an exponential fit function for a certain temperature range (Richter and Coll 2002).

If several thermal channels are available, iterative temperature/emissivity separation (TES) methods can be applied (Gillespie et al. 1998; Dash et al. 2002; Young, Johnson, and Hackwell 2002). As an example, in the normalized emissivity method (NEM), the surface temperature is calculated for all channels with a constant user-defined emissivity, and for each pixel the channel with the highest temperature is finally selected, because it is closest to the kinetic surface temperature. If the assumed start emissivity is correct, the true kinetic temperature will be obtained; otherwise, the result will have a small absolute temperature error. Afterward, the emissivities are calculated for each channel. In the adjusted NEM (Coll et al. 2001), the start emissivity is not constant but depends on the surface cover (vegetation, soil, sand, or water), which is determined by the reflective bands. Therefore, a closer match with the actual pixel-dependent emissivity can be expected and, as a consequence, a higher temperature accuracy.

Surface and air temperature are among the key parameters of weather and climate. Together with the factor of water, they determine plant growth, crop yield, carbon uptake by vegetation, evapotranspiration, and energy balance and influence the hydrological cycle (Carlson et al. 1981; Friedl 2002). Therefore, even a single thermal channel added to an instrument with reflective bands will distinctly broaden the range of applications. Additionally, multispectral or hyperspectral thermal bands allow an evaluation of the emissivity spectrum, which contains material-specific diagnostic features (Vincent 1975; Salisbury and D'Aria 1992).

7.3 Atmospheric Correction for Multispectral Instruments (Solar Region)

Typical multispectral instruments have a small number (smaller than 10) of broad bands in the solar or reflective region. Aerosol retrieval using Equation 7.3 and the radiative transfer in Equation 7.1 requires at least a channel in the red spectrum (around 650 nm) and a SWIR1 (1.6 μm) or SWIR2 (2.2 μm). Since the channels are placed in atmospheric window regions, they are only marginally influenced by the atmospheric water vapor column, and as water vapor cannot be retrieved from those channels, a climatologic average value or data from nearby weather stations or from other satellites has to be used. For sensors with only three or four VNIR bands, an empirical aerosol retrieval algorithm has been published that can be used for scenes containing dark vegetation areas (Richter, Schläpfer, and Müller 2006).

The proposed European Space Agency (ESA) Sentinel-2 instrument (ESA 2007) has 13 bands of different spatial resolutions (10, 20, and 60 m) with AC channels at 60 m (443, 940, and 1375 nm). The swath width is 290 km. Sentinel-2 will provide enhanced-quality continuity with existing missions of SPOT and Landsat. The launch is planned for 2013. The blue, red, and SWIR2 bands allow aerosol retrieval and the NIR bands (865, 940 nm) allow water vapor retrieval. In addition, thin cirrus can be detected with the 1375-nm band. As this sensor exceeds the typical number of bands of multispectral instruments and can perform image-based aerosol and water vapor retrieval, it is sometimes called "superspectral."

7.4 Combined Atmospheric and Topographic Correction

A large part of the land surface of the Earth is occupied by mountains. In these areas, there is a strong influence of the topography on the signal recorded by optical remote sensing instruments, that is, for the same surface cover, slopes oriented away from and toward the sun will appear darker and brighter, respectively, if compared to a horizontal geometry. This behavior causes problems for subsequent classification and thematic evaluation. Therefore, a combined atmospheric and topographic correction has to be performed in rugged terrain. A number of topographic correction techniques have been developed to eliminate or at least reduce the topographic influence (Teillet, Guindon, and Goodenough 1982; Riano et al. 2003; Richter, Kellenberger, and Kaufmann 2009).

All proposed methods rely on a digital elevation model (DEM) of the scene to describe the topography. If θ_n, θ_s, ϕ_s, and ϕ_n denote solar zenith angle, terrain slope, solar azimuth, and topographic azimuth, respectively, the local solar illumination angle β can be obtained from the topographic slope, aspect angles, and the solar geometry

$$\cos\beta(x,y) = \cos\theta_s \cos\theta_n(x,y) + \sin\theta_s \sin\theta_n \cos\{\phi_s - \phi_n(x,y)\} \tag{7.5}$$

where x, y indicate the pixel coordinates in an image that depends on the terrain slope θ_n and aspect ϕ_n. If ρ_T and ρ_H denote the reflectance of an inclined (terrain) and a horizontal surface, respectively, then the Lambertian method of topographic normalization is defined as

$$\rho_H = \rho_T \frac{\cos\theta_s}{\cos\beta} \tag{7.6}$$

For a low illumination, that is, small values of $\cos\beta$, the corrected reflectance is too large and the corresponding parts of an image are overcorrected. In the case of topographic shadow, $\cos\beta$ tends to 0 and ρ_H to infinity. All methods can be applied to surface reflectance (after AC) or the apparent or top-of-atmosphere (TOA) reflectance, that is, $\rho = \pi L/E\cos\theta_s$ (L = at-sensor radiance, E = extraterrestrial solar irradiance).

The Minnaert method uses an exponent K for the term ($\cos\theta_s/\cos\beta$) where K usually ranges between 0 and 1, which is derived from the image data on a per-channel basis. The third technique (C normalization) also belongs to a class of non-Lambertian methods. It uses an additive term c in the numerator and denominator of the cosine functions of

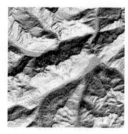

FIGURE 7.5
(See color insert following page 426.) Example of a combined atmospheric and topographic correction of a SPOT-5 scene from a part of the Swiss Alps. Left to right: Original SPOT-5 scene (color coding for red, green, and blue bands is 1650, 840, and 660 nm, respectively), illumination map, and combined atmospheric and topographic correction. (From Richter, R. et al., *Rem Sens*, 1, 2009. With permission.)

Equation 7.6 to avoid very high values in the topographic correction. This term c accounts for the diffuse radiance component, and it is calculated on a per-channel basis, evaluating a statistical regression of image data (see the study by Riano et al. [2003] for details). The fourth approach (Shepherd and Dymond 2003) calculates horizontal reflectance as a function of solar and terrain angles (θ_s, β) and also depends on the sensor view angles on a flat terrain (θ_v) and an inclined terrain (β_v). This algorithm includes an additive geometric term in the denominator of Equation 7.6 to avoid an overcorrection in faintly illuminated areas.

A fifth method (Richter, Kellenberger, and Kaufmann 2009) is a modified Minnaert approach that differs from the standard Minnaert method by employing a set of empirical rules for determining the threshold solar illumination angle, and a criterion to prevent overcorrection. The problem is that no method achieves the best ranking in all situations.

Figure 7.5 presents an example of a combined atmospheric and topographic correction for a SPOT-5 scene from a part of the Swiss Alps (dated September 21, 2005). Elevations range between 1200 and 3000 m, and the scene contains steep slopes of up to 58°. The corrected scene was processed according to the fifth method, and most topographic features are compensated well in the result.

7.5 Nonstandard Atmospheric Conditions (Haze, Cirrus, Cloud Shadow)

Conditions such as haze, cirrus, and cloud shadow comprise situations with boundary layer haze of varying optical thickness, with cirrus and scattered clouds, and with cloud shadow. These cases pose special scene-dependent problems that are difficult or impossible to solve with RT codes. Usually, simplifications have to be made, for example, single scattering has to be assumed because the cloud geometry is unknown or too complex. Nevertheless, significant progress has been achieved during the last decade. Figures 7.6 and 7.7 present two examples of haze removal and deshadowing of satellite imagery.

In Figure 7.6, haze was removed for an ALOS AVNIR-2 scene (http://www.alos-restec.jp, accessed 16/09/2010) with the assumption that it is an additive signal component to the ground-reflected radiance and normal path radiance in the haze-free regions. Multiple scattering effects were neglected. Haze is subtracted with the so-called haze optimized

FIGURE 7.6
Haze removal of ALOS AVNIR-2 imagery from northern Germany (dated April 16, 2007; band 1 at 463 nm). Left: Original subscene. Right: surface reflectance after haze removal.

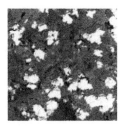

FIGURE 7.7
(See color insert following page 426.) Deshadowing of SPOT-5 imagery (dated May 22, 2005; color coding of red, green, and blue bands is 830, 660, and 555 nm, respectively). Left and right: Original and deshadowed scene, respectively.

transform (HOT; Zhang, Guindon, and Cihlar 2002), followed by an AC to obtain surface reflectance.

Figure 7.7 (left) shows a SPOT-5 subscene from Romania that contains a large percentage of scattered clouds. The right part of the figure is the result of applying a deshadowing algorithm on the original subscene (Richter and Muller 2005). Obviously, a lot of features that appear hidden in the original scene can be recognized in the deshadowed result.

7.6 Atmospheric Correction Codes for Land

A brief survey on commercially available AC codes for land imagery is included in Gao et al. (2009). Among the most popular algorithms are ACORN (ImSpec LLC, http://www. imspec.com, Palmdale, CA), FLAASH (developed by Spectral Sciences, Inc., MA; distributed by ITT Visual Information Solutions, CO, http://www.ittvis.com), ISDAS (Canada Centre for Remote Sensing, Quebec, Canada, http://www.ccrs.nrcan.gc.ca), and ATCOR (German Aerospace Center [DLR], Cologne, Germany; distributed by ReSe company, Langeggweg, Switzerland, http://www.rese.ch). Table 7.2 summarizes the main features of these codes. In the table, a plus sign indicates that the corresponding feature is supported, whereas a minus sign marks that the capability is missing. Most features are supported by all codes; however, processing of thermal band imagery can be done only with ATCOR. In addition, topographic correction is supported only by ISDAS and ATCOR.

TABLE 7.2

Comparison of Popular AC Codes

Feature	ACORN	FLAASH	ISDAS	ATCOR
Multispectral instruments	+	+	+	+
Hyperspectral instruments	+	+	+	+
Adjacency correction	−	+	+	+
Water vapor retrieval	+	+	+	+
Haze removal	−	−	−	+
Spectral polishing	−	+	−	+
Spectral smile correction	+	−	−	+
Thermal region: Surface temperature, emissivity	−	−	−	+
Rugged terrain: DEM topographic correction	−	−	+	+

Note: A plus sign indicates that the corresponding feature is supported, whereas a minus sign indicates the capability is missing.

7.7 Open Challenges

Atmospheric and topographic correction algorithms will continue to be improved in the future. Enhanced processing of hyperspectral imagery will benefit from an increase in the accuracy of RT models, particularly concerning scattering in the blue spectral region and updates of molecular absorption parameters. In addition, the sensor signal-to-noise ratio, radiometric calibration accuracy, and stability are likely to be improved. An open concern is the question of the most accurate solar irradiance database. The Committee on Earth Observation Satellites (CEOS; http://www.ceos.org, accessed 15/09/2010) recommends the Thuillier database, whereas others approve of the new Kurucz (1997) database, which is the default used in MODTRAN4. Although the solar constant, that is, irradiance integrated over the whole spectral range, is known with an accuracy of about 1%, much larger discrepancies exist for the spectral irradiance, depending on the spectral resolution. Figure 7.8 presents relative differences between the Thuillier (Thuillier et al. 2003) and the new Kurucz spectra for bandwidths of 3 and 10 nm. There are large differences between these sources, especially in the blue part of the spectrum. These discrepancies can probably be resolved within a few years when updated and more accurate measurements become available. Another problem is that the Thuillier database ends at 2.4 μm, whereas a number of hyperspectral instruments have channels up to 2.5 μm.

However, a number of challenges will probably persist for many years, especially for fully automated processing environments. Examples include the difficult cases of nonstandard atmospheric conditions, that is, removal of boundary layer haze of varying thickness and deshadowing of cloud shadow regions, especially under geometrically complex situations with scattered clouds at different altitude layers or a combination of haze, cloud, and shadow regions. Additionally, topographic correction techniques need to be improved, as there is no acknowledged method that works best in all mountainous regions of the Earth under all surface-cover conditions and seasons. This means that AC will remain an exciting research topic for a long time.

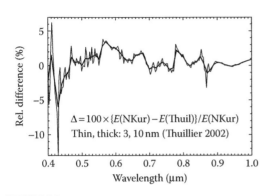

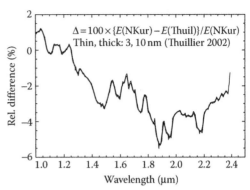

FIGURE 7.8
Comparison of the relative differences between new Kurucz and Thuillier irradiance.

References

Berk, A., L. S. Bernstein, G. P. Anderson, P. K. Acharya, D. C. Robertson, J. H. Chetwynd, and S. M. Adler-Golden. 1998. MODTRAN cloud and multiple scattering upgrades with application to AVIRIS. *Rem Sens Environ* 65:367–75.

Carlson, T., J. Dodd, S. Benjamin, and J. Coope. 1981. Satellite estimation of the surface energy balance, moisture availability, and thermal inertia. *J Appl Meteorol* 20:67–87.

Carrere, V., and J. E. Conel. 1993. Recovery of atmospheric water vapor total column abundance from imaging spectrometer data around 940 nm—sensitivity analysis and applications to airborne visible/infrared imaging spectrometer (AVIRIS) data. *Rem Sens Environ* 44:179–204.

Coll, C., V. Caselles, E. Rubio, F. Sospreda, and E. Valor. 2001. Temperature and emissivity separation from calibrated data of the digital airborne imaging spectrometer. *Rem Sens Environ* 76:250–9.

Dash, P., F.-M. Götsche, F.-S. Oleson, and H. Fischer. 2002. Land surface temperature and emissivity estimation from passive sensor data: Theory and practice—current trends. *Int J Rem Sens* 23:2563–94.

Dave, J. V. 1980. Effect of atmospheric conditions on remote sensing of a surface nonhomogeneity. *Photogramm Eng Rem Sensing* 46:1173–80.

ESA. 2007. GMES Sentinel-2 mission requirements document. http://esamultimedia.esa.int/docs/GMESSentinel-2_MRD.pdf, accessed 15/09/2010.

Friedl, M. A. 2002. Forward and inverse modelling of land surface energy balance using surface temperature measurements. *Rem Sens Environ* 79:344–54.

Gao, B.-C., and A. F. H. Goetz. 1990. Column atmospheric water vapour and vegetation liquid water retrieval from airborne imaging spectrometer data. *J Geophys Res* 95(D4):3549–64.

Gao, B.-C., M. J. Montes, C. O. Davis, and A. F. H. Goetz. 2009. Atmospheric correction algorithms for hyperspectral remote sensing data of land and ocean. *Rem Sens Environ* 113:S17–24.

Gao, B.-C., P. Yang, W. Han, R.-R. Li, and W. J. Wiscombe. 2002. An algorithm using visible and 1.38 µm channels to retrieve cirrus cloud reflectances from aircraft and satellite data. *IEEE Trans Geosci Rem Sens* 40:1659–68.

Gillespie, A., S. Rokugawa, T. Matsunaga, J. S. Cothern, S. Hook, and A. B. Kahle. 1998. A temperature and emissivity separation algorithm for advanced spaceborne thermal emission and reflection radiometer (ASTER). *IEEE Trans Geosci Rem Sens* 36:1113–26.

Guanter, L., L. Alonso, and J. Moreno. 2005. A method for the surface reflectance retrieval from PROBA/CHRIS data over land: Application to ESA SPARC campaigns. *IEEE Trans Geosci Rem Sens* 43:2908–17.

Hook, S. J., J. J. Myers, K. J. Thome, M. Fitzgerald, and A. B. Kahle. 2001. The MODIS/ASTER airborne simulator (MASTER)—a new instrument for earth science studies. *Rem Sens Environ* 76:93–102.

Kaufman, Y. J., A. Wald, L. A. Remer, B.-C. Gao, R. R. Li, and L. Flynn. 1997. The MODIS 2.1 mum channel—correlation with visible reflectance for use in remote sensing of aerosol. *IEEE Trans Geosci Rem Sens* 35:1286–98.

Kurucz, R. L. 1995. The solar spectrum: Atlases and line identifications. *Astron Soc Pac Conf Ser* 81:17–31.

Mayer, B., and A. Kyling. 2005. Technical note: The libRadtran software package for radiative transfer calculations – description and examples of use. *Atmos Chem Phys* 5:1855–77.

Mouroulis, P., R. O. Green, and T. G. Chrien. 2000. Design of pushbroom imaging spectrometers for optimum recovery of spectroscopic and spatial information. *Appl Opt* 39:2210–20.

Mushkin, A., L. K. Balick, and A. R. Gillespie. 2005. Extending surface temperature and emissivity retrieval to the mid-infrared (3–5 µm) using the multispectral thermal imager (MTI). *Rem Sens Environ* 98:141–51.

Riano, D., E. Chuvieco, J. Salas, I. Aguado. 2003. Assessment of different topographic corrections in Landsat TM data for mapping vegetation types. *IEEE Trans Geosci Rem Sens* 41:1056–61.

Richter, R., M. Bachmann, W. Dorigo, and A. Müller. 2006. Influence of the adjacency effect on ground reflectance measurements. *IEEE Geosci Rem Sens Lett* 3:565–9.

Richter, R., and C. Coll. 2002. Bandpass-resampling effects for the retrieval of surface emissivity. *Appl Opt* 41:3523–29.

Richter, R., T. Kellenberger, and H. Kaufmann. 2009. Comparison of topographic correction methods. *Rem Sens* 1:184–96.

Richter, R., and A. Müller. 2005. De-shadowing of satellite/airborne imagery. *Int J Rem Sens* 26:3137–48.

Richter, R., D. Schläpfer, and A. Müller. 2006. An automatic atmospheric correction algorithm for visible/NIR imagery. *Int J Rem Sens* 27:2077–85.

Salisbury, J. W., and D. M. D'Aria. 1992. Emissivity of terrestrial materials in the 8–14 µm atmospheric window. *Rem Sens Environ* 42:83–106.

Schläpfer, D., C. C. Borel, J. Keller, and K. I. Itten. 1998. Atmospheric precorrected differential absorption technique to retrieve columnar water vapor. *Rem Sens Environ* 65:353–66.

Shepherd, J. D., and J. R. Dymond. 2003. Correcting satellite imagery for the variance of reflectance and illumination with topography. *Int J Rem Sens* 24:3503–14.

Stamnes, K., S. C. Tsay, W. J. Wiscombe, and K. Jayaweera. 1988. Numerically stable algorithm for discrete-ordinate-method radiative transfer in multiple scattering and emitting media. *Appl Opt* 27:2502–9.

Teillet, P. M., B. Guindon, and D. G. Goodenough. 1982. On the slope-aspect correction of multispectral scanner data. *Int J Rem Sens* 8:84–106.

Thuillier, G., M. Herse, D. Labs et al. 2003. The solar spectral irradiance from 200 to 2400 nm as measured by the SOLSPEC spectrometer from the ATLAS and EURECA missions. *Solar Phy* 214:1–22.

Vermote, E. F., D. Tanre, J. L. Deuze, M. Herman, and J. J. Morcrette. 1997. Second simulation of the satellite signal in the solar spectrum: 6S: An overview. *IEEE Trans Geosci Rem Sens* 35:675–86.

Vincent, R. K. 1975. The potential role of thermal infrared multispectral scanners in geological remote sensing. *Proc IEEE* 63:137–47.

Young, S. J., B. R. Johnson, and J. A. Hackwell. 2002. An in-scene method for atmospheric compensation of thermal hyperspectral data. *J Geophys Res* 107(D24):4774–93.

Zhang, Y., B. Guindon, and J. Cihlar. 2002. An image transform to characterize and compensate for spatial variations in thin cloud contamination of Landsat images. *Rem Sens Environ* 82:173–87.

8

Three-Dimensional Geometric Correction of Earth Observation Satellite Data

Thierry Toutin

CONTENTS

8.1 Introduction

Why orthorectify Earth observation (EO) satellite data? Any EO data, regardless of whether they are acquired by a scanner or a frame camera aboard a satellite, or by a photographic system in an aircraft or any other platform/sensor combination, will have various geometric distortions, depending on the manner in which the data are acquired. This problem is inherent in remote sensing, as we attempt to accurately represent the three-dimensional (3D) surface of Earth as a two-dimensional (2D) image. Consequently, raw images contain such significant geometric distortions that they cannot be used directly with geographic information system (GIS)–ready products. Thus, multisource data integration (raster and vector) for geomatics applications requires geometric and radiometric processing adapted to the nature and characteristics of the data in order to keep the best information from each image in the composite orthorectified products.

The processing of multisource data can be based on the concept of "terrain-geocoded images," a term originally invented in Canada for defining value-added products (Guertin and Shaw 1981). Photogrammetrists, however, prefer the term "orthoimage" when referring to the unit of terrain-geocoded data, where all distortions including those of the relief are corrected. To integrate different data under this concept, each raw image must be separately converted to an orthoimage so that each component orthoimage of the data set can be registered, compared, combined, and so on, not only pixel-by-pixel but also with cartographic vector data in a GIS.

Why does the geometric correction process seem more important today than in the past? In 1972, the impact of geometric distortions was quite negligible for different reasons:

- The images, such as those from a Landsat multispectral scanner (Landsat-MSS), were nadir viewing, and the resolution was coarse (around 80–100 m).
- The products resulting from the image processing were analog on paper.
- The interpretation of the final products was performed visually.
- The fusion and integration of multisource and multiformat data did not exist at that time.

Today, the impacts of geometric distortions, although they are similar to the ones in the past, are less negligible because of the following factors:

- The images are off-nadir viewing, and the resolution is fine (submeter level).
- The products resulting from image processing are fully digital products.
- The interpretation of the final products is realized on the computer.
- The fusion of multisource images (different platforms and sensors) is in general use.
- The integration of multiformat data (raster/vector) is a general tendency in geomatics.

One must admit that the new EO data, their method and processing, the resulting processed data, and their analysis and interpretation introduced new needs and requirements for geometric corrections, due to a drastic evolution accompanied by large scientific and technology improvements between the two periods. Even if the literature is quite abundant mainly in terms of books and peer-reviewed articles (an exhaustive list is given in the references section), it is important to update the problems and the solutions recently adopted for geometrically correcting remote sensing images with the latest developments and research studies from around the world. This chapter will then address the following concepts:

- The sources of geometric distortions and deformations with different categorizations (Section 8.2)
- The modeling of these distortions with different 2D/3D physical/empirical models and mathematical functions (Section 8.3)
- The 3D geometric correction method and algorithms with their processing steps and errors (Section 8.4)

Comparisons between the models and mathematical functions, their applicability, and their performance on different types of images (frame camera, visible and infrared [VIR] oscillating or pushbroom scanners, and side-looking antenna radar [SLAR] or synthetic aperture radar [SAR] sensors at high, medium, or low resolutions) are also addressed. The errors with their propagation from the input data to the final results are also evaluated through the full processing steps.

8.2 Sources of Geometric Distortions

Each EO image acquisition system produces unique geometric distortions in its raw images and consequently the geometry of these images in its own local coordinate system does not correspond to the terrain and to the user's specific map projection. Obviously, the geometric distortions vary considerably with different factors, such as the platform (airborne and satellite), the sensor (VIR and SAR; total field of view [FOV], low to high resolution), and the associated scanner (whiskbroom, pushbroom, frame, etc.). However, it is possible to make general categorizations of these distortions.

The sources of distortions (Table 8.1) can be grouped into two broad categories: (1) the "observer" or the acquisition system (platform, imaging sensor, and other measuring instruments, such as gyroscope and stellar sensors) and (2) the "observed" (atmosphere and Earth). In addition to these distortions, deformations related to map projections have to be taken into account because the terrain and most GIS end-user applications are generally represented and performed respectively in a topographic map space and not in a referenced ellipsoid. Figures 8.1 and 8.2 illustrate the geometry of acquisition and the quasi-polar elliptical orbit approximation of remote sensing satellites around the Earth, respectively. The map deformations are logically included in the distortions of the observed.

Previous studies made a second-level categorization into low-, medium-, and high-frequency distortions (Friedmann et al. 1983), where frequency is determined or compared

TABLE 8.1

Description of Error Sources for the Two Categories, the Observer and the Observed, with Different Subcategories

Category	Subcategory	Description of Error Sources
The observer or the acquisition system	Platform (spaceborne or airborne)	Variation of the movement; variation in platform attitude (low to high frequencies)
	Sensor (VIR, SAR, or HR images)	Variation in sensor mechanics (scan rate, scanning velocity, etc.); lens distortions, viewing/look angles; panoramic effect with the FOV
	Measuring instruments	Time variations or drift; clock synchronicity
The observed	Atmosphere	Refraction and turbulence
	Earth	Curvature, rotation, topographic effect
	Map	Geoid to ellipsoid, ellipsoid to map

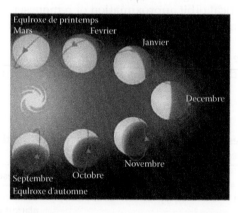

FIGURE 8.1
(See color insert following page 426.) Geometry of viewing of a satellite scanner in orbit around the Earth. (Courtesy and copyright Serge Riazanoff, VisioTerra, 2009.)

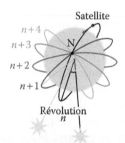

FIGURE 8.2
(See color insert following page 426.) Near-Earth, quasi-circular, quasi-polar, sun-synchronous orbit for EO satellites. The different revolutions around the poles with a constant illumination angle (top) showing the same illumination condition all the year (bottom). (Courtesy and copyright Serge Riazanoff, VisioTerra, 2009.)

with the image acquisition time. Examples of low-, medium-, and high-frequency distortions are those arising from orbit variations, Earth rotation, and local topographic effects, respectively. Although this categorization was suitable in the 1980s when there were very few remote sensing systems, today, with so many different acquisition systems, it is no longer acceptable because it differs with each acquisition system. For example, attitude variations are a high-frequency distortion for QuickBird or airborne pushbroom scanner, a medium-frequency distortion for System Pour l'Observation de la Terre (high resolution in the visible SPOT-HRV) and Landsat Enhanced Thematic Mapper (Landsat-ETM+), and a low-frequency distortion for Landsat-MSS, but not a distortion for a medium resolution imaging spectrometer (MERIS).

The geometric distortions and their error sources given in Table 8.1 are deterministic and predictable and generally well understood. Some of these distortions, especially those related to instrumentation, are systematic and generally corrected at ground receiving stations or by the image vendors. Other distortions are not taken into account and corrected because they are specific to each acquisition time and location; further, information on the atmosphere is rarely available. Such distortions are also geometrically negligible for low- to medium-resolution images.

8.2.1 Distortions Related to the Platform

Some basic information on satellite orbits and celestial mechanics are useful to better understand platform-related distortions. The EO satellites obey the celestial mechanical laws defined by Newton and Kepler for an unperturbed trajectory (Keplerian orbit) and by Gauss and Lagrange for a perturbed trajectory (osculatory orbit; Escobal 1965; Centre National d'Études Spatiales 1980). A number of perturbations (due to Earth gravity and surface irregularities, atmospheric drag, etc.) slowly change the Keplerian orbit based on the two-body attraction of Newton's law into an osculatory orbit (Centre National d'Études Spatiales 1980). Information on orbits is often needed, and different orbital models can be used depending on their utility and required accuracy (Bakker 2000):

- To calculate the satellite location on its osculatory orbit in order to compute the Earth coordinates of scanned pixels, requiring high accuracy (submeters) over a small time frame (seconds)

- To predict when the satellite will pass over a specific area, requiring low accuracy (kilometers) but over a long time frame (days)

Many orbital models have been developed since 1960 using the same mechanical laws with Gaussian/Lagrangian equations; the differences between the orbital models are mainly in the number and types of perturbations and the techniques to integrate them. As defined and adapted by the North American Aerospace Defense Command, simplified general perturbations (SGPs), SGP4, and the most accurate SGP8 are the orbital models to be used for low- and near-Earth satellites (orbital period less than 225 minutes and altitude less than 6000 km). Most, if not all, of the civilian EO spacecrafts have near-Earth, retrograde, quasi-circular, quasi-polar, geosynchronous, and sun-synchronous orbits (Figure 8.2).

Near-Earth orbits (altitude more than 300 km) are high enough to reduce the atmospheric drag. Retrograde orbits with 90°–180° inclination are westward-launched orbits, which require extra fuel to compensate for the Earth's rotation, but they provide the only solution for obtaining sun-synchronous orbits. Quasi-circular orbits avoiding large changes in altitude enable images with similar scales to be acquired, which is desirable for EO. Quasi-polar orbits with 90°–100° inclination enable sensors to image the entire Earth, including most of the poles. Because geosynchronous orbits have a repeating ground track, they have an orbital period that is an integer multiple of the

Earth's sidereal rotation period. This integer multiple is called the "repeat cycle." The special case of a geosynchronous orbit that is circular and directly above the equator is called a "geostationary orbit." The satellite track on the ground, also called the "path," is kept fixed within certain limits related to orbit maintenance accuracy. Paths are artificially divided into squared scenes at regular intervals, generating rows. An example of the path–row system is the World Reference System of Landsat satellites. The main advantage is that the satellite follows a fixed pattern on the Earth, which is desirable for operational EO systems. Sun-synchronous orbits enable satellites to pass overhead at the same local solar time and thus to acquire images with almost identical illumination or lighting conditions (Figure 8.2). Although the comparison of multidate images is easier with a sun-synchronous orbit, this approach has some disadvantages, especially as variations in illumination reveal different structural details. In addition, sun-synchronous orbits require retrograde orbits and strict relationships between orbital parameters (mainly inclination and height), which must be preserved during the satellite's lifetime.

The distortions associated with near-Earth satellites are mainly due to the interaction between the platform and Earth (Earth's gravity, shape, and movement, generating a quasi-elliptic movement; Escobal 1965; Centre National d'Études Spatiales 1980; Light et al. 1980). The satellite has six degrees of freedom in space, which can be determined by (Toutin 1983; Kim and Dowman 2006) knowing any one of the following sets of details:

- The satellite's position relative to Earth's center (three parameters: X, Y, Z) and orientation (a solid angle)
- Its position/velocity (left part of Figure 8.3) while the attitude is roll, pitch, yaw, or a specific kind of Euler angles (right part of Figure 8.3)

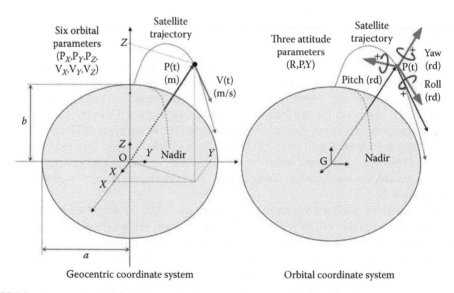

FIGURE 8.3
Orbital (left) and attitude (right) parameters defining the orbit and the satellite in its orbit as a function of time. (Courtesy and copyright Serge Riazanoff, VisioTerra, 2009.)

- Its osculatory orbit with six parameters relative to the instantaneous orbit (semi-major axis, eccentricity, inclination, ascending-node longitude, perigee argument, and mean anomaly; Figure 8.4)

$$\vec{Z}_{yaw} = \frac{\vec{P}(t)}{\left\|\vec{P}(t)\right\|}; \quad \vec{X}_{pitch} = \frac{\vec{V}(t) \wedge \vec{Z}_{yaw}}{\left\|\vec{V}(t) \wedge \vec{Z}_{yaw}\right\|}; \quad \vec{Y}_{roll} = \vec{Z}_{yaw} \wedge \vec{X}_{pitch}$$

Depending on the acquisition time (used as the seventh parameter) and the size of the image, all these parameters are time dependent due to orbital perturbations and thus generate a range of nonsystematic distortions. All these nonsystematic and time-dependent distortions are not predictable and must be evaluated for each image from satellite tracking data or ground-control information or both. Some effects of these distortions include the following:

- Platform altitude variation (z axis) is the most critical among position parameters. It can change the pixel spacing in the across-track direction, whereas X and Y parameters generate only a translation in the sensed ground area. However, the altitude variation of the satellite (around 8–10 m/s for near-Earth orbits at around 800-km altitude) is not too significant over the time required to acquire a full scene (5–10 seconds for high- to medium-resolution satellite sensors). It only represents 1/8000 relative pixel variation. For SPOT-5 panchromatic super mode (12,000 pixels with 2.5-m spacing), it generates 0.3-m variation in pixel spacing, which cumulatively generates an error around 2 m at the edge of the image.

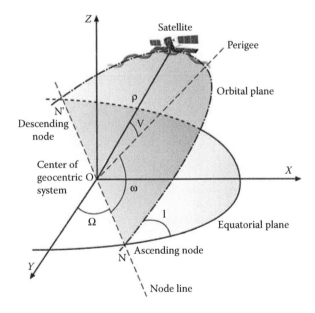

FIGURE 8.4
Description of a satellite osculatory orbit and its approximation by a Keplerian ellipse. X, Y, Z are the position coordinates in a geocentric frame reference system, I is the orbit inclination, Ω is the longitude of the ascending node (N), ω is the argument of the perigee (P), ($\omega + v$) is the argument of the satellite, and ρ is the distance between the Earth's center (O) and the satellite.

- Platform velocity variations can change the line spacing or create line gaps/over-laps. Variations in spacecraft velocity only cause distortions in the along-track direction.

- Platform altitude variation (Figure 8.5) can change the orientation and the shape of VIR images (but it does not affect SAR image geometry). The roll is the rotation around the flight vector (*x* axis; right part of Figure 8.3), hence in a "wing down" direction, its variation causes lateral shifts and scale changes in the across-track direction. The pitch is the vertical rotation of the platform, in the "nose up" plane, and its variation results in scan-line spacing changes. The yaw is the rotation around the vertical axis and its variation changes the orientation of each scanned line, resulting in a skew between the scan lines. Variations in platform altitude can be severe in terms of location on the ground (absolute offset of hundreds of meters) over the time required to scan a full scene (5–10 seconds), and are signifi-cant when the variation is sudden over a few lines (relative offset of tens of meters within milliseconds).

8.2.2 Distortions Related to Sensors

Sensor-related distortions include the following:

- Calibration parameters uncertainties arise from lens distortions (pertaining to focal length, principal point, detector position, and decentering, as well as radial/tangential distortions) and the instantaneous FOV (IFOV) for VIR sensors, or the range gate delay (timing) for SAR sensors. The uncertainties in focal length change

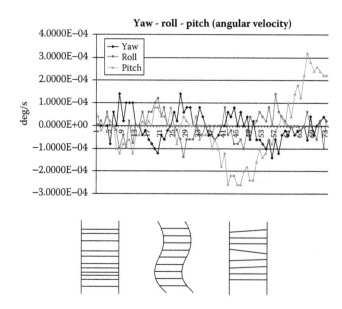

FIGURE 8.5
Example of System Pour l'Observation de la Terre attitude variations (top), and its impact on image geometry (bottom): pitch (left), roll (center), and yaw (right).

the pixel size and those in the principal point cause a systematic shift of the scan lines and of the full image. Lens distortion causes imaged positions to be distorted along radial/tangential lines from the principal point, which depends on the detector position in the focal plane. This last distortion is already corrected for preprocessed/georeferenced images (e.g., IKONOS), but it has to be corrected for high-resolution (HR) sensors (below 10-m resolution). Information is sometimes included in the metadata (e.g., SPOT-5).

- The panoramic distortion in combination with the curvature of Earth and topographic relief changes the ground pixel sampling along the column direction. Panoramic distortion occurs for large ±30°-FOV sensors (MERIS, Moderate Res-olution Imaging Spectroradiometer [MODIS], etc.) As the sensor scans across each line, the distance from the sensor to the ground increases further away from the center of the swath, and the sensor scans a larger area as it moves closer to the edges. The further away from nadir an object, the larger the compression. This effect results in the compression of image features at points away from the nadir, and this distortion is called "tangential scale" distortion. On the other hand, with large off-nadir viewing systems (such as an agile HR sensor with an FOV of few degrees), the area covered by the IFOV for a large off-nadir image is larger than that covered for a nadir image. However, pixel spacing in the column direction is almost constant (due to small FOV) for a given off-nadir viewing. This distortion is called an "off-nadir viewing" distortion.

8.2.3 Distortions Related to the Earth

Earth-related distortions include the following:

- The rotation of the Earth generates latitude-dependent displacements from east to west between scanned lines. The eastward rotation of the Earth, during a satellite orbit, causes the sweep of scanning systems to cover an area slightly to the west of each previous scan, which cumulates as the number of scan lines accumulates. The resultant imagery is thus skewed westward across the image. The amount of linear rotation varies with the scanning time: It is longer for Landsat than for SPOT. This is known as "skew" distortion.

- Earth's curvature creates variation of ground spacing in the along-track direction, which increases with the distance from the ground nadir point. Its effect is related to both the FOV and the off-nadir viewing angle, which increases with the distance from the nadir point.

- Earth's topographic relief generates a parallax in the viewing direction.

Because the extracted information from the EO satellite data are afterward integrated with multisource multiformat data, generally in GIS, they have to be finally projected into the user's map reference system. The deformations associated with map projection include the approximation of the geoid by a reference ellipsoid (Figure 8.6) and the projection of the reference ellipsoid on a plane surface. In general, a conformal projection, which preserves angles, is used in mapping, such as the universal transverse mercator projection with its latitudinal (B–Y) and 3°-longitudinal (1–60) zones (Figure 8.7).

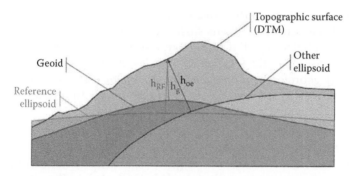

FIGURE 8.6
The approximation of the geoid by ellipsoids. (Courtesy and copyright Serge Riazanoff, VisioTerra, 2009.)

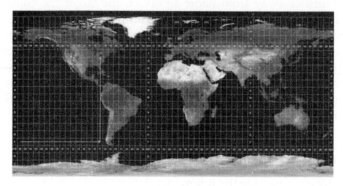

FIGURE 8.7
(See color insert following page 426.) Example of a cylindrical conformal projection: the universal transverse mercator projection with its 3° longitudinal (1–60) zones and latitudinal (A–Z) zones.

8.3 Geometric Modeling of Distortions

All geometric distortions other than the ones discussed in Sections 8.2.1 through 8.2.3 require models and mathematical functions to perform geometric corrections of imagery, either with 2D/3D empirical models (such as 2D/3D polynomial or 3D rational functions [RFs]) or with rigorous 2D/3D physical and deterministic models. With 2D/3D physical models, which reflect the physical reality of the viewing geometry (platform, sensor, Earth, and sometimes map projection), geometric correction can be performed either step by step with a mathematical function for each distortion/deformation or simultaneously with a "combined" mathematical function. The step-by-step solution is generally applied at the ground receiving station when the image distributors sell value-added products (georeferenced, map-oriented, or geocoded), whereas end users generally use and prefer the combined solution.

8.3.1 Two-Dimensional/Three-Dimensional Empirical Models

The 2D/3D empirical models can be used when the parameters of the acquisition systems or a rigorous 3D physical model are not available. Since they do not reflect the source of distortions described in Section 8.2, such models do not require a priori information on any component of the total system (platform, sensor, Earth, and map projection).

Such empirical models are based on different mathematical functions:

- Two-dimensional polynomial functions, such as

$$P_{2D}(XY) = \sum_{i=0}^{m} \sum_{j=0}^{n} a_{ij} X^i Y^j \tag{8.1}$$

- Three-dimensional polynomial functions, such as

$$P_{3D}(XYZ) = \sum_{i=0}^{m} \sum_{j=0}^{n} \sum^{p} a_{ijk} X^i Y^j Z^k \tag{8.2}$$

- Three-dimensional RFs, such as

$$R_{3D}(XYZ) = \frac{\displaystyle\sum_{i=0}^{m} \sum_{j=0}^{n} \sum_{k=0}^{p} a_{ijk} X^i Y^j Z^k}{\displaystyle\sum_{i=0}^{m} \sum_{j=0}^{n} \sum_{k=0}^{p} b_{ijk} X^i Y^j Z^k} \tag{8.3}$$

where X, Y, and Z are the terrain or cartographic coordinates, i, j, and k are integer increments, and m, n, and p are integer values, generally between 0 and 3, with $m = n$ (=p) being the order of the polynomial functions, generally 3.

Each 2D first-, second-, and third-order polynomial function will then have 3, 6, and 10 unknown terms. Each 3D first-, second-, and third-order polynomial function will then have 4, 10, and 20 unknown terms. The 2D/3D first-order polynomial functions are also called "affine transformations." Each 3D first-, second-, and third-order RF will have 8, 20, and 40 unknown terms, named the rational polynomial coefficients (RPCs). In fact, the 3D RFs are extensions of the collinearity equations (Section 8.3.2), which are equivalent to 3D first-order RFs. Depending on the imaging geometry in each axis (flight and scan), the order of polynomial functions (numerator and denominator for RFs) can be different and/ or specific terms, such as XZ for 2D or XZ, YZ^2, or Z^3 for 3D, can be differently dropped from the polynomial functions when these terms cannot be related to any physical element of the image acquisition geometry. Then these "intelligent" polynomial functions reflect better the geometry in both axes and reduce overparameterization and the correlation between terms. Okamoto (1981, 1988) has already applied this reduction of terms for one-dimensional central perspective photographs and line scanners, respectively.

8.3.1.1 Two-Dimensional Polynomial Functions

Because the 2D polynomial functions, with their formulation, are well known and documented since the 1970s (Wong 1975; Billingsley 1983), only a few characteristics are given here. Polynomial functions of the first order (6 terms) allow for only correcting a translation in both axes, a rotation, a scaling in both axes, and an obliquity. Polynomial functions of the second order (12 terms) allow for correction of, in addition to the previous parameters, torsion and convexity in both axes. Polynomial functions of the third order (20 terms) allow for correction of the same distortions as a second-order polynomial function along with other distortions, which do not necessarily correspond to any physical reality of the image acquisition system. In fact, previous research studies have demonstrated that third-order polynomial functions introduce errors in relative pixel positioning in orthoimages, such as Landsat-Thematic Mapper (TM) or SPOT-HRV (Caloz and Collet 2001), as well as in geocoding and integrating multisensor images, such as SPOT-HRV and airborne SAR (Toutin 1995a).

Since 2D polynomial functions do not reflect the sources of distortion during image formation and do not correct for terrain relief distortions, they are limited to images with few or small distortions, such as nadir viewing images, systematically corrected images, and/or small images over flat terrain (Bannari et al. 1995). Since these functions correct for local distortions at the ground control point (GCP) location, they are very sensitive to input errors and hence GCPs have to be numerous and regularly distributed (de Leeuw, Veugen, and van Stokkom 1988). Consequently, these functions should not be used when precise geometric positioning is required for multisource/multiformat data integration and in high relief areas.

The 2D polynomial functions, as the simplest solution, were mainly used until the 1980s on images, whose systematic distortions, excluding the relief, had already been corrected by the image providers. As reported by Wong (1975), 2D fourth-order polynomial functions are theoretically valid for low-resolution Earth Resources Technology Satellite (ERTS)-1 imagery to approximate a rigorous 2D physical model (Kratky 1971). Extensions to conformal/orthogonal polynomial functions (Wong 1975; de Leeuw, Veugen, and van Stokkom 1988) and surface spline functions (Goshtasby 1988) were also used for Landsat-MSS. As mentioned in Section 8.1, good geometric accuracy was not a key point in the analysis of analog images, for which 2D polynomial functions could be appropriated. More recently, simple affine and projective functions applied to IKONOS Geo images (Hanley and Fraser 2001) or hybrid and projective functions applied to IKONOS Geo and Indian Remote Sensing (IRS)-1C images (Valadan Zoej et al. 2002) achieved good results because the images were acquired with near-nadir viewing angles over a flat terrain. However, although it is now known that 2D polynomial functions are not suitable regardless of the image type and size as well as the terrain relief, some users still apply them, apparently without knowing their implications for subsequent processing operations and the resulting digital products.

8.3.1.2 Three-Dimensional Polynomial Functions

3D polynomial functions are an extension of 2D polynomial functions, and are obtained by adding Z terms related to the third dimension of the terrain. However, they are prone to the same problems as any empirical function, except for the relief, that is, they are applicable to small images, they need numerous regularly distributed GCPs, they correct locally at GCPs, they are very sensitive to errors, and they have a lack of robustness and consistencies in an operational environment. Their use should thus be limited to small images or to systematically corrected images, in which all distortions except the relief have been precorrected. For these reasons, second-order conformal polynomial functions have been primarily used in aerial photogrammetry during the 1960s (Baetslé 1966; Schut 1966). Due to the larger size of satellite images, they have been mainly used with georeferenced images: SPOT-HRV (level 1 and 2) using first-order functions (Baltsavias and Stallmann 1992; Okamoto et al. 1998), and SPOT-HRV (level 1B) and Landsat-TM (level bulk or georeferenced) using second-order functions (Palà and Pons 1995). More recently, first-order affine functions were applied to IKONOS Geo products (Ahn, Cho, and Jeon 2001; Fraser, Hanley, and Yamakawa 2002; Fraser, Baltsavias, and Gruen 2002; Jacobsen 2002; Vassilopoulou et al. 2002). The terms related to terrain elevation in the 3D polynomial function could be reduced to a_iZ for VIR images and to a_iZ and a_jZ^2 for SAR images, whatever the order of the polynomial functions used. The main reason is that there is no physical interrelation between the X and Z or the Y and Z directions for most of the sensors used.

In fact, Kratky (1971, 1989) has already used third- or fourth-order 3D polynomial functions with this term reduction to approximate the 2D or 3D physical models developed for ERTS or SPOT raw images, respectively. The main reason why he developed his SPOT 3D polynomial model was that the real-time computation for implementing his specific physical model solution was not feasible on a stereo workstation. He would certainly not have done this approximation with the higher-performance computers now available. More recently, tests were also performed using Kratky's polynomial functions with IKONOS Geo products acquired with near-nadir viewing over high relief areas (Kersten et al. 2000; Vassilopoulou et al. 2002). The second study (Vassilopoulou et al. 2002) evaluated orthoimage errors over the GCPs, and 1–2-m errors were achieved depending on their number, definition, and image measurement accuracy; however, this statistical evaluation is biased, as it used checked data applied in the geometric correction process. However, as previously mentioned in this section, the 3D affine transformation, also evaluated in the second study (Vassilopoulou et al. 2002), gave the same results as fourth-order polynomial functions, but with much less GCPs.

8.3.1.3 Three-Dimensional Rational Functions

Although they were occasionally used during the 1980s (Okamoto 1981, 1988), interest in 3D RFs was recently renewed (Li 1998) among the civilian photogrammetric and remote sensing communities with the launch of the first civilian HR IKONOS sensor in 1999 (Grodecki 2001; Fraser, Dial, and Grodecki 2006; Shaker 2007; Fraser and Ravanbakhsh 2009; Tong, Liu, and Weng 2009). Since sensor and orbit parameters are not included in the metadata, 3D rational function models (RFMs) could be an alternative to 3D physical models. The 3D RFMs can be used in two approaches (Madani 1999):

1. To approximate an already-solved existing 3D physical model (terrain independent)
2. To normally compute the RPCs of all the polynomial RFs with GCPs (terrain dependent)

8.3.1.3.1 Terrain-Independent Rational Function Model Approach

The first approach, inappropriately called terrain independent because the process still requests some GCPs to remove RFM bias (explained in this section), is performed in two steps. A 3D regular grid of the imaged terrain is first defined, and the image coordinates of the 3D grid ground points are computed using an already-solved existing 3D physical model. These grid points and their 3D ground and 2D image coordinates are then used as GCPs to resolve the 3D RFs and compute the RPCS of polynomial functions. There are some disadvantages to RFs (Madani 1999):

- The inability to model local distortions (such as high-frequency variations with VIR sensors or SAR sensors)
- A limitation in the image size
- The difficulty in the interpretation of RPCs due to the lack of physical meaning

- A potential failure to zero denominator
- A potential correlation between the RPCs of RFs

A different strategy can be adopted to reduce these limitations. To solve the inability to model local distortions, RFs should be applied to georeferenced data with systematic distortions corrected (such as IKONOS Geo images) rather than raw data with no geometric distortions corrected (such as QuickBird-2 or SPOT-5). To reduce the limitation in image size, the image itself could be subdivided into subimages with separate 3D RFs required for each subimage (Yang 2001). This results in much more geometric and also radiometric processing. To reduce the disadvantage of potential failure and correlation, the nonsignificant and/or high correlated RPCs can be eliminated to avoid zero crossing and instability of RFs depending on image geometry (Dowman and Dolloff 2000). To overcome some of these problems, a "universal real-time image geometry model" based on RFs has been developed (OGC 1999). It is a dynamic RF of variable orders, whose RPCs can be either chosen as a function of the sensor geometry or eliminated using an iterative procedure (Robertson 2003). This reduction of RPCs is then similar to the orientation theory developed in the 1980s (Okamoto 1981, 1988). In addition, when the denominator functions are omitted, RFs become simple 3D polynomial functions.

Dowman and Dolloff (2000) addressed the advantages of a universal real-time image geometry model, such as universality, confidentiality, efficiency, and ability for information transfer, as well as its disadvantages, such as loss of accuracy, numerical instability of the solution (due to overparameterization, correlation, interpolation errors), failure for highly distorted imagery, uncertainty (there is no relation to physical perturbations), and complexity (in defining the functions and number of GCPs). Some of these advantages or disadvantages are also related to 3D polynomial functions, as mentioned in Section 8.3.1.3.

Image vendors, government agencies that do not want to deliver satellite/sensor information with the image, and commercial photogrammetric workstation suppliers are the main users of this first approach. Image vendors thus provide with the image all the parameters of 3D RFs. Consequently, the users can theoretically process the images without GCP for generating orthoimages with digital elevation model (DEM). This approach is adopted by different image resellers around the world, which provide third-order RPCs with small-FOV HR images: GeoEye with IKONOS and GeoEye-1 images (Grodecki 2001; Grodecki and Dial 2003), DigitalGlobe and MacDonald, Dettwiler, and Associates (MDA) with QuickBird-2 and WorldView-2 images (Hargreaves and Roberston 2001; Robertson 2003), the Indian Space Research Organization (ISRO) with Cartosat-1, and, recently, MDA with Radarsat-2. This first approach was also tested under specific circumstances in an academic environment with aerial photographs, SPOT, Earth Resources Observation System (EROS)-A, Formosat-2, and QuickBird images by computing parameters of first- to third-order RFs from already-solved 3D physical models (Tao and Hu 2001; Chen, Teo, and Liu 2006; Bianconi et al. 2008).

The application of the first approach using vendor-provided third-order RPCs has become more popular these last years with HR images because of their small FOV, for example, IKONOS, QuickBird, Cartosat-1, WorldView, GeoEye, and Radarsat-2 (Fraser, Hanley, and Yamakawa 2002; Fraser, Baltsavias, and Gruen 2002; Tao and Hu 2002; Robertson 2003; Tao, Hu, and Jiang 2004; Fraser and Ravanbakhsh 2009; Toutin and Omari 2011). Applications using stereo images for digital surface model (DSM) generation were also performed (Lehner, Müller, and Reinartz 2005) with accuracy comparisons of stereo-extracted DSMs using empirical (RFM second approach) and physical (collinearity) models (Toutin 2006a). However, biases or random errors still exist after applying the RFs, and the results need to be postprocessed. The original RPCs can be refined with linear equations using additional accurate user GCPs (Lee et al. 2002) or using 2D polynomial functions (Fraser, Hanley, and Yamakawa 2002; Fraser, Baltsavias, and Gruen 2002; Di, Ma, and Li 2003; Tao, Hu, and Jiang 2004; Noguchi et al. 2004; Fraser and Hanley 2005; Wang et al 2005; Tong, Liu, and Weng 2010). For the last solution, 1 or 2 GCPs are used for zero-order 2D polynomial functions (bidirectional shift) and 6–10 GCPs for first- or second-order 2D polynomial functions to compute their parameters with a least-squares adjustment process. The use of these GCPs in RFM postprocessing is the reason why this approach is inappropriately called terrain independent. Later, Toutin (2006a) showed that first- or second-order polynomial functions that refine the IKONOS RFM do not significantly improve the final accuracy when compared to just a bias compensation (zero order). The main reason is that there is no more systematic geometric distortion in georeferenced IKONOS images except the relief. On the other hand, experiments for refining the RFM of raw QuickBird images, which display more severe geometric distortions, were not so coherent between different studies:

- Noguchi et al. (2004) demonstrated that a bias with a time-dependent drift (partial first-order functions with two parameters) has to be used for correcting some "unexplained" systematic errors.
- Fraser and Hanley (2005) reversed the findings of Noguchi et al. (2004), stating that just a bias refinement (zero-order) is enough because the time-dependent drift did not correct systematic errors.
- Cheng, Smith, and Sutton (2005), Toutin (2006a), and Tong, Liu, and Weng (2010) demonstrated that full first-order functions (with six parameters) have at least to be used.
- Tong, Liu, and Weng (2010) found the existence of potential high-order error signals in vendor-provided RPCs that can be corrected with second-order functions and a large number (around 20) of GCPs.

A likely explanation for these contradictions for QuickBird RFM refinement is mainly related to the RFM dependency on local distortions with "level-1A" images and the terrain relief. The study sites of Cheng and his coauthors (2005) and Toutin (2006a) were 1000 and 450 m in elevation range, respectively (first-order polynomial refinement), whereas the study site of Noguchi et al. (2004) was 240 m (shift and time-dependent drift refinement) and that of Fraser and Hanley (2005) was only 50 m (shift refinement). No indication of the terrain relief in Shanghai, China, was given in the study by Tong, Liu, and Weng (2010).

On the other hand, some operational studies using 3D RFs with different HR images (level-1A EROS-A1, IKONOS Geo, level-1A QuickBird-2) showed inferior and less consistent results (Kristóf, Csató, and Ritter 2002) than previous ones conducted in university environments. Some inconsistencies and errors with the IKONOS orthoimages generated from RFs were not explained (Davis and Wang 2001), and these errors did not appear when the 3D physical model was used. Tao and Hu (2002) achieved 2.2-m horizontal accuracy with almost 7-m bias when processing stereo IKONOS images using the first-approach RF method. Kristóf, Csató, and Ritter (2002) and Kim and Muller (2002) obtained 5-m random errors computed on precise independent check points (ICPs); larger errors away from the GCPs were also reported (Petrie 2002).

8.3.1.3.2 Terrain-Dependent Rational Function Model Approach

The second approach, called the terrain-dependent approach, can be performed by end users with the same processing method as with polynomial functions. Since there are 40 and 80 RPCs for the four second- and third-order polynomial functions, a minimum of 20 and 40 GCPs, respectively, are required to resolve 3D RFs. However, the RFs do not model the physical reality of the image acquisition geometry, and they are sensitive to input errors, similar to 2D/3D polynomial functions. Since RFs, similar to 2D/3D polynomial functions, mainly correct locally at GCP locations with some remaining distortions between GCPs (Petrie 2002), many more GCPs will be required to reduce their error propagation in an operational environment. A piecewise approach as described in Section 8.3.1.3 (Yang 2001) should also be used for large images (SPOT, Landsat, IRS, Radarsat), which will increase the number of GCPs proportionally to the number of subimages, making the method inadequate in an operational environment.

Some academic studies conducted in well-controlled research environments demonstrated the feasibility of using medium-resolution to HR images: level-1B SPOT and/or Landsat-TM georeferenced images (Okamoto et al. 1998; Tao and Hu 2001), Radarsat-SAR fine-mode ground-range image (Dowmann and Dolloff 2000), and IKONOS Geo images (Fraser, Hanley, and Yamakawa 2002; Fraser, Baltsavias, and Gruen 2002; Tao and Hu 2002). All the results were presented with georeferenced images (all systematic distortions corrected) acquired over a flat/hilly terrain, or over high relief areas with almost-nadir viewing angles, and consequently with almost no geometric distortions. In fact, the RFM solution is highly dependent on actual terrain relief, and on the number, accuracy, and distribution of GCPs (Tao and Hu 2002). In addition, the results for IKONOS images were not significantly better than using simple 3D first-order polynomial functions (translation–rotation scaling corrections in addition to the relief correction; Fraser, Hanley, and Yamakawa 2002; Fraser, Baltsavias, and Gruen 2002; Vassilopoulou et al. 2002) due to the fact that IKONOS Geo images are already corrected for all geometric distortions except the relief. However, some of these studies and results cannot be easily extrapolated because the conditions of experimentation were not properly and completely defined, such as the geometric characteristics of images used, level of geometric correction already applied to images, source, accuracy, distribution, and number of GCPs, source, accuracy, distribution, and number of checked data, and the size of the terrain and its relief. Since this approach is not efficient, it is no longer used.

8.3.2 Two-Dimensional/Three-Dimensional Physical Models

The 2D/3D physical functions used to perform the geometric correction differ from one another, depending on the sensor, platform, and the sensor's image acquisition geometry (Figure 8.8):

Camera　　　　**Pushbroom**　　　　**Wiskbroom**　　　　**Radar**
(instantaneous acquisition)　(CCD line acquisition)　(mechanical sweeping acquisition)　(side-looking acquisition)

FIGURE 8.8
Image acquisition geometry of different satellite sensors. (Courtesy and copyright Serge Riazanoff, VisioTerra, 2009.)

- Array camera systems (for instantaneous acquisition), such as photogrammetric cameras, metric cameras (MC), or large format cameras (LFC)
- Mechanical rotating or sweeping mirrors, such as Landsat-MSS, TM, or ETM+
- Pushbroom scanners (for line acquisition), such as MERIS, SPOT-HRV/high resolution in geometry (HRG), and IRS-1C/D
- Agile scanners (for line acquisition), such as IKONOS, QuickBird, WorldView
- The SAR sensors (for side-looking acquisition), such as Environmental Satellite (ENVISAT), Radarsat-1/2, CosmoSkyMed, Terra-SAR

Although each sensor has its own unique characteristics, one can draw generalities for the development of 2D/3D physical models in order to fully correct all distortions described in Section 8.2. The physical model should mathematically model all distortions of the platform (position, velocity, and attitude for VIR sensors), sensor (lens, viewing/look angles, and panoramic effect), Earth (ellipsoid and relief for 3D), and cartographic projection. The geometric correction process can address each distortion one-by-one, either step-by-step or simultaneously. In fact, it is better to consider the total geometry of viewing (platform + sensor + Earth + map), because some of the distortions are correlated and have the same type of impact on the ground. It is theoretically more precise to compute only one combined parameter rather than each component of this combined parameter separately; this also avoids overparameterization and correlation between terms.

Some examples of combined parameters include the following:

- The "orientation" of the image is a combination of the platform heading due to orbital inclination, yaw of the platform, and convergence of the meridian.
- The "scale factor" in the along-track direction is a combination of the velocity, altitude, and pitch of the platform, the detection signal time of the sensor, and a component of the Earth's rotation in the along-track direction.
- The "leveling angle" in the across-track direction is a combination of platform roll, the viewing angle, orientation of the sensor, Earth's curvature, etc.

Considerable research has been carried out to develop robust and rigorous 3D physical models that describe the acquisition geometry related to different types of images (VIR and SAR images; low-, medium-, and high-resolution images) and of platforms (spaceborne and airborne). The 2D physical model was developed for ERTS imagery (Kratky 1971), and 3D physical models were developed for the following:

- Low-/medium-resolution VIR satellite images (Bähr 1976; Masson d'Autume 1979; Konecny 1979; Sawada et al. 1981; Khizhnichenko 1982; Friedmann et al. 1983; Guichard 1983; Toutin 1983; Salamonowicz 1986; Konecny, Kruck, and Lohmann 1986; Gugan 1987; Konecny et al. 1987; Kratky 1987; Shu 1987; Paderes, Mikhail, and Fagerman 1989; Westin 1990; Novak 1992; Robertson et al. 1992; Ackermann et al. 1995; Sylvander et al. 2000; Westin 2000)
- High-resolution VIR satellite images (Gopala Krishna et al. 1996; Jacobsen 1997; Cheng and Toutin 1998; Toutin and Cheng 2000; Bouillon et al. 2002, 2006; Chen and Teo 2002; Hargreaves and Robertson 2001; Toutin 2003a; Westin and Forsgren 2002)

- The SAR satellite images (Rosenfield 1968; Gracie et al. 1970; Leberl 1978; Wong, Orth, and Friedmann 1981; Curlander 1982; Naraghi, Stromberg, and Daily 1983; Guindon and Adair 1992; Toutin and Carbonneau 1992; Tannous and Pikeroen 1994; Toutin and Chénier 2009)
- Airborne VIR images (Derenyi and Konecny 1966; Konecny 1976; Gibson 1984; Ebner and Muller 1986; Hoffman and Muller 1988)
- Airborne SLAR/SAR images (La Prade 1963; Rosenfield, 1968; Gracie et al. 1970; Derenyi 1970; Konecny 1970; Leberl 1972; Hoogeboom, Binnenkade, and Veugen 1984; Toutin, Carbonneau, and St-Laurent 1992)

The 2D physical model for ERTS by Kratky (1971) took into consideration and mathematically modeled, in a step-by-step manner, the effects of scanner geometry, panoramic effect, Earth's rotation, satellite circular orbit and attitude, nonuniform scan rate, and map projection to finalize with a dual simple equation (one for each axis), which mathematically integrated all the previous error equations.

The general starting points of other research studies in deriving the mathematical functions of the 3D physical model are, generally, as follows:

- The well-known collinearity condition and equations (Bonneval 1972; Wong 1980) for VIR images are given by

$$x = (-f)\frac{m_{11}(X - X_0) + m_{12}(Y - Y_0) + m_{13}(Z - Z_0)}{m_{31}(X - X_0) + m_{32}(Y - Y_0) + m_{33}(Z - Z_0)} \tag{8.4}$$

$$y = (-f)\frac{m_{21}(X - X_0) + m_{22}(Y - Y_0) + m_{23}(Z - Z_0)}{m_{31}(X - X_0) + m_{32}(Y - Y_0) + m_{33}(Z - Z_0)} \tag{8.5}$$

where (x,y) are the image coordinates; (X,Y,Z) are the map coordinates, (X_0,Y_0,Z_0) are the projection center coordinates, $-f$ is the focal length of the VIR sensor, and $[m_{ij}]$ are the nine elements of the orthogonal 3-rotation matrix.

- The Doppler and range equations for radar images are as follows:

$$f = \frac{2(\vec{V_S} - \vec{V_P}) \cdot (\vec{S} - \vec{P})}{\lambda|\vec{S} - \vec{P}|} \tag{8.6}$$

$$r = |\vec{S} - \vec{P}| \tag{8.7}$$

where f is the Doppler value, r is the range distance, $\vec{S}$ and $\vec{V_S}$ are the sensor position and velocity, $\vec{P}$ and $\vec{V_P}$ are the target point position and velocity on the ground, and λ is the radar wavelength.

It should be noted that collinearity equations were adapted as radargrammetric equations to process radar images (Leberl 1972, 1990) and later as an integrated and unified mathematical equation to process multisensor (VIR or radar) images (Toutin 1995b).

The collinearity equations are valid for an instantaneous image or scan-line acquisition, such as photogrammetric cameras (LFC, MC), and VIR scanner sensors (used aboard SPOT, Landsat), and the Doppler-range equations are valid for a SAR scan line. However, since the parameters of neighboring scan lines of scanners are highly correlated, it is possible to link the exposure centers and rotation angles of different scan lines to integrate supplemental information, such as either of the following:

- The ephemeris and attitude data using the laws of celestial mechanics (Figures 8.3 and 8.4) for satellite images
- The global positioning system (GPS) and inertial navigation system (INS) data for airborne images

The integration of different distortions and the mathematical derivation of equations for different sensors are outside the scope of this chapter. They are described for photogrammetric cameras in the works of Bonneval (1972) or Wong (1980), for scanner images in those of Leberl (1972), Konecny (1976), or de Masson d'Autume (1979), for ERTS/Landsat images in the works of Kratky (1971), Bähr (1976), or Salamonowicz (1986), for pushbroom scanners, such as SPOT, in those of Guichard (1983), Toutin (1983), Konecny, Kruck, and Lohmann (1986), and Konecny et al. (1987), and for SAR data in the works of Leberl (1978) or Curlander (1982).

8.4 Methods, Processing, and Errors

Whatever the mathematical functions used, the geometric correction method and processing steps are more or less the same. The processing steps are as follows (Figure 8.9):

- Acquisition of image(s) and "preprocessing of metadata"
- Acquisition of the ground points (control/check/pass) with image coordinates and map coordinates X, Y, (Z)
- Computation of the unknown terms of the mathematical functions used for the geometric correction model for one or more images
- Image(s) rectification with or without DEM

The main differences in the processing steps between physical and empirical models are denoted in italic style, and between 2D and 3D models in bold style. The metadata are useless for empirical models because the models do not reflect the geometry of viewing, whereas the Z-elevation coordinates for GCPs and DEM are of no use for 2D empirical models.

8.4.1 Acquisition of Images and Metadata

With VIR images, different types of image data with different levels of preprocessing can be obtained, but different image providers unfortunately use a range of terminology to denominate the same type of image data. Terminology should be standardized, mainly for the convenience of end users.

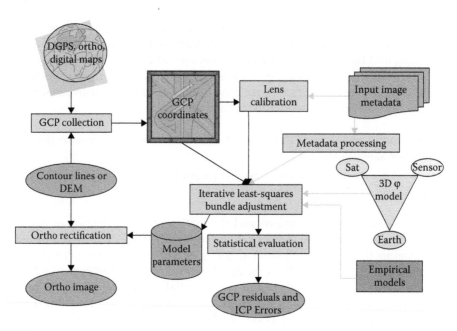

FIGURE 8.9
(See color insert following page 426.) Description of the geometric correction method and its processing steps. The ellipse symbols denote input/output data and the box symbols denote processes.

- Raw images with only normalization and calibration of the detectors (e.g., level 1A for SPOT, EROS, and Formosat, level 1B1 for Advanced Land-Observing Satellite [ALOS], basic for QuickBird/WorldView, single look complex for Radarsat) without any geometric correction are satellite-track oriented. For SAR data, this level corresponds to the slant-range geometry. In addition, the full metadata related to sensor, satellite (ephemeris and attitude), and image are provided.

- Georeferenced images (e.g., level 1B for SPOT, 1SYS for Cartosat, standard for QuickBird/WorldView, 1G for Landsat-ETM+, SAR georeferenced fine/SAR georeferenced extrafine [SGF/SGX] for Radarsat) corrected for systematic distortions due to sensor, platform, and the Earth's rotation and curvature are satellite-track oriented. For SAR data, this level corresponds to the ground-range geometry. Generally, only a few metadata related to sensor and satellite are provided; some of this metadata are related to level-1B processing.

- Map-oriented images, also called geocoded images, (e.g., level 2A for SPOT, Geo standard for IKONOS, 1B2 for ALOS, systematically geocoded/precision geocoded [SSG/SPG] for Radarsat) corrected for the same distortions as georeferenced images are oriented toward the north. For SAR data, this level corresponds to the ground-range geometry. Generally, very few metadata related to sensor and satellite are provided; most of this metadata are related to level-2A processing and ellipsoid/map characteristics.

8.4.1.1 Raw Level-1A Images

For the sake of understanding, the easiest terminology defined for SPOT images are used. The raw level-1A images are preferred by photogrammetrists because the 3D physical models derived from collinearity equations are well known and well developed, and are easily used in soft-copy workstations. Since different 3D physical models are largely available for such VIR images, raw level-1A-type images should now be favored by the remote sensing community as well. Specific software to read and preprocess the appropriate metadata (ephemeris, attitude, sensor, and image characteristics) have to be realized for each image sensor according to the 3D physical model used. Using laws of celestial mechanics and Lagrangian equations (Escobal 1965; Centre National d'Études Spatiales 1980; Light et al. 1980), the ephemeris (position and velocity; Figure 8.3 [left]) can be transformed into specific osculatory orbital parameters (Figure 8.4) to reduce time-related effects (Toutin 1983). Since the Lagrangian equations take into account the variation of the Earth's gravitational potential to link the different positions of the satellite during image formation, this method is more accurate and robust than using a constant ellipse with second-order time-dependent polynomial functions (Guichard 1983; Toutin 1983; Tannous and Pikeroen 1994; Bannari et al. 1995). This statement is more applicable when "long-strip" images from the same orbit are used with low-resolution images (Robertson et al. 1992; Sylvander et al. 2000; Westin 2000), with multiimage path processing (Toutin 1985; Sakaino et al. 2000) or with a block bundle adjustment method using single-sensor images (Veillet 1991; Campagne 2000; Kornus, Lehner, and Schroeder 2000; Cantou 2002; Toutin 2003b, c, d, e), and when using multisensor images (Toutin 2004a, 2006b).

The 3D physical models were also applied to HR airborne images (Konecny 1976; Gibson 1984; Ebner and Muller 1986) and HR spaceborne images, such as those from push-broom scanners IRS–1C/D (Gopala Krishna et al. 1996; Jacobsen 1997; Cheng and Toutin 1998), the asynchronous scanner from EROS (Chen and Teo 2002; Westin and Forsgren 2002; Bianconi et al. 2008), and SPOT-5 scanners (Bouillon et al. 2002, 2006; Toutin 2004c, 2006c) and the new HR images from, for example, QuickBird-2, Formosat-2 (Hargreaves and Robertson 2001; Toutin 2004b, c; Chen, Teo, and Liu 2006; Bianconi et al. 2008), and others for achieving subpixel accuracy. On the other hand, results using the RFM second approach were also published with raw HR small-FOV images (QuickBird-2, Cartosat-1, Formosat-2, etc.; Noguchi et al. 2004; Cheng, Smith, and Sutton 2005; Fraser and Hanley 2005; Lehner, Müller, and Reinartz 2005; Chen, Teo, and Liu 2006; Toutin 2006a), but not with large-FOV images due to their inability to model high-frequency distortions inherent in raw level-1A large-swath or long-strip images (Madani 1999; Dowman and Dolloff 2000).

8.4.1.2 *Georeferenced Level-1B Images*

Since they have been systematically corrected and georeferenced, level-1B images retain just the terrain elevation distortion, in addition to a rotation–translation related to the map reference system. A 3D first-order polynomial model with Z-elevation parameters can thus be efficient for this approach, depending on the requested final accuracy. For scanners with across-track viewing capability, only the Z-elevation parameter in the X equation is useful. The second-order polynomial models could also be used (Palà and Pons 1995) for correcting some residual errors of 1B processing. Possible solutions to overcome the empirical model approximation are either the conversion of 1B images back to 1A images using metadata and reverse transformation (Al-Roussan et al. 1997) or the "reshaping and resizing" of 1B images to the raw imagery format (Valadan Zoej and Petrie 1998). This 1B geometric modeling can be mathematically combined with normal level-1A 3D physical models to avoid multiple image resampling. Although this mathematical procedure used for 1B images works better than empirical models, it is recommended that raw images with rigorous 3D physical models (collinearity equations) be directly used.

8.4.1.3 Map-Oriented Level-2A Images

Similar to level-1B images, map-oriented images (level 2A) also retain elevation distortion, but image lines and columns are not related to sensor-viewing and satellite directions. A 3D first-order polynomial model with Z-elevation parameters in both axes can thus be efficient for this approach, depending on the requested final accuracy. For level 1B, second-order empirical models (polynomial or rational) can be used for correcting some residual errors of the 2A processing, but it is generally no longer possible to convert back the 2A image with the reverse transformation. Due to the fact that IKONOS Geo images are already corrected for all systematic geometric distortions except terrain relief, the following 2D/3D empirical models were recently applied to achieve pixel accuracy or better:

- The 2D first-order polynomial and RF functions on flat terrain (Hanley and Fraser 2001)

- The 3D first-order polynomial functions (Ahn, Cho, and Jeon 2001; Fraser, Hanley, and Yamakawa 2002; Fraser, Baltsavias, and Gruen 2002; Vassilopoulou et al. 2002)

- The 3D fourth-order polynomial functions (Kersten et al. 2000; Vassilopoulou et al. 2002)

- The 3D third-order RF functions using the first approach (described in Section 8.3.1.3.1) with parameters computed from Space Imaging's (Thornton, CO) camera model (Grodecki 2001; Fraser, Hanley, and Yamakawa 2002; Fraser, Baltsavias, and Gruen 2002; Tao and Hu 2002)

- The 3D third-order RF functions using the first approach (described in Section 8.3.1.3.1) with parameters provided with IKONOS Geo images and using GCPs either to remove bias (Fraser, Hanley, and Yamakawa 2002) or to improve the original RF parameters (Lee et al. 2002)

Although the results are in the order of pixel accuracy, they are generally achieved in an academic environment or by the image providers using images or subimages acquired over flat/hilly terrain. Only few results were published in high relief terrain (Kersten et al. 2000; Vassilopoulou et al. 2002). Conversely, other academic or operational studies using the first or second RF approach obtained larger errors of few (2–5) pixels (Davis and Wang 2001; Kristóf, Csató, and Ritter 2002; Kim and Muller 2002; Petrie 2002; Tao and Hu 2002).

The 3D physical model has been approximated and developed for IKONOS Geo images using basic information such as metadata and the laws of celestial mechanics (Toutin and Cheng 2000). Even in the approximated form, this 3D physical model ("using a global geometry and adjustment") has proven to be robust and to achieve consistent results over different study sites and environments (urban, semirural, and rural; Europe, North America, and South America), different relief (flat to high), and different cartographic data (differential GPS [DGPS], orthophotos, digital topographic maps, DEM; Toutin 2003a). This 3D physical model has been used in different operational applications, such as in digital image base map generation (Davis and Wang 2001), urban management (Hoffmann et al. 2001; McCarthy, Cheng, and Toutin 2001; Meinel and Reder 2001; Ganas, Lagios, and Tzannetos 2002), and land resources management (Kristóf, Csató, and Ritter 2002; Toutin 2003b).

8.4.1.4 Synthetic Aperture Radar Images

SAR images are standard products in slant- or ground-range presentations. They are generated digitally during postprocessing from raw signal SAR data (Doppler frequency, time delay). Errors present in the input parameters related to the image geometry model will propagate through to the image data. These include errors in the estimation of slant range and Doppler frequency, and also errors related to the satellite's ephemeris and the ellipsoid. Assuming the presence of some geometric error residuals, the parameters of a 3D physical model reflect these residuals. As mentioned in Section 8.3.2, the 3D physical model starts generally either from traditional Doppler and range equations (Curlander 1982), from the equations of radargrammetry (Konecny 1970; Leberl 1978, 1990), or from generalized equations (Leberl 1972; de Masson d'Autume 1979; Toutin 1995b). Due to the large elevation distortions in SAR images, 2D polynomial models cannot be used even in rolling topography (Toutin 1995a) or to extract planimetric features (de Sève, Toutin, and Desjardins 1996). Further, since different 3D SAR physical models are largely available, few attempts have been made to apply 3D polynomial or RF empirical models to SAR images (spaceborne or airborne images). Dowman and Dolloff (2000) present preliminary results with one Radarsat-1 SAR fine-mode image, but the conditions of experimentation (study site and terrain relief, cartographic data and accuracy, type and approach of RFs) are not described. More recently, third-order RFMs with some term reduction are provided with Radarsat-2, whatever the mode (ultrafine to ScanSAR), beam (20°–40°), or format (slant, ground) considered. According to MDA, their accuracies are similar to that of the physical model used for their computation, but with larger errors with ScanSAR and seep angles (Robertson 2009, pers. comm.). Without independent evaluation of these SAR RFMs, extrapolation to other SAR data should be carefully evaluated (Toutin and Omari 2011).

8.4.2 Acquisition of Ground Control Points

Whatever the VIR and/or SAR geometric model used, even when the RF terrain-independent approach is used to remove the bias or refine RF parameters, some GCPs have to be acquired to compute or refine the parameters of the mathematical functions in order to obtain a cartographic standard accuracy. Generally, an iterative least-squares adjustment process is applied when more GCPs than the minimum number required by the model (as a function of unknown parameters) are used. The number of GCPs used is a function of different conditions: the method of collection, sensor type and resolution, image spacing, geometric model, study site, physical environment, GCP definition and accuracy, and final expected accuracy. Figure 8.10 shows examples of well-defined GCPs and tools to extract their image coordinates. If GCPs are determined a priori without any knowledge of the images to be processed, 50% of the points may be rejected. If GCPs are determined a posteriori with knowledge of the images to be processed, the reject factor will be smaller (20%–30%). Consequently, all the aspects of GCP collection do not have to be considered separately, but rather as a whole to avoid too large discrepancies in the accuracies of these different aspects. For example, a DGPS survey is too good to process Landsat data in mountainous study sites, and on the other hand, road intersections and topographic maps of scale 1:50,000 are not good enough to process QuickBird images if you expect a high final pixel accuracy. The weakest aspect in the GCP collection, which is of course different for each study site and image, will thus be the major source of error in error propagation and the overall error budget of the mathematical model computation.

FIGURE 8.10
Examples of well-defined ground control points (GCPs) and tools for image pointing and for extracting their image coordinates.

Since empirical models do not reflect the geometry of viewing and do not filter errors, many more GCPs than the theoretical minimum are required to reduce the propagation of input errors in geometric models. When the cartographic data accuracy and/or the positioning accuracy are in the same order of magnitude as the sensor resolution, twice as many GCPs is a minimum requirement; around 20, 40, or 80 GCPs should then be acquired for second-order 2D polynomial, 3D polynomial, or 3D terrain-dependent RF models, respectively. The third-order models obviously require more GCPs, mainly the RFs. Further, in order to ensure robustness and consistency in an operational environment, it is safer to collect more than twice the required minimum mentioned previously in this section. It could then be a restriction on the use of such empirical models in an operational environment. However, when using 3D first-approach RF models with the already-computed RPCs provided by an image vendor, only a few (1–10) GCPs are needed to remove the errors with 2D polynomial functions or to refine the RF parameters. When more than one image is processed, each image requires its own GCPs and the geometric models are generally computed separately, that is, there is no relative orientation or link between adjacent images because RFMs were computed independently by the image providers. However, some block adjustment can be performed with RFs (Dial and Grodecki 2002; Fraser, Hanley, and Yamakawa 2002; Grodecki and Dial 2003). Because empirical models are sensitive to GCP distribution and number, GCPs should be spread over the full image(s) in planimetry and also in the elevation range for the 3D models to avoid large errors between GCPs. It is also better to have medium-accurate GCPs (lakes, tracks, ridges) than no GCP at the tops of mountains. If the image is larger than the study site, it is recommended to reduce the GCP collection to the study site area because the empirical models correct only locally.

With 3D physical models, fewer GCPs (1–6) are required per image. When more than one image is processed, a spatiotriangulation method with 3D block-bundle adjustment can be used to process all the images together (VIR and SAR). This enables users to drastically reduce the number of GCPs for the block with the use of tie points (TPs; Veillet 1991; Belgued et al. 2000; Campagne 2000; Kornus, Lehner, and Schroeder 2000; Sakaino et al. 2000; Cantou 2002; Toutin 2003b, c, d, e). When the map and positioning accuracy is of the same order of magnitude as the sensor resolution, twice (or a little less) the theoretical minimum is recommended. When the accuracy is lower, the number should be increased depending also on the final expected accuracy (Savopol et al. 1994). Since more confidence, consistency, and robustness can be expected with physical models (due to global image processing and filtering of input errors) than with empirical models, it is not necessary to increase the number of GCPs in operational environments. The GCPs should preferably be spread at the border of the image(s) to avoid extrapolation in planimetry, and it is also preferable to cover the full elevation range of the terrain (lowest and highest elevations). Contrary to empirical models, it is not necessary for physical models to have a regular distribution in the planimetric and elevation ranges. Since physical models correct globally, the GCP collection has to be performed in the full image size even if the study site is smaller. First, it will be easier to find GCPs over the full image than over a subarea, and second, more homogeneity is thus obtained in the different area of the image.

The GCP cartographic coordinates can be obtained from GPS, air photo surveys, paper or digital maps, GIS, orthorectified photos or images, chip databases, and so on, depending on the requested accuracy of the input/output data. The cartographic coordinates obtained from these sources have drastically different accuracies: from better than 0.2 m with DGPS to 25–50 m with 1:50,000 paper maps, certainly the most common GCP source used around the world. Consequently, with lower accuracy, more GCPs must be used (Savopol et al. 1994). Image coordinates are obtained interactively on the screen or automatically using a GCP chip database and image correlation tools. When multiple images with overlapping coverage are processed, image coordinates are obtained in stereoscopy (the best solution) or simultaneously in "double monoscopy" because some workstations do not have full stereoscopic capabilities for multisensor images. The double monoscopy image measurements will then create artificial X- and Y-parallaxes (few pixels) between the images, and the parallax errors will propagate through the bundle adjustment (relative and absolute orientations). The error propagation is larger with SAR images than with VIR images due to a lower image measurement accuracy (1–2 pixels vs. 1/3–1/2 pixel), and not only increases with smaller intersection angles but also with shallower same-side SAR look angles (Toutin 1998, 1999). Consequently, when possible, true stereoscopic image measurements using human depth perception, which enables a better relative correspondence of the GCP between images and a better absolute positioning on the ground, should be used.

8.4.3 Geometric Model Computation

When more than one image (VIR or SAR) is processed over large study sites (Figure 8.11), a spatiotriangulation process based on a block adjustment can be first applied to simultaneously compute all geometric models (Figure 8.9). The spatiotriangulation method has been applied using 3D physical models to different VIR/SAR/HR data, acquired either from a single sensor (Veillet 1991; Belgued et al. 2000; Campagne 2000; Kornus, Lehner, and Schroeder 2000; Sakaino et al. 2000; Cantou 2002; Toutin 2003b, c, d, e) or from multiple

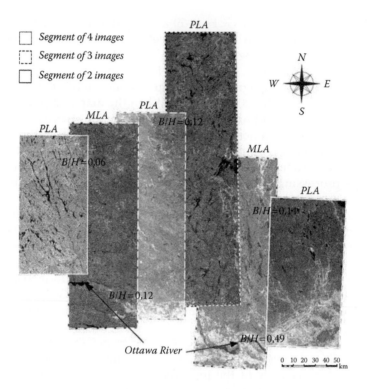

FIGURE 8.11

Image block of 17 level-1A panchromatic/multiband System Pour l'Observation de la Terre high geometric resolution (SPOT-HRG) images over Gatineau Hills, Canada (300 × 240 km) generated from six strips of two to four images. (SPOT Image © CNES 1991–1998.)

sensors (Toutin, 2004a, 2006c). Spatiotriangulation was also applied using 3D RF models to HR optical data acquired only from a single sensor (Dial and Grodecki 2002; Fraser, Hanley, and Yamakawa 2002; Grodecki and Dial 2003). Figure 8.11 shows an example of a block formed with 17 level-1A panchromatic/multiband SPOT HRG images (300 × 240 km) acquired over Gatineau Hills, Canada, generated from six strips of two to four images (Toutin 2003e).

All model parameters of each image/strip are determined by a common least-squares adjustment so that individual models are properly tied in and the entire block is optimally oriented in relation to the GCPs. With the spatiotriangulation process, the same number of GCPs is theoretically needed to adjust a single image, an image strip, or a block. However, some TPs between the adjacent images have to be used to link the images or strips or both. The elevation of TPs (ETPs) must be added when the intersection geometry of the adjacent images is weak, such as with intersection angles less than 15°–20° (Toutin 2003b, c, d, e). There are a number of advantages to the spatiotriangulation process:

- The reduction of the number of GCPs
- A better relative accuracy between images
- A more homogeneous and precise mosaic over large areas
- A homogeneous GCP network for future geometric processing

Whatever the number of images (spatiotriangulation or single image) and the geometric models (physical or empirical) used, each GCP contributes to two observation equations: (1) an equation in X and (2) an equation in Y. The observation equations per image are used to esta-blish the error equations for GCPs, TPs, and ETPs. Each group of error equations can be weighted as a function of the accuracy of the image and cartographic data. The normal equations are then derived and resolved with the unknowns computed. In addition, for the 3D physical models, conditions or constraints on osculatory orbital parameters or other parameters (GPS/INS) can be added in the adjustment to take into account the knowledge and the accuracy of the ephemeris or other data, when available. These conditions and con-straints thus prevent the adjustment from diverging, and they also filter the input errors.

Since there are always redundant observations to reduce the input error propagation in geometric models, a least-squares adjustment is generally used. When the mathemati-cal equations are nonlinear, which is the case for physical and second- and higher-order empirical models, some means of linearization (series expansions or Taylor's series) must be used. A set of approximate values for the unknown parameters in the equations must thus be initialized:

- To zero for empirical models, because they do not reflect the image acquisition geometry
- From the osculatory orbital/flight and sensor parameters of each image for physi-cal models

More information on least-squares methods applied to geomatics data can be obtained from the studies of Mikhail (1976) and Wong (1980). The results of this processing step are as follows:

- The parameter values for the geometric model used for each image
- The residuals in X and Y directions (and the Z direction if more than one image is processed) for each GCP/ETP/TP and their root-mean-square (RMS) residuals
- The errors and bias in X and Y directions (and the Z direction if more than one image is processed) for each ICP if any, and their RMS errors
- The computed cartographic coordinates for each point, including ETPs and TPs

When more GCPs than the theoretically required minimum are used, the GCP residuals reflect the modeling accuracy, whereas the ICP RMS errors reflect the final accuracy, tak-ing into account ICP accuracy. As mentioned in Section 8.4.2, this final accuracy is mainly dependent on the geometric model and the number of GCPs used versus their cartographic and image coordinates. When ICPs are not accurate, their errors are included in the com-puted RMS errors; consequently, the final internal accuracy of the modeling will be better than these RMS errors.

When no ICP is available, GCP RMS residuals can be carefully used as an approxima-tion of the final accuracy, only when using physical models. However, the fact that RMS residuals can be small with empirical models does not necessarily mean a good accuracy because these models correct locally at GCPs and the least-squares adjustment minimizes residuals at GCPs. Errors are still present among GCPs (Davis and Wang 2001; Petrie 2002). On the other hand, by using overabundant GCPs with physical models, the input data errors (image measurement or map or both) do not propagate through the physical models but are mainly reflected in the GCP residuals due to a global adjustment. Consequently, it is thus "normal and safe" with 3D physical models to obtain RMS residuals in the same order of magnitude as the GCP accuracy, and the physical model by itself will be more accurate. In other words, the internal accuracy of images will be better than the RMS residuals. In contrast to empirical methods, which are sensitive to GCP number and spatial distribution (including their elevation), 3D physical models are not affected by these factors because they precisely retain the complete viewing geometry, given that there is no extrapolation in planimetry and also in elevation.

8.4.4 Digital Elevation Model Generation from Stereo Images

When two images are acquired over the same site from two different viewpoints, it is possible to reconstruct the terrain relief and to generate digital terrain models (DTMs). Because the sensor did not image the bald Earth but the top of feature surfaces, DEMs are in fact DSMs, which include the height, or a part, of natural and human-made surfaces (trees, houses, fences, etc.; Figure 8.12). Principally, two image-matching methods can be used to extract the elevation parallax for generating DSMs: (1) the computer-assisted (visual) method or (2) the automatic method. These two methods can of course be integrated to take into account the strength of each one.

Computer-assisted visual matching is an extension of the traditional photogrammetric method to extract elevation data (contour lines) on a stereoplotter. It then requires full stereoscopic capabilities to generate the online 3D reconstruction of the stereo model and to capture in real time the 3D planimetric and elevation features. For elevation, spot elevations, contour lines, or irregular grid, DEM can be generated. Stereoscopic viewing is realized on a computer screen using a system of optics. The stereo images are separated spatially, radiometrically, or temporally. Spatial separation is achieved by the use of two monitors or a split screen and an optical system using mirror or convex lenses or both. Radiometric separation is achieved by anaglyphic or polarization techniques with colored or polarized lens, respectively. Temporal separation is achieved by an alternate display of the two images and using special synchronized lenses (Walker and Petrie 1996).

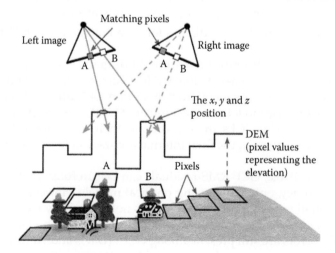

FIGURE 8.12
Digital elevation models from stereoscopic images acquired from two different viewpoints. The heights of natural or human-made surfaces are included in the elevation.

To obtain true 3D performance in a stereo workstation, the images are resampled into an epipolar or a quasi-epipolar geometry, in which only the X-parallax related to the elevation is retained (de Masson d'Autumne 1979; Baker and Binford 1981). Another solution to control the image positioning from the raw imagery is to automatically follow the dynamic change by cancelling the Y-parallax using the previously computed stereo model (Toutin et al. 1993; Toutin and Beaudoin 1995). In the same way as with a conventional stereoplotter, the operator cancels the X-parallax by fusing the two floating marks (one per image) on the ground. The system then measures the bidimensional parallax between the images for each point, and computes the X, Y, and Z cartographic coordinates using 3D intersection. The visual matching then combines in the brain a geometric aspect (fusing the floating marks together) and a radiometric aspect (fusing the floating marks on the corresponding image point). Some automatic tasks (displacement of the image or cursor, prediction of the corresponding image point position) are added.

However, computer-assisted visual matching, principally used with paper-format images and analytical stereo workstations, is a long and expensive process to derive DEM. When using digital images, automated image matching can thus be used. Since image matching has been a lively research topic for the last 30 years, an enormous body of research work and literature exists on the image matching of different EO sensors.

Most of the research studies on satellite image matching are based on Marr's research (1982) at the Massachusetts Institute of Technology (MIT), dealing with the modeling of human vision. If a computer program can be realized to see things as a human would, then the algorithm must have some basis in human visual processing. The stereo disparity is based on the following two "correct" assumptions about the real world (Marr and Poggio 1977): (1) a point of a surface has a unique position in space at any one time and (2) matter is cohesive. The first generation of image matching processes based on these assumptions is the gray-level image matching process. Gray-level matching between two images really implies that the radiometric intensity data from one image, representing a particular element of the real world, must be matched with the intensity data from the second image, representing the same real-world element.

Although satellite images of the real world represented by gray levels is not like a random-dot stereogram (which is easily matchable), gray-level matching has been widely studied and applied to remote sensing data. Most of the matching systems operate on reference and search windows. For each position in the search window, a match value is computed from the gray-level values in the reference window. The local maximum of all the match values computed in the search window is the good spatial position of the searched point. The match value can be computed with the normalized cross-correlation coefficient, sum of mean normalized absolute differences, stochastic sign change, or outer minimal number estimator methods. The first is considered to be the most accurate (Leberl et al. 1994) computation method and is largely used with remote sensing images. Leberl et al. also noticed that matching errors were smaller with SPOT images and digitized aerial photographs than with SAR images. The last two match-value computation methods have rarely or never been used by the remote sensing community.

Another solution to the problem of matching, introduced by Förstner (1982), is the least-squares approach, minimizing the squares of the image gray-level differences in an iterative process. This method makes possible the use of well-known mathematical tools and the estimation of error. Rosenholm (1986) found that the more-complicated least-squares method applied on simulated SPOT images did not give any significant improvement when compared with the cross-correlation coefficient method. However, this least-squares method seems to be more accurate with real SPOT data (Day and Muller 1988).

The notion of least-squares matching in the object domain (groundel) rather than in the image domain (pixel) was later introduced by Helava (1988). Predicted image densities, corresponding to each groundel, are mathematically computed with known geometric and radiometric image parameters, and matched to the original ones. The uncertainty in the parameters of a particular groundel is resolved by the least-squares method. An advantage of this approach is that more than two images from the same or different sensors can be used to make the least-squares solution meaningful; a disadvantage is the inability to correctly model the groundel attributes for each image. Due to these reasons, this matching technique is mainly used with air photos, since more than two images overlap the same ground area and their geometry and radiometry are better controlled.

Since one of Marr's assumptions was either missing or incorrectly implemented in gray-level matching (mainly with images of the real world), Marr developed a second generation of image matching: feature-based matching (Marr and Hildreth 1980). The same element of the real world may look considerably different in remote sensing images acquired at different times and with different geometries between the sensor, illumination, and terrain. Instead, the edges in the images reflect the true structures (Cooper, Friedman, and Wood 1987). Although feature-based matching has not become very popular among the remote sensing community with satellite data, some applications have been realized with simulated SPOT and real Landsat-TM (Cooper, Friedman, and Wood 1987). The DEM results were not as good as those obtained by Simard and Slaney (1986) with Landsat-TM stereo pair using gray-level matching. Hähn and Förstner (1988) also found that the least-squares matching method is more accurate than feature-based matching, which is converse to Marr's theoretical prediction. Later, Schneider and Hahn (1995) tested the two methods to extract TPs on Modular Opto-electronic Multispectral Stereo Scanner (MOMS-2/D2) stereo images. Their results in planimetry and elevation were twice as accurate with intensity-based matching than with feature-based matching.

Hybrid approaches using multiprimitive multi-image matching can thus achieve better and faster results by combining gray-level matching and feature-based matching with a hierarchical multiscale algorithm, and also with computer-assisted visual matching. An example is the algorithm and software, satellite image precision processing (SAT-PP), developed by the Institute of Geodesy and Photogrammetry of the Swiss Federal Institute of Technology, Zurich (ETH Zurich), Switzerland, for multisensor data (Eisenbeiss et al. 2005; Figure 8.13). The feature-based approach may produce good results for identified features, but it produces no elevation at intermediate points. They can then be used as seed points for gray-level matching. Another hybrid approach is to generate gradient amplitude images in a first step with gray-level values derived from the original stereo images instead of gradient images with only binary edge values. In the second step, any gray-level matching technique can be used on these preprocessed images (Paillou and Gelautz 1999). The linear gradient operator can be designed to be optimal to remove noise (if any) and to enhance edges. No attempt has been made with VIR images.

Although computer-assisted visual matching is a long process, it has been proven to be very accurate with photos or different satellite VIR data (Leberl et al. 1994; Raggam et al. 1994; Dorrer et al. 1995; Toutin and Beaudoin 1995). It can thus be used to eliminate blunders, to fill mismatched areas, or in areas where automated image matching gives errors larger than 1 pixel (about 10% for SPOT and 15% for digitized photographs; Leberl et al. 1994). It can also be used to generate seed points for automated matching.

Other developments have been realized and tested principally for airborne or close-range stereo images, but rarely with satellite images, such as the global approach, scale space algorithms, relational matching, consideration of break lines, and multiple image

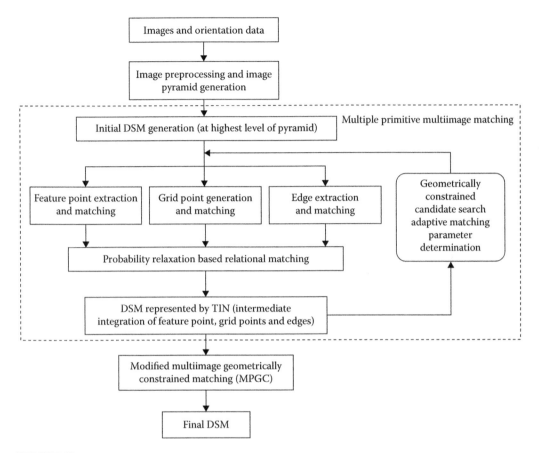

FIGURE 8.13
Overview of the multiprimitive, multiimage matching method employed in satellite image precision processing (SAT-PP) software package developed by the Institute of Geodesy and Photogrammetry of the Swiss Federal Institute of Technology Zurich (ETH Zurich), Switzerland. (From Baltsavias, E., L. Zhang, and H. Eisenbeiss. 2005. DSM generation and interior orientation determination of ikonos images using a test field in Switzerland. In *International Society of Photogrammetry, Remote Sensing Workshop "High-Resolution Earth Imaging for Geospatial Information"*, May 17–20. CD-ROM. With permission.)

primitives. Some other research studies using the recognition of corresponding structures (Della Ventura et al. 1990) or of uniform regions (Petit-Frère 1992; Abbasi-Dezfouli and Freeman 1996), a moment-based approach with fine-invariant features (Flusser and Suk 1994), or a wavelet transform approach (Djamdji and Bijaoui 1995) were performed. They were only used to extract well-defined GCPs for image registration between different spaceborne VIR images.

More development must be done to integrate these solutions for generating seed points for gray-level matching. Some apparent contradictions should also be considered in future research studies, such as

- The theoretical prediction of Marr (1982) that feature-based matching is better than gray-level matching versus better experimental results with gray-level matching than with feature-based matching

- The theoretical automated image matching error (much better than one pixel) versus the experimental results (one and more pixels, depending on the data)

- The so-called superiority of computer matching over visual matching versus the experimental results

Overall, studies conducted until now confirm our earlier statement in this section that image matching has been a lively research topic for the past 30 years, but only time will tell whether it will remain so for the next 30 years.

Whatever the matching method and the strategy adopted, there is always a need for postprocessing the extracted elevation data, for example, to remove blunders, fill the mismatched areas, correct for vegetation cover, and smooth the DEM. Different methods can be used depending on the capability of the (stereo) workstation: manual, automatic, or interactive. A blunder-removal function is needed to remove any artifacts or noise when an elevation value is drastically different from its neighbors. These functions generally use existing filters based on statistical computation (mean, standard deviation). Some functions tend to remove small noisy areas, whereas inversely, some tend to increase failed areas on the rationale that the pixels surrounded by failed pixels tend to have a high probability of being noisy. These functions are well adapted to be performed automatically.

To fill the mismatched and the noisy areas once they are detected, interpolation functions are used to replace the mismatched values by interpolating from good elevation values of the edges of the failed areas. Standard interpolation functions (bilinear, distance-weighted), which can be performed automatically, are adequate for small areas (less than 200 pixels). For larger areas, an operator should interactively stereo extract seed points to fill the mismatched areas of the raw DEM. Another solution is to first transform the DEM into a triangular irregular network (TIN) and then display it over the stereo pair in the stereo workstation. The operator can then edit the appropriate vertex of triangles to better fit the shape of the TIN with his or her 3D perception of the terrain relief. In addition, the operator can extract some specific geomorphologic features (mountain crests, thalwegs, lake shorelines), which can be integrated to generally reduce the largest errors at the lowest and highest elevations in the DEM. Using human 3D perception to edit DEM is thus advantageous since it produces a more coherent and consistent terrain relief reconstruction.

Forested areas also have to be edited for vegetation cover, depending on the relation between sensor resolution, the expected DEM accuracy, and canopy height. An automatic classification or an interactive stereo extraction or both can delimit the different forested areas and measure their canopy height. This information is then used to reduce the elevations at the ground level. Finally, an appropriate method of filtering must also be applied to smooth the "pit and hummock" pattern of the DEM, while preserving the sharp breaks in slopes. Filtering improves the relative DEM accuracy or the relationship between neighboring values, whereas the absolute DEM accuracy appears to be controlled by the generation method, system, and software (Giles and Franklin 1996). Unfortunately, only a few research studies and scientific results have been devoted to and published on the postprocessing step. Most of the time, stereo workstation manufacturers develop their own methods and tools to achieve this last, but not least, step of DEM generation.

8.4.5 "Orthorectification"

The last step of the geometric processing is image rectification with DEM (Figure 8.14). To orthorectify the original image into a map image, there are two processing operations:

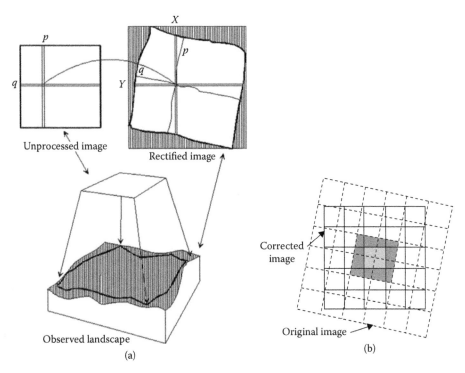

FIGURE 8.14
Image rectification to project the original image to the ground reference system: the geometric (a) and radio-metric (b) operations.

1. A geometric operation to compute the cell coordinates in the original image for each map image cell (Figure 8.14a)
2. A radiometric operation to compute the intensity value or digital number (DN) of the map-image cell as a function of the intensity values of original image cells that surround the previously computed position of the map image cell (Figure 8.14b)

8.4.5.1 Geometric Operation

The geometric operation requires the two equations of the geometric model with the previously computed unknown parameters, and sometimes elevation information. Since the 2D models do not use elevation information, the accuracy of the resulting rectified image will depend on the image viewing/look angle and the terrain relief. On the other hand, 3D models take into account the elevation distortion and a DEM is thus needed to create accurate orthorectified images. This rectification should then be called an orthorectification. But if no DEM is available, different altitude levels can be input for different parts of the image (a kind of "rough" DEM) in order to minimize this elevation distortion. It is then important to have a quantitative evaluation of the DEM impact on the rectification/orthorectification process, both in terms of elevation accuracy for the positioning accuracy and grid spacing for the level of details. This last aspect is more important with HR images because a poor grid spacing when compared with the image spacing could generate artifacts for linear features (wiggly roads or edges).

Figures 8.15 and 8.16 give the relationship between DEM accuracy (including interpolation in the grid), and the viewing and look angles with the resulting positioning error on VIR and SAR orthoimages, respectively. These curves were mathematically computed with the elevation distortion parameters of a 3D physical model (Toutin 1995b). However, they could also be used as an approximation for other 3D physical and empirical models. One of the advantages of these curves is that they can be used to find any third parameter when two others are known. It can be useful not only for the quantitative evaluation of

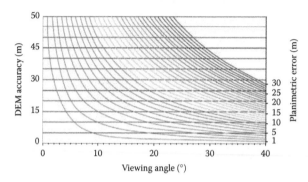

FIGURE 8.15
Relationship between the digital elevation model (DEM) accuracy (in meters), the viewing angle (in degrees) of the visible and infrared (VIR) image, and the resulting positioning error (in meters) generated on the orthoimage. (From Toutin, T., *EARSeL J Adv Remote Sens*, 4, 2, 1995b.)

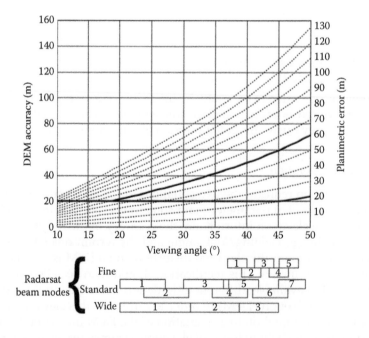

FIGURE 8.16
Relationship between the digital elevation model (DEM) accuracy (in meters), the look angle (in degrees) of the synthetic aperture radar (SAR) image, and the resulting positioning error (in meters) generated on the SAR orthoimage. The different boxes at the bottom represent the range of look angles for each Radarsat beam mode. (From Toutin, T., *Journal canadien de télédétection*, 24, 1998.)

the orthorectification but also to forecast the appropriate input data, DEM, or the viewing/look angles, depending on the objectives of the project.

For example (Figure 8.15), with a SPOT image acquired with a viewing angle of 10° and having a 45-m accurate DEM, the error generated on the orthoimage is 9 m. Inversely, if a 4-m final positioning accuracy for the orthoimage is required and there is a 10-m accurate DEM, the VIR image should be acquired with a viewing angle less than 20°. The same error evaluation can be applied to SAR data using the curves given in Figure 8.16. As another example, if positioning errors of 60 and 20 m on standard-1 (S1) and fine-5 (F5) orthoimages, respectively, are required, a 20-m elevation error, which includes the DEM accuracy and the interpolation into the DEM, is thus sufficient. For HR images (spaceborne or airborne), the surface heights (buildings, forest, hedges) should be either included in the DTM to generate a DSM or taken into account in the overall elevation error. In addition, an inappropriate DEM in terms of grid spacing can generate artifacts with HR images acquired with large viewing angles, principally over high relief areas (Zhang, Tao, and Mercer 2001).

Finally, for any map coordinates (X, Y), with the Z-elevation parameter extracted from a DEM when 3D models are used, the original image coordinates (column and line) are computed from the two resolved equations of the model. However, the computed image coordinates of the map image will not be directly overlaid on a pixel center of the original image; in other words, the column and line computed values will be rarely, if ever, integer values.

8.4.5.2 Radiometric Operation

Since the computed coordinate values in the original image are not integers, one must compute the DN to be assigned to the map image cell. In order to compute the DN to be assigned to the map image cell, the radiometric operation uses a resampling kernel applied to original image cells: either the DN of the closest cell (called "nearest neighbor resampling"), or a specific interpolation or deconvolution algorithm using the DNs of surrounding cells. In the first case, the radiometry of the original image and the image spectral signatures are not altered, but the visual quality of the image is degraded. In addition to radiometric degradation, a geometric error of up to half a pixel is introduced. This can cause a disjointed appearance in the map image. If these visual and geometric degradations are acceptable to the end user, it can be an advantageous solution.

In the second case, different interpolation or deconvolution algorithms (bilinear interpolation or sinusoidal function) can be applied. The bilinear interpolation takes into account the four cells surrounding the cell. The final DN is then computed either from two successive linear interpolations in line and column using the DNs of the two surrounding cells in each direction or in one linear interpolation using the DNs of the four surrounding cells. The DNs are weighted as a function of the cell distance from the computed coordinate values. Due to the weighting function, this interpolation creates a smoothing in the final map image.

The theoretically ideal deconvolution function is the $\sin(x)/x$ function. As this $\sin(x)/x$ function has an infinite domain, it cannot be exactly computed. Instead, it can be represented by a piecewise cubic function, such as the well-known cubic convolution. The cubic convolution then computes third-order polynomial functions using a 4×4 cell window. The DNs are first computed successively in the four-column and -line directions, and the final DN is the arithmetic mean of these DNs. This cubic convolution does not smooth, but enhances and generates some contrast in the map image (Kalman 1985).

Due to technological improvements in computers these last years, the sin(x)/x function can now be directly applied as a deconvolution function with different window sizes (generally, 8×8 or 16×16). The computation time with the 16×16-cell window can be 40–80 times more than the computation time required for nearest neighbor resampling. The final image is, of course, sharper, with more details on features.

All these interpolation or deconvolution functions can be applied to VIR and SAR images. However, they are geometric resampling kernels, which are not very well adapted to SAR images. Instead, for SAR images it is better to use statistical functions based on the characteristics of the radar used, such as existing adaptive filters using local statistics (Lee 1980; Lopes et al. 1993; Touzi 2002). Combining the filtering process with the resampling process also avoids multiple radiometric processing and transformation, which largely degrades the image content and its interpretation (Toutin 1995b).

Since interpolation or deconvolution functions transform the DNs and then alter the radiometry of the original image, problems may be encountered in subsequent spectral signature or pattern recognition analysis. Consequently, any process based on image radiometry should be performed before using the interpolation or deconvolution algorithms.

Figures 8.17 and 8.18 are examples of the application of different resampling kernels to WorldView-1 panchromatic mode image and Radarsat-2 SAR ultrafine-mode (U2) ground-range image, respectively, during their orthorectification process with DEM. Subimages (around 200×200 pixels) were resampled with a factor of three to better illustrate the variations among the resampling kernels. The WorldView and Radarsat resampled image

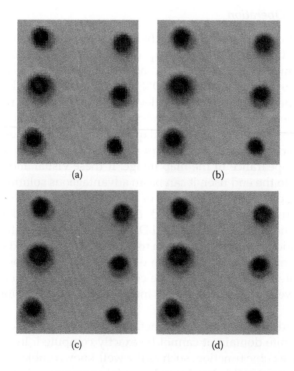

(a) (b)

(c) (d)

FIGURE 8.17
Examples of geometric resampling kernels applied to WorldView-2 panchromatic mode image during the orthorectification process with digital elevation model: The subimages are 193×219 pixels with 0.15-m spacing. Letters (a) through (d) refer to different geometric resampling kernels (nearest neighbor, bilinear, cubic convolution, and sin(x)/x with 16×16 window, respectively). (WorldView-1 Image © and courtesy Digital Globe, 2009.)

pixels are then 0.15 and 0.5 m, respectively. Letters (a) through (d) refer to different geometric resampling kernels (nearest neighbor, bilinear, cubic convolution, and $\sin(x)/x$ with 16×16 window, respectively), and letters (e) and (f) refer to statistical adaptive SAR filters (enhanced Lee and gamma filters with 5×5 window, respectively). For both VIR and SAR images, the nearest neighbor resampling kernel (a) generates "blocky" images, whereas the bilinear resampling kernel (b) generates fuzzy images or out-of-focus images. One can also see the step gradient, which is reduced with the kernel sequence (from a to d). The best results are obtained with sinusoidal resampling kernels (c and d), but the true sinusoidal function (d) generates sharper features that represent the original circle shapes extremely well (Figure 8.17). The $\sin(x)/x$ function with the 16×16 window kernel should thus be favored for optical images during orthorectification, although the processing time is longer for this kernel.

On the other hand, with the Radarsat-2 SAR image (Figure 8.18), all the white dots (corresponding to houses) change from square shapes (a) to round shapes (b, c, d), which do not correspond with the original geometry of square houses. In addition, worm-shaped artifacts (typically from large oversampling) are generated in homogeneous flat areas around the houses and in the "black" streets; these artifacts become more evident with the sequence of (a) through (d) resampling kernels. Consequently, the three last geometric resampling kernels (bilinear, cubic convolution, and $\sin(x)/x$) should not be used with SAR images. The two adaptive filters (e and f) not only give a better image appearance

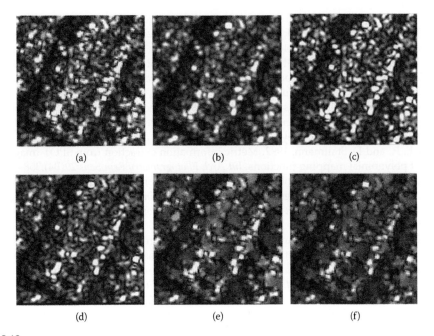

(a) (b) (c)

(d) (e) (f)

FIGURE 8.18
Examples of geometric/statistical resampling kernels applied to Radarsat-2 synthetic aperture radar (SAR) ultrafine-mode (U2) image during the orthorectification process with digital elevation model (DEM). The subimages are 265×262 pixels with 0.5-m spacing. Letters (a) through (d) refer to geometric resampling kernels (nearest neighbor, bilinear, cubic convolution, $\sin(x)/x$ with 16×16 window, respectively), and letters (e) and (f) refer to statistical adaptive SAR filters (enhanced Lee and gamma filters with 5×5 window, respectively). (Radarsat-2 Data © MacDonald, Dettwiler, and Associates Ltd. (2008)—All Rights Reserved and Courtesy of Canadian Space Agency.)

than all the geometric resampling kernels due to the fact that the SAR speckle is filtered at the same time, but also keep the original feature shapes (even the neighboring houses are better separated and discriminated) without generating worm-shaped artifacts. These statistical kernels are better adapted for SAR images, even with six-time oversampling of SAR resolution (3 m), and they should always be favored during orthorectification.

References

Abbasi-Dezfouli, M., and T. G. Freeman. 1996. Stereo-images registration based on uniform patches. In *International Archives of Photogrammetry and Remote Sensing*, Vienna, Austria, July 9–18, vol. 31(B2), 101–6. Vienna: Austrian Society for Surveying and Geoinformation.

Ackermann, F., D. Fritsch, M. Hahn, F. Schneider, and V. Tsingas. 1995. Automatic generation of digital terrain models with MOMS-02/D2 data. In *Proceedings of the MOMS-02 Symposium*, Köln, Germany, July 5–7, 79–86. Paris, France: EARSeL.

Ahn, C. -H., S. -I. Cho, and J. C. Jeon. 2001. Orthorectification software applicable for IKONOS high-resolution images: GeoPixel-Ortho. In *Proceedings of IGARSS*, Sydney, Australia, July 9–13, 555–7. Piscataway, NJ: IEEE.

Al-Roussan, N., P. Cheng, G. Petrie, T. Toutin, and M. J. Valadan Zoej. 1997. Automated DEM extraction and ortho-image generation from SPOT level-1B imagery. *Photogramm Eng Rem Sensing* 63:965–74.

Baetslé, P. L. 1966. Conformal transformations in three dimensions. *Photogramm Eng* 32:816–24.

Bähr, H. P. 1976. Geometrische Modelle für Abtasteraufzeichnungen von Erkundungdsatlliten. *Bildmessung und Luftbildwesen* 44:198–202.

Baker, H. H., and T. O. Binford. 1981. Depth from edge and intensity based stereo. In *Proceedings of the Seventh International Joint Conference on Artificial Intelligence*, Vancouver, B. C., Canada, 631–6. San Diego: Martin Kaufman Publishers.

Bakker, W. 2000. Satellite and sensor systems for environmental monitoring. In *Encyclopedia of Analytical Chemistry: Applications, Theory and Instrumentation*. ed. R. A. Meyers, vol. 10, 8693–746. Chichester, UK: John Wiley & Sons.

Baltsavias, E. P., and D. Stallmann. 1992. Metric information extraction from SPOT images and the role of polynomial mapping functions. *Int Arch Photogrametry Rem Sens* 29(B4):358–64.

Baltsavias, E., L. Zhang, and H. Eisenbeiss. 2005. DSM generation and interior orientation determination of ikonos images using a test field in Switzerland. In *International Society of Photogrammetry, Remote Sensing Workshop "High-Resolution Earth Imaging for Geospatial Information,"* May 17–20. CD-ROM.

Bannari, A., D. Morin, G. B. Bénié, and F. J. Bonn. 1995. A theoretical review of different mathematical models of geometric corrections applied to remote sensing images. *Rem Sens Rev* 13:27–47.

Belgued, Y., S. Goze, J. -P. Planès, and P. Marthon. 2000. Geometrical block adjustment of multisensor radar images. In *Proceedings of EARSeL Workshop: Fusion of Earth Data*, Sophia-Antipolis, France, January 26–28 2000, 11–16. Nice, France: SEE GréCA/EARSeL.

Bianconi, M., M. Crespi, F. Fratarcangeli, F. Giannone, and F. Pierlice. 2008. A new strategy for rational polynomial coefficients generation. In *Proceedings of the EARSeL Joint Workshop*, C. Jürgens, ed. Bochum, Germany, March 5–7, 21–28.

Billingsley, F. C. 1983. Data processing and reprocessing. In *Manual of Remote Sensing*, vol. 1, 2nd ed., ed. R. N. Colwell, 719–22. Falls Church, VA: Sheridan Press.

Bonneval, H. 1972. Levés topographiques par photogrammétrie aérienne, dans *Photogrammétrie générale: Tome 3*, Collection scientifique de l'Institut Géographique National. Paris, France: Eyrolles Editeur.

Bouillon, A., E. Breton, F. de Lussy, and R. Gachet. 2002. SPOT5 HRG and HRS first in-flight geometric quality results. In *Proceedings of SPIE, Vol. 4881A Sensors, system, and Next Generation Satellites VII*, Agia Pelagia, Crete, Greece, September 22–27. Bellingham, WA: SPIE.

Caloz, R., and C. Collet. 2001. Transformations géométriques. In *Précis de télédétection, Volume 3: Traitements numériques d'images de télédétection.* 76–105. Ste Foy, Québec, Canada: Presse de l'Université du Québec.

Campagne, P. 2000. Apport de la spatio-triangulation pour la caractérisation de la qualité image géométrique. *Bulletin de la Société Française de Photogrammétrie et de Télédétection* 159:66–26.

Cantou, P. 2002. French Guiana Mapped using ERS-1 Radar Imagery. Web site of SPOT-Image, Toulouse, France, http://www.spotimage.fr/home/appli/apcarto/guiamap/welcome.htm/ (accessed April 20, 2003).

Centre National d'Études Spatiales (CNES). 1980. Le mouvement du véhicule spatial en orbite. Toulouse, France: CNES.

Chen, L. -C., and T. -A. Teo. 2002. Rigorous generation of orthophotos from EROS-A high resolution satellite images. In *International Archives of Photogrametry and Remote Sensing and Spatial Information Sciences*, Ottawa, Canada, July 8–12, vol. 34(B4), 620–5. Ottawa, Ontario: Natural Resources Canada.

Chen, L. -C., T. -A. Teo, and C. -L. Liu. 2006. The geometrical comparison of RSM and RFM for FORMOSAT-2 satellite images. *Photogramm Eng Rem Sensing* 72(5):573–9.

Cheng, P., D. Smith, and S. Sutton. 2005. Mapping of QuickBird images: Improvement in accuracy since release of first QuickBird. *GeoInformatics* 8:50–2.

Cheng, P., and T. Toutin. 1998. Unlocking the potential for IRS-1C data. *Earth Observation Magazine* 7(3):24–6.

Cooper, P. R., D. E. Friedman, and S. A. Wood. 1987. The automatic generation of digital terrain models from satellite images by stereo. *Acta Astronaut* 15(3):171–80.

Curlander, J. C. 1982. Location of spaceborne SAR imagery. *IEEE Trans Geosci Rem Sens* 22:106–12.

Davis, C. H. and X. Wang. 2001. Planimetric accuracy of IKONOS 1-m panchromatic image products. In *Proceedings of the ASPRS Annual Conference*, St. Louis, Missouri, April 23–27. Bethesda, MD: ASPRS.

Day, T., and J. -P. A. Muller. 1988. Quality assessment of digital elevation models produced by automatic stereo-matchers from SPOT image pairs. *Photogramm Record* 12(72):797–808.

Della Ventura, A., A. Rampini, and R. Schettini. 1990. Image Registration by Recognition of Corresponding Structures. *IEEE Trans Geosci Rem Sens* 28(3):305–314.

Derenyi, E. E. 1970. "An Exploratory Investigation into the Relative Orientation of Continuous Strip Imagery," Research Report No. 8, University of New Brunswick, Canada.

Derenyi, E. E., and G. Konecny. 1966. Infrared scan geometry. *Photogramm Eng* 32:773–8

de Leeuw, A. J., L. M. M. Veugen, and H. T. C. van Stokkom. 1988. Geometric correction of remotely sensed imagery using ground control points and orthogonal polynomials. *Int J Rem Sens* 9:1751–9.

de Masson d'Autume, G. 1979. Le traitement géométrique des images de télédétection. *Bulletin de la Société Française de Photogrammétrie et de Télédétection* 73–74:5–16.

de Sève, D., T. Toutin, and R. et Desjardins. 1996. Evaluation de deux méthodes de corrections géométriques d'images Landsat-TM et ERS-1 RAS dans une étude de linéaments géologiques. *Int J Rem Sens* 17:131–42.

Di, K., R. Ma, and R. Li. 2003. Geometric processing of ikonos stereo imagery for coastal applications. *Photogramm Eng Rem Sensing* 69(8):873–9.

Dial, G., and J. Grodecki. 2002. Block adjustment with rational polynomial camera models. In *Proceedings of the ACSM-ASPRS Annual Conference/XXII FIG International Congress*, Washington D. C., April 19–26. Bethesda, MD: ASPRS.

Djamdji, J. -P., and A. Bijaoui. 1995. Disparity analysis: a wavelet transform approach. *IEEE Trans Geosci Rem Sens* 33(1):67–76.

Dorrer, E., W. Maier, and V. Uffenkamp. 1995. Stereo Compilation of MOMS-02 Scenes on the Analytical Plotter. In *Proceedings of the MOMS-02 Symposium*, Köln, Germany, July 5–7, 95–110.

Dowman I., and J. Dolloff. 2000. An evaluation of rational function for photogrammetric restitution. In *International Archives of Photogrammetry and Remote Sensing*, Amsterdam, The Netherlands, July 16–23, vol. 33(B3), 254–66. GITC, Amsterdam, The Netherlands.

Ebner, H., and F. Muller. 1986. Processing of digital three line imagery using a generalized model for combined point determination. In *International Archives of Photogrametry and Remote Sensing*, Rovamieni, Finland, August 19–22, vol. 26(B3), 212–22. Helsinki, Finland: ISPRS.

Escobal, P. R. 1965. *Methods of Orbit Determination*, 479. Malabar, FL: Krieger Publishing Company.

Flusser, J., and T. Suk. 1994. A moment-based approach to registration of images with affine geometric distortions. *IEEE Trans Geosci Rem Sens* 32(2):382–7.

Förstner, W. 1982. On the geometric precision of digital correlation. In *International Archives of Photogrammetry*, Helsingfors, Finland, vol. 24(B3), 176–89. Helsinki, Finland: ISPRS.

Fraser, C. S., E. Baltsavias, and A. Gruen. 2002. Processing of IKONOS imagery for sub-metre 3D positioning and building extraction. *ISPRS J Photogramm Rem Sens* 56:177–94.

Fraser, C. S., G. Dial, and J. Grodecki. 2006. Sensor orientation via RPCs. *ISPRS J Photogramm Rem Sens* 60(3):182–94.

Fraser, C. S., and H. B. Hanley. 2005. Bias-compensated RPCs for sensor orientation of high-resolution satellite imagery. *Photogramm Eng Rem Sensing* 71(8):909–15.

Fraser, C. S., H. B. Hanley, and T. Yamakawa. 2002. Three-dimensional geopositioning accuracy of IKONOS imagery. *Photogramm Record* 17:465–79.

Fraser, C. S., and M. Ravanbakhsh. 2009. Georeferencing accuracy of GeoEye-1 imagery. *Photogramm Eng Rem Sensing* 75(6):634–8.

Friedmann, D. E., J. P. Friedel, K. L. Magnusses, K. Kwok, and S. Richardson. 1983. Multiple scene precision rectification of spaceborne imagery with few control points. *Photogramm Eng Rem Sensing* 49:1657–67.

Ganas, A., E. Lagios, and N. Tzannetos. 2002. An investigation into the spatial accuracy of the IKONOS-2 orthoimagery within an urban environment. *Int J Rem Sens* 23:3513–9.

Gibson, J. 1984. Processing stereo imagery from line imagers. In *Proceedings of 9th Canadian Symposium on Remote Sensing*, St. John's, Newfoundland, Canada, August 14–17, 471–88. Ottawa, Canada: Canadian Society for Remote Sensing.

Giles, P. T., and S. E. Franklin. 1996. Comparison of derivative topographic surfaces of a DEM generated from stereoscopic SPOT images with field measurements. *Photogramm Eng Rem Sens* 62(10):1165–71.

Gopala Krishna, B., B. Kartikeyan, K. V. Iyer, R. Mitra, and P. K. Srivastava. 1996. Digital photogrammetric workstation for topographic map updating using IRS-1C stereo imagery. In *International Archives of Photogrammetry and Remote Sensing*, Vienna, Austria, July 9–18, vol. 31(B4), 481–5. Vienna: Austrian Society for Surveying and Geoinformation.

Goshtasby, A. 1988. Registration of images with geometric distortions. *IEEE Trans Geosci Rem Sens* 26:60–4.

Gracie, G., J. W. Bricker, R. K. Brewer, and R. A. Johnson. 1970. "Stereo Radar Analysis". Report No. FTR-1339-1, U.S. Engineer Topographic Laboratory, Ft. Belvoir, VA.

Grodecki, J. 2001. IKONOS stereo feature extraction—RPC approach. In *Proceedings of the ASPRS Annual Conference*, held in St Louis, Missouri, April 23–27. Bethesda, MD: ASPRS.

Grodecki, J., and G. Dial. 2003. Block adjustment of high-resolution satellite images described by rational polynomials. *Photogramm Eng Rem Sens* 69(1):59–68.

Guertin, F., and E. Shaw. 1981. Definition and potential of geocoded satellite imagery products. In *Proceedings of the 7th Canadian Symposium on Remote Sensing*, Winnipeg, Canada, September 8–11, 384–94. Manitoba: Manitoba Remote Sensing Centre.

Gugan, D. J. 1987. Practical aspects of topographic mapping from SPOT imagery. *Photogramm Record* 12:349–55.

Guichard, H. 1983. Etude théorique de la précision dans l'exploitation cartographique d'un satellite à défilement: application à SPOT. *Bulletin de la Société Française de Photogrammétrie et de Télédétection* 90:15–26.

Guindon, B., and M. Adair. 1992. Analytic formulation of spaceborne SAR image geocoding and value-added products generation procedures using digital elevation data. *Can J Rem Sens* 18:2–12.

Hanley, H. B., and C. S. Fraser. 2001. Geopositioning accuracy of IKONOS imagery: Indications from two dimensional transformations. *Photogramm Record* 17:317–29.

Hähn, M., and W. Förstner. 1988., The applicability of feature-based and a least squares matching algorithm for DEM acquisition. *Int Arch Photogramm Rem Sens* 27(B9):III137–50.

Hargreaves, D., and B. Robertson. 2001. Review of QuickBird-1/2 and OrbView-3/4 products from MacDonald Dettwiler processing systems. In *Proceedings of the ASPRS Annual Conference*, St Louis, Missouri, April 23–27. Bethesda, MD: ASPRS.

Helava, U. V. 1988. Object space least squares correlation. In *Proceedings of the ACSM-ASPRS Annual Convention*, St. Louis, Missouri, vol. 3, 46–55. Bethesda, MD: ACSM/ASPRS.

Hoffmann, C., K. Steinnocher, M. Kasanko, T. Toutin, and P. Cheng. 2001. Urban mapping with high resolution satellite imagery: IKONOS and IRS data put to the test. *GeoInformatics* 4(10):34–7.

Hoffman, O., and F. Muller. 1988. Combined point determination using digital data of three line scanner systems. In *International Archives of Photogrammetry and Remote Sensing*, Kyoto, Japan, July 3–9, vol. 27(B3), 567–77. Helsinki, Finland: ISPRS.

Hoogeboom, P., P. Binnenkade, and L. M. M. Veugen. 1984. An algorithm for radiomeric and geometric correction of digital SLAR data. *IEEE Trans Geosci Rem Sens* 22:570–6.

Jacobsen, K. 1997. Calibration of IRS-1C pan camera. In *ISPRS Workshop on Sensors and Mapping from Space*, Hannover, Germany, September 29–October 2, 163–70. Bonn, Germany: German Society for Photogrammetry and Remote Sensing/ISPRS.

Jacobsen, K. 2002. Generation of orthophotos with carterra geo images without orientation information. In *Proceedings of the ACSM-ASPRS Annual Conference/XXII FIG International Congress*, Washington DC., April 19–26. Bethesda, MD: ASPRS.

Kalman, L. S. 1985. Comparison of cubic-convolution interpolation and least-squares restoration for resampling landsat-MSS imagery. In *Proceedings of the 51st Annual ASP-ASCM Convention: "Theodolite to Satellite,"* Washington D.C., March 10–15, 546–56. Falls Church, VA: ASP.

Kersten, T., E. Baltsavias, M. Schwarz, and I. Leiss. 2000. Ikonos-2 Cartera Geo—Erste geometrische Genauigkeitsuntersuchungen in der Schweiz mit hochaufgeloesten Satellitendaten. *Vermessung, Photogrammetrie, Kulturtechnik* 8:490–7.

Khizhnichenko, V. I. 1982. Co-ordinates transformation when geometrically correcting Earth space scanner images. [In Russian.] *Earth Explor From Space* 5:96–103.

Kim, T., and I. Dowman. 2006. Comparison of two physical sensor models for satellite images: Position-rotation model and orbit-attitude model. *The Photogrametric Rec* 21(114):110–23.

Kim, J. R., and J. -P. Muller. 2002. 3D reconstruction from very high resolution satellite stereo and its application to object identification. In *International Archives of Photogrametry and Remote Sensing and Spatial Information Sciences*, Ottawa, Canada, July 8–12, vol. 34(B4), 637–43. Ottawa, Ontario: Natural Resources Canada.

Konecny, G. 1970. Metric problems in remote sensing. *ITC Publ Ser A* 50:152–77.

Konecny, G. 1976. Mathematische Modelle und Verfahren zur geometrischen Auswertung von Zeilenabtaster-Aufnahmen. *Bildmessung und Luftbildwesen* 44:188–97.

Konecny, G. 1979. Methods and possibilities for digital differential rectification. *Photogramm Eng Rem Sens* 45:727–34.

Konecny, G., E. Kruck, and P. Lohmann. 1986. Ein universeller Ansatz für die geometrischen Auswertung von CCD-Zeilenabtasteraufnahmen. *Bildmessung und Luftbildwesen* 54:139–46.

Konecny, G., P. Lohmann, H. Engel, and E. Kruck. 1987. Evaluation of SPOT imagery on analytical instruments. *Photogramm Eng Rem Sens* 53:1223–30.

Kornus, W., M. Lehner, and M. Schroeder. 2000. Geometric in-flight calibration by block adjustment using MOMS-2P 3-line-imagery of three intersecting stereo-strips. *Bulletin de la Société Française de Photogrammétrie et de Télédétection* 159:42–54.

Kratky, W. 1971. Precision processing of ERTS imagery. In *Proceedings of ASP-ACSM Fall Convention*, San Francisco, CA, September 7–11, 481–514. Falls Church, VA: ASP.

Kratky, W. 1987. Rigorous stereophotogrammetric treatment of SPOT images. In *Proceedings of the International Conference on SPOT-1: Image utilization, assessment, and results*. Paris, France, November, 1281–88. Toulouse, France: CEPADUES Editions.

Kratky, W. 1989. On-line aspects of stereophotogrammetric processing of SPOT images. *Photogramm Eng Rem Sens* 55:311–6.

Kristóf, D., É. Csató, and D. Ritter. 2002. Application of high-resolution satellite images in forestry and habitat mapping – evaluation of IKONOS images through a hungarian case study. In *International Archives of Photogrametry and Remote Sensing and Spatial Information Sciences*, Ottawa, Canada July 8–12, vol. 34(B4), 602–7. Ottawa, Ontario: Natural Resources Canada.

La Prade, G. L. 1963. An analytical and experimental study of stereo for radar. *Photogramm Eng* 29:294–300.

Leberl, F. W. 1972. On model formation with remote sensing imagery. *Österreichiches Zeitschrift für Vermessungswesen* 2:43–61.

Leberl, F. W. 1978. *Satellite Radargrammetric*, vol. 239, Serie C. Munich, Germany: Deutsche Geodactische Kommission.

Leberl, F. W. 1990. *Radargrammetric Image Processing*. Norwood, MA: Artech House.

Leberl, F., K. Maurice, J. K. Thomas, and M. Millot. 1994. Automated radar image matching experiment. *ISPRS J Photogramm Rem Sens* 49(3):19–33.

Lee, J. S. 1980. Digital Image enhancement and noise filtering by use of local statistics. *IEEE Trans Pattern Anal Mach Intell* 2:165–8.

Lee, J. -B., Y. Huh, B. Seo, and Y. Kim. 2002. Improvement the positional accuracy of the 3D terrain data extracted From IKONOS-2 satellite imagery. In *International Archives of Photogrammetry and Remote Sensing*, Graz, Austria, September 9–13, vol. 34(B3), B142–B145. Institute for Computer Graphics and Vision, Graz, Austria.

Lehner, M., R. Müller, and P. Reinartz. 2005. DSM and orthoimages from QuickBird and IKONOS data using rational polynomial functions. In *International Archives of Photogrammetry and Remote Sensing*, Hannover, Germany, May 17–20, vol. 36.

Li, R. 1998. Potential of high-resolution satellite imagery for national mapping products. *Photogramm Eng Rem Sens* 64(12):1165–70.

Light, D. L., D. Brown, A. Colvocoresses, F. Doyle, M. Davies, A. Ellasal, J. Junkins, J. Manent, A. McKenney, R. Undrejka, and G. Wood. 1980. Satellite photogrammetry. In *Manual of Photogrammetry*, 4th ed, Chapter XVII, ed. C. C. Slama, 883–977. Falls Church, VA: ASP Publishers.

Lopes, A., E. Nezry, R. Touzi, and H. Laur. 1993. Structure detection and statistical adaptive speckle filtering in SAR images. *Int J Rem Sens* 14:1735–58.

Madani, M. 1999. Real-time sensor-independent positioning by rational functions. In *Proceedings of ISPRS Workshop on Direct Versus Indirect Methods of Sensor Orientation*, Barcelona, Spain, November 25–26, 64–75. Helsinki, Finland: ISPRS.

Marr, D. and T. Poggio. 1977. A Computation of Stereo Disparity. In *Proceedings of the Royal Society of London*, B194:283–287.

Marr, D. 1982. *Vision: A Computational Investigation into the Human Representation and Processing of Visual Information*. San Francisco: W.H. Freeman and Co.

Marr, D., and E. Hildreth. 1980. A theory of edge detection. *Proc R Soc Lond* B207:187–217.

McCarthy, F., P. Cheng, and T. Toutin. 2001. Case study of using IKONOS imagery in small municipalities. *Earth Observation Mag* 10(11):13–6.

Meinel, G., and J. Reder. 2001. IKONOS Satellitenbilddaten - ein erster Erfahrungsbericht. *Kartograp hische Nachrichten* 51:40–6.

Mikhail, E. M. 1976. *Observations and Least Squares*. New York: Harper & Row Publishers.

Naraghi, M., W. Stromberg, and M. Daily. 1983. Geometric rectification of radar imagery using digital elevation models. *Photogramm Eng Rem Sens* 49:195–59.

Noguchi, M., C. S. Fraser, T. Nakamura, T. Shimono, and S. Oki. 2004. Accuracy assessment of QuickBird stereo imagery. *Photogramm Rec* 19(106):128–37.

Novak, K. 1992. Rectification of digital imagery. *Photogramm Eng Rem Sens* 58:339–44.

OGC. 1999. *The Open GISTM Abstract Specifications: The Earth Imagery Case*, vol. 7. http://www.opengis.org/techno/specs/htm/(accessed September 24, 2009).

Okamoto, A. 1981. Orientation and construction of models. Part III: Mathematical basis of the orientation problem of one-dimensional central perspective photographs. *Photogramm Eng Rem Sens* 47:1739–52.

Okamoto, A. 1988. Orientation theory of CCD line-scanner images. In *International Archives of Photogrammetry and Remote Sensing*, Kyoto, Japan, July 3–9, vol. 27(B3), 609–17. Helsinki, Finland: ISPRS.

Okamoto, A., C. Fraser, S. Hattori, H. Hasegawa, and T. Ono. 1998. An alternative approach to the triangulation of SPOT imagery. In *International Archives of Photogrammetry and Remote Sensing*, Stuttgart, Germany, September 7–10, vol. 32(B4), 457–62. German Society for Photogrammetry and Remote Sensing, Bonn, Germany.

Paderes, F. C., E. M. Mikhail, and J. A. Fagerman. 1989. Batch and on-line evaluation of stereo SPOT imagery. In *Proceedings of the ASPRS - ACSM Convention*, vol. 3, 31–40. Baltimore, MD. Bethesda, MD: ASPRS.

Palà, V., and X. Pons. 1995. Incorporation of relief in polynomial-based geometric corrections. *Photogramm Eng Rem Sens* 61:935–44.

Petit-Frère, J. 1992. Prise en compte des différences photométriques entre images dans les techniques de stéréorestitution. In *International Archives of Photogrammetry and Remote Sensing*, Washington DC, August 2–14, vol. 29(B4), 392–8. Bethesda, MD: ASPRS.

Petrie, G. 2002. The ACSM-ASPRS Conference: A report on the Washington Meeting. *GeoInformatics* 5(6):42–43.

Raggam, J., A. Almer, and D. Strobl. 1994. A Combination of SAR and Optical Line Scanner Imager for Stereoscopic Extraction of 3-D Data. *ISPRS Journal of Photogrammetry and Remote Sensing* 49(4):11–21.

Robertson, B. 2003. Rigorous geometric modeling and correction of QuickBird imagery. In *Proceedings of the International Geoscience and Remote Sensing IGARSS 2003*, Toulouse, France, July 2003, 21–5. Toulouse, France: Centre national d'Etudes Spatiales (CNES).

Robertson, B., A. Erickson, J. Friedel, B. Guindon, T. Fisher, R. Brown, P. Teillet, M. D'iorio, J. Cihlar, and A. Sanz. 1992. GeoComp, a NOAA AVHRR geocoding and compositing system. In *International Archives of Photogrammetry and Remote Sensing*, Wahshington DC, August 3–14, vol. 24(B2), 223–8. Bethesda, MD: ASPRS.

Rosenfield, G. H. 1968. Stereo radar techniques. *Photogramm Eng* 34:586–94.

Rosenholm, D. 1986. Numerical accuracy of automatic parallax measurement of simulated SPOT images. *Can J Rem Sens* 12(2):103–13.

Sakaino, S., H. Suzuki, P. Cheng, and T. Toutin. 2000. Updating maps of Kazakhstan using stitched SPOT images. *Earth Observation Mag* 9(3):11–13.

Salamonowicz, P. H. 1986. Satellite orientation and position for geometric correction of scanner imagery. *Photogramm Eng Rem Sens* 52:491–9.

Savopol, F., A. Leclerc, T. Toutin, and Y. Carbonneau. 1994. La correction géométrique d'images satellitaires pour la Base nationale de données topographiques. *Geomatica, été* 48:193–207.

Sawada, N., M. Kikode, H. Shinoda, H. Asada, M. Iwanaga, S. Watanabe, and K. Mori. 1981. An analytic correction method for satellite MSS geometric distortion. *Photogramm Eng Rem Sens* 47:1195–203.

Schneider, F., and M. Hahn. 1995. Automatic DEM generation using MOMS-02/D2 image dat. In *Proceedings of the Photogrammetric Week '95*, Stuttgart, Germany, September 11–15, 85–94. Heidelberg, Germany: Wichmann.

Shaker, A. 2007. Satellite sensor modeling and 3D geo-positioning using empirical models. *Int J Appl Earth Observation Geoinformation* 10(3):282–95.

Shu, N. 1987. Restitution géométrique des images spatiales par la méthode de l'équation de colinéarité. *Bulletin de la Société Française de Photogrammétrie et de Télédétection* 105:27–40.

Schut, G. H. 1966. Conformal transformations and polynomials. *Photogramm Eng* 32:826–9.

Simard, R., and R. Slaney. 1986. Digital terrain model and image integration for geologic interpretation. In *Proceedings of the 5th Thematic Conference on Remote Sensing for Exploration Geology*, Reno, Nevada, September 29–October 2, 49–60. Ann Arbor, MI: ERIM.

Sylvander, S., P. Henry, C. Bastien-Thiery, F. Meunier, and D. Fuster. 2000. Vegetation geometrical image quality. *Bulletin de la Société Française de Photogrammétrie et de Télédétection* 159:59–65.

Tannous, I., and B. Pikeroen. 1994. Parametric modelling of spaceborne SAR image geometry. Application: SEASAT/SPOT image registration. *Photogramm Eng Rem Sens* 60:755–66.

Tao, V., and Y. Hu. 2001. A comprehensive study of the rational function model for photogrammetric processing. *Photogramm Eng Rem Sens* 67:1347–57.

Tao, V., and Y. Hu. 2002. 3D reconstruction methods based on the rational function model. *Photogramm Eng Rem Sens* 68:705–14.

Tao, C. V., Y. Hu, and W. Jiang. 2004. Photogrammetric exploitation of IKONOS imagery for mapping application. *Int J Rem Sens* 25(14):2833–53.

Tong, X. H., S. J. Liu, and Q. H. Weng. 2009. Geometric processing of QuickBird images for land use mapping—a case study in Shanghai, China. *IEEE J Sel Top Appl Earth Observations Rem Sens* 2(2):61–6.

Tong, X. H., S. J. Liu, and Q. H. Weng. 2010. Bias-corrected RPCs for high accuracy positioning of QuickBird images. *ISPRS J Photogramm Eng Rem Sens* 65(2):218–26.

Toutin, Th., Cl. Nolette, Y. Carbonneau, P. -A. Gagnon et al. 1993. Stéréo-restitution interactive des données SPOT: description d'un nouveau système. *Journal canadien de télédétection* 19(2):146–151.

Toutin, T. 1983. Analyse mathématique des possibilités cartographiques du satellite SPOT. In *Mémoire du diplôme d'Etudes Approfondies*, 1–74. Ecole Nationale des Sciences Géodésiques Saint-Mandé, France.

Toutin, T. 1985. Analyse mathématique des possibilités cartographiques de système SPOT. In *Thèse de Docteur-Ingénieur*, 163. École Nationale des Sciences Géodésiques, St-Mandé, France: IGN.

Toutin, T. 1995a. Intégration de données multi-sources: comparaison de méthodes géométriques et radiométriques. *Int J Rem Sens* 16:2795–811.

Toutin, T. 1995b. Multi-source data integration with an integrated and unified geometric modelling. *EARSeL J Adv Rem Sens* 4(2):118–29.

Toutin, T. 1998. Evaluation de la précision géométrique des images de RADARSAT. *Journal canadien de télédétection* 24:80–8.

Toutin, T. 1999. Error tracking of radargrammetric DEM from RADARSAT images. *IEEE-Trans Geosci Rem Sens* 37:2227–38.

Toutin, T. 2003a. Error tracking in IKONOS geometric processing using a 3D parametric modelling. *Photogramm Eng Rem Sens* 69:43–51.

Toutin, T. 2003b. Block bundle adjustment of Ikonos in-track images. *Int J Rem Sens* 24:851–7.

Toutin, T. 2003c. Block bundle adjustment of Landsat7 ETM+ images over mountainous areas. *Photogramm Eng Rem Sens* 69:1341–9.

Toutin, T. 2003d. Path processing and block adjustment with RADARSAT-1 SAR images. *IEEE-Trans Geosci Rem Sens* 41:2320–8.

Toutin T. 2003e. Compensation par segment et bloc d'images panchromatiques et multibandes de SPOT. *Journal canadien de télédétection* 29(1):36–42.

Toutin, T. 2004a. Spatiotriangulation with multi-sensor VIR/SAR images. *IEEE Trans Geosci Rem Sens* 42(10):2096–103.

Toutin, T. 2004b. DSM generation and evaluation from QuickBird stereo imagery with 3D physical modelling. *Int J of Rem Sens* 25(22):5181–93.

Toutin, T. 2004c. Comparison of stereo-extracted DTM from different high-resolution sensors: SPOT-5, EROS-A, IKONOS-II, and QuickBird. *IEEE Trans Geosci Rem Sens* 42(10):2121–29.

Toutin, T. 2006a. Comparison of DSMs generated from stereo HR images using 3D physical or empirical models. *Photogramm Eng Remote Sens* 72(5):597–604.

Toutin, T. 2006b. Generation of DSM from SPOT-5 in-track HRS and across-track HRG stereo data using spatiotriangulation and autocalibration. *ISPRS J Photogramm Eng Remote Sens* 60(3):170–81.

Toutin, T. 2006c. Spatiotriangulation with multisensor HR stereo-images. *IEEE Trans Geosci Remote Sens* 44(2):456–62.

Toutin, T., and Y. Carbonneau. 1992. MOS and SEASAT image geometric corrections. *IEEE Trans Geosci Remote Sens* 30:603–9.

Toutin, T., Y. Carbonneau, and L. St. Laurent. 1992. An integrated method to rectify airborne radar imagery using DEM. *Photogramm Eng Remote Sens* 58:417–22.

Toutin, T., and P. Cheng. 2000. Demystification of IKONOS. *Earth Observation Mag* 9(7):17–21.

Toutin, T., and R. Chénier. 2009. 3-D radargrammetric modeling of RADARSAT-2 ultrafine mode: Preliminary results of the geometric calibration. *IEEE-GRSL* 6(2):282–6.

Touzi, R. 2002. A review of speckle filtering in the context of estimation theory. *IEEE Trans Geosci Remote Sens* 40:2392–404.

Valadan Zoej, M. J., A. Mansourian, B. Mojaradi, and S. Sadeghian. 2002. 2D Geometric correction of IKONOS imagery using genetic algorithm. *Int Arch Photogrametry Remote Sens Spatial Inf Sci* 34(B4).

Valadan Zoej, M. J., and G. Petrie. 1998. Mathematical modelling and accuracy testing of SPOT Level-1B stereo-pairs. *Photogramm Rec* 16:67–82.

Vassilopoulou, S., L. Hurni, V. Dietrich, E. Baltsavias, M. Pateraki, E. Lagios, and I. Parcharidis. 2002. Orthophoto generation using IKONOS imagery and high-resolution DEM: A case study on volcanic hazard monitoring of Nisyros Island (Greece). *ISPRS J Photogramm Remote Sens* 57:24–38.

Veillet, I. 1991. Triangulation spatiale de blocs d'images SPOT. In *Thèse de Doctorat*, 101. Paris, France: Observatoire de Paris. École Nationale des Sciences Géodésiques, St-Mandé, France: IGN.

Walker, A. S., and G. Petrie. 1996. Digital photogrammetric workstation. In *International Archives of Photogrammetry and Remote Sensing*, Vienna, Austria, July 9–18, vol. 3(B2), 384–95. Vienna: Austrian Society for Surveying and Geoinformation.

Wang, J., K. Di, R. Ma, and R. Li. 2005. Evaluation and improvement of geo-positioning accuracy of IKONOS stereo imagery. *J Surv Eng* 131(2):35–42.

Westin, T. 1990. Precision rectification of SPOT imagery. *Photogramm Eng Remote Sens* 56:247–53.

Westin, T. 2000. Geometric modelling of imagery from the MSU-SK conical scanner. *Bulletin de la Société Française de Photogrammétrie et de Télédétection* 159:55–8.

Westin, T., and J. Forsgren. 2002. Orthorectification of EROS-A1 Images. ImageSat International Web site, http://www.imagesatintl.com/customersupport/techarticles/Orthorectification_EROSA1Images.pdf/(accessed April 20, 2003).

Wiesel, J. W. 1984. Image rectification and registration. In *International Archives of Photogrammetry and Remote Sensing*, Rio de Janeiro, Brazil, June 18–29, vol. 25(A3b), 1120–29. Helsinki, Finland: ISPRS.

Wong, K. W. 1975. Geometric and cartographic accuracy of ERTS-1 imagery. *Photogramm Eng Remote Sens* 41:621–35.

Wong, K. W. 1980. Basic mathematics of photogrammetry. In *Manual of Photogrammetry* 4th ed, Chapter II, ed. C. C. Slama, 37–101. Falls Church, VA: ASP Publishers.

Wong, F., R. Orth, and D. Friedmann. 1981. The use of digital terrain model in the rectification of satellite-borne imagery. In *Proceedings of the 15th International Symposium on Remote Sensing of Environment*, Ann Arbor, Michigan, May 11–15, 653–62.

Yang, X. 2001. Piece-wise linear rational function approximation in digital photogrammetry. In *Proceedings of the ASPRS Annual Conference*, St Louis, Missouri, April 23–27. Bethesda, MD: ASPRS.

Zhang Y., V. Tao, and B. Mercer. 2001. Assessment of the influences of satellite viewing angle, Earth curvature, relief height, geographical location and DEM characteristics on planimetric displacement of high-resolution satellite imagery. In *Proceedings of the ASPRS Annual Conference*, St Louis, Missouri, April 23–27. Bethesda, MD: ASPRS.

9

Remote Sensing Image Classification

Dengsheng Lu, Qihao Weng, Emilio Moran, Guiying Li, and Scott Hetrick

CONTENTS

9.1 Introduction

The classification of remotely sensed data has long attracted the attention of the remote sensing community because classification results are fundamental sources for many environmental and socioeconomic applications. Scientists and practitioners have made great efforts in developing advanced classification approaches and techniques for improving classification accuracy (Gong and Howarth 1992; Kontoes et al. 1993; Foody 1996; San Miguel-Ayanz and Biging 1997; Aplin, Atkinson, and Curran 1999; Stuckens, Coppin, and Bauer 2000; Franklin et al. 2002; Pal and Mather 2003; Gallego 2004; Lu and Weng 2007; Blaschke 2010; Ghimire, Rogan, and Miller 2010). However, classifying remotely sensed data into a thematic map remains a challenge because many factors, such as the complexity of the landscape under investigation, the availability of reference data, the selected remotely sensed data, image-processing and image classification approaches, and the analyst's experiences, may affect classification accuracy. Many uncertainties or errors may be introduced into the classification results; thus, much effort should be devoted to the identification of these major factors in the image classification process and then to improving them. This chapter provides a brief overview of the major steps involved in the process of image classification, discusses the potential techniques for improving land-cover classification performance, and provides a case study for land use/cover classification in a moist tropical region of the Brazilian Amazon with Landsat thematic mapper (TM) imagery.

9.2 Overview of Image Classification Procedure

Classification of remotely sensed imagery is a complex process and requires the consideration of many factors. Figure 9.1 illustrates the major steps of an image classification procedure. Sections 9.2.1 through 9.2.8 provide brief descriptions for each step.

9.2.1 Nature of Remote Sensing Image Classification

Before implementing image classification for a specific study area, it is very important to clearly define the research problems that need to be solved, the objectives, and the location and size of the study area (Jensen 2005). In particular, clearly understanding the needs of the end user is critical. It is helpful to list some questions, such as the following: What is the detailed classification system and what are the most interesting land covers? What is the accuracy for each land cover or overall accuracy? What is the minimum mapping unit? What previous research work has been done and how can one maintain compatibility with it? What data sources are available and what data are required? What are the time, cost, and labor constraints? These questions directly affect the selection of remotely sensed data, selection of classification algorithms, and design of a classification procedure for a specific purpose.

9.2.2 Determination of a Classification System and Selection of Training Samples

A suitable classification system is a prerequisite for successful classification. In general, a classification system is designed based on the user's needs, the spatial resolution of the remotely sensed data, compatibility with previous work, available image-processing

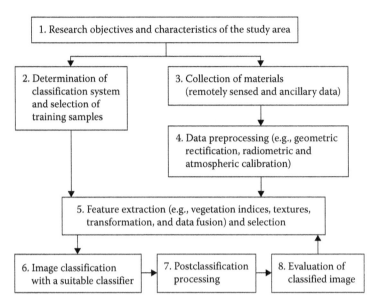

FIGURE 9.1
Major steps involved in the image classification procedure.

and classification algorithms, and time constraints. Such a system should be informative, exhaustive, and separable (Landgrebe 2003; Jensen 2005). In many cases, a hierarchical classification system is adopted to take different conditions into account.

A sufficient number of training samples and their representativeness are critical for image classifications (Hubert-Moy et al. 2001; Chen and Stow 2002; Landgrebe 2003; Mather 2004). Training samples are usually collected from fieldwork or from fine spatial resolution aerial photographs and satellite images. Different collection strategies, such as single pixel, seed, and polygon, may be used, but they will influence classification results, especially for classifications with fine spatial resolution image data (Chen and Stow 2002). When the landscape under investigation is complex and heterogeneous, selection of a sufficient number of training samples becomes difficult. This problem becomes complicated if medium or coarse spatial resolution data are used for classification, because a large volume of mixed pixels may occur. Therefore, selection of training samples must consider the spatial resolution of the remote sensing data being used, the availability of ground reference data, and the complexity of the landscapes under investigation.

9.2.3 Selection of Remotely Sensed Data

Remotely sensed data have different spatial, radiometric, spectral, and temporal resolutions. Understanding the strengths and weaknesses of different types of sensor data is essential for selecting suitable remotely sensed data for image classification. Some previous literature has reviewed the characteristics of major types of remote sensing data (Barnsley 1999; Estes and Loveland 1999; Althausen 2002; Lefsky and Cohen 2003). The selection of suitable remotely sensed data requires considering such factors as the needs of the end user, the scale and characteristics of the study area, available image data and their characteristics, cost and time constraints, and the analyst's experience in using the selected images. The end user's need determines the nature of classification and the scale of the study area, thus affecting the selection of remotely sensed data. In general, at a local level, a fine-scale classification system is needed, thus high spatial resolution data such as IKONOS and QuickBird data are helpful. At a regional scale, medium spatial resolution data such as those from Landsat TM and Terra Advanced Spaceborne Thermal Emission and Reflection Radiometer (ASTER) are the most frequently used data. At a continental or global scale, coarse spatial resolution data such as Advanced Very High Resolution Radiometer (AVHRR), Moderate Resolution Imaging Spectroradiometer (MODIS), and System Pour l'Observation de la Terre (SPOT) vegetation data are preferable.

Atmospheric condition is another important factor that influences the selection of remote sensing data. The frequent cloudy conditions in moist tropical regions are often an obstacle for capturing high-quality optical sensor data. Therefore, different kinds of radar data may serve as an important supplementary data source. Since multiple sources of sensor data are now readily available, image analysts have more choices to select suitable remotely sensed data for a specific study. In this situation, monetary cost is often an important factor affecting the selection of remotely sensed data.

9.2.4 Image Preprocessing

Image preprocessing may include the examination of image quality, geometric rectification, and radiometric and atmospheric calibration. If different ancillary data are used, data conversions among different sources or formats and quality evaluation of these data are necessary before they can be incorporated into a classification procedure. The examination of original images to see any remote sensing system–induced radiometric errors is necessary before the data are used for further processing. Accurate geometric rectification or image registration of remotely sensed data is a prerequisite for combining different source data in a classification process.

If a single-date image is used for classification, atmospheric correction may not be required (Song et al. 2001). However, when multitemporal or multisensor data are used, atmospheric calibration is mandatory. This is especially true when multisensor data, such as TM and SPOT or TM and radar are integrated for an image classification. A variety of methods, ranging from simple relative calibration to the dark-object subtraction (DOS) method and complex physically based models (e.g., second simulation of the satellite signal in the solar spectrum [6S]), have been developed for radiometric and atmospheric correction (Markham and Barker 1987; Gilabert, Conese, and Maselli 1994; Chavez 1996; Stefan and Itten 1997; Vermote et al. 1997; Tokola, Löfman, and Erkkilä 1999; Heo and FitzHugh 2000; Song et al. 2001; Du, Teillet, and Cihlar 2002; Lu et al. 2002; McGovern et al. 2002; Canty, Nielsen, and Schmidt 2004; Hadjimitsis, Clayton, and Hope 2004; Chander, Markham, and Helder 2009). Topographic correction is important if the study area is located in rugged or mountainous regions (Teillet, Guindon, and Goodenough 1982; Civco 1989; Colby 1991; Meyer et al. 1993; Richter 1997; Gu and Gillespie 1998; Hale and Rock 2003; Lu et al. 2008a). A detailed description of atmospheric and topographic correction is beyond the scope of this chapter. Interested readers may check the references cited in this section to identify a suitable approach for a specific study.

9.2.5 Feature Extraction and Selection

Selecting suitable variables is a critical step for successfully performing an image classification. Many potential variables may be used in image classification, including spectral signatures, vegetation indices, transformed images, textural or contextual information, multitemporal images, multisensor images, and ancillary data. Because of the different capabilities of these variables in land-cover separability, the use of too many variables in a classification procedure may decrease classification accuracy (Price, Guo, and Stiles 2002). It is important to select only those variables that are most useful in separating land-cover or vegetation classes, especially when hyperspectral or multisource data are employed. Many approaches, such as principal component analysis, minimum noise fraction transform, discriminant analysis, decision boundary feature extraction, nonparametric weighted feature extraction, wavelet transform, and spectral mixture analysis (Myint 2001; Okin et al. 2001; Rashed et al. 2001; Asner and Heidebrecht 2002; Lobell et al. 2002; Neville et al. 2003; Landgrebe 2003; Platt and Goetz 2004), may be used for feature extraction, in order to reduce the data redundancy inherent in remotely sensed data or to extract specific land-cover information.

Optimal selection of spectral bands for image classification has been extensively discussed in the literature (Mausel, Kramber, and Lee 1990; Landgrebe 2003). Graphic analysis (e.g., bar graph spectral plots, cospectral mean vector plots, two-dimensional feature space plot, and ellipse plots) and statistical methods (e.g., average divergence, transformed divergence, Bhattacharyya distance, and Jeffreys–Matusita distance) have been used to identify optimal subsets of bands (Jensen 2005). In practice, divergence-related algorithms based on training samples are often used to evaluate class separability and select optimal bands.

9.2.6 Selection of a Suitable Classification Algorithm

In recent years, many advanced classification approaches, such as artificial neural networks, decision trees, fuzzy sets, and expert systems, have been widely applied in image classification. Cihlar (2000) discussed the status and research priorities of land-cover mapping for large areas. Franklin and Wulder (2002) assessed land-cover classification approaches with medium spatial resolution remotely sensed data. Published works by Tso and Mather (2001) and Landgrebe (2003) specifically focused on image-processing approaches and classification algorithms. In general, image classification approaches can be grouped into different categories, such as supervised versus unsupervised, parametric versus nonparametric, hard versus soft (fuzzy) classification, per-pixel, subpixel, and per-field (Lu and Weng 2007). There are many different classification methods available. For the sake of convenience, Lu and Weng (2007) grouped classification approaches as per-pixel, subpixel, per-field, contextual, and knowledge-based approaches, and a combination approach of multiple classifiers, and described the major advanced classification approaches that have appeared in the recent literature. In practice, many factors, such as the spatial resolution of the remotely sensed data, different data sources, classification systems, and the availability of classification software, must be taken into account when selecting a classification method for use. If the classification is based on spectral signatures, parametric classification algorithms such as maximum likelihood are often used; otherwise, if multisource data are used, nonparametric classification algorithms such as the decision tree and neural network are commonly used. Spatial resolution is an important factor affecting the selection of a suitable classification method. For example, high spectral variation within the same land-cover class in high spatial and radiometric resolution images such as those from QuickBird and IKONOS often results in poor classification accuracy when a traditional per-pixel classifier is used. In this circumstance, per-field or object-oriented classification algorithms outperform per-pixel classifiers (Thomas, Hendrix, and Congalton 2003; Benz et al. 2004; Jensen 2005; Stow et al. 2007;

Mallinis et al. 2008; Zhou, Troy, and Grove 2008). For medium and coarse spatial resolution data, however, spectral information is a more important attribute than spatial information because of the loss of spatial information. Since mixed pixels create a problem in medium- and coarse-resolution imagery, per-pixel classifiers have repeated difficulties in dealing with them. Subpixel-based classification methods can provide better area estimation than per-pixel-based methods (Lu and Weng 2006).

9.2.7 Postclassification Processing

Research has indicated that postclassification processing is an important step in improving the quality of classifications (Harris and Ventura 1995; Murai and Omatu 1997; Stefanov, Ramsey, and Christensen 2001; Lu and Weng 2004). Its roles include the recoding of land use/cover classes, removal of "salt-and-pepper" effects, and modification of the classified image using ancillary data or expert knowledge. Traditional per-pixel classifiers based on spectral signatures often lead to salt-and-pepper effects in classification maps due to the complexity of the landscape. Thus, a majority filter is often applied to reduce noise. Also, ancillary data are often used to modify the classification image based on established expert rules. For example, forest distribution in mountainous areas is related to elevation, slope, and aspects. Data describing terrain characteristics can be used to modify classification results based on the knowledge of specific vegetation classes and topographic factors. In urban areas, housing or population density is related to urban land-use distribution patterns, and such data can be used to correct some classification confusions between commercial and high-intensity residential areas or between recreational grass and crops (Lu and Weng 2006). As more and more ancillary data, such as digital elevation models (DEMs) and soil, roads, population, and economic data become available, geographic information systems (GIS) techniques will play an important role in managing these ancillary data and in modifying the classification results using the established knowledge or relationships between land cover and these ancillary data.

9.2.8 Evaluation of Classification Performance

The evaluation of classification results is an important process in the classification procedure. Different approaches may be employed, ranging from a qualitative evaluation based on expert knowledge to a quantitative accuracy assessment based on sampling strategies. A classification accuracy assessment generally includes three basic components: (1) sampling design, (2) response design, and (3) estimation and analysis procedures (Stehman and Czaplewski 1998). The error matrix approach is one of the most widely used in accuracy assessment (Foody 2002). In order to properly generate an error matrix, one must consider the following factors: reference data collection, classification scheme, sampling scheme, spatial autocorrelation, and sample size and sample unit (Congalton and Plourde 2002). After the generation of an error matrix, other important accuracy assessment elements, such as overall accuracy, omission error, commission error, and kappa coefficient, can be derived (Congalton and Mead 1983; Hudson and Ramm 1987; Congalton 1991; Janssen and van der Wel 1994; Kalkhan, Reich, and Czaplewski 1997; Stehman 1996; Smits, Dellepiane, and Schowengerdt 1999; Congalton and Plourde 2002; Foody 2002, 2004; Congalton and Green 2008). In particular, kappa analysis is recognized as a powerful method for analyzing a single error matrix and for comparing the differences among various error matrices (Congalton 1991; Smits, Dellepiane, and Schowengerdt 1999; Foody 2004). Many authors, such as Congalton (1991),

Janssen and van der Wel (1994), Smits, Dellepiane, and Schowengerdt (1999), Foody (2002), and Congalton and Green (2008), have reviewed the methods for classification accuracy assessment.

9.3 Overview of Major Techniques for Improving Classification Performance

Different remotely sensed data will have variations in spatial, spectral, radiometric, and temporal resolutions, as well as differences in polarization. Making full use of these characteristics is an effective way of improving classification accuracy (Lu and Weng 2005; Lu et al. 2008b). Generally speaking, spectral response is the most important information used for land-cover classification. As high spatial resolution data become readily available, textural and contextual information become significant in image classification (Lu et al. 2008b). This section discusses some major techniques used for improving the performance of land-cover classification.

9.3.1 Use of Spatial Information

The spatial resolution of an image determines the level of detail that can be observed on the Earth's surface, and spatial information plays an important part in improving land use/cover classification accuracy, especially when high spatial resolution images such as IKONOS and QuickBird images are employed (Sugumaran, Zerr, and Prato 2002; Goetz et al. 2003; Herold, Liu, and Clarke 2003; Hurtt et al. 2003; van der Sande, de Jong, and de Roo 2003; Xu et al. 2003; Zhang and Wang 2003; Wang et al. 2004; Stow et al. 2007; Mallinis et al. 2008; Zhou, Troy, and Grove 2008). A major advantage of these fine spatial resolution images is that such data greatly reduce the mixed-pixel problem, and there is the potential to extract much more detailed information on land-cover structures from these data than from medium or coarse spatial resolution data. However, some new problems associated with fine spatial resolution image data emerge, notably the shadows caused by topography, tall buildings, or trees, and the high spectral variation within the same land-cover class. These challenges may lower classification accuracy if classifiers cannot effectively handle them (Irons et al. 1985; Cushnie 1987). The huge amount of data storage capacity and severe shadow problems in fine spatial resolution images leads to challenges in selecting suitable image-processing approaches and classification algorithms. Spatial information may be used in different ways, such as in contextual-based or object-oriented classification approaches, or textural images (Blaschke 2010; Ghimire, Rogan, and Miller 2010).

9.3.2 Integration of Different Sensor Data

Images from different sensors may contain distinctive features in reflecting land-cover surfaces. Data fusion or integration of multisensor data takes advantage of the strengths of distinct image data for improving visual interpretation and quantitative analysis. Many methods have been developed to integrate spectral and spatial information (Gong 1994; Dai and Khorram 1998; Pohl and van Genderen 1998; Chen and Stow 2003; Ulfarsson, Benediktsson, and Sveinsson2003; Lu et al. 2008b; Amarsaikhan et al. 2010; Ehlers et al. 2010). Solberg, Taxt, and Jain (1996) broadly divided data fusion methods into four categories: (1) statistical,

(2) fuzzy logic, (3) evidential reasoning, and (4) neural network. Pohl and van Genderen (1998) reviewed data fusion methods, including color-related techniques (e.g., color composite, intensity, hue, and saturation [IHS], and luminance and chrominance), statistical/ numerical methods (e.g., arithmetic combination, principal component analysis, high-pass filtering, regression variable substitution, canonical variable substitution, component substitution, and wavelets transforms), and various combinations of these methods. A recent review paper by Zhang (2010) further overviewed multisource data fusion techniques and discussed their trends. Li, Li, and Gong (2010) discussed the measures based on multivariate statistical analysis to evaluate the quality of data fusion results. In general, data fusion involves two major procedures: (1) geometric coregistration of two data sets and (2) mixture of spectral and spatial information contents to generate a new data set that contains the enhanced information from both data sets. Accurate registration between the two data sets is extremely important for precisely extracting information contents from both data sets, especially for line features such as roads and rivers. Radiometric and atmospheric calibrations are also needed before multisensor data are merged.

9.3.3 Use of Multitemporal Data

Temporal resolution refers to the time interval in which a satellite revisits the same location. A higher temporal resolution provides better opportunities to capture high-quality images. This is particularly useful for areas such as moist tropical regions, where adverse atmospheric conditions regularly occur. The use of remotely sensed data collected over different seasons has proven useful in improving classification accuracy, especially for crop and vegetation classification (Brisco and Brown 1995; Wolter et al. 1995; Lunetta and Balogh 1999; Oetter et al. 2000; Liu, Takamura, and Takeuchi 2002; Guerschman et al. 2003). For example, Lunetta and Balogh (1999) compared single- and two-date Landsat-5 TM images (spring leaf-on and fall leaf-off images) for wetland mapping in Maryland and Delaware, and found that multitemporal images provided better classification accuracies than single-date imagery by itself. An overall classification accuracy of 88% was achieved from multitemporal images, compared with 69% from single-date imagery.

9.3.4 Use of Ancillary Data

Ancillary data, such as topography, soils, roads, and census data, may be combined with remotely sensed data to improve classification performance. Harris and Ventura (1995) and Williams (2001) suggested that ancillary data may be used to enhance image classification in three ways: (1) preclassification stratification, (2) classifier modification, and (3) postclassification sorting. Since land-cover distribution is related to topography, topographic data have proven to be valuable in improving land-cover classification accuracy in mountainous regions (Janssen, Jaarsma, and van der Linden 1990; Meyer et al. 1993; Franklin, Connery, and Williams 1994), and topographic data are useful at all three stages of image classification as (1) a stratification tool in preclassification, (2) an additional channel during classification, and (3) a smoothing means in postclassification (Senoo et al. 1990; Maselli et al. 2000). In urban studies, DEM data are rarely used to aid image classification due to the fact that urban regions are often located in relatively flat areas. Instead, data related to human systems such as population distribution and road density are frequently incorporated in urban classifications (Mesev 1998; Epstein, Payne, and Kramer 2002; Zhang et al. 2002; Lu and Weng 2006). As discussed in Section 9.2.7, GIS techniques play an important role in the effective use of ancillary data in improving land use/cover classification performance.

9.4 Case Study for Land-Cover Classification with Landsat Thematic Mapper Imagery

The previous sections have briefly reviewed major steps for image classification and potential measures for improving classification accuracy. The following section provides a case study in the moist tropical region of Brazil for showing how combination of remote sensing-derived variables and original spectral bands improved classification performance.

9.4.1 Research Problem and Objective

Landsat TM imagery is the most common data source for land-cover classification, and much previous research has explored methods to improve classification performance, including the use of advanced classification options such as neural network, extraction and classification of homogeneous objects (ECHO), object-oriented classifiers, decision tree classifier, and subpixel-based methods (Lu et al. 2004a, Lu and Weng 2007; Blaschke 2010). However, the role of vegetation indices and textural images in improving land-cover classification performance is still poorly understood, in particular in moist tropical regions such as the Brazilian Amazon. Therefore, we selected Altamira, Pará state, Brazil, as a case study to explore the role of vegetation indices and textural images in improving vegetation classification performance.

9.4.2 Study Area

Altamira is located along the Trans-Amazonian Highway (BR-230) in the northern Brazilian state of Pará. The city of Altamira lies on the Xingu River at the eastern edge of the study area (see Figure 9.2). In the 1950s, an effort was made to attract colonists from northeastern

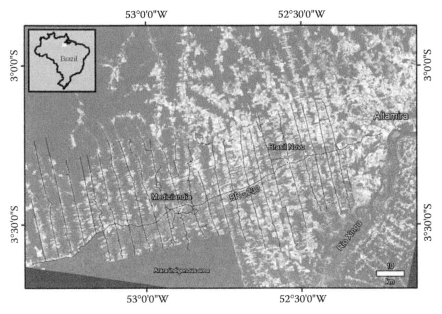

FIGURE 9.2
Altamira of Para state, Brazil, was selected as the area for the case study.

Brazil, who came and settled along streams as far as 20 km from the city center. With the construction of the Trans-Amazonian Highway in 1970, this population and older caboclo settlers from earlier rubber boom eras claimed land along the new highway and legalized their land claims. Early settlement was driven by geopolitical goals of settling in the northern region of Brazil and by political economic policies aimed at shifting production of staples like rice, corn, and beans from the southernmost Brazilian states to the northern region. The uplands have a somewhat rolling topography, with highest elevation measuring approximately 350 m. Floodplains along the Xingu are flat, with the lowest elevation measuring approximately 10 m. Nutrient-rich alfisols and infertile ultisols and oxisols are found in the uplands of this area. The overall soil quality of this area is above-average fertility for Amazonia. The dominant native types of vegetation are mature moist forest and liana forest. Major deforestation in the area, began in 1972, which was concurrent with the construction of the Trans-Amazonian Highway (Moran 1981). Deforestation has led to a complex composition of different vegetation types in this area, such as different secondary succession stages, pasture, and agroforestry (Moran et al. 1994; Moran, Brondízio, and Mausel 1994; Moran and Brondízio 1998). Annual rainfall in Altamira is approximately 2000 mm and is concentrated during the period from late October through early June; the dry period occurs between June and September. The average temperature is about 26°C (Tucker, Brondízio, and Moran 1998).

9.4.3 Methods

After the research problems were clearly identified, research objectives were defined, and the study area was selected, the next step was to design a feasible classification procedure, which may include reference data collection for use as training samples, development of suitable variables from the selected remote sensing data, selection of a suitable classification algorithm, and evaluation of the classified image.

9.4.3.1 Data Collection and Preprocessing

Sample plots for different land covers, especially for different stages of secondary succession and pasture, were collected during the summer of 2009 in the Altamira study area. Prior to fieldwork, candidate sample locations of complex vegetation areas were identified in the laboratory. In each sample area, the locations of different vegetation-cover types were recorded using a global positioning system (GPS) device, and detailed written descriptions and photographs of vegetation stand structures (e.g., height, canopy cover, species composition) were recorded. Sketch-map forms were used in conjunction with small field maps showing the candidate sample locations on A4 paper to note the spatial extent and patch shape of vegetation-cover types in the area surrounding the GPS point. Following the fieldwork, GPS points and field data were edited and processed using GIS and remote sensing software to create representative area of interest (AOI) polygons to be used for image classification. The AOI polygons were created by identifying areas of uniform pixel reflectance in an approximate 3 × 3 pixel window size on the Landsat TM imagery. A land-cover classification system was designed based on our research objectives, compatibility with our previous research work (Mausel et al. 1993; Moran et al. 1994; Moran, Brondízio, and Mausel 1994; Moran and Brondízio 1998) and field surveys. The land-cover classification system included three forest classes (upland, flooding, and liana), three succession stages (initial, intermediate, and advanced stages, or SS1, SS2, and SS3), pasture, and four nonvegetated classes (water, wetland, urban, and burn scars).

A Landsat-5 TM image acquired on July 2, 2008 was geometrically registered to a previously corrected Landsat TM image with a geometric error of less than half a pixel. The nearest-neighbor resampling algorithm was used to resample the TM imagery to a pixel size of 30 × 30 m. An improved image-based DOS model was used to perform radiometric and atmospheric correction (Chavez 1996; Lu et al. 2002; Chander, Markham, and Helder 2009). The gain and offset for each band and solar elevation angle were obtained from the image header file. The path radiance was identified based on clear water for each band.

9.4.3.2 Selection of Suitable Vegetation Indices

Many vegetation indices have been used for different purposes, such as estimation of biophysical parameters (Bannari et al. 1995; McDonald, Gemmell, and Lewis 1998). Lu et al. (2004b) examined the relationships between vegetation indices and forest stand structure attributes such as biomass, volume, and average stand diameter in different biophysical conditions in the Brazilian Amazon. In this research, they found that vegetation indices with TM band 5 had higher correlation coefficients than those without band 5, such as normalized difference vegetation index (NDVI), in study areas like Altamira with complex forest stand structure. Therefore, in this research, different vegetation indices, including band 5, were designed, as well as other indices as summarized in Table 9.1. In order to identify suitable vegetation indices for improving vegetation classification performance, training sample plots for different vegetation types based on field surveys were selected for conducting separability analysis with the transformed divergence algorithm (Mausel, Kramber, and Lee 1990; Landgrebe 2003). Individual vegetation indices and a combination of two or more indices were explored. When different combinations of two or more indices

TABLE 9.1

Vegetation Indices Used in Research

Sl. No.	Vegetation Index	Equation
1	TC1	$0.304\text{TM1} + 0.279\text{TM2} + 0.474\text{TM3} + 0.559\text{TM4} + 0.508\text{TM5} + 0.186\text{TM7}$
2	TC2	$-0.285\text{TM1} - 0.244\text{TM2} - 0.544\text{TM3} + 0.704\text{TM4} + 0.084\text{TM5} - 0.180\text{TM7}$
3	TC3	$0.151\text{TM1} + 0.197\text{TM2} + 0.328\text{TM3} + 0.341\text{TM4} - 0.711\text{TM5} - 0.457\text{TM7}$
4	ASVI	$((2\text{NIR}+1) - \sqrt{(2\text{NIR}+1)^2 - 8(\text{NIR} - 2\text{RED}+\text{BLUE})})/2$
5	MSAVI	$((2\text{NIR}+1) - \sqrt{(2\text{NIR}+1)^2 - 8(\text{NIR} - 2\text{RED})})/2$
6	ND4_2	$(\text{TM4} - \text{TM2})/(\text{TM4} + \text{TM2})$
7	ND4_25	$(\text{TM4} - \text{TM2} - \text{TM5})/(\text{TM4} + \text{TM2} + \text{TM5})$
8	ND42_53	$(\text{TM4} + \text{TM2} - \text{TM5} - \text{TM3})/(\text{TM4} + \text{TM2} + \text{TM5} + \text{TM3})$
9	ND42_57	$(\text{TM4} + \text{TM2} - \text{TM5} - \text{TM7})/(\text{TM4} + \text{TM2} + \text{TM5} + \text{TM7})$
10	ND4_35	$(\text{TM4} - \text{TM3} - \text{TM5})/(\text{TM4} + \text{TM3} + \text{TM5})$
11	ND45_23	$(\text{TM4} + \text{TM5} - \text{TM2} - \text{TM3})/(\text{TM4} + \text{TM5} + \text{TM2} + \text{TM3})$
12	ND4_57	$(2 \times \text{TM4} - \text{TM5} - \text{TM7})/(\text{TM4} + \text{TM5} + \text{TM7})$
13	NDVI	$(\text{TM4} - \text{TM3})/(\text{TM4} + \text{TM3})$
14	NDWI	$(\text{TM4} - \text{TM5})/\text{TM4} + \text{TM5})$

Note: ND = normalized difference; ASVI = atmospheric and soil vegetation index; MSAVI = modified soil adjusted vegetation index; TC = tasseled-cap transform. NIR, RED, and BLUE represent near-infrared, red, and blue band in TM image, that is, TM bands 4, 3 and 1. The ND number represents the TM spectral band.

are tested, standard deviation and correlation coefficients are used to determine the best combination of vegetation indices according to the following equation:

$$\text{Best combination} = \sum_{i=1}^{n} \text{STD}_i \left/ \left| \sum_{j=1}^{n} R_{ij} \right| \right.$$

(9.1)

where STD_i is the standard deviation of the vegetation index image i, R_{ij} is the correlation coefficient between two vegetation index images i and j, and n is the number of vegetation index images.

9.4.3.3 Selection of Suitable Textural Images

Many texture measures have been developed and textural images have proven useful in improving land-cover classification accuracy (Haralick, Shanmugam, and Dinstein 1973; Kashyap, Chellappa, and Khotanzad 1982; Marceau et al. 1990; Augusteijn, Clemens, and Shaw 1995; Shaban and Dikshit 2001; Chen, Stow, and Gong 2004; Lu et al. 2008b). Of the many texture measures, gray-level co-occurrence matrix (GLCM)-based textural images have been extensively used in image classification (Marceau et al. 1990; Lu et al. 2008b). Lu (2005) explored the roles of textural images in biomass estimation and found that textural images based on variance with TM band 2 and a window size of 9 × 9 had a significant relationship with biomass. In another study, Lu and his colleagues (Lu et al. 2008b) explored textural images in vegetation classification and found that textural images based on entropy, second moment, dissimilarity, and contrast, with window sizes of 7 × 7 or 9 × 9, exhibit better performance. Therefore, in our research, GLCM-based texture measures such as variance, homogeneity, contrast, dissimilarity, and entropy were explored with a window size of 9 × 9 and Landsat TM bands 2, 3, 4, 5, and 7. Separability analysis with transformed divergence based on selected training sample plots of different vegetation classes was used for the selection of a potential single textural image or a combination of two or more textural images. The analysis of correlation and standard deviation of each textural image was used to identify the best combination according to Equation 9.1.

9.4.3.4 Land-Cover Classification

Maximum likelihood classification (MLC) is the most common parametric classifier that assumes normal or near-normal spectral distribution for each feature of interest and an equal prior probability among the classes. This classifier is based on the probability that a pixel belongs to a particular class. It takes the variability of classes into account by using the covariance matrix. A detailed description of MLC can be found in many textbooks (e.g., Richards and Jia 1999; Lillesand and Kiefer 2000; Jensen 2005). In our research, MLC was used to conduct land-cover classification based on different scenarios, in order to explore the roles of vegetation indices and textural images in improving land-cover, especially vegetation classification in the moist tropical region. The scenarios included the consideration of six TM spectral bands, a combination of spectral and vegetation indices, a combination of spectral and textural images, and a combination of spectral indices, vegetation indices, and textural images. These classification results were analyzed based on accuracy assessment.

9.4.3.5 Accuracy Assessment

Accuracy assessment is often required for a land-cover classification. A common method for accuracy assessment involves the use of an error matrix, for which the literature has provided the meanings of and calculation methods for overall accuracy, producer's accuracy, user's accuracy, and kappa coefficient (Congalton 1991; Smits, Dellepiane, and Schowengerdt 1999; Foody 2002; Congalton and Green 2008). In this study, a total of 338 test sample plots were used for accuracy assessment. An error matrix was developed for each classification scenario, and then producer's accuracy and user's accuracy for each class and overall accuracy and kappa coefficient for each scenario were calculated based on the corresponding error matrix.

9.4.4 Results

This section provides the analysis of the identified vegetation indices and textural images and compared the classified results with MLC based on different scenarios.

9.4.4.1 Identification of Vegetation Indices and Textural Images

Since the classification of vegetation is especially difficult in our research, the selection of vegetation indices or textural images is essential to enhance vegetation separability, especially for different types of forest and secondary succession stages. Therefore, three forest types (upland forest, flooding forest, and liana forest), three succession stages (initial, intermediate, and advanced succession stages, or SS1, SS2, and SS3), and pasture were selected. The separability analysis indicated that the best single vegetation index includes ND4-25, TC2 (TC stands for tasseled cap), ND42-53, ND4-35, and TC3, and the best single textural images are from the dissimilarity on TM bands 2 or 3 (TM2-DIS, TM3-DIS), contrast on TM band 2 (TM2-CON), and homogeneity on TM bands 2 or 3 (TM2-HOM or TM3-HOM). However, no single individual vegetation index or textural image could separate the vegetation types. According to the separability analysis and the best combination model, a combination of two vegetation indices or two textural images provided the best results for vegetation separability. Three or more vegetation indices or textural images did not significantly improve vegetation separability; a similar conclusion was reached in our previous research (Lu et al. 2008b). Therefore, the best combination for two vegetation indices is TC2 and ND42-57, and the best combination for two best textural images is TM2-DIS and TM4-DIS (dissimilarity based on TM bands 2 and 4). Figure 9.3 provides the comparison of TM spectral bands, two selected vegetation indices, and two textural images, showing the different features for vegetation types, especially the textural images.

9.4.4.2 Comparison of Classification Results

The comparison of accuracy assessment among different scenarios (see Table 9.2) indicated that although the incorporation of vegetation indices into spectral bands has a limited role in improving vegetation classification performance, it is helpful in improving the extraction and separability of pasture, water, and urban land covers; in contrast, the incorporation of textural images into spectral bands was valuable for improving vegetation classification performance, especially for upland forest, flooding forest, and intermediate and advanced succession classes. This research indicates that the incorporation of both

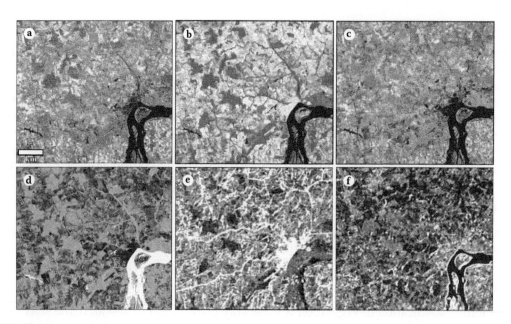

FIGURE 9.3

A comparison of thematic mapper (TM) bands 4 and 5, two vegetation indices, and two textural images. (a) and (b) TM bands 4 and 5; (c) and (d) the second component from tasseled cap transformation and the vegetation index based on bands 4, 2, 5, and 7; and (e) and (f) textural images based on dissimilarity on band 2 and band 4 and a window size of 9 × 9 pixels.

TABLE 9.2

Comparison of Accuracy Assessment Results with MLC among Different Scenarios

Land-Cover Types	6SB		6SB and 2VI		6SB and 2TX		6SB and 2VI and 2TX	
	PA	UA	PA	UA	PA	UA	PA	UA
Upland forest	37.04	95.24	24.07	92.86	66.67	78.26	66.67	78.26
Flooding forest	93.75	50.00	100.00	41.03	100.00	66.67	100.00	69.57
Liana forest	95.45	66.67	95.45	63.64	81.82	66.67	84.09	69.81
SS1	84.00	61.76	80.00	64.52	92.00	57.50	92.00	58.97
SS2	67.86	90.48	67.86	86.36	78.57	95.65	82.14	92.00
SS3	89.66	74.29	86.21	75.76	79.31	85.19	86.21	89.29
Pasture	83.33	94.83	86.36	95.00	75.76	96.15	77.27	98.08
Water	68.18	100.00	95.45	100.00	72.73	100.00	95.45	100.00
Nonvegetated wetland	53.85	100.00	69.23	90.00	69.23	100.00	53.85	87.50
Urban	100.00	71.05	100.00	100.00	100.00	79.41	100.00	100.00
Burn scars	100.00	87.50	92.86	86.67	92.86	100.00	100.00	87.50
Overall accuracy		77.22		77.51		80.18		82.84
Kappa coefficient		0.7446		0.7485		0.7770		0.8071

6SB represents TM six spectral bands; 6SB and 2VI represent the combination of six spectral bands and two vegetation indices; 6SB and 2TX represent the combination of six spectral bands and two textural images; and 6SB and 2VI and 2TX represent the combination of six spectral bands, two vegetation indices, and two textural images. PA and UA represent producer's accuracy and user's accuracy.

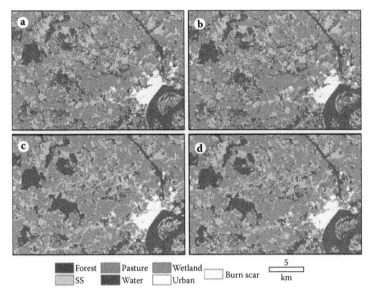

Forest Pasture Wetland
SS Water Urban Burn scar

FIGURE 9.4
(See color insert following page 426.) Comparison of classification results among different scenarios with the maximum likelihood classifier: (a) six Thematic Mapper spectral bands, (b) combination of spectral bands and two vegetation indices, (c) combination of spectral bands and two textural images, and (d) combination of spectral bands, two vegetation indices, and two textural images.

vegetation indices and textural images into spectral bands provides the best classification performance. Figure 9.4 provides a comparison of classification results among the four scenarios. It indicates that the use of textural images can reduce the salt-and-pepper effect in the classification image, which is often produced with the per-pixel-based classification method.

9.4.5 Summary of the Case Study

This study indicates the importance of textural images in improving vegetation classification accuracies. A critical step is to identify suitable textural images that can provide the best separability for specified classes. For the selection of a single textural image, one can select the textural image with the highest separability, but for the selection of two or more textural images, a method based on comparing the standard deviation and correlation coefficients between the images provides an easy way to identify a suitable combination.

9.5 Final Remarks

Image classification has made great progress over the past decades in the following three areas: (1) development and use of advanced classification algorithms, such as subpixel, per-field, and knowledge-based classification algorithms; (2) use of multiple remote sensing features, including spectral, spatial, multitemporal, and multisensor information; and (3) incorporation of ancillary data into classification procedures, including such data as topographic, soils, roads, and census data. Spectral features are the most important information required for image classification. As spatial resolution increases, how to effectively use the spatial information inherent in the image becomes an important question to be considered. Thus, object-, texture-, or contextual-based methods have attracted increased attention (Lam 2008; Blaschke 2010; Ghimire, Rogan, and Miller 2010). Classification approaches may vary with different types of remote sensing data. In high spatial resolution data such as those from IKONOS and QuickBird, the high spectral variation within the same landcover class poses a challenge. A combination of spectral and textural information and the use of per-field or object-oriented classification algorithms can reduce this problem. For medium and coarse spatial resolution data, mixed pixels are a problem, resulting in poor area estimation for classified images when per-pixel classifiers are used. Thus, subpixel features from spectral mixture analysis or fuzzy membership have been used in image classification. Moreover, image data have been integrated with ancillary data as another means for enhancing image classification in which GIS plays an important role. When multisource data are used in a classification, parametric classification algorithms such as MLC are typically not appropriate. Advanced nonparametric classifiers, such as neural network, decision tree, and evidential reasoning, or the knowledge-based approach appear to be the most appropriate choices.

The success of an image classification depends on many factors. The availability of high-quality remotely sensed imagery and ancillary data, design of a proper classification procedure, and skills and experiences of the analyst are most important. For a particular study, it is often difficult to identify the best classifier due to a lack of guidelines for classifier selection and the unavailability of suitable classification algorithms at hand. Comparative studies of different classifiers are thus frequently conducted. Moreover, the combination of different classification approaches has been shown to be helpful for improving classification accuracy. Future research is necessary to develop guidelines for the applicability and capability of major classification algorithms.

Acknowledgments

The authors wish to thank the National Institute of Child Health and Human Development at the National Institutes of Health (grant #R01 HD035811) and the National Science Foundation (grant #BCS 0850615) for their support for this research.

References

Althausen, J. D. 2002. What remote sensing system should be used to collect the data? In *Manual of Geospatial Science and Technology*, ed. J. D. Bossler, J. R. Jensen, R. B. McMaster and C. Rizos, 276. New York: Taylor & Francis.

Amarsaikhan, D. et al. 2010. Fusing high-resolution SAR and optical imagery for improved urban land cover study and classification. *Int J Image Data Fusion* 1:83.

Aplin, P., P. M. Atkinson, and P. J. Curran. 1999. Per-field classification of land use using the forthcoming very fine spatial resolution satellite sensors: problems and potential solutions. In *Advances in Remote Sensing and GIS Analysis*, ed. P. M. Atkinson and N. J. Tate, 219. New York: John Wiley & Sons.

Asner, G. P., and K. B. Heidebrecht. 2002. Spectral unmixing of vegetation, soil and dry carbon cover in arid regions: Comparing multispectral and hyperspectral observations. *Int J Remote Sens* 23:3939.

Augusteijn, M. F., L. E. Clemens, and K. A. Shaw. 1995. Performance evaluation of texture measures for ground cover identification in satellite images by means of a neural network classifier. *IEEE Trans Geosci Remote Sens* 33:616.

Bannari, A. et al. 1995. A review of vegetation indices. *Remote Sens Rev* 13:95.

Barnsley, M. J. 1999. Digital remote sensing data and their characteristics. In *Geographical Information Systems: Principles, Techniques, Applications, and Management*, 2nd ed., ed. P. Longley, M. Goodchild, D. J. Maguire and D. W. Rhind, 451. New York: John Wiley & Sons.

Benz, U. C. et al. 2004. Multi-resolution, object-oriented fuzzy analysis of remote sensing data for GIS-ready information. *ISPRS J Photogramm Remote Sens* 58:239.

Blaschke, T. 2010. Object based image analysis for remote sensing. *ISPRS J Photogramm Remote Sens* 65:2.

Brisco, B., and R. J. Brown. 1995. Multidate SAR/TM synergism for crop classification in Western Canada. *Photogramm Eng Remote Sens* 61:1009.

Canty, M. J., A. A. Nielsen, and M. Schmidt. 2004. Automatic radiometric normalization of multitemporal satellite imagery. *Remote Sens Environ* 91:441.

Chander, G., B. L. Markham, and D. L. Helder. 2009. Summary of current radiometric calibration coefficients for Landsat MSS, TM, ETM+, and EO-1 ALI sensors. *Remote Sens Environ* 113:893.

Chavez Jr., P. S. 1996. Image-based atmospheric corrections—revisited and improved. *Photogramm Eng Remote Sens* 62:1025.

Chen, D., and D. A. Stow. 2002. The effect of training strategies on supervised classification at different spatial resolution. *Photogramm Eng Remote Sens* 68:1155.

Chen, D., and D. A. Stow. 2003. Strategies for integrating information from multiple spatial resolutions into land-use/land-cover classification routines. *Photogramm Eng Remote Sens* 69:1279.

Chen, D., D. A. Stow, and P. Gong. 2004. Examining the effect of spatial resolution and texture window size on classification accuracy: An urban environment case. *Int J Remote Sens* 25:2177.

Cihlar, J. 2000. Land cover mapping of large areas from satellites: Status and research priorities. *Int J Remote Sens* 21:1093.

Civco, D. L. 1989. Topographic normalization of Landsat Thematic Mapper digital imagery. *Photogramm Eng Remote Sens* 55:1303.

Colby, J. D. 1991. Topographic normalization in rugged terrain. *Photogramm Eng Remote Sens* 57:531.

Congalton, R. G. 1991. A review of assessing the accuracy of classification of remotely sensed data. *Remote Sens Environ* 37:35.

Congalton, R. G., and K. Green. 2008. *Assessing the Accuracy of Remotely Sensed Data: Principles and Practice*, 2nd ed, 183. Boca Raton, FL: CRC Press/Taylor & Francis Group.

Congalton, R. G., and R. A. Mead. 1983. A quantitative method to test for consistency and correctness in photo interpretation. *Photogramm Eng Remote Sens* 49:69.

Congalton, R. G., and L. Plourde. 2002. Quality assurance and accuracy assessment of information derived from remotely sensed data. In *Manual of Geospatial Science and Technology*, ed. J. Bossler, 349. London: Taylor & Francis.

Cushnie, J. L. 1987. The interactive effect of spatial resolution and degree of internal variability within land-cover types on classification accuracies. *Int J Remote Sens* 8:15.

Dai, X., and S. Khorram. 1998. A hierarchical methodology framework for multisource data fusion in vegetation classification. *Int J Remote Sens* 19:3697.

Du, Y., P. M. Teillet, and J. Cihlar. 2002. Radiometric normalization of multitemporal high-resolution satellite images with quality control for land cover change detection. *Remote Sens Environ* 82:123.

Ehlers, M. et al. 2010. Multisensor image fusion for pansharpening in remote sensing. *Int J Image Data Fusion* 1:25.

Epstein, J., K. Payne, and E. Kramer. 2002. Techniques for mapping suburban sprawl. *Photogramm Eng Remote Sens* 68:913.

Estes, J. E., and T. R. Loveland. 1999. Characteristics, sources, and management of remotely-sensed data. In *Geographical Information Systems: Principles, Techniques, Applications, and Management,* 2nd ed., ed. P. Longley, M. Goodchild, D. J. Maguire, and D. W. Rhind, 667. New York: John Wiley & Sons.

Foody, G. M. 1996. Approaches for the production and evaluation of fuzzy land cover classification from remotely-sensed data. *Int J Remote Sens* 17:1317.

Foody, G. M. 2002. Status of land cover classification accuracy assessment. *Remote Sens Environ* 80:185.

Foody, G. M. 2004. Thematic map comparison: evaluating the statistical significance of differences in classification accuracy. *Photogramm Eng Remote Sens* 70:627.

Franklin, S. E., D. R. Connery, and J. A. Williams. 1994. Classification of alpine vegetation using Landsat Thematic Mapper, SPOT HRV and DEM data. *Can J Remote Sens* 20:49.

Franklin, S. E. et al. 2002. Evidential reasoning with Landsat TM, DEM and GIS data for land cover classification in support of grizzly bear habitat mapping. *Int J Remote Sens* 23:4633.

Franklin, S. E., and M. A. Wulder. 2002. Remote sensing methods in medium spatial resolution satellite data land cover classification of large areas. *Prog Phys Geog* 26:173.

Gallego, F. J. 2004. Remote sensing and land cover area estimation. *Int J Remote Sens* 25:3019.

Ghimire, B., J. Rogan, and J. Miller. 2010. Contextual land cover classification: incorporating spatial dependence in land cover classification models using random forests and the Getis statistic. *Remote Sens Lett* 1:45.

Gilabert, M. A., C. Conese, and F. Maselli. 1994. An atmospheric correction method for the automatic retrieval of surface reflectance from TM images. *Int J Remote Sens* 15:2065.

Goetz, S. J. et al. 2003. IKONOS imagery for resource management: tree cover, impervious surfaces, and riparian buffer analyses in the mid-Atlantic region. *Remote Sens Environ* 88:195.

Gong, P. 1994. Integrated analysis of spatial data from multiple sources: An overview. *Can J Remote Sens* 20:349.

Gong, P., and P. J. Howarth. 1992. Frequency-based contextual classification and gray-level vector reduction for land-use identification. *Photogramm Eng Remote Sens* 58:423.

Gu, D., and A. Gillespie. 1998. Topographic normalization of Landsat TM images of forest based on subpixel sun-canopy-sensor geometry. *Remote Sens Environ* 64:166.

Guerschman, J. P. et al. 2003. Land cover classification in the Argentine Pampas using multitemporal Landsat TM data. *Int J Remote Sens* 24:3381.

Hadjimitsis, D. G., C. R. I. Clayton, and V. S. Hope. 2004. An assessment of the effectiveness of atmospheric correction algorithms through the remote sensing of some reservoirs. *Int J Remote Sens* 25:3651.

Hale, S. R., and B. N. Rock. 2003. Impacts of topographic normalization on land-cover classification accuracy. *Photogramm Eng Remote Sens* 69:785.

Haralick, R. M., K. Shanmugam, and I. Dinstein. 1973. Textural features for image classification. *IEEE Trans Syst Man Cybern* SMC-3:610.

Harris, P. M., and S. J. Ventura. 1995. The integration of geographic data with remotely sensed imagery to improve classification in an urban area. *Photogramm Eng Remote Sens* 61:993.

Heo, J., and T. W. FitzHugh. 2000. A standardized radiometric normalization method for change detection using remotely sensed imagery. *Photogramm Eng Remote Sens* 66:173.

Herold, M., X. Liu, and K. C. Clarke. 2003. Spatial metrics and image texture for mapping urban land use. *Photogramm Eng Remote Sens* 69:991.

Hubert-Moy, L. et al. 2001. A comparison of parametric classification procedures of remotely sensed data applied on different landscape units. *Remote Sens Environ* 75:174.

Hudson, W. D., and C. W. Ramm. 1987. Correct formulation of the Kappa coefficient of agreement. *Photogramm Eng Remote Sens* 53:421.

Hurtt, G. et al. 2003. IKONOS imagery for the large scale biosphere—atmosphere experiment in Amazonia (LBA). *Remote Sens Environ* 88:111.

Irons, J. R. et al. 1985. The effects of spatial resolution on the classification of Thematic Mapper data. *Int J Remote Sens* 6:1385–1403.

Janssen, L. F., M. N. Jaarsma, and E. T. M. van der Linden. 1990. Integrating topographic data with remote sensing for land-cover classification. *Photogramm Eng Remote Sens* 56:1503.

Janssen, L. F. J., and F. J. M. van der Wel. 1994. Accuracy assessment of satellite derived land-cover data: A review. *Photogramm Eng Remote Sens* 60:419.

Jensen, J. R. 2005. *Introductory Digital Image Processing: A Remote Sensing Perspective*. 3rd ed., 526. Upper Saddle River, NJ: Prentice Hall.

Kalkhan, M. A., R. M. Reich, and R. L. Czaplewski. 1997. Variance estimates and confidence intervals for the Kappa measure of classification accuracy. *Can J Remote Sens* 23:210.

Kashyap, R. L., R. Chellappa, and A. Khotanzad. 1982. Texture classification using features derived from random field models. *Pattern Recognit Lett* 1:43.

Kontoes, C. et al. 1993. An experimental system for the integration of GIS data in knowledge-based image analysis for remote sensing of agriculture. *Int J Geogr Inf Syst* 7:247.

Lam, N. S. 2008. Methodologies for mapping land cover/land use and its change. In *Advances in Land Remote Sensing: System, Modeling, Inversion and Application*, ed. S. Liang, 341. New York: Springer.

Landgrebe, D. A. 2003. *Signal Theory Methods in Multispectral Remote Sensing*. 508. Hoboken, NJ: Wiley.

Lefsky, M. A., and W. B. Cohen. 2003. Selection of remotely sensed data. In *Remote Sensing of Forest Environments: Concepts and Case Studies*, ed. M. A. Wulder and S. E. Franklin, 13. Boston, MA: Kluwer Academic.

Li, S., Z. Li, and J. Gong. 2010. Multivariate statistical analysis of measures for assessing the quality of image fusion. *Int J Image Data Fusion* 1:47.

Lillesand, T. M., and R. W. Kiefer. 2000. *Remote Sensing and Image Interpretation*. 4th ed., 724. New York: John Wiley & Sons.

Liu, Q. J., T. Takamura, and N. Takeuchi. 2002. Mapping of boreal vegetation of a temperate mountain in China by multitemporal Landsat TM imagery. *Int J Remote Sens* 23:3385.

Lobell, D. B. et al. 2002. View angle effects on canopy reflectance and spectral mixture analysis of coniferous forests using AVIRIS. *Int J Remote Sens* 23:2247.

Lu, D. 2005. Aboveground biomass estimation using Landsat TM data in the Brazilian Amazon. *Int J Remote Sens* 26:2509.

Lu, D., and Q. Weng. 2004. Spectral mixture analysis of the urban landscapes in Indianapolis with Landsat ETM+ imagery. *Photogramm Eng Remote Sens* 70:1053.

Lu, D., and Q. Weng. 2005. Urban classification using full spectral information of Landsat ETM+ imagery in Marion County, Indiana. *Photogramm Eng Remote Sens* 71:1275.

Lu, D., and Q. Weng. 2006. Use of impervious surface in urban land use classification. *Remote Sens Environ* 102:146.

Lu, D., and Q. Weng. 2007. A survey of image classification methods and techniques for improving classification performance. *Int J Remote Sens* 28:823.

Lu, D. et al. 2002. Assessment of atmospheric correction methods for Landsat TM data applicable to Amazon basin LBA research. *Int J Remote Sens* 23:2651.

Lu, D. et al. 2004a. Comparison of land-cover classification methods in the Brazilian Amazon basin. *Photogramm Eng Remote Sens* 70:723.

Lu, D. et al. 2004b. Relationships between forest stand parameters and Landsat Thematic Mapper spectral responses in the Brazilian Amazon basin. *For Ecol Manage* 198:149.

Lu, D. et al. 2008a. Pixel-based Minnaert correction method for reducing topographic effects on the Landsat 7 ETM+ image. *Photogramm Eng Remote Sens* 74:1343.

Lu, D. et al. 2008b. A comparative study of Landsat TM and SPOT HRG images for vegetation classification in the Brazilian Amazon. *Photogramm Eng Remote Sens* 70:311.

Lunetta, R. S., and M. E. Balogh. 1999. Application of multi-temporal Landsat 5 TM imagery for wetland identification. *Photogramm Eng Remote Sens* 65:1303.

Mallinis, G. et al. 2008. Object-based classification using QuickBird imagery for delineating forest vegetation polygons in a Mediterranean test site. *ISPRS J Photogramm Remote Sens* 63:237.

Marceau, D. J. et al. 1990. Evaluation of the grey-level co-occurrence matrix method for land-cover classification using SPOT imagery. *IEEE Trans Geosci Remote Sens* 28:513.

Markham, B. L., and J. L. Barker. 1987. Thematic Mapper bandpass solar exoatmospheric irradiances. *Int J Remote Sens* 8:517.

Maselli, F. et al. 2000. Classification of Mediterranean vegetation by TM and ancillary data for the evaluation of fire risk. *Int J Remote Sens* 21:3303.

Mather, P. M. 2004. *Computer Processing of Remotely-Sensed Images: An Introduction.* 3rd ed. Chichester, UK: John Wiley & Sons Ltd.

Mausel, P. W., W. J. Kramber, and J. K. Lee. 1990. Optimum band selection for supervised classification of multispectral data. *Photogramm Eng Remote Sens* 56:55.

Mausel, P. W. et al. 1993. Spectral identification of succession stages following deforestation in the Amazon. *Geocarto Int* 8:61.

McDonald, A. J., F. M. Gemmell, and P. E. Lewis. 1998. Investigation of the utility of spectral vegetation indices for determining information on coniferous forests. *Remote Sens Environ* 66:250.

McGovern, E. A. et al. 2002. The radiometric normalization of multitemporal Thematic Mapper imagery of the midlands of Ireland: A case study. *Int J Remote Sens* 23:751.

Mesev, V. 1998. The use of census data in urban image classification. *Photogramm Eng Remote Sens* 64:431.

Meyer, P. et al. 1993. Radiometric corrections of topographically induced effects on Landsat TM data in alpine environment. *ISPRS J Photogramm Remote Sens* 48:17.

Moran, E. F. 1981. *Developing the Amazon.* Bloomington, IN: Indiana University Press.

Moran, E. F., and E. S. Brondízio. 1998. Land-use change after deforestation in Amazônia. In *People and Pixels: Linking Remote Sensing and Social Science,* ed. D. Liverman, E. F. Moran, R. R. Rindfuss and P. C. Stern, 94. Washington, DC: National Academy Press.

Moran, E. F., E. S. Brondízio, and P. Mausel. 1994. Secondary succession. *Res Explor* 10:458.

Moran, E. F. et al. 1994. Integrating Amazonian vegetation, land use, and satellite data. *Bioscience* 44:329.

Murai, H., and S. Omatu. 1997. Remote sensing image analysis using a neural network and knowledge-based processing. *Int J Remote Sens* 18:811.

Myint, S. W. 2001. A robust texture analysis and classification approach for urban land-use and land-cover feature discrimination. *Geocarto Int* 16:27.

Neville, R. A. et al. 2003. Spectral unmixing of hyperspectral imagery for mineral exploration: Comparison of results from SFSI and AVIRIS. *Can J Remote Sens* 29:99.

Oetter, D. R. et al. 2000. Land cover mapping in an agricultural setting using multiseasonal Thematic Mapper data. *Remote Sens Environ* 76:139.

Okin, G. S. et al. 2001. Practical limits on hyperspectral vegetation discrimination in arid and semiarid environments. *Remote Sens Environ* 77:212.

Pal, M., and P. M. Mather. 2003. An assessment of the effectiveness of decision tree methods for land cover classification. *Remote Sens Environ* 86:554.

Platt, R. V., and A. F. H. Goetz. 2004. A comparison of AVIRIS and Landsat for land use classification at the urban fringe. *Photogramm Eng Remote Sens* 70:813.

Pohl, C., and J. L. van Genderen. 1998. Multisensor image fusion in remote sensing: Concepts, methods, and applications. *Int J Remote Sens* 19:823.

Price, K. P., X. Guo, and J. M. Stiles. 2002. Optimal Landsat TM band combinations and vegetation indices for discrimination of six grassland types in Eastern Kansas. *Int J Remote Sens* 23:5031.

Rashed, T. et al. 2001. Revealing the anatomy of cities through spectral mixture analysis of multispectral satellite imagery: A case study of the Greater Cairo region, Egypt. *Geocarto Int* 16:5.

Richards, J. A., and X. Jia. 1999. *Remote Sensing Digital Image Analysis: An Introduction.* 3rd ed., 363. Berlin, Germany: Springer-Verlag.

Richter, R. 1997. Correction of atmospheric and topographic effects for high spatial resolution satellite imagery. *Int J Remote Sens* 18:1099.

San Miguel-Ayanz, J., and G. S. Biging. 1997. Comparison of single-stage and multi-stage classification approaches for cover type mapping with TM and SPOT data. *Remote Sens Environ* 59:92.

Senoo, T. et al. 1990. Improvement of forest type classification by SPOT HRV with 20 m mesh DTM. *Int J Remote Sens* 11:1011.

Shaban, M. A., and O. Dikshit. 2001. Improvement of classification in urban areas by the use of textural features: The case study of Lucknow city, Uttar Pradesh. *Int J Remote Sens* 22:565.

Smits, P. C., S. G. Dellepiane, and R. A. Schowengerdt. 1999. Quality assessment of image classification algorithms for land-cover mapping: A review and a proposal for a cost-based approach. *Int J Remote Sens* 20:1461.

Solberg, A. H. S., T. Taxt, and A. K. Jain. 1996. A Markov random field model for classification of multisource satellite imagery. *IEEE Trans Geosci Remote Sens* 34:100.

Song, C. et al. 2001. Classification and change detection using Landsat TM data: When and how to correct atmospheric effect. *Remote Sens Environ* 75:230.

Stefan, S., and K. I. Itten. 1997. A physically-based model to correct atmospheric and illumination effects in optical satellite data of rugged terrain. *IEEE Trans Geosci Remote Sens* 35:708.

Stefanov, W. L., M. S. Ramsey, and P. R. Christensen. 2001. Monitoring urban land cover change: An expert system approach to land cover classification of semiarid to arid urban centers. *Remote Sens Environ* 77:173.

Stehman, S. V. 1996. Estimating the Kappa coefficient and its variance under stratified random sampling. *Photogramm Eng Remote Sens* 62:401.

Stehman, S. V., and R. L. Czaplewski. 1998. Design and analysis for thematic map accuracy assessment: Fundamental principles. *Remote Sens Environ* 64:331.

Stow, D. et al. 2007. Object-based classification of residential land use within Accra, Ghana based on QuickBird satellite data. *Int J Remote Sens* 28:5167.

Stuckens, J., P. R. Coppin, and M. E. Bauer. 2000. Integrating contextual information with per-pixel classification for improved land cover classification. *Remote Sens Environ* 71:282.

Sugumaran, R., D. Zerr, and T. Prato. 2002. Improved urban land cover mapping using multitemporal IKONOS images for local government planning. *Can J Remote Sens* 28:90.

Teillet, P. M., B. Guindon, and D. G. Goodenough. 1982. On the slope-aspect correction of multispectral scanner data. *Can J Remote Sens* 8:84.

Thomas, N., C. Hendrix, and R. G. Congalton. 2003. A comparison of urban mapping methods using high-resolution digital imagery. *Photogramm Eng Remote Sens* 69:963.

Tokola, T., S. Löfman, and A. Erkkilä. 1999. Relative calibration of multitemporal Landsat data for forest cover change detection. *Remote Sens Environ* 68:1.

Tso, B., and P. M. Mather. 2001. *Classification Methods for Remotely Sensed Data*. New York: Taylor & Francis.

Tucker, J. M., E. S. Brondízio, and E. F. Moran. 1998. Rates of forest regrowth in Eastern Amazônia: a comparison of Altamira and Bragantina regions, Pará State, Brazil. *Interciencia* 23:64.

Ulfarsson, M. O., J. A. Benediktsson, and J. R. Sveinsson. 2003. Data fusion and feature extraction in the wavelet domain. *Int J Remote Sens* 24:3933.

van der Sande, C. J., S. M. de Jong, and A. P. J. de Roo. 2003. A segmentation and classification approach of IKONOS-2 imagery for land cover mapping to assist flood risk and flood damage assessment. *Int J Appl Earth Obs Geoinf* 4:217.

Vermote, E. et al. 1997. Second simulation of the satellite signal in the solar spectrum, 6S: An overview. *IEEE Trans Geosci Remote Sens* 35:675.

Wang, L. et al. 2004. Comparison of IKONOS and QuickBird images for mapping mangrove species on the Caribbean coast of panama. *Remote Sens Environ* 91:432.

Williams, J. 2001. *GIS Processing of Geocoded Satellite Data*. 327. Chichester, UK: Springer and Praxis Publishing.

Wolter, P. T. et al. 1995. Improved forest classification in the northern lake states using multi-temporal Landsat imagery. *Photogramm Eng Remote Sens* 61:1129.

Xu, B. et al. 2003. Comparison of gray-level reduction and different texture spectrum encoding methods for land-use classification using a panchromatic IKONOS image. *Photogramm Eng Remote Sens* 69:529.

Zhang, J. 2010. Multisource remote sensing data fusion: Status and trends. *Int J Image Data Fusion* 1:5.

Zhang, Q., and J. Wang. 2003. A rule-based urban land use inferring method for fine-resolution multispectral imagery. *Can J Remote Sens* 29:1.

Zhang, Q. et al. 2002. Urban built-up land change detection with road density and spectral information from multitemporal Landsat TM data. *Int J Remote Sens* 23:3057.

Zhou, W., A. Troy, and J. M. Grove. 2008. Object-based land cover classification and change analysis in the Baltimore metropolitan area using multi-temporal high resolution remote sensing data. *Sensors* 8:1613.

10

Object-Based Image Analysis for Vegetation Mapping and Monitoring

Thomas Blaschke, Kasper Johansen, and Dirk Tiede

CONTENTS

10.1 Introduction

Environmental monitoring requirements, conservation goals, spatial planning enforcement, and ecosystem-oriented natural resources management, to name just a few drivers, lend considerable urgency to the development of operational solutions that can extract tangible information from remote sensing data. The "workhorses" of satellite data generation, such as the Landsat and System Pour l'Observation de la Terre (SPOT) satellites or the Advanced Spaceborne Thermal Emission and Reflection Radiometer (ASTER) and Moderate Resolution Imaging Spectroradiometer (MODIS) instruments, have become important in global and regional studies of biodiversity, nature conservation, food security, deforestation impact, desertification monitoring, and other application fields. With the increasing spatial resolution of the "1-m generation" of IKONOS (launched in 1999), QuickBird (2001), and OrbView (2003) sensors and the increasing availability of airborne optical digital high spatial resolution imaging sensors and laser scanners, new application fields that had previously been the domain of analog airborne remote sensing can now be tackled with increasing flexibility and precision, at reduced costs and for remote areas not previously accessible. In late 2007, the first commercial satellite with a spatial resolution of less than half a meter (WorldView-1; 0.44-m panchromatic) became operational, followed by multispectral sensors such as GeoEye-1 (2008) and WorldView-2 (2009), providing pan-sharpened subhalf-meter resolution image data. At present, we see security applications, vehicle detection, urban mapping, and vegetation assessment applications developing rapidly, in terms of both number and sophistication. By simplification and generalization, we can distinguish two major trends: (1) an increasing amount of data is being produced in an ever-broadening range of spatial, spectral, radiometric, and temporal resolutions, including the aforementioned high spatial resolutions; and (2) national and supranational programs and systems are being developed for regular or on-demand vegetation surveys.

Suitable remotely sensed image data for mapping vegetation cover and properties at different spatial scales (from global to local) are becoming increasingly available. There is an extensive body of work covering the usefulness of these image data and the associated methods for vegetation mapping and monitoring (Coppin et al. 2004; Lu et al. 2004). However, mapping and monitoring of biophysical vegetation parameters and extraction of stand parameters such as height, age, and foliage projective cover on a regular and cost-effective basis had not been possible until recently due to a lack of data with sufficient spatial resolution (Congalton et al. 2002; Gergel et al. 2007; Kayitakire, Hamel, and Defourny 2006). The availability of data from high spatial resolution sensors such as the IKONOS, QuickBird, GeoEye-1, and WorldView-2 satellite sensors and airborne multispectral, hyperspectral, and light detection and ranging (lidar) sensors has opened up new opportunities for the development of operational mapping and monitoring of small features such as individual tree crowns and narrow riparian zones (Hurtt et al. 2003). The capacity to map these small features and related vegetation structural parameters has improved over the last decade through the use of object-based image analysis (OBIA).

This chapter assesses the potential of OBIA for vegetation mapping and monitoring. The acronyms OBIA and GEOBIA, which stands for geospatial OBIA, are both used herein interchangeably. The chapter summarizes trends in vegetation remote sensing and reflects very briefly on underlying concepts such as image segmentation, which is much older than the popularized commercial software of today. The goal of this chapter is to provide a review of OBIA applications for remote sensing of vegetation. Sections 10.2.1 through 10.2.4 provide a background on the use of OBIA for remote sensing of vegetation and information on how the main remote sensing systems used for vegetation/forestry applications differ from one another. The intent is not to review the sensors, but to focus on OBIA methods used for vegetation studies, which are also demonstrated through two separate case studies in this chapter.

10.2 Object-Based Image Analysis for Remotely Sensed Vegetation Mapping

The objective of OBIA is to develop and apply theory, methods, and tools for replicating and improving human interpretation of remotely sensed image data in an automated manner. OBIA consists of image segmentation, that is, clustering of pixels into homogenous objects, and subsequent classification or labeling of the objects, and modelling based on the characteristics of objects (Johansen, Bartolo and Phinn, 2010).

10.2.1 Object-Based Image Analysis

Having identified an increasing dissatisfaction with pixel-by-pixel image analysis, Blaschke and Strobl (2001) raised the provocative question "What's wrong with pixels?" Although this critique was not new (see Blaschke and Strobl 2001; Burnett and Blaschke 2003; Blaschke, Burnett, and Pekkarinen 2004 for a more thorough discussion), they observed a hype in applications "beyond pixels." A common denominator of these applications was, and still is, that they are built on image segmentation, that is, the partitioning of an image into meaningful geographically based objects (see also Hay et al. 2003; Benz et al. 2004; Liu et al. 2006; Hay et al. 2005; Lang and Blaschke 2006; Lang 2008; Hay and Castilla 2008; Blaschke, Lang, and Hay 2008). Image segmentation is not at all a new concept (Haralick and Shapiro 1985; Pal and Pal 1993); it has its roots in industrial image processing and was not used extensively in geospatial applications during the 1980s and 1990s (Blaschke, Burnett, and Pekkarinen 2004).

It is widely agreed (Blaschke and Strobl 2001; Hay et al. 2003; Burnett and Blaschke 2003; Flanders, Hall-Beyer, and Pereverzoff 2003; Benz et al. 2004; Blaschke, Burnett, and Pekkarinen 2004; Zhang et al. 2005; Liu et al. 2006; Lang 2008; Hay and Castilla 2008) that OBIA builds on older segmentation, edge-detection, feature extraction, and classification concepts that have been used in remote sensing image analysis for decades (Kettig and Landgrebe 1976; Haralick and Shapiro 1985; Pal and Pal 1993; Hay, Niemann, and McLean 1996; Ryherd and Woodcock 1996). Its emergence has nevertheless provided a bridge between the spatial concepts applied in multiscale landscape analysis (Hay et al. 2001; Wu and David 2002; Burnett and Blaschke 2003; Laliberte et al. 2004), geographic information systems (GISs), and the synergy between image-objects and their radiometric characteristics and analyses in Earth observation satellite data (Blaschke, Burnett, and Pekkarinen 2004; Langanke, Burnett, and Lang 2007; Laliberte, Fredrickson, and Rango 2007; Stow et al. 2008; Tiede, Lang, and Hoffmann 2008; Trias-Sanz, Stamon, and Louchet 2008; Weinke, Lang, and Preiner 2008).

Uses for segmentation methods outside remote sensing are legion (Pal and Pal 1993). Within remote sensing applications, segmentation algorithms are numerous, and the number has been rapidly increasing over the past few years (Blaschke, Burnett, and Pekkarinen 2004; Neubert, Herold, and Meinel 2008). Image segmentation from an algorithmic perspective is generally divided into four categories: (1) point based, (2) edge based, (3) region based, and (4) combined (for technical details of segmentation techniques, readers are referred to the study by Pal and Pal [1993]). As known to most readers, image segmentation is not a remote sensing–specific concept. Rather, many algorithms originate from industrial image processing. Blaschke, Burnett, and Pekkarinen (2004) specifically reviewed various segmentation algorithms for remote sensing applications. Segmentation provides the building blocks of OBIA (Hay and Castilla 2008; Lang 2008). Segments are regions that are generated by one or more criteria of homogeneity in one or more dimensions (of a feature space), respectively. Thus, segments have additional spectral information compared to single pixels (e.g., mean values per band, median values, minimum and maximum values, mean ratios, variance); but even greater advantage than the diversification of spectral value descriptions of objects is the additional spatial information for objects (Blaschke and Strobl 2001; Flanders, Hall-Beyer, and Pereverzoff 2003; Benz et al. 2004; Hay and Castilla 2008). It has been frequently claimed that this spatial dimension (distances, neighborhood, topologies, etc.) is crucial to OBIA methods, and that this is a major reason for the marked increase in the use of segmentation-based methods in recent times compared to the use of image segmentation in remote sensing during the 1980s and 1990s (Hay et al. 2003; Benz et al. 2004; Blaschke, Burnett, and Pekkarinen 2004; Liu et al. 2006). It is this additional information and the reduction of feature reflectance variation at the object level that makes object-based feature extraction and conversion of image data sets into thematic map products so unique.

10.2.2 Object-Based Image Analysis and Increasing Spatial Resolution

The OBIA approach is tied in with high spatial resolution situations. In an image, such a situation may occur if the pixels are significantly smaller than the objects under consideration (Blaschke 2010; Strahler 1986). Only then is regionalization of the pixels into groups of the pixels and finally objects useful and needed. In a high spatial resolution image, the specific advantages of the OBIA approach can be deployed, although regionalization approaches have also been applied to other situations, for example, to Landsat images. Recent studies have also utilized OBIA methods for medium or coarse spatial resolution data (Dorren, Maier, and Seijmonsbergen 2003; Duveiller et al. 2008; Myint et al. 2008).

A central task of image segmentation is the production of a set of nonoverlapping segments (polygons). Before performing OBIA, this step is separated from the classification process (Blaschke 2010). The problem, though, is scale. Scale is a "window of perception" (Marceau 1999), and we typically end up with several scales in imagery, if the spatial resolution is finer than the size of the objects of interest (Figure 10.1). A segmentation algorithm is used with the expectation that it will divide the image into relatively homogeneous and semantically significant groups of pixels. Burnett and Blaschke (2003) called these groups "object candidates," which must be recognized by further processing steps and must be transformed into meaningful objects. It is well known that semantically significant regions are found in an image at different spatial scales of analysis (Hay et al. 2001; Hay et al. 2003), and OBIA is inextricably linked to multiscale analysis concepts (Burnett and Blaschke 2003; Benz et al. 2004; Laliberte et al. 2004; Lang 2008; Hay and Castilla 2008), even if single levels are targeted for specific applications (Lang and Langanke 2006; Lang 2008; Weinke, Lang, and Preiner 2008). Burnett and Blaschke (2003) called this OBIA concept "multiscale segmentation/object relationship modeling" (MSS/ORM). Lang and Langanke (2006) developed an iterative one-level representation (OLR), and Tiede, Lang, and Hoffmann (2008) applied the OLR concept convincingly to airborne lidar data for tree crown segmentation (as did many other research groups, e.g., Brennan and Webster 2006; Bunting and Lucas 2006). Weinke, Lang, and Preiner (2008) empirically applied and evaluated both these OBIA concepts, and found pros and cons for each approach. For high spatial resolution image data (pixels <5 m), we can discriminate fields or forest stands at coarse scales, whereas at finer spatial scales, we can discriminate individual trees or plants (Figure 10.1). Parameters and thresholds in a typical single-scale segmentation algorithm must therefore be tuned to the correct scale for analysis. It is, however, often not possible

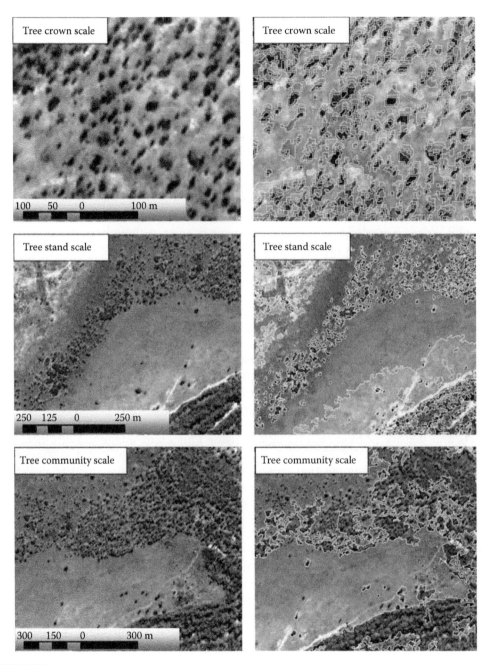

FIGURE 10.1
(See color insert following page 426.) Different spatial scales of observations in high spatial resolution QuickBird image data shown as a false-color composite. The three spatial scales show individual tree crowns and stands and associated feature segmentation within an Australian tropical savanna landscape as well as a tree community segmentation level showing riparian vegetation, savanna woodlands, and rangelands. This figure illustrates the multiscale concept by creating multiple scales of segmentation through successive grouping of image pixels into homogeneous image objects, providing a more intuitive and hierarchical partitioning of the image results, which cannot effectively be achieved in per-pixel approaches.

to determine the correct spatial scale of analysis in advance, as different kinds of images require different scales of analysis. Further, significant objects may appear at different spatial scales of analysis in the same image.

It should be clearly stated that much of the work referred to as OBIA originated around the software known as "eCognition" (Flanders et al. 2003; Benz et al. 2004). Further, only few "early" OBIA developers used the term "object based." Some authors used "object oriented" (Blaschke and Hay 2001; Benz et al. 2004), and some of these later switched to object based (with or without a hyphen), whereas some authors still use the term object oriented. It has so far been assumed that most authors prefer to use the term "based" since "oriented" may be too closely related to the object-oriented programming paradigm (see Hay and Castilla 2008, for a deeper discussion, and Blaschke 2010, for an overview of the different usages). Interestingly, there is not much critique on OBIA. In order to avoid potential flaws or too much optimism in this chapter, we performed a systematic search in the ISI Web of Knowledge and by using Google Scholar. Although hardly any critique was found, Blaschke (2010, p. 12) identified a "'technopositivistic' tendency" in the early OBIA literature. The only article we found that attests a poor performance in a comparison study is a recent publication by Mas, Gao, and Pacheco (2010). These authors analyzed the sensitivities of 85 landscape metrics to different classification methods and parameters for a Landsat image. Images classified based on pixel-based methods were smoothed using different methods (majority fitering, sieving, and clumping). Not surprisingly, for the spatial resolution of Landsat, this study found a poor performance of the OBIA method.

The idea of incorporating contextual information in the classification of remote sensing images can be traced back to the 1970s (Kettig and Landgrebe 1976), although the importance of incorporating texture increases with increasing spatial resolution (Kayitakire, Hamel, and Defourny 2006). What might have been considered a relatively homogeneous forest patch using Landsat imagery may be differentiated further to include structural aspects and density or biomass estimations using QuickBird imagery. With increasing spatial resolution comes the ability to map in more detail vegetation cover, forest structure, forest function, species composition, crown closure, stand height, stem density, vegetation age and volume, and other structural biophysical parameters (Wulder et al. 2004).

10.2.3 Remote Sensing of Vegetation

The mapping and monitoring of vegetation conditions has been one of the most important objectives of remote sensing since the advent of remote sensing technology (Lillesand, Kiefer, and Chipman 2008). Aerial photography was the sole source of information for vegetation mapping prior to 1972, when the first Landsat satellite was launched. The mapping and monitoring of vegetation properties has been revolutionized by satellite remote sensing and multispectral imaging systems, which became commercially available during the 1970s. The number of methods used in image analysis and classification is legion. These methods may be grouped roughly into an investigation of vegetation classes and ecosystems with a focus on classification and delineation of extents of units classified as homogeneous concerning a given property (Baltzer 2001; Foody 2003; Lefsky, Cohen, and Spies 2001; Lucas et al. 2000; Roberts et al. 2002), and biophysical parameter studies (Wulder 1998; Shoshany 2000; Turner, Ollinger, and Kimball 2004; Boyd and Danson 2005). For example, Wulder (1998) summarized some potential image processing methods that may be useful for the estimation of forest structural biomass estimation parameters. Treitz and Howarth (1999) reviewed hyperspectral remote sensing for biophysical parameter estimation in the forest ecosystem. Asner, Hicke, and Lobell (2003) summarized the per-pixel analysis of forest structure using vegetation indices, spectral mixture analysis, and canopy-reflectance modeling. Regarding remote sensing applications, we may summarize that in addition to the experimental, hypothetical, and case study styles of investigations, we witness more and more the use of remote sensing within vegetation studies with increasing modes of operational usage (Boyd and Danson 2005; Boyd 2009).

Vegetation mapping and monitoring has benefited significantly from the more readily available high spatial resolution image data, which have proven essential for vegetation and forest inventories at local to regional spatial scales. Higher spatial resolution may not necessarily result in improved mapping accuracies (Harvey and Hill 2001; Wang, Sousa, and Gong 2004), as the accuracy also depends on the size of the features to be mapped (Johansen and Phinn 2006; Gergel et al. 2007). Although increased spatial resolution provides opportunities for detecting small features and for mapping objects of interest in great detail, it also creates a variety of challenges in image analysis due to the variability of reflectance values within features of interest, for example, sunlit and shaded parts within one tree crown (Aplin, Atkinson, and Curran 1999; Cochrane 2000; Goetz et al. 2003; Sawaya et al. 2003). The suitability of high spatial resolution image data for detailed vegetation mapping and monitoring in various environments is supported by the ability to scale-up mapping results derived at high spatial resolutions.

Most optical sensors are only capable of providing detailed information on the horizontal distribution and not the vertical distribution of woody vegetation and forests. In many areas of the world, active microwave sensing is the only operational option. Radar data are routinely used for acquiring remotely sensed data within given time frameworks because radar systems can collect Earth feature data irrespective of weather and light conditions. Due to this unique feature of radar data compared with optical sensor data, radar data have been used extensively in many fields, including forest-cover identification and mapping, discrimination of forest compartments and forest types, and estimation of forest stand parameters. A large number of research papers have proven the potential of radar data for ecological applications, including aboveground biomass estimation (for an overview, readers are referred to Wulder 1998; Santos et al. 2002; Lu 2006).

Airborne lidar sensors derive information on the elevation and reflectance of terrain and vegetation from a pulse or continuous-wave laser fired from an airborne transmitter, for which the transmitter's position is precisely and accurately measured. Processing of the reflected lidar signal provides an accurate measure of the distance between the transmitter and the reflecting surfaces based on the time of travel of the pulse and the position of the sensor (Lefsky et al. 2002). Since 2004, full-waveform lidar has become commercially available, allowing the complete waveform of the backscattered signal echo to be recorded (Mallet and Bretar 2009). Both discrete-return and full-waveform lidar data can provide detailed information on the heights of canopy and understory surfaces with vertical and horizontal accuracies within a few centimeters. There are several published studies that suggest lidar data can be accurately employed for mapping and monitoring vegetation condition. Some of them have focused on the detection of individual tree crowns (Brandtberg et al. 2003; Persson, Holmgren, and Soderman 2002) and the estimation of tree heights, crown dimensions, and vertical structure (Anderson et al. 2006; Zimble et al. 2003).

With the improvement of analysis techniques and OBIA software capacity along with the increasing amount of commercially available high spatial resolution image and lidar data, mapping capabilities are expected to grow rapidly in the future in terms of both accuracy and the amount of biophysical vegetation properties that can be successfully mapped. A number of remotely sensed data studies of vegetation have used OBIA. These studies are reviewed in Sections 10.2.4.1 through 10.2.4.5.

10.2.4 State-of-the-Art Object-Based Image Analysis of Vegetation

In general, the number of OBIA publications is growing rapidly (Blaschke 2010) as is—more specifically—the utilization of OBIA for vegetation analysis and classification. Over the last few years, the number of empirical studies published in peer-reviewed journals reflects the improvements that OBIA offers over per-pixel analyses (Blaschke 2010). Whereas per-pixel image analysis takes into account only spectral reflectance and texture calculated through the use of moving square windows, OBIA includes information on feature shape, context, neighborhood, and multiple spatial scales.

10.2.4.1 Vegetation Inventory and Classification

Yu et al. (2006) carried out a comprehensive vegetation inventory for protected seashore areas in northern California using high spatial resolution airborne image data and ancillary topographic data, and found object-based approaches more suitable than pixel-based approaches for vegetation mapping, as they overcame the problem of the salt-and-pepper effects found in pixel-based classification. Dorren, Maier, and Seijmonsbergen (2003) favored an OBIA approach rather than a pixel-based analysis to discriminate broad-scale forest-cover types from Landsat image and digital elevation model (DEM) data of a mountainous area in Austria. Yan et al. (2006) compared per-pixel and OBIA classifications for land-cover mapping in a coal fire area in Inner Mongolia, and found the differences in accuracy, expressed in terms of proportions of correctly allocated pixels, to be statistically significant. They concluded that the thematic mapping result using an OBIA approach gave a much higher accuracy than that obtained using the per-pixel approach.

10.2.4.2 Change Detection

Im, Jensen, and Tullis (2008) compared three different change detection techniques, based on object/neighborhood correlation, image analysis, and image segmentation, with two different per-pixel approaches, and found that object-based change classifications were superior (kappa up to 90%) compared to the other change detection results (kappa from 80% to 85%). Johansen et al. (2010) compared QuickBird-based change detection maps of different vegetation types derived from object-based and per-pixel inputs used in three change detection techniques (postclassification comparison, image differencing, and tasseled cap transformation) and found the object-based inputs to provide more accurate change detection results in all cases. Desclée, Bogaert, and Defourny (2006) proved the effectiveness of object-based change detection by detecting forestland-cover changes in deciduous and coniferous stands (>90% detection accuracy) from three System Pour l'Observation de la Terre (SPOT) images covering an 1800-km² study area in east Belgium over a 10-year period. Duveiller et al. (2008) investigated land-cover change by combining a systematic regional sampling scheme based on high spatial resolution imagery with object-based, unsupervised, classification techniques for a multidate segmentation, to obtain objects with similar land-cover change trajectories, which were then classified by unsupervised procedures. This approach was applied to the Congo River basin to accurately estimate deforestation at regional, national, and landscape levels. Krause et al. (2004) integrated Landsat and ASTER data, aerial photographs, and point data obtained by fieldwork. They assessed temporal–spatial changes on a mangrove peninsula in northern Brazil and the adjacent rural socioeconomic impact area, as well as the nature of the mangrove structure.

Structural change was analyzed, and the authors were able to differentiate between strong and weak patterns in the mangrove ecosystem, suggesting different management measures and monitoring at hierarchical scales. For mangroves on the Caribbean coast of Panama, Wang, Sousa, and Gong (2004) were able to enhance spectral separability among mangrove species by using objects as the basic spatial units, as opposed to pixels. Another example of an efficient OBIA-based analysis of a mangrove ecosystem is described by Conchedda, Durieux and Mayaux (2008).

10.2.4.3 High Spatial Resolution Optical Data

Chubey, Franklin, and Wulder (2006) used OBIA to derive forest inventory parameters from IKONOS image data of a 77-km^2 study area in Alberta, Canada, and achieved the best relationships between field- and image-derived discrete land-cover types, species composition, and crown closure. Radoux and Defourny (2007) used high spatial resolution satellite images and OBIA methods to produce large-scale maps and quantitative information about the accuracy and precision of delineated boundaries for forest management using IKONOS and SPOT-5 image data. They found that tree shade and the interaction of stand patterns and sensor viewing angles produced a positive bias along forest/nonforest boundaries. For a highly fragmented forest landscape on southern Vancouver Island, Canada, Hay et al. (2005) proved how segments corresponded cognitively to individual tree crowns, ranging up to forest stands, using segmentation, object-specific analysis, and object-specific upscaling. Gergel et al. (2007) distinguished forest structural classes in riparian forests in British Columbia, Canada, for riparian restoration planning using QuickBird image data, and achieved accuracies ranging from 70% to 90% for most classes. Bunting and Lucas (2006) delineated tree crowns using seed identification and a region-growing algorithm within mixed-species forests of complex structure in central-east Queensland, Australia, based on 1-m airborne Compact Airborne Spectrographic Imager (CASI) hyperspectral data, and achieved mapping accuracies of greater than 70%. Mallinis et al. (2008) performed a multiscale, object-based analysis of a QuickBird satellite image to delineate forest vegetation polygons in a natural forest in northern Greece and found the inclusion of texture important; they also found that the use of classification trees yielded better results than the nearest-neighbor algorithm. Johansen et al. (2007) mapped the vegetation structure of Vancouver Island and discriminated structural stages in vegetation for riparian and adjacent forested ecosystems, using various texture parameters for a QuickBird image including co-occurrence contrast, dissimilarity, and homogeneity texture measures. An OBIA classification resulted in a very detailed map of vegetation structural classes, with an overall accuracy of 79%.

10.2.4.4 Light Detection and Ranging Data

Due to the high spatial resolution of lidar data, OBIA is increasingly used for both urban applications and delineating artificial objects, as well as for natural or near-natural objects. For instance, Xie, Roberts, and Johnson (2008) used an object-based geographic image retrieval approach for detecting invasive, exotic Australian pine in southern Florida using high spatial resolution orthophotos and lidar data. A moderate retrieval performance was achieved, with the lidar data proving to be most useful. Maier, Tiede, and Dorren (2008) incorporated very detailed information from lidar-derived canopy surface models and found that single and multilayered stands could be correctly distinguished in 82% of the sample plots. Also, stands with many small gaps and few but large gaps could be discriminated. Pascual et al. (2008) presented a two-stage approach for characterizing the structure of *Pinus sylvestris* stands in forests of central Spain. Building on the delineation of forest stands and a digital canopy height model derived from lidar data, they investigated forest structure types.

10.2.4.5 Incorporating Nonspectral Information

Weiers et al. (2004), Bock et al. (2005), Lathrop, Montesano, and Haag (2006), and Diaz-Varela et al. (2008) demonstrated the usefulness of OBIA methods for habitat mapping tasks. Whereas Weiers et al. (2004) and Bock et al. (2005) used time-series analysis of Landsat Thematic Mapper (TM)/Enhanced TM (ETM+) image data for parts of northern Germany, Lathrop, Montesano, and Haag (2006) assessed sea grass on New Jersey's Atlantic coast using high spatial resolution airborne image data. Diaz-Varela et al. (2008) mapped highly heterogeneous landscapes of northern Spain from Landsat TM image data. Wiseman, Kort, and Walker (2009) successfully identified and quantified 93 out of 97 shelterbelts across the Canadian Prairie provinces using multispectral reflectance, shape, texture, and other relational properties in comparison with 1:40,000 scale orthophoto interpretation. Spectral reflectance, variance, and shape parameters were combined to differentiate species compositions for six shelterbelts. Addink, de Jong, and Pebesma (2007) demonstrated with airborne hyperspectral image data, in a very detailed study with 243 field plots, that the accuracy of vegetation parameters, aboveground biomass, and leaf area index (LAI) in southern France was higher for object-based analysis than for per-pixel analysis and that object size affects prediction accuracy. Stow et al. (2008) could differentiate changes in "true shrubs" and "subshrubs" within coastal sage scrub vegetation communities in California with an overall accuracy of 83% using high spatial resolution airborne image data, and they proved that patterns of shrub distribution were more related to anthropogenic disturbance than to a long drought. Su et al. (2008) used OBIA methods to improve texture analysis based on both segmented image objects and moving windows across a QuickBird image scene, and co-occurrence matrix (gray-level co-occurrence matrix [GLCM]) textural features (homogeneity, contrast, angular second moment, and entropy) were calculated. Single additional features, such as Moran's I, were able to improve the user's and producer's accuracies by up to 16% for shrub- and grasslands. A comparison of results between spectral and textural–spatial information indicated that textural and spatial information can be used to improve the object-based classification of vegetation in urban areas using high spatial resolution imagery. Luscier et al. (2006) precisely evaluated an OBIA method based on digital photographs of vegetation to objectively quantify the percentage ground cover of grasses, forbs, shrubs, litter, and bare ground within 90 plots of 2 × 2 m. The observed differences between true cover and OBIA results ranged from 1% to 4% for each category. Ivits et al. (2005) analyzed landscape patterns for 96 sampling plots in Switzerland, based on object-derived patch indices for land-use intensities ranging from old-growth forests to intensive agricultural landscapes. Landscape patterns could be quantified on the basis of merged Landsat ETM–Indian Remote Sensing (IRS), QuickBird, and aerial photographic data. Gitas, Mitri, and Ventura (2004) mapped burned areas on the Spanish Mediterranean coast from National Oceanic and Atmospheric Administration (NOAA) Advanced Very High Resolution Radiometer (AVHRR) image data using object-based classification and achieved 90% spatial agreement with a digital fire perimeter map. These are just some examples of an increasing body of peer-reviewed literature on OBIA. For the sake of completeness, we should mention that OBIA methods include ways to incorporate various kinds of auxiliary information such as elevation, cadastre, bioclimate data, soil information, road networks, and transportation networks, to name just a few. In information-rich societies, we may regard remote sensing as only one out of many sources of information. Within spatial data infrastructures (SDIs), many examples exist that prove the potential for joint remote sensing/GIS applications. This is one of the basic ideas of the theoretical framework described by Burnett and Blaschke (2003), briefly outlined in Section 10.4.

10.3 Case Studies of Object-Based Image Analysis for Mapping Vegetation

To demonstrate how OBIA can be used to deliver environmental management–ready vegetation information, two case studies involving innovative approaches are presented in this section. The two case studies focus on automating the mapping process for vegetation feature extraction, integrating field and image data, and mapping small features, that is, individual tree crowns, over large spatial extents.

10.3.1 Case Study 1: Mapping Plant Projective Cover within Riparian Zones Using Object-Based Image Analysis and Light Detection and Ranging Data

The first case study focusing on the mapping of riparian plant projective cover (PPC) from LiDAR data using OBIA will first provide an introduction on riparian zone and PPC mapping and research objective. Subsequent sections will present the methods used and the OBIA results of riparian zone extent and PPC mapping.

10.3.1.1 Introduction

Riparian zones, or areas bordering streams, are found along rivers and creeks, and extend outward to the limits of historic flooding (Naiman and Decamps 1997). Plant project cover (PPC) is an important parameter to map in riparian zones, as it provides information on a number of riparian zone functions. The PPC value affects the amount of light reaching the streams, which in turn regulates water temperature and the level of algal growth. The PPC measurements can also provide an estimate of riparian zone fragmentation. The PPC value has been linked to bank stability, as tree roots provide a stabilizing effect on stream banks (Bennett and Simon 2004). Finally, PPC can directly influence riparian zone and in-stream wildlife habitats, as riparian zones provide unique microclimatic conditions, refuge from fires, and in-stream woody debris for important habitats (Land and Water Australia 2002). The objective of this study was to use lidar data to map riparian zone extent, then estimate PPC within the riparian zone and validate the PPC results. The work was conducted in central Queensland, Australia. Whereas PPC was calculated at the pixel level, OBIA was used for mapping riparian zone extent and validating the PPC results.

10.3.1.2 Methods

This methods section will first present the study area followed by methods used to produce the LiDAR derived products and the OBIA of riparian zone extent and PPC with riparian zones. Finally the validation approach based on field data will be explained.

FIGURE 10.2
(See color insert following page 426.) Location and field photos of the Mimosa Creek savanna woodland riparian study site in central Queensland.

10.3.1.2.1 Study Area

The study area was located in central Queensland along Mimosa Creek in a subtropical savanna woodland area (24°31.S; 149°46.E; Figure 10.2). Although the riparian zone remains relatively intact, the surrounding woody vegetation has been extensively cleared for agriculture and grazing. However, patches of remnant woodland vegetation remain, and regrowth is common. The study area receives 600–700 mm of rain, mainly between October and March. The stream and riparian zone widths of Mimosa Creek were in most cases between 10 and 30 m and 15 and 80 m, respectively. Riparian vegetation ranged from open to very dense canopy with varying amounts of subcanopy vegetation.

10.3.1.2.2 Production of Standard Light Detection and Ranging Products

Lidar data were captured by the Leica ALS50-II sensor on July 15, 2007 along Mimosa Creek, 6 weeks after the field acquisition campaign. The lidar data were captured approximately 1000 m above terrain, and consisted of 4 returns and a point density of 4 points/m². The lidar returns were classified as ground or nonground by the data provider using proprietary software.

The following four lidar products were produced for use in OBIA: (1) digital terrain model (DTM), (2) terrain slope, (3) fractional cover count, and (4) canopy height model (Figure 10.3). The DTM was produced at a pixel size of 0.5 m using an inverse distance-weighted interpolation of returns classified as ground hits. From this DTM, the rate of change in horizontal and vertical directions was calculated to produce a terrain slope layer. Fractional cover count, defined as 1 minus the gap fraction probability, was calculated from the proportion of counts from first returns 2 m above the ground level

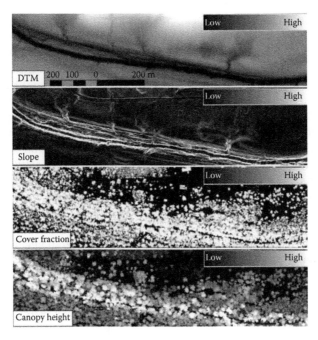

FIGURE 10.3
Light detection and ranging–derived raster products, including a digital terrain model, terrain slope, fractional cover counts, and a layer showing the maximum height of first returns within each pixel, used as input bands for object-based image analysis.

within 2.4 × 2.4 m bins. The height threshold of 2 m above the ground was also used in the field for measuring PPC. A detailed explanation of calculating PPC from fractional cover counts can be found in the study by Armston et al. (2009). The height of all first returns above the ground was calculated by subtracting the ground elevation from the first return elevation to obtain a representation of the top of the canopy (Suarez et al. 2005). These lidar-derived raster products were used for OBIA to derive maps of riparian zone extent and PPC.

10.3.1.2.3 Mapping Riparian Zone Extent

To map PPC within the riparian zone, the riparian zone extent had to be mapped. As the streambed defines the interior edge of the riparian zone, the streambed extent was first classified. The external perimeter of the riparian zone was mapped from a combination of geomorphic and vegetation characteristics. The Definiens Developer 7 software (Definiens 2007; Munich, Germany) was used for mapping the streambed and the riparian zone extent.

The segmentation process first split the image up into small square objects using chessboard segmentation. As streambeds are normally located at the lowest elevation in the landscape, the small square objects with the lowest minimum DTM values were first identified. As streambeds are generally flat and surrounded by steep stream banks, objects in contact with the minimum-value DTM object were continuously fused as long as they did not exceed a terrain slope of 8° and a DTM height difference of more than 1.5 m compared to the minimum DTM values. This produced one large object, with the majority of the area belonging to the streambed. This area was segmented once again using chessboard segmentation, and context, slope, and elevation information was used to refine the area belonging to the streambed to establish the internal borders of the riparian zone.

The mapping of the riparian zone extent was carried out using both vegetation and geomorphic information from the lidar-derived PPC, canopy height model, DTM, and terrain slope layers (Figure 10.4a). The segmentation and classification processes required some prior knowledge, which in this case was obtained through fieldwork measurements to ensure accurate mapping. A chessboard segmentation producing small square objects was initially used to set conditions for mapping potential riparian vegetation if the objects were within 100 m from the streambed and had a PPC value >40% and a tree height >8 m (Figure 10.4b). The thresholds were based on field measurements. Then, areas next to the streambed with steep slopes of >10° were classified as banks. The stream banks can be considered part of the riparian zone even without the presence of vegetation (Figure 10.4c). Objects enclosed by potential riparian vegetation and stream banks were classified as gaps and included as potential riparian vegetation (Figure 10.4d and e). After merging potential riparian vegetation, banks, and gaps, those objects that did not border the streambed were omitted (Figure 10.4e and f). Elevation differences between the streambed and the external perimeter of the riparian zone provided very useful information for mapping the riparian zone extent to ensure the riparian zones do

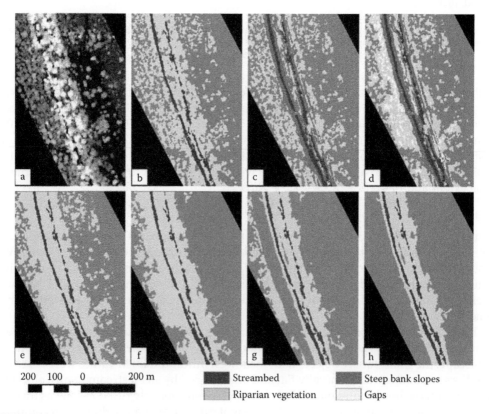

▬ Streambed		▬ Steep bank slopes
▬ Riparian vegetation		▫ Gaps

200   100   0   200 m

FIGURE 10.4
(See color insert following page 426.) Object-based image analysis steps for mapping the extent of the riparian zone from light detection and ranging data ((a) through (h)).

not extend into nonriparian areas in hilly landscapes. Based on field observations, a DTM value of 5 m above the streambed was set as the maximum elevation for riparian zones within a distance of 100 m from the streambed (Figure 10.4g). Potential riparian zone objects were then merged and omitted if they were not in contact with the streambed. As the external riparian zone perimeter is often defined based on the vegetation and canopy extent (Naiman and Decamps 1997), a region-growing algorithm was applied to grow the extent of the riparian zone if PPC >70%, the distance to the streambed was <100 m, and the elevation difference between the streambed and the riparian zone perimeter was <7 m (Figure 10.4h).

10.3.1.2.4 Plant Projective Cover Mapping

The PPC was mapped at the pixel level using the lidar-derived fractional cover count layer and an existing power function (Armston et al. 2009):

$$PPC = 1 - P_{gap}^{0.6447}$$

where P_{gap} is 1 minus the lidar cover fraction. The riparian zone extent map was used to mask the PPC layer.

10.3.1.2.5 Field Data Acquisition and Validation

Field data were obtained at 11 sites within the area covered by the lidar data. At each site, the widths of the riparian zone and the streambed were measured with a tape measure. Digital vertical photos were taken of the canopy cover using a lens with a focal length of 35 mm, at 5-m intervals along 5 parallel transects at each site located perpendicular to the stream. These transects started at the edge of the streambed and ended at least 10 m beyond the external perimeter of the riparian zone. The field photos were analyzed and converted to a measure of PPC using the approach by van Gardingen et al. (1999). The global positioning system (GPS) readings averaged for 20 minutes at the start and end of each transect line, and were used for georeferencing.

As the field photos covered slightly different areas of the canopy because of the varying heights of the lower parts of the canopy and in order to take into account slight geometric offsets between field and lidar data, a pixel-based validation approach was deemed inappropriate. An object-based approach was found most suitable for the integration of field and lidar data. To keep objects homogenous and thereby avoid the averaging out of variance of pixel values, the PPC layer was used for a multiresolution segmentation with a maximum focus on color as opposed to shape. The object sizes were approximated to single tree crowns, clusters of tree crowns, and individual gaps. This preserved extreme values from areas with very dense and sparse canopy cover. The PPC values derived from field photos taken within the corresponding objects were averaged and related to the mean PPC value of the respective object. Field-derived photo points located within 2 m of an object edge were omitted to take into account geometric offsets between field and lidar data and the uncertainty of the areal coverage of the canopy within the field photos.

10.3.1.3 Results and Discussion

The width of the riparian zone, and hence the riparian zone extent, was accurately mapped from the lidar data as PPC; tree height and geomorphic characteristics such as bank slope and elevation differences provided useful information in addition to contextual

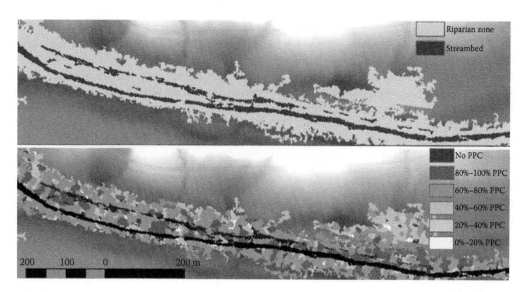

FIGURE 10.5
(See color insert following page 426.) Mapped riparian zone extent and plant project cover overlaying the digital terrain model.

class-related information (R^2 = 0.99, root-mean-square error (RMSE) = 3.2 m, and n = 10; Figure 10.5). However, knowledge from fieldwork regarding the maximum riparian zone width, stream bank slope, and elevation differences between the streambed and the external perimeter of the riparian zone enabled calibration of the parameters of the developed rule set. The PPC within the riparian zone could be mapped using the OBIA-derived riparian zone extent as a mask and the algorithm developed by Armston et al. (2009) for the conversion of lidar-derived fractional cover counts to PPC (Figure 10.5). A major advantage of the OBIA approach was the use of contextual information in relation to the mapped streambed. Knowing the location and the extent of the streambed enabled the use of features such as distance to the streambed, relative border of the streambed, and elevation difference in relation to the streambed for mapping the riparian zone extent. The initial segmentation into small square objects and the subsequent merging and region-growing algorithms were useful and enabled the integration of context information. The object-based approach using lidar data for mapping riparian zone extent and validating predicted PPC within the riparian zone was robust for the study area examined and reduced the effects of slight misregistrations between field and image data, which often complicate the integration of pixel-based analysis of high spatial resolution image and field data.

The matchup of single photo points with single pixels or windows of pixels, for example, 3 × 3 pixels, within imaged areas exhibiting large PPC variations resulted in a relationship with larger RMSE values between field and lidar data. The validation of PPC using independent PPC measurements showed a better relationship between field and lidar-derived PPC when assessed at the object level (Figure 10.6). This was because of the averaging of field measurements and pixels into homogenous objects rather than using square windows of pixels not taking into account PPC homogeneity. The use of windows of 15 × 15 m averaged out many of the extreme PPC values, as no consideration was given to the PPC homogeneity within the windows. As a small-scale parameter of 10 was used

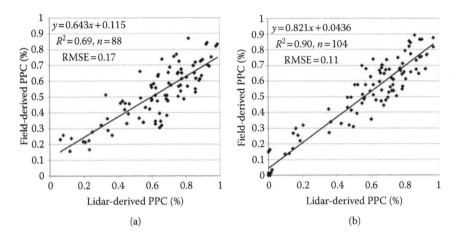

FIGURE 10.6
Relationship between field and light detection and ranging–derived plant projective cover assessed at (a) 225-m^2 square plots and (b) object-derived plots. (From Johansen, K. et al. 2010b. *Ecol Indic* 10(4). With permission.)

for the multiresolution segmentation of the PPC layer with emphasis on color, that is, PPC values, areas with similar PPC values were merged. The integration of field data (averaged within each object) and PPC objects reduced issues related to slight geometric offsets between field and image data. Also, as the canopy photos corresponded to areas of different sizes of the canopy, because of the varying heights of the lower parts of the photographed canopy cover, the object-based approach was found to be more suitable. These results indicate that object-based techniques may be used for model inversion for extracting key riparian biophysical parameters from high spatial resolution image and field data.

10.3.1.4 Summary

Object-based image analysis proved to be a powerful tool for mapping riparian zone extent and integrating high spatial resolution image and field data. Contextual information proved to be essential for mapping riparian zone extent from lidar-derived raster products. The fine detail of high spatial resolution lidar and field data was more appropriately integrated at the object level to preserve the full range of image layer values and to account for potential geometric offsets between the data sets. The developed approach may be used in other woodland riparian areas having similar contextual landscape and stream characteristics with minor adjustments required for slope, elevation, PPC, and riparian zone width input parameters based on field measurements.

10.3.2 Case Study 2: Tree Crown Extraction in a Low Mountain Range Area from UltraCamX-Derived Surface Models Using Object-Based Image Analysis and Grid Computing Techniques

The second case study focuses on the identification of single trees and the delineation of tree crowns, based on UltraCamX-derived surface models and the application of grid computing techniques for specific high data volume–processing.

10.3.2.1 Introduction

The second case study focuses on the identification of single trees. In the last decade, several algorithms were developed to extract individual tree parameters from high spatial resolution data supporting forest inventories (see Brandtberg et al. 2003; Persson, Holmgren, and Soderman 2002). Through the availability of high spatial resolution digital surface models (DSMs)—at the moment primarily captured by airborne laser scanning (lidar)—forest inventories of the future will be increasingly based on such data sets. Biophysical structural parameters such as tree density, tree height, crown width, and plant projective cover can be automatically extracted from high spatial resolution DSM data over large spatial extents. Various research applications already exist in this area, and especially the combination of multispectral imagery and DSM data is considered very promising for future forest inventories. An overview of applications for automated forest parameter extraction is given by Koch and Dees (2008), and Mallet and Bretar (2009). Since acquisition of laser-scanning data for DMS creation is still expensive and complex for short-term monitoring duties, for example, for yearly bark beetle monitoring (Wermelinger 2004), the aim of this study was to extract individual tree crowns from DSM data, which was calculated using airborne high spatial resolution UltraCamX (Vexcel Imaging GmbH, Graz, Austria) stereo image data. In this study, trees that were taller than 2 m were considered, and their height and crown width were derived. Coniferous and deciduous species were differentiated based on the spectral reflectance information of the imagery. Moreover, grid computing techniques were applied to cope with the large amount of data and, in this case, the computationally intensive OBIA.

10.3.2.2 Methods

This methods section gives an overview about the study area and the data sets used followed by the description of the developed algorithm for single tree crown extraction and the field data based validation approach.

10.3.2.2.1 Study Area and Data

The study area comprises almost 14 km^2 of forested area in the federal state of Upper Austria, Austria (Figure 10.7a). It is a low mountain forest area dominated by spruce (*Picea abies*) and beech (*Fagus sylvatica*) stands. Other tree species such as firs, sycamores, Douglas firs, and alders cover less than 7% of the area.

In this case study, data from different sources were combined: airborne multispectral UltraCamX stereo image data, a DSM derived from these data, representing the Earth's surface including features such as vegetation, buildings, and bridges, and an already existing DEM, representing ground surface topography, derived from lidar data (Figure 10.7c). Because of limited ground surface topographic variation, the DEM was deemed suitable to use for several monitoring cycles to normalize the DSM (nDSM) derived from the UltraCamX stereo data, by subtracting the DEM from the DSM. This means that the time frame between different airborne lidar acquisitions can be expanded for our study area and compensated by UltraCamX data, with the benefit of capturing not only up-to-date DSM data but also current optical imagery.

The UltraCamX frames were acquired with 80%/60% overlap in October 2008, resulting in digital infrared orthophotos with a pixel size of 0.125 m. From these data, a DSM was produced by Forest Mapping Management (FMM) of Austria, with the same spatial resolution. Additionally, a DEM based on lidar data from April 2007 with a pixel size of 1 m was provided by the federal state of Upper Austria, which was used to normalize

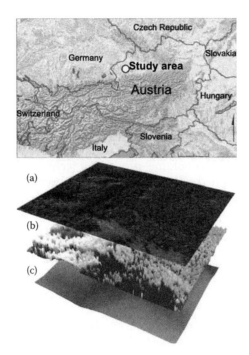

FIGURE 10.7
(See color insert following page 426.) Location of the study area in the Austrian state of Upper Austria. Data sets used in the case study are (a) UltraCamX digital infrared orthophotos, (b) normalized digital surface model derived from the UltraCamX stereo imagery, and (c) an existing light detection and ranging–based digital elevation model.

the surface model (nDSM), that is, to get real vegetation heights. The height accuracy of the lidar data in the forested area was between ±20 and ±50 cm, positional accuracy was ±30 cm, and point density was 1.7 points/m^2 (DORIS 2009).

10.3.2.2.2 Algorithm Development

For individual tree crown extraction, an object-based algorithm written in Cognition Network Language (CNL; in the Definiens developer software) by Tiede and Hoffmann (2006) was adapted to the very high spatial resolution nDSM data set. The algorithm starts from local nDSM maxima as seed points and delineates individual tree crowns based on underlying height values and height-value changes. This innovative approach uses a region-controlled extraction of local maxima as well as region-controlled parameters in the rule set. Regions were initially delineated at a coarser spatial scale to represent stand-like units with similar height and vegetation structure. In this case, four different regions representing different average stand heights were distinguished. This a priori information controlled the tree crown extraction rule sets for every region, aiming to adapt the algorithm to the regional forest structure types. The core delineation process was performed in two steps: (1) The regions were broken down into pixels ("pixel-sized objects") within a region's boundary. (2) From these pixel-sized objects, tree crowns were built in a region-growing manner using local maxima as seed points, that is, the following parameters were automatically adapted depending on the particular region (see the study by Tiede, Lang, and Hoffmann [2008]): The search radius for the local-maximum method was automatically adapted for each region depending on the average height; a stopping criterion for the region-growing process depending on the underlying nDSM data was adapted, if the candidate objects that were taken into account exceeded a certain height difference, assuming that in this case the crown edge has been reached; and a maximum crown width to avoid uncontrolled growth of tree crowns in case a local maximum was not detected correctly, preventing a merging of objects with other potential tree crowns. This last parameter was also dependent on the height information given by the initially delineated regions. In the last step, the resulting tree crowns were separated into coniferous or deciduous trees, based on their spectral reflectance properties represented in the orthophoto data.

Because of the large data volume (nDSM >10 GB; approximately 900 million pixels), there was a need for specific high data volume–processing techniques to be applied. Grid computing techniques were applied within the eCognition Server (developed by Definiens) environment to automatically split the data set into 65 tiles, which were then distributed for processing among different computers (Figure 10.8a). The same rule set was applied to each tile, and the tiled results were subsequently merged. This process allowed the processing of large data sets and significantly decreased the processing time. However, stitching of the tiles required postprocessing of the results in order to remove errors introduced at the boundaries of the different tiles. Examples of these errors include double crowns because of biased local-maxima calculations or half-delineated crowns due to the breaking off of the region-growing algorithm, which can occur at the border of the tiles if a crown is divided. Although the crown representation in nDSM data yields only one local maximum, the division of the crown due to the tiling process can potentially bias the local-maximum search. For each of the divided crown representations, a maximum is found, and the region-growing algorithm uses each maximum as a seed point but breaks off at the image tile border (Figure 10.8b). A Visual Basic for Applications (VBA) routine in ArcGIS

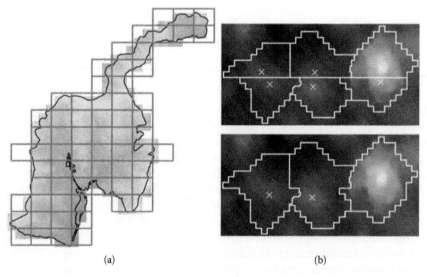

(a)   (b)

FIGURE 10.8
Tiling of normalized digital surface model (nDSM) using the eCognition Server: (a) Tiling of the nDSM into 65 parts for applying grid computing techniques, and (b) automated postprocessing of extracted tree crowns and crown maxima at the border of the tiles. Double crowns were merged and double local maxima were removed. (From Tiede, D., A. Osberger, and H. Novak. Automatisierte Baumextraktion mit höchstaufgelösten Oberflächenmodellen abgeleitet aus UltraCamX-Daten. In *Angewandte Geoinformatik 2009*, ed. J. Strobl, T. Blaschke and G. Griesebner, 2009. Wichmann Verlag, Heidelberg. With permission.)

was programmed to postprocess the results, that is, by removing multiple local maxima within the same tree crown by keeping the tallest point, and subsequently merging split tree crowns that were originally delineated as two separate crowns in two different tiles.

10.3.2.2.3 Validation

Quantitative validation of the tree crown extraction results was conducted by a forest expert using field measurements and classical forest inventories in the study area (Austrian Forest Inventory), which is one of the most intensive national forest-monitoring systems in Europe. From circular plots of 100 m², relevant tree parameters, such as tree height, tree species, and forest stand structure, were measured. The position of each plot was measured with a GPS receiver from the center of the circular plots.

10.3.2.3 Results

A total of approximately 380,000 tree crowns with heights above 2 m were automatically extracted, and almost the same amount of tree crowns were delineated (Figure 10.9). The relatively few exceptions were mainly dead trees or trees with no distinct crown. Calculation time, without pre- and postprocessing, comprised 20 hours by usage of three standard personal computers. The required processing time can be reduced further if the number of computers used is increased. The development of the rule set was more time consuming, but through the use of normalized surface height data, the transferability of the rule set to other images or areas was improved. Rule sets relying on spectral information generally require modification of thresholds and membership functions between different images, because of differences in seasonality, time of image acquisition, and atmospheric effects. The only part of the algorithm relying on spectral information was the differentiation of different species after the tree crown extraction process.

10.3.2.3.1 Single Tree Extraction

Field validation showed that the automated tree crown extraction results depended on the height of the individual tree crown. In Table 10.1, the validation results for different stand height classes are visualized together with the number of GPS-measured validation plots and the average tree detection rate per stand height class. In stands with average tree heights of 14–18 m, the average detection rate was 64%, whereas stands with an average height >26 m had detection rates over 90%. Problems were encountered in stands with complex structures, where several individual tree crowns were counted as one tree. The opposite situation, that is, identification of more crowns than the number of trees present, occurred for some deciduous trees and trees with distinct within-canopy foliage clumping (double crowns), where two or more local maxima per tree were detected. The latter case can be considered a general methodological problem utilizing local-maximum-based algorithms.

In coniferous stands with an average height >18 m, comparison with on-ground forest inventories revealed results that are suitable for use in operational mapping environments without postclassification corrections. In mixed stands, results depended on the proportion of different species, type of species, and vertical structure of the stands.

10.3.2.3.2 Tree Height Derivation

The extracted heights of the trees showed a higher accuracy than measurements for classical forest inventories. Comparisons between automated and manual height estimations

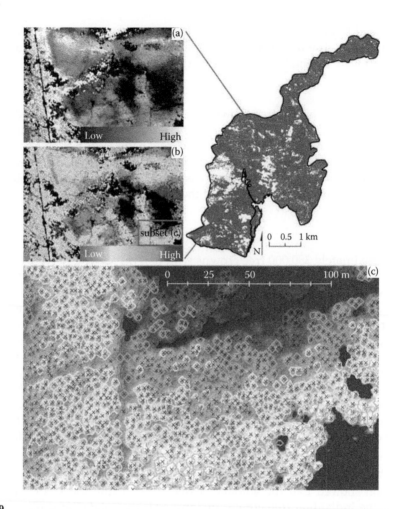

FIGURE 10.9
(See color insert following page 426.) Results of individual tree crown extraction and delineation for the whole study area (right) and subsets showing (a) the normalized digital surface model, (b) the overlaid tree crowns, and (c) the tree crowns with the extracted local maxima (color coded according to the extracted height values).

TABLE 10.1

Sample-Based Validation of Individual Tree Crown Extraction

Average Stand Height (m)	Number of GPS Sample Points	Tree Detection Rate (%)[a]
<14	3	52
14–18	11	64
18–22	14	77
22–26	12	86
26–30	9	94

[a] Percentage of all correctly extracted trees within a radius of 5.64 m around each GPS sample point.

in the field revealed much better results for the automated derivation of mean tree crown height and tree height (local-maximum extraction) values.

10.3.2.3.3 Tree Species

The differentiation of different tree species was hampered by the low solar illumination angle at the time of acquiring the UltraCamX data in October 2008. Therefore only a differentiation between coniferous and deciduous trees was performed, based on the normalized difference vegetation index (NDVI) of the imagery. In the sample plots, no errors were observed for this differentiation.

10.3.2.4 Discussion and Outlook

Individual tree crown parameters extracted from the UltraCamX data and the derived DSM could be obtained in a fast and accurate manner using automated OBIA methods. The results offer possibilities for more cost-efficient forest-monitoring tasks in the future. The available data sets were acquired in October 2008, and yielded a high degree of shadows. It is likely that the results could be improved using data acquired between May and August. It turned out that the acquisition date did not influence the part of the workflow that used lidar data, although it hampered a species differentiation based on optical data.

For even-aged forests, the proposed method is able to deliver relevant stand and tree crown parameters. Although the tree crown detection in evenly structured stands is troubled, the delineation of stand structures based on individual trees is another advantage compared to the manual estimations used in forest inventories. Also, an automated derivation of growing stock under the consideration of the respective tree crown size could lead to more objective or comprehensible growing stock estimations than forest taxation in the field.

Compared to classical forest inventories (in western and central Europe), this approach reveals a possibility to produce faster and more accurate results, mainly through the reduction of manual measurement expenditure. Object-based approaches are in this case bridging the gap between remote sensing and GIS. The results are GIS-ready and parameterized information about single trees and can directly be fed into forest inventory databases and GISs. Future research will focus on the transferability (Walker and Blaschke 2008) of the approach to areas that are dominated by deciduous trees, and also on a more general comparison of parameters derived from airborne lidar data–based DSMs and high-resolution optical data. It should be emphasized that the comparisons of lidar data and UltraCamX data in this context address only the derived products (DSM and DEM), not the sensors, generally. Lidar data with multiple returns allow measurements of vertical structural parameters from a single data set, whereas the calculation of a DSM from stereo UltraCamX data, as used in this case study, offers continuous optical data at very high spatial resolutions. On the contrary, achieving a similarly high spatial resolution and vertical accuracy with lidar data, would require a very high point density during lidar data acquisition.

10.4 Summary and Discussion

The case studies described in this chapter demonstrate how OBIA can be used to improve vegetation mapping through the use of not only spectral and textural information but also geometric (width or size of an object) and contextual information. These capabilities significantly improve the ability to accurately map vegetation biophysical parameters. For high spatial resolution vegetation mapping applications, OBIA can improve the integration of field and image data for validation and potentially also for calibration and model development. The case studies also demonstrate the suitability of OBIA for feature extraction. Feature extraction from high spatial resolution image data has previously been hampered by the large spectral reflectance variability of individual features at the pixel level, for example, for the extraction of individual tree crowns. Through the use of objects, we may overcome this limitation by reducing the reflectance variability of single pixels. The case studies prove that there is great potential for automating the feature extraction process for vegetation studies over large spatial extents, which will undoubtedly be a major focus of future research.

Applications of OBIA are developing rapidly. Several books on OBIA/GEOBIA have been published, and several special issues of scientific journals were recently devoted to this topic. One of the most recent trends is for OBIA methods to become part of dedicated workflows and converge with mainstream GIS applications (Baatz, Hoffmann, and Willhauck 2008). Blaschke (2010) concludes that this rapidly increasing body of scientific literature conveys a sense of optimism that OBIA methods generate multiscale geospatial information, which is tempered by some disquiet that the increasingly complex classification rule sets and workflows raise at least as many research questions as they resolve.

There is a realization that the higher spatial resolution and detection detail available using improved optical instruments, such as radar, lidar, and even sonar (sound navigation ranging; Lucieer 2008), create problems with the "traditional" approach to land-use/landcover mapping. The OBIA approach supports attempts to overcome a purely descriptive categorization of the spectral characteristics of pixels, and paves the way for a combined use of spectral and spatial (contextual) information toward developing conditioned information (Lang 2008).

Recently, OBIA research has been directed more toward the automation of image processing. As a consequence of the rapidly increasing proliferation of high spatial resolution imagery and the improved access to this type of imagery, more and more research is now concerned with automatic object delineation. Automated object recognition is certainly an end goal, but at present, it is mainly achieved in a stepwise manner, either with strongly interlinked procedures building workflows or with clear breaks in these workflows. In both cases, the necessary steps involve addressing various multiscale instances of related objects within a single image (e.g., individual tree crowns, tree clusters, stands, and forests). An increasing number of research articles deal with object and feature recognition and feature extraction. Still, although intrinsically tied to OBIA, for the majority of applications we can note that they are not an end in themselves.

We pointed to the fact that when dealing with high spatial resolution imagery, the question of scale or addressing the right scale gains importance. As mentioned in Section 10.2.2, Burnett and Blaschke (2003) have developed a methodology to derive objects at several levels simultaneously and to utilize this information in a classification. They called it MSS/ORM. One of the underlying ideas is to provide a methodological basis for a seamlessly integrated GIS/remote sensing analysis environment. In this respect, the multiscale approach is powerful when addressing area metrics, shape metrics, topological relationships ("borders to," "is embedded in," "is surrounded by"), and hierarchical relationships ("is subobject of," "is super object of"). Lang and Langanke (2006) have, however, convincingly shown that for specific cases, an OLR might be sufficient and more straightforward than MSS/ORM. In either case, the delineation of relatively homogeneous areas is the basic method, and the common denominator of various realizations of OBIA is the objective to derive "meaningful objects." Since the appropriate scale of observation is a function of the type of environment and the type of information that is being sought (Woodcock and Harward 1992; Marceau 1999; Hay et al. 2001), the selection of scale is very important and is a hot research topic of OBIA. Recently, Dragut, Tiede, and Levick (in press) developed a tool called "estimation of scale parameter" (ESP), which builds on the idea of local variance (LV) of object heterogeneity within a scene and allows one to iteratively generate image-objects at multiple scale levels in a bottom-up approach, while calculating the LV for each scale. Variation in heterogeneity is explored by evaluating LV plotted against the corresponding scale. It is hoped that one of the major obstacles of OBIA, namely, the selection of the right scale, may be overcome with such tools. In this case, thresholds in the rates of change of LV may indicate scale levels at which the image can be segmented in an appropriate manner, relative to the data properties at the scene level.

10.5 Future Research

Today, the bottleneck of large data volume throughput can be overcome by the use of server technologies. As this technology has become accessible for government departments and private agencies only within the last few years, it is likely to have major future implications on the OBIA of vegetation for large-area mapping relative to spatial resolution. In this chapter, we described two case studies with large data volumes. The Austrian study area was spread over 14 km² with a pixel size of 12.5 cm. This resulted in 900 million pixels. Such large amounts of data were impossible to process a few years ago. Despite the enormous progress in processing power, we point out that as the spatial extent, and hence the data volume, for OBIA increases, more conditions will need to be fulfilled in order to accurately map vegetation parameters over large areas. We conclude that more complex rule sets with increasing transferability are required, and multiscale analyses are often mandatory, which require (multi-)scale concepts. The transferability of rule sets may be improved through the use and integration of existing geographic information from spatial data infrastructures. The integration of the temporal domain through the time-series analysis of vegetation dynamics may also provide additional information for integration into rule sets, but the ontological and epistemological foundation of OBIA is still in its infancy.

An increasing number of OBIA subdisciplines are likely to develop in the future through further developments in software, available image and spatial data, and increasing research in this area. From this review of OBIA and the two case studies, we conclude that OBIA has major advantages for vegetation mapping because of the high levels of reflectance variability within individual vegetation features. The addition of shape and contextual information improves the capability to map vegetation features and structural parameters. However, because of the complexity of vegetation reflectance characteristics and vegetation dynamics, many challenges still remain for OBIA-based vegetation studies. We believe that with the continuing human impact on vegetation and predicted climate change, there is an urgent need to explore this new mapping discipline in order to take full advantage of its potential and to build up further capacities for detailed operational mapping and monitoring of vegetation through the use of OBIA.

References

Addink, E. A., S. M. de Jong, and E. J. Pebesma. 2007. The importance of scale in object-based mapping of vegetation parameters with hyperspectral imagery. *Photogramm Eng Remote Sens* 73(8):905–12.

Anderson, J., M. E. Martin, M. -L. Smith, R. O. Dubayah, M. A. Hofton, P. Hyde, B. E. Peterson, J. B. Blair, and R. G. Knox. 2006. The use of waveform lidar to measure northern temperate mixed conifer and deciduous forest structure in New Hampshire. *Remote Sens Environ* 105:248–61.

Aplin, P., P. M. Atkinson, and P. Curran. 1999. Fine spatial resolution simulated satellite sensor imagery for land-cover mapping in the United Kingdom. *Remote Sens Environ* 68:206–16.

Armston, J., R. Denham, T. Danaher, P. Scarth, and T. Moffiet. 2009. Prediction and validation of foliage projective cover from Landsat-5 TM and Landsat-7 ETM+ imagery for Queensland, Australia. *J Appl Remote Sens* 3:1–28.

Asner, G. P., J. A. Hicke, and D. B. Lobell. 2003. Per-pixel analysis of forest structure: Vegetation indices, spectral mixture analysis and canopy reflectance modeling. In *Remote Sensing of Forest Environments: Concepts and Case Studies*, ed. M. A. Wulder and S. E. Franklin, 209–54. Boston: Kluwer Academic.

Baatz, M., C. Hoffmann, and G. Willhauck. 2008. Progressing from object-based to object-oriented image analysis. In *Object Based Image Analysis*, ed. T. Blaschke, S. Lang and G. Hay. New York: Springer.

Baltzer, H. 2001. Forest mapping and monitoring with interferometric synthetic aperture radar (InSAR). *Prog Phys Geog* 25:159–77.

Bennett, S. J., and A. Simon. 2004. *Riparian Vegetation and Fluvial Geomorphology.* Washington D.C.: American Geophysical Union.

Benz, U. C., P. Hofmann, G. Willhauck, I. Lingenfelder, and M. Heynen. 2004. Multi-resolution, object-oriented fuzzy analysis of remote sensing data for GIS-ready information. *ISPRS J Photogramm Remote Sens* 58(3–4):239–58.

Blaschke, T. 2010. Object based image analysis for remote sensing. *ISPRS Int J Photogramm Remote Sens* 65(1):2–16.

Blaschke, T., C. Burnett, and A. Pekkarinen. 2004. New contextual approaches using image segmentation for object-based classification. In *Remote Sensing Image Analysis: Including the Spatial Domain*, ed. F. De Meer and S. de Jong, 211–36. Dordrecht: Kluver Academic Publishers.

Blaschke, T. and G. J. Hay. 2001. Object-oriented image analysis and scale-space: Theory and methods for modeling and evaluating multi-scale landscape structure. *Int Arch Photogramm Remote Sens* 34:22–9.

Blaschke, T., S. Lang, and G. J. Hay, eds. 2008. *Object Based Image Analysis*, 817. New York: Springer.

Blaschke, T. and J. Strobl. 2001. What's wrong with pixels? Some recent developments interfacing remote sensing and GIS. *GIS – Z Geoinf Sys* 14(6):12–7.

Bock, M., P. Xofis, J. Mitchley, G. Rossner, and M. Wissen. 2005. Object-oriented methods for habitat mapping at multiple scales: Case studies from northern Germany and Wye Downs, UK. *J Nat Conserv* 13(2–3):75–89.

Boyd, D. S. 2009. Remote sensing in physical geography: A twenty-first-century perspective. *Prog Phys Geog* 33(4):451–6.

Boyd, D. S., and F. M. Danson. 2005. Satellite remote sensing of forest resources: Three decades of research development. *Prog Phys Geog* 29:1–26.

Brandtberg, T., T. A. Warner, R. E. Landerberger, and J. B. McGraw. 2003. Detection and analysis of individual leaf-off tree crowns in small footprint, high sampling density lidar data from the eastern deciduous forest in North America. *Remote Sens Environ* 58:290–303.

Brennan, R., and T. L. Webster. 2006. Object-oriented land-cover classification of lidar-derived surfaces. *Can J Remote Sens* 32(2):162–72.

Bunting, P., and R. Lucas. 2006. The delineation of tree crowns in Australian mixed species forests using hyperspectral Compact Airborne Spectrographic Imager (CASI) data. *Remote Sens Environ* 101(2):230–48.

Burnett, C., and T. Blaschke. 2003. A multi-scale segmentation/object relationship modelling methodology for landscape analysis. *Ecol Modell* 168(3):233–49.

Castilla, G., G. J. Hay, and J. R. Ruiz. 2008. Size-constrained region merging (SCRM): An automated delineation tool for assisted photointerpretation. *Photogramm Eng Remote Sens* 74(4):409–19.

Chubey, M. S., S. E. Franklin, and M. A. Wulder. 2006. Object-based analysis of IKONOS-2 imagery for extraction of forest inventory parameters. *Photogramm Eng Remote Sens* 72(4):383–94.

Cochrane, M. A. 2000. Using vegetation reflectance variability for species level classification of hyperspectral data. *Int J Remote Sens* 21(10):2075–87.

Congalton, R. G., Birch, K., Jones, R., and Schriever, J. 2002. Evaluating remotely sensed techniques for mapping riparian vegetation. *Computers and Electronics in Agriculture,* 37:113–126.

Coppin, P., I. Jonckheere, K. Nackaerts, and B. Muys, 2004. Digital change detection methods in ecosystem monitoring: A review. *Int J Remote Sens* 25(9):1565–96.

Definiens. 2007. *Definiens Developer 7: User Guide,* document version 7.0.1.872, 497. Munich, Germany: Definiens AG.

Desclée, B., P. Bogaert, and P. Defourny. 2006. Forest change detection by statistical object-based method. *Remote Sens Environ* 102(1–2):1–11.

Diaz-Varela, R. A., P. Ramil Rego, S. C. Iglesias, and C. Muñoz Sobrino. 2008. Automatic habitat classification methods based on satellite images: A practical assessment in the NW Iberia coastal mountains. *Environ Monit Assess* 144(1–3):229–50.

DORIS (Digitales Oberösterreichisches Raum-Informations-System). 2009. Metainformation zu Digitales Geländemodell. http://doris.ooe.gv.at/geoinformation/metadata/pdf/dhm.pdf (5 October 2009)

Dorren, L. K., B. Maier, and A. C. Seijmonsbergen. 2003. Improved Landsat-based forest mapping in steep mountainous terrain using object-based classification. *For Ecol Manage* 183(1–3):31–46.

Dragut, L., D. Tiede, and S. Levick. 2010. ESP: A tool to estimate scale parameter for multiresolution image segmentation of remotely sensed data. *Int J Geog Inf Sci* 24(6):859–871.

Duveiller, G., P. Defourny, B. Desclée, and P. Mayaux. 2008. Deforestation in central Africa: Estimates at regional, national and landscape levels by advanced processing of systematically distributed Landsat extracts. *Remote Sens Environ* 112(5):1969–81.

Flanders, D., M. Hall-Beyer, and J. Pereverzoff. 2003. Preliminary evaluation of eCognition object-based software for cut block delineation and feature extraction. *Can J Remote Sens* 29(4):441–52.

Foody, G. M. 2003. Remote sensing of tropical forest environments: Towards the monitoring of environmental resources for sustainable development. *Int J Remote Sens* 24:4035–46.

Gergel, S. E., Y. Stange, N. C. Coops, K. Johansen, and K. R. Kirby. 2007. What is the value of a good map? An example using high spatial resolution imagery to aid riparian restoration. *Ecosystems* 10(5):688–702.

Gitas, I. Z., G. H. Mitri, and G. Ventura. 2004. Object-based image classification for burned area mapping of Creus Cape, Spain, using NOAA-AVHRR imagery. *Remote Sens Environ* 92:409–13.

Goetz, S. J., R. K. Wright, A. J. Smith, E. Zinecker, and E. Schaub. 2003. IKONOS imagery for resource management: Tree cover, impervious surfaces, and riparian buffer analyses in the mid-Atlantic region. *Remote Sens Environ* 88:195–208.

Haralick, R. M., and L. Shapiro. 1985. Survey: Image segmentation techniques. *Comput Vision, Graphics, and Image Proc* 29:100–32.

Harvey, K. R., and G. J. E. Hill. 2001. Vegetation mapping of a tropical freshwater swamp in the Northern Territory, Australia: A comparison of aerial photography, Landsat TM and SPOT satellite imagery. *Int J Remote Sens* 22(15):2911–25.

Hay, G. J., T. Blaschke, D. J. Marceau, and A. Bouchard. 2003. A comparison of three image-object methods for the multiscale analysis of landscape structure. *ISPRS J Photogramm Remote Sens* 57(5–6):327–45.

Hay, G. J., and G. Castilla. 2008. Geographic object-based image analysis (GEOBIA): A new name for a new discipline. In *Object-Based Image Analysis*, ed. T. Blaschke, S. Lang and G. Hay, 93–112. New York: Springer.

Hay, G. J., G. Castilla, M. A. Wulder, and J. R. Ruiz. 2005. An automated object-based approach for the multiscale image segmentation of forest scenes. *Int J Appl Earth Obs Geoinf* 7(4):339–59.

Hay, G. J., D. J. Marceau, P. Dube, and A. Bouchard. 2001. A multiscale framework for landscape analysis: object-specific analysis and upscaling. *Landsc Ecol* 16(6):471–90.

Hay, G. J., K. O. Niemann, and G. F. McLean. 1996. An object-specific image-texture analysis of H-resolution forest imagery. *Remote Sens Environ* 55(2):108–22.

Hurtt, G., X. Xiao, M. Keller, M. Palace, G. P. Asner, and R. Braswell. 2003. IKONOS imagery for the large scale biosphere–atmosphere experiment in Amazonia (LBA). *Remote Sens Environ* 88:111–27.

Im, J., J. R. Jensen, and J. A. Tullis. 2008. Object-based change detection using correlation image analysis and image segmentation. *Int J Remote Sens* 29(2):399–423.

Ivits, E., B. Koch, T. Blaschke, M. Jochum, and P. Adler. 2005. Landscape structure assessment with image grey-values and object-based classification at three spatial resolutions. *Int J Remote Sens* 26(4):2975–93.

Johansen, K., L. A. Arroyo, J. Armston, S. Phinn, and C. Witte. 2010. Mapping riparian condition indicators in a sub-tropical savanna environment from discrete-return LiDAR data using object-based image analysis. *Ecol Indic* 10(4):796–807.

Johansen, K., L. A. Arroyo, S. Phinn, and C. Witte. 2010. Comparison of geo-object based and pixel-based change detection of riparian environments using high spatial resolution multi-spectral imagery. *Photogramm Eng Remote Sens* 76(2):123–36.

Johansen, K., Bartolo, R., and Phinn, S. 2010. Special feature – Geospatial object based image analysis. *Journal of Spatial Science* 55(1):3–7.

Johansen, K., N. C. Coops, S. E. Gergel, and J. Stange. 2007. Application of high spatial resolution satellite imagery for riparian and forest ecosystem classification. *Remote Sens Environ* 110(1):29–44.

Johansen, K., and S. Phinn. 2006. Linking riparian vegetation structure in Australian tropical savannas to ecosystem health indicators: Semi-variogram analysis of high spatial resolution satellite imagery. *Can J Remote Sens* 32(3):228–43.

Kayitakire, F., C. Hamel, and P. Defourny. 2006. Retrieving forest structure variables based on image texture analysis and IKONOS-2 imagery. *Remote Sens Environ* 102(3–4):390–401.

Kettig, R., and D. Landgrebe. 1976. Classification of multispectral image data by extraction and classification of homogeneous objects. *IEEE Trans on Geosci Electron* 14(1):19–26.

Koch, B., and M. Dees. 2008. Forestry applications. In *Advances in Photogrammetry, Remote Sensing and Spatial Information Sciences*. 2008 ISPRS congress book, ed. Z. Li, J. Chen and E. Baltsavias, 439–68. London: Taylor & Francis.

Krause, G., M. Bock, S. Weiers, and G. Braun. 2004. Mapping land-cover and mangrove structures with remote sensing techniques: A contribution to a synoptic GIS in support of coastal management in north Brazil. *Environ Manage* 34(3):429–40.

Laliberte, A. S., E. L. Fredrickson, and A. Rango. 2007. Combining decision trees with hierarchical object-oriented image analysis for mapping arid rangelands. *Photogramm Eng Remote Sens* 73(2):197–207.

Laliberte, A. S., A. Rango, K. M. Havstad, J. F. Paris, R. F. Beck, R. McNeely, and A. L. Gonzalez. 2004. Object-oriented image analysis for mapping shrub encroachment from 1937 to 2003 in southern New Mexico. *Remote Sens Environ* 93(1–2):198–210.

Land and Water Australia (2002). *River Landscapes: Fact Sheet*. Canberra, Australia: Land and Water Australia.

Lang, S. 2008. Object-based image analysis for remote sensing applications: Modeling reality—dealing with complexity. In *Object-Based Image Analysis*, ed. T. Blaschke, S. Lang and G. J. Hay, 1–25. New York: Springer.

Lang, S., and T. Blaschke. 2006. Bridging remote sensing and GIS: What are the main supporting pillars? *Int Arch Photogramm Remote Sens Spatial Inf Sci* XXXVI-4/C42.

Lang, S., and T. Langanke. 2006. Object-based mapping and object-relationship modeling for land use classes and habitats. *Photogramm Fernerkundung Geoinf* 10(1):5–18.

Langanke, T., C. Burnett, and S. Lang. 2007. Assessing the mire conservation status of a raised bog site in Salzburg using object-based monitoring and structural analysis. *Landscape Urban Plann* 79(2):160–9.

Lathrop, R. G., P. Montesano, and S. Haag. 2006. A multi-scale segmentation approach to mapping sea grass habitats using airborne digital camera imagery. *Photogramm Eng Remote Sens* 72(5):665–75.

Lefsky, M. A., W. B. Cohen, G. G. Parker, and D. J. Harding. 2002. Lidar remote sensing for ecosystem studies. *BioScience* 52:19–30.

Lefsky, M. A., W. B. Cohen, and T. A. Spies. 2001. An evaluation of alternate remote sensing products for forest inventory, monitoring, and mapping of Douglas fir forests in western Oregon. *Can J For Res* 31:78–87.

Lillesand, T. M., R. W. Kiefer, and J. W. Chipman. 2008. *Remote Sensing and Image Interpretation*. 6th ed. New York: Wiley.

Liu, Y., M. Li, L. Mao, and F. Xu. 2006. Review of remotely sensed imagery classification patterns based on object oriented image analysis. *Chin Geog Sci* 16(3):282–8.

Lu, D. 2006. The potential and challenge of remote sensing-based biomass estimation. *Int J Remote Sens* 27(7):1297–328.

Lu, D., P. Mausel, E. Brondizio, and E. Moran. 2004. Change detection techniques. *Int J Remote Sens* 25:2365–407.

Lucas, R. M., K. M. Honza, P. J. Curran, G. M. Foody, R. Milne, T. Brown, and S. Amaral. 2000. Mapping the regional extent of tropical forest regeneration stages in the Brazilian Legal Amazon using NOAA AVHRR data. *Int J Remote Sens* 21:2855–81.

Lucieer, V. L. 2008. Object-oriented classification of sidescan sonar data for mapping benthic marine habitats. *Int J Remote Sens* 29(3):905–21.

Luscier, J. D., W. L. Thompson, J. M. Wilson, B. E. Gorham, L. D. Dragut. 2006. Using digital photographs and object- based image analysis to estimate percent ground cover in vegetation plots. *Front Ecol Environ* 4(8):408–13.

Maier, B., D. Tiede, and I. Dorren. 2008. Characterising mountain forest structure using landscape metrics on LiDAR-based canopy surface models. In *Object-Based Image Analysis*, ed. T. Blaschke, S. Lang and G. J. Hay, 625–44. New York: Springer.

Mallet, C., and F. Bretar. 2009. Full-waveform topographic lidar: State-of-the-art. *ISPRS J Photogramm Remote Sens* 64(1):1–16.

Mallinis, G., N. Koutsias, M. Tsakiri-Strati, and M. Karteris. 2008. Object-based classification using Quickbird imagery for delineating forest vegetation polygons in a Mediterranean test site. *ISPRS J Photogramm Remote Sens* 63(2):237–50.

Marceau, D. 1999. The scale issue in the social and natural sciences. *Can J Remote Sens* 25(4):347–56.

Mas, J. -F., Y. Gao, and J. A. N. Pacheco. 2010. Sensitivity of landscape pattern metrics to classification approaches. *For Ecol Manage* 259:1215–24.

Myint, S. W., M. Yuan, R. S. Cerveny, and C. P. Giri. 2008. Comparison of remote sensing image processing techniques to identify tornado damage areas from Landsat TM data. *Sensors* 8(2):1128–56.

Naiman, R. J., and H. Decamps. 1997. The ecology of interfaces: Riparian zones. *Annu Rev Ecol Syst* 28:621–58.

Neubert, M., H. Herold, and G. Meinel. 2008. Assessing image segmentation quality: Concepts, methods and application. In *Object-Based Image Analysis*, ed. T. Blaschke, S. Lang and G. J. Hay, 760–84. New York: Springer.

Pal, R., and K. Pal. 1993. A review on image segmentation techniques. *Pattern Recognit* 26(9):1277–94.

Pascual, C., A. García-Abril, L. G. García-Montero, S. Martín-Fernández, and W. B. Cohen. 2008. Object-based semi-automatic approach for forest structure characterization using lidar data in heterogeneous *Pinus sylvestris* stands. *For Ecol Manage* 255(11):3677–85.

Persson, A., J. Holmgren, and U. Soderman. 2002. Detecting and measuring individual trees using an airborne laser scanner. *Photogramm Eng Remote Sens* 28:1–8.

Radoux, J., and P. Defourny. 2007. A quantitative assessment of boundaries in automated forest stand delineation using very high resolution imagery. *Remote Sens Environ* 110(4):468–75.

Roberts, D. A., I. Numata, K. Holmes, G. Batista, T. Krug, A. Monteiro, B. Powell, and O. A. Chadwick. 2002. Large area mapping of land-cover change in Rondonia using decision tree classifiers. *J Geophys Res* 107(D20):8073 LBA 40–1 to 40–18.

Ryherd, S., and C. E. Woodcock. 1996. Combining spectral and texture data in the segmentation of remotely sensed images. *Photogramm Eng Remote Sens* 62(2):181–94.

Santos, J. R., M. S. Pardi LaCruz, L. S. Araujo, and M. Keil. 2002. Savanna and tropical rainforest biomass estimation and spatialization using JERS-1 data. *Int J Remote Sens* 23:1217–29.

Sawaya, K. E., L. G. Olmanson, N. J. Heinert, P. L. Brezonik, and M. E. Bauer. 2003. Extending satellite remote sensing to local scales: Land and water resource monitoring using high-resolution imagery. *Remote Sens Environ* 88(1–2):144–56.

Shoshany, M. 2000. Satellite remote sensing of natural Mediterranean vegetation: A review within an ecological context. *Prog Phys Geog* 24:153–78.

Stow, D., Y. Hamada, L. Coulter, and Z. Anguelova. 2008. Monitoring shrubland habitat changes through object-based change identification with airborne multispectral imagery. *Remote Sens Environ* 112(3):1051–61.

Strahler, A. H. 1986. On the nature of models in remote sensing. *Remote Sensing of Environment* 20:121–139.

Su, W., J. Li, Y. Chen, Z. Liu, J. Zhang, T. M. Low, I. Suppiah, and S. A. M. Hashim. 2008. Textural and local spatial statistics for the object-oriented classification of urban areas using high resolution imagery. *Int J Remote Sens* 29(11):3105–17.

Suarez, J. C., C. Ontiveros, S. Smith, and S. Snape. 2005. Use of airborne LiDAR and aerial photography in the estimation of individual tree heights in forestry. *Comput Geosci* 31:253–62.

Tiede, D., and C. Hoffmann. 2006. Process oriented object-based algorithms for single tree detection using laser scanning data. *EARSeL-Proceedings of the Workshop on 3D Remote Sensing in Forestry* 14th–15th Feb 2006, Vienna, 151–6.

Tiede, D., S. Lang, and C. Hoffmann. 2008. Domain-specific class modelling for one-level representation of single trees. In *Object-Based Image Analysis*, ed. T. Blaschke, S. Lang and G. Hay, 133–51. New York: Springer.

Tiede, D., A. Osberger, and H. Novak. 2009. Automated tree extraction using very high resolution Digital Surface Models from UltraCamX-data. In *Applied Geoinformatics 2009*, ed. J. Strobl, T. Blaschke and G. Griesebner, 55–60. Wichmann, Heidelberg.

Trias-Sanz, R., G. Stamon, and J. Louchet. 2008. Using colour, texture, and hierarchical segmentation for high-resolution remote sensing. *ISPRS J Photogramm Remote Sens* 63(2):156–68.

Treitz, P. M., and Howarth, P. J. Hyperspectral remote sensing for estimating biophysical parameters of forest ecosystems. *Progress in Physical Geography* 23(3):359–390.

Turner, D. P., S. V. Ollinger, and J. S. Kimball. 2004. Integrating remote sensing and ecosystem process models for landscape- to regional-scale analysis of the carbon cycle. *BioScience* 54:573–84.

van Gardingen, P. R., G. E. Jackson, S. Hernandez-Daumas, G. Russel, and L. Sharp. 1999. Leaf area index estimates obtained for clumped canopies using hemispherical photography. *Agric For Meteorol* 94:243–57.

Walker, J. S., and T. Blaschke. 2008. Object-based land-cover classification for the Phoenix metropolitan area: Optimization vs. transportability. *Int J Remote Sens* 29(7):2021–40.

Wang, L., W. P. Sousa, and P. Gong. 2004. Integration of object-based and pixel-based classification for mapping mangroves with IKONOS imagery. *Int J Remote Sens* 25(24):5655–68.

Weiers, S., M. Bock, M. Wissen, and G. Rossner. 2004. Mapping and indicator approaches for the assessment of habitats at different scales using remote sensing and GIS methods. *Landscape and Urban Plann* 67(1–4):43–65.

Weinke, E., S. Lang, and M. Preiner. 2008. Strategies for semi-automated habitat delineation and spatial change assessment in an Alpine environment. In *Object-Based Image Analysis*, ed. T. Blaschke, S. Lang and G. J. Hay, 711–32. New York: Springer.

Wermelinger, B. 2004. Ecology and management of the spruce bark beetle *Ips typographus*—a review of recent research. *For Ecol Manage* 202(1–3):67–82.

Wiseman, G., J. Kort, and D. Walker. 2009. Quantification of shelterbelt characteristics using high-resolution imagery. *Agric Ecosyst Environ* 131(1–2):111–7.

Woodcock, C., and V. J. Harward. 1992. Nested-hierarchical scene models and image segmentation. *Int J Remote Sens* 13(16):3167–87.

Wu, J., and J. L. David. 2002. A spatially explicit hierarchical approach to modeling complex ecological systems: Theory and applications. *Ecol Modell* 153(1–2):7–26.

Wulder, M. 1998. Optical remote-sensing techniques for the assessment of forest inventory and biophysical parameters. *Prog Phys Geog* 22(4):449–76.

Wulder, M. A., R. J. Hall, N. Coops, and S. Franklin. 2004. High spatial resolution remotely sensed data for ecosystem characterization. *BioScience* 54:511–21.

Xie, Z., C. Roberts, and B. Johnson. 2008. Object-based target search using remotely sensed data: A case study in detecting invasive exotic Australian pine in South Florida. *ISPRS J Photogramm Remote Sens* 63(6):647–60.

Yan, G., J. -F. Mas, B. H. P. Maathuis, Z. Xiangmin, and P. M. Van Dijk. 2006. Comparison of pixel-based and object-oriented image classification approaches—a case study in a coal fire area, Wuda, Inner Mongolia, China. *Int J Remote Sens* 27(18):4039–55.

Yu, Q., P. Gong, N. Chinton, G. Biging, M. Kelly, and D. Schirokauer. 2006. Object-based detailed vegetation classification with airborne high spatial resolution remote sensing imagery. *Photogramm Eng Remote Sens* 72(7):799–811.

Zhang, Q.F., M. Molenaar, K. Tempfli, and W. Shi. 2005. Quality assessment for geo-spatial objects derived from remotely sensed data. *Int J Remote Sens* 26(14):2953–74.

Zimble, D. A., D. L. Evans, G. C. Carlson, R. C. Parker, S. C. Grado, and P. D. Gerard. 2003. Characterizing vertical forest structure using small-footprint airborne LiDAR. *Remote Sens Environ* 87:171–82.

11

Land-Use and Land-Cover Change Detection

Dengsheng Lu, Emilio Moran, Scott Hetrick, and Guiying Li

CONTENTS

11.1 Introduction

Change detection is the process of identifying differences in the state of an object or phenomenon by observing it at different times (Singh 1989). Timely and accurate change detection of Earth's surface features provides the foundation for a better understanding of the relationships and interactions between human and natural phenomena in order to better manage and use resources. The advantages of repetitive data acquisition, its synoptic view, and a digital format suitable for computer processing have made remotely sensed data the major data sources for different change detection applications during the past decades (Lu et al. 2004; Kennedy et al. 2009). In general, change detection involves the application of multitemporal data sets to quantitatively analyze the temporal effects of the phenomena of interest. Good change detection research should provide the following information: area change and rate of change, spatial distribution of changed types, change trajectories of land-cover types, and accuracy assessment of change detection results (Lu et al. 2004).

The objective of change detection is to compare spatial representation of two points in time by controlling all variances caused by differences in variables that are not of interest and to measure changes caused by differences in the variables of interest (Green, Kempka, and Lackey 1994). The basic premise in using remotely sensed data for change detection is that changes in the objects of interest will result in changes in reflectance values or local textures that are separable from changes caused by other factors, such as differences in atmospheric conditions, illumination, viewing angles, and soil moisture (Deer 1995). In practice, many factors, such as the quality of image registration, the quality of atmospheric correction or normalization between multitemporal images, the complexity of the landscape and topography under investigation, the analyst's skill and experience, and the selected change detection methods, can affect change detection results (Lu et al. 2004; Jensen 2005). Errors and uncertainties may come from different steps taken, such as in image preprocessing and selection of the change detection algorithm. It is important to understand the major steps in implementing the change detection procedure and to reduce errors or uncertainties in each step. Different authors have often arrived at different and sometimes controversial conclusions about which change detection techniques are most effective. Therefore, in this chapter, we describe the major steps used in the change detection procedure and provide a case study showing how to use change detection techniques to solve practical problems.

11.2 Overview of Change Detection Procedure

There are two categories of changes: changes between classes and changes within classes. A change between classes is a conversion of land cover from one category to a completely different category, for example through deforestation or urbanization. A change within classes is a modification of the condition of the land-cover type within the same category, for example through selective logging (Lu et al. 2004). The change detection procedure can be based on per-pixel, subpixel, or object-oriented methods, which require different image processing and change detection algorithms. The research objectives, remote sensing data used, and the geographical size of the study area can affect the design of the change detection procedure, including the use of different image processing methods and change detection techniques (Lu et al. 2004; Jensen 2005). In general, at a local scale, object-oriented methods are useful for reducing the spectral variation within the same land cover when very high spatial resolution images such as QuickBird or IKONOS are available. At a regional scale, per-pixel-based methods are often used when medium spatial resolution images such as Landsat Thematic Mapper (TM) images are available. For a national and global scale at which coarse spatial resolution images such as Moderate Resolution Imaging Spectroradiometer (MODIS) and Advanced Very High Resolution Radiometer (AVHRR) images are available, subpixel-based change detection techniques may provide better results than per-pixel-based techniques due to the mixed-pixel problem. However, in practice, per-pixel-based techniques are still the most common methods for land-cover change detection at different scales because the image processing techniques and change detection algorithms are mainly based on per-pixel data analysis. Therefore, this section mainly focuses on the per-pixel-based change detection procedure. Figure 11.1 illustrates the major steps and corresponding contents for

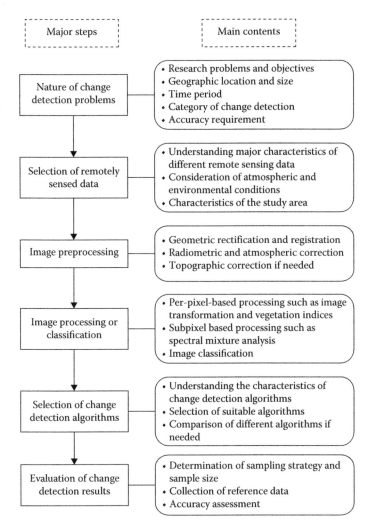

FIGURE 11.1
Major steps and corresponding main contents for a remote sensing-based land-cover change detection procedure.

the change detection procedure. The following subsections provide a brief description for each step.

11.2.1 Nature of Change Detection Problems

Before change detection is conducted in a specific study area, it is very important to clearly define the research problems that need solving, the objectives, and the location and size of the study area (Jensen 2005). These issues directly affect the selection of remotely sensed data and the selection of change detection algorithms. It is helpful to list some questions, for example, the kinds of change detection contents that are required: binary change and nonchange information, detailed "from-to" change trajectories, or the detection of continuous change. What is the accuracy overall and for each change detection trajectory? How large is the study area? What is the change detection period? What kinds of remote sensing data and/or ancillary data are available? Based on these questions, one can identify specific research questions and objectives. After that, one can design a change detection procedure suitable for the specific study area and purpose.

11.2.2 Data Collection

Many remotely sensed data generated from both airborne and spaceborne sensors with different spatial, radiometric, spectral, and temporal resolutions, are available. In order to select suitable data sets for a specific study, it is important to understand the strengths and weaknesses of different types of sensor data. Some previous literature has reviewed the characteristics of the major types of remote sensing data (Barnsley 1999; Estes and Loveland 1999; Althausen 2002; Lefsky and Cohen 2003). For example, Barnsley (1999) and Lefsky and Cohen (2003) summarized the characteristics of different remote sensing data in spectral, radiometric, spatial and temporal resolutions, polarization, and angularity. The user's need, scale and characteristics of the study area, availability of various image data and their characteristics, cost and time constraints, and the analyst's experience in using the selected image may be the most important factors affecting the selection of remotely sensed data for a specific study area (Lu et al. 2004; Lu and Weng 2007). In general, at a local level, a fine-scale land-cover change scheme is required; thus, high spatial resolution data such as IKONOS, QuickBird, and SPOT 5 HRG (High-Resolution Geometric) data are helpful. At a regional scale, medium spatial resolution data such as Landsat TM and Terra Advanced Spaceborne Thermal Emission and Reflection Radiometer (ASTER) are the most frequently used data sources. On a continental or global scale, coarse spatial resolution data such as AVHRR, MODIS, and SPOT VEGETATION are preferable.

For a successful implementation of a change detection analysis using remotely sensed data, careful considerations of the remote sensor system, environmental characteristics, and image processing methods are important. The temporal, spatial, spectral, and radiometric resolutions of remotely sensed data have a significant impact on the success of a remote sensing change detection project. The important environmental factors include atmospheric conditions, soil moisture conditions, and phenological characteristics (Weber 2001; Jensen 2005). When selecting remote sensing data for change detection applications, it is important to use the same sensor, radiometric, and spatial resolution data with anniversary or very near anniversary acquisition dates in order to eliminate the effects of external sources, such as sun angle, seasonal, and phenological differences. However, in a specific study, selection of the same sensor data may be difficult, especially in moist tropical regions due to often cloudy conditions. Thus, the use of different sensor data for change detection is required (Lu et al. 2008a). Detailed descriptions about the considerations of remote sensing systems and environmental characteristics before implementing a change detection study are available in previous literature, such as Coppin and Bauer (1996), Biging et al. (1999), and Jensen (2005).

11.2.3 Image Preprocessing

Before implementing a change detection analysis, the following two conditions should be satisfied: (1) precise coregistration between multitemporal images and (2) precise radiometric and atmospheric calibration or normalization between multitemporal images. The importance of accurate geometric registration of multitemporal images is obvious, because largely spurious results of change detection are produced if there is misregistration (Townshend et al. 1992; Dai and Khorram 1998; Stow 1999; Verbyla and Boles 2000;

Carvalho et al. 2001; Stow and Chen 2002). Many textbooks have detailed the description of image-to-map rectification or image-to-image registration (e.g., Jensen 2005).

The same invariant objects could have different spectral signatures in different acquisition-date images due to different sun elevation angles and azimuth angles, vegetation phenological conditions, soil moisture, and atmospheric conditions. In such a case, the conversion of digital numbers to radiance or surface reflectance is a requirement for quantitative analyses of multitemporal images. A variety of methods, such as relative calibration, dark object subtraction, and 6S (second simulation of the satellite signal in the solar spectrum) have been developed for radiometric and atmospheric normalization or correction (Markham and Barker 1987; Gilabert, Conese, and Maselli 1994; Chavez 1996; Stefan and Itten 1997; Vermote et al. 1997; Tokola, Lofman, and Erkkila 1999; Heo and FitzHugh 2000; Yang and Lo 2000; Song et al. 2001; Du, Teillet, and Cihlar 2002; Lu et al. 2002; McGovern et al. 2002; Vicente-Serrano, Perez-Cabello, and Lasanta 2008; Chander, Markham, and Helder 2009). If the study area is rugged or mountainous, topographic correction may be necessary. More detailed information about topographic correction is available in Teillet, Guindon, and Goodenough (1982), Civco (1989), Colby (1991), Meyer et al. (1993), and Lu et al. (2008b).

11.2.4 Image Processing and Classification

Change detection can be conducted using spectral bands or derived images such as vegetation indices and transformed images. For example, much previous research has indicated the usefulness of the visible red-band images in change detection analysis (Jensen and Toll 1982; Fung 1990; Chavez and Mackinnon 1994; Lu et al. 2005) because vegetation has low reflectance, but impervious surfaces or soils have high reflectance in this band. However, a single band cannot reflect all changed information due to the complex landscapes. For many situations, use of transformed images or vegetation indices can be more effective in extracting the differences of changed features than single spectral bands (Lu et al. 2005). Image transformation is often used to reduce data redundancy and the number of image channels so that the information contents are concentrated in a few transformed images (Jensen 2005). Different techniques have been developed to transform the multispectral data into a new data set. Principal component analysis (PCA), tasseled cap, minimum noise fraction, wavelet transform, and spectral mixture analysis (Myint 2001; Okin et al. 2001; Rashed et al. 2001; Asner and Heidebrecht 2002; Lobell et al. 2002; Neville et al. 2003; Landgrebe 2003; Platt and Goetz 2004; Lu et al. 2008a) are among the most commonly used techniques. Vegetation indices are recommended for removing the variability, which is caused by canopy geometry, soil background, sun view angles, and atmospheric conditions when measuring biophysical properties (Elvidge and Chen 1995; Blackburn and Steele 1999). Many vegetation indices have thus been developed and applied to biophysical parameter studies (Anderson and Hanson 1992; Anderson, Hanson, and Haas 1993; Eastwood et al. 1997; Mutanga and Skidmore 2004). Vegetation indices are often used for land-cover change detection, especially for vegetation change.

For many applications, detailed land-cover change trajectories are required; thus, classification-based change detection techniques, such as postclassification comparison, are often used. Scientists and practitioners have made great efforts to develop advanced classification approaches and techniques for improving classification accuracy (Gong and Howarth 1992; Kontoes et al. 1993; Foody 1996; San Miguel-Ayanz and Biging 1997; Aplin, Atkinson, and Curran 1999; Stuckens, Coppin, and Bauer 2000; Franklin et al. 2002; Pal and Mather 2003; Gallego 2004; Lu and Weng 2007; Blaschke 2010; Ghimire, Rogan, and Miller 2010).

Much previous literature has been specifically concerned with image classification (Tso and Mather 2001; Landgrebe 2003). Lu and Weng (2007) provided a comprehensive up-to-date review of classification approaches and techniques. Since classification results are the required data sets for detecting detailed land-use and land-cover "from-to" change trajectories, high classification accuracies are critical for change detection because individual classification accuracy will affect the change detection accuracy.

11.2.5 Selection of Change Detection Algorithms

For a given research purpose, when the remotely sensed data and study areas are identified, selection of an appropriate change detection method has considerable significance in producing a high-quality change detection product. Some techniques, such as image differencing, can only provide change or nonchange information, whereas other techniques, such as postclassification comparison, can provide a complete matrix of change directions. In general, change detection techniques can be roughly grouped into two categories: (1) those detecting binary change or nonchange information, such as using image differencing, image ratioing, vegetation index differencing, and PCA and (2) those detecting detailed "from-to" change trajectory, such as using the postclassification comparison and hybrid change detection methods (Lu et al. 2004). Previous literature has reviewed many change detection techniques (Singh 1989; Coppin and Bauer 1996; Yuan, Elvidge, and Lunetta 1998; Serpico and Bruzzone 1999; Coppin et al. 2004; Lu et al. 2004; Jensen 2005; Kennedy et al. 2009). Lu et al. (2004) grouped change detection methods into seven categories: (1) algebra, (2) transformation, (3) classification, (4) advanced models, (5) geographic information system approaches, (6) visual analysis, and (7) other approaches, and summarized the major characteristics, advantages, and disadvantages for the selected techniques. Due to the importance of monitoring changes among Earth's surface features, the research of change detection techniques has long been an active topic, and new techniques are constantly appearing.

When implementing change or nonchange detection, one critical step is to select appropriate thresholds in both tails of the histogram representing the changed areas (Singh 1989). Two methods are often used for the selection of thresholds (Singh 1989; Yool, Makaio, and Watts 1997): (1) interactive procedure or manual trial-and-error procedure—an analyst interactively adjusts the thresholds and evaluates the resulting image until satisfied; and (2) statistical measures—selection of a suitable standard deviation from the mean. The two disadvantages of the threshold technique are (1) the resulting differences may include external influences caused by atmospheric conditions, sun angles, soil moistures, and phenological differences in addition to true land-cover change; and (2) the threshold is highly subjective and scene dependent, depending on the analyst's familiarity with the study area and skill (Lu et al. 2004, 2005). When implementing the detailed "from-to" change detection, the results are mainly dependent on the classification accuracy for each date being analyzed (Jensen 2005). In other words, classification errors from the individual-date images will affect the final change detection accuracy. The critical step is to develop an accurate classification image for each date.

In practice, an analyst often selects several methods to implement change detection in a study area and then compares and identifies the best result through accuracy assessment (Muchoney and Haack 1994; Michener and Houhoulis 1997; Macleod and Congalton 1998; Yuan and Elvidge 1998; Mas 1999; Dhakal et al. 2002; Lu et al. 2005). Although a large number of change detection applications have been implemented and different change detection techniques have been tested, the question of which method is best suited for a specific study area remains unanswered. No single method is suitable for all cases. The method selected depends on an analyst's knowledge of the change detection methods and skills in handling remote sensing data, the image data used, and characteristics of the study areas.

11.2.6 Evaluation of Change Detection Results

Accuracy assessment for change detection is particularly difficult due to problems in collecting reliable temporal field-based data sets; thus, much previous research on change detection did not provide a quantitative analysis of the research results. Standard accuracy assessment techniques have been developed mainly for single-date remotely sensed data. Previous literature has provided the meanings and calculation methods for these elements (Congalton, Oderwald, and Mead 1983; Congalton 1991; Janssen and van der Wel 1994; Kalkhan, Reich, and Czaplewski 1997; Biging et al. 1999; Smits, Dellepiane, and Schowengerdt 1999; Congalton and Plourde 2002; Foody 2002; Congalton and Green 2008). The error matrix-based accuracy assessment method is the most common and valuable method for the evaluation of change detection results. In addition, some new methods have been developed to analyze the accuracy of change detection (Morisette and Khorram 2000; Lowell 2001). Morisette and Khorram (2000) used "accuracy assessment curves" to analyze satellite-based change detection, and Lowell (2001) developed an area-based accuracy assessment method for the analysis of change maps. A monograph titled "Accuracy assessment of remotely sensed-derived change detection," edited by Siamak Khorram (Biging et al. 1999), is specifically focused on the accuracy assessment of land-cover change detection. This monograph describes the issues affecting the accuracy assessment of land-cover change detection, identifies the factors of a remote sensing processing system that affect accuracy assessment, presents a sampling design to estimate the elements of the error matrix efficiently, illustrates possible applications, and gives recommendations for accuracy assessment of change detection.

11.3 Case Study for Detecting Urbanization with Multitemporal Landsat TM Images

Section 11.2 briefly overviewed the major steps used in the change detection procedure. This section provides a case study for showing how to conduct a change detection for examining urban expansion based on multitemporal TM images in a complex urban-rural landscape.

11.3.1 Research Problem and Objective

Digital change detection in urban environments is a challenge due to three characteristics unique to urban areas: (1) urban land-use and land-cover changes usually account for a small proportion of the study area and are scattered in different locations; (2) impervious surfaces and similar spectral features between impervious surfaces and other nonvegetation land covers are complex; and (3) the spatial resolution of remotely sensed imagery is limited. Although many change detection techniques, such as PCA, image differencing, and postclassification comparison, can be applied to urban land-use and land-cover change detection (Singh 1989; Coppin and Bauer 1996; Coppin et al. 2004; Lu et al. 2004;

Kennedy et al. 2009), the detection results are often poor, especially in urban–rural frontiers. Therefore, this research aims to develop a change detection procedure suitable for detecting urbanization in a complex urban–rural frontier, based on the comparison of extracted impervious surface data sets from multitemporal Landsat TM images.

11.3.2 Description of the Study Area

Lucas do Rio Verde (hereafter called simply Lucas) in Mato Grosso state, Brazil, has a relatively short history and small urban extent. It was established in the early 1980s (Figure 11.2), and has experienced rapid urbanization. This region is connected to the city of Santarém, a river port on the Amazon River, and to the heart of the soybean growing area at the city of Cuiabá by highway BR-163, which runs through the county. The economic base of Lucas is large-scale agriculture, including the production of soy, cotton, rice, and corn, as well as poultry and swine. The county is at the epicenter of soybean production in Brazil, and it is expected to grow in population threefold in the next 10 years. Because it is, at present, a relatively small town, yet has complex urban–rural spatial patterns derived from its highly capitalized agricultural base, large silos and warehouses, and planned urban growth, Lucas is an ideal site for exploring techniques for detecting urbanization with remote sensing data.

11.3.3 Methods

After the research objectives were clearly defined, the next step was to select suitable remote sensing data and to design a feasible procedure for implementing change detection.

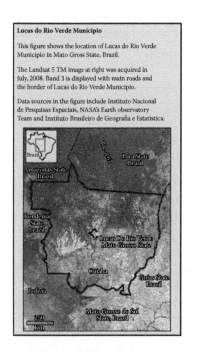
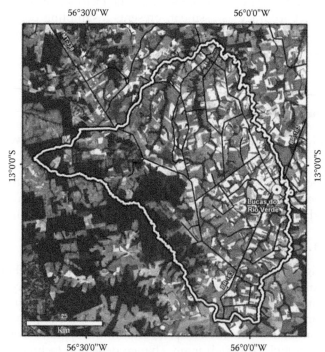

FIGURE 11.2
Study area—Lucas do Rio Verde Municipio, Mato Grosso state, Brazil.

11.3.3.1 Data Collection and Preprocessing

Landsat images acquired on June 21, 1984, June 6, 1996, and May 22, 2008 were used in this research. Radiometric and atmospheric calibration was conducted using the image-based dark-object subtraction (DOS) method. The DOS model is an image-based procedure that standardizes imagery for the effects caused by solar zenith angle, solar radiance, and atmospheric scattering (Lu et al. 2002; Chander, Markham, and Helder 2009). The equations used for Landsat TM image calibration are

$$R_\lambda = \frac{PI \times D \times (L_\lambda - L_{\lambda.haze})}{[Esun_\lambda \times \cos(\theta)]} \tag{11.1}$$

$$L_\lambda = gain_\lambda \times DN_\lambda + bias_\lambda \tag{11.2}$$

where L_λ is the apparent at-satellite radiance for spectral band λ, DN_λ is the digital number of spectral band λ, R_λ is the calibrated reflectance, $L_{\lambda.haze}$ is path radiance, $Esun_\lambda$ is exo atmospheric solar irradiance, D is the distance between the Earth and sun, and θ is the sun zenith angle. The path radiance for each band is identified based on the analysis of water bodies and shades in the images. The $gain_\lambda$ and $bias_\lambda$ are the radiometric gain and bias corresponding to spectral band λ, and they are often provided in an image header file or metadata file or calculated from maximal and minimal spectral radiance values (Lu et al. 2002). All TM images were geometrically coregistered into the UTM projection with geometric errors of less than one pixel, so that all images have the same coordinate system. The nearest neighbor resampling technique was used to resample the Landsat TM images into a pixel size of 30 m × 30 m during image-to-image registration.

11.3.3.2 Mapping of Impervious Surface Distribution

Per-pixel impervious surface mapping is often based on the image classification of spectral signatures (Shaban and Dikshit 2001; Dougherty et al. 2004; Jennings, Jarnagin, and Ebert 2004), but the spectral confusion between impervious surfaces and other land covers often results in a poor classification performance in the urban landscape (Lu and Weng 2005), especially in a complex urban–rural frontier. This research developed a method based on the combination of filtering images and unsupervised classification of Landsat spectral signatures for mapping per-pixel impervious surface distribution. The fact that the red-band image in Landsat TM has high spectral values for impervious surfaces, but has low spectral values for vegetation and water or wetlands provides the potential for rapidly mapping impervious surfaces. The minimum and maximum filters with a window size of 3 × 3 pixels were separately applied to the Landsat red-band image. The image differencing between maximum and minimum filtering images was used to highlight linear features (mainly roads) and other impervious surfaces. Examining the difference image indicated that a threshold value of 13 can be used to extract the impervious surface image. The spectral signature of the initial impervious surface image was then extracted and was further classified into 60 clusters using an unsupervised classification method to refine the impervious surface image by removing the nonimpervious surface pixels. Finally, manual editing of the impervious surface image was conducted to make sure that all impervious surfaces, especially in urban regions, were extracted. The final impervious surface image was overlain on the TM color composite to visually examine the quality of the impervious surface results to assure that all urban area and major roads were properly extracted. The same method was applied to all three dates of TM imagery to generate a time series of impervious surface images.

11.3.3.3 Detection of Urbanization

Many change detection methods may be used for land-cover change detection (Lu et al. 2004), but most of them are not suitable for the detection of urbanization due to the unique characteristics of the urban landscape. Therefore, the change detection of urbanization in this research is based on the comparison of extracted impervious surface images in order to eliminate the impacts of spectral confusion between impervious surfaces and other land covers, such as between dark impervious surfaces and water or wetland, and between bright impervious surfaces and bare soils or harvested fields. Two methods were used in this research. The first method was to produce a color composite by assigning the 2008, 1996, and 1984 impervious surface images as red, green, and blue, for visual interpretation of impervious surface change. Another method was to produce the change detection result based on a comparison of extracted impervious surface images pixel by pixel. The total impervious surface area change was also calculated.

11.3.3.4 Evaluation of Urbanization Results

Accuracy assessment of change detection results is an important part of the change detection procedure for understanding the reliability and confidence in the results. In this research, quantitative assessment of the change detection was difficult due to the lack of high spatial resolution images or field survey data for Landsat TM imagery in 1996 and 1984. Therefore, the evaluation of change detection results was based on a cross-comparison between the TM color composite and urbanization images. No quantitative evaluation was conducted.

11.3.4 Results

Evaluation of the per-pixel impervious surface image based on overlaying it with the TM color composite indicated that a combination of filtering images and unsupervised classification methods developed in this research can effectively extract the pixel-based impervious surface image in a complex urban–rural frontier. Figure 11.3 shows where impervious surface change occurred between the TM acquisition dates. The impervious surface images of 2008, 1996, and 1984 were assigned as red, green, and blue in the color composite; thus, red indicates that impervious surfaces increased between 1996 and 2008, and yellow indicates that the impervious surface increased between 1984 and 1996. This figure shows that the major impervious surface increase between 1984 and 1996 was in central Lucas because it was established in the early 1980s, and then, the impervious surface rapidly increased in the north, northwest, and south parts of town, and more roads were constructed after 1996.

In per-pixel-based results, each extracted impervious surface pixel is assumed to be 100% impervious surface. Thus, the total impervious surface area for this study area can be calculated by multiplying the total pixel number of impervious surfaces and the TM pixel size (30 m by 30 m). This research indicates that the total impervious surface area in 1984 only accounted for 0.24% of the total study area, which gradually increased to 0.43% in 1996 and to 1.29% in 2008, implying rapid urbanization rate during the change detection periods.

11.3.5 Summary of the Case Study

The per-pixel-based method for mapping impervious surface distribution and monitoring its change is valuable for visual interpretation of urbanization. The method, based on the combination of filtering image differencing and unsupervised classification, can be successfully

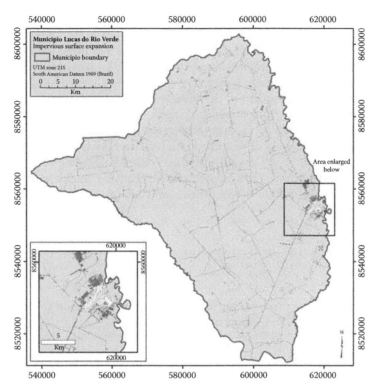

FIGURE 11.3
(See color insert following page 426.) Color composite of three dates of impervious surface distribution in 2008, 1996, and 1984 by assigning them as red, green, and blue, illustrating the spatial distribution and patterns of impervious surface changes.

used to map impervious surface distribution in the complex urban–rural frontier, which is often difficult for traditional classification methods. In addition, the detection of urbanization based on the extracted impervious surface images can eliminate the impacts of environmental conditions on remote sensing surface reflectance, which often results in a different reflectance for the same land covers. However, the areal extent of impervious surfaces is overestimated significantly, especially in the urban–rural frontier due to the mixed-pixel problem in Landsat TM images (Wu and Murray 2003; Lu and Weng 2006). From the view of area calculation of urbanization, fractional impervious surface distribution based on subpixel-based method, such as spectral mixture analysis, must be developed (Lu and Weng 2006).

11.4 Final Remarks

Change detection has long been an active research topic, and many techniques have been developed in recent decades. The availability of more and more different types of sensor data and different ancillary data, along with a need for more detailed and accurate change detection information, provides new challenges for developing suitable change detection techniques for specific purposes. Change detection is a comprehensive procedure that requires careful design of different steps, including the statement of research problems and objectives, data collection, preprocessing, selection of suitable change detection algorithms, and evaluation of the results. Errors or uncertainties may emerge from any of the different steps, thus affecting the change detection results. Understanding the relationships between the change detection stages, identifying the weakest links in the image processing chain, and then devoting efforts to improving them are keys to a successful change detection project. In addition, the designed change detection procedure should carefully take the spectral and spatial resolutions of the data, polarization, and angle features into account. Previous research on change detection is based mainly on per-pixel comparison. As high spatial resolution images such as QuickBird and WorldView become readily available, object-based or texture-based change detection methods may provide new insights and have recently attracted increasing attention (Lam 2008; Zhou, Troy, and Grove 2008; Blaschke 2010; Wu, Yang, and Lishman 2010).

Acknowledgments

The authors thank the National Institute of Child Health and Human Development at NIH (grant # R01 HD035811) for the support of this research.

References

Althausen, J. D. 2002. What remote sensing system should be used to collect the data? In *Manual of Geospatial Science and Technology*, ed. J. D. Bossler, J. R. Jensen, R. B. McMaster, and C. Rizos, 276. New York: Taylor & Francis.

Anderson, G. L., and J. D. Hanson. 1992. Evaluating handheld radiometer derived vegetation indices for estimating above ground biomass. *Geocarto Int* 7:71.

Anderson, G. L., J. D. Hanson, and R. H. Haas. 1993. Evaluating Landsat Thematic Mapper derived vegetation indices for estimating aboveground biomass on semiarid rangelands. *Rem Sens Environ* 45:165.

Aplin, P., P. M. Atkinson, and P. J. Curran. 1999. Per-field classification of land use using the forthcoming very fine spatial resolution satellite sensors: Problems and potential solutions. In *Advances in Remote Sensing and GIS Analysis*, ed. P. M. Atkinson and N. J. Tate, 219. New York: John Wiley & Sons.

Asner, G. P., and K. B. Heidebrecht. 2002. Spectral unmixing of vegetation, soil and dry carbon cover in arid regions: Comparing multispectral and hyperspectral observations. *Int J Rem Sens* 23:3939.

Barnsley, M. J. 1999. Digital remote sensing data and their characteristics. In *Geographical Information Systems: Principles, Techniques, Applications, and Management*, 2nd ed, ed. P. Longley, M. Goodchild, D. J. Maguire, and D. W. Rhind, 451. New York: John Wiley & Sons.

Biging, G. S. et al. 1999. "Accuracy assessment of remote sensing-detected change detection," Monograph Series, ed. S. Khorram. Maryland: American Society for Photogrammetry and Remote Sensing.

Blackburn, G. A., and C. M. Steele. 1999. Towards the remote sensing of Matorral vegetation physiology: Relationships between spectral reflectance, pigment, and biophysical characteristics of semiarid bushland canopies. *Rem Sens Environ* 70:278.

Blaschke, T. 2010. Object based image analysis for remote sensing. *ISPRS J Photogramm Rem Sens* 65:2.

Carvalho, L. M. T. et al. 2001. Digital change detection with the aid of multiresolution wavelet analysis. *Int J Rem Sens* 22:3871.

Chavez Jr., P. S. 1996. Image-based atmospheric corrections—revisited and improved. *Photogramm Eng Rem Sens* 62:1025.

Chavez Jr., P. S., and D. J. Mackinnon. 1994. Automatic detection of vegetation changes in the southwestern United States using remotely sensed images. *Photogramm Eng Rem Sens* 60:571.

Chander, G., B. L. Markham, and D. L. Helder. 2009. Summary of current radiometric calibration coefficients for Landsat MSS, TM, ETM+, and EO-1 ALI sensors. *Rem Sens Environ* 113:893.

Civco, D. L. 1989. Topographic normalization of Landsat Thematic Mapper digital imagery. *Photogramm Eng and Rem Sens* 55:1303.

Colby, J. D. 1991. Topographic normalization in rugged terrain. *Photogramm Eng Rem Sens* 57:531.

Congalton, R. G., R. G. Oderwald, and R. A. Mead. 1983. Assessing Landsat classification accuracy using discrete multivariate analysis statistical techniques. *Photogramm Eng Rem Sens* 49:1671.

Congalton, R. G. 1991. A review of assessing the accuracy of classification of remotely sensed data. *Rem Sens Environ* 37:35.

Congalton, R. G., and L. Plourde. 2002. Quality assurance and accuracy assessment of information derived from remotely sensed data. In *Manual of Geospatial Science and Technology*, ed. J. Bossler, 349. London: Taylor & Francis.

Congalton, R. G., and K. Green. 2008. *Assessing the Accuracy of Remotely Sensed Data: Principles and Practice*, 2nd ed, 183. Boca Raton: CRC Press/Taylor & Francis Group.

Coppin, P. R., and M. E. Bauer. 1996. Digital change detection in forest ecosystems with remote sensing imagery. *Rem Sens Rev* 13:207.

Coppin, P. et al. 2004. Digital change detection methods in ecosystem monitoring: A review. *Int J Rem Sens* 25:1565.

Dai, X. L., and S. Khorram. 1998. The effects of image misregistration on the accuracy of remotely sensed change detection. *IEEE Trans Geosci Rem Sens* 36:1566.

Deer, P. J. 1995. Digital change detection techniques: Civilian and military applications, *International Symposium on Spectral Sensing Research 1995 Report*. Greenbelt, MD: Goddard Space Flight Center.

Dhakal, A. S. et al. 2002. Detection of areas associated with flood and erosion caused by a heavy rainfall using multitemporal Landsat TM data. *Photogramm Eng Rem Sens* 68:233.

Dougherty, M. et al. 2004. Evaluation of impervious surface estimates in a rapidly urbanizing watershed. *Photogramm Eng Rem Sens* 70:1275.

Du, Y., P. M. Teillet, and J. Cihlar. 2002. Radiometric normalization of multitemporal high-resolution satellite images with quality control for land cover change detection. *Rem Sens Environ* 82:123.

Eastwood, J. A. et al. 1997. The reliability of vegetation indices for monitoring saltmarsh vegetation cover. *Int J Rem Sens* 18:3901.

Elvidge, C. D., and Z. Chen. 1995. Comparison of broad-band and narrow-band red and near-infrared vegetation indices. *Rem Sens Environ* 54:38.

Estes, J. E., and T. R. Loveland. 1999. Characteristics, sources, and management of remotely-sensed data. In *Geographical Information Systems: Principles, Techniques, Applications, and Management*, 2nd ed., eds. P. Longley, M. Goodchild, D. J. Maguire, and D. W. Rhind, 667. New York: John Wiley & Sons.

Foody, G. M. 1996. Approaches for the production and evaluation of fuzzy land cover classification from remotely-sensed data. *Int J Rem Sens* 17:1317.

Foody, G. M. 2002. Status of land cover classification accuracy assessment. *Rem Sens Environ* 80:185.

Franklin, S. E. et al. 2002. Evidential reasoning with Landsat TM, DEM and GIS data for land cover classification in support of grizzly bear habitat mapping. *Int J Rem Sens* 23:4633.

Fung, T. 1990. An assessment of TM imagery for land-cover change detection. *IEEE Trans Geosci Rem Sens* 28:681.

Gallego, F. J. 2004. Remote sensing and land-cover area estimation. *Int J Rem Sens* 25:3019.

Ghimire, B., J. Rogan, and J. Miller. 2010. Contextual land cover classification: Incorporating spatial dependence in land cover classification models using random forests and the Getis statistic. *Rem Sens Lett* 1:45.

Gilabert, M. A., C. Conese, and F. Maselli. 1994. An atmospheric correction method for the automatic retrieval of surface reflectance from TM images. *Int J Rem Sens* 15:2065.

Gong, P., and P. J. Howarth. 1992. Frequency-based contextual classification and gray-level vector reduction for land-use identification. *Photogramm Eng Rem Sens* 58:423.

Green, K., D. Kempka, and L. Lackey. 1994. Using remote sensing to detect and monitor land-cover and land-use change. *Photogramm Eng Rem Sens* 60:331.

Heo, J., and T. W. FitzHugh. 2000. A standardized radiometric normalization method for change detection using remotely sensed imagery. *Photogramm Eng Rem Sens* 66:173.

Janssen, L. F. J., and F. J. M. van der Wel. 1994. Accuracy assessment of satellite derived land-cover data: A review. *Photogramm Eng Rem Sens* 60:419.

Jennings, D. B., S. T. Jarnagin, and C. W. Ebert. 2004. A modeling approach for estimating watershed impervious surface area from national land cover data 92. *Photogramm Eng Rem Sens* 70:1295.

Jensen, J. R., and D. L. Toll. 1982. Detecting residential land use development at the urban fringe. *Photogramm Eng Rem Sens* 48:629.

Jensen, J. R. 2005. *Introductory Digital Image Processing: A Remote Sensing Perspective*, 3rd ed., 526. Upper Saddle River, NJ: Prentice Hall.

Kalkhan, M. A., R. M. Reich, and R. L. Czaplewski. 1997. Variance estimates and confidence intervals for the Kappa measure of classification accuracy. *Can J Rem Sens* 23:210.

Kennedy, R. E. et al. 2009. Remote sensing change detection tools for natural resource managers: Understanding concepts and tradeoffs in the design of landscape monitoring projects. *Rem Sens Environ* 113:1382.

Kontoes, C. et al. 1993. An experimental system for the integration of GIS data in knowledge-based image analysis for remote sensing of agriculture. *Int J Geogr Inf Syst* 7:247.

Lam, N. S. 2008. Methodologies for mapping land cover/land use and its change. In *Advances in Land Remote Sensing: System, Modeling, Inversion and Application*, ed. S. Liang, 341. New York: Spinger.

Landgrebe, D. A. 2003. *Signal Theory Methods in Multispectral Remote Sensing*, 508. Hoboken, NJ: Wiley.

Lefsky, M. A., and W. B. Cohen. 2003. Selection of remotely sensed data. In *Remote Sensing of Forest Environments: Concepts and Case Studies*, eds. M. A. Wulder, and S. E. Franklin, 13. Boston: Kluwer Academic.

Lobell, D. B. et al. 2002. View angle effects on canopy reflectance and spectral mixture analysis of coniferous forests using AVIRIS. *Int J Rem Sens* 23:2247.

Lowell, K., 2001. An area-based accuracy assessment methodology for digital change maps. *Int J Rem Sens* 22:3571.

Lu, D. et al. 2002. Assessment of atmospheric correction methods for Landsat TM data applicable to Amazon basin LBA research. *Int J Rem Sens* 23:2651.

Lu, D. et al. 2004. Change detection techniques. *Int J Rem Sens* 25:2365.

Lu, D. et al. 2005. Land-cover binary change detection methods for use in the moist tropical region of the Amazon: A comparative study. *Int J Rem Sens* 26:101.

Lu, D., and Q. Weng. 2005. Urban classification using full spectral information of Landsat ETM+ imagery in Marion County. *Indiana Photogramm Eng Rem Sens* 71:1275.

Lu, D., and Q. Weng. 2006. Use of impervious surface in urban land use classification. *Rem Sens Environ* 102:146.

Lu, D., and Q. Weng. 2007. A survey of image classification methods and techniques for improving classification performance. *Int J Rem Sens* 28:823.

Lu, D., M. Batistella, and E. Moran. 2008a. Integration of Landsat TM and SPOT HRG images for vegetation change detection in the Brazilian Amazon. *Photogramm Eng and Rem Sens* 70:421.

Lu, D. et al. 2008b. Pixel-based Minnaert correction method for reducing topographic effects on the Landsat 7 ETM+ image. *Photogramm Eng Rem Sens* 74:1343.

Macleod, R. D., and R. G. Congalton. 1998. A quantitative comparison of change-detection algorithms for monitoring eelgrass from remotely sensed data. *Photogramm Eng Rem Sens* 64:207.

Markham, B. L., and J. L. Barker. 1987. Thematic Mapper bandpass solar exoatmospheric irradiances. *Int J Rem Sens* 8:517.

Mas, J. F. 1999. Monitoring land-cover changes: A comparison of change detection techniques. *Int J Rem Sens* 20:139.

McGovern, E. A. et al. 2002. The radiometric normalization of multitemporal Thematic Mapper imagery of the midlands of Ireland—a case study. *Int J Rem Sens* 23:751.

Meyer, P. et al. 1993. Radiometric corrections of topographically induced effects on Landsat TM data in alpine environment. *ISPRS J Photogramm Rem Sens* 48:17.

Michener, W. K., and P. F. Honhoulis. 1997. Detection of vegetation associated with extensive flooding in a forested ecosystem. *Photogramm Eng Rem Sens* 63:1363.

Morisette, J. T., and S. Khorram. 2000. Accuracy assessment curves for satellite-based change detection. *Photogramm Eng Rem Sens* 66:875.

Muchoney, D. M., and B. N. Haack. 1994. Change detection for monitoring forest defoliation. *Photogramm Eng Rem Sens* 60:1243.

Mutanga, O., and A. K. Skidmore. 2004. Narrow band vegetation indices overcome the saturation problem in biomass estimation. *Int J Rem Sens* 25:3999.

Myint, S. W. 2001. A robust texture analysis and classification approach for urban land-use and land-cover feature discrimination. *Geocarto Int* 16:27.

Neville, R. A. et al. 2003. Spectral unmixing of hyperspectral imagery for mineral exploration: Comparison of results from SFSI and AVIRIS. *Can J Rem Sens* 29:99.

Okin, G. S. et al. 2001. Practical limits on hyperspectral vegetation discrimination in arid and semi-arid environments. *Rem Sens Environ* 77:212.

Pal, M., and P. M. Mather. 2003. An assessment of the effectiveness of decision tree methods for land cover classification. *Rem Sens Environ* 86:554.

Platt, R. V., and A. F. H. Goetz. 2004. A comparison of AVIRIS and Landsat for land use classification at the urban fringe. *Photogramm Eng Rem Sens* 70:813.

Rashed, T. et al. 2001. Revealing the anatomy of cities through spectral mixture analysis of multispectral satellite imagery: A case study of the Greater Cairo region. *Egypt Geocarto Int* 16:5.

San Miguel-Ayanz, J., and G. S. Biging. 1997. Comparison of single-stage and multi-stage classification approaches for cover type mapping with TM and SPOT data. *Rem Sens Environ* 59:92.

Serpico, S. B., and L. Bruzzone. 1999. Change detection. In *Information Processing for Remote Sensing*, ed. C. H. Chen, 319. Singapore: World Scientific Publishing.

Shaban, M. A., and O. Dikshit. 2001. Improvement of classification in urban areas by the use of textural features: The case study of Lucknow city. *Uttar Pradesh Int J Rem Sens* 22:565.

Singh, A. 1989. Digital change detection techniques using remotely sensed data. *Int J Rem Sens* 10:989.

Smits, P.C., S. G. Dellepiane, and R. A. Schowengerdt. 1999. Quality assessment of image classification algorithms for land-cover mapping: A review and a proposal for a cost-based approach. *Int J Rem Sens* 20:1461.

Song, C. et al. 2001. Classification and change detection using Landsat TM data: When and how to correct atmospheric effect. *Rem Sens Environ* 75:230.

Stefan, S., and K. I. Itten. 1997. A physically-based model to correct atmospheric and illumination effects in optical satellite data of rugged terrain. *IEEE Trans Geosci Rem Sens* 35:708.

Stow, D. A. 1999. Reducing the effects of misregistration on pixel-level change detection. *Int J Rem Sens* 20:2477.

Stow, D. A., and D. M. Chen. 2002. Sensitivity of multitemporal NOAA AVHRR data of an urbanizing region to land-use/land-cover change and misregistration. *Rem Sens Environ* 80:297.

Stuckens, J., P. R. Coppin, and M. E. Bauer. 2000. Integrating contextual information with per-pixel classification for improved land cover classification. *Rem Sens Environ* 71:282.

Teillet, P. M., B. Guindon, and D. G. Goodenough. 1982. On the slope-aspect correction of multispectral scanner data. *Can J Rem Sens* 8:84.

Tokola, T., S. Löfman, and A. Erkkilä. 1999. Relative calibration of multitemporal Landsat data for forest cover change detection. *Rem Sens Environ* 68:1.

Townshend, J. R. G. et al. 1992. The effect of image misregistration on the detection of vegetation change. *IEEE Trans Geosci Rem Sens* 30:1054.

Tso, B., and P. M. Mather. 2001. *Classification Methods for Remotely Sensed Data.* New York: Taylor & Francis.

Verbyla, D. L., and S. H. Boles. 2000. Bias in land cover change estimates due to misregistration. *Int J Rem Sens* 21:3553.

Vermote, E. et al. 1997. Second simulation of the satellite signal in the solar spectrum, 6S: An overview. *IEEE Trans Geosci Rem Sens* 35:675.

Vicente-Serrano, S. M., F. Pérez-Cabello, and T. Lasanta. 2008. Assessment of radiometric correction techniques in analyzing vegetation variability and change using time series of Landsat images. *Rem Sens Environ* 112:3916.

Weber, K. T. 2001. A method to incorporate phenology into land cover change analysis. *J Range Manage* 54:A1.

Wu, C., and A. T. Murray. 2003. Estimating impervious surface distribution by spectral mixture analysis. *Rem Sens Environ* 84:493.

Wu, X., F. Yang, and R. Lishman. 2010. Land cover change detection using texture analysis. *J Comp Sci* 6:92.

Yang, X., and C. P. Lo. 2000. Relative radiometric normalization performance for change detection from multi-date satellite images. *Photogramm Eng Rem Sens* 66:967.

Yool, S. R., M. J. Makaio, and J. M. Watts. 1997. Techniques for computer-assisted mapping of rangeland change. *J Range Manage* 50:307.

Yuan, D., and C. Elvidge. 1998. NALC land cover change detection pilot study: Washington D.C. area experiments. *Rem Sens Environ* 66:166.

Yuan, D., C. D. Elvidge, and R. S. Lunetta. 1998. Survey of multispectral methods for land cover change analysis. In *Remote Sensing Change Detection: Environmental Monitoring Methods and Applications*, eds. R. S. Lunetta and C. D. Elvidge, 21. Chelsea, MI: Ann Arbor Press.

Zhou, W., A. Troy, and J. M. Grove. 2008. Object-based land cover classification and change analysis in the Baltimore metropolitan area using multi-temporal high resolution remote sensing data. *Sensors* 8:1613.

Section III

Environmental Applications-Vegetation

12

Remote Sensing of Ecosystem Structure and Function

Alfredo R. Huete and Edward P. Glenn

CONTENTS

12.1 Introduction

Earth-observing remote sensing technologies are becoming widely adopted within the resource management, ecosystem sciences, and sustainable development communities. Satellite data offer unprecedented capabilities to capture the spatial and temporal detail of ecosystem properties at regional to global scales, and remote sensing tools are now employed in characterizing ecosystem structure and biologic properties and in monitoring ecosystem health, seasonal dynamics, and functional processes. The moderate resolution Landsat and coarser resolution National Oceanic and Atmospheric Administration (NOAA) Advanced Very High Resolution Radiometer (AVHRR) programs have provided systematic time-series observations since the early and late 1970s, respectively, for ecosystem and agricultural productivity assessments, land-cover mapping, vegetation-climate studies, drought monitoring, biodiversity and habitat loss, invasive species, and many other ecological applications.

Many global change advancements in vegetation responses to climate variability, including the impacts of the El Niño-Southern Oscillation events, can be attributed to the development and consistency of these programs (Tucker et al. 1991; Myneni et al. 1997).

Earth's terrestrial surface is covered by vegetation canopies consisting of diverse structural and functional land-cover types and ecosystems. The relationship between the solar energy incident at the surface and the spectral composition of the reflected energy provides a wealth of information about the biogeochemical nature (pigments, leaf chemistry, soil mineralogy), moisture status, and physical and structural characteristics of the surface (canopy height, leaf area, vegetation physiognomy, soil roughness). Remotely sensed data in the spectral, spatial, and temporal domains further reveal information about surface processes, including photosynthesis, evapotranspiration, land surface functioning, and ecosystem disturbance (Running 2006). In this chapter, we review many of the advancements made in the remote sensing of ecosystem structures, processes, and functions.

12.2 State and Composition of the Surface

Vegetated canopies are assemblages of plant species with grass-, shrub-, or tree-like structures overlying fresh and decaying litter and soils with mineral and organic properties. The species-specific leaf elements vary in morphology, specific leaf area, and color. Canopy biophysical and biochemical variables relevant to the characterization of vegetation states include vegetation fraction (F_v), leaf area index (LAI), fraction of absorbed photosynthetically active radiation (f_{APAR}), chlorophyll content, and canopy water content, all of which are important ecosystem parameters of interest for biogeochemical, productivity, and climate models. It is an enormous task to collect reliable field data of these properties with sufficient spatial and temporal coverage for use in ecosystem analysis. In many cases, remote sensing measurements, with their synoptic and repetitive sampling capabilities, may be the most efficient and only method to assess widespread ecosystem properties (Kerr and Ostrevsky 2003).

12.2.1 Biogeochemistry

Narrowband hyperspectral reflectance signatures are commonly used to find the biogeochemistry of landscapes and retrieve quantitative information about canopy chemistry, water content, and soil composition (Ustin et al. 2004; Thenkabail et al. 2000; Ben-Dor et al. 2008), and many studies are aimed at finding the right, optimal combination of narrow waveband reflectance data to extract such information.

Spectral signatures of leaf, soil, litter, and woody samples are obtained by laboratory and field spectroradiometer measurements. Leaf spectra can also be measured in situ with a leaf clip apparatus that is attached to field spectrometers or with an integrating sphere, in which reflectance, transmittance, and absorptance spectra are obtained (Carter and Knapp 2001). Spectral signature libraries of soils, rocks, leaves, and plants are available, for example, the USGS digital spectral library (http://speclab.cr.usgs.gov/spectral-lib.html) and the ASTER spectral library (http://speclib.jpl.nasa.gov). Leaf spectra can also be generated from simulation models, such as the PROSPECT model (Jacquemoud 1990).

The reflectance properties of a leaf vary with pigment content, dry matter organic compounds, water, and leaf structural characteristics (leaf shape, specific leaf area). Leaf spectra are therefore highly variable, with species, leaf age, stress, and health providing opportunities to remotely sense physiological responses to growth conditions and environmental change. Many comprehensive reviews on leaf optics are available in the literature (Gates et al. 1965; Knipling 1970; Colwell 1974).

An example of a healthy spectral signature of a green leaf is shown in Figure 12.1. There are three main spectral domains: (1) the visible (VIS, 0.4–0.7 μm), (2) the near-infrared (NIR, 0.7–1.1 μm), and (3) the shortwave-infrared (SWIR, 1.3–2.5 μm) regions. Very low reflectances in the VIS region are associated with the strong absorbing capacity of leaf pigments, mainly chlorophyll a, chlorophyll b, carotenoids, and xanthophylls, with the chlorophyll pigments accounting for 60–75% of the energy absorbed (Gates et al. 1965; Figure 12.2).

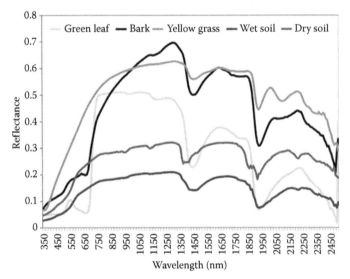

FIGURE 12.1
Spectral reflectance signatures of green and senesced (yellow) leaves, dry and wet soil, and woody bark.

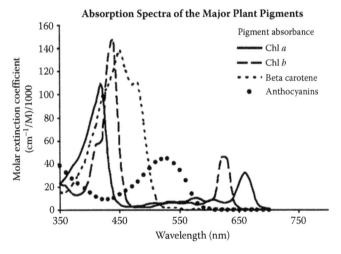

FIGURE 12.2
Absorption spectra of dominant plant pigments (From Blackburn, G. A., *J Exp Bot*, 58, 855, 2007. With permission.)

A green reflectance peak is evident at 0.55 μm, owing to slightly less absorptance, and is responsible for the green appearance of healthy leaves. Chlorophyll content at the leaf level is directly inferred using hyperspectral indices and leaf- and plant-canopy simulation models (Broge and Leblanc 2001; Zarco-Tejada et al. 2004). Carotenoid content provides complementary information on vegetation physiological status. Thus, the spatial and temporal dynamics of plant pigments are key indicators of ecosystem properties, given the strong coupling of pigments to plant physiological status and stress (Blackburn 2007). Because most of the energy in the 0.4–0.7 μm range is apparently absorbed by pigments to drive photosynthesis, the term "photosynthetically active radiation" (PAR) is often used to describe radiation in the VIS part of the spectrum.

Beyond the highly absorbing red region is a sharp "red-edge" transition region at around 0.70–0.78 μm, in which the absorbing leaves become highly reflective and transmissive in NIR wavelengths. Spectral reflectance curves in the NIR are sensitive to leaf morphology and leaf structural properties, particularly the spongy mesophyll layer in which NIR reflectance increases with an increasing number of cell layers and intercellular spaces. Several secondary water absorption features at 0.96, 1.1, and 1.24 μm can be used to assess leaf water content. As a result of NIR sensitivity to leaf structural properties, the NIR spectral region is very useful in plant biodiversity studies and in discriminating among plant species and leaf shape (needle-leaf, broadleaf, grass) that are often not discriminable in the VIS spectrum.

The NIR spectral domain also has a strong transition region between 1.1 and 1.3 μm, where reflectances decrease sharply and become highly absorbing in the SWIR, attributed to strong absorption by leaf water. Thus, variations in leaf water content have a large effect on reflectances in the SWIR as well as on portions of the NIR, with important water absorption features present at 1240, 1640, and 2100 nm, resulting in negative relationships between reflectances at these wavelengths and leaf water content. However, such variations are commonly leaf-type and plant-species dependent.

12.2.2 Leaf Stress

Plant physiological stress due to drought, disease, and nutrient deficiencies alters leaf spectral signatures throughout the spectrum; however, many studies have found very consistent stress-related changes occurring in the VIS wavelengths (400–720 nm; Carter 1994). In general, reflectance increases in the red, green–yellow, and red-edge spectral regions with decreasing chlorophyll amounts, and it also increases consistently with plant stress at wavelengths near 700 nm, causing the slope of the red edge to shift toward the blue spectrum. The concentrations of other pigments, such as carotenoids, are more persistent with stress and remain high enough in stressed leaves that there is very little change in absorption in the blue range (400–500 nm) compared to healthy leaves (Gitelson et al. 2003).

Beyond the red edge, in the NIR and SWIR spectral regions, reflectances become independent of chlorophyll, and changes in reflectances occur with leaf anatomy or water content changes in response to stress. Leaf reflectances in the NIR generally decrease with stress, relative to healthy leaves, due to deterioration of leaf cellular structure (Knipling 1970). In the SWIR region, several studies have shown increases in leaf reflectances associated with plant stress (Gausman and Allen 1973; Tucker 1980), providing useful information to infer canopy moisture content and soil moisture status in the plant root zone.

12.2.3 Soil, Litter, and Woody Components

Soil spectral reflectance signatures have been studied in laboratory and field measurements and have been found to be primarily a function of absorption features associated with primary and secondary minerals (iron oxides, clays, carbonates, salts), organic matter and litter content, as well as soil structure and surface moisture status (Baumgardner et al. 1985). Overall, soil spectral reflectances increase with increasing wavelengths across the VIS and NIR as a result of strong Fe–O charge transfer absorption in the blue region (Figure 12.1). Soils have distinct features caused by vibrational processes in the SWIR, which include two broad water absorption bands at 1.4 and 1.9 μm. Minerals with OH, CO_3 (e.g., calcite), and SO_4 (e.g., gypsum) exhibit vibrational features in the 1.8–2.5 μm region, whereas silicate layers with OH absorb near 1.4 and 2.2 μm (Mulders 1987). However, strong mineral absorption spectra are rarely found in soil samples, due to their highly heterogeneous mineral mixtures and strong soil particle coatings of secondary minerals (FeO) and organic (humic) molecules. As a result, Stoner and Baumgardner (1981) reported only five basic soil spectral reflectance curve shapes determined by the relative presence and amounts of iron and organic compounds.

The spectral signature of a soil is further modified by surface structural properties (particle size, roughness) and optical geometric factors (sun view geometry). Reflectance decreases with increasing surface roughness because coarse aggregates contain a lot of interaggregate spaces and "light traps." Roughness will alter the spectral signature of a soil and the inferences made of its soil color and mineralogy because the shorter wavelengths are most affected. The bidirectional reflectance distribution function (BRDF) describes the manner in which surfaces scatter radiation across all sun-surface-sensor view geometries and can be used, through models, to derive geometric descriptors, such as size, shape, and orientation of surface "roughness" elements, of a soil (Irons et al. 1992). Kimes et al. (2000) further suggested that the relationships between spectral bands are most useful in assessing soil properties.

Quantitative methods for extracting information about physical and biochemical characteristics of soils have been developed and recognized in the field of soil spectroscopy. As an example, Ben-Dor and Bannin (1994) used a VIS and NIR analysis scheme to predict a wide variety of chemical constituents of soils, including $CaCO_3$, Fe_2O_3, Al_2O_3, SiO_2, free iron oxides, and K_2O, from fine-resolution spectra of arid and semiarid soils. Soil spectral properties and their use in soil applications and mapping are reviewed comprehensively by Baumgardner et al. (1985), Ben-Dor et al. (2008), and Anderson and Croft (2009).

Soil spectral reflectance curves also decrease with wetting, by as much as 40–60%, with the strongest decrease occurring in the water-sensitive SWIR spectral region (Figure 12.1). The reflectance ratio between dry and wet soils further varies with soil type, having larger ratios in forest soils with low organic matter, relative to darker, organic-rich grassland soils (Baumgardner et al. 1985). A fundamental limitation of optical remote sensing is that only soil moisture at the surface and near-surface soil moisture changes more quickly than soil moisture at greater depths, making it very difficult to infer soil wetness below a few centimeters.

The presence of plant litter and its stage of decomposition will alter soil spectral reflectance signatures (Stoner and Baumgardner 1981). There are also nonphotosynthetic vegetation (NPV) components within a canopy that influence canopy spectra. Wood and standing litter have distinctive narrowband lignocellulose absorption features in the SWIR that allow for their discrimination (Elvidge 1990; Asner et al. 2000). As in assessments of plant stress and chlorophyll, there are many soil and NPV absorption features that cannot be resolved with the use of broadband sensors and require high spectral resolution sampling.

12.3 Ecosystem Structure

A complete characterization of landscape ecological properties requires information on both canopy biogeochemical and structural properties. At the canopy scale, reflectances become an integrated signal of many leaves and species, NPV elements, and soil organized in various canopy structural assemblages. The structural makeup of a canopy will include leaf angle distribution (LAD), leaf clumpiness, plant physiognomy (shrub/tree/grass), plant height, and crown diameter. There are complex and dynamic patterns of light interactions with the photosynthetic elements of a canopy, resulting in sunlit and shaded or diffuse surfaces, dependent on the sun position and atmospheric conditions. This complicates the interpretation and extrapolation of laboratory and field spectra to canopy- and ecosystem-scale measurements.

Despite the structural complexities and mixing of spectral signatures, there remain many common features between leaf- and canopy-scale spectra, especially the high spectral reflectance contrasts commonly observed across the VIS, NIR, and SWIR spectral regions. For example, the greater the contrast between the red and the NIR spectral regions, the greater the amount and/or vigor of vegetation (Figure 12.3). Thus, information contained in leaf, NPV, and soil spectra will manifest itself at the canopy scale, although with some important modifications. Spectral reflectance signatures at the canopy and landscape scales are collected with field- and tower-mounted spectroradiometers, as well as with airborne (e.g., AVIRIS) and the EO-1 Hyperion spaceborne imaging sensors (Asner 1998; Ustin et al. 2004; Shimabukuro et al. 1994).

12.3.1 Spectral Measures

The spectral behavior of vegetation canopies in the VIS, NIR, and SWIR regions forms the theoretical basis for the design and development of various empirical and physical-based methods to retrieve biophysical variables, including the use of hyperspectral techniques, spectral indices, radiative transfer model inversions, and BRDF approaches. Vegetation indices (VIs) are optical measures of canopy *greenness*, a composite property of total canopy chlorophyll, leaf area, canopy cover, and structure. VIs are robust and seamless

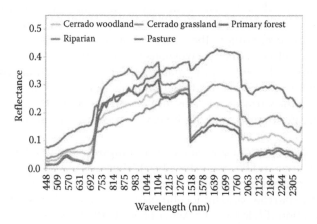

FIGURE 12.3
(See color insert following page 426.) Canopy level spectral reflectance signatures measured by an EO-1 Hyperion sensor over Araguaia National Park, Brazil.

biophysical measures, computed the same across all pixels in time and space regardless of biome type, land-cover condition, and soil type. VIs are typically used as proxies to characterize ecosystem states and processes and have become indispensable tools in many ecological applications, including productivity and phenology studies, vegetation-climate interactions, land-cover classification, land-use change detection, drought monitoring, biodiversity, habitat loss, and public health.

A VI measures the contrast between the chlorophyll-absorbing red spectral region and nonabsorbing leaf NIR reflectance signal to quantify the amount and vigor of vegetation. There are a variety of ways in which these two bands may be combined to estimate greenness, and this has resulted in a multitude of VI equations from spectral band ratios, normalized differences, linear band combinations to optimized band combinations (Gobron et al. 2000; Huete 1988; Tucker 1979). For an effective VI, its capability to capture essential biophysical phenomena with adequate fidelity and minimal external influences (atmosphere and soil), and its global extension in time and space should be considered.

The normalized difference VI (NDVI) is a functional variant of the simple ratio (SR = ρ_N/ρ_R) that provides greenness values normalized between -1 and $+1$:

$$\text{NDVI} = \frac{(\text{SR}-1)}{(\text{SR}+1)} = \frac{(\rho_N - \rho_R)}{(\rho_N + \rho_R)} \tag{12.1}$$

where ρ_N and ρ_R are reflectances in the NIR and red bands, respectively. In NIR–red space, NDVI depicts vegetation isolines (i.e., lines of constant NDVI value) with increasing slopes but with constant zero intercepts (Figure 12.4). An important advantage of ratio-based indices is their ability to produce stable values by normalizing many extraneous sources of noise. This normalizing ability was particularly useful in the early AVHRR era, in which data quality problems due to sensor instrument noise and atmosphere contamination were particularly prevalent. The NDVI also presents some disadvantages in landscape studies, related to the nonlinearity of ratios, sensitivity to soil background, and saturation at moderate-to-high vegetation densities. Canopies overlying darker soils have much higher NDVI values (up to twofold) than equivalent canopies underlain by bright soils, particularly at moderate vegetation covers (~50% and LAI of 1; Huete 1988).

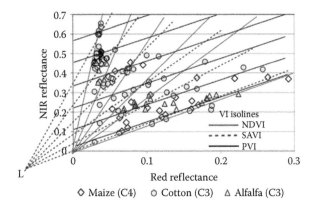

FIGURE 12.4
Concept of vegetation indices and isolines in NIR–red reflectance space. The isolines depict the domain of equal vegetation indices values, and the spectral data points correspond to observed measurements over different crop types and soils (note that the three VIs yield inconsistent interpretations of the vegetation canopy spectra).

The "soil-line" concept was formulated by Richardson and Wiegand (1977) as a way to separate soil-induced optical variations from those of vegetation. The perpendicular vegetation index (PVI) measured the vector distance of a vegetated pixel in NIR–red spectral space, orthogonal to the soil line, as a quantitative measure of vegetation amount in a pixel (Figure 12.4). This provided a vegetation measure referenced to the spectral variations of the underlying soils. The tasseled cap concept expands on this 2-band concept to 4 and 6 bands (Landsat MSS and TM, respectively) and allows one to separate other pixel features besides vegetation (Healey et al. 2005).

A first-order Beer's law approximation of soil-vegetation radiative transfer allows the influence of the backscattered soil signal to be effectively removed, as in the soil adjusted vegetation index (SAVI):

$$\text{SAVI} = (1+L)\,(\rho_N - \rho_R)/(\,L + \rho_N + \rho_R\,) \tag{12.2}$$

where L is the soil adjustment factor. The parameter L shifts the vegetation isolines in NIR–red space away from the origin and has a value of 0.5 for the global case (Figure 12.4; Huete, 1988). The enhanced vegetation index (EVI) is an optimized combination of blue, red, and NIR bands, developed as a Moderate Resolution Imaging Spectroradiometer (MODIS) satellite product and designed to extract canopy greenness, independent of underlying soil backgrounds and atmospheric aerosol variations (Figure 12.5). The EVI gains its heritage from the SAVI and the atmospherically resistant vegetation index (ARVI; Kaufman and Tanré 1992; Huete et al. 2002)

$$\text{EVI} = 2.5(\rho_N - \rho_R)/(L + \rho_N + C_1\rho_R - C_2\rho_B) \tag{12.3}$$

where ρ_N, ρ_R, ρ_B are reflectances in the NIR, red, and blue bands, respectively, L is the canopy background adjustment factor, and C_1 and C_2 are the aerosol resistance weights. Although MODIS surface reflectance data is atmospherically corrected, the aerosol resistance term stabilizes residual aerosol noise after atmospheric correction, such as from

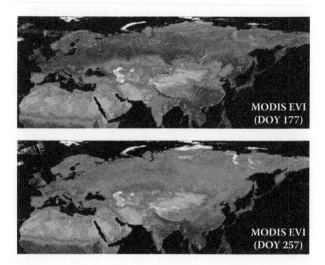

FIGURE 12.5
Example of MODIS-enhanced vegetation index spatial and temporal variations over Europe and Asia for composite day of year 177 (June) and 257 (September). The darker colors represent maximum levels of greenness.

over- and undercorrected pixels. The coefficients of the EVI equation in the standard MODIS products are $L = 1$, $C_1 = 6$ and $C_2 = 7.5$.

There are many optical measures and techniques that are used to specifically exploit hyperspectral remote sensing, including higher-order derivative analysis of spectral signatures, band ratioing and continuum removal, lignocellulose absorption indices, mixture modeling (Asner and Lobell 2000), and plant physiological indices such as the photochemical reflectance index (PRI; Gamon, Serrano, and Surfus 1997) and the "green" chlorophyll index (Gitelson et al. 2005). Spectral mixture analysis is a linear unmixing technique that decomposes spectral measurements into a set of unique reflecting features, such as green vegetation, NPV, soils, and shade (Smith et al. 1990; Roberts et al. 1993). The basis of a linear mixing scheme is that the measured spectral response is equal to the weighted sum of multiple reflecting spectral features:

$$d_{ik} = \sum_{j=1}^{n} r_{ij} c_{jk} + \varepsilon \qquad (12.4)$$

where d_{ik} is the measured spectral response of spectral mixture k in waveband i, n is the number of independent reflecting components in the mixture, r_{ij} is the response of component j in waveband i, c_{jk} is the relative contribution of component j in spectral mixture k, and ε is the residual error. Spectral unmixing has been effectively applied to AVIRIS (224 bands) hyperspectral imagery to extract soils information from vegetated canopies (Palacios-Orueta et al. 1999). Asner and Lobell (2000) developed a SWIR-based endmember approach to separate and generate green vegetation, NPV, and soil component images. Mixing models based on hyperspectral imaging data have also been used to study land-cover and land-use changes in the Amazon (Roberts et al. 2003; Shimabukuro et al. 1994) and in soil and precipitation influences in Hawaiian forests (Asner et al. 2005). Some shortcomings of linear mixing models are that prior identification of important endmembers may be required, and there may be nonlinear mixing influences present within pixels.

12.3.2 Biophysical Ecological Variables

There are both horizontal and vertical aspects of canopy structure of importance to their ecological characterization. LAI, or foliage density, is defined as the one-sided area of leaves in a canopy projected on the ground and integrates all leaf area from the top to bottom of a canopy. Vegetation fraction (F_v) represents the ground surface covered with vegetation and only considers the horizontal properties of a canopy. Various ecological applications and models will require estimates of one or both of these canopy variables, and some models require separate estimates of LAI and F_v, such as in the calculation of roughness lengths for turbulent energy transfer (Glenn et al. 2008). The chlorophyll and water contents of a canopy are also important in analyzing canopy ecophysiological functioning, and the f_{APAR} is a fundamental canopy parameter to understand radiation dynamics and canopy processes. Some thorough reviews on remote sensing of ecological properties can be found in Glenn et al. (2008), Kerr and Ostrovsky (2003), and Pettorelli et al. (2005).

In deriving spectral relationships with specific biophysical variables, the strong covariation among the variables is an important limitation during vegetation canopy development. Thus, as a woody or herbaceous canopy grows, F_v, LAI, total chlorophyll and water contents, as well as phenology (leaf age) and leaf structure (LAD), may change simultaneously, rendering it difficult to isolate spectral variations attributed to specific vegetation biophysical components. These covarying factors become integrated in spectral–biophysical relationships, resulting in strong local-scale empirical relationships, which become weaker when extended to other sites with different canopy structures. Thus, for example, an NDVI–F_v empirical relationship may also include variations in spectral reflectance due to changes in LAI, leaf age, and chlorophyll status. One method to circumvent this problem is to use canopy radiative transfer models, in which individual canopy factors can be varied while keeping the other factors constant. Such models are useful to assess spectral sensitivities to the individual canopy factors and to categorize structural assemblages into standard conditions with implicit assumptions, for use in canopy radiative transfer inversions and lookup tables (LUT; Jacquemoud et al. 1995).

12.3.2.1 Vegetation Fraction (Fᵥ)

Vegetation fraction, F_v, is an important ecosystem variable useful to assess Et, rainfall interception, and turbulent energy transfer parameters. In general, NDVI is most commonly used to assess F_v, with both linear and nonlinear relationships reported (Carlson and Ripley 1997). Leprieur et al. (2000) found a curvilinear regression between the F_v and NDVI over a precipitation gradient in the south Sahel and found most of the uncertainties in the relationship to be associated with the degree of clumping from the woody vegetation (shrubs and trees). Others have shown distinctly linear NDVI–F_v relationships to be nonlinear with a change in soil background type for the same canopy (Huete et al. 1985; Bausch 1993). Montandon and Small (2008) and Jiang et al. (2006) concluded that the NDVI may not be suitable to infer F_v because of its nonlinearity and scaling problems, which are most pronounced over moderately vegetated canopies underlain by darker soil backgrounds or standing water.

To minimize the nonlinear effects, satellite-based F_v products are sometimes derived through linear and normalized combinations of high (vegetation) and low (soil) NDVI values within a scene or biome type (Zeng et al. 2000; Gutman and Ignatov 1998). This is done by scaling the NDVI from bare soil to dense vegetation for F_v retrieval

$$F_v = \frac{(NDVI - NDVI_0)}{(NDVI_\infty - NDVI_0)} \tag{12.5}$$

where $NDVI_0$ and $NDVI_\infty$ are the signals from bare soil and dense green vegetation, respectively. However, Glenn et al. (2008) showed that different plant species will yield different F_v–NDVI relationships because they have different VI values at dense cover, associated with differences in chlorophyll content and canopy architecture, particularly clump LAI values. Furthermore, such relationships may also be sensor dependent, as AVHRR NDVI values over dense vegetation are lower than those of MODIS NDVI due to the narrower spectral bandwidths and more complete atmospheric correction (aerosols, water vapor) in MODIS NDVI.

There have been several studies aiming to linearize the NDVI equation itself, resulting in improvements to VI–F_v relationships (e.g., Jiang and Huete 2010). The PVI, tasseled cap greenness, and decomposed vegetation fraction component that is derived from spectral mixture modeling also provide measures of F_v that are much more linear than ratio methods, such as NDVI (Anderson et al. 2010; Ustin et al. 2004; Roberts et al. 1993).

12.3.2.2 Leaf Area Index

LAI (m^2/m^2) is an important canopy variable for many ecosystem processes and model studies. LAI is difficult to measure in the field and optical and direct harvesting methods frequently produce different results; therefore, there is much interest in developing remote sensing methods for LAI estimations. Spectral VIs are generally sensitive to LAI for relatively low LAI values (Bégué 1993; Asner et al. 2003), with increasingly weak and nonlinear relationships at LAI beyond 2 or 3, after which the NDVI becomes saturated (Baret and Guyot 1991).

LAI–VI relationships strongly vary with vegetation type, phenology period, and satellite sensor. Significant but nonlinear relationships have been found between Landsat NDVI with field-measured LAI values in coniferous forest sites in Siberia; however, poorer relationships were found using coarser resolution MODIS NDVI (Chen et al. 2005). Lower correlations are reported in broadleaf forest canopies compared with needle-leaf stands, possibly a result of NDVI saturation (Fassnacht et al. 1997). For example, in a beech deciduous forest in Europe, NDVI–LAI relationships varied across different phenology periods, with the worst correlations occurring in periods of maximum LAI owing to NDVI saturation (Wang et al. 2005). In larger-scale field validation campaigns conducted over a series of biomes, Cohen et al. (2003) reported only weak correlations between field-measured LAI and several satellite products, including a MODIS LAI product based on three-dimensional (3D) canopy radiative transfer modeling and generated LUT (Myneni et al. 2002). The canopy model uses generalized attributes for each of six biome structural types, including leaf optical properties, F_v, plant LAI, and soil background as input variables to the model. The product works well over areas that match the generalized biome types; however, in other ecosystems, MODIS LAI overpredicts ground LAI by a factor of 2 (Leuning et al. 2005), thus, empirical approaches are more common in the accurate estimation of LAI at a local scale.

Whereas the NDVI is very sensitive to the red band that has a low canopy optical penetration depth, other indices show a greater sensitivity on the NIR that senses to greater canopy optical depths. These include the PVI, SAVI, and EVI, all exhibiting a higher LAI sensitivity (Fensholt et al. 2004; Huete et al. 2002). Houborg and Soegaard (2004) found MODIS EVI was able to accurately describe the variation in green biomass over agriculture areas in Denmark, up to a maximum green LAI of 5 ($r^2 = 0.91$). Thus, the EVI may better depict biophysical canopy structural variations and be less prone to saturation in high-biomass (LAI) areas.

12.3.2.3 Canopy Chlorophyll Content

Chlorophyll content at the canopy level is a key property of plant communities, and its assessment from satellite data is relevant to studies of ecosystem productivity, CO_2 fluxes, and vegetation stress (Gitelson et al. 2003; Gitelson et al. 2006; Blackburn 2007). Hyperspectral indices that are sensitive to leaf chlorophyll include the first derivative of the red-edge slope at 700–740 nm and various hyperspectral (narrowband) versions of VIs. In the intense chlorophyll absorption range at 670–680 nm, chlorophyll sensitivity is very low in moderate-to-high LAI conditions, and relatively large amounts of chlorophyll must be lost from the leaves before a significant optical difference occurs. Increased chlorophyll sensitivity can be attained by shifting away from this intense absorption region, such as at wavelengths near 690–720 nm (beginning of the red edge) as well as in the green–yellow spectrum (Carter and Knapp 2001). Although spectral absorption in the red is normally correlated with chlorophyll content, other factors need to be considered at the canopy scale, including canopy architecture, chlorophyll distribution within the canopy, LAI, and absorption associated with soil background, all of which confound the direct retrieval of canopy chlorophyll contents.

Fundamentally, VIs are more directly related to the chlorophyll content or greenness of the canopy, rather than leaf area or vegetation ground cover, as plants with different chlorophyll contents will also have different VI values for the same F_v or LAI. In general, total canopy chlorophyll content tends to be curvilinearly related with VIs because surface leaves intercept more light than leaves deeper in the canopy (Gitelson et al. 2006), and the VI is also subject to red-band chlorophyll saturation problems. Most satellite sensors acquire red reflectances from broad bandwidths, which are inadequate for the remote sensing of vegetation biochemical properties and are unable to capture fine chlorophyll spectral features and their variation with plant stress and leaf phenology (Broge and Leblanc 2001).

Canopy chlorophyll activity decreases with stress and disturbance, affecting reflectances in the VIS spectrum. Stress is also associated with a displacement in reflectances or shift in the "red-edge" reflectances toward shorter wavelengths. This is sometimes called "blue shift" and is readily quantified by the first derivative of the red-edge inflection point. Blue shift has been observed with plant nutrients and mineral stress, as well as with plants affected by heavy metal contamination (Rock et al. 1986), and corresponds strongly to lower leaf chlorophyll concentrations.

12.3.2.4 Canopy Water Content and Soil Moisture

Knowledge of the water status of a vegetation canopy can provide valuable information on vegetation drought status and soil moisture conditions. Soil moisture further acts as an integrator of the amount and occurrence of precipitation events, influencing the partitioning of surface available energy into sensible and latent heat fluxes. Despite its importance, there has been limited success in implementing remotely sensed soil moisture observations at appropriate time and space scales needed for current hydrology, climate, and biogeochemical models. Accurate assessments of soil moisture are difficult due to complex land-cover conditions and very large soil moisture variability across landscapes, with unknown scale dependencies.

Vegetation water indices (WIs) employing 1240-, 1640-, or 2100-nm wavelengths in lieu of the red band used in VIs have recently been used as independent vegetation measures related to moisture condition rather than chlorophyll amount (Zarco-Tejada, Rueda, and Ustin 2003; Figure 12.6). Although VIs have also been correlated with vegetation water content, they are physiologically related to canopy chlorophyll content, and several studies have found WIs to be more effective in mapping canopy water content, such as the Soil Moisture Experiments 2002 (SMEX02) campaign, in which the NDVI was found to be saturated while the WIs remained sensitive to increasing amounts of green vegetation (Jackson et al. 2004). Ceccato et al. (2001) concluded that VIs were less suitable for retrieving canopy water contents because the relationship between chlorophyll and water is species specific. Furthermore, decreases in chlorophyll content do not always imply a decrease in water content.

The assessment of vegetation water content or equivalent water thickness $(g \cdot H_2O/cm^2$ leaf area) is obtained by combining NIR and SWIR reflectance values. The use of two or more bands helps minimize variations in leaf internal structure and leaf dry matter content as well as in canopy geometry, shadowing, and soil surface moisture that also influence SWIR reflectance (Ceccato et al. 2001). An example of a WI is the moisture stress index, calculated as the simple ratio between SWIR (1600 nm) and NIR (820 nm) spectral reflectances, which has been successfully used to derive leaf scale functions of water content (Hunt and Rock 1989).

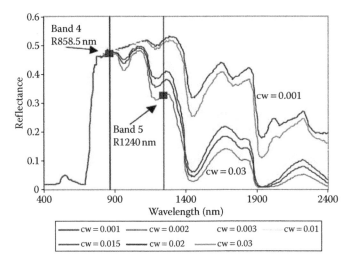

FIGURE 12.6
Spectral reflectance variations in the NIR and SWIR spectral regions for various leaf model simulated water contents. MODIS bands 4 and 5 are depicted and used in formulation of the NDWI. (From *Remote Sensing of Environment*, 85, Zarco-Tejada, P. J., Rueda, C. A., and Ustin, S. L., Water content estimation in vegetation with MODIS reflectance data and model inversion methods, 109. Copyright (2003), with permission from Elsevier.)

Hardisky, Smart, and Klemas (1983) developed the normalized difference infrared index (NDII), contrasting the NIR with SWIR (1600 nm) wavelengths:

$$\text{NDII} = \frac{(\rho_{\text{NIR}} - \rho_{\text{SWIR}})}{(\rho_{\text{NIR}} + \rho_{\text{SWIR}})} \tag{12.6}$$

They found that this index was strongly correlated with canopy water content. Several variants of this index are now used in a wide range of studies using high spectral resolution as well as broadband reflectances. Xiao, Zhang et al. (2004) and Xiao, Hollinger et al. (2004) used the normalized difference between the NIR and SWIR bands (1580–1750 nm) from the SPOT-4 VEGETATION (VGT) and MODIS sensors as measures of land surface moisture status and named this the land surface water index (LSWI). Xiao et al. (2005) found satellite-derived LSWI to be sensitive to seasonal fluctuations in canopy or leaf water contents in Amazon tropical forests.

Fensholt and Sandholt (2003) formulated the shortwave-infrared water stress index (SIWSI) from daily MODIS NIR and SWIR (1628–1652 nm) reflectances:

$$\text{SIWSI} = \frac{(\rho_{\text{SWIR}} - \rho_{\text{NIR}})}{(\rho_{\text{SWIR}} + \rho_{\text{NIR}})} \tag{12.7}$$

They reported high correlations between SIWSI and soil moisture in the root zone in a Sahel vegetation study in Senegal. The normalized difference water index (NDWI) uses NIR reflectances at 860 and 1240 nm wavelengths:

$$\text{NDWI} = \frac{(\rho_{860\text{nm}} - \rho_{1240\text{nm}})}{(\rho_{860\text{nm}} + \rho_{1240\text{nm}})} \tag{12.8}$$

with the liquid water absorption feature at 1240 nm enhanced by the high NIR scattering in the leaf (Gao 1996). Whereas the SWIR region responds to both vegetation and soil surface moisture, the 1240-nm water-absorbing region has been shown to respond to canopy moisture status only. This formulation was applied to MODIS bands 5 (1230–1250 nm) and 2 (841–876 nm) and was found to be a strong indicator of canopy water content during the growing season in the Sahel (Fensholt and Sandholt 2003). However, it was found that in dry years the vegetation cover was too dry to provide information on canopy water content, suggesting that a minimum, threshold vegetation amount must be present for detection with WIs.

The growth of vegetated ecosystems is largely sustained by the water availability in the soil, with the subsoil supplying water to plants long after the surface has dried out. There are other indirect methods to derive information on the moisture status of soil and the canopy it supports, involving the use of thermal and optical remote sensing data (Moran et al.; Nemani and Running 1989). For example, the water deficit index measures variations in canopy temperatures associated with subsoil (root zone) and soil moisture conditions. As soil moisture declines, canopy temperatures tend to increase due to decreases in transpiration. This is depicted in a Vegetation Index-Surface minus air temperature scatterplot (trapezoid shaped), which enables us to separate and quantify variations in canopy temperature attributed to vegetation cover from those due to changes in soil moisture status (Figure 12.7). In some cases, differences between canopy and air temperatures are used, which enables a direct coupling of the trapezoid end points to Penman–Monteith energy balance equations.

12.3.2.5 Fraction of Photosynthetically Active Radiation Absorbed

The fraction of f_{APAR} by a vegetation canopy is related to primary productivity as a function of a light-use efficiency (LUE) coefficient defining the amount of carbon fixed per unit radiation intercepted (Blackburn 2007):

$$f_{APAR} = APAR/PAR \qquad (12.9)$$

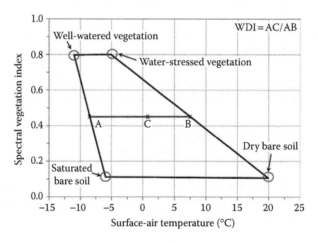

FIGURE 12.7
Illustration of the water deficit index (WDI) trapezoid as used in open canopies (From Moran, M.S. et al., *Remote Sens Environ*, 49, 246, 1994. With permission.)

where APAR is the absorbed photosynthetically active radiation and PAR is the incident photosynthetically active radiation, which encompasses the radiation in the VIS portion of the spectrum.

Monteith and Unsworth (1990) showed that in theory the simple ratio of NIR to red reflectance, and by extension other VIs, is a unique function of the f_{APAR}. The NDVI has been found to be linearly correlated with f_{APAR} in field and modeling studies, across several biome types, and its integral over the growing season has been correlated with ecosystem net primary production (Asrar et al. 1984). However, such relationships are generally site- and phenology-dependent and become increasingly nonlinear in canopies underlain by darker soil backgrounds (Sellers 1987). In a Kalahari field campaign, known as SAFARI, field-measured f_{APAR} was found to be strongly linearly related with MODIS NDVI, and reported as, $f_{APAR} = 0.88 \times NDVI + 0.03$; however, this varied distinctly between seasonal green-up and dry-down periods, with slopes ranging from 0.96 to 0.44, respectively (Huemmrich et al. 2005). In a multibiome MODIS study across North America, Sims et al. (2006) empirically derived a linear "green" f_{APAR}–NDVI relationship across different species as $f_{APAR} = 1.24 \times NDVI - 0.168$.

A significant proportion, up to 40%, of incident PAR may be absorbed by nonphotosynthetic elements of a canopy, such as woody or senesced plant material (Asner et al. 2000). Using MODIS data and a coupled leaf-canopy radiative transfer model (PROSAIL-2), Zhang et al. (2005) separated f_{APAR} into canopy, leaf, and chlorophyll components and found MODIS NDVI best approximated total canopy f_{APAR}, while MODIS EVI was more closely coupled with f_{APAR} associated with chlorophyll. They noted that only chlorophyll f_{APAR} is used for photosynthesis and therefore useful in quantifying primary production. This was also demonstrated by Xiao, Zhang et al. (2004) and Xiao, Hollinger et al. (2004) in studies on broadleaf and needle-leaf forests with seasonal MODIS data, suggesting a close relationship of EVI with the chlorophyll content of the canopy.

12.3.2.6 Surface Roughness

Laser altimetry, or lidar (light detection and ranging), is an emerging remote sensing technology that directly measures a 3D representation of the surface structure and roughness and maps the 3D spatial patterns of vegetated canopies (Figure 12.8). The basic measurement made by a lidar sensor is the distance between the sensor and the target surface, determined by the elapsed time between the emission of a laser pulse and the arrival of

FIGURE 12.8
(See color insert following page 426.) Example of airborne lidar for mapping the three-dimensional properties of canopy surfaces (Adapted with permission from Macmillan Publishers Ltd., Tollefson, J., *Nature* 461:1048, copyright 2009.)

the reflection of that pulse (Lefsky et al. 2002). Lidar can increase the accuracy of biophysical retrievals by directly measuring the 3D structure of plant canopies as well as subcanopy topography, thereby producing highly accurate estimates of vegetation height, cover, and landscape stability (Anderson and Croft 2009). There have been many recent studies integrating airborne lidar with hyperspectral imagery for forest disturbance, selective logging, and the spread of invasive species and associated biochemical alterations (Asner and Knapp 2008; Tollefson 2009).

The BRDF has also been used to characterize the structure and spatial heterogeneity of landscapes. The BRDF specifies the reflectance behavior of a surface and scattering as a function of view and illumination angles for a given wavelength and is a fundamental and intrinsic property governing the reflectance behavior of a surface. The anisotropy of a BRDF signal varies with the biophysical properties of the canopy (F_v, LAI, LAD), specular and diffuse reflection of leaves, and different types of shadows cast by the surface. Even a relatively bare surface soil exhibits a range of BRDF behaviors associated with soil microrelief, compactness, and smoothness, resulting in corresponding variations in specular and diffuse reflection (Anderson and Croft 2009).

In vegetated canopies, surface and volume scattering components are in the BRDF signal and gap-driven BRDF signals from forested canopies. BRDF profiles can be measured with field goniometer systems such as the PARABOLA or obtained from wide viewing and multiangular satellite sensors, as in MODIS and MISR (Diner et al. 1991). The BRDF properties of a canopy will influence the reflectances derived from satellite sensors with important implications for the interpretation of time-series composited satellite products (Shuai et al. 2008). Generally, sensor view geometries in the forward scatter direction will detect more canopy shade and will be darker relative to the backscatter measurement that views more of the sunlit portion of canopies.

12.3.3 Pedosphere

Soils are an integral part of an ecosystem and its functioning, and they function as the Earth's geomembrane, regulating biogeochemical and hydrologic cycles of matter and energy within terrestrial ecosystems. The soil body is sometimes referred to as the *pedosphere* and exhibits great spatial variability as a result of the interactions of climate, topography, parent material, and organisms acting on it over time. With the majority of the Earth's terrestrial surface classified as "open canopies," there is an appreciable soil surface signal present in most satellite imagery, providing information about soil degradation and erosional processes, salinity, crust formation, organic matters, soil moisture, and texture.

EO-1 Hyperion and airborne AVIRIS imagery have been used to derive soil properties such as clays and carbonates that contribute to soil stability, and are therefore useful in quantifying vulnerability to erosion. Erosion alters the biochemical composition of the surface as well as its structure, and therefore, spectral signatures will vary with different stages of erosion (Ben-Dor et al. 2009). Hyperion data have been used in assessments of soil carbon and salinity, and ASTER data have been used in land degradation applications, gully erosion studies, and salinity detection (Anderson and Croft 2009).

Many studies have demonstrated the possibility of inferring information about a soil from the spectral signatures of the overlying vegetation. This technique is known as geobotany and relies on the detection of anomalies in the vegetation cover to obtain information on soil and geologic characteristics. Subsurface soil properties cannot be detected with air- and spaceborne remote sensing measures; however, *pedotransfer* functions are increasingly being developed to derive soil properties beneath the surface. In one study, fiber-optic methods inserted into a soil profile provided soil spectral information at varying depths, which enabled the inference of subsoil properties from soil surface spectral measurements (Ben-Dor et al. 2009).

Most remote sensing techniques used to derive soil information, however, remain undeveloped, and there are currently no operational remote sensing algorithms available for deriving soil-related data products. As noted by Ben-Dor et al. (2008), the adoption of imaging spectroscopy techniques remains a new frontier in soil science, with advances limited by the lack of availability of operational hyperspectral sensors. Nevertheless, remote sensing remains the only viable technique to map, monitor, and manage soils.

12.4 Ecosystem Functioning

Remote sensing studies of ecosystem function consider biogeochemical and water exchanges within the soil–plant–atmosphere system, light absorption and ecosystem productivity, and environmental controls and stress factors that impact vegetation physiological function. Ecosystem metabolic processes include plant photosynthesis and transpiration, primary productivity and decomposition, carbon and water cycles, and phenology. In contrast to field-based measurements of vegetation function, remote sensing provides measures of ecological processes across a diversity of spatial and temporal scales (Kerr and Ostrovsky 2003).

Gross primary production (GPP) is the amount of carbon fixed during photosynthesis with units of mass of carbon per unit area per time (g-C $\cdot$ m^{-2} $\cdot$ d^{-1}). GPP is related to the amount of PAR absorbed by green vegetation multiplied by LUE; that is, the efficiency with which the absorbed light is used in photosynthesis (Monteith and Unsworth 1990):

$$GPP = LUE \cdot APAR = LUE \cdot f_{APAR} \cdot PAR \qquad (12.10)$$

where GPP and APAR have the same molar units (e.g., μmol $\cdot$ m^{-2} $\cdot$ d^{-1}) and f_{APAR} is derived through empirical VI relationships, typically through its relationship with NDVI (Baret and Guyot 1991; Goetz et al. 1999; Goward and Huemmrich 1992; Gamon et al. 1995).

LUE depicts the ability of an ecosystem to convert energy into biomass, that is, the efficiency of carbon captured by a canopy, and it can be expressed as the ratio of GPP and APAR. LUE values can vary considerably across vegetation types and in response to the environmental controls of water, temperature, and light, and hence are quite difficult to derive. Many ecosystem models estimate LUE$_{max}$ using LUT for each vegetation type and then adjust these values downward on the basis of environmental stress factors derived from available meteorologic information (Turner et al. 2006). The net primary production and photosynthesis (NPP/PSN) products are based on the Biome-BGC (BioGeochemical) terrestrial ecosystems model. These products combine remotely sensed vegetation structural parameters, LAI, and f_{APAR} with meteorologic data sets to derive photosynthesis rates, turbulent energy, and mass fluxes, and subsequently GPP (Figure 12.9; Running et al. 2004).

MODIS EVI and LSWI have also been incorporated into LUE-productivity models, such as the vegetation photosynthesis model (VPM; Xiao, Zhang et al. 2004) and the vegetation photosynthesis and respiration model (Mahadevan et al. 2008), yielding tower-calibrated predictions of GPP and gross or net ecosystem exchange across evergreen and deciduous

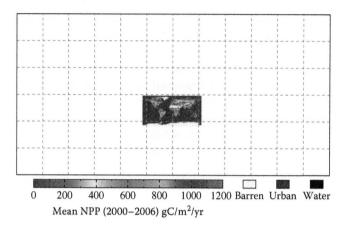

0   200   400   600   800   1000   1200 Barren Urban Water

Mean NPP (2000–2006) gC/m²/yr

FIGURE 12.9
(See color insert following page 426.) MODIS annual NPP global product averaged for the years 2000–2006 (From Running, S., Numerical Terradynamic Simulation Group. http://www.ntsg.edu. Accessed September 2010. With permission.)

forest sites in temperate North America and in seasonally moist tropical evergreen forest in the Amazon (Xiao, Zhang et al. 2004; Xiao, Hollinger et al. 2004; Xiao et al. 2005).

$$GPP = (LUE_{max} \cdot T \cdot W \cdot P) \cdot f_{APAR} \cdot PAR \quad (12.11)$$

where f_{APAR} is derived from EVI, and LUE_{max} is multiplied by temperature (T), moisture (W), and phenology (P) scalars. LSWI is used to adjust the scalars with respect to changes in canopy and soil moisture status. Potter et al. (2007) input MODI EVI data into the NASA CASA (Carnegie Ames Stanford Approach) model and found modeled outputs enabled prediction of peak growing season uptake rates of CO_2 in irrigated croplands and in moist temperate forests.

Several researchers have proposed the use of narrowband spectral indices as a direct measurement of LUE to avoid land-cover generalizations and reduce uncertainties in modeling efforts. The PRI is a normalized difference ratio of two narrow wavelengths, one at ~531 nm and a second reference band at 570 nm, that provides a *scaled* LUE (Gamon, Serrano, and Surfus 1997). Reflectance variations at 531 nm have been associated with the conversion of xanthophyll pigments from an epoxidized (violaxanthin) state to a de-epoxidized (antheraxanthin and zeaxanthin) state as a way of dissipating excess light energy to protect the photosynthetic apparatus (Drolet et al. 2008). Xanthophyll pigment composition is therefore closely related to the photosynthetic LUE (Blackburn 2007).

PRI has been computed from MODIS ocean color narrowband reflectances and is found to be well correlated with daily net primary productivity in a temperate deciduous forest (Drolet et al. 2005). Other narrowband spectral techniques are also being studied for discerning photosynthetic efficiency and LUE, including certain spectroscopic approaches and "red-edge" reflectance indices that have performed better than PRI (Gitelson et al. 2009). Chlorophyll fluorescence (ChlF) is another indicator of photosynthetic function that can be extracted from high-resolution reflectance spectra at canopy and ecosystem scales. Overall, more research is needed in the use of narrowband hyperspectral data to better understand photosynthetic efficiency at the ecosystem level and its role in limiting carbon uptake.

Remote sensing has also been used to directly generate independent measures of vegetation processes without considering LUE or f_{APAR}. Rahman et al. (2005) used MODIS EVI to directly estimate GPP, rather than attempt to derive LUE and estimate f_{APAR}. They analyzed 10 AmeriFlux tower sites over a wide range of biome types and found that MODIS EVI alone was able to estimate GPP with relatively high accuracy and without the need for meteorologic data or direct estimation of LUE (Figure 12.10a). Sims et al. (2006) further noted that the relationship between tower GPP fluxes and MODIS EVI was stronger for deciduous forests and weaker for evergreen sites; however, the EVI was able to estimate GPP with relatively high accuracy across many sites and without directly considering LUE, thus simplifying carbon balance models over most vegetation types. Yang et al. (2007) developed a continental-scale measure of GPP by integrating MODIS EVI and AmeriFlux data using an inductive machine learning technique called support vector machines (SVM).

Strong linear relationships between MODIS EVI and tower-calibrated GPP measurements of seasonal carbon fluxes were also found in high-biomass tropical forests in the

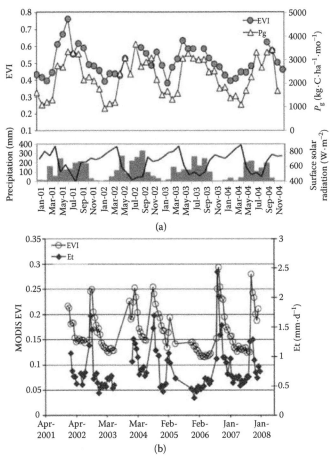

(a)

(b)

FIGURE 12.10
Relationships of MODIS greenness with Fluxnet tower measures of (a) GPP and (b) Et at MaeKlong and La Paz sites, respectively. (From *Agric for Meteorol*, 148, Huete, A.R. et al., Multiple site tower flux and remote sensing comparisons of tropical forest dynamics in Monsoon Asia, 748. Copyright (2008), with permission from Elsevier; From Huete, A.R. et al., Applications of the MODIS sensor for monitoring terrestrial land cover, ed. J. Maas, Instituto Nacional de Ecologia, Centre de Investigaciones en Geografia Ambiental, Mexico, 2010. With permission.)

Amazon and Southeast Asia (Huete et al. 2006; Huete et al. 2008). With this seasonal synchrony of EVI with tower GPP, Ichii et al. (2007) constrained the Biome-BGC ecosystem model with MODIS satellite EVI values and mapped the spatial variability in the rooting depths of forest trees over the Amazon, improving the assessments of carbon and water and energy cycles in tropical forests.

VIs are also used to estimate canopy foliage properties for use in surface energy balance and soil–vegetation–atmosphere transfer models and also in combination with ground data to directly estimate Et. Combined remote sensing and in situ tower flux measurements have also yielded close relationships with water fluxes (Figure 12.10b). With the aid of an SVM model, Yang et al. (2006) derived continental-scale estimates of Et by combining MODIS data with eddy covariance flux tower measurements and found MODIS EVI to be the most important explanatory factor in their estimates of Et (root mean square of $0.62 \cdot mm \cdot d^{-1}$). Nagler et al. (2005) reported seasonal Et measurements at flux tower sites in semi-arid upland grass, shrub, and riparian communities in Arizona and New Mexico to be strongly correlated with MODIS EVI ($r = 0.80$–0.94). Seasonal VI profiles tend to track tower Et fairly well when transpiration dominates water fluxes and soil evaporation is minimal, such as in riparian areas and during interstorm periods in upland sites, because the VI values would not be able to detect soil evaporation contributions to total canopy Et following precipitation events (Glenn et al. 2007).

The high correlation between VIs and tower fluxes in so many different ecosystems may seem surprising because, in theory, CO_2 and water exchanges are controlled in part by stomatal resistance and related not just to canopy properties but also to environmental variables (PAR, soil moisture, air temperature, vapor pressure deficit, wind), which can vary considerably over short periods. However, as reviewed by Glenn et al. (2008), ecological processes tend to adjust plant characteristics (e.g., foliage density) over periods of weeks or months to match the capacity of the environment to support photosynthesis and maximize growth. This is known as the resource optimization theory (Field et al. 1995), which treats photosynthesis and plant production as integrators of resource availability. Leaves are expensive to produce and maintain, and so when plants undergo water or nutrient stress, or are exposed to unfavorable conditions, they reduce their leaf area to use resources more efficiently.

Monteith and Unsworth (1990) noted that VIs can legitimately be used to estimate the rate of processes that depend on radiation absorbed by the leaves (R_a), such as GPP and transpiration (T), whereas the relationship of LAI or F_v to R_a is strongly nonlinear and depends on leaf architecture and spectral properties. Although there is often a lack of 1:1 correspondence between VIs and canopy attributes such as LAI or F_v, this does not compromise the utility of VIs in predicting physiological processes such as transpiration and photosynthesis, which are primarily driven by light absorbed by leaves (LAI) at the top of canopy (Glenn et al. 2008). Sellers (1987) similarly concluded that VIs may be more indicative of biophysical processes, such as GPP and T, and less reliable predictors of LAI and F_v. In summary, remote sensing is most commonly used to generate maps of vegetation "state" properties (LAI, F_v, f_{APAR}) that are then utilized in productivity and hydrology models, along with meteorologic parameters. On the other hand, remote sensing can also be used to directly provide independent estimates of water and carbon fluxes, which can then be used to constrain models in both spatial and temporal domains.

There remain many challenges on how best to integrate spatially extensive satellite data with local tower measures from multiple sites for regional scaling, modeling, and predicting vegetation processes in response to climate variability (Running et al. 1999). High temporal frequency and coarse resolution satellite data, such as MODIS, are critical to achieving greater sensitivity to seasonal landscape patterns; however, such data may be too coarse, spatially, to adequately define heterogeneous landscapes. Li et al. (2008) found that the source area influencing the tower measurement, in heterogeneous semi-arid landscapes, was poorly resolved by the size of the satellite pixels. This affected the interpretation of spatial variations of land surface fluxes over heterogeneous areas and the evaluation of the performance of the land surface model. Other challenges include the need to assess autotrophic and heterotrophic ecosystem respiration and the efficiency of conversion of assimilate into growth. To improve remote sensing of ecosystem functioning, both spectral and temporal data are needed, and in the hyperspectral domain, more knowledge is needed on key spectral bands diagnostic of physiological function, water status, and carbon exchange.

12.5 Phenology

Phenology is the study of the timing of recurring biological events, which can include the timing of budbreak, flowering, pollination, leaf flushing and extension, maturity, and senescence within canopies. Phenology is an integrative science for exploring vegetation responses to environmental controls, and as such, represents a critical biological response to climate change (Schwartz, Ahas, and Aasa 2006; Morisette et al. 2009). Variations and shifts in phenology influence biogeochemical processes, photosynthesis, water cycling, soil moisture depletion, and canopy physiology. Thus, there is much interest in characterizing seasonal cycles within and across ecosystems and understanding the impact of changes in phenology on ecosystem functioning. We can further suggest that an accurate representation of the seasonal dynamics of ecosystem functioning is a prerequisite to driving ecosystem productivity models and predicting future interannual trends and changes resulting from climate change and land-use impacts.

Remote sensing enables the synoptic and repetitive monitoring of vegetation for investigating landscape phenology at regional and continental scales. High temporal frequency satellite time-series data, such as MODIS, AVHRR, and SPOT-VGT, have been extensively used for long-term analysis of biome-dependent responses to climate warming (Zhang et al. 2004). Remote sensing provides spatial and spectral depictions of phenological-event timing, event value, direction and rates of change, and integrated time between events (Figure 12.11; White et al. 2009; Zhang et al. 2003; Jenerette, Scott, and Huete 2010). Most often, VIs are used to depict vegetation seasonal activity, and there is currently a 30+ year record of AVHRR NDVI data available for phenology studies at 5–8 km spatial resolution. MODIS has a 10+ year satellite phenology product derived from EVI and available at 0.250–1 km resolutions (Zhang et al. 2003; Liang and Schwartz 2009). An example of a maximum greenness date phenology map derived from MODIS EVI is depicted in Figure 12.12).

The seasonality of ecosystem metabolism in temperate forests is dominated by an active growing season that transitions into a dormant season in which metabolism substantially slows. However, in tropical forests, seasonality is not so obvious, and controls on phenology are not well understood. For example, recent MODIS phenology studies in tropical areas have found positive *greening* vegetation responses to seasonal drying (Huete et al. 2006; Myneni et al. 2007) and interannual drought (Saleska et al. 2007), which suggest these tropical forests are light-limited and respond to increased dry-period light availability. Tropical rainforests present challenges to remote sensing due to the near ubiquitous

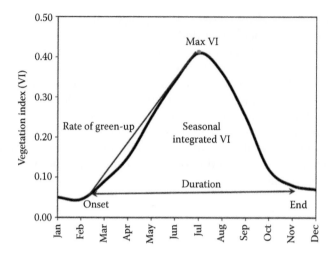

FIGURE 12.11
Diagram depicting phenology metrics derived from remote sensing.

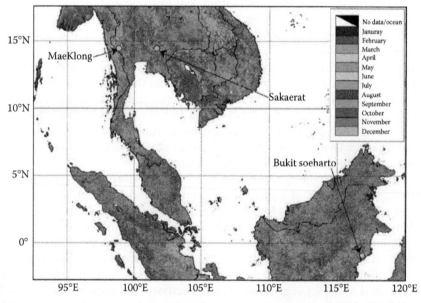

FIGURE 12.12
(See color insert following page 426.) Maximum greenness date phenology image derived from MODIS over Southeast Asia. (From *Agric and Forest Meteorol*, 148, Huete, A.R. et al., Multiple site tower flux and remote sensing comparisons of tropical forest dynamics in Monsoon Asia, 748. Copyright (2008), with permission from Elsevier.)

presence of clouds and high LAI levels that cause saturation in the photosynthetically active and chlorophyll-sensitive spectral region. Other ecosystems also present challenges in remote sensing of assessments of phenology. Savannas are very complex due to multiple physiognomies of mixed woody-herbaceous systems (shrub, grass, and trees), each with phenologies responding to different environmental controls (Ratana, Huete, and Ferreira 2005). Arid regions exhibit high-frequency rainfall pulse events that are spatially heterogeneous and have weak vegetation spectral signals (Jenerette, Scott, and

Huete 2010; Okin et al. 2001). Similarly, northern latitude regions have very strong and highly variable snow background signals that complicate the interpretation of remote sensing of phenology.

There are many factors that confuse the assessment of ecosystem phenology with remote sensing data. Foremost is the issue of clouds, aerosols, and view/sun angle effects, which contaminate the remotely sensed signal (Los et al. 2005). Despite various compositing and filtering schemes, coarse resolution satellite data remain noisy and need to be carefully examined for both systematic and random errors. During humid periods, cloud cover restricts the availability of good quality satellite data, whereas in the dry season, aerosols from biomass burning may pose problems. As a result, different satellite products from the AVHRR and MODIS instruments show variable and inconsistent seasonal patterns over tropical rainforests, with some products showing canopy drying in the dry season while other products show greening and a positive response to drought. Large inconsistencies have been reported in cross-comparisons of satellite products (including MODIS, SPOT-VGT, and AVHRR) for tropical evergreen broadleaf forests (Garrigues et al. 2008), and inconsistencies have been reported among MODIS products and field LAI observations (Doughty and Goulden 2008).

12.6 Conclusions

In this chapter we have highlighted some important advancements in the assessments and studies of ecosystem structure and functioning from space. Remotely sensed measures of green foliage density and vegetation dynamics are powerful tools for assessing the physiological status of vegetation and for monitoring ecosystem processes related to light absorption, in particular canopy photosynthesis, primary production, phenological greening and browning, and plant transpiration. However, there exist important trade-offs and compromises in characterizing ecosystems from space related to the spatial, spectral, and temporal capabilities of the imaging sensors. Multiple sensor systems with appropriate combinations of spectral, spatial, and temporal resolutions are needed to improve the remote characterization of ecosystem structure and function.

High temporal frequency measurements are important in obtaining the sufficient cloud-free data necessary to sample phenological variations. Even in arid regions, cloud-free satellite imagery may be difficult to acquire during the short growing season due to the sporadic rainfall and clouds that accompany vegetation activity. However, high temporal frequency satellite data are spatially too coarse to adequately resolve complex ecosystems and highly heterogeneous landscapes and fragmentation patterns associated with disturbance and land-use activities.

Spatial patterns of ecosystem variability are best defined at fine (<1 m) and moderate (<30 m) resolutions. There is normally an enormous mismatch between leaf- and species-level ecological variables and satellite spatial resolutions that render it difficult to validate satellite-derived products. Lastly, most satellite sensors measure ecosystem characteristics and monitor phenological events from broadband reflectances that are limited in their ability to capture fine chlorophyll and biochemical spectral variability corresponding to multispecies canopies, leaf age spectral variations, and variable plant stress responses. Hyperspectral remote sensing measurements add spectral fidelity to extract important biochemical canopy features of ecosystems that can potentially be useful in characterizing species and functional types (Sanchez-Azofeifa and Castro 2006).

Hyperspectral remote sensing combined with high temporal frequency satellite measurements provide powerful monitoring tools for characterizing landscape phenology and ecosystem processes. Hyperspectral imagery provides fine spectral and spatial detail that improves detection of land-cover types and physiognomies, while coarser resolution sensors, such as MODIS, provide frequent temporal measurements for obtaining more complete phenology profiles. The important advancements being made in surface sensor networks, such as Fluxnet, will further offer well-calibrated, time-series-based in situ data sets, for a more thorough validation and characterization of satellite ecology products. Finally, the upcoming potential launches of new hyperspectral missions, such as Hyperspectral Infrared Imager (HyspIRI) and the NPOESS Visible/Infrared Imager Radiometer Suite (VIIRS), as well as lidar missions and hyperspatial sensors, such as the recently launched DigitalGlobe's WorldView-2, will greatly expand ecosystem studies and provide new opportunities and challenges for multisensor data fusion and the scaling and extension of leaf physiologic processes and phenology from species and ecosystem to regional and global scales.

References

Anderson, K., and H. Croft. 2009. Remote sensing of soil surface properties. *Prog Phys Geog* 33:457.

Anderson, L., Y. Shimabukuro, S. Almeida, and A. R. Huete. 2010. Use of fraction images for monitoring intra-annual phenology of different vegetation physiognomies in Amazonia. *Int J Remote Sens.*

Asner, G. P. 1998. Biophysical and biochemical sources of variability in canopy reflectance. *Remote Sens Environ* 64:234.

Asner, G. P., K. M. Carlson et al. 2005. Substrate age and precipitation effects on Hawaiian forest canopies from spaceborne imaging spectroscopy. *Remote Sens Environ* 98:457.

Asner, G. P., and D. E. Knapp. 2008. Invasive species detection in Hawaiian rainforests using airborne imaging spectroscopy and lidar. *Remote Sens Environ* 112:1942.

Asner, G. P., and D. B. Lobell. 2000. A biogeophysical approach for automated SWIR unmixing of soils and vegetation. *Remote Sens Environ* 74:99.

Asner, G. P., J. M. O. Scurlock et al. 2003. Global synthesis of leaf area index observations: Implications for ecological and remote sensing studies. *Global Ecol Biogeogr* 12:191.

Asner, G. P., C. A. Wessman et al. 2000. Impact of tissue, canopy, and landscape factors on the hyperspectral reflectance variability of arid ecosystems. *Remote Sens Environ* 74:69.

Asrar, G., M. Fuchs et al. 1984. Estimating absorbed photosynthetic radiation and leaf area index from spectral reflectance in wheat. *Agron J* 76:300.

Baret, F., and G. Guyot. 1991. Potentials and limits of vegetation indices for LAI and APAR assessment. *Remote Sens Environ* 35:161.

Baumgardner, M. F., L. F. Silva, L. L. Biehl, and E. R. Stoner. 1985. Reflectance properties of soils. *Adv Agron* 38:1.

Bausch, W. C. 1993. Soil background effects on reflectance-based crop coefficients for corn. *Remote Sens Environ* 46:213.

Bégué, A. 1993. Leaf area index, intercepted photosynthetically active radiation, and spectral vegetation indices: A sensitivity analysis for regular-clumped canopies. *Remote Sens Environ* 46:45.

Ben-Dor, E., and A. Bannin. 1994. Visible and near-infrared (0.4–1.1 um) analysis of arid and semi-arid soils. *Remote Sens Environ* 48:261.

Ben-Dor, E., S. Chabrillat, J. A. M. Dematte et al. 2009. Using imaging spectroscopy to study soil properties. *Remote Sens Environ* 113(S1):S289–310, doi:10.1016/j.rse.2008.09.019.

Ben-Dor, E., R. G. Taylor et al. 2008. Imaging spectrometery for soil applications. *Adv Agron* 97:321.

Blackburn, G. A. 2007. Hyperspectral remote sensing of plant pigments. *J Exp Bot* 58:855.

Broge, N. H., and E. Leblanc. 2001. Comparing prediction power and stability of broadband and hyperspectral vegetation indices for estimation of green leaf area index and canopy chlorophyll density. *Remote Sens Environ* 76:156.

Carlson, T. N., and D. A. Ripley. 1997. On the relation between NDVI, fractional vegetation cover and leaf area index. *Remote Sens Environ* 62:241.

Carter, G. A. 1994. Ratios of leaf reflectances in narrow wavebands as indicators of plant stress. *Int J Remote Sens* 15:697.

Carter, G. A., and A. K. Knapp. 2001. Leaf optical properties in higher plants: Linking spectral characteristics to stress and chlorophyll concentration. *Am J Bot* 88:677.

Ceccato, P., S. Flasse et al. 2001. Detecting vegetation leaf water content using reflectance in the optical domain. *Remote Sens Environ* 77:22.

Chen, X., L. Vierling et al. 2005. Monitoring boreal forest leaf area index across a Siberian burn chronosequence: A MODIS validation study. *Int J Remote Sens* 26:5433.

Cohen, W. B., T. K. Maiersperger et al. 2003. Comparisons of land cover and LAI estimates derived from ETM+ and MODIS for four sites in North America: A quality assessment of 2000/2001 provisional MODIS products. *Remote Sens Environ* 88:233.

Colwell, J. E. 1974. Vegetation canopy reflectance. *Remote Sens Environ* 3:175.

Diner, D., C. J. Bruegge, J. V. Martonchik et al. 1991. A multi-angle image spectroradiometer for terrestrial remote sensing with the Earth Observation System. *Int J Imaging Syst Tech* 3:92.

Doughty, C. E., and M. L. Goulden. 2008. Seasonal patterns of tropical forest leaf area index and CO_2 exchange. *J Geophys Res* 113:G00B06, doi:10.1029/2007JG000590.

Drolet, G. G., K. F. Huemmrich, F. G. Hall et al. 2005. A MODIS-derived photochemical reflectance index to detect inter-annual variations in the photosynthetic light-use efficiency of a boreal deciduous forest. *Remote Sens Environ* 98:212.

Drolet, G. G., E. M. Middleton et al. 2008. Regional mapping of gross light-use efficiency using MODIS spectral indices. *Remote Sens Environ* 112:3064.

Elvidge, C. D. 1990. Visible and near-infrared reflectance characteristics of dry plant materials. *Int J Remote Sens* 11:1775.

Fassnacht, K. S., S. T. Gower, M. D. MacKenzie et al. 1997. Estimating the leaf area index of north central Wisconsin forests using the Landsat Thematic Mapper. *Remote Sens Environ* 61:229.

Fensholt, R., and I. Sandholt. 2003. Derivation of a shortwave infrared water stress index from MODIS near- and shortwave infrared data in a semiarid environment. *Remote Sens Environ* 87:111.

Fensholt, R., I. Sandholt et al. 2004. Evaluation of MODIS LAI, fAPAR and the relation between fAPAR and NDVI in a semi-arid environment using in situ measurements. *Remote Sens Environ* 91:490.

Field, C. B., J. T. Randerson et al. 1995. Global net primary production: Combining ecology and remote sensing. *Remote Sens Environ* 51:74.

Gamon, J. A., C. B. Field et al. 1995. Relationship between NDVI, canopy structure, and photosynthesis in three californian vegetation types. *Ecol Appl* 5:28.

Gamon, J. A., L. Serrano, and J. S. Surfus. 1997. The photochemical reflectance index: An optical indicator of photosynthetic radiation use efficiency across species, functional types, and nutrient levels. *Oecologia* 112:492.

Garrigues, S., R. Lacaze, F. Baret et al. 2008. Validation and intercomparison of global Leaf Area Index products derived from remote sensing data. *J Geophys Res* 113, doi:10.1029/2007JG000635.

Gao, B. C. 1996. NDWI. A normalized difference water index for remote sensing of vegetation liquid water from space. *Remote Sens Environ* 58:257.

Gates, D. M., H. J. Keegan et al. 1965. Spectral properties of plants. *Appl Opt* 4:11.

Gausman, H. W., and W. A. Allen. 1973. Optical parameters of leaves of 30 plant species. *Plant Physiol* 52:57.

Gitelson, A. A., O. B. Chivkunova, and M. N. Merzlyak. 2009. Nondestructive estimation of anthocyanins and chlorophylls in anthocyanic leaves. *Am J Bot* 96:1861.

Gitelson, A. A., Y. Gritz et al. 2003. Relationships between leaf chlorophyll content and spectral reflectance and algorithms for non-destructive chlorophyll assessment in higher plant leaves. *J Plant Physiol* 160:271.

Gitelson, A. A., A. Viña, V. Ciganda et al. 2005. Remote estimation of canopy chlorophyll content in crops. *Geophys Res Lett* 32:L08403, doi:10.1029/2005GL022688.

Gitelson, A. A., A. Viña, S. B. Verma et al. 2006. Relationship between gross primary production and chlorophyll content in crops: Implications for the synoptic monitoring of vegetation productivity. *J Geophys Res* 111:D08S11, doi:10.1029/2005JD006017.

Glenn, E. P., A. R. Huete et al. 2007. Integrating remote sensing and ground methods to estimate evapotranspiration. *Crit Rev Plant Sci* 26:139.

Glenn, E. P., A. R. Huete, P. L. Nagler et al. 2008. Relationship between remotely-sensed vegetation indices, canopy attributes and plant physiological processes: What vegetation indices can and cannot tell us about the landscape. *Sensors* 8:2136.

Gobron, N., B. Pinty, M. M. Verstraete et al. 2000. Advanced vegetation indices optimized for upcoming sensors: Design, performance, and applications. *IEEE Trans Geosci Remote Sens* 38:2489.

Goetz, S. J., S. D. Prince et al. 1999. Satellite remote sensing of primary production: An improved production efficiency modeling approach. *Ecol Modell* 122(3):239–255.

Goward, S. N., and K. F. Huemmrich. 1992. Vegetation canopy PAR absorptance and the normalized difference vegetation index: An assessment using the SAIL model. *Remote Sens Environ* 39:119.

Gutman, G., and A. Ignatov. 1998. The derivation of the green vegetation fraction from NOAA/AVHRR data for use in numerical weather prediction models. *Int J Remote Sens* 19:1533.

Hardisky, M. A., R. M. Smart, and V. Klemas. 1983. Seasonal spectral characteristics and above ground biomass of the tidal marsh plant Spartina alterniflora. *Photogramm Eng Remote Sens* 49:85.

Healey, S. P., W. B. Cohen, Y. Zhiqiang et al. 2005. Comparison of tasseled cap-based Landsat data structures for use in forest disturbance detection. *Remote Sens Environ* 97:301.

Houborg, R. M., and H. Soegaard. 2004. Regional simulation of ecosystem CO2 and water vapor exchange for agricultural land using NOAA AVHRR and Terra MODIS satellite data, application to Zealand, Denmark. *Remote Sens Environ* 93:150.

Huemmrich, K. F., J. L. Privette et al. 2005. Time-series validation of MODIS land biophysical products in a Kalahari woodland, Africa. *Int J Remote Sens* 26:4381.

Huete, A. R. 1988. A soil-adjusted vegetation index (SAVI). *Remote Sens Environ* 25:295.

Huete, A., K. Didan et al. 2002. Overview of the radiometric and biophysical performance of the MODIS vegetation indices. *Remote Sens Environ* 83:195.

Huete, A. R., K. Didan, Y. E. Shimabukuro et al. 2006. Amazon rainforests green-up with sunlight in dry season. *Geophys Res Lett* 33:L06405, doi:10.1029/2005GL025583.

Huete, A. R., R. D. Jackson et al. 1985. Spectral response of a plant canopy with different soil backgrounds. *Remote Sens Environ* 17:37.

Huete, A. R., N. Restrepo-Coupe et al. 2008. Multiple site tower flux and remote sensing comparisons of tropical forest dynamics in monsoon Asia. *Agric For Meteorol* 148:748.

Huete, A. R., R. Solano-Barajas, and N. Restrepo-Coupe. 2010. Indices de vegetacion y procesos ecosistemicos. In *Aplicaciones Del Sensor MODIS Para El Monitoreo De Cubiertas Terrestres*, ed. J. Maas. Mexico: Instituto Nacional de Ecología, Centro de Investigaciones en Geografía Ambiental. UNAM2010.

Hunt, E. R., and B. N. Rock. 1989. Detection of changes in leaf water content using near- and middle-infrared reflectances. *Remote Sens Environ* 30:43.

Ichii, K., H. Hashimoto, M. A. White et al. 2007. Constraining rooting depths in tropical rainforests using satellite data and ecosystem modeling for accurate simulation of gross primary production seasonality. *Global Change Biol* 13:67.

Irons, J. R., G. Campbell, J. M. Norman et al. 1992. Prediction and measurement of soil bidirectional reflectance. *IEEE Trans Geosci Remote Sens* 30:249.

Jackson, T. J., D. Chen et al. 2004. Vegetation water content mapping using Landsat data derived normalized difference water index for corn and soybeans. *Remote Sens Environ* 92:475.

Jacquemoud, S. 1990. PROSPECT: A model to leaf optical properties spectra. *Remote Sens Environ* 34:74.

Jacquemoud, S., F. Baret et al. 1995. Extraction of vegetation biophysical parameters by inversion of the PROSPECT+SAIL models on sugar beet canopy reflectance data: Application to TM and AVIRIS sensors. *Remote Sens Environ* 52:163.

Jenerette, G. D., R. L. Scott, and A. R. Huete. 2010. Functional differences between summer and winter season rain assessed with MODIS-derived phenology in a semi-arid region. *J Veg Sci* 21:16.

Jiang, Z., and A. R. Huete. 2010. Linearization of NDVI based on its relationship with vegetation fraction. *Photogramm Eng Remote Sens* 76(8):965–975.

Jiang, Z., A. R. Huete et al. 2006. Analysis of NDVI and scaled difference vegetation index retrievals of vegetation fraction. *Remote Sens Environ* 101:366.

Kaufman, Y., and D. Tanré. 1992. Atmospherically resistant vegetation index (ARVI) for EOS MODIS. *IEEE Trans Geosci Remote Sens* 30:261.

Kerr, J. T., and M. Ostrovsky. 2003. From space to species: Ecological applications for remote sensing. *Trends Ecol Evol* 18:299.

Kimes, D., Y. Knyazikhin et al. 2000. Inversion methods for physically based models. *Remote Sens Rev* 18:381.

Knipling, E. B. 1970. Physical and physiological basis for the reflectance of visible and near-infrared radiation from vegetation. *Remote Sens Environ* 1:155.

Lefsky, M. A., W. B. Cohen, G. G. Parker et al. 2002. Lidar remote sensing for ecosystem studies. *Bioscience* 52:19.

Leprieur, C., Y. H. Kerr et al. 2000. Monitoring vegetation cover across semi-arid regions: Comparison of remote observations from various scales. *Int J Remote Sens* 21:281.

Leuning, R., H. A. Cleugh, S. J. Zegelin et al. 2005. Carbon and water fluxes over a temperate Eucalyptus forest and a tropical wet/dry savanna in Australia: Measurements and comparion with MODIS remote sensing estimates. *Agric For Meteorol* 129:151.

Li, F., W. P. Kustas, M. C. Anderson et al. 2008. Effect of remote sensing spatial resolution on interpreting tower-based flux observations. *Remote Sens Environ* 112:337.

Liang, L., and M. Schwartz. 2009. Landscape phenology: An integrative approach to seasonal vegetation dynamics. *Landsc Ecol* 24:465.

Los, S. O., P. R. J. North, W. M. F. Grey, and M. J. Barnsley. 2005. A method to convert AVHRR normalized difference vegetation index time series to a standard viewing and illumination geometry. *Remote Sens Environ* 99:400.

Mahadevan, P., S. C. Wofsy, D. M. Matross et al. 2008. A satellite-based biosphere parameterization for net ecosystem CO_2 exchange: Vegetation photosynthesis and respiration model (VPRM). *Global Biogeochem Cycles* 22:GB2005, doi:2010.1029/2006GB002735.

Montandon, L. M., and E. E. Small. 2008. The impact of soil reflectance on the quantification of the green vegetation fraction from NDVI. *Remote Sens Environ* 112(4):1835–1845, doi: 10.1016/j.rse.2007.09.007.

Monteith, J. L., and M. H. Unsworth. 1990. *Principles of Environmental Physics*. London: Arnold.

Moran, M. S., T. R. Clarke et al. 1994. Estimating crop water deficit using the relation between surface-air temperature and spectral vegetation index. *Remote Sens Environ* 49:246.

Morisette, J. T., A. D. Richardson, A. K. Knapp et al. 2009. Learning the rhythm of the seasons in the face of global change: Phenological research in the 21st Century. *Front Ecol Environ* 7:253.

Mulders, M. A. 1987. *Remote Sensing in Soil Science*. Amsterdam: Elsevier.

Myneni, R. B., S. Hoffman et al. 2002. Global products of vegetation leaf area and fraction absorbed PAR from year one of MODIS data. *Remote Sens Environ* 83:214.

Myneni, R. B., C. D. Keeling et al. 1997. Increased plant growth in the northern high latitudes from 1981 to 1991. *Nature* 386:698.

Myneni, R. B., W. Yang, R. R. Nemani et al. 2007. Large seasonal swings in leaf area of Amazon rainforests. *Proc Nat Acad Sci* 104:4820.

Nagler, P. L., J. Cleverly et al. 2005. Predicting riparian evapotranspiration from MODIS vegetation indices and meteorological data. *Remote Sens Environ* 94:17.

Nemani, R. R., and S. W. Running. 1989. Estimation of regional surface resistance to evapotranspiration from NDVI and thermal-IR AVHRR data. *J Appl Meteorol* 28:276.

Numerical Terradynamic Simulation Group. http://www.ntsg.edu (accessed in September 2010).

Okin, G. S., D. Roberts, B. Murray et al. 2001. Practical limits on hyperspectral vegetation discrimination in arid and semiarid environments. *Remote Sens Environ* 77:212.

Palacios-Orueta, A., J. E. Pinzon et al. 1999. Remote sensing of soil properties in the Santa Monica Mountains II. hierarchical foreground and background analysis. *Remote Sens Environ* 68:138.

Pettorelli, N., J. O. Vik et al. 2005. Using the satellite-derived NDVI to assess ecological responses to environmental change. *Trends in Ecol Evol* 20:503.

Potter, C., S. Klooster et al. 2007. Terrestrial carbon sinks for the United States predicted from MODIS satellite data and ecosystem modeling. *Earth Interact* 11:1.

Rahman, A., D. Sims, V. Cordova et al. 2005. Potential of MODIS EVI and surface temperature for directly estimating per-pixel ecosystem C fluxes. *Geophys Res Lett* 32(19):L19404, doi.10.1029/2005GL024127.

Ratana, P., A. R. Huete, and L. Ferreira. 2005. Analysis of cerrado physiognomies and conversion in the MODIS seasonal–temporal domain. *Earth Interact* 9:1.

Richardson, A. J., and C. L. Wiegand. 1977. Distinguishing vegetation from soil background information. *Photogramm Eng Remote Sens* 43:1541.

Roberts, D. A., M. Keller et al. 2003. Studies of land-cover, land-use, and biophysical properties of vegetation in the large scale biosphere atmosphere experiment in Amazônia. *Remote Sens Environ* 87:377.

Roberts, D., G. M. Smith et al. 1993. Green vegetation, non-photosynthetic vegetation and soils in AVIRIS data. *Remote Sens Environ* 44:255.

Rock, B. N., J. E. Vogelmann et al. 1986. Remote detection of forest damage. *Bioscience* 36:439.

Running, S. W. 2006. Is global warming causing more, larger wildfires? *Science* 313:927.

Running, S. W., D. D. Baldocchi et al. 1999. A global terrestrial monitoring network integrating tower fluxes, flask sampling, ecosystem modeling and EOS satellite data. *Remote Sens Environ* 70:108.

Running, S. W., R. R. Nemani, F. A. Heinsch et al. 2004. A continuous satellite-derived measure of global terrestrial primary production. *Bioscience* 54:547.

Saleska, S. R., K. Didan, A. R. Huete et al. 2007. Amazon forests green-up during 2005 drought. *Science* 318:612.

Sanchez-Azofeifa, G. A., and K. Castro. 2006. Canopy observations on the hyperspectral properties of a community of tropical dry forest lianas and their host trees. *Int J Remote Sens* 27:2101.

Schwartz, M. D., R. Ahas, and A. Aasa. 2006. Onset of spring starting earlier across the Northern Hemisphere. *Global Change Biol* 12:343.

Sellers, P. J. 1987. Canopy reflectance, photosynthesis and transpiration: II. The role of biophysics in the linearity of their interdependence. *Remote Sens Environ* 21:143.

Shimabukuro, Y. E., B. N. Holben et al. 1994. Fraction images derived from NOAA-AVHRR data for studying the deforestation in the Brazilian Amazon. *Int J Remote Sens* 15:517.

Shuai, Y., C. B. Schaaf, A. H. Strahler et al. 2008. Quality assessment of BRDF/albedo retrievals in MODIS operational system. *Geophys Res Lett* 35:L05407, doi:10.1029/2007GL032568.

Sims, D. A., A. F. Rahman, V. D. Cordova et al. 2006. On the use of MODIS EVI to assess gross primary productivity of North American ecosystems. *J Geophys Res* 111(4):G04015, doi:04010.01029/02006JG000162.

Smith, M. O., S. L. Ustin et al. 1990. Vegetation in deserts: I. A regional measure of abundance from multispectral images. *Remote Sens Environ* 31:1.

Stoner, E. R., and M. F. Baumgardner. 1981. Characteristic variations in reflectance of surface soils. *Soil Sci Soc Am J* 45:1161.

Thenkabail, P. S., R. B. Smith et al. 2000. Hyperspectral vegetation indices and their relationships with agricultural crop characteristics. *Remote Sens Environ* 71:158.

Tollefson, J. 2009. Climate: Counting carbon in the Amazon. *Nature* 461:1048.

Tucker, C. J. 1979. Red and photographic infrared linear combinations for monitoring vegetation. *Remote Sens Environ* 8:127.

Tucker, C. J. 1980. Remote sensing of leaf water content in the near infrared. *Remote Sens Environ* 10:23.

Tucker, C. J., H. E. Dregne et al. 1991. Expansion and contraction of the Sahara desert from 1980 to 1990. *Science* 253:299.

Turner, D. P., W. D. Ritts et al. 2006. Evaluation of MODIS NPP and GPP products across multiple biomes. *Remote Sens Environ* 102:282.

Ustin, S. L., D. A. Roberts et al. 2004. Using imaging spectroscopy to study ecosystem processes and properties. *BioScience* 54:523.

Wang, Q., S. Adiku et al. 2005. On the relationship of NDVI with leaf area index in a deciduous forest site. *Remote Sens Environ* 94:244.

White, M. A., K. M. De Beurs, K. Didan et al. 2009. Intercomparison, interpretation, and assessment of spring phenology in North America estimated from remote sensing for 1982–2006. *Global Change Biol* 15(10):2335–2359.

Xiao, X., D. Hollinger, J. Aber et al. 2004. Satellite-based modeling of gross primary production in an evergreen needle-leaf forest. *Remote Sens Environ* 89:519.

Xiao, X., Q. Zhang et al. 2004. Modeling gross primary production of temperate deciduous broadleaf forest using satellite images and climate data. *Remote Sens Environ* 91:256.

Xiao, X., Q. Zhang et al. 2005. Satellite-based modeling of gross primary production in a seasonally moist tropical evergreen forest. *Remote Sens Environ* 94:105.

Yang, F., K. Ichii et al. 2007. Developing a continental-scale measure of gross primary production by combining MODIS and AmeriFlux data through support vector machine approach. *Remote Sens Environ* 110:109.

Yang, F., M. A. White et al. 2006. Prediction of continental-scale evapotranspiration by combining MODIS and AmeriFlux data through support vector machine. *IEEE Trans Geosci Remote Sens* 44:3452.

Zarco-Tejada, P., J. R. Miller et al. 2004. Hyperspectral indices and model simulation for chlorophyll estimation in open-canopy tree crops. *Remote Sens Environ* 90:463.

Zarco-Tejada, P. J., C. A. Rueda, and S. L. Ustin. 2003. Water content estimation in vegetation with MODIS reflectance data and model inversion methods. *Remote Sens Environ* 85:109.

Zeng, X., R. E. Dickinson et al. 2000. Derivation and evaluation of global 1-km fractional vegetation cover data for land modeling. *J Appl Meteorol* 39:826.

Zhang, X., M. A. Friedl, C. B. Schaaf et al. 2003. Monitoring vegetation phenology using MODIS. *Remote Sens Environ* 84:471.

Zhang, X., M. A. Friedl, C. B. Schaaf et al. 2004. The footprint of urban climates on vegetation phenology. *Geophys Res Lett* 31:L12209.

Zhang, Q., X. Xiao et al. 2005. Estimating light absorption by chlorophyll, leaf and canopy in a deciduous broadleaf forest using MODIS data and a radiative transfer model. *Remote Sens Environ* 99:357.

13

Remote Sensing of Live Fuel Moisture

Stephen R. Yool

CONTENTS

13.1 Introduction and Background

Pyrogeography (i.e., the geography of fire) is a multidimensional subfield of geography concerned with the study of the complex space–time interactions between fire and people. Humans have always been the keepers of the flame; thus, pyrogeography is, true to its geographic roots, integrative, spanning the physical, biological, and social sciences (Yool 2009). In this chapter, we consider that humans can use remote sensing technology as a tool to understand and monitor how climate dynamics mediate planetary pyrogeography and to manage fuels under changing climate conditions. We believe that such information is key to the long-term security of human systems and ecosystems.

13.1.1 Problem Statement

Climate warming and earlier springs have increased fire activity significantly in the fuel-laden forests of the western United States (Westerling et al. 2006). Coincidental expansion of the wildland–urban intermix has increased the probability of ignitions in these fire-starved forests. Ignition probabilities are mediated chiefly by variations in the interannual precipitation and topography (slope, elevation, and aspect), which affect live and dead fuel moistures. The diversity of microclimates within rugged-montane landscapes challenges conventional applications of the normalized difference vegetation index (NDVI) as a proxy for live fuel moistures, limiting fire management to broad-scale fire seasons (i.e., large areal extents over intervals of months). In this chapter, we explore the complex space–time variability in live fuel moistures. Our chief aim is to review current research on remote sensing for monitoring live fuel moistures, present a new monitoring metric, and use a case study and field data for validation.

13.1.2 Chapter Organization and Preview

In this chapter, we use NDVI time-series data from the 1-km resolution Advanced Very High Resolution Radiometer (AVHRR) to introduce the fuel moisture stress index (FMSI). We also used the free online U.S. Geological Survey AVHRR maximum value composite (MVC) product, which is compiled every 10–14 days. The FMSI is the inverse z-score of the multitemporal NDVI MVC; thus, it has a nominal spatial resolution of 1 km^2. We show that the FMSI predicts the length of the fire season for each 1-km^2 area with a temporal resolution of about 2 weeks. We begin with a review of live fuel moisture metrics, starting with the NDVI. We discuss NDVI advantages, summarize the debate regarding the NDVI, and then describe the shortwave and thermal infrared spectra as other useful moisture metrics. In Section 13.2, "Data and Methods," we introduce FMSI development and implementation. Validation appears in Section 13.3, Results and Discussion, followed by the summary and conclusions in Section 13.4.

13.1.3 Live Fuel Moisture Metrics: The NDVI

The NDVI, which was developed in the early 1970s, is an established metric for vegetation conditions and is the most well-known, well-used, and well-debated metric in remote sensing science. The NDVI is produced by dividing the difference with the sum of the reflectances from the near infrared (NIR) and red spectral bands (Rouse et al. 1974):

$$NDVI = (\rho_{NIR} - \rho_{red})/(\rho_{NIR} + \rho_{red}) \tag{13.1}$$

producing values from –1 to 1 (positive values indicate greater "greenness" and negative values indicate lesser "greenness"). The NDVI is not a direct measure of vegetation moisture content, but it measures photosynthetic activity and cell turgor, both of which have a strong relationship with internal plant moisture. The internal structure characteristics of hydrated plant leaves have been shown, for example, to reflect NIR radiation strongly, but as the vegetation loses moisture, the NIR reflectance decreases, due possibly to the difference between the index of refraction of air and water occupying the space within these cells (Chuvieco et al 2002). As the vegetation cures (dries), red reflectance rises and NIR reflectance falls, decreasing the NDVI.

13.1.4 NDVI Advantages

There are multiple advantages to adopting the NDVI for live fuel moisture modeling, and particularly the NDVI from the AVHRR. AVHRR "sees" the world twice daily and provides a nearly complete stream of biweekly maximum value composites (MVC) data that span 1989 (start of MVC data record) to the present. Although we detected no obvious radiometric bias in the AVHRR time series selected for our protocols (i.e., no drifting of the mean NDVI), we should note that the quality of the NDVI time series can be limited by the orbital drift of the platform (Trishchenko, Cihlar, and Li 2002). AVHRR NDVI pixels also have a comparatively coarse nominal spatial resolution (1 km^2), limiting analyses to broad landscape-scale dynamics. The sacrifice of fine spatial resolution, however, favors access to landscape-scale fuel conditions and also opens data accessibility; small file sizes facilitate image processing on less powerful computers. Moreover, we believe that AVHRR is useful to monitor fire hazards broadly because climate affects fuel moisture variations at broad spatial scales.

13.1.5 NDVI Debate

There is noteworthy debate about the merits and demerits of the NDVI as a proxy for the vegetation moisture content. Leblon (2001) provides a comprehensive overview of studies using the NDVI for live fuel moisture monitoring. Multiple studies have confirmed a functional relationship between the NDVI and the moisture condition of grasses, shrubs, and forest understory species (Paltridge and Barber 1988; Chladil and Nunez 1995; Alonso et al. 1996; Deshayes et al. 1998; Burgan, Klaver, and Klaver 1998). However, the relationship between the NDVI and reflectance from canopies of conifer forests has not been well characterized (Hardy and Burgan 1999). This is likely due to the number of complex factors contributing to the spectral response of each pixel within a mixed conifer forest, including diverse species and morphology, as well as the obscuring of the understory by the overstory signal, which may have a very different moisture condition (Eidenshink, Burgan, and Hass 1990; Hardy and Burgan 1999; Leblon 2001). One potential solution to control such variability, which we feature in this chapter, is to measure the relative NDVI flux of each conifer canopy over time; the 1-km grid cell serves to "self reference" each canopy's characteristics (e.g., species, morphology), isolating distinctive canopy characteristics through time.

13.1.6 NDVI Distortions

One particular difficulty in applying the NDVI to fuel moisture monitoring is the distortion of the signal by rock and bare soil when biomass density varies significantly across time within a given pixel. Consider two scenarios, A and B, shown in Figure 13.1, which illustrates a highly schematized canopy as viewed by a sensor over three seasons. Scenario A, season 1, shows a healthy tree that becomes a moisture-stressed tree in season 2 and then returns to health again in season 3. In scenario A, the resulting NDVI signal accurately quantifies the rise and fall of stress within the constant population. Scenario B, season 1, shows a single healthy tree. In season 2, a number of other healthy plants have sprouted to fill the scene. In season 3, a drought causes stress in this new larger population. But the NDVI signal indicates that season 3 is in less stress than season 1, because the new stressed foliage filling in the canopy results in a higher NDVI value than the unvegetated soil that surrounds the single lush plant in season 1 of scenario B. Techniques have been suggested

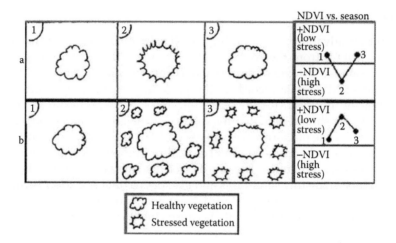

FIGURE 13.1
The population variability problem. Soil and background signal can distort the NDVI signal when biomass density changes within a pixel.

to handle the signal distortion by bare soil, including the noteworthy soil-adjusted vegetation index (SAVI) and its derivatives (Huete 1988). SAVI requires an estimate of soil background cover to be most effective. The NDVI MVC lacks the spectral resolution for SAVI and at 1 km the NDVI MVC product averages the dominant surface covers; it is as such the most practical, economical, and understandable product for fire researchers and managers.

13.1.7 Relative Greenness: An NDVI-Based Indicator of Live Fuel Moisture Stress

The relative greenness (RG; Burgan and Hartford 1993) and departure from average (DA; Hartford and Burgan 1994) indices were developed to produce a measure of fire potential in live fuels that is made comparable from one pixel to another by relating current NDVI observations to a historical range or mean. RG compares NDVI maximums and minimums for the same period (i.e., from 1989 to the start of the MVC data record) to the current NDVI value:

$$RG = (NDVI - \text{min since } 1989)/(\text{max since } 1989 - \text{min since } 1989) \qquad (13.2)$$

DA divides the current NDVI value by the mean for the same period since 1989:

$$DA = NDVI/\text{mean NDVI since } 1989 \qquad (13.3)$$

In addition to indicating high or low values that are uniquely meaningful to a given pixel, calculating relative values over the same periods for multiple years distinguishes changes in NDVI, which are not due to annual phenological patterns. RG and DA related well to the 1994 wildfires in the northern Rocky Mountains, but Leblon (2001) notes that the critical fire hazard threshold values are difficult to define for both indices.

13.1.8 NDVI Alternatives

Previous work established the importance of the thermal infrared, which when combined with the NDVI data produced a negative slope that characterized the range of vegetation water content (Nemani and Running 1989; Sandholt, Rasmussen, and Andersen 2002; Dupigny-Giroux and Lewis 1999). Dupigny-Giroux and Lewis (1999) found good relationships between plant moisture, NDVI, and surface temperature using Landsat, but reported that these relationships did not hold for coarser resolution AVHRR data. Ceccato et al. (2001) described why most surrogates for vegetation water content (e.g., vegetation stress indices, curing and chlorophyll content) cannot retrieve the water content at the leaf level, using instead a combination of shortwave infrared (SWIR) and NIR. Bowyer and Danson (2004) reported related work that the leaf area index, fractional vegetation cover, and solar zenith angle influence the longer SWIR wavelengths more than the shorter SWIR wavelengths and NIR, concluding that useful empirical relationships between the vegetation moisture content and remotely sensed vegetation indices are possible. Although we had to rule out the SWIR because the AVHRR carries no SWIR bands, we conducted experiments to test the relationships between AVHRR NDVI and thermal infrared bands. Our results for the AVHRR were consistent with those reported for AVHRR by Dupigny-Giroux and Lewis (1999). Because we wanted our protocols to be as portable as possible, we took the conservative position of using the NDVI as the exclusive metric for fuel moisture stress.

13.2 Data and Methods

13.2.1 Study Regions

Fire season length depends in part on moisture variations in live vegetative fuels (grasses, shrubs, trees) as mediated by the precipitation, temperature, topography, and winds. We will feature the data from three montane regions in southeastern Arizona: (1) the Catalina-Rincon Mountains, (2) the Chiricahua Mountains, and (3) the Huachuca Mountains (Figure 13.2). Southeastern Arizona is characterized by a classic basin and range topography, with numerous mountain ranges or sky islands studding a relatively level plain that rises from about 914 m at the base of the Catalina-Rincons to about 1524 m at the base of the Huachucas. The Catalina-Rincon site spans the largest range of elevation of the three regions, whereas the Chiricahuas and Huachucas have higher base elevations and give rise to higher peaks.

The annual precipitation pattern in this region is bimodal, with high intensity "monsoon" rains occurring in July and August, when moisture from the tropical regions of Mexico and the Gulf of California is drawn into Arizona. Longer duration precipitation is supplied by fronts and synoptic scale disturbances during the winter months of December and January, giving rise to snow occurring at higher elevations. Arid conditions may persist from March through July at lower elevations. This arid fore-summer cures fuels, which are especially susceptible to ignition by lightning strikes early in the monsoon season.

The large range of elevation and rugged terrain found over these regions supports a rich biodiversity. Vegetation type "transitions" from desert scrub or grassland at the base of sky islands up to forests of mixed conifer and subalpine fir. As far back as the nineteenth century, the concept of "life zones" was developed by C. Hart Merriam, who identified climate/vegetation relationships in the San Francisco Peaks in northern Arizona (Merriam 1890).

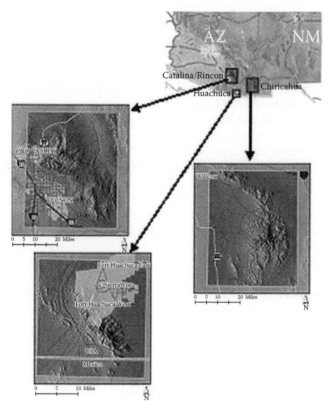

FIGURE 13.2
Location of three study sites in southeastern Arizona. The Chiricahuas are located just west of the Arizona/ New Mexico border. These shaded relief maps highlight mountains, urban areas, roads, and rivers.

Forrest Shreve documented relationships between vegetation and climate with changing elevation and topography in the Catalinas (Shreve 1915). Later workers confirmed that major vegetation communities form a sequence of zones where the elevation and aspect create favorable microclimates (Brown 1994; Whittaker and Niering 1964, 1968).

13.2.2 Fuel Moisture Stress Index

We used the U.S. Geological Survey NDVI MVC product to develop the FMSI. The maximum NDVI is selected from each (roughly) two week period of daily observations, producing an MVC image for each pixel over this interval. The effects of clouds, directional and off-nadir viewing effects, atmospheric attenuation, sun angle, and shadow effects are minimized in MVC images (Holben 1986). The NDVI MVC is rescaled from 0 to 200 (Eidenshink 1992). We noted that the MVC data are not error free. Hence, they must be screened for undetected clouds and missing data records.

We transformed NDVI values into z-scores to measure the live fuel moisture stress relative to a given pixel's unique history (Yool 2001):

$$\mathrm{FMSI}(x) = -\frac{(x-\mu)}{\sigma} \tag{13.4}$$

where x is the NDVI value of a pixel for a given date, μ is the mean value for this pixel for the same AVHRR biweekly period in all years since 1989, and σ is the standard deviation for this pixel for the same period in all years since 1989. We extended our work through 2006 only due to time and data processing constraints, but note that the updated NDVI MVC product is available free from the U.S. Geological Survey. The period "drilldown" relates each period to the same period in all other years. This transformation produces the FMSI—the inverse of the standardized NDVI—scaled so that high FMSI values represent high-moisture stress.

To measure fuel condition variations throughout the fire season of each year, we produced an FMSI time series from the AVHRR NDVI MVC imagery for three sites of the study, which included an FMSI image for each biweekly period. Positive values of the FMSI indicate above-average moisture stress and negative values indicate below-average moisture stress. We compared the FMSI time series during the fire season with the fire records for our sites.

13.2.3 Defining the Fire Season across a Heterogeneous Landscape

Parameter elevations on independent slopes model (PRISM) data were used to show low- and high-elevation correlations between precipitation, temperature, and FMSI for each 1-km grid cell over southern Arizona's April–June fire season (Crimmins 2003, pers. comm.). PRISM data are continuous, digital grid estimates of monthly, yearly, and event-based climatic parameters. These data sets are used worldwide as the highest-quality spatial climate data. The climate data had a 4-km spatial resolution, but were autocorrelated sufficiently to be downscaled to 1 km for this study (Crimmins 2003, pers. comm.). We used a 1500-m threshold to represent the transition between grass and scrub brush versus the upland Madrean evergreen woodlands. We believed that the grass and brush fuels responded to rain or drought at shorter time scales; woody upland fuels, by contrast, can rely more on stored soil moisture. Results indicated that the average FMSI for April through June in elevations below 1500 m correlated significantly (R^2 value >0.8 at 95% confidence) with the preceding winter (December through March) mean precipitation anomaly and also to current (April through June) mean temperature anomaly. In elevations above 1500 m (i.e., the limit of the Lower Sonoran Life Zone [Merriam 1890]), however, no significant correlation was found between the FMSI averaged over the months April through June and the preceding years (Crimmins 2003, pers. comm.).

13.3 Results and Discussion

13.3.1 Comparing FMSI Time Series to Legacy Observations

Forrest Shreve observed nearly a century ago that the arid fore-summer in and around the Catalinas varied in length from 16 weeks in the lower bajada to 6 weeks at 8000 feet. At low elevations, frost is unlikely beyond the end of February (Figure 13.3), but at high elevations, frost can occur as late as the end of May. The bimodal precipitation pattern does not change with elevation, but a slow progression of warming temperatures clearly proceeds upward in elevation as the spring weeks pass. FMSI trends are consistent with Shreve's findings.

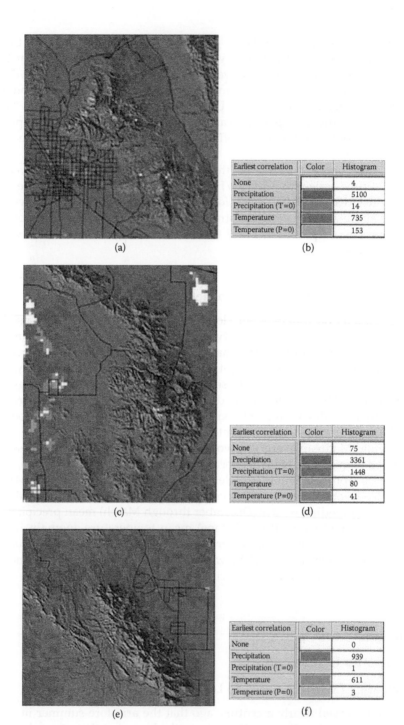

FIGURE 13.3
(See color insert following page 426.) (a;c;e) Earliest correlating climate drivers across the three sites. Precipitation is most often the earliest correlating climate driver in low elevations. (b;d;f) Color-coding in this figure indicates (1) pixels where FMSI did not correlate significantly to either driver, (2) pixels where FMSI correlated first to preceding winter precipitation, (3) pixels where FMSI correlated only to precipitation, (4) pixels where FMSI correlated first to temperatures and (5) pixels that correlated only to temperature.

13.3.2 Comparing the FMSI to the Fire Record

Does the incidence of wildfire expand to increasingly higher elevations as the arid fore-summer progresses (i.e., Shreve [1915] and the FMSI consistent)? To answer this question, we show the fire record and corresponding box plots of these fires (Figures 13.4a and b) and the FMSI (Figure 13.4c). (The Coronado National Forest provided geographic information system files on the fire activity in the Catalina-Rincon Mountains for the study period of record.)

Box plots provide insight to the statistical distribution of values in data by graphically depicting median (location), spread (dispersion), skewness, and outliers. The horizontal line contained within the box marks the median value (second quartile) of the data. The box contains the middle 50% of the data points. The difference between the upper and lower quartiles is known as the interquartile range, and is a measure of the variability of the data.

Figure 13.4a plots the number of wildfires >10 acres (~4 hectares). Wildfires in the three study sites occur at lower elevations starting in March and continue all the way through to July, but occurs only at higher elevations later in the season. Figure 13.4b, derived from Figure 13.4a, strongly suggests that the season of the most critical fuel moisture stress and peak vulnerability to ignition begins at different times across these landscapes, depending on various local factors (e.g., wind patterns and aspect), but it appears mediated chiefly by elevation. Area burned increased monthly (as the curing progressed to higher elevations and put more fuels in play), peaking in June, then falling in July with the arrival of monsoonal rains. Figure 13.4c depicts the FMSI representation in terms of the length of the fire season. The FMSI agrees with the elevation basis of curing revealed by actual fires.

13.3.3 Shreve Was Correct

We concluded that the initial poor correlations between the FMSI and the two climate variables in elevations above 1500 m arose because we assumed the same range of time for the fire season, April through June, for every pixel in the three sky island sites. But Shreve (1915) was correct. Before an average fire season, the FMSI can be calculated and a representative fire season composite map can be produced. Hence, the length of the fire season for each pixel in the study site must be determined.

13.3.4 Determining Length of the Fire Season for Each Pixel

We determined for every pixel in each of the three sites the strongest correlations between the FMSI over all possible ranges of biweekly periods between March and July, and precipitation in the preceding winter months (December through March). We also determined the strongest correlations between the FMSI over all possible ranges of biweekly periods between March and July, and the temperature over all possible months between February and July that precede or are concurrent with each range of the FMSI periods tested. Figure 13.5 summarizes the results.

The results showed significant (95% confidence) correlations between the FMSI and one or both climate variables for most pixels in each of the three sites. However, correlating ranges of time are often too short to be considered reasonable fire season approximations, and correlations between the FMSI and the two climate drivers often exist over different ranges of biweekly periods. Precipitation correlations can extend from early March through July, and some number of precipitation correlations end before the earliest of monsoon rains, which is feasible. Nearly all temperature-correlating FMSI period ranges are concentrated between May and June, and some continue through July.

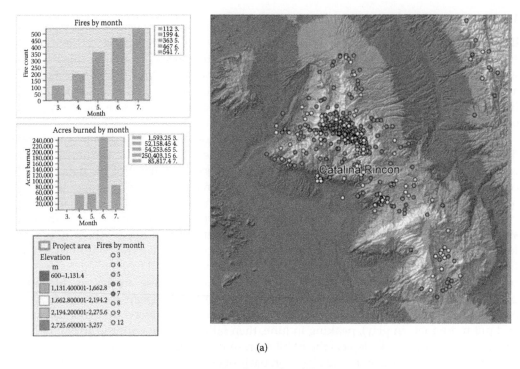

(a)

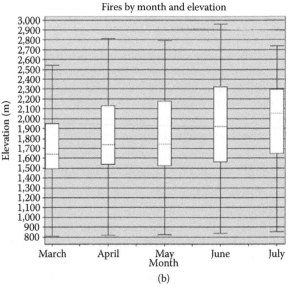

(b)

FIGURE 13.4
(See color insert following page 426.) (a) Actual fires for the period of record covered by the FMSI. Earliest fires (e.g., March) occur at lower elevations in the Catalina-Rincons. The FMSI shows lower elevations remain vulnerable as the fire season "progresses" in elevation. By July, all elevations are vulnerable; this is evident in Figure 13.4b, which shows July fires span high and low elevations; and in Figure 13.4c, where the FMSI shows the early, sustained vulnerability of low-elevation fuels. (b) Box plot showing fire counts by elevation per month. The medians are the dotted lines within the box. Each box contains 50% of the values. The "whiskers" denote minimum and maximum values. Median fire counts tend to increase in elevation by month. (c) The length of fire season (LOFS) as determined by the FMSI, which shows live fuels cure later at higher elevations; fire season thus is later at higher elevations, and this is consistent with Figures 13.4a and 13.4b.

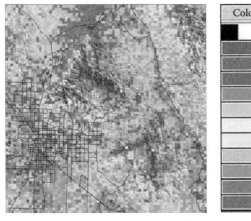

Color	LOFS(wks)	Histogram
	No corr	4
	2	189
	4	461
	6	513
	8	610
	10	566
	12	728
	14	988
	16	880
	18	683
	20	274
	22	110

(c)

FIGURE 13.4 *(Continued)*

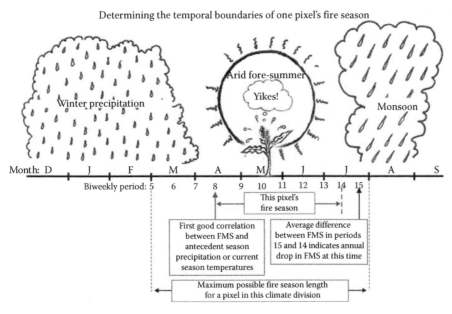

FIGURE 13.5
For this pixel, the earliest significant correlation occurs in period 8, indicating the beginning of the fire season. The average difference between FMSI in periods 14 and 15 indicates FMSI characteristically drops in period 15 each year, indicating monsoonal moistures usually begins alleviating fuel moisture stress by this time. Thus, period 14 is selected to be the last period of this pixel's fire season.

When the FMSI begins to correlate with one or both climate drivers, we interpret this to mean that moisture resources have begun to be tapped significantly, and vegetation has entered its annual season of arid fore-summer survival, with increased vulnerability to combustion. We designated the FMSI period associated with the earliest significant correlation to either climate variable as the first period in a given pixel's fire season. Correlations to the second climate variable, when they occur later in the season, could indicate an exhaustion stage. The earliest correlating climate driver is most often the pre-cipitation, except in upper elevations and urbanized areas, where temperature is often the earliest correlating climate driver. The maps in Figure 13.3 show the earliest correlating climate driver for each of the three sites.

The end of the fire season should be tied closely to the arrival of the monsoon rains (typi-cally between periods 13 and 15). We designated the last period of each pixel's fire season as that period between 13 and 15 that precedes the period when the average (taken over all years in the dataset) FMSI difference between periods is negative and stays negative through period 15. This average drop in FMSI suggests that monsoonal moisture is usu-ally effective by this time for this pixel. When no drop in value is detected, period 15 is assumed to be the end of the season. Lack of a drop in the FMSI could possibly be due to a characteristic longer lag in vegetative response to monsoon moisture and characteristi-cally later rains in these locations. Figure 13.6 illustrates how one pixel's fire season would be modeled.

13.3.5 Annual Fire Season Fuel Moisture Stress Maps

Using the model described previously, we determined the fire season temporal bound-aries for each pixel in each study site. To characterize a year's fire season over a study site, the FMSI for that year was averaged over the range of fire season biweekly periods

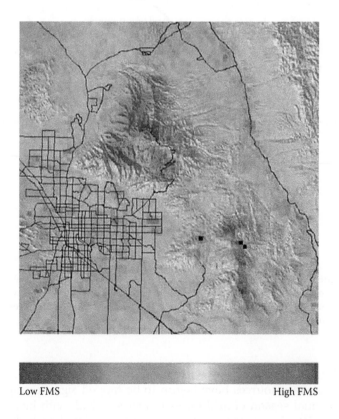

Low FMS   High FMS

FIGURE 13.6
(See color insert following page 426.) Fire season summary map for the Catalina-Rincon site for the year 2004, clearly showing fire scars from the previous year's Aspen fire in the Catalinas and the smaller Helens II fire in the Rincons.

unique to each pixel in the given site. As shown in Figure 13.5, the fire season summary of each year for the given pixel would be found by averaging the FMSI for each year over periods 8 through 14. The fire season summary is a composite image in which pixel values represent their unique fire season means for the given year and is visualized in Figure 13.6, which shows the fire season summary map of the Catalina-Rincon site for the year 2004. Figure 13.7 shows the chronological sequence for all years for all three sites.

Fire seasons begin earliest in the Catalina region, and the distribution of fire season range here is the most diverse of the three regions, which may be attributable to the Catalina's considerable variation in land cover and elevation. Fire seasons begin later in the slightly higher-elevation Chiricahua region and much later in the Huachuca region, which is the highest-elevation region of the three study regions.

The population variability problem is very evident in fire season summary maps following fires. Scars from the Aspen and Bullock fires, which occurred in the Catalina-Rincons in 2003, are still visible in the 2006 Catalina fire season summary (see Figure 13.7). In the years following a fire, there will be markedly less photosynthetic activity. Depending on the degree of the burn, however, there may be lifeless biomass or no biomass in a given location. As in the years following severe drought, interpretation of the FMSI in recently burned locations can be ambiguous because the FMSI does not distinguish between absent biomass and dead biomass, which presents a much greater fire hazard. Fire history would establish whether the FMSI is responding to live fuels under stress or to recent burns.

Figure 13.7 shows the fire season summaries for all three sites from 1989 to 2006. This chronological sequence reveals comparable trends in fuel moisture stress patterns among the three regions, suggesting synoptic climate forcing of live fuel moistures. Fuel moisture stress appears most severe in 1989–1990, 1996, 2002–2004, and 2006. The years with the least stress are 1993 and 1998. In general, the Huachucas appear to have had more stress, whereas the Catalinas appear to have had slightly less stress, than the other two regions.

Fire season summaries in chronological sequence for the three southeastern Arizona study sites

The figure below shows 15–18 years of fire season fuel moisture stress in three sky island regions of southeastern Arizona. Each column represents a year and each row represents a site.

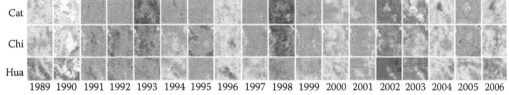

1989 1990 1991 1992 1993 1994 1995 1996 1997 1998 1999 2000 2001 2002 2003 2004 2005 2006

Notes:
– Because satellite imagery was not available for all periods spanning March through July, the following fire season summaries are approximations:
  – Catalina: 1990, 1999, and 2003
  – Chiricahua: 1990, 1991, and 1999
  – Huachuca: 1990
– White pixels indicate nondata.

FIGURE 13.7
(See color insert following page 426.) Fire season summaries in chronological sequence.

13.3.6 Visualizing Links between the Average Fire Season FMSI and Climate Drivers

We plotted fire season summaries on a grid to show the links between the FMSI and climate factors (Figure 13.8). Monthly means were converted to z-scores so they could be compared to FMSI results, also z-scores. Climatologies, over which z-scores were calculated, were determined by examining every possible range of years to find the range that yields the highest count/mean of significant correlations between the FMSI and the given climate variable.

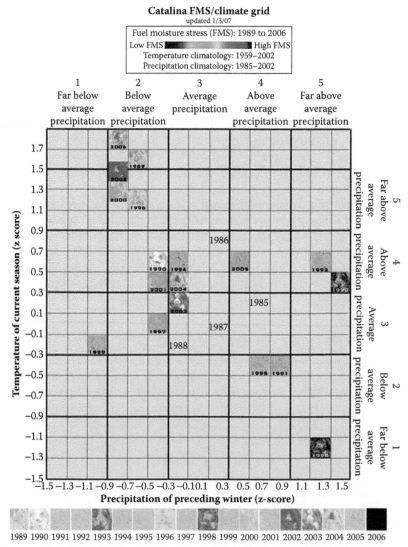

FIGURE 13.8

(See color insert following page 426.) The FMSI/climate grid for the Catalina-Rincon study site. Fire season summaries for 18 years are plotted using precipitation and temperature z-scores for each year.

The precipitation coordinate of each year is simply the mean z-score calculated over the immediately preceding December through March period for Arizona climate division 7; thus, precipitation coordinates in the FMSI/climate grids are the same for all three study regions. However, the values represented by the temperature axis in the FMSI/climate grids are weighted mean z-scores based on the ranges of months over which temperature correlated strongest to the FMSI over the pixels of the given study site, making the FMSI/climate grid for each study region unique. The FMSI/climate grid for the Catalina site is shown in Figure 13.8.

Orange–red-colored fire season summaries cluster in the upper left (hot and dry) corners of FMSI/climate grids, whereas blue-colored fire season summaries gravitate into the lower right (cool and wet) corners of FMSI/climate grids. No years appear in the lower left (cool and dry) or upper right (hot and wet) corners of FMSI/climate grids, describing spatiotemporal patterns since 1989.

The noteworthy anomaly on all three FMSI/climate grids is the 2003 fire season summary, with "hot" colors that contrast sharply with nearby fire season summaries in the average or above-average regions. We verified with personnel at the Arizona Remote Sensing Center that the sudden increase in the FMSI in the fire seasons of 2002 and 2003 was not caused by any known inconsistency in sensor calibration or imagery preprocessing between 1989 and 2003. Our conclusion, therefore, is that the 2002 and 2003 fire season summaries accurately reported unusually severe FMSI. However, if conditions in 2003 were not as severe as in 2000, why does the 2003 fire season summary show overall higher FMSI than the 2000 fire season summary? The simplest assumption is that FMSI is mediated by more than just antecedent winter rains and fire season temperatures; we speculate here that strong seasonal winds in the region are responsible for the high FMSI values observed in 2003. Through anecdotal evidence, we are aware that strong winds fanned the 2003 Aspen fire out of control. Wind thus plays such a strong role in fire behavior that it cannot be discounted as a possible explanation for this anomaly.

Is it possible that severe conditions occurring over a number of years in a row pushed vegetation populations within these landscapes over a threshold of fuel moisture stress "criticality"? A sequence of severe years culminating in the 2002 fire season may have forced large amounts of the vegetation in these landscapes beyond their survival capacity, such that they could not revive in the next spring despite the moister and cooler conditions of 2003. This is yet another example of the population variability problem. Is FMSI high in 2003 because there is more bare dirt between living plants or because there is more dead biomass still standing from the previous harsh year? More field work would be required to answer these questions.

13.4 Summary and Conclusions

We used a standardized AVHRR NDVI time series to derive FMSI to characterize the vegetation moisture stress for the highly variable landscapes of three different montane regions in southeastern Arizona. The algorithm was developed to search (since 1989 data onset) for the strongest correlations between measured fuel moisture stress, winter precipitation from December through March, and temperature from March through July.

These correlations suggest a significant relationship between the FMSI and climate drivers. Results also support Shreve's observations that the arid fore-summer begins progressively later as elevations increase. Dependency of vegetation survival on the preceding winter precipitation and current temperature conditions is often evident early at lower elevations, but not until June or later at the highest elevations. In addition to the elevation, other microclimate factors, such as aspect, vegetation type, soil type, and so on, impact the fuel moisture stress in any given location.

We produced a composite map representing annual fire season fuel moisture stress for each year by averaging the FMSI over each pixel's unique fire season. We plotted these fire season summary maps against the winter precipitation and current fore-summer temperatures, depicting patterns in fuel moisture stress between 1989 and 2006. This is a short span of time, climatologically speaking, but as fire season summaries are added with each new year, longer-term climate patterns could emerge within the FMSI/climate grid. Such modeling efforts with current and new systems will bring us closer to understanding fire behavior and will empower humans to manage planetary fuel resources effectively.

Acknowledgments

The author wishes to thank Susan Taunton for designing and implementing the research described in this chapter. Gary Christopherson assisted with select graphics.

References

Alonso, M., A. Camarasa, E. Chuvieco, D. Cocero, L. A. Kyun, M. P. Martin, and F. J. Salas. 1996. Estimating temporal dynamics of fuel moisture of Mediterranean species from NOAA-AVHRR data. *EARSel Adv Rem Sens* 4(4):9–24.

Bowyer, P., and F. M. Danson. 2004. Sensitivity of spectral reflectance to variation in live fuel moisture content at leaf and canopy level. *Rem Sens Environ*, 92:297–308.

Brown, D. E., ed. 1994. *Biotic Communities: Southwestern United States and Northwestern Mexico.* Salt Lake City, UT: University of Utah Press.

Burgan, R. E., and R.A. Hartford. 1993. Monitoring vegetation greenness with satellite data. General Technical Report INT-297, United States Dept. of Agriculture, Forest Service, Intermountain Research Station, Ogden, Utah.

Burgan, R. E., R. W. Klaver, and J. M. Klaver. 1998. Fuel models and fire potential from satellite and surface observations. *Int J Wildland Fire* 8(3):159–70.

Ceccato, P., S. Flasse, S. Tarantola, S. Jacquemoud, and J. M. Grégoire. 2001. Detecting vegetation leaf water content using reflectance in the optical domain. *Rem Sens Environ*, 77:22–33.

Chladil, M. A., and M. Nunez. 1995. Assessing grassland moisture and biomass in Tasmania: The application of remote sensing and empirical models for a cloudy environment. *Int J Wildland Fire* 5:165–71.

Chuvieco, E., D. Riaño, I. Aguado, and D. Cocero. 2002. Estimation of fuel moisture content from multitemporal analysis of Landsat Thematic Mapper reflectance data: Applications in fire danger assessment. *Int J Rem Sens* 23(11):2145–62.

Deshayes, M., E. Chuvieco, D. Cocero, M. Karteris, N. Koutsias, and N. Stach. 1998. Evaluation of different NOAA-AVHRR derived indices for fuel moisture content estimation: Interest for short term fire risk assessment. In *III International Conference on Forest Fire Research—Fourteenth Conference on Fire and Forest Meteorology*, ed. D. X. Viegas, 1149–67. Coimbra, Portugal: ADAI.

Dupigny-Giroux, L. A., and J. Lewis. 1999. A moisture index for surface characterization over a semi-arid area. *Photogramm Eng & Rem Sens*, 65(8):937–45.

Eidenshink, J. C. 1992. The 1990 conterminous U.S. AVHRR data set. *Photogramm Eng Rem Sens* 58(6):809–13.

Eidenshink, J. C., R. E. Burgan, and R. H. Hass. 1990. Monitoring fire fuels condition by using time series composites of advanced very high resolution radiometer (AVHRR) data. In *Proceedings of the Second International Symposium on Advanced Technology in Natural Resource Management*, Washington, DC, 68–82.

Hardy, C. C., and R. E. Burgan. 1999. Evaluation of NDVI for monitoring live moisture in three vegetation types of the western U.S. *Photogramm Eng & Rem Sens* 65:603–10.

Hartford, R. A., and R. E. Burgan. 1994. Vegetation condition and fire occurrence: A remote sensing connection. In *Proceedings of the Interior West Fire Council Meeting and Symposium*, Coeur d'Alene, ID, November 1–3, 1994.

Holben, B. N. 1986. Characteristics of maximum-value composite images from temporal AVHRR data. *Int J Rem Sens* 7(11):1417–34.

Huete, A. R. 1988. A soil-adjusted vegetation index (SAVI). *Rem Sens Environ* 25:295–309.

Leblon, B. 2001. Forest wildfire hazard monitoring using remote sensing: A review. *Rem Sens Rev* 20:1–43.

Merriam, C. H. 1890. Results of a biological survey of the San Francisco mountains region and desert of the Little Colorado in Arizona. United States Dept. of Agriculture, Forest Service, North American Fauna 3, Washington, DC.

Nemani, R. R., and S. W. Running. 1989. Estimation of regional surface resistance to evapotranspiration from NDVI and thermal-IR AVHRR data. *J Appl Meteorol Climatol* 28:276–84.

Paltridge, G. W., and J. Barber. 1988. Monitoring grassland dryness and fire potential in Australia with NOAA-AVHRR data. *Rem Sens Environ* 25:381–94.

Rouse, J. W., J. A. Haas, J. A. Schell, and D. W. Deering. 1974. Monitoring vegetation systems in the Great Plains with ERTS. In *Proceedings of the Third Earth Resources Technology Satellite-1 Symposium*, SP-351, 310–17. Greenbelt, MD: NASA.

Sandholt, I., K. Rasmussen, and J. Andersen. 2002. A simple interpretation of the surface temperature/vegetation index space for assessment of surface moisture status. *Rem Sens Environ*, 79:213–24.

Shreve, F. 1915. *The Vegetation of a Desert Mountain Range as Conditioned by Climatic Factors*. Washington, DC: Carnegie Institution of Washington.

Trishchenko, A. P., J. Cihlar, and Z. Q. Li. 2002. Effects of spectral response function on surface reflectance and NDVI measured with moderate resolution satellite sensors. *Rem Sens Environ*, 81:1–18.

Westerling, A. L., H. G. Hidalgo, D. R. Cayan, and T. W. Swetnam. 2006. Warming and earlier spring increase western U.S. forest wildfire activity. *Science* 313(5789):940–3.

Whittaker, R. H., and W. A. Niering. 1964. Vegetation of the Santa Catalina Mountains, Arizona: I. Species distribution and floristic relations on the north slope. *J Arizona Acad Sci* 3:9–34.

Whittaker, R. H., and W. A. Niering. 1968. Vegetation of the Santa Catalina Mountains, Arizona: III. Ecological classification and distribution of species. *J Arizona Acad Sci* 5:3–21.

Yool, S. R. 2001. Enhancing fire scar anomalies in AVHRR NDVI time series data. *GeoCarto Int* 16(1):5–12.

Yool, S. R. 2009. Pyrogeography. In *Sage Encyclopedia of Geography*. London: Sage.

14

Forest Change Analysis Using Time-Series Landsat Observations

Chengquan Huang

CONTENTS

14.1 Introduction

Forest change is highly related to many surface processes of the Earth. Covering about 40% of the ice-free land surface, forests contain nearly 80% of the total carbon estimated to be in the terrestrial aboveground biosphere (Waring and Running 1998). Forest disturbance and recovery processes are major mechanisms that determine the carbon residence time in the terrestrial biosphere and the net carbon flux between the biosphere and the atmosphere (Law et al. 2004; Hirsch et al. 2004). By shaping the landscape pattern of forest age and structure (Peterken 2001), forest change can affect land hydrology, climate, and biogeochemical processes (Band 1993; Sahin and Hall 1996; Giambelluca et al. 2000), and it has complex but often adverse impacts on biological conservation by threatening the habitats of endangered species (DeFries et al. 2005; Zartman 2003; e.g., Kinnaird et al. 2003). Knowledge of forest disturbance history and recovery is therefore necessary in order to advance many earth science applications.

Reliable characterization of forest change requires dense time-series observations at fine spatial resolutions. Spatially, many disturbances caused by human activities are at hectare or subhectare levels, while damage caused by fires and other natural disturbances can range from a few hectares to hundreds of thousands of hectares or more. Temporally, they can be abrupt or gradual processes, depending on the nature of disturbances. Harvest, fire, and storm typically result in abrupt changes, while changes due to insects and diseases can last several years or longer. However, forest growth is always a gradual process. Establishment of a forest stand always takes time, and trees continue to grow after a stand is established. Proper characterization of such gradual processes requires temporally dense observations. Furthermore, although many disturbances often result in abrupt spectral changes that are relatively easy to detect using satellite images acquired before and immediately after each disturbance, the spectral change signals often become obscured and eventually undetectable as trees grow back following those disturbances (Huang, Goward, Masek et al. 2009; Figure 14.1). As a result, forest change products derived using temporally sparse observations typically have considerable omission errors (Lunetta et al. 2004; Masek et al. 2008).

A collection of Landsat images acquired through current and previous Landsat missions (Goward et al. 2006), referred to as the Landsat record in this chapter, provides a unique data source for reconstructing forest change history. With the earliest Landsat images acquired in 1972, this record allows for assembling temporally dense image stacks

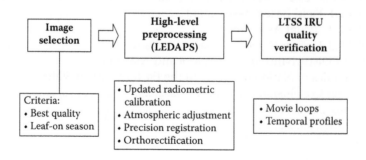

FIGURE 14.1
Major steps for developing a Landsat time-series stack (LTSS). (From Huang, C. et al. *Int J Digital Earth*, 2, 3, 2009. With permission.)

for assessing forest change in many areas (Huang, Goward, Masek et al. 2009). The fine spatial resolutions of Landsat images provide the spatial details necessary for characterizing many of the changes arising due to both natural and anthropogenic disturbances (Townshend and Justice 1988). Two major steps are involved in forest analysis using temporally dense Landsat image stacks: (1) development of Landsat time-series stacks (LTSS) and (2) change analysis using LTSS. The LTSS should consist of high-quality images that can be readily analyzed for accurate detection of land-cover change and related phenomena, a status called imagery-ready-to-use (IRU; Goward 2006). Change analysis using LTSS can be achieved in several ways. One is to divide each LTSS into a sequence of image pairs and analyze each pair using any of the existing bitemporal change detection techniques (Coppin et al. 2004; Lu et al. 2004; Singh 1989). Obviously, such an approach would be extremely inefficient. Furthermore, bitemporal techniques cannot take advantage of the rich temporal information contained in the LTSS, which is particularly useful for characterizing land-cover and change processes. Although algorithms capable of analyzing three or more images simultaneously have also been developed (Lunetta et al. 2004; Cohen et al. 2002; Coppin and Bauer 1996), most of them suffer shortcomings similar to those mentioned above. Analyzing all images simultaneously in an LTSS is an alternative to change analysis using LTSS. Two existing change detection algorithms belong to this approach. One is a trajectory-based change detection algorithm developed by Kennedy, Choen, and Schroeder (2007) and the other is a vegetation change tracker (VCT) algorithm developed by Huang et al. (2010). The latter was developed specifically for use with LTSS. LTSS development and the VCT algorithm have been described separately in previous publications (Huang, Goward, Masek et al. 2009; Huang et al. 2010). This chapter intends to provide a coherent description of the entire LTSS-VCT approach, along with summaries of assessments of the disturbance products derived using this approach and two applications.

14.2 Landsat Time-Series Stack Development

An LTSS is defined as a temporal sequence of Landsat images acquired at a nominal temporal interval for an area defined by a path/row tile of the World Reference System (WRS). These images should have an IRU quality, which is defined as follows: They are the best available images, and they have high levels of geolocation accuracy and radiometric integrity. The temporal frequency of observations in an LTSS is driven by data availability and the temporal characteristics of the changes to be detected. For forest change analysis, annual or seasonal observations are desirable, but if such frequent observations are not available, LTSS with biennial or longer temporal intervals may also be used, especially in high-latitude regions where trees grow slowly. However, due to limited data availability and cloud contamination, the actual temporal intervals between consecutive acquisitions in an LTSS can be different from the nominal interval of that LTSS (example acquisition dates of some LTSS can be found in the study by Huang, Goward, Schleeweis et al. 2009; Table 14.1).

The process for developing an LTSS comprises an image selection protocol, automated high-level preprocessing algorithms, and IRU quality-verification procedures (Figure 14.1). The high-level preprocessing algorithms include updated radiometric calibration for Landsat-5 images, atmospheric adjustment to surface reflectance, precision registration, and

TABLE 14.1

Standard Deviation Values (Reflectance in Percentage) Used in Equations 14.4 and 14.5

Band 1	Band 2	Band 3	Band 4	Band 5	Band 7
0.80	0.582	0.617	3.575	1.214	0.768

Source: Huang, C. et al. *Remote Sens Environ,* 114, 1, 2010. With permission.
Note: These values were the average of those derived using images acquired in different years from different places.

orthorectification. Here, the term "high level" is used to differentiate these algorithms from the standard correction algorithms for Landsat images (Landsat Project Science Office 2000). These high-level preprocessing algorithms have been implemented as fully automated routines in the Landsat Ecosystem Disturbance Adaptive Processing System (LEDAPS) to facilitate batch job processing.

14.2.1 Image Selection

The main purpose of image selection is to identify high-quality Landsat acquisitions that are needed to constitute an LTSS. Because Thematic Mapper (TM) and Enhanced Thematic Mapper+ (ETM+; except the panchromatic band) images have almost identical spatial and spectral characteristics, they are used interchangeably in the LTSS, as in many other land-cover change analyses (e.g., Vicente-Serrano, Pérez-Cabello, and Lasanta 2008; Lo and Yang 2002). The following two issues need to be considered in image selection:

1. A selected image must be acquired during the leaf-on season. Images acquired outside this temporal window are generally not suitable for forest change analysis, because leaf-off deciduous forests can be spectrally confused with disturbed forest land. For each WRS path/row, the leaf-on season can be defined based on vegetation phenology derived using Moderate Resolution Imaging Spectroradiometer (MODIS) and Advanced Very High Resolution Radiometer (AVHRR) measurements (Schwartz, Reed, and White 2002; Zhang et al. 2003), which includes June to mid-September for most areas in the conterminous United States. This criterion can be relaxed to include May and October for the southern United States.

2. In order to maximize the proportion of usable pixels within each selected image, it should have minimal or no quality problems arising from instrument errors or from cloud and shadow contamination.

14.2.2 Updated Radiometric Calibration

Radiometric calibration is part of the U.S. Geological Survey (USGS) process for producing standard level-1 Landsat imagery (Landsat Project Science Office 2000). For Landsat-7 images, the ETM+ sensor has been monitored since its launch (Markham et al. 2004). As such, the conversion of ETM+ level-1 radiometry to at-sensor radiance is a simple matter of applying the rescaling gains and biases from the ETM+ header file to the imagery. However, for Landsat-5 images, there have been many revisions to the calibration coefficients (Chander et al. 2004; Chander and Markham 2003; Markham and Barker 1986; Markham et al. 2004). Prior to May 2003, internal calibration (IC) lamps measurements were used to determine the gain coefficient. During the 1990s, it became increasingly apparent that variations in the IC-derived gain values reflected a combination of real changes in sensor calibration (i.e., detector sensitivity and filter properties) and changes in the IC lamps themselves.

From May 2003 to April 2007, following the Landsat-7/Landsat-5 underfly experiment in April 1999, a cross-calibration between ETM+ and TM was established (Chander et al. 2004; Teillet et al. 2004), which was used to determine the gain value for Landsat-5 images. After April 2007, further investigations using invariant ground targets in North Africa suggested that the initial lookup table (LUT) had an error for the first part of the Landsat-5 history (about 1985–1992). Instead, records of at-sensor radiance from these sites suggested a more gradual decay in gain throughout the mission life. As such, Landst-5 imagery processed after April 2007 used a revised LUT (Chander et al. 2004).

The production date of a Landsat-5 image needs to be used to determine the calibration that was originally applied to that image, which can then be "undone" by applying the reciprocal of the gain, and then the most recent LUT can be applied. Conversion to top-of-atmosphere (TOA) reflectance is then performed for the reflective bands using the scene-specific solar geometry, and sensor-specific bandpasses convolved with the CHKUR exoatmospheric irradiances from MODTRAN-4 (Landsat Project Science Office 2000; Markham and Barker 1986).

For the thermal band, the raw digital number is converted to TOA (apparent) temperature using the standard approach provided by Markham and Barker (1986) for TM images and by the Landsat Project Science Office (2000) for ETM+ images. For Landsat-5 images, however, a radiance correction to the calibration published in late 2007 needs to be considered (Barsi et al. 2007).

14.2.3 Atmospheric Adjustment to Surface Reflectance

The LEDAPS atmospheric adjustment algorithm was designed to calculate surface reflectance by compensating for atmospheric scattering and absorption effects on the TOA reflectance (Masek et al. 2006). The basic assumptions of this algorithm are that the target is Lambertian and infinite and the gaseous absorption and particle scattering in the atmosphere can be decoupled.

Developed based on a similar method used for MODIS and AVHRR (Vermote, El Saleous, and Justice 2002), this approach uses the 6S (second simulation of a satellite signal in the solar spectrum) radiative transfer code to compute the transmission, intrinsic reflectance, and spherical albedo for relevant atmospheric constituents, including gases, ozone, water vapor, and aerosols (Vermote et al. 1997).

Ozone concentration was derived from the Total Ozone Mapping Spectrometer (TOMS) aboard the Nimbus-7, Meteor-3, and Earth Probe platforms as well as from the National Oceanic and Atmospheric Administration's (NOAA) Tiros Operational Vertical Sounder ozone data when TOMS data was not available. Column water vapor was taken from NOAA National Centers for Environmental Prediction (NCEP) reanalysis data (available at http://dss.ucar.edu/datasets/ds090.0). Digital topography (1-km GTOPO30) and NCEP surface pressure data were used to adjust Rayleigh scattering to local conditions. Aerosol optical thickness was directly derived from the Landsat image using the dark, dense vegetation method of Kaufman et al. (1997).

14.2.4 Precision Registration and Orthorectification

Raw satellite images usually contain significant geometric distortions arising due to a range of sources, including platform- and instrument-related sources, as well as those due to the Earth's curvature, rotation, and topography (Toutin 2003). Beginning from late 2008, the USGS decided to distribute terrain-corrected (L1T) images as the standard Landsat imagery product. In general, no additional geometric correction is necessary for those L1T images because their geolocation errors are typically less than 1 pixel. Unfortunately, as of the writing of this chapter, this USGS standard has not been adopted by international ground-receiving stations. The standard systematic correction (L1G) products distributed by those stations can have geolocation errors of 500 m or more. This results primarily from uncontrolled orbital drift. Further analysis of image relief displacement as a result of topography has shown that at swath edges geolocation is in error by about 120 m per kilometer of elevation. Therefore, the LEDAPS precision registration and orthorectification algorithm are still needed for images obtained from international ground-receiving stations. Details of this algorithm have been provided by Gao, Masek, and Wolfe (2009).

14.2.5 Landsat Time-Series Stacks Imagery-Ready-to-Use Quality Verification

Prior to its use in downstream applications, each developed LTSS needs to be verified to determine whether the processed images have geometric or radiometric artifacts, which can result from the following:

- Unidentified quality problems with the input images
- Unknown bugs that may exist in the LEDAPS preprocessing algorithms
- Incorrect inputs regarding the geometry or radiometry of the concerned images

If artifacts are found in some LTSS images, they need to be fixed, or the images contaminated by those artifacts need to be excluded from downstream change analysis to avoid resulting in spurious changes. The verification procedures include a quick visualization approach and a spectral–temporal profile method. Prior to verification, the images of each LTSS need to be clipped so that they have exactly the same spatial domain.

14.2.5.1 Image Clipping

Due to difficulties in orbital control, satellite orbits can shift slightly among repeat passes. As a result, several images for a single WRS location are not necessarily congruent to the same geographic region (i.e., they do not overlay on top of each other exactly). Therefore, pixels near the edge of a WRS tile can have valid values on some dates but not on other dates. Temporal analysis of such incomplete observations is difficult. To avoid this problem, a common area mask is used to exclude such observations from being analyzed. For each LTSS, this mask is defined as the maximal geographic area where all image acquisitions have valid pixel values (except the missing data area caused by missing scan lines that may exist in certain acquisitions), and all images of the LTSS are clipped using this mask.

14.2.5.2 Visual Verification

Visual inspection is a simple yet effective method for verifying the quality of the LTSS images. By flipping the images from one date to another, an experienced image analyst can quickly identify inconsistencies among the images, which are often indicators of geometric or radiometric artifacts, and can gain first-hand knowledge of the change processes to be analyzed later. To facilitate quick visualization of the LTSS images, the clipped images are converted to the JPEG format (note that other visualization-ready formats can be used in the place of the JPEG format here), which are then assembled to create a movie loop. For each LTSS, a single-stretching method is used during the conversion in order to create comparable color tones among the images. After testing with different stretching methods, it was found that a single-stretching method per band could produce satisfactory visual effects for images acquired in most areas consisting primarily of closed or near-closed canopy forests. The general linear stretching equation is

$$\text{out\_value} = \frac{\text{in\_value} - \text{refl\_min}}{\text{refl\_max} - \text{refl\_min}} \times 255 \tag{14.1}$$

where in_value and out_value are the input surface reflectance (%) and output stretched values, respectively. The single set of refl_min and refl_max applicable to most vegetated areas are given in Table 14.2.

In semiarid areas, some partially vegetated areas may appear saturated in bands 2 and 3 when stretched using Equation 14.1. To avoid this problem, the following nonlinear stretching method is used for those two bands for images acquired in such areas (band 4 is stretched using Equation 14.2):

$$\text{Out\_value} = 13 \times \text{In\_value} - 0.15 \times \text{In\_value}^2 \tag{14.2}$$

14.2.5.3 Spectral–Temporal Profile

Spectral–temporal profiles can provide a more quantitative assessment of the radiometric consistency among the images within each LTSS. Such profiles are created using the spectral values of targets that are considered relatively stable over time. Because the primary use of the LTSS assembled here is forest change analysis, the "stable targets" in this context refer to conifer stands that do not have visual signs of being disturbed during the entire observing period of each LTSS. For each LTSS, a few examples of such stands are identified, and the average spectral values of those stands are calculated for each acquisition date. The values for all dates are then plotted as a function of the acquisition date. Figure 14.2 shows that the TOA reflectance values can vary greatly, mostly due to changing atmospheric conditions from year to year. Most of those variations are removed or greatly reduced by performing atmospheric corrections.

TABLE 14.2

Parameters Used in Equation 14.1 for Stretching the Surface Reflectance Images to Create JPEG Images for Use in the Movie Loop

	Refl_min (%)	Refl_max (%)
Band 2	0	15
Band 3	0	15
Band 4	2	50

Source: Huang, C. et al. *Int J Digital Earth*, 2, 3, 2009. With permission.

Note: Only standard false color images were created using the TM/ETM+ bands 4, 3, and 2 shown in red, green, and blue colors.

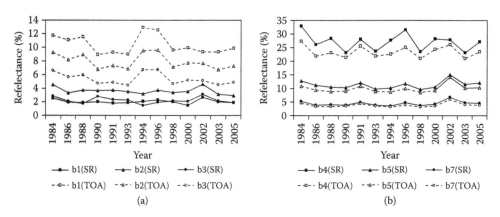

FIGURE 14.2

A comparison of the temporal variability of the top-of-atmospheric (TOA) reflectance and the LEDAPS surface reflectance (SR) for (a) bands 1–3 and (b) bands 4, 5, and 7 for conifer stands within the WRS path 16/row 36 LTSS that did not experience major disturbances. The LEDAPS atmospheric adjustment substantially reduced the temporal variability in the visible bands.

14.3 Vegetation Change Tracker Algorithm

14.3.1 Overview of the Algorithm

The VCT algorithm is developed based on the following spectral–temporal properties of forest, disturbance, and postdisturbance recovery processes:

- Due to light absorption by green vegetation and canopy shadowing, forest is one of the darkest vegetated surfaces in satellite images acquired during the leaf-on growing season in visible and some shortwave infrared bands (Colwell 1974; Goward, Huemmrich, and Waring 1994; Huemmrich and Goward 1997; Kauth and Thomas 1976).

- During the mid-growing season, undisturbed forests typically maintain relatively stable spectral signatures over many years, while most nonforest land-cover types have more spectral variability, both seasonally and interannually.

- Most forest disturbance events result in a sudden reduction or removal of forest canopy cover and woody biomass and are often manifested by abrupt spectral changes.

- Depending on the nature of a disturbance, the resultant change signal in the spectral data can last several years or longer. This is because reestablishment of a new forest stand due to a disturbance takes time, or no forest stand will be reestablished if that disturbance results in a conversion from a forest to a nonforest land-cover type.

The algorithm consists of two major steps: individual image masking and normalization and time-series analysis (Figure 14.3).

Note that because VCT uses the spectral–temporal information as recorded to detect forest disturbances in an LTSS, the detected disturbances are not necessarily limited to those defined in forestry or ecology. In VCT, a disturbance refers to any event that can result in significant reduction or removal of forest canopy cover and woody biomass, including

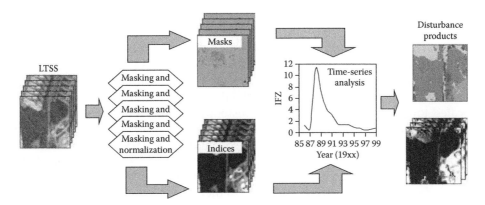

FIGURE 14.3
Overall data flow and processes of the vegetation change tracker algorithm. (From Huang, C. et al. *Remote Sens Environ*, 114, 1, 2010. With permission.)

harvest, selective logging, tree reduction for fuel treatment or other purposes, and damages due to fire, storm, insects, or diseases, although not all these events can be detected reliably by the current version of the VCT. Throughout this chapter, recovery, regrowth, and regeneration are used interchangeably, referring to the recovery process of a forest stand from a nonstand replacement disturbance, or the reestablishment of a new forest stand from a stand-clearing disturbance.

14.3.2 Individual Image Masking and Normalization

In this step, each image is analyzed individually to create initial masks for water, cloud, and shadow and to normalize the image using known forest samples. This step has the following major processes: creation of a land–water mask, identification of forest samples, calculation of forest indices, and masking of cloud and cloud shadow.

14.3.2.1 Land–Water Masking

Based on the known spectral properties of typical water bodies (Jensen 1996), a pixel is flagged as a water pixel if it has a low reflectance value in the shortwave infrared band (band 5) and satisfies at least one of the following two conditions:

1. It has a decreasing trend of reflectance values from the visible to the infrared bands.
2. It has a low normalized difference vegetation index (NDVI) value, where NDVI is calculated using the reflectance value of the red (R_{red}) and near-infrared (R_{NIR}) bands:

$$\text{NDVI} = \frac{R_{NIR} - R_{red}}{R_{NIR} + R_{red}} \tag{14.3}$$

14.3.2.2 Identification of Forest Samples

Although the LTSS images have been corrected to achieve high levels of radiometric integrity, VCT uses forest samples to further normalize image radiometry and to calculate forest likelihood measures. Such forest samples are identified based on the known spectral properties of forest. Specifically, dense, mature forests typically appear dark and green in a true color composite imagery and are among the most easily distinguishable features in remote sensing imagery (Dodge and Bryant 1976). As such, some of them can be identified reliably using histograms created from local image windows (e.g., 5 × 5 km). Because forest pixels are typically the darkest vegetated pixels, they are generally located toward the lower end of each histogram. When a local image window has a significant portion of forest pixels, those pixels form a peak called a "forest peak" in the histogram. In the absence of water, dark soil, and other dark nonvegetated surfaces, which are masked out using appropriately defined NDVI and brightness threshold values, forest pixels are delineated using threshold values defined by the forest peak. Huang et al. (2008) described this approach in detail in their study.

14.3.2.3 Calculation of Forest Indices

The identified forest samples are used to calculate a number of indices that are indicative of the likelihood of each pixel being a forest pixel. Suppose the mean and standard deviation of the band i spectral values of forest samples within an image are $\bar{b}_i$ and SD_i, respectively, then, for any pixel with a band i value of b_i, a forest z-score (FZ_i) value for the band can be calculated as follows:

$$FZ_i = \frac{b_i - \bar{b}}{SD_i} \tag{14.4}$$

For multispectral satellite images, an integrated forest z-score (IFZ) value for that pixel is defined by integrating FZ_i over the spectral bands, as follows:

$$IFZ = \sqrt{\frac{1}{NB} \sum_{l=1}^{NB} (FZ_i)^2} \tag{14.5}$$

where NB is the number of bands used. For Landsat TM and ETM+ images, bands 3, 5, and 7 are used to calculate the IFZ. Bands 1 and 2 are not used because they are highly correlated with band 3. The near-infrared band is not included in the IFZ calculation because (1) it is less sensitive to logging and other nonfire disturbances than the other spectral bands and (2) spectral changes in this band do not always correlate with disturbance events.

A major problem with using SD_i calculated from forest samples within each individual image is that the value can vary greatly as a function of the forest type composition in that Landsat image. The SD_i calculated this way will be low for images consisting of forest pixels that are spectrally similar but can be very high for images consisting of both open canopy forests with bright backgrounds and closed canopy forests. Such a dependency of the SD_i and hence the IFZ on the composition of forest types within each image makes it difficult to develop generic change detection algorithms for use over a wide range of forest biomes. To mitigate this problem, the average standard deviation values derived using images acquired in different years from different places of the United States are used in Equations 14.4 and 14.5 (Table 14.2). The FZ_i and IFZ indices calculated using Equations 14.4 and 14.5 have a number of appealing properties:

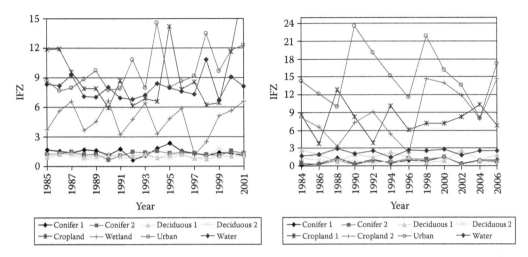

FIGURE 14.4
IFZ values of different land-cover types in eastern Virginia (left, WRS path 15/row 34) and Oregon (right, WRS path 45/row 29) show that deciduous and conifer forests have IFZ values that are generally below 3 and are different from those of other land-cover types (except for some water). (From Huang, C. et al. *Remote Sens Environ*, 114, 1, 2010. With permission.)

- IFZ is an inverse measure of the likelihood of a pixel being a forest pixel. Pixels having a low IFZ value near 0 are close to the spectral center of forest samples, while those having high IFZ values are likely nonforest pixels (Figure 14.4).

- Assuming forest pixels have a normal distribution in the spectral space, FZ_i could be directly related to the probability of a pixel being a forest pixel using the standardized normal distribution table (SDST; Davis 1986). As the root mean square of FZ_i, IFZ can be interpreted similarly. Specifically, over 99% of forest pixels likely have IFZ values less than 3. Although in reality forests may not have a rigorous normal distribution, and the standard deviation values used here are not calculated from the image of interest, such an approximate probability interpretation makes it possible to define probability-based threshold values that might be applicable to images acquired on different dates over different locations.

- While deciduous and coniferous forests often have different spectral characteristics, during the growing season they have similar IFZ values that are substantially more stable over time and are mostly lower than those of nonforest land-cover types (Figure 14.4). This observation makes it possible to detect forest changes using the IFZ index without knowing the forest type, although the differences between the IFZ values of different forest types can be greater than those shown in Figure 14.4 (see also Section 14.3.3.2).

In addition to the FZ_i and IFZ indices, the VCT also calculates a normalized burn ratio index (NBRI):

$$\text{NBRI} = \frac{R_{\text{NIR}} - R_7}{R_{\text{NIR}} + R_7} \tag{14.6}$$

where R_{NIR} and R_7 are reflectance in the near-infrared band (band 4) and band 7, respectively. NBRI is correlated with field-measured burn severity indices (Chen et al. 2008; Escuin, Navarro, and Fernandez 2008) and is used to improve the detection of fire disturbance events using VCT.

14.3.2.4 Cloud and Shadow Masking

Although the major goal of image selection in LTSS development is to minimize cloud contamination due to frequent cloudy conditions in many areas, some LTSS images inevitably contain cloudy pixels. Cloudy pixels generally have high brightness values and low greenness values. If unflagged, most likely they will be mapped as nonforest regardless of the actual surface conditions beneath the clouds. For forest change analysis, unflagged clouds over forests likely will be mapped as forest disturbance. Cloud shadow over forests may also be mapped as disturbance, because as the spectral signature of forests under shadow can be quite different from that of sunlit forests. The cloud masking algorithm used in the VCT is based on the observation that clouds generally appear bright in reflective bands and cold in thermal bands and can be separated from cloud-free observations using threshold values defined by a set of linear boundaries in a spectral-temperature space. Once a cloud patch is flagged, its height is calculated using its temperature and a normal lapse rate (Smithson, Addison, and Atkinson 2008). The shadow location of the cloud is then predicted according to solar illumination geometry and the calculated cloud height; the dark pixels at or near the predicted shadow location are flagged as actual shadow. Details on this cloud and shadow algorithm have been provided by Huang et al. (in press).

14.3.3 Time-Series Analysis

After the masking and normalization steps are completed for all images in an LTSS, temporal interpolation is used to derive interpolated values for the pixels flagged as cloud, shadow, or another bad observation. The resultant masks and indices are then used to determine change and no-change classes and to derive a suite of attributes to characterize the changes that are mapped.

14.3.3.1 Temporal Interpolation

Pixels contaminated by cloud or cloud shadow should not be used in change analysis because they could result in spurious changes. However, ignoring such pixels in the analysis will result in holes in the derived change products. To avoid this problem, VCT uses temporal interpolation to derive interpolated values for pixels flagged as cloud or shadow. Specifically, for each pixel masked as cloud or cloud shadow in a particular year i, the temporally nearest noncloud and nonshadow observations acquired before (p) and after (n) year i are used to calculate its value as follows:

$$x_i = x_p + (i-p) \times \frac{x_n - x_p}{n-p} \tag{14.7}$$

where x is any of the indices calculated in Section 14.3.2. If no noncloud, nonshadow observation can be found in the years before (or after) the current acquisition year, then the value for the current year is set to that of the temporally nearest noncloud, nonshadow observation acquired after (or before) year i. For each pixel flagged as cloud or shadow, temporal interpolation is applied to all the indices calculated in Section 14.3.2.

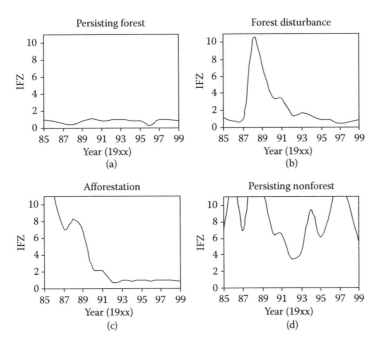

FIGURE 14.5
Typical IFZ temporal profiles of major forest cover change processes (a–c) and nonforest (d) that are used to characterize different change processes. (From Huang, C. et al. *Remote Sens Environ*, 114, 1, 2010. With permission.)

14.3.3.2 Determination of Change and No-Change Classes

Time-series analysis of forest cover and change is based mainly on the physical interpretation of the IFZ. Because the IFZ measures the likelihood of a pixel being a forest pixel, its value should change in response to forest change. Figure 14.5 shows typical temporal profiles of the IFZ for major land-cover and forest change processes. For persisting forest land where no major disturbance occurred during the years being monitored (throughout this chapter the term "persisting" indicates that the cover type of a pixel remained the same during the entire observing period), the IFZ value is low and relatively stable throughout the monitoring period (Figures 14.4 and 14.5a). During any year a sharp increase in the IFZ value indicates the occurrence of a disturbance in that year. A sequence of gradually decreasing IFZ values following that disturbance represents the regeneration process of a new forest stand (Figure 14.5b). Conversion from nonforest to forest (afforestation) or regeneration of a forest stand from a disturbance that occurred before the first LTSS acquisition is documented by a gradual decrease in the IFZ from high values to the level of undisturbed forests (Figure 14.5c). While certain crops may be spectrally similar to forest and can have low IFZ values during certain seasons, their IFZ values likely will fluctuate greatly as surface conditions change from one year to another due to harvest and crop rotation (Figures 14.4 and 14.5d).

Based on these distinctive IFZ temporal profiles of different land-cover and forest change processes, decision rules are used to identify persisting land-cover types and to detect disturbances in a sequence of steps (Figure 14.6).

Step 1: *Persisting water*—Pixels identified as water by the water masks created in Section 14.3.2.1 for all acquisition years are classified as persisting water. Because some water bodies may be turbid during some seasons and may not be masked as water during

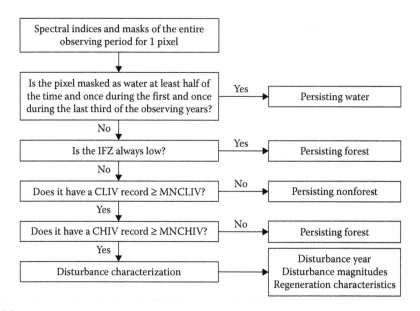

FIGURE 14.6
Major steps and decision rules used by the vegetation change tracker to determine persisting land-cover types and forest disturbance classes. (From Huang, C. et al. *Remote Sens Environ*, 114, 1, 2010. With permission.)

that season, any pixel that is masked as water at least half of the time and is also masked as water at least once during the first third of the observing period and once during the last third of the observing period is also classified as persisting water.

Step 2: *Persisting forest*—Pixels not classified as persisting water in step 1 are further analyzed in this step. Persisting forest pixels are characterized by having low IFZ values throughout the entire observing period. Based on the approximate probability interpretation of the IFZ (see Section 14.3.2.3), most forests with closed or near-closed canopy cover should have IFZ values of less than a threshold value of 3. In the next two steps, this threshold value is used to separate low and high IFZ values.

Step 3: *Persisting nonforest*—Pixels not classified as persisting forest in step 2 are further analyzed in this step. While most persisting nonforest pixels have high and often temporally variable IFZ values, some of them can be spectrally similar to certain forest pixels and can have low IFZ values during a particular season of a year. The likelihood of most nonforest pixels to have consecutive low IFZ values (CLIV), however, is low (Figure 14.4). Therefore, if a pixel has a CLIV record, it was likely a forest pixel at least during the years when it had the CLIV record. Such a pixel is referred to as a *once-forested pixel*. Obviously, the longer the CLIV record a pixel has, the more likely the pixel was a once-forested pixel and less likely a persisting nonforest pixel. In VCT, a minimum number of consecutive low IFZ values (MNCLIV) is used to determine whether a pixel is a persisting nonforest pixel or was once forested. A pixel is classified as a persisting nonforest pixel if its longest CLIV record is less than the MNCLIV threshold value.

Step 4: *Disturbance detection*—A pixel not classified as any of the three persisting land-cover classes in the previous steps should be once-forested and should have consecutive low IFZ values during the years when the pixel remained forested. A disturbance typically results in a sharp increase in the IFZ values (Figure 14.5b).

Unfortunately, an increase in the IFZ can also result from noisy observations, including unflagged cloud, shadow, and instrument- or processing-related errors, because these noisy observations typically have high IFZ values. With each LTSS consisting of carefully selected Landsat images (Huang, Goward, Masek, et al. 2009), the likelihood of a pixel having unflagged data quality problems in consecutive acquisition years should be low. Therefore, if a pixel remained forested before and after a noisy observation, the noisy observation most likely will result in a spike in the IFZ temporal profile, that is, a high IFZ value preceded and immediately followed by low IFZ values, not consecutive high IFZ values. On the other hand, most disturbances, especially those leading to significant losses of forest canopy and live biomass, likely will result in consecutive high IFZ values (CHIV).

A conversion from forest to a nonforest land-cover type typically should result in nonforest signals (i.e., mostly high IFZ values) in the years following that disturbance event. For a disturbance followed by forest regeneration, including reestablishment of urban trees in areas converted from forest to an urban environment, the IFZ should remain high until the young trees grow to a stage such that they spectrally look like forest.

Therefore, VCT uses the CHIV record following an IFZ hike to determine whether the increase was caused by a noisy observation or a disturbance. Only an IFZ hike followed by a CHIV record at least as long as a predefined minimum number of consecutive high IFZ values (MNCHIV) is mapped as a disturbance.

Note that open canopy forests with bright backgrounds typically have IFZ values much higher than those of closed canopy forests and likely will be classified as persisting nonforest using the IFZ threshold value of 3 as defined in step 2. To minimize this problem, for images consisting of both closed and open canopy forest types, steps 2 through 4 are performed twice. In the first iteration, the initial IFZ threshold value of 3 is used to characterize forests and disturbances for areas having closed canopy forests. Pixels classified as persisting nonforest in the first iteration are reanalyzed in the second iteration, during which the IFZ threshold value in step 2 is relaxed for better characterization of forest and disturbances for areas having open canopy forests. Based on extensive examination of various sparse forests in the semi-arid western United States, the IFZ threshold value is set to 6.5 in the second iteration.

Some fires, especially understory fires, do not always result in high IFZ values. The VCT uses the NBRI to detect fire pixels. Because fires typically result in low NBRI values (Escuin, Navarro, and Fernandez 2008), they are detected by searching for significant decreases in the NBRI temporal profile.

For disturbances that occurred at the beginning of the time series, there may not be a CLIV record that satisfies the MNCLIV criterion. Likewise, disturbances that occurred at the end of the time series will not have a CHIV record that satisfies the MNCHIV criterion. Therefore, use of the MNCLIV and MNCHIV threshold values will not allow detection of such disturbances. To alleviate this problem, the MNCLIV is relaxed for disturbances that occurred at the beginning of the time series, and the MNCHIV criteria is relaxed for disturbances that occurred at the end of the time series. If a pixel reaches step 4 but its longest CHIV record is shorter than the MNCHIV and the CHIV record is not at the end of the time series, it is classified as a persisting forest.

14.3.3.3 *Disturbance Characterization*

For each disturbance detected by the VCT, a disturbance year and several disturbance magnitude measures are calculated to characterize the disturbance. In addition, two attributes are used to summarize the IFZ profile that follows the disturbance, which may be useful for characterizing the regeneration process that may follow each detected disturbance. These attributes are defined in Figure 14.7.

Disturbance year: For a typical forest disturbance, the IFZ value increases sharply following a CLIV record (Figure 14.5b). While the disturbance occurs somewhere between the acquisition dates of two consecutive images that exhibit a sharp increase, the disturbance year is defined by the acquisition year of the second image. Regeneration from a disturbance that occurred before the first LTSS acquisition is indicated by an initially high IFZ value, which decreases gradually in subsequent years and remains low for several consecutive years (Figure 14.5c). This category of disturbance is called *preseries disturbance.*

Disturbance magnitude measures: Disturbance magnitude refers to the spectral change resulting from a disturbance. For the multispectral Landsat images, different disturbance magnitude measures can be calculated using different spectral bands or indices. The VCT calculates three disturbance magnitudes, with the first calculated using the IFZ, the second using the NDVI, and the third using the NBRI. Figure 14.7 shows the calculation of the IFZ disturbance magnitude. Note that other spectral indices, such as the band-specific FZ_i values or the tasseled cap indices (Crist and Cicone 1984; Huang et al. 2002), can also be used to calculate disturbance magnitudes. Further studies are needed to determine the disturbance magnitude measures that can better characterize the nature and intensity of detected disturbances.

Regeneration characteristics: If forest regeneration occurred after a disturbance, the regeneration process is tracked by a *regeneration curve* (Figure 14.7). This curve refers to the portion of the IFZ profile from the disturbance year to a target measurement year later than the disturbance year, or to the year before the next disturbance if that disturbance occurred before the target measurement year. Further

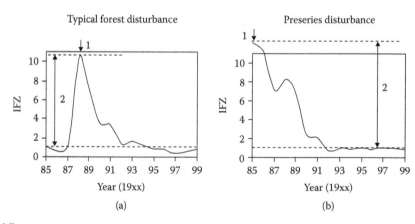

FIGURE 14.7
Disturbance year (1), disturbance magnitude (2), and the regeneration curve for (a) a typical forest disturbance and (b) a preseries disturbance. The regeneration curve is the part of the IFZ profile for the period from the disturbance year (1) to a target measurement years later than the disturbance year, or to the year before the next disturbance if that disturbance occurred before the target measurement year. The disturbance magnitude is calculated as the difference between the two dashed lines representing the IFZ value at the disturbance year and the average IFZ value of consecutive low IFZ values. (From Huang, C. et al. *Remote Sens Environ*, 114, 1, 2010. With permission.)

studies are needed to evaluate how this curve tracks the growing process of forests in terms of biomass accumulation and height growth.

Note that in regions like the southeastern United States where certain tree species grow fast enough to allow more than one forest harvest during the observing period of an LTSS, some fields may experience more than one disturbance. In such cases, the VCT detects all disturbances and calculates the above-described attributes for each detected disturbance.

14.4 Algorithm Assessment and Example Applications

The LTSS-VCT approach has been used in many areas of the United States, including Mississippi (Li et al. 2009b), Alabama (Li et al. 2009a), and 30 locations across the United States where LTSS have been assembled through the North American Forest Dynamics (NAFD) project (Goward et al. 2008; Huang, Goward, Masek et al. 2009). The disturbance maps derived at selected locations have been evaluated both qualitatively and quantitatively. This section briefly describes the assessments of the disturbance products and summarizes two applications of the LTSS-VCT approach, one for Mississippi and Alabama and the other for seven national forests (NFs) in the eastern United States (Figure 14.8). More details on the assessments of disturbance products are provided by Huang et al. (2010) and Thomas et al. (forthcoming), and detailed descriptions of these two applications have been provided by Li et al. (2009a,b) and Huang, Goward, Schleewis, et al. (2009).

14.4.1 Products Assessment

The comprehensive validation of the entire suite of VCT products as described in Section 14.3.3.3 was found to be extremely challenging. Linking the disturbance magnitude measures to changes in biomass or other biophysical variables requires pre- and post-disturbance measurements obtained using methods that would allow reliable retrieval of those variables. Similarly, linking the regeneration curve to vegetation biophysical changes associated with regeneration processes requires multitemporal reference data sets on those biophysical variables. Such reference data sets will likely be scarce, especially for older disturbances that occurred in the 1990s and 1980s, although their availability has yet to be better understood. Therefore, current efforts have focused on the disturbance year product, including limited field assessment, visual assessment, and design-based accuracy assessment to evaluate VCT disturbance products.

14.4.1.1 Field-Based Assessment

The disturbance year maps have been assessed through limited field trips conducted in Virginia (path 15/row 34, October 2005), Mississippi and Alabama (path 21/row 37 and path 21/row 39, May 2007), and Oregon (path 45/row 49, July 2007). Ground data collected through these field trips were generally inadequate for assessing all disturbance classes. However, they allowed a better understanding of the nature of the mapped disturbances and the processes that occurred following those disturbances. Field evidence was easy to find for recent disturbance events that resulted in complete or near-complete removal of the forest canopy, including harvest and urban development. Older disturbances that occurred years before the field trips and were followed by regeneration of new forest stands were often evidenced by the existence of young, even-aged forests, and the disturbance

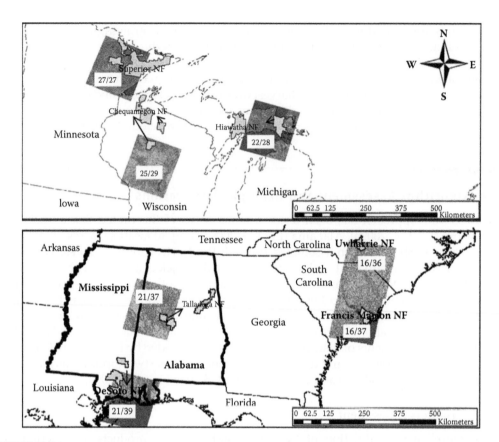

FIGURE 14.8
Locations of Mississippi and Alabama as well as the seven national forests that were selected as the study areas in the two example applications of the LTSS-VCT approach described in this chapter. The NFs are shown in gray polygons. The images (with the WRS path/row shown as two numbers separated by a slash) in the background are disturbance maps produced using the LTSS-VCT approach. (From Huang, C. et al. *Remote Sens Environ*, 113, 7, 2009. With permission.)

year was roughly reflected by the height of the regenerating trees. Likely due to vigorous growth of understory vegetation in Virginia, Mississippi, and Alabama, evidence of less-intensive disturbances such as storm damage, insect or disease defoliation, and selective logging was difficult to find in these areas. Due to slow vegetation growth under dry environmental conditions on the eastern side of the Cascades in Oregon, many stumps left from recent and old (likely >5 years) selective loggings were found during the field trip, and some selective loggings were mapped successfully by the VCT.

14.4.1.2 Visual Assessment

Because the spectral change signals of most forest disturbances can be identified reliably by experienced image analysts (Huang et al. 2008; Masek et al. 2008), especially when images acquired immediately before and after the occurrence of those disturbances are available (Cohen et al. 1998), visual inspection of the disturbances mapped by VCT against the input Landsat images can provide an immediate and still reliable way to evaluate those disturbances (Figure 14.9). Based on this observation, the VCT disturbance year maps were

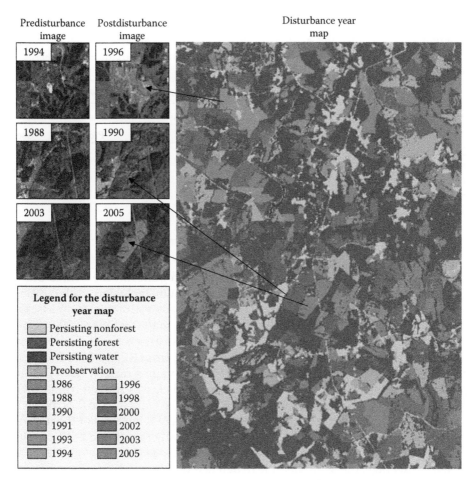

FIGURE 14.9
(See color insert following page 426.) Visual validation of three mapped disturbances using pre- and post-disturbance Landsat images. The disturbance year map was selected from a 17.1 × 11.4 km area in the Uwharrie national forest located in North Carolina (WRS path 16/row 36). The size of each Landsat image chip shown to the left is 2.85 × 2.85 km. (From Huang, C. et al. *Remote Sens Environ*, 113, 7, 2009. With permission.)

evaluated qualitatively and quantitatively. Qualitative assessments included visual inspection of most of the maps generated by the VCT. When suspicious changes were noticed, the pre- and postdisturbance Landsat images were inspected to determine whether those changes were real. These qualitative assessments revealed that most of the disturbance maps were quite reasonable. Here, a "reasonable" disturbance map was defined as follows: the map had minimum speckles; for human disturbance events such as harvest and logging, the mapped disturbance polygons had regular shapes or linear features that were often the results of human activities and for natural disturbances such as fire and storm, the disturbance patches typically had irregular shapes.

14.4.1.3 Design-Based Accuracy Assessment

To obtain quantitative estimates of the accuracies of the disturbance year maps produced by the VCT, a design-based accuracy assessment was conducted over six sites selected

TABLE 14.3

Overall Accuracy, Kappa, and Average Producers' and Users' Accuracy Values of the VCT Disturbance Year Products Assessed for All Land Cover and Disturbance Year Classes Seen in Those Products

WRS Path/ Row	Overall Accuracy	Kappa	Average Accuracy of Individual Classes			
			All Classes		Disturbance Classes	
			Producers'	Users'	Producers'	Users'
12/31	0.85	0.76	0.67	0.67	0.63	0.63
15/34	0.80	0.75	0.67	0.78	0.62	0.77
21/37	0.78	0.74	0.64	0.79	0.59	0.79
27/27	0.77	0.67	0.64	0.80	0.60	0.79
37/34	0.86	0.43	0.31	0.55	0.24	0.52
45/29	0.84	0.73	0.57	0.72	0.52	0.70

to represent different forest biomes and disturbance regimes in the United States. For each site, reference samples were selected with known inclusion probabilities, and those probabilities were considered in deriving accuracy estimates. Such assessments would allow unbiased inference on the accuracy of a map (Stehman 2000). Table 14.3 summarizes the accuracy estimates derived from these assessments. It shows that the disturbance year maps had overall accuracies ranging from 0.77 to 0.86. Except for the southern Utah site (WRS path 37/row 34), the kappa values ranged from 0.67 to 0.76. The producer's and user's accuracies averaged over all classes ranged from 0.57 to 0.67 and 0.67 to 0.80, respectively, and ranged from 0.52 to 0.63 and 0.63 to 0.79, respectively, when averaged over the disturbance classes. The average accuracies of the disturbance classes indicate that although those classes were typically rare (up to 1–3% of total area per disturbance year) as compared with no-change classes (Masek et al. 2008; Lunetta et al. 2004), on average, the VCT was able to detect more than half the disturbances with relatively low levels (i.e., 21–37% for five validation sites) of false alarm.

14.4.2 Forest Disturbance in Mississippi and Alabama

Mississippi and Alabama are located next to each other in the deep south of the United States, having a total land area of 125,443 km^2 and 135,775 km^2, respectively. Eighteen WRS path/row tiles are required to cover these two states. Both states are heavily forested, and forestry is a vital component of their economy. Forest management activities and hurricanes and storms along the Gulf coast are the major drivers of forest change. To quantify the rate of forest change in the two states, an LTSS consisting of approximately one image every 2 years from 1984 to 2007 was assembled for each of 18 WRS path/row tiles; the images were then analyzed using the VCT to produce disturbance products and to calculate forest fragmentation metrics (Li et al. 2009a, b). A wall-to-wall disturbance map showing the most recent disturbances was produced for each state by mosaicking the maps at the WRS path/row tile level (Figure 14.10). The results revealed that the two states had widespread disturbances in some coastal areas in recent years, most of which were likely the result of Hurricane Katrina and other tropical storms. Most of the disturbances mapped in inland areas were stand-clearing harvest. The two states had roughly the same level of disturbance rates with similar trends (Figure 14.11). The average annual

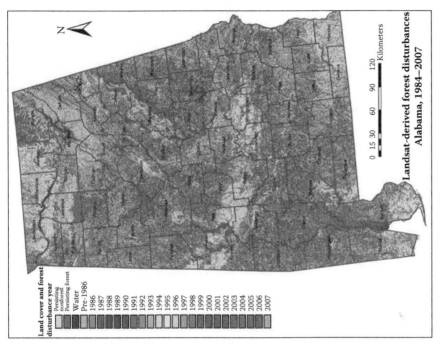

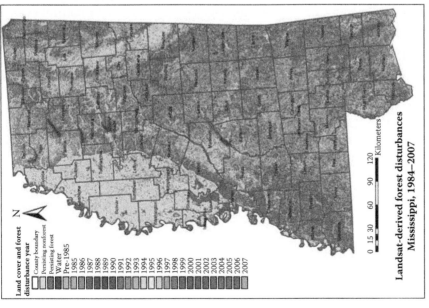

FIGURE 14.10

(See color insert following page 426.) Disturbance year map derived using the LTSS-VCT approach for Mississippi (left) and Alabama (right).

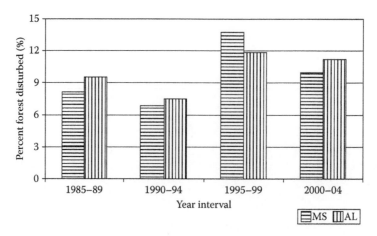

FIGURE 14.11

5-year cumulative disturbance rates (percent forest area disturbed) derived using the LTSS-VCT approach in Mississippi and Alabama.

disturbance rate was 1.98% or 1592 km^2 for Mississippi, each year from 1985 to 2004, and for Alabama, 2.02% or 1970 km^2 for the same period.

14.4.3 Dynamics of National Forests in the Eastern United States

The NFs in the United States are managed for multiple purposes, including outdoor recreation, rangeland, timber, watershed, and wildlife and fish (USDA 2007). They are subject to disturbances arising from various management activities and natural events such as fire, storms, insects, and diseases. Continuous monitoring of forest changes arising from such disturbances is essential for assessing the conditions of the NFs and the effectiveness of management approaches. The sample areas selected through the NAFD project (Goward et al. 2008; Huang, Goward, Masek, et al. 2009) covered or intersected with seven NFs in the eastern United States, including the De Soto National Forest in Mississippi, Talladega National Forest in Alabama, Francis Marion National Forest in South Carolina, Uwharrie National Forest in North Carolina, Chequamegon National Forest in Wisconsin, Hiawatha National Forest in Michigan, and the Superior National Forest in Minnesota (Figure 14.8). The disturbance maps produced by NAFD project using biennial LTSS allowed an assessment of these NFs and their surrounding areas (Huang, Goward, Schleeweis, et al. 2009). Specifically, the results showed that each of the seven NFs consisted of 90% or more forest land. During the observing period of 1984–2006, about 30–45% of the land pixels in four NFs in the southeastern United States and 10–20% in three NFs in the northern United States were disturbed at least once. For each NF, three buffer zones, defined at 0–5 km, 5–10 km, and 10–15 km from the boundary of the NF, generally had lower percentages of forest land than that within the NF, and the proportion of disturbed forest in the buffer zones were considerably higher than that within the NF. Temporally, the annual disturbance rates varied considerably both within the boundary and in the three buffer zones of each NF. Except for the Uwharrie National Forest, where no obvious trend was found as to whether the NF experienced higher or lower disturbance rates than its buffer zones, the disturbance rates within the other NFs were generally lower than in their buffer zones during most of the years of the observing period of each LTSS.

14.5 Summary and Conclusions

The Landsat record provides a unique data source for understanding the dynamics of land cover and the related surface properties for the decades, dating back to the early 1970s. This chapter presents an approach for reconstructing forest disturbance history over the last few decades using the Landsat record. In this approach, LTSS consisting of a dense time series of IRU quality Landsat observations are produced using streamlined algorithms and procedures, and forest changes are mapped with known disturbance year using the VCT algorithm. This approach has been used to produce disturbance products for many areas in the United States. Two applications of this approach in Mississippi and Alabama and in seven NFs in the eastern United States were summarized in this chapter. Visual assessments of the disturbance year products derived using this LTSS-VCT approach revealed that most of them were reasonably reliable. Design-based accuracy assessment revealed that overall accuracies of around 80% were achieved for disturbances mapped at individual disturbance year level. Average user's and producer's accuracies of the disturbance classes were around 70% and 60% for five of the six validation sites, respectively, suggesting that although forest disturbances were typically rare as compared with no-change classes, on average the VCT was able to detect more than half of those disturbances with relatively low levels of false alarms. Field assessment revealed that VCT was able to detect most stand-clearing disturbance events, including harvest, fire, and urban development, while some non-stand-clearing events such as thinning and selective logging were also mapped in the western United States.

In addition to the disturbance year products for characterizing the occurrence of disturbances, using spectral indices the VCT algorithm also calculates several change magnitude measures and tracks postdisturbance processes. Validation or calibration of these measures and indices requires a time series of ground measurements or other types of reference data sets that match the LTSS acquisitions temporally. Existing reference data sets will not likely be adequate for this purpose, although their availability has yet to be better understood. Obtaining a time series of reference data sets suitable for calibrating or validating data products derived using dense satellite observations should be one of the major goals in planning future reference data collection efforts.

The ability to reconstruct forest disturbance history using the LTSS-VCT approach for a given area depends on the availability of a long-term satellite data record consisting of quality, temporally frequent acquisitions for that area. Based on knowledge gained through the NAFD project and an in-depth analysis of the USGS Landsat archive (Goward et al. 2006), most areas in the United States have some Landsat images for use with the LTSS-VCT. An inventory of Landsat holdings at international ground-receiving stations will be needed in order to determine the feasibility of assembling an LTSS at a specific temporal interval for regions outside the United States. To ensure that long-term records of global forest disturbance history can be reconstructed in the future, it is necessary to develop the satellite capabilities today that will allow acquisition of adequate Landsat or Landsat-class images for making at least one cloud-free composite during the peak growing season of every year.

Acknowledgments

Development of the LTSS-VCT approach was made possible through funding support from NASA's Terrestrial Ecology, Carbon Cycle Science, and Applied Sciences programs, the U.S. Geological Survey, and the LANDFIRE project, which was sponsored by the intergovernmental Wildland Fire Leadership Council of the United States. The NAFD study contributes to the North American Carbon Program. The author wishes to thank Drs. Samuel Goward, Zhiliang Zhu, Jeffrey Masek, and his other collaborators at the University of Maryland, the U.S. Geological Survey Earth Resources Observation and Science (EROS) Center, the NASA Goddard Space Flight Center, and the U.S. Forest Service for their support of the studies described in this chapter.

References

Band, L. E. 1993. Effect of land surface representation on forest water and carbon budgets. *J Contam Hydrol* 150:749–72.

Barsi, J. A., S. J. Hook., J. R. Schott, N. G. Raqueno, and B. L. Markham. 2007. Landsat-5 thematic mapper thermal band calibration update. *IEEE Geosci Remote Sens Lett* 4(4):552–5.

Chander, G., D. L. Helder, B. L. Markham, J. D. Dewald, E. Kaita, K. J. Thome, E. Micijevic, and T. A. Ruggles. 2004. Landsat-5 TM reflective-band absolute radiometric calibration. *IEEE Trans Geosci Remote Sens* 42(12):2747–60.

Chander, G., and B. Markham. 2003. Revised Landsat-5 TM radiometric calibration procedures and postcalibration dynamic ranges. *IEEE Trans Geosci Remote Sens* 41(11):2674–7.

Chen, X., Z. Zhu, D. Ohlen, C. Huang, and H. Shi. 2008. Use of multiple spectral indices to estimate burn severity in the Black Hills of South Dakota. Paper read at Pecora 17: The Future of Land Imaging ... Going Operational, Denver, Colorado, November 18–20, 2008.

Cohen, W. B., T. K. Maiersperger, M. Fiorella, T. A. Spies, R. J. Alig, and D. R. Oetter. 2002. Characterizing 23 years (1972–95) of stand replacement disturbance in western Oregon forests with Landsat imagery. *Ecosystems* 5(2):122–37.

Cohen, W. B., F. Maria, B. John, H. Eileen, and A. Karen. 1998. An efficient and accurate method for mapping forest clearcuts in the Pacific Northwest using Landsat imagery. *Photogramm Eng Remote Sens* 64(4):293–300.

Colwell, J. E. 1974. Vegetation canopy reflectance. *Remote Sens Environ* 3:174–83.

Coppin, P., and M. Bauer. 1996. Digital change detection in forested ecosystems with remote sensing imagery. *Remote Sens Rev* 13:234–7.

Coppin, P., E. Lambin, I. Jonckheere, K. Nackaerts, and B. Muys. 2004. Digital change detection methods in ecosystem monitoring: A review. *Int J Remote Sens* 25(9):1565–96.

Crist, E. P., and R. C. Cicone. 1984. A physically-based transformation of Thematic Mapper data—the TM Tasseled Cap. *IEEE Trans Geosci Remote Sens* GE–22(3):256–63.

Davis, J. C. 1986. *Statistics and data analysis in geology.* New York: John Wiley & Sons.

DeFries, R., A. Hansen, A. C. Newton, and M. C. Hansen. 2005. Increasing isolation of protected areas in tropical forests over the past twenty years. *Ecol Appl* 15(1):19–26.

Dodge, A., and E. Bryant. 1976. Forest type mapping with satellite data. *J For* 74:23–40.

Escuin, S., R. Navarro, and P. Fernandez. 2008. Fire severity assessment by using NBR (Normalized Burn Ratio) and NDVI (Normalized Difference Vegetation Index) derived from LANDSAT TM/ETM images. *Int J Remote Sens* 29(4):1053–73.

Gao, F., J. Masek, and R. Wolfe. 2009. An automated registration and orthorectification package for Landsat and Landsat-like data processing. *J Appl Remote Sens* 3,033515, doi: 10.1117/1.3104620: doi: 10.1117/1.3104620.

Giambelluca, T. W., M. A. Nullet, A. D. Ziegler, and L. Tran. 2000. Latent and sensible energy flux over deforested land surfaces in the eastern Amazon and northern Thailand. *J Trop Geogr* 21(2):107–30.

Goward, S. N. 2006. Future of land remote sensing: What is needed. Paper read at Joint Agency Commercial Imagery Evaluation (JACIE) Workshop, Patuxent National Wildlife Research Center, Laurel, MD, March 14–16, 2006.

Goward, S. N., K. F. Huemmrich, and R. H. Waring. 1994. Visible-near infrared spectral reflectance of landscape components in western Oregon. *Remote Sens Environ* 47:190–203.

Goward, S., J. Irons, S. Franks, T. Arvidson, D. Williams, and J. Faundeen. 2006. Historical record of Landsat global coverage: Mission operations, NSLRSDA, and international cooperator stations. *Photogramm Eng Remote Sens* 72(10):1155–69.

Goward, S. N., G. M. Jeffrey, C. Warren, M. Gretchen, J. C. George, H. Sean, H. Richard et al. 2008. Forest disturbance and North American carbon flux. *EOS Trans Am Geophys Union* 89(11):105–6.

Hirsch, A. I., S. L. William, A. H. Richard, A. S. Neal, and D. W. Joseph. 2004. The net carbon flux due to deforestation and forest re-growth in the Brazilian Amazon: Analysis using a process-based model. *Glob Chang Biol* 10(5):908–24.

Huang, C., W. Bruce, H. Collin, Y. Limin, and Z. Gregory. 2002. Derivation of a tasseled cap transformation based on Landsat 7 at-satellite reflectance. *Int J Remote Sens* 23(8):1741–8.

Huang, C., S. N. Goward, J. G. Masek, G. Feng, E. F. Vermote, N. Thomas, K. Schleeweis et al. 2009. Development of time series stacks of Landsat images for reconstructing forest disturbance history. *Int J Digital Earth* 2(3):195–218.

Huang, C., S. N. Goward, J. G. Masek, N. Thomas, Z. Zhu, and J. E. Vogelmann. 2010. An automated approach for reconstructing recent forest disturbance history using dense Landsat time series stacks. *Remote Sens Environ* 114(1):183–98.

Huang, C., S. N. Goward, K. Schleeweis, N. Thomas, J. G. Masek, and Z. Zhu. 2009. Dynamics of national forests assessed using the Landsat record: Case studies in eastern U.S. *Remote Sens Environ* 113(7):1430–42.

Huang, C., S. Kuan, K. Sunghee, J. R. G. Townshend, D. Paul, M. Jeffrey, and S. N. Goward. 2008. Use of a dark object concept and support vector machines to automate forest cover change analysis. *Remote Sens Environ* 112:970–85.

Huang, C., N. Thomas, S. N. Goward, M. Jeffrey, Z. Zhu, J. R. G. Townshend, and J. E. Vogelmann. In press. Automated masking of cloud and cloud shadow for forest change analysis. *Int J Remote Sens* doi:10.1080/01431160903369642.

Huemmrich, K. F., and S. N. Goward. 1997. Vegetation canopy PAR absorptance and NDVI: An assessment for ten tree species with the SAIL model. *Remote Sens Environ* 61:254–69.

Jensen, J. R. 1996. *Introductory Digital Image Processing: A Remote Sensing Perspective.* Englewood Cliffs, NJ: Prentice-Hall.

Kaufman, Y. J., A. E. Wald, L. A. Remer, B.-C. Gao, R.-R. Li, and L. Flynn. 1997. The MODIS 2.1 mm Channel—Correlation with visible reflectance for use in remote sensing of aerosol. *IEEE Trans Geosci Remote Sens* 35:1286–98.

Kauth, R. J., and G. S. Thomas. 1976. The tasseled cap—a graphic description of the spectral-temporal development of agricultural crops as seen in Landdat. Paper read at Proceedings on the Symposium on Machine Processing of Remotely Sensed Data, West Lafayette, Indiana, June 29–July 1, 1976.

Kennedy, R. E., W. B. Cohen, and T. A. Schroeder. 2007. Trajectory-based change detection for automated characterization of forest disturbance dynamics. *Remote Sens Environ* 110(3):370–86.

Kinnaird, M. F., E. W. Sanderson, T. G. O'Brien, H. T. Wibisono, and W. Gillian. 2003. Deforestation trends in a tropical landscape and implications for endangered large mammals. *Conserv Biol* 17(1):245–57.

Landsat Project Science Office. 2000. *Landsat 7 Science Data User's Handbook*. National Aeronautics and Space Administration. http://landsathandbook.gsfc.nasa.gov/handbook/handbook_toc.html (accessed on March 31, 2008).

Law, B. E., S. Van Tuyl, W. D. Ritts, W. B. Cohen, D. Turner, J. Campbell, and O. J. Sun. 2004. Disturbance and climate effects on carbon stocks and fluxes across western Oregon, USA. *Glob Chang Biol* 10(9):1429–44.

Li, M., C. Huang, Z. Zhu, H. Shi, H. Lu, and S. Peng. 2009a. Assessing rates of forest change and fragmentation in Alabama, USA, using the vegetation change tracker model. *For Ecol Manage* 257(6):1480–8.

Li, M., C. Huang, Z. Zhu, H. Shi, H. Lu, and S. Peng. 2009b Use of remote sensing coupled with a vegetation change tracker model to assess rates of forest change and fragmentation in Mississippi, USA. *Int J Remote Sens* 30(24):6559–74.

Lo, C. P., and X. Yang. 2002. Drivers of land-use/land-cover changes and dynamic modeling for the Atlanta, Georgia metropolitan area. *Photogramm Eng Remote Sens* 68(10):1073–82.

Lu, D., P. Mausel, E. Brondízio, and E. Moran. 2004. Change detection techniques. *Int J Remote Sens* 25(12):2365–407.

Lunetta, R. S., D. M. Johnson, J. G. Lyon, and J. Crotwell. 2004. Impacts of imagery temporal frequency on land-cover change detection monitoring. *Remote Sens Environ* 89(4):444–54.

Markham, B. L., and J. L. Barker. 1986. Landsat MSS and TM postcalibration dynamic ranges, exoatmospheric reflectances and at-satellite temperatures. *EOSAT Landsat Tech Note* 1:3–8.

Markham, B. L., J. C. Storey, D. L. Williams, and J. R. Irons. 2004. Landsat sensor performance: History and current status. *IEEE Trans Geosci Remote Sens* 42(12):2691–4.

Masek, J. G., C. Huang, W. Cohen, J. Kutler, F. Hall, and R. E. Wolfe. 2008. Mapping North American forest disturbance from a decadal Landsat record: Methodology and initial results. *Remote Sens Environ* 112:2914–26.

Masek, J. G., E. F. Vermote, N. E. Saleous, R. Wolfe, F. G. Hall, K. F. Huemmrich, F. Gao, J. Kutler, and L. Teng-Kui. 2006. A Landsat surface reflectance dataset for North America, 1990–2000. *IEEE Geosci Remote Sens Lett* 3(1):68–72.

Peterken, G. F. 2001. Structural dynamics of forest stands and natural processes. In *The Forest Handbook*, ed. J. Evans. Malden, MA: Blackwell Science.

Sahin, V., and M. J. Hall. 1996. The effects of afforestation and deforestation on water yields. *J Hydrol* 178:293–309.

Schwartz, M. D., B. C. Reed, and M. A. White. 2002. Assessing satellite-derived start-of-season measures in the conterminous USA. *Int J Climat* 22(14):1793–805.

Singh, A. 1989. Digital change detection techniques using remotely-sensed data. *Int J Remote Sens* 10(6):989–1003.

Smithson, P., K. Addison, and K. Atkinson. 2008. *Fundamentals of the Physical Environment*, 4th ed. London: Routledge.

Stehman, S. V. 2000. Practical implications of design-based sampling inference for thematic map accuracy assessment. *Remote Sens Environ* 72:35–45.

Teillet, P. M., D. L. Helder, T. A. Ruggles, R. Landry, F. J. Ahern, N. J. Higgs, J. A. Barsi, et al. 2004. A definitive calibration record for the Landsat-5 thematic mapper anchored to the Landsat-7 radiometric scale. *Can J Remote Sens* 30(4):631–43.

Thomas, N., C. Huang, S. N. Goward, S. Powell, K. Rishmawi, K. Schleeweis, and H. Adrienne. Forthcoming. Assessment of NAFD forest change products derived from Landsat Time Series Stacks (LTSS). *Remote Sens Environ*.

Toutin, T. 2003. Geometric correction of remotely sensed images. In *Methods and Applications for Remote Sensing of Forests: Concepts and Case Studies*, ed. M. Wulder and S. Franklin. Kluwer Academic Publishers, Boston, MA.

Townshend, J. R. G., and C. O. Justice. 1988. Selecting the spatial resolution of satellite sensors required for global monitoring of land transformations. *Int J Remote Sens* 9:187–236.

USDA. 2007. *USDA Forest Service Strategic Plan FY 2007–2012*. United States Department of Agriculture Forest Service, Washington, DC.

Vermote, E. F., N. Z. El Saleous, and C. O. Justice. 2002. Atmospheric correction of MODIS data in the visible to middle infrared: First results. *Remote Sens Environ* 83(1–2):97–111.

Vermote, E. F., D. Tanre, J. L. Deuze, M. Herman, and J. J. Morcrette. 1997. Second simulation of the satellite signal in the solar spectrum, 6S: An overview. *IEEE Trans Geosci Remote Sens* 35:675–86.

Vicente-Serrano, S. M., F. Pérez-Cabello, and T. Lasanta. 2008. Assessment of radiometric correction techniques in analyzing vegetation variability and change using time series of Landsat images. *Remote Sens Environ* 112(10):3916–34.

Waring, R. H., and S. W. Running. 1998. *Forest Ecosystems: Analysis at Multiple Scales*, 2nd ed. New York: Academic Press.

Zartman, C. E. 2003. Habitat fragmentation impacts on epiphyllous bryophyte communities in central Amazonia. *Ecology* 84(4):948–54.

Zhang, X., A. H. Strahler, J. C. F. Hodges, F. Gao, B. C. Reed, A. Huete, M. A. Friedl, and C. B. Schaaf. 2003. Monitoring vegetation phenology using MODIS. *Remote Sens Environ* 84(3):471–75.

15

Satellite-Based Modeling of Gross Primary Production of Terrestrial Ecosystems

Xiangming Xiao, Huimin Yan, Joshua Kalfas, and Qingyuan Zhang

CONTENTS

15.1 Introduction

Plant photosynthesis occurs within the chloroplasts of plant leaves and is composed of two distinct processes: (1) light absorption, that is, chlorophyll absorbs photosynthetically active radiation (PAR, mostly visible spectrum) from sunlight; and (2) carbon fixation, that is, the absorbed energy is then used to combine water and CO_2 to produce sugar (Figure 15.1). Plant photosynthesis is well understood at the chloroplast and leaf levels through direct measurements by instruments (Taiz and Zeiger 2002). However, there is no direct instrument-based measurement of plant photosynthesis at

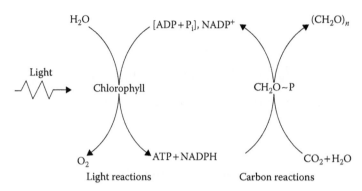

FIGURE 15.1
Schematic diagram of plant photosynthesis that illustrates the light absorption process and the carbon fixation process.

the canopy and landscape scales, and how to scale up photosynthesis from individual leaves to the canopy and landscape is still a challenging and hotly debated topic. At the canopy and landscape scales, photosynthesis is often termed gross primary production (GPP).

Application of the eddy covariance technique to measure the net ecosystem exchange (NEE) of CO_2 between terrestrial ecosystems and the atmosphere dates back to 1974 (Shaw et al. 1974). In 1990, the first year-long continuous CO_2 flux measurements using the eddy covariance technique were conducted at the Harvard forest site in Massachusetts (Wofsy et al. 1993). CO_2 flux tower sites provide integrated CO_2 flux measurements over footprints with sizes and shapes (linear dimensions typically ranging from hundreds of meters to several kilometers) that vary with the tower height, canopy physical characteristics, and wind velocity (Baldocchi et al. 1996). Continuous measurements of the CO_2 NEE between terrestrial ecosystems and the atmosphere through the eddy covariance technique have allowed for more detailed study of ecosystem respiration and GPP at ecosystem and landscape scales (Wofsy et al. 1993). NEE between the terrestrial ecosystem and the atmosphere, as measured at a half-hourly frequency throughout a year, is the difference between the GPP and the ecosystem respiration (R_e):

$$NEE = GPP - R_e \qquad (15.1)$$

Since the early 1990s, more than 600 eddy flux tower sites have been established, covering all major biome types in the world. It is important to note that the footprint sizes of CO_2 eddy flux towers are comparable with the spatial resolution of several major satellite observation platforms (e.g., Moderate Resolution Imaging Spectroradiometer [MODIS], SPOT-4/Vegetation). Therefore, several studies have compared the dynamics of satellite-derived vegetation indices with CO_2 fluxes from flux towers, with a goal to establish a linkage between the ecosystem metabolism (CO_2 flux) and satellite-based observations of vegetation dynamics (Xiao et al. 2004). Due to the changes in climate, soils, land use, and management, however, there are still great uncertainties in estimating seasonal dynamics and spatial variation of GPP at the canopy and landscape scales.

Satellite-based optical remote sensing platforms (e.g., Landsat, Advanced Very High Resolution Radiometer [AVHRR], SPOT-4/Vegetation, and MODIS) provide frequent observations of the land surface of the entire Earth, and the radiometric values recorded by the optical sensors are associated with the biophysical and biochemical properties of vegetation and soils. As the mathematical transformations are calculated using different spectral bands (e.g., red and near-infrared [NIR]), vegetation indices have been widely used to track vegetation dynamics at the land surface. For example, the normalized difference vegetation index (NDVI), which is calculated from the red and NIR bands of the National Oceanic and Atmospheric Administration (NOAA) AVHRR sensors, is now the longest time-series data record for vegetation study (Myneni et al. 1997; White et al. 2005). NDVI is calculated as follows:

$$NDVI = \frac{\rho_{NIR} - \rho_{red}}{\rho_{NIR} + \rho_{red}} \tag{15.2}$$

In the early 1970s, a satellite-based production efficiency model was first proposed to estimate the net primary production (NPP) using photosynthetically active radiation absorbed (APAR) by the vegetation canopy ($APAR_{canopy}$) and the radiation use efficiency (Monteith 1972, 1977). Since then, a number of models driven by satellite images have been developed to estimate the GPP and NPP (Potter et al. 1993; Field et al. 1995; Prince and Goward 1995; Running et al. 1994; Running et al. 2004; Xiao et al. 2004; Sims et al. 2008; Sims et al. 2006). Satellite-based models of GPP were largely founded on the concept of light-use efficiency (LUE). Depending upon their approaches to estimating APAR for photosynthesis, these production efficiency models (PEMs) can be grouped into two categories based on how they calculate light absorption for photosynthesis: (1) those using the fraction of photosynthetically active radiation (FPAR) absorbed by vegetation canopy ($FPAR_{canopy}$); and (2) those using the FPAR absorbed by chlorophyll ($FPAR_{chl}$). This chapter aims to provide a brief review of satellite-based PEMs and to highlight the major differences between these two approaches ($FPAR_{canopy}$ and $FPAR_{chl}$). The following discussion is composed of the following: (1) the concepts of leaf area index (LAI), $FPAR_{canopy}$, and $APAR_{canopy}$ and a brief introduction of two PEMs built upon the concept of $FPAR_{canopy}$, (2) the concept of chlorophyll, $FPAR_{chl}$, and $APAR_{chl}$, (3) a detailed description of the vegetation photosynthesis model (VPM) that is built upon the concept of $FPAR_{chl}$, and (4) a case study of VPM simulation results from a cropland site.

15.2 Leaf Area Index, APARcanopy, and FPARcanopy

LAI, $APAR_{canopy}$, and $FPAR_{canopy}$ have all been a focus of both the ecology and the remote sensing communities over the past few decades. A number of remote sensing studies have been conducted to develop quantitative relationships between the NDVI and LAI, and between the NDVI and $FPAR_{canopy}$ (Prince and Goward 1995; Ruimy et al. 1999). Approaches based on the NDVI-LAI and NDVI-$FPAR_{canopy}$ relationships have been the dominant paradigm at the crossroads of the fields of remote sensing science and ecology, for example, satellite-based PEMs (Figure 15.2).

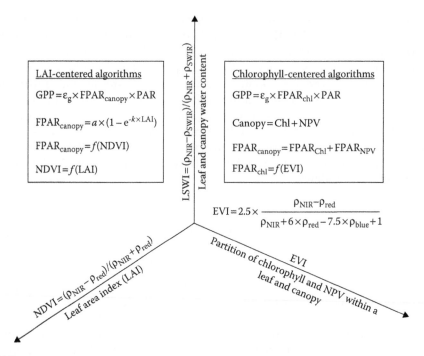

FIGURE 15.2
A simple comparison between two paradigms of production efficiency models. GPP = gross primary production; NDVI = normalized difference vegetation index; FPAR = fraction of photosynthetically active radiation; PAR = photosynthetically active radiation; EVI = enhanced vegetation index; LSWI = land surface water index; Chl = chlorophyll; NPV = nonphotosynthetic vegetation; NIR = near-infrared; and SWIR = shortwave infrared.

A number of satellite-based PEMs use the concept of $FPAR_{canopy}$ to estimate the GPP and NPP (Potter et al. 1993; Field et al. 1995; Prince and Goward 1995; Running et al. 2004). GPP is calculated as follows:

$$GPP = \varepsilon_g \times FPAR_{canopy} \times PAR \tag{15.3}$$

where ε_g is the LUE for photosynthesis or GPP. Brief descriptions of two models are provided in Sections 15.2.1 and 15.2.2.

15.2.1 Global Production Efficiency Model

The global production efficiency model (GLO-PEM) estimates both the GPP and NPP based on the production efficiency approach (see Equation 15.3). It has several linked components that describe the processes of canopy radiation absorption, utilization, autotrophic respiration, and the regulation of these processes by environmental factors (Prince and Goward 1995; Goetz et al. 2000). The GLO-PEM uses NDVI to estimate $FPAR_{canopy}$ (see Goward and Huemmrich 1992 for more details):

$$FPAR_{canopy} = 1.08 \times NDVI - 0.08 \tag{15.4}$$

In the GLO-PEM, ε_g is estimated through a modeling approach based on plant physiological principles (Prince and Goward 1995). Plant photosynthesis depends on both the capacity of the photosynthetic enzymes to assimilate CO_2 (Collatz et al. 1991; Farquhar et al. 1980) and the stomatal conductance of CO_2 from the atmosphere into the intercellular spaces (Harley et al. 1992). These two processes are affected by environmental factors, such as air temperature, water vapor pressure deficit, soil moisture, and atmospheric CO_2 concentration. Detailed descriptions of approaches for modeling ε_g have been provided in many earlier publications (Prince and Goward 1995; Collatz et al. 1991; Goetz and Prince 1998; Collatz et al. 1992; Goetz and Prince 1999).

15.2.2 MODIS Daily Photosynthesis Model

The photosynthesis (PSN) model uses Equation 15.3 to estimate GPP, but ε_g and $FPAR_{canopy}$ are derived using different methods (Running et al. 2004; Running et al. 1999; Running et al. 2000). $FPAR_{canopy}$ is produced as a part of the MOD15 (LAI and FPAR) product suite. In MOD17, a set of biome-specific maximum LUE parameters is extracted from the biome properties lookup table (Running et al. 2000).

$$\varepsilon_g = \varepsilon_0 \times T_{scalar} \times W_{scalar} \qquad (15.5)$$

where ε_0 is the maximum LUE, T_{scalar} is estimated as a function of daily minimum temperature, and W_{scalar} is estimated as a function of daylight average water vapor pressure deficit. In this approach, biome is defined according to the MODIS land-cover product (MOD12) (Running et al. 2004; Running et al. 2000; Friedl et al. 2002).

15.3 Chlorophyll, Light Absorption by Chlorophyll, and $FPAR_{chl}$

From the biochemical perspective, vegetation canopies are composed of chlorophyll (chl) and nonphotosynthetic vegetation (NPV). The latter includes both canopy-level (e.g., stem, senescent leaves) and leaf-level (e.g., cell walls, vein, and other pigments) materials. Therefore, $FPAR_{canopy}$ should be partitioned into $FPAR_{chl}$ and FPAR absorbed by NPV ($FPAR_{NPV}$) (Xiao et al. 2004a,b; Xiao et al. 2005a).

$$Canopy = chlorophyll + NPV \qquad (15.6)$$

$$FPAR_{canopy} = FPAR_{chl} + FPAR_{NPV} \qquad (15.7)$$

How much difference is there between $FPAR_{canopy}$ and $FPAR_{chl}$ in a vegetation canopy? Does the difference between $FPAR_{canopy}$ and $FPAR_{chl}$ change over the plant growing season? Using a radiative transfer model (PROSAIL2) and daily MODIS data, results from temperate deciduous forests (Zhang et al. 2005; Zhang et al. 2006) have shown that $FPAR_{canopy}$ is significantly larger than $FPAR_{chl}$, and the difference between $FPAR_{canopy}$ and $FPAR_{chl}$ changes as much as 30%–40% over the plant growing season (Figure 15.3).

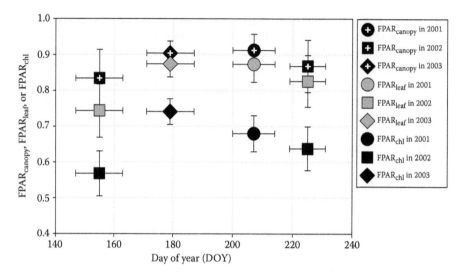

FIGURE 15.3
A comparison of the fractions of photosynthetically active radiation absorbed by canopy, leaf, and chlorophyll (FPAR$_{canopy}$, FPAR$_{leaf}$, and FPAR$_{chl}$ respectively), as illustrated in a deciduous broadleaf forest at the Harvard forest site, Massachusetts (see Zhang et al. 2005 for more details).

As shown in Figure 15.1, photosynthesis starts with light absorption by leaf chlorophyll. Only the PAR absorbed by chlorophyll (product of PAR × FPAR$_{chl}$) is responsible for photosynthesis or GPP. Based on the conceptual partitioning of chlorophyll and NPV within a leaf and canopy, the VPM was developed for estimating GPP over the photosynthetically active period of vegetation (Xiao et al. 2004a). The VPM is briefly described as follows:

$$GPP = \varepsilon_g \times FPAR_{chl} \times PAR \qquad (15.8)$$

This biochemical approach, based on the chlorophyll–FPAR$_{chl}$ relationship, is currently an emerging paradigm in the field of remote-sensing–based PEMs, and other additional models have been developed using the FPAR$_{chl}$ concept (Sims et al. 2008; Sims et al. 2006; Mahadevan et al. 2008). Figure 15.2 summarizes the major differences between FPAR$_{canopy}$ and FPAR$_{chl}$ approaches in estimating light absorption and GPP.

15.4 Detailed Description of the Vegetation Photosynthesis Model

15.4.1 Model Input Data

15.4.1.1 Satellite Data

The satellite-based VPM uses two vegetation indices as input data: the enhanced vegetation index (EVI) and the land surface water index (LSWI). These vegetation indices differ from the widely used NDVI (Equation 15.2). NDVI is often applied in PEMs to estimate the vegetation productivity of terrestrial ecosystems (Field et al. 1995; Prince and Goward 1995; Nemani et al. 2003). It is known that NDVI suffers several limitations, including sensitivity to atmospheric conditions, sensitivity to soil background (e.g., soil moisture), and saturation of index values in multilayered and closed canopies (Xiao et al. 2004). EVI directly adjusts the reflectance in the red band as a function of the reflectance in the blue band, accounting for residual atmospheric contamination (e.g., aerosols), variable soil, and canopy background reflectance (Huete et al. 1997):

$$\text{EVI} = \frac{G(\rho_{\text{NIR}} - \rho_{\text{red}})}{\rho_{\text{NIR}} + (C_1\rho_{\text{red}} - C_2\rho_{\text{blue}}) + L} \tag{15.9}$$

where G is 2.5, C_1 is 6, C_2 is 7.5, and L is 1, and ρ_{NIR}, ρ_{red}, and ρ_{blue} are land surface reflectances of the NIR, red, and blue bands, respectively.

Because the shortwave infrared (SWIR) spectral band is sensitive to vegetation water content and soil moisture, a combination of NIR and SWIR bands has been used to derive water-sensitive vegetation indices (Xiao et al. 2004b; Ceccato et al. 2002a,b; Ceccato et al. 2001). LSWI is calculated as the normalized difference between NIR and SWIR spectral bands (Xiao et al. 2002):

$$\text{LSWI} = \frac{\rho_{\text{NIR}} - \rho_{\text{SWIR}}}{\rho_{\text{NIR}} + \rho_{\text{SWIR}}} \tag{15.10}$$

where ρ_{NIR} and ρ_{SWIR} are reflectances of the NIR and the SWIR band, respectively.

Satellite images from two advanced optical sensors (vegetation onboard SPOT-4 satellite and MODIS onboard Terra satellite) have blue, red, NIR, and SWIR bands, which allow the calculation of EVI and LSWI indices. EVI and LSWI have now been used widely to characterize the growing conditions of vegetation (Zhang et al. 2003; Boles et al. 2004).

15.4.1.2 Climate Data

The climate input data sets for the VPM include daily minimum temperature (°C), daily maximum temperature (°C), and the daily sum of PAR (mol/day). The daily climate data come from either *in situ* measurements (e.g., CO_2 flux tower sites, weather stations) or climate model simulations (e.g., NCEP Reanalysis climate data), depending upon the availability of climate data (Zhang et al. 2009; Zhao et al. 2005; Raich et al. 1991).

15.4.2 Estimation of Vegetation Photosynthesis Model Parameters

15.4.2.1 Light Absorption by Chlorophyll

In the VPM, FPAR_{chl} within the photosynthetically active period of vegetation is estimated as a linear function of EVI, and the coefficient, a, is set to be 1.0 (Xiao et al. 2004a,b):

$$\text{FPAR}_{\text{chl}} = \alpha \times \text{EVI} \tag{15.11}$$

15.4.2.2 Effect of Temperature on Gross Primary Production

Temperature affects photosynthesis; there are a number of ways to estimate the effect of temperature on photosynthesis (T_{scalar}). In the VPM, T_{scalar} is estimated at each time step using the equation developed for the terrestrial ecosystem model (Raich et al. 1991):

$$T_{scalar} = \frac{(T - T_{min})(T - T_{max})}{[(T - T_{min})(T - T_{max})] - (T - T_{opt})^2} \tag{15.12}$$

where T_{min}, T_{max}, and T_{opt} are the minimum, maximum, and optimum temperatures for photosynthetic activities, respectively. If air temperature falls below T_{min}, T_{scalar} is set to be zero. The values of the T_{min}, T_{max}, and T_{opt} parameters vary with vegetation types.

15.4.2.3 Effect of Water on Gross Primary Production

W_{scalar}, the effect of water on plant photosynthesis, has been estimated as a function of soil moisture and/or water vapor pressure deficit in a number of PEMs (Field et al. 1995; Prince and Goward 1995; Running et al. 2000). For instance, in the Carnegie Ames Stanford Approach (CASA) model, soil moisture was estimated using a one-layer bucket model (Malmstrom et al. 1997). Soil moisture represents water supply to the leaves and canopy, and water vapor pressure deficit represents evaporative demand in the atmosphere. The leaf and canopy water content is largely determined by dynamic changes of both the soil moisture and water vapor pressure deficit.

The availability of time-series data of NIR and SWIR bands from the new generation of advanced optical sensors (e.g., variable geometry turbocharger, MODIS) offers opportunities for quantifying the canopy water content at large spatial scales through both the vegetation indices approach (Ceccato et al. 2002a) and the radiative transfer modeling approach (Zarco-Tejada et al. 2003). Vegetation indices that are based on NIR and SWIR bands are sensitive to changes in equivalent water thickness (g/cm^2) at the leaf and canopy levels (Ceccato et al. 2002a,b; Ceccato et al. 2001; Hunt and Rock 1989). As a first-order approximation, the VPM uses a satellite-derived water index to estimate the seasonal dynamics of W_{scalar}:

$$W_{scalar} = \frac{1 + LSWI}{1 + LSWI_{max}} \tag{15.13}$$

where $LSWI_{max}$ is the maximum LSWI value within the plant growing season for individual pixels.

15.4.2.4 Effect of Leaf Age and Phenology on Gross Primary Production

The leaf age affects the seasonal patterns of photosynthetic capacity and NEE in deciduous forest (Wilson et al. 2001). Turner et al. (Turner et al. 2003) compared daily LUE from four CO_2 flux tower sites: an agriculture field, a tall grass prairie, a deciduous broadleaf forest, and a boreal forest. Their results suggested that parameters on cloudiness and the phenological status of vegetation should be included in modeling vegetation primary production. In the VPM, P_{scalar} is used to account for the effect of the leaf age on photosynthesis at the canopy level. Calculation of P_{scalar} is dependent upon leaf longevity (deciduous versus evergreen). For a canopy that is dominated by leaves with a life expectancy of 1 year (one growing season, e.g., deciduous trees and shrubs), P_{scalar} is calculated at two different phases:

$$P_{scalar} = \frac{1 + LSWI}{2} \text{ (From bud-burst to complete leaf expansion)} \tag{15.14}$$

$$P_{scalar} = 1 \text{ (After complete leaf expansion)} \tag{15.15}$$

Evergreen trees and shrubs have a green canopy throughout the year, because foliage is retained for several years. The canopy of evergreen forests is thus composed of green leaves of various ages. To deal with different age classes in evergreen forest canopies, fixed turnover rates of foliage of evergreen forests at the canopy level have been used in some process-based ecosystem models (Aber and Federer 1992; Law et al. 2000). For evergreen forests, we simply assume $P_{scalar} = 1$ (Xiao et al. 2004b); we also assume this for tundra, grassland, and cropland (e.g., wheat) vegetation, which have new leaves emerging through most of the plant growing season (Li et al. 2007).

15.4.2.5 Maximum LUE

LUE (ε_g) is affected by temperature, water, and leaf phenology:

$$\varepsilon_g = \varepsilon_0 \times T_{scalar} \times W_{scalar} \times P_{scalar} \tag{15.16}$$

where ε_0 is the apparent quantum yield or the maximum LUE (μmol CO_2/μmol PPFD), and T_{scalar}, W_{scalar}, and P_{scalar} are the scalars for the effects of temperature, water, and leaf phenology on the LUE of vegetation, respectively. A full description of the VPM is given elsewhere (Xiao et al. 2004b; Xiao et al. 2005a).

The maximum LUE (ε_0) for individual vegetation types can be estimated from the nonlinear analysis of the observed half-hourly NEE and incident PAR data from eddy covariance flux tower sites. In the VPM, the ecosystem-level ε_0 values vary with vegetation types. The Michaelis–Menten function (Equation 15.17) is used to estimate the ε_0 values of individual vegetation types; half-hourly NEE and PAR data for weekly to 10-day periods within the peak period of the plant growing season (e.g., from July to August) are used:

$$NEE = \frac{\alpha \times PPFD \times GPP_{max}}{\alpha \times PPFD + GPP_{max}} - R_e \tag{15.17}$$

where α is the maximum LUE or apparent quantum yield (as photosynthetic photon flux density [PPFD] approaches 0), GPP_{max} is the maximum gross ecosystem exchange, and R_e

is the ecosystem respiration. The estimated α value is used as an estimate of the ε_0 parameter in the VPM.

15.4.3 Model Evaluation

Evaluating GPP estimates at the canopy level is a challenging task. Recent progress in partitioning the observed NEE data into GPP and R_e makes it possible to directly evaluate GPP estimates from various models. Daily GPP and R_e flux data at individual flux sites are generated from half-hourly NEE flux data by the CO_2 eddy covariance flux community (Baldocchi et al. 2001; Mizoguchi et al. 2009).

15.5 Case Study Estimating Gross Primary Production of C$_4$ Maize Cropland Using the Vegetation Photosynthesis Model

The VPM has been evaluated for and applied to several major biome types, including tropical rainforests (Xiao et al. 2005b), temperate deciduous broadleaf forests (Xiao et al. 2004a; Wu et al. 2009), evergreen needle-leaf forests (Xiao et al. 2004b; Xiao et al. 2005a), alpine tundra (Li et al. 2007), grassland (Wu et al. 2008), and winter wheat–maize croplands (Yan et al. 2009). Here, we present a case study of C$_4$ maize cropland to illustrate the model simulation.

The Rosemount G21 site is located at the University of Minnesota's Rosemount Research and Outreach Center, approximately 25 km south of St. Paul, MN (Baker and Griffis 2005). The site has silty loam soil with a surface layer of high organic carbon content. It has a temperate continental climate, and the plant growing season begins in May and ends in October. In 2005, maize was planted at this site, and no irrigation occurred during the entire growing season.

In this case study, MODIS 8-day Land Surface Reflectance (MOD09A1) data sets were downloaded from the EROS Data Center of the U.S. Geological Survey (http://www.edc.usgs.gov/). The MODIS sensor onboard the National Aeronautics and Space Administration (NASA) Terra satellite has 36 spectral bands. Seven spectral bands are primarily designed for the study of vegetation and land surface: blue (459–479 nm), green (545–565 nm), red (620–670 nm), NIR (841–875 nm, 1230–1250 nm), and SWIR (1628–1652 nm, 2105–2155 nm). The reflectance values of four spectral bands (blue, red, NIR [841–875 nm], and SWIR [1628–1652 nm]) during 2005 were used to calculate vegetation indices (NDVI, EVI, and LSWI). Time series of vegetation indices for one MODIS pixel, within which an eddy covariance flux tower is located, were used for the VPM simulation. In this case, the extent of the flux tower's footprint (<250 m radius) is approximately the same as one MODIS pixel (a 500-m square).

We obtained level-4 data for the Rosemount G21 site from the AmeriFlux network Web site (http://public.ornl.gov/ameriflux/), including climate data (air temperature, PAR) and CO_2 flux data. Level-4 data contain gap-filled and u*-filtered NEE data, with GPP and R_e calculated at hourly, daily, weekly, and monthly intervals. Flags regarding the quality of the original and gap-filled data were also provided within the data set.

Figure 15.4 shows the seasonal dynamics of PAR, mean air temperature, precipitation, three vegetation indices (NDVI, EVI, and LSWI), and GPP for 2005. The seasonal dynamics

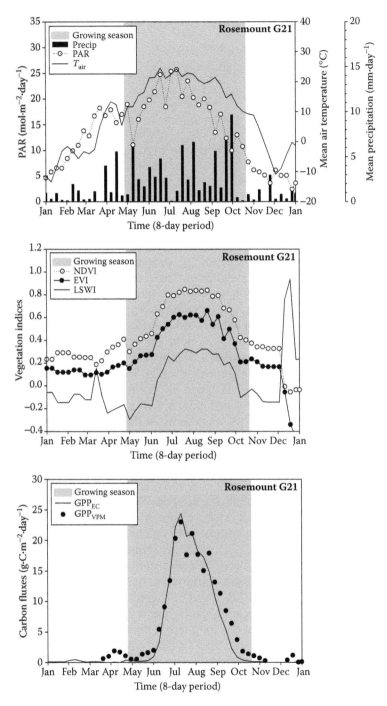

FIGURE 15.4
Seasonal dynamics of (a) mean air temperature (°C), photosynthetically active radiation (mol PPFD day^{-1}), and precipitation; (b) normalized difference vegetation index (NDVI), enhanced vegetation index (EVI), and land surface water index (LSWI); and (c) vegetation photosynthesis model (VPM)-predicted gross primary production (GPP, g C/m^2/day) and estimated GPP from the CO_2 eddy covariance (EC) flux tower data at the Rosemount G21 flux tower site, Minnesota, in 2005. Maize was cultivated in 2005. In the VPM simulation, the model parameters are $\varepsilon_0 = 1.5$ g C/mol PPFD, $LSWI_{max} = 0.37$, $T_{min} = 0°C$, $T_{opt} = 27°C$, and $T_{max} = 48°C$.

of GPP predicted by the VPM agree reasonably well with GPP estimated by partitioning the NEE data of the flux tower site.

15.6 Summary

$FPAR_{canopy}$ can be conceptually partitioned into $FPAR_{chl}$ and $FPAR_{NPV}$, but quantifying both leaf chlorophyll content and $FPAR_{chl}$ across various terrestrial biomes over time is a challenging task. It will require (1) extensive field measurements of chlorophyll content at the leaf, canopy, and landscape levels; (2) improved radiative transfer models that couple both leaf-level biochemical properties (e.g., chlorophyll, other pigments, and dry matter) and canopy-level biophysical properties (e.g., the plant area index, stem fraction, and LAI); and (3) high-quality satellite images (Zhang et al. 2009).

A comparison between $FPAR_{chl}$ and $FPAR_{canopy}$ is an important topic for both the remote sensing and ecosystem modeling communities. The concept of $FPAR_{canopy}$ has been widely accepted in scientific literature and is useful for estimating light absorption by vegetation canopy. A radiative transfer model was applied to generate standard products of $FPAR_{canopy}$ and the LAI, using MODIS data (Myneni et al. 2002). Seasonal dynamics of resultant $FPAR_{canopy}$ data was often compared with the field-measured LAI, and the impact of $FPAR_{canopy}$ data on the seasonal dynamics of GPP in the context of an input data set for the PSN model was also assessed (Turner et al. 2006; Heinsch et al. 2006). Note that the amount of PAR absorbed by canopy ($APAR_{canopy} = FPAR_{canopy} \times PAR$) is often much larger than the amount of PAR absorbed by leaf chlorophyll ($APAR_{chl} = FPAR_{chl} \times PAR$). When $APAR_{canopy}$ is used in a GPP model, its uncertainty in light absorption for photosynthesis is likely to propagate into GPP estimates. In recent years, several new models have used $FPAR_{chl}$ or EVI to estimate light absorption by chlorophyll and GPP (Sims et al. 2008; Sims et al. 2006; Yan et al. 2009); they are more consistent with the light absorption process of photosynthesis at the chlorophyll level (Figure 15.1). Therefore, research efforts for retrieving the leaf chlorophyll content and $FPAR_{chl}$ are critically needed for reducing uncertainty in modeling GPP. The terrestrial ecosystem science and the remote sensing communities need to undertake regular measurements of both biochemical (chlorophyll, nitrogen, and $FPAR_{chl}$) and structural (LAI, $FPAR_{canopy}$) variables across the leaf, canopy, and landscape levels and develop data sets of the chlorophyll content over land ecosystems. Furthermore, because a large portion of leaf nitrogen is within leaf chloroplasts, developing quantitative relationships among chlorophyll, $FPAR_{chl}$, and nitrogen could have significant implication for reducing the uncertainty in estimating the GPP and the carbon cycle.

Validation of satellite-based PEMs is a long-term effort that requires coordination from both the remote sensing community and the CO_2 eddy covariance flux community. At present, there are more than 600 eddy covariance flux tower sites operating across various biomes of the world. These sites cover different types of land uses and management practices, as well as different stages of disturbance and recovery. The eddy flux community needs to (1) partition half-hourly NEE data into GPP and ecosystem respiration; (2) provide the GPP data to users in a timely fashion; and (3) quantify LUE in a consistent approach, using observed half-hourly NEE and PAR data and the nonlinear Michaelis–Menten equation (Equation 15.17). As there are a number of methods for estimating the GPP from the observed NEE data, additional effort is needed to reduce the error of GPP estimates from the NEE data using a method consistent across flux tower sites (Richardson et al. 2006; Falge et al. 2002).

Acknowledgments

This study was supported by research grants from the NASA EOS Data Analysis Program (NNX09AE93G), the Interdisciplinary Science program (NAG5-11160, NAG5-10135) and the Land-Cover and Land-Use Change Program (NN-H-04-Z-YS-005-N, and NNG05GH80G). The site-specific climate and CO_2 flux data for the Rosemount G21 CO_2 flux tower site were provided by the AmeriFlux network (http://public.ornl.gov/ameriflux/) and the site's principal investigators, Drs. Timothy Griffis and John Baker. We thank our reviewers for their efforts and insightful reviews of the earlier version of the manuscript.

References

Aber, J. D., and C. A. Federer. 1992. A generalized, lumped-parameter model of photosynthesis, evapotranspiration and net primary production in temperate and boreal forest ecosystems. *Oecologia* 92:463–74.

Baker, J. M., and T. J. Griffis. 2005. Examining strategies to improve the carbon balance of corn/soybean agriculture using eddy covariance and mass balance techniques. *Agric For Meteorol* 128:163–77.

Baldocchi, D. et al. 1996. Strategies for measuring and modelling carbon dioxide and water vapour fluxes over terrestrial ecosystems. *Global Change Biol* 2:159–68.

Baldocchi, D. et al. 2001. FLUXNET: A new tool to study the temporal and spatial variability of ecosystem-scale carbon dioxide, water vapor, and energy flux densities. *Bull Am Meteorol Soc* 82:2415–34.

Boles, S. H. et al. 2004. Land cover characterization of temperate East Asia using multi-temporal VEGETATION sensor data. *Rem Sens Environ* 90:477–89.

Ceccato, P. et al. 2001. Detecting vegetation leaf water content using reflectance in the optical domain. *Rem Sens Environ* 77:22–33.

Ceccato, P. et al. 2002a. Designing a spectral index to estimate vegetation water content from remote sensing data: Part 1—theoretical approach. *Rem Sens Environ* 82:188–97.

Ceccato, P. et al. 2002b. Designing a spectral index to estimate vegetation water content from remote sensing data: Part 2—validation and applications. *Rem Sens Environ* 82:198–207.

Collatz, G. J. et al. 1991. Physiological and environmental regulation of stomatal conductance, photosynthesis and transpiration: A model that includes a laminar boundary layer. *Agric For Meteorol* 54:107–36.

Collatz, G. J. et al. 1992. Coupled photosynthesis-stomatal conductance model for leaves of C_4 plants. *Aust J Plant Physiol* 19:519–38.

Falge, E. et al. 2002. Seasonality of ecosystem respiration and gross primary production as derived from FLUXNET measurements. *Agric For Meteorol* 113:53–74.

Farquhar, G. et al. 1980. A biochemical model of photosynthetic CO_2 assimilation in leaves of C_3 species. *Planta* 149:78–90.

Field, C. B. et al. 1995. Global net primary production: Combining ecology and remote-sensing. *Rem Sens Environ* 51:74–88.

Friedl, M. A. et al. 2002. Global land cover mapping from MODIS: Algorithms and early results. *Rem Sens Environ* 83:287–302.

Goetz, S. J., and S. D. Prince. 1998. Variability in carbon exchange and light utilization among boreal forest stands: Implications for remote sensing of net primary production. *Can J For Res* 28:375–89.

Goetz, S. J., and S. D. Prince. 1999. Modelling terrestrial carbon exchange and storage: Evidence and implications of functional convergence in light-use efficiency. *Adv Ecol Res* 28:57–92.

Goetz, S. J. et al. 2000. Interannual variability of global terrestrial primary production: Results of a model driven with satellite observations. *J Geophys Res Atmos* 105:20077–91.

Goward, S. N., and K. F. Huemmrich. 1992. Vegetation canopy PAR absorptance and the normalized difference vegetation index: An assessment using the SAIL model. *Rem Sens Environ* 39:119–40.

Harley, P. C. et al. 1992. Theoretical considerations when estimating the mesophyll conductance to CO_2 flux by analysis of the response of photosynthesis to CO_2. *Plant Physiol* 98:1429–36.

Heinsch, F. A. et al. 2006. Evaluation of remote sensing based terrestrial productivity from MODIS using regional tower eddy flux network observations. *IEEE Trans Geosci Rem Sens* 44:1908–25.

Huete, A. R. et al. 1997. A comparison of vegetation indices over a global set of TM images for EOS-MODIS. *Rem Sens Environ* 59:440–51.

Hunt, E. R., and B. N. Rock. 1989. Detection of changes in leaf water-content using near-infrared and middle-infrared reflectances. *Rem Sens Environ* 30:43–54.

Law, B. E. et al. 2000. Measurements of gross and net ecosystem productivity and water vapour exchange of a Pinus ponderosa ecosystem, and an evaluation of two generalized models. *Global Change Biol* 6:155–68.

Li, Z. Q. et al. 2007. Modeling gross primary production of alpine ecosystems in the Tibetan Plateau using MODIS images and climate data. *Rem Sens Environ* 107:510–9.

Mahadevan, P. et al. 2008. A satellite-based biosphere parameterization for net ecosystem CO_2 exchange: Vegetation photosynthesis and respiration model (VPRM). *Global Biogeochem Cycles* 22:GB2005, doi:10.1029/2006GB002735.

Malmstrom, C. M. et al. 1997. Interannual variation in global-scale net primary production: Testing model estimates. *Global Biogeochem Cycles* 11:367–92.

Mizoguchi, Y. et al. 2009. A review of tower flux observation sites in Asia. *J For Res* 14:1–9.

Monteith, J. L. 1972. Solar radiation and productivity in tropical ecosystems. *J Appl Ecol* 9:747–66.

Monteith, J. L. 1977. Climate and efficiency of crop production in Britain. *Philos Trans R Soc Lond B Biol Sci* 281:277–94.

Myneni, R. B. et al. 1997. Increased plant growth in the northern high latitudes from 1981 to 1991. *Nature* 386:698–702.

Myneni, R. B. et al. 2002. Global products of vegetation leaf area and fraction absorbed PAR from year one of MODIS data. *Rem Sens Environ* 83:214–31.

Nemani, R. R. et al. 2003. Climate-driven increases in global terrestrial net primary production from 1982 to 1999. *Science* 300:1560–3.

Potter, C. S. et al. 1993. Terrestrial ecosystem production: A process model-based on global satellite and surface data. *Global Biogeochem Cycles* 7:811–41.

Prince, S. D., and S. N. Goward. 1995. Global primary production: A remote sensing approach. *J Biogeogr* 22:815–35.

Raich, J. W. et al. 1991. Potential net primary productivity in South-America: Application of a global-model. *Ecol Appl* 1:399–429.

Richardson, A. D. et al. 2006. A multi-site analysis of random error in tower-based measurements of carbon and energy fluxes. *Agric For Meteorol* 136:1–18.

Ruimy, A. et al. 1999. Comparing global models of terrestrial net primary productivity (NPP): Analysis of differences in light absorption and light-use efficiency. *Global Change Biol* 5:56–64.

Running, S. W. et al. 1994. Terrestrial remote-sensing science and algorithms planned for Eos Modis. *Int J Rem Sens* 15:3587–620.

Running, S. W. et al. 1999. MODIS daily photosynthesis (PSN) and annual net primary production (NPP) product (MOD17). In *Algorithm Theorectical Basis Document, Version 3.0.* htpp://modis.gsfc.nasa.gov/1999 (accessed on September 16, 2010).

Running, S. W. et al. 2000. Global terrestrial gross and net primary productivity from the earth obse-rving system. In *Methods in Ecosystem Science*, ed. O. E. Sala, R. B. Jackson, H. A. Mooney, and R. W. Howarth, 44–57. New York: Springer Verlag.

Running, S. W. et al. 2004. A continuous satellite-derived measure of global terrestrial primary pro-duction. *Bioscience* 54:547–60.

Shaw, R. H. et al. 1974. Some observations of turbulence and turbulent transport within and above plant canopies. *Boundary-Layer Meteorol* 5:429–49.

Sims, D. A. et al. 2006. On the use of MODIS EVI to assess gross primary productivity of North American ecosystems. *J Geophys Res Biogeosciences* 111:G04015, doi:10.1029/2006JG000162.

Sims, D. A. et al. 2008. A new model of gross primary productivity for North American ecosystems based solely on the enhanced vegetation index and land surface temperature from MODIS. *Rem Sens Environ* 112:1633–46.

Taiz, L., and E. Zeiger. 2002. *Plant Physiology*, 3rd ed. Sunderland, MA: Sinauer Associates, Inc.

Turner, D. P. et al. 2003. A cross-biome comparison of daily light use efficiency for gross primary production. *Global Change Biol* 9:383–95.

Turner, D. P. et al. 2006. Assessing interannual variation in MODIS-based estimates of gross primary production. *IEEE Trans Geosci Rem Sens* 44:1899–907.

White, M. A. et al. 2005. A global framework for monitoring phenological responses to climate change. *Geophys Res Lett* 32:L04705, doi:10.1029/2004GL021961.

Wilson, K. B. et al. 2001. Leaf age affects the seasonal patterns of photosynthetic capacity and net ecosystem exchange of carbon in a deciduous forest. *Plant Cell Environ* 24:571–83.

Wofsy, S. C. et al. 1993. Net exchange of CO_2 in a mid-latitude forest. *Science* 260:1314–7.

Wu, J. B. et al. 2009. Estimation of the gross primary production of an old-growth temperate mixed forest using eddy covariance and remote sensing. *Int J Rem Sens* 30:463–79.

Wu, W. X. et al. 2008. Modeling gross primary production of a temperate grassland ecosystem in Inner Mongolia, China, using MODIS imagery and climate data. *Sci China Ser D Earth Sci* 51:1501–12.

Xiao, X. et al. 2004a. Modeling gross primary production of a deciduous broadleaf forest using satel-lite images and climate data. *Rem Sens Environ* 91:256–70.

Xiao, X. et al. 2004b. Satellite-based modeling of gross primary production in an evergreen needleleaf forest. *Rem Sens Environ* 89:519–34.

Xiao, X. M. et al. 2002. Characterization of forest types in Northeastern China, using multi-temporal SPOT-4 VEGETATION sensor data. *Rem Sens Environ* 82:335–48.

Xiao, X. M. et al. 2004. Modeling gross primary production of temperate deciduous broadleaf forest using satellite images and climate data. *Rem Sens Environ* 91:256–70.

Xiao, X. M. et al. 2005a. Modeling gross primary production of an evergreen needleleaf forest using MODIS and climate data. *Ecol Appl* 15:954–69.

Xiao, X. M. et al. 2005b. Satellite-based modeling of gross primary production in a seasonally moist tropical evergreen forest. *Rem Sens Environ* 94:105–22.

Yan, H. M. et al. 2009. Modeling gross primary productivity for winter wheat-maize double crop-ping system using MODIS time series and CO_2 eddy flux tower data. *Agric Ecosyst Environ* 129:391–400.

Zarco-Tejada, P. J. et al. 2003. Water content estimation in vegetation with MODIS reflectance data and model inversion methods. *Rem Sens Environ* 85:109–24.

Zhang, J. H. et al. 2009. Satellite-based estimation of evapotranspiration of an old-growth temperate mixed forest. *Agric For Meteorol* 149:976–84.

Zhang, Q. Y. et al. 2005. Estimating light absorption by chlorophyll, leaf and canopy in a decidu-ous broadleaf forest using MODIS data and a radiative transfer model. *Rem Sens Environ* 99:357–71.

Zhang, Q. Y. et al. 2006. Characterization of seasonal variation of forest canopy in a temperate decid-uous broadleaf forest, using daily MODIS data. *Rem Sens Environ* 105:189–203.

Zhang, Q. Y. et al. 2009. Can a satellite-derived estimate of the fraction of PAR absorbed by chlorophyll (FAPAR(chl)) improve predictions of light-use efficiency and ecosystem photosynthesis for a boreal aspen forest? *Rem Sens Environ* 113:880–8.

Zhang, X. Y. et al. 2003. Monitoring vegetation phenology using MODIS. *Rem Sens Environ* 84:471–5.

Zhao, M. S. et al. 2005. Improvements of the MODIS terrestrial gross and net primary production global data set. *Rem Sens Environ* 95:164–76.

16

Global Croplands and Their Water Use from Remote Sensing and Nonremote Sensing Perspectives

Prasad S. Thenkabail, Munir A. Hanjra, Venkateswarlu Dheeravath, and Muralikrishna Gumma

CONTENTS

16.1 Introduction

Croplands are the largest user of water worldwide. Much of the water is used for food production, making global croplands and their water use important to world food security. Of all the water used by humans, croplands consume an overwhelming proportion (60%–90%; Thenkabail et al. 2009a; Thenkabail et al. 2009b; Thenkabail et al. 2009c). There are two types of cropland water use (Falkenmark and Rockström 2006): (1) green water use by rain-fed croplands from the unsaturated soil zone and (2) blue water use by irrigated croplands from rivers, reservoirs, lakes, and from saturated zones or groundwater aquifers and rain over irrigated croplands (Hoff et al. 2010; Rockström et al. 2008). However, alternative demands for land water use are increasing steeply due to urbanization (Deyong et al. 2009), industrialization (Liu et al. 2005), environmental flows and ecosystem services (Gordon, Finlayson, and Falkenmark 2009), maintaining water quality (and associated health of the populations), and recreational and other demands (Gordon, Peterson, and Bennett 2008).

Global cropland maps and water-use assessments are identified as one of the most important variables for all areas of societal benefits in the 10-year implementation plan of the Group on Earth Observation (Herold et al. 2008). Croplands nearly stagnated in the last decade. Especially, irrigated croplands that were at the center of the "green revolution" and currently produce about 45% of the world's food from just 25% of cropland areas has stopped expanding (Thenkabail et al. 2009a). Yet the demand for food and water continues to increase rapidly. The world population is growing at nearly 100 million per year and is expected to reach 9.15 billion by 2050 (UNDP 2009; Table 16.1). Irrigated croplands that increased rapidly between the 1970s and 2000s have almost stagnated, and the era of "green revolution" (increases in crop per unit area and cropping intensity) seems nearly over while the populations of emerging markets are consuming more calories per capita, croplands are diverted to biofuel crops, and the much anticipated "blue revolution" (growing "more crop per drop") has not taken off. As a result, the world is facing a food crisis that has not been experienced since the 1970s. Investments in agricultural research and development have fallen in all developing countries and even in Organization for Economic Cooperation and Development (OECD) member countries, and the era of cheap food has come to an end as food prices are likely to continue rising until 2015 and even beyond. The current trends in world population growth, food demand, food prices, global cropland, and water competition in agriculture pose new challenges to world food security (Khan and Hanjra 2009).

Increasing the water use and cropland area for producing more food are not easy options. Alternative demands for water in other fields as well as uncertainties in water availability associated with climate change rule out such options. Increasing cropland areas is also not

TABLE 16.1

World Population (in Thousands) under All Variants, 1950–2050

Year	Medium Variant	High Variant	Low Variant	Constant-Fertility Variant
1950	2,529,346	2,529,346	2,529,346	2,529,346
1955	2,763,453	2,763,453	2,763,453	2,763,453
1960	3,023,358	3,023,358	3,023,358	3,023,358
1965	3,331,670	3,331,670	3,331,670	3,331,670
1970	3,685,777	3,685,777	3,685,777	3,685,777
1975	4,061,317	4,061,317	4,061,317	4,061,317
1980	4,437,609	4,437,609	4,437,609	4,437,609
1985	4,846,247	4,846,247	4,846,247	4,846,247
1990	5,290,452	5,290,452	5,290,452	5,290,452
1995	5,713,073	5,713,073	5,713,073	5,713,073
2000	6,115,367	6,115,367	6,115,367	6,115,367
2005	6,512,276	6,512,276	6,512,276	6,512,276
2010	6,908,688	6,908,689	6,908,687	6,908,688
2015	7,302,186	7,369,003	7,235,360	7,342,730
2020	7,674,833	7,850,649	7,498,821	7,798,900
2025	8,011,533	8,324,226	7,698,240	8,264,771
2030	8,308,895	8,762,174	7,855,775	8,741,186
2035	8,570,570	9,181,935	7,966,536	9,241,316
2040	8,801,196	9,606,206	8,024,592	9,782,041
2045	8,996,344	10,037,286	8,022,171	10,374,956
2050	9,149,984	10,461,086	7,958,779	11,030,273

Source: UNDP. 2009. *Population Division of the Department of Economic and Social Affairs of the United Nations Secretariat.* World Population Prospects: The 2008 Revision, http://esa.un.org/unpp (October 31, 2009).

an option for a number of reasons. For instance, land-use and land-cover (LULC) changes, particularly deforestation for crop production, are shown to have a stronger impact on ecosystem carbon budgets than the projected climate change scenarios (Tan et al. 2009). At current levels, agricultural croplands already account for 50% of methane (CH_4) and 60% of nitrous oxide (N_2O; Zou et al. 2009). Agricultural N_2O emissions are projected to increase by 35–60% by 2030 due to increased chemical and manure nitrogen inputs (IPCC 2007). Irrigated rice paddies are a major source of atmospheric CH_4 (Maraseni, Mushtaq, and Maroulis 2009; USEPA 2006). Conversion of natural ecosystems to croplands will also mean a loss of the already endangered or threatened flora and fauna. Inclusion of agriculture in any carbon trading scheme will exert pressure on food crops and reallocate water from crops with higher emissions to perennial crops or forest plantation grown for carbon sequestration.

The above factors raise the following central questions:

- How can the world population be fed without unsustainably increasing cropland areas and/or water allocation for food production?
- What strategies can save water from existing cropland areas so as to create "new water" (saved water from agriculture) for alternative uses?

In order to answer the above questions, we need to determine existing cropland areas and their water use. This can help better understand and plan for food production in the coming decades by sustainable allocations of cropland areas and their water-use patterns. Thereby, the goal of this chapter is to determine and discuss the current state-of-art on (1) global cropland area maps and statistics and (2) global cropland water use.

This chapter discusses the following: Global croplands and their water use are determined by remote sensing and nonremote sensing approaches, with emphasis on the former (Section 16.2). Sources of uncertainty in the areas and limitations of existing cropland maps are discussed in Sections 16.3 and 16.4. Estimates of global croplands water use are provided at a country-by-country level using a uniform and systematic framework to allow comparisons across countries (Section 16.5) as well as from regional and local studies using various remote sensing products (Section 16.6). The main conclusions and policy implications on global cropland and their water use for future food security are discussed in Section 16.7.

16.2 Global Croplands (Rain-Fed and Irrigated) Using Remote Sensing and Nonremote Sensing Approaches

Global cropland (irrigated and rain-fed) areas increased from 265 Mha in 1700 to about 1.5 Mha in 2000. Major studies on cropland area (Goldewijk et al. 2009; Portmann, Siebert, and Döll 2009; Ramankutty et al. 2008; Siebert and Döll 2009; Thenkabail et al. 2009a; Thenkabail et al. 2009c) estimated the total croplands (irrigated and rain-fed) to be about 1.5 billion hectares for the nominal year 2000. Thus, by 2000, agriculture covered about 10% of the world's terrestrial surface (148,940,000 km²).

Global cropland mapping is possible by integrating agricultural statistics and census data from national systems and spatial mapping technologies involving geographic information systems (Ramankutty et al. 2008; Figure 16.1). The availability of advanced remote sensing imagery (Table 16.2) along with secondary data and recent advances in data access,

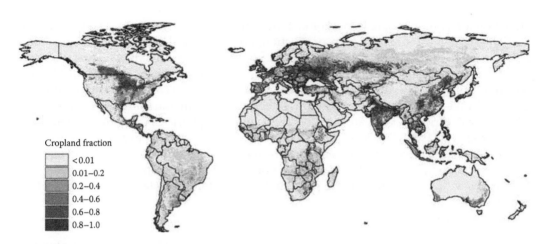

FIGURE 16.1
(See color insert following page 426.) Global cropland map at nominal 5-minutes (0.083333 decimal degrees) resolution using national statistics and geospatial techniques for the nominal year 2000. Total area of croplands is 1.47 billion hectares. (Adapted from Ramankutty, N. et al. *Global Biogeochem Cycles*, 22, 2008.)

quality, processing, and delivery have made possible the estimation of croplands based on remote sensing at the global level (Thenkabail et al. 2009a; Thenkabail et al. 2009c; Figure 16.2).

16.2.1 Remote Sensing Advances in Global Cropland Mapping

The specific remote sensing advances enabling global cropland mapping and generation of their statistics include factors such as free access to well-calibrated and guaranteed data such as Landsat and Moderate Resolution Imaging Spectrometer (MODIS), frequent temporal coverage of data such as MODIS backed by high-resolution Landsat data, free access to high-quality secondary data such as long-term precipitation, evapotranspiration, surface temperature, soils, and global digital elevation map (GDEM), global coverage of data, Web-access and broad bandwidth, and advances in computer technology and data processing.

These advances have enabled better estimates of global cropland and water use at the country level, using unified and systematic frameworks. For instance, Ramankutty et al. (2008) estimated global croplands to be 1.54 billion hectares for the nominal year 2000. Thenkabail et al. (2009a) and Thenkabail et al. (2009c) also obtained a similar estimate (1.53 billion hectares; Table 16.3). However, Portmann, Siebert, and Doll's (2009) estimates were lower (1.3 billion hectares; Table 16.3). A country-by-country comparison of cropland areas for 197 countries showed a very high correlation with an R^2 value of 0.89 (Figure 16.3) between the studies Ramankutty et al. (2008), Thenkabail et al. (2009a), and Thenkabail et al. (2009c).

16.2.2 Global Irrigated Croplands Using Remote Sensing and Nonremote Sensing Approaches

Irrigated croplands use about 80% of all blue water (water in rivers, reservoirs, lakes, aquifer groundwater, and direct rainfall) used by humans. About 45% of the total food in the world is produced from about 25% of the irrigated croplands (399 million hectares;

TABLE 16.2

Satellite Sensor Data Characteristics

Sensor	Spatial (m)	Spectral (#)	Radiometric (bit)	Band Range (μm)	Band Widths (μm)	Irradiance (Wm⁻²sr⁻¹μm⁻¹)	Data Points (# per hectares)	Frequency of Revisit (d)
Coarse-Resolution Sensors								
AVHRR	1000	4	11	0.58–0.68	0.10	1390	0.01	daily
				0.725–1.1	0.375	1410		
				3.55–3.93	0.38	1510		
				10.30–10.95	0.65	0		
				10.95–11.65	0.7	0		
MODIS	250, 500, 1000	36/7	12	0.62–0.67	0.05	1528.2	0.16, 0.04, 0.01	daily
				0.84–0.876	0.036	974.3	0.16, 0.04, 0.01	
				0.459–0.479	0.02	2053		
				0.545–0.565	0.02	1719.8		
				1.23–1.25	0.02	447.4		
				1.63–1.65	0.02	227.4		
				2.11–2.16	0.05	86.7		
Multispectral Sensors								
Landsat-1, 2, 3 MSS	56 × 79	4	6	0.5–0.6	0.1	1970	2.26	16
				0.6–0.7	0.1	1843		
				0.7–0.8	0.1	1555		
				0.8–1.1	0.3	1047		
Landsat-4, 5 TM	30	7	8	0.45–0.52	0.07	1970	11.1	16
				0.52–0.60	0.80	1843		
				0.63–0.69	0.60	1555		
				0.76–0.90	0.14	1047		
				1.55–1.74	0.19	227.1		
				10.4–12.5	2.10	0		
				2.08–2.35	0.25	80.53		

(Continued)

TABLE 16.2 (*Continued*)

Satellite Sensor Data Characteristics

Sensor	Spatial (m)	Spectral (#)	Radiometric (bit)	Band Range (µm)	Band Widths (µm)	Irradiance (W m⁻²sr⁻¹µm⁻¹)	Data Points (# per hectares)	Frequency of Revisit (d)
Multispectral Sensors								
Landsat-7 ETM+	30	8	8	0.45–0.52	0.65	1970	44.4, 11.1	16
				0.52–0.60	0.80	1843		
				0.63–0.69	0.60	1555		
				0.50–0.75	0.150	1047		
				0.75–0.90	0.200	227.1		
				10.0–12.5	2.5	0		
				1.75–1.55	0.2	1368		
				0.52–0.90	0.38	1352.71		
				(panchromatic)				
ASTER	15, 30, 90	15	812	0.52–0.63	0.11	1846.9	44.4, 11.1, 1.23	16
				0.63–0.69	0.06	1546.0		
				0.76–0.86	0.1	1117.6		
				0.76–0.86	0.1	1117.6		
				1.60–1.70	0.1	232.5		
				2.145–2.185	0.04	80.32		
				2.185–2.225	0.04	74.96		
				2.235–2.285	0.05	69.20		
				2.295–2.365	0.07	59.82		
				2.360–2.430	0.07	57.32		
				8.125–8.475	0.35	0		
				8.475–8.825	0.35	0		
				8.925–9.275	0.35	0		
				10.25–10.95	0.7	0		
				10.95–11.65	0.7	0		
ALI	30	10	12	0.048–0.69 (p)	0.64	1747.8600	11.1	16
				0.433–0.453	0.20	1849.5		
				0.450–0.515	0.65	1985.0714		
				0.425–0.605	0.80	1732.1765		
				0.633–0.690	0.57	1485.2308		
				0.775–0.805	0.30	1134.2857		
				0.845–0.890	0.45	948.36364		
				1.200–1.300	1.00	439.61905		
				1.550–1.750	2.00	223.39024		
				2.080–2.350	2.70	78.072727		

SPOT-1	2.5–20	15	16	0.50–0.59	0.09	1858	1600, 25	3–5
-2				0.61–0.68	0.07	1575		
-3				0.79–0.89	0.1	1047		
-4				1.5–1.75	0.25	234		
				0.51–0.73 (p)	0.22	1773		
IRS-1C	23.5	15	8	0.52–0.59	0.07	1851.1	18.1	16
				0.62–0.68	0.06	1583.8		
				0.77–0.86	0.09	1102.5		
				1.55–1.70	0.15	240.4		
				0.5–0.75 (p)	0.25	1627.1		
IRS-1	23.5	15	8	0.52–0.59	0.07	1852.1	18.1	16
				0.62–0.68	0.06	1577.38		
				0.77–0.86	0.09	1096.7		
				1.55–1.70	0.15	240.4		
				0.5–0.75 (p)	0.25	1603.9		
IRS-P6-AWiFS	56	4	10	0.52–0.59	0.07	1857.7	3.19	16
				0.62–0.68	0.06	1556.4		
				0.77–0.86	0.09	1082.4		
				1.55–1.70	0.15	239.84		
CBERS-2	20 m pan		11	0.51–0.73	0.22	1934.03	25, 25	
-3B	20 m MS			0.45–0.52	0.07	1787.10	400, 25	
-3	5 m pan			0.52–0.59	0.07	1587.97		
-4	20 m MS			0.63–0.69	0.06	1069.21		
				0.77–0.89	0.12	1664.3		
Hyperspectral Sensor								
Hyperion	30	196[a]	16	196 effective calibrated bands VNIR (bands 8–57) 427.55–925.85 nm SWIR (bands 79–224) 932.72–2395.53 nm	10 nm wide (approx.) for all 196 bands	See data in Neckel and Labs (1984). Plot it and obtain values for Hyperion bands.	11.1	16

(Continued)

TABLE 16.2 (*Continued*)

Satellite Sensor Data Characteristics

Sensor	Spatial (m)	Spectral (#)	Radiometric (bit)	Band Range (µm)	Band Widths (µm)	Irradiance (W m^{-2}sr^{-1}µm^{-1})	Data Points (# per hectares)	Frequency of Revisit (d)
Hyperspatial Sensor								
IKONOS	1–4	4	11	0.445–0.516	0.71	1930.9	10000, 625	5
				0.506–0.595	0.89	1854.8		
				0.632–0.698	0.66	1156.5		
				0.757–0.853	0.96	1156.9		
QUICKBIRD	0.61–2.44	4	11	0.45–0.52	0.07	1381.79	14872, 625	5
				0.52–0.60	0.08	1924.59		
				0.63–0.69	0.06	1843.08		
				0.76–0.89	0.13	81574.77		
RESOURSESAT	5.8	3	10	0.52–0.59	0.07	1853.6	33.64	24
				0.62–0.68	0.06	1581.6		
				0.77–0.86	0.09	1114.3		
RAPID EYE-A -E	6.5	5	12	0.44–0.51	0.07	1979.33	236.7	1–2
				0.52–0.59	0.07	1752.33		
				0.63–0.68	0.05	1499.18		
				0.69–0.73	0.04	1343.67		
				0.77–0.89	0.12	1039.88		
WORLDVIEW	0.55	1	11	0.45–0.51	0.06	1996.77	40000	1.7–5.9
FORMOSAT-2	2–8	5	11	0.45–0.52	0.07	1974.93	2500, 156.25	Daily
				0.52–0.60	0.08	1743.12		
				0.63–0.69	0.06	1485.23		
				0.76–0.90	0.14	1041.28		
				0.45–0.90 (p)	0.45	1450		

KOMPSAT-2	1–4	5	10	0.5–0.9	0.4	1379.46	10000, 625	3–28
				0.45–0.52	0.07	1974.93		
				0.52–0.6	0.08	1743.12		
				0.63–0.59	0.04	1485.23		
				0.76–0.90	0.14	1041.28		

Note: FORMOSAT is a Taiwanis satellite operated by Taiwanis National Space Organization NSPO; data marketed by SPOT. Hyperion is first spaceborne Hyperspectral Sensor Onboard Earth Observing-1(EO-1). IKONOS is a high-resolution satellite operated by GeoEye. KOMFOSAT is a Korean multipurpose satellite; data marketed by SPOT image. RESOURSESAT is a satellite launched by India. QUICKBIRD is a satellite from DigitalGlobe, a private company in the United States. RAPID EYE-A/E is a satellite constellation from Rapideye, a German company.

a   Adapted from Melesse, M. et al. 2007. *Sens J* 7:3209–41. http://www.mdpi.org/sensors/papers/s7123209.pdf.

b   Of 242 bands, 196 are unique and calibrated. These are band 8 (427.55 nm) to band 57 (925.85 nm) that are acquired by a visible and near-infrared (VNIR) sensor; and band 79 (932.72 nm) to band 224 (2395.53 nm) that are acquired by a shortwave infrared (SWIR) sensor.

ASTER = Advanced Spaceborne Thermal Emission and Reflection Radiometer; ALI = Advanced Land Imager; AVHRR = Advanced Very High Resolution Radiometer; CBERS-2 = China–Brazil Earth Resources Satellite; IRS-1C/D-LISS = Indian Remote Sensing Satellite/Linear Imaging Self Scanner; IRS-P6-AWiFS = Indian Remote Sensing Satellite/Advanced Wide Field Sensor; Landsat-1, 2, 3 MSS = multispectral scanner; Landsat-4, 5 TM = Thematic Mapper; Landsat-7 ETM+ = Enhanced Thematic Mapper plus; MODIS = Moderate Imaging Spectral Radiometer; SPOT = Satellites Pour l'Observation de la Terre or Earth-observing satellites; SWIR = short-wave infrared sensor; VNIR = visible near-infrared sensor.

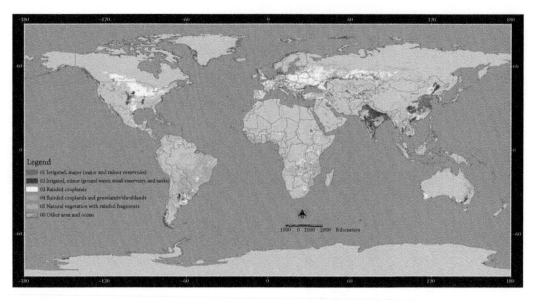

FIGURE 16.2
(See color insert following page 426.) Global cropland map at nominal 1-km resolution using remote sensing for the nominal year 2000. Total cropland area was determined to be 1.53 billion hectares, of which 399 Mha was irrigated area. Because irrigated areas often had more than one crop per year, the total annualized irrigated area was 467 Mha. (Adapted from Thenkabail, P.S. et al. *Rem Sens,* 1, 2009b. http://www.mdpi.com/2072-4292/1/2/50; Thenkabail, P.S. et al. *Remote Sensing of Global Croplands for Food Security,* CRC Press/Taylor & Francis, Boca Raton, FL, 2009c.)

Thenkabail et al. 2009a). This denotes that the importance of irrigated cropland for water and food security is very high.

There are also two premier global irrigated area maps (GIAMs) and statistics. These are (1) Thenkabail et al. (2009a) and Thenkabail et al. (2009c); and (2) Portmann, Siebert, and Döll (2009) and Siebert and Döll (2008, 2009), which is revised from their previous work (Siebert, Hoogeveen, and Frenken 2006). These products are also referred to as (1) GIAM of the International Water Management Institute's (IWMI), which is based on coarse-resolution remote sensing (Thenkabail et al. 2007; Thenkabail et al. 2009a; Thenkabail et al. 2009c; Thenkabail et al. 2005); and (2) global map of irrigated areas (GMIA) of the Food and Agricultural Organization of the United Nations and the University of Frankfurt (FAO/UF), which is based on national statistics (Siebert, Hoogeveen, and Frenken 2006). The irrigated and rain-fed cropland data, products, and state-of-the-art methods are disseminated through a dedicated Web portal (http://www.iwmigiam.org), a comprehensive book (Thenkabail et al. 2009c) with contributions from 53 authors worldwide, and several peer-reviewed publications (Thenkabail et al. 2009a; Thenkabail et al. 2009b).

These two premier GIAMs and statistics have provided better estimates of irrigated cropland worldwide. Portmann, Siebert, and Doll (2009) estimated "area equipped for irrigation" (but not necessarily irrigated, as some areas may be left fallow depending on water availability) in the world to be 312 Mha for the nominal year 2000 (Table 16.3). An equivalent definition of the phrase "area equipped for irrigation" is the total area available for irrigation (TAAI) or net irrigated area (Thenkabail et al. 2009a). The TAAI of the world for the year 2000 was 399 Mha. Thenkabail et al. (2009a) also determined annualized irrigated

TABLE 16.3

Global Irrigated and Rain-Fed Cropland Areas for the Nominal Year 2000 Based on Three Different Studies

Rank Based on Total IWMI GIAM Cropland Areas # A1	Country Name A2	Croplands: Irrigated — Total Area Available for Irrigation or Net Irrigated Areas, Hectares — Thenkabail et al.(2009a), Thenkabail et al.(2009b) A3	Croplands: Irrigated — Annualized Irrigated Areas or Gross Irrigated Areas, Hectares — Thenkabail et al.(2009a), Thenkabail et al.(2009b) A4	Croplands: Irrigated — Percentage of Total Global Annualized Irrigated Areas, % — Thenkabail et al.(2009a), Thenkabail et al.(2009b) A5	Croplands: Rain-Fed — Total Rain-Fed Cropland Areas, Hectares — Thenkabail et al.(2009a), Biradar et al.(2008) A6	Croplands: Rain-Fed — Percentage of Total Global Rain-Fed Cropland Areas, % — Thenkabail et al.(2009a), Biradar et al.(2008) A7	Croplands (Irrigated & Rain-Fed) — Total NET Cropland Areas (Irrigated TAAI & Rain-Fed), Hectares — Thenkabail et al.(2009a) A8	Croplands (Irrigated & Rain-Fed) — Total GROSS Cropland Areas (AIA & Rain-Fed), Hectares — Thenkabail et al.(2009a) A9	Croplands (Irrigated & Rain-Fed) — Percentage of Total Global GROSS Cropland Areas, % — Thenkabail et al.(2009a) A10	Croplands: Rain-Fed & Irrigated — Harvested Area of Rain-Fed & Irrigated Crops, Hectares\yr — Rammkutty and Foley (1998) A11	Croplands: Irrigated — Harvested Area of Irrigated Crops, Hectares\yr — Portmann et al.(submitted); applied in Siebert and Döll (2009) A12	Croplands: Rain-Fed — Harvested Area of Rain-Fed Crops, Hectares\yr — Portmann et al.(submitted); applied in Siebert and Döll (2009) A13	Croplands: Rain-Fed & Irrigated — Harvested Area of Rain-Fed & Irrigated Crops, Hectares\yr — Portmann et al.(submitted); applied in Siebert and Döll (2009) A14
1	China	111,988,772	151,802,086	32.52	91635702	8.10	203624474	243437788	14.33	147070700	85655000	82691500	168346500
2	India	101,234,893	132,253,854	28.34	48824269	4.31	150059162	181078123	10.66	171696820	68724900	115719000	184443900
3	USA	28,045,478	24,309,188	5.21	133571602	11.80	161617080	157880790	9.30	183979540	20548500	111394000	131942500
5	Russia	13,886,856	11,203,530	2.40	114788560	10.14	128675416	125992090	7.42	126892130	3772920	75288900	79061820
14	Brazil	4,195,118	4,085,844	0.88	87408556	7.72	91603674	91494400	5.39	51341076	2820970	47144500	49965470
6	Argentina	9,304,258	8,766,412	1.88	34318900	3.03	43623158	43085312	2.54	34010544	1352380	29024400	30376780
11	Australia	11,865,244	5,373,409	1.15	36758302	3.25	48623546	42131711	2.48	30030778	2384300	21219600	23603900
9	Kazakhstan	7,227,718	6,469,685	1.39	31722986	2.80	38950704	38192671	2.25	23507754	1804750	14085200	15889950
20	Canada	2,658,297	2,874,252	0.62	34944402	3.09	37602699	37818654	2.23	42773136	707056	34353900	35060956
26	Ukraine	2,995,578	2,381,799	0.51	28290153	2.50	31285731	30671952	1.81	36282376	1005120	26733700	27738820
16	Indonesia	3,172,879	3,322,443	0.71	17573608	1.55	20746487	20896051	1.23	54709968	7108330	24425300	31533630
21	France	2,399,518	2,687,153	0.58	17648821	1.56	20048339	20335974	1.20	19494778	1708020	16226300	17934320
4	Pakistan	14,036,151	15,959,342	3.42	3642557	0.32	17678708	19601899	1.15	23634900	19344800	3471930	22816730
18	Spain	3,421,724	3,025,823	0.65	15392046	1.36	18813770	18417869	1.08	18712148	3423510	11499900	14923410
7	Thailand	6,610,586	7,397,368	1.59	9931747	0.88	16542333	17329115	1.02	17151778	6187300	11514700	17702000
164	Zambia	779	536	0.00	16677106	1.47	16677885	16677642	0.98	5338720	55387	1071070	1126457
107	Tanzania	47,022	46,998	0.01	16410652	1.45	16457674	16457650	0.97	5477548	227000	5641460	5868460
15	Mexico	3,854,673	3,608,730	0.77	12497923	1.10	16352596	16106653	0.95	38267104	5958090	11246700	17204790
124	Congo, Dem. Rep.	21,833	20,375	0.00	15815336	1.40	15837169	15835711	0.93		7771	6061650	6069421
56	Poland	351,514	454,111	0.10	14424037	1.27	14775551	14878148	0.88	14790640	83292	12150000	12233292
103	Mozambique	56,415	60,742	0.01	13726544	1.21	13782959	13787286	0.81	4435618	40063	3093520	3133583
112	Angola	23,316	34,158	0.01	13454118	1.19	13477434	13488276	0.79	3404562	42000	2123380	2165380

(Continued)

TABLE 16.3 (Continued)

Global Irrigated and Rain-Fed Cropland Areas for the Nominal Year 2000 Based on Three Different Studies

| Rank Based on Total IWMI GIAM Cropland Areas # | Country Name | Croplands: Irrigated — Total Area Available for Irrigation or Net Irrigated Areas, Hectares (Thenkabail et al. (2009a), Thenkabail et al. (2009b)) | Croplands: Irrigated — Annualized Irrigated Areas or Gross Irrigated Areas, Hectares (Thenkabail et al. (2009a), Thenkabail et al. (2009b)) | Croplands: Irrigated — Percentage of Total Global Annualized Irrigated Areas, % (Thenkabail et al. (2009b)) | Croplands: Rain-Fed — Total Rain-Fed Cropland Areas, Hectares (Thenkabail et al. (2009a), Biradar et al. (2008)) | Croplands: Rain-Fed — Percentage of Total Global Rain-Fed Cropland Areas, % (Thenkabail et al. (2009a), Biradar et al. (2008)) | Croplands: (Irrigated & Rain-Fed) — Total NET Cropland Areas (Irrigated TAAI & Rain-Fed), Hectares (Thenkabail et al. (2009a)) | Croplands: (Irrigated & Rain-Fed) — Total GROSS Cropland Areas (AIA & Rain-Fed), Hectares (Thenkabail et al. (2009a)) | Croplands: (Irrigated & Rain-Fed) — Percentage of Total Global GROSS Cropland Areas, % (Thenkabail et al. (2009a)) | Croplands: Rain-Fed & Irrigated — Harvested Area of Rain-Fed & Irrigated Crops, Hectares\yr (Ramnkutty and Foley (1998)) | Croplands: Irrigated — Harvested Area of Irrigated Crops, Hectares\yr (Portmann et al. (submitted); applied in Siebert and Döll (2009)) | Croplands: Rain-Fed — Harvested Area of Rain-Fed Crops, Hectares\yr (Portmann et al. (submitted); applied in Siebert and Döll (2009)) | Croplands: Rain-Fed & Irrigated — Harvested Area of Rain-Fed & Irrigated Crops, Hectares\yr (Portmann et al. (submitted); applied in Siebert and Döll (2009)) |
A1	A2	A3	A4	A5	A6	A7	A8	A9	A10	A11	A12	A13	A14
10	Myanmar (Burma)	4,452,997	6,306,671	1.35	6257996	0.55	10710993	12564667	0.74	10997484	2263060	10890200	13153260
32	Turkey	1,753,382	1,577,313	0.34	10603366	0.94	12356748	12180679	0.72	22992722	3476000	17142500	20618500
19	Germany	2,197,697	3,001,674	0.64	8998878	0.80	11196575	12000552	0.71	12543913	266827	12099400	12366227
102	Belarus	84,088	60,926	0.01	10968114	0.97	11052202	11029040	0.65	5872892	115000	6050970	6165970
43	South Africa	821,040	828,491	0.18	10097803	0.89	10918843	10926294	0.64	15444497	1664300	5768150	7432450
13	Vietnam	4,384,022	4,949,533	1.06	5967528	0.53	10351550	10917061	0.64	0	5228400	6450190	11678590
74	Ethiopia	184,239	162,808	0.04	10564343	0.93	10748582	10727151	0.63	11042325	410557	7810690	8221247
70	Nigeria	197,909	216,154	0.05	9572289	0.85	9770698	9788943	0.58	36278520	164000	36852800	37016800
30	Sudan	1,737,118	1,930,592	0.41	7816063	0.69	9553181	9746655	0.57	17520156	1208110	10245300	11453410
8	Bangladesh	5,235,050	7,166,028	1.54	2536292	0.22	7771342	9702320	0.57	9707694	6431080	8571160	15002240
28	Romania	2,375,239	2,049,888	0.44	7563254	0.67	9938493	9613142	0.57	10143706	422724	9404020	9826744
31	Philippines	1,542,629	1,789,108	0.38	7479645	0.66	9022274	9268753	0.55	10265201	2067000	11243600	13310600
22	Italy	2,829,523	2,644,140	0.57	6436452	0.57	9265975	9080592	0.53	8788428	2670360	6671640	9342000
73	Bolivia	214,091	163,036	0.04	8803829	0.78	9017920	8966865	0.53	3155506	127001	2201000	2328001
146	Zimbabwe	4,744	3,533	0.00	8781932	0.78	8786676	8785465	0.52	3544333	202816	2057260	2260076
12	Uzbekistan	3,601,487	5,295,515	1.14	2821987	0.25	6423474	8117502	0.48	5312912	3819100	1144480	4963580
24	Iran	2,623,336	2,488,558	0.53	5509694	0.49	8133030	7998252	0.47	14774583	7296520	5899260	13195780
37	United Kingdom	970,733	1,060,204	0.23	5014629	0.44	5985362	6074833	0.36	7140452	183461	5674230	5857691
83	Kenya	85,401	104,527	0.02	5944333	0.53	6029734	6048860	0.36	5094582	76813	4199470	4276283
53	Colombia	546,186	592,495	0.13	5359287	0.47	5905473	5951782	0.35	3660984	645000	3017570	3662570
25	Japan	2,525,096	2,468,596	0.53	3428667	0.30	5953763	5897263	0.35	3771969	2167230	2189900	4357130
120	Paraguay	28,582	25,029	0.01	5538996	0.49	5567578	5564025	0.33	2964544	54000	4861040	4915040

94	Madagascar	72,359	75,156	0.02	5345476	0.47	5417835	0.32	5420632	3723170	1105690	1287310	2393000
66	Malaysia	258,766	274,565	0.06	5042468	0.45	5301234	0.31	5317033	8588696	501606	5310880	5812486
60	Peru	355,956	374,954	0.08	4846774	0.43	5202730	0.31	5221728	4382453	1109000	1494450	2603620
85	Cote d'Ivoire	95,138	101,890	0.02	4998420	0.44	5081162	0.30	5087914	7050554	41618	6355950	6397568
113	Uganda	30,017	30,586	0.01	5012869	0.44	5042886	0.30	5043455	8363950	2330	6288240	6290570
41	Cambodia	736,318	938,441	0.20	3868166	0.34	4604484	0.28	4806607	4115292	336992	2098280	2435272
36	Morocco	1,045,119	1,153,817	0.25	3603724	0.32	4648843	0.28	4757541	9755383	1468600	5362090	6830690
33	Nepal	1,251,988	1,477,303	0.32	3131060	0.28	4383048	0.27	4608363	2892683	1257980	2950220	4208200
72	Hungary	241,714	186,221	0.04	4358475	0.39	4600189	0.27	4544696	4793370	103764	4786070	4889834
38	Bulgaria	1,301,804	1,012,064	0.22	3416518	0.30	4718322	0.26	4428582	3678408	50898	3164360	3215258
46	Venezuela	894,880	807,078	0.17	3256971	0.29	4151851	0.24	4064049	8062588	491000	1251740	1742740
23	Iraq	2,220,024	2,626,564	0.56	1356711	0.12	3576735	0.23	3983275	5958466	2439000	36974	2475974
34	Chile	1,514,922	1,445,230	0.31	2412213	0.21	3927135	0.23	3857443	2457098	897274	831023	1728297
49	Czech Republic	518,036	701,727	0.15	3068209	0.27	3586245	0.22	3769936	3141046	16554	2593460	2610014
61	Uruguay	381,403	360,055	0.08	3354348	0.30	3735751	0.22	3714403	1621174	216979	551795	768774
42	Afghanistan	1,008,138	923,490	0.20	2748082	0.24	3756220	0.22	3671572	8355130	1912920	1427810	3340730
78	Algeria	144,349	136,946	0.03	3520819	0.31	3665168	0.22	3657765	6174593	570447	2908470	3478917
27	Korea, Dem. Rep.	1,467,262	2,053,625	0.44	1598207	0.14	3065469	0.22	3651832	2858257	1278000	1557170	2835170
17	Egypt	2,144,099	3,292,726	0.71	281590	0.02	2425689	0.21	3574316	2494046	6027120	1199970	7227090
48	Greece	907,739	766,678	0.16	2757498	0.24	3665237	0.21	3524176	2742187	1237970	2060890	3298860
35	Korea, Rep.	1,192,469	1,313,755	0.28	1928760	0.17	3121229	0.19	3242515	1858862	875415	1290420	2165835
142	Botswana	5,417	4,278	0.00	3198620	0.00	3204037	0.19	3202898	778086	620	166867	167487
67	Serbia	171,939	234,348	0.05	2947604	0.05	3119543	0.19	3181952	3181952	60071	3099970	3160041
65	Ecuador	288,581	281,166	0.06	2844430	0.06	3133011	0.18	3125596	2705382	686000	1792480	2478480
63	Portugal	358,865	313,908	0.07	2756177	0.07	3115042	0.18	3070085	2853130	638947	1727700	2366647
96	Ghana	60,647	71,764	0.02	2716954	0.02	2776954	0.16	2788071	6248967	17138	5140830	5157968
47	Kyrgyzstan	700,876	770,274	0.17	1990967	0.17	2691843	0.16	2761241	1159904	1140610	341595	1482205
110	Lithuania	57,272	41,591	0.01	2651512	0.01	2708784	0.16	2693103	3734526	4416	2369650	2374066
51	Cuba	486,898	637,159	0.14	2007424	0.14	2494322	0.16	2644583	4963997	822225	1440130	2262355
106	Cameroon	52,694	52,128	0.01	2591767	0.01	2644461	0.16	2643895	7437664	45079	3243950	3289029
29	Turkmenistan	1,522,372	1,999,984	0.43	542979	0.43	2065351	0.15	2542963	1878684	1402830	473466	1876296
59	Mongolia	422,332	376,378	0.08	2136984	0.08	2559316	0.15	2513362	1823339	57300	252370	309670
62	Guinea	302,633	320,350	0.07	2190800	0.07	2493433	0.15	2511150	1691191	20386	1856450	1876836
158	Central African Republic	1,155	1,086	0.00	2393214	0.00	2394369	0.14	2394300	2129445	69	776393	776462
179	Congo	0	0	0.00	2386480	0.00	2386480	0.14	2386480	733403	2000	211392	213392
64	Senegal	211,416	290,572	0.06	1980242	0.06	2191658	0.13	2270814	2475760	83904	1693600	1777504
45	Sri Lanka	948,029	809,579	0.17	1439246	0.17	2387275	0.13	2248825	2046457	731700	1344560	2076260
44	Azerbaijan	835,627	821,980	0.18	1398784	0.12	2234411	0.13	2220764	2054320	730129	734327	1464456
99	Mali	56,355	65,879	0.01	2051073	0.18	2107428	0.12	2116952	5015301	180317	2424560	2604877

(Continued)

TABLE 16.3 (*Continued*)

Global Irrigated and Rain-Fed Cropland Areas for the Nominal Year 2000 Based on Three Different Studies

Rank Based on Total IWMI GIAM Cropland Areas # A1	Country Name A2	Croplands: Irrigated — Total Area Available for Irrigation or Net Irrigated Areas, Hectares A3	Croplands: Irrigated — Annualized Irrigated Areas or Gross Irrigated Areas, Hectares A4	Croplands: Irrigated — Percentage of Total Global Annualized Irrigated Areas, % A5	Croplands: Rain-Fed — Total Rain-Fed Cropland Areas, Hectares A6	Croplands: Rain-Fed — Percentage of Total Global Rain-Fed Cropland Areas, % A7	Croplands (Irrigated & Rain-Fed) — Total NET Cropland Areas (Irrigated TAAI & Rain-Fed), Hectares A8	Croplands (Irrigated & Rain-Fed) — Total GROSS Cropland Areas (AIA & Rain-Fed), Hectares A9	Croplands (Irrigated & Rain-Fed) — Percentage of Total Global GROSS Cropland Areas, % A10	Croplands: Rain-Fed & Irrigated — Ramnkutty and Foley (1998) — Harvested Area of Rain-Fed & Irrigated Crops, Hectares\yr A11	Croplands: Irrigated — Portmann et al. (submitted); applied in Siebert and Döll (2009) — Harvested Area of Irrigated Crops, Hectares\yr A12	Croplands: Rain-Fed — Portmann et al. (submitted); applied in Siebert and Döll (2009) — Harvested Area of Rain-Fed Crops, Hectares\yr A13	Croplands: Rain-Fed & Irrigated — Portmann et al. (submitted); applied in Siebert and Döll (2009) — Harvested Area of Rain-Fed & Irrigated Crops, Hectares\yr A14
137	Latvia	12,683	7,325	0.00	2040565	0.18	2053248	2047890	0.12	1838370	833	917421	918254
128	Burkina Faso	15,663	14,660	0.00	2025961	0.18	2041624	2040621	0.12	3914244	20233	3483640	3503873
81	Laos	105,585	107,734	0.02	1917269	0.17	2022854	2025003	0.12	1137818	354642	637728	992370
149	Malawi	3,293	2,794	0.00	1996142	0.18	1999435	1998936	0.12	1440040	56515	1541930	1598445
87	Austria	116,456	98,551	0.02	1822194	0.16	1938650	1920745	0.11	1569707	41076	1331250	1372326
50	Taiwan, Province of China	499,043	677,877	0.15	1111947	0.10	1610990	1789824	0.11		588798	363675	952473
93	Slovakia	109,904	75,488	0.02	1690815	0.15	1800719	1766303	0.10	1339293	104560	1324930	1429490
68	Moldova	294,070	22,9433	0.05	1533012	0.14	1827082	1762445	0.10	2046245	256377	1530200	1786577
57	Tajikistan	383,243	44,9153	0.10	1190392	0.11	1573635	1639545	0.10	1091111	637213	347061	984274
55	Belgium	324,796	507,430	0.11	1101425	0.10	1426221	1608855	0.09		10378	416387	426765
192	Papua New Guinea	0	0	0.00	1607752	0.14	1607752	1607752	0.09	2024525	0	923169	923169
77	New Zealand	125,390	141,686	0.03	1459699	0.13	1585089	1601385	0.09	660818	383236	372708	755944
168	Liberia	237	300	0.00	1598806	0.14	1599043	1599106	0.09	379859	2100	399802	401902
109	Croatia	35,202	44,630	0.01	1551680	0.14	1586882	1596310	0.09	1992810	5000	1138000	1143000
58	Somalia	372,476	403,574	0.09	1189487	0.11	1561963	1593061	0.09	1093790	206000	340769	546769
39	Netherlands	870,243	1,011,340	0.22	564102	0.05	1433345	1575442	0.09	833890	153650	737499	891149
89	Guatemala	69,373	91,313	0.02	1440867	0.13	1510240	1532180	0.09	2630816	139788	1479390	1619178
40	Denmark	1,164,705	979,539	0.21	517719	0.05	1681824	1496658	0.09	3197374	204071	2450940	2655011
52	Syria	566,990	596,263	0.13	879249	0.08	1446239	1475512	0.09	4837218	1507870	3160120	4667990
92	Honduras	70,584	77,729	0.02	1384346	0.12	1454930	1462075	0.09	1623373	100000	852399	952399
86	Tunisia	109,144	100,647	0.02	1284882	0.11	1394026	1385529	0.08	2444246	367000	1690130	2057130

115	Sierra Leone	21,807	29,037	0.01	1336205	0.12	1358012	1365242	0.08	582745	30000	477979	507979
130	Bosnia and Herzegovina	10,766	14,203	0.00	1303620	0.12	1314386	1317823	0.08	995513	3000	633413	636413
122	Nicaragua	16,439	22,720	0.01	1241957	0.11	1258396	1264677	0.07	2542479	75222	867862	943084
97	Sweden	83,918	71,108	0.02	1040821	0.09	1124739	1111929	0.07	2804305	53440	2227520	2280960
69	Albania	223,777	225,864	0.05	864549	0.08	1088326	1090413	0.06	677557	180000	264223	444223
181	Gabon	0	0	0.00	1084861	0.10	1084861	1084861	0.06	315822	8450	205735	214185
108	Panama	49,069	45,048	0.01	1037572	0.09	1086641	1082620	0.06	826045	30811	278973	309784
129	Estonia	24,637	14,476	0.00	1052562	0.09	1077199	1067038	0.06	897736	600	765029	765629
116	Chad	25,234	27,698	0.01	925287	0.08	950521	952985	0.06	3831578	26804	2082790	2109594
76	Georgia	128,538	146,141	0.03	796878	0.07	925416	943019	0.06	1138420	196702	547279	743981
79	Macedonia	169,843	131,620	0.03	695920	0.06	865763	827540	0.05	717538	42500	360056	402556
136	Burundi	11,793	8,490	0.00	798743	0.07	810536	807233	0.05	588748	20130	1125320	1145450
95	Finland	125,307	71,961	0.02	721148	0.06	846455	793109	0.05	2447813	20000	1991230	2011230
126	Costa Rica	12,628	15,791	0.00	772096	0.07	784724	787887	0.05	697421	123030	349366	472396
100	Rwanda	80,067	64,806	0.01	710557	0.06	790624	775363	0.05	777961	5500	1349640	1355140
121	Togo	21,727	23,843	0.01	725130	0.06	746857	748973	0.04	3045446	2557	1376440	1378997
111	Switzerland	29,523	36,976	0.01	690849	0.06	720372	727825	0.04	374221	14500	428373	442873
141	Niger	4,129	4,317	0.00	703697	0.06	707826	708014	0.04	13904998	96125	10238200	10334325
127	Benin	15,173	15,415	0.00	668742	0.06	683915	684157	0.04	2880695	2823	1787270	1790093
135	Namibia	10,526	9,303	0.00	672697	0.06	683223	682000	0.04	819824	8806	185874	194680
185	Ireland	0	0	0.00	630766	0.06	630766	630766	0.04	1348463	1100	586198	587298
90	Dominican Republic	70,876	79,648	0.02	550415	0.05	621291	630063	0.04	1667190	220000	687649	907649
71	Libya	230,656	210,022	0.05	412158	0.04	642814	622180	0.04	805312	316000	357915	673915
54	Saudi Arabia	678,677	551,066	0.12	63518	0.01	742195	614584	0.04	2148407	1280720	122034	1402754
145	Lesotho	5,675	3,681	0.00	571627	0.05	577302	575308	0.03	417547	203	188366	188569
88	Swaziland	149,274	97,004	0.02	446942	0.04	596216	543946	0.03	128573	45482	134479	179961
165	Slovenia	439	510	0.00	542220	0.05	542659	542730	0.03	254712	10324	183915	194239
104	Haiti	50,848	53,903	0.01	486161	0.04	537009	540064	0.03	1567588	89000	1048170	1137170
80	Armenia	106,695	118,324	0.03	415591	0.04	522286	533915	0.03	410979	172806	333254	506060
155	Norway	2,072	1,453	0.00	485336	0.04	487408	486789	0.03	569607	36200	604876	641076
131	Montenegro	10,331	13,908	0.00	364360	0.03	374691	378268	0.02		2109	347634	349743
75	Guinea Bissau	108,042	155,389	0.03	191959	0.02	300001	347348	0.02	546406	8562	369965	378527
134	El Salvador	11,592	10,401	0.00	294667	0.03	306259	305068	0.02	1036444	50710	738223	788933
84	Guyana	96,276	102,930	0.02	184027	0.02	280303	286957	0.02	438246	178029	53756	231785
91	Yemen	91,688	79,188	0.02	206310	0.02	297998	285498	0.02	8690437	399668	661291	1060959
101	Gambia	39,872	63,415	0.01	197119	0.02	236991	260534	0.02	540746	0	0	0
132	Eritrea	17,017	13,776	0.00	232850	0.02	249867	246626	0.01	523471	5969	501042	507011
147	Belize	3,887	3,510	0.00	228077	0.02	231964	231587	0.01	102505	3000	82274	85274
143	East Timor	3,800	4,061	0.00	222063	0.02	225863	226124	0.01		7000	179410	186410

(Continued)

TABLE 16.3 (*Continued*)

Global Irrigated and Rain-Fed Cropland Areas for the Nominal Year 2000 Based on Three Different Studies

Rank Based on Total IWMI GIAM Cropland Areas # A1	Country Name A2	Croplands: Irrigated — Total Area Available for Irrigation or Net Irrigated Areas, Hectares — Thenkabail et al. (2009a), Thenkabail et al. (2009b) A3	Croplands: Irrigated — Annualized Irrigated Areas or Gross Irrigated Areas, Hectares — Thenkabail et al. (2009a), Thenkabail et al. (2009b) A4	Croplands: Irrigated — Percentage of Total Global Annualized Irrigated Areas, % — Thenkabail et al. (2009a), Thenkabail et al. (2009b) A5	Croplands: Rain-Fed — Total Rain-Fed Cropland Areas, Hectares — Thenkabail et al. (2009), Biradar et al. (2008) A6	Croplands: Rain-Fed — Percentage of Total Global Rain-Fed Cropland Areas, % — Thenkabail et al. (2009), Biradar et al. (2008) A7	Croplands (Irrigated & Rain-Fed) — Total NET Cropland Areas (Irrigated TAAI & Rain-Fed), Hectares — Thenkabail et al. (2009) A8	Croplands (Irrigated & Rain-Fed) — Total GROSS Cropland Areas (AIA & Rain-Fed), Hectares — Thenkabail et al. (2009) A9	Croplands (Irrigated & Rain-Fed) — Percentage of Total Global GROSS Cropland Areas, % — Thenkabail et al. (2009) A10	Croplands: Rain-Fed & Irrigated — Harvested Area of Rain-Fed & Irrigated Crops, Hectares\yr — Ramnkutty and Foley (1998) A11	Croplands: Irrigated — Harvested Area of Irrigated Crops, Hectares\yr — Portmann et al. (submitted); applied in Siebert and Döll (2009) A12	Croplands: Rain-Fed — Harvested Area of Rain-Fed Crops, Hectares\yr — Portmann et al. (submitted); applied in Siebert and Döll (2009) A13	Croplands: Rain-Fed & Irrigated — Harvested Area of Rain-Fed & Irrigated Crops, Hectares\yr — Portmann et al. (submitted); applied in Siebert and Döll (2009) A14
82	Israel	99,806	104,542	0.02	101665	0.01	201471	206207	0.01	407057	184072	76797	260869
157	Bhutan	997	1,396	0.00	154068	0.01	155065	155464	0.01	226391	43507	77788	121295
139	Cyprus	7,099	4,863	0.00	148942	0.01	156041	153805	0.01		36210	5342	41552
119	Lebanon	24,747	25,268	0.01	126708	0.01	151455	151976	0.01	264201	139292	137695	276987
133	Puerto Rico	11,964	11,253	0.00	138701	0.01	150665	149954	0.01		17465	49918	67382
105	Jordan	72,717	52,541	0.01	75548	0.01	148265	128089	0.01	595772	100105	42477	142582
125	Mauritania	15,124	20,036	0.00	100469	0.01	115593	120505	0.01	745161	23084	272066	295150
187	Luxembourg	0	0	0.00	111041	0.01	111041	111041	0.01	1051585	24	34221	34245
123	Suriname	19,845	20,774	0.00	88593	0.01	108438	109367	0.01	77794	51180	9400	60580
114	Oman	17,853	30,145	0.01	66449	0.01	84302	96594	0.01	98737	72461	3143	75604
160	Brunei	799	1,002	0.00	86509	0.01	87308	87511	0.01	57747	1000	15028	16028
98	United Arab Emirates	93,810	70,603	0.02	0	0.00	93810	70603	0.00	180144	204951	19453	224404
140	Jamaica	4,881	4,556	0.00	61139	0.01	66020	65695	0.00	3801523	24666	171191	195857
154	West Bank	1,612	1,542	0.00	64136	0.01	65748	65678	0.00		0	0	0
150	Equatorial Guinea	2,812	2,644	0.00	57260	0.01	60072	59904	0.00	130423	0	127208	127208
117	Qatar	38,509	27,596	0.01	0	0.00	38509	27596	0.00	0	9544	0	9544
118	Kuwait	37,333	26,753	0.01	0	0.00	37333	26753	0.00	15704	8509	719	9229
148	French Guyana	2,860	2,822	0.00	18869	0.00	21729	21691	0.00	15752	6007	9658	15665
153	Trinidad and Tobago	1,859	1,720	0.00	8590	0.00	10449	10310	0.00	167262	3600	57308	60908
195	Singapore	0	0	0.00	8941	0.00	8941	8941	0.00	102300	0	405	405
138	Gaza Strip	5,909	6,790	0.00	693	0.00	6602	7483	0.00		0	0	0

#	Country												
175	Andorra	0	0	0.00	5776	0.00	5776	5776	0.00	20558	150	0	150
186	Liechtenstein	0	0	0.00	4839	0.00	4839	4839	0.00		0	0	0
144	Mauritius	5,312	3,910	0.00	0	0.00	5312	3910	0.00		20919	67663	88582
162	San Marino	1,102	797	0.00	2060	0.00	3162	2857	0.00		0	1957	1957
151	Antigua and Barbuda	2,270	2,468	0.00	0	0.00	2270	2468	0.00		130	3478	3608
152	Guadeloupe	1,894	2,022	0.00	0	0.00	1894	2022	0.00		5697	20111	25808
156	St. Kitts and Nevis	1,650	1,445	0.00	0	0.00	1650	1445	0.00		18	7831	7849
159	Virgin Islands	827	1,015	0.00	0	0.00	827	1015	0.00	1516774	185	0	185
161	Reunion	651	846	0.00	0	0.00	651	846	0.00		7584	46469	54053
163	Djibouti	905	587	0.00	0	0.00	905	587	0.00	1092	388	707	1095
166	Comoros	241	417	0.00	0	0.00	241	417	0.00		85	75698	75783
167	Anguilla	489	404	0.00	0	0.00	489	404	0.00		0	0	0
173	Monaco	73	53	0.00	132	0.00	205	185	0.00		0	0	0
169	Turks and Caicos Islands	214	170	0.00	0	0.00	214	170	0.00		0	0	0
170	Montserrat	69	115	0.00	0	0.00	69	115	0.00		0	554	554
171	St. Pierre and Miquelon	70	59	0.00	0	0.00	70	59	0.00		0	0	0
172	Cayman Islands	66	55	0.00	0	0.00	66	55	0.00		0	440	440
174	Seychelles	66	44	0.00	0	0.00	66	44	0.00		224	0	224
176	Bahrain	0	0	0.00	0	0.00	0	0	0.00		3113	488	3601
177	Barbados	0	0	0.00	0	0.00	0	0	0.00		1000	13175	14175
178	Cape Verde	0	0	0.00	0	0.00	0	0	0.00		2578	46964	49542
180	Fiji	0	0	0.00	0	0.00	0	0	0.00		3000	154135	157135
183	Grenada	0	0	0.00	0	0.00	0	0	0.00		219	15440	15659
184	Guam	0	0	0.00	0	0.00	0	0	0.00		312	13715	14028
188	Malta	0	0	0.00	0	0.00	0	0	0.00		3540	12330	15870
189	Martinique	0	0	0.00	0	0.00	0	0	0.00		6730	15595	22325
190	Northern Marianna Islands	0	0	0.00	0	0.00	0	0	0.00		60	0	60
193	Pitcairn Islands	0	0	0.00	0	0.00	0	0	0.00		0	0	0
194	Sao Tome and Principe	0	0	0.00	0	0.00	0	0	0.00		9700	15415	25115
196	St. Lucia	0	0	0.00	0	0.00	0	0	0.00		297	24167	24464
197	St. Vincent and the Grenadines	0	0	0.00	0	0.00	0	0	0.00		0	17261	17261
198	Vatican City	0	0	0.00	0	0.00	0	0	0.00	3525306	0	0	0
	TOTAL	398,526,952	466,757,677	100	1,131,552,272	100	1,530,079,224	1,598,309,949	94	1,537,977,307	312,384,543	992,349,053	1,304,733,596

area (AIA or gross irrigated area) taking into account the cropping intensity. The AIA of the world in 2000 was 467 Mha (Table 16.3). A country-by-country comparison of irrigated cropland areas for 197 countries (Figure 16.4) showed very high correlation with R^2 value of 0.94 between the studies by Portmann, Siebert, and Döll (2009), Siebert and Döll (2009), and Thenkabail et al. (2009a).

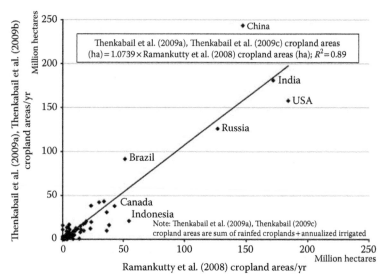

FIGURE 16.3
Comparison of cropland (rain-fed and irrigated) areas of 197 countries by two methods: the primarily nonremote sensing approach of Ramankutty et al. (2008) versus the remote sensing approach of Thenkabail, P.S. et al. (2009a) and Thenkabail, P.S. et al. (2009c).

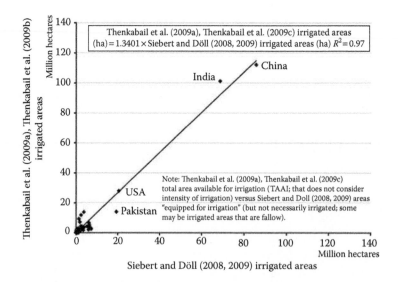

FIGURE 16.4
Comparison of irrigated cropland areas of 197 countries by two methods: the primarily nonremote sensing approach of Siebert and Döll (2008, 2009) versus the primarily remote sensing approach of Thenkabail, P.S. et al. (2009a) and Thenkabail, P.S. et al. (2009c).

16.2.3 Global Rain-Fed Croplands Using Remote Sensing and Nonremote Sensing Approaches

Rain-fed croplands are the major user of rainfall and green water. About 70% of all precipitation is stored as green water (unsaturated zone of soils). Rain-fed croplands depend on this water for crop growth and food production. Recent estimates show that rain-fed croplands produce about 55% of the world's food from 75% (1.13 billion hectares) of the cropland areas (Thenkabail et al. 2009c). Rain-fed croplands, even though they are far less productive than irrigated areas, are the main source of livelihood for subsistence farmers and are the key focus of future crop and water productivity increases (Falkenmark and Rockström 2006). They are also considered environmentally friendly, given the problems of salinization and land degradation in irrigated cropland areas (Khan and Hanjra 2008).

16.3 Uncertainty in Global Cropland Areas

There are substantial differences in cropland areas when compared at a country-by-country level (Table 16.3, Figures 16.3 and 16.4). The differences (Dheeravath et al. in press; Thenkabail et al. 2009b) were in both traditional approaches (Portmann, Siebert, and Döll 2009; Ramankutty et al. 2008; Siebert and Döll 2008, 2009) and remote sensing (Thenkabail et al. 2009a; Thenkabail et al. 2009c). The main causes of differences in areas reported in various studies can be attributed to, but not limited to, reluctance of agencies at national and state levels to furnish census data on irrigated areas in view of their institutional interests in sharing of water and water data; reporting of large volumes of census data with inadequate statistical analysis; subjectivity involved in observation-based data collection processes; inadequate accounting of irrigated areas, especially minor irrigation from groundwater, in national statistics; definition issues involved in mapping using remote sensing as well as national statistics; difficulties in arriving at precise estimates of area fractions using remote sensing; difficulties in separating irrigated croplands from rain-fed croplands; and imagery resolution in remote sensing (Ramankutty et al. 2008 versus; Thenkabail et al. 2009a; Thenkabail et al. 2009c).

16.4 Limitations of Existing Global Cropland Maps and Statistics

The various coarser-resolution cropland area maps discussed in Section 16.2 have many limitations. These include the following:

- Absence of precise spatial location of these cropland areas
- Uncertainties in differentiating irrigated areas from rain-fed areas
- Absence of crop types or varieties, cropping intensities, and sources of irrigation

Addressing such limitations remains crucial for comprehensive water-use assessments and food security analysis. The errors of omissions in coarser-resolution data are high as a result of its inability to capture fragmented, smaller patches of croplands accurately. In addition, the coarser-resolution data adds significant errors of commissions. As a result of the large pixel size, it can at times map patchy noncropland areas that surround croplands as cropland areas. In either case, the need for finer spatial resolution products to resolve the issue is a must (Wardlow and Egbert 2008).

This will require scientists to map global cropland areas at a higher resolution (e.g., Landsat 30-m resolution in fusion with MODIS 250-m time series) or ideally a new finer-resolution product. Such a product should (1) define more precisely the actual area and the spatial distribution of cropland areas; (2) develop methods and techniques for consistent and unbiased estimates of irrigated versus rain-fed cropland areas over space and time for the entire world; (3) elaborate on the extent of multiple irrigated and rain-fed cropping over a year, particularly in Brazil (Galford et al. 2008) and Asia (Xiao et al. 2006), where two or even three crops may be grown in 1 year, but where cropping intensities are not accurately known or recorded in secondary statistics; and (4) account for irrigation source. This will be a significant advance because irrigated and rain-fed cropping intensity and their crop types have a huge influence on the quantum of water used by crops and associated indicators of agricultural productivity, crop diversification, and food security. Accounting for irrigation sources can help identify the hot spots where surface or groundwater could be overexploited and thus enhance the resilience of cropland and water resources to climate change and associated pressure points.

16.5 Cropland Water Use: A Global Perspective

Globally, only about 6% (3,091 km³/y) to 7% (3,798 km³/y) of the available renewable water (54,695 km³/y) is currently withdrawn (but not necessarily used) by irrigated croplands (Table 16.4). However, a country-by-country assessment (Table 16.5) shows water withdrawal of more than 100% (meaning importation or groundwater mining) in countries such as United Arab Emirates, Saudi Arabia, and Libya; more than 50% in countries such as Jordan, Cyprus, Yemen, Israel, and Tunisia, 20–50% in countries such as China, India, Uzbekistan, Iran, and Iraq; less than 10% in countries such as the United States, Italy, and Greece; and less than 1% in countries such as the Democratic Republic of Congo, Canada, and Brazil. This implies a highly uneven spatial distribution in water availability and water withdrawal for irrigation around the world. Further, such spatial variabilities are also very high within countries such as the United States, Australia, and China. Intracountry

TABLE 16.4

Water Use by Croplands for the Nominal Year 2000

Blue Water Use by Irrigated Crops (km³/yr)	Green Water Use by Irrigated Crops (km³/yr)	Green Water Use by Rainfed Crops (km³/yr)	Total Water Use by Irrigated and Rainfed Crops (km³/yr)	Reference
1180	919	4586	6685	Siebert and Döll (2009)
1800	-	5000	6800	Falkenmark and Rockström (2006)
			7500	Postel (1998)

TABLE 16.5

Global Water Withdrawal and Water Use by Irrigated Croplands: A Country-by-Country Assessment

Rank Based on Total IWMI GIAM Cropland Areas # A1	Country Name A2	Water: Renewable Glieck et al. (2009) Annual Renewable Water Resource km³/y A22	Water: Withdrawal for Irrigation Wisser et al. (2008) Total Water Withdrawal for Irrigation Based on IWMI GIAM Irrigated Areas km³/y A26	Water: Withdrawal for Irrigation Wisser et al. (2008) Total Water Withdrawal for Irrigation Based on FAO/UF V4.0 Irrigated Areas km³/y A27	Water: Requirement (ET) for Irrigation Siebert and Döll (2009, 2008) Blue Water Requirement for Irrigation Based on Irrigated Areas of FAO/UF V4.0 km³/y A28	Water Green: Requirement (ET) for Irrigation Siebert and Döll (2009, 2008) Green Water Availability over Irrigated Areas Based on FAO/UF V4.0 km³/y A29	Water Blue: Requirement (ET) for Irrigation Siebert and Döll (2009, 2008) Total (Blue & Green) Water Requirement for Irrigation Based on FAO/UF V4.0 km³/y A30
1	China	2830	755	606	147	257	404
2	India	1908	1694	844	287	175	462
3	USA	3069	122.3	141.2	139.1	79.1	218.3
5	Russia	4498	71.3	17.1	11.6	13.4	25.0
14	Brazil	8233	28.1	14.2	8.3	18.0	26.4
6	Argentina	814.0	47.4	11.1	5.8	5.7	11.5
11	Australia	398.0	25.0	12.8	13.6	10.9	24.5
9	Kazakhstan	109.6	40.0	12.7	8.9	3.2	12.1
20	Canada	3300.0	5.1	2.2	2.7	2.3	5.1
26	Ukraine	139.5	8.5	11.4	3.5	3.6	7.1
16	Indonesia	2838.0	46.3	53.4	13.6	43.2	56.8
21	France	189.0	5.6	4.9	3.2	6.0	9.2
4	Pakistan	233.8	136.2	414.7	117.0	19.3	136.3
18	Spain	111.1	13.7	13.1	18.6	9.2	27.8
7	Thailand	409.9	123.9	124.3	19.1	30.8	49.9
164	Zambia	105.2	0.0	0.1	0.4	0.2	0.6
107	Tanzania	91.0	0.2	0.8	1.0	0.8	1.8

(Continued)

TABLE 16.5 (*Continued*)

Global Water Withdrawal and Water Use by Irrigated Croplands: A Country-by-Country Assessment

| Rank Based on Total IWMI GIAM Cropland Areas # | Country Name | Water: Renewable Glieck et al. (2009) Annual Renewable Water Resource km³/y | Water: Withdrawal for Irrigation Wisser et al. (2008) Total Water Withdrawal for Irrigation Based on IWMI GIAM Irrigated Areas km³/y | Water: Withdrawal for Irrigation Wisser et al. (2008) Total Water Withdrawal for Irrigation Based on FAO/UF V4.0 Irrigated Areas km³/y | Water: Requirement (ET) for Irrigation Siebert and Döll (2009, 2008) Blue Water Requirement for Irrigation Based on Irrigated Areas of FAO/UF V4.0 km³/y | Water Green: Requirement (ET) for Irrigation Siebert and Döll (2009, 2008) Green Water Availability over Irrigated Areas Based on FAO/UF V4.0 km³/y | Water Blue: Requirement (ET) for Irrigation Siebert and Döll (2009, 2008) Total (Blue & Green) Water Requirement for Irrigation Based on FAO/UF V4.0 km³/y |
A1	A2	A22	A26	A27	A28	A29	A30
15	Mexico	457.2	36.9	32.0	26.8	24.2	51.0
124	Congo, Dem. Rep.	1283.0	0.0	0.0	0.0	0.1	0.1
56	Poland	63.1	0.7	0.3	0.1	0.3	0.4
103	Mozambique	216.0	0.2	0.8	0.2	0.2	0.4
112	Angola	184.0	0.2	0.2	0.2	0.1	0.3
10	Myanmar (Burma)	1045.6	40.0	25.3	5.9	6.5	12.4
32	Turkey	234.0	14.6	25.0	14.6	6.6	21.3
19	Germany	188.0	4.5	1.9	0.2	0.9	1.1
102	Belarus	58.0	0.0	0.0	0.2	0.5	0.6
43	South Africa	50.0	3.1	8.3	8.8	6.5	15.3
13	Vietnam	891.2	78.9	70.0	7.4	24.7	32.1
74	Ethiopia	110.0	1.7	1.9	1.2	1.4	2.6
70	Nigeria	286.2	2.3	1.5	0.9	0.6	1.5
30	Sudan	154.0	17.7	11.2	10.1	2.9	13.0
8	Bangladesh	1210.6	65.0	83.3	18.7	24.3	43.0
28	Romania	42.3	8.1	11.9	0.9	1.3	2.2
31	Philippines	479.0	17.6	25.1	3.8	10.4	14.2
22	Italy	175.0	8.7	8.3	6.5	8.8	15.2

73	Bolivia	622.5	2.2	0.4	0.4	0.3	0.8
146	Zimbabwe	20.0	0.0	0.5	0.9	0.7	1.6
12	Uzbekistan	72.2	32.9	35.5	24.1	5.5	29.7
24	Iran	137.5	32.7	64.2	40.8	10.7	51.5
37	United Kingdom	160.6	2.1	0.3	0.2	0.6	0.9
83	Kenya	30.2	0.6	0.4	0.5	0.3	0.8
53	Colombia	2132.0	3.3	10.6	1.0	3.7	4.6
25	Japan	430.0	21.1	31.4	1.7	9.1	10.7
120	Paraguay	336.0	0.1	0.4	0.1	0.5	0.6
94	Madagascar	337.0	0.4	5.7	2.8	4.4	7.2
66	Malaysia	580.0	1.3	2.8	0.6	2.8	3.5
60	Peru	1913.0	3.7	11.5	5.1	1.4	6.5
85	Cote d'Ivoire	81.0	0.0	0.0	0.1	0.3	0.5
113	Uganda	66.0	0.1	0.0	0.0	0.0	0.0
41	Cambodia	476.1	19.5	10.6	1.0	1.8	2.8
36	Morocco	29.0	5.5	4.1	9.0	2.2	11.2
33	Nepal	210.2	14.2	16.1	4.2	4.0	8.2
72	Hungary	120.0	0.8	1.4	0.2	0.4	0.6
38	Bulgaria	19.4	4.9	4.2	0.1	0.1	0.2
46	Venezuela	1233.2	7.0	8.1	2.0	3.4	5.4
23	Iraq	96.4	42.0	44.6	20.9	2.1	23.0
34	Chile	922.0	7.1	3.6	3.0	2.1	5.1
49	Czech Republic	16.0	0.8	0.1	0.0	0.1	0.1
61	Uruguay	139.0	4.1	1.7	0.6	1.0	1.6
42	Afghanistan	65.0	9.8	25.1	9.4	2.6	12.0
78	Algeria	14.3	0.8	1.1	4.3	1.0	5.2
27	Korea, Dem. Rep.	77.1	8.5	7.9	1.1	5.6	6.7
17	Egypt	86.8	19.0	37.8	46.9	0.8	47.7
48	Greece	72.0	3.9	5.8	6.9	2.5	9.4
35	Korea, Rep.	69.7	5.3	5.7	0.8	4.0	4.9
142	Botswana	14.7	0.0	0.0	0.0	0.0	0.0
67	Serbia	0.0	0.0	0.0	0.2	0.2	0.4

(Continued)

TABLE 16.5 (*Continued*)

Global Water Withdrawal and Water Use by Irrigated Croplands: A Country-by-Country Assessment

Rank Based on Total IWMI GIAM Cropland Areas # A1	Country Name A2	Water: Renewable Glieck et al. (2009) Annual Renewable Water Resource km³/y A22	Water: Withdrawal for Irrigation Wisser et al. (2008) Total Water Withdrawal for Irrigation Based on IWMI GIAM Irrigated Areas km³/y A26	Water: Withdrawal for Irrigation Wisser et al. (2008) Total Water Withdrawal for Irrigation Based on FAO/UF V4.0 Irrigated Areas km³/y A27	Water: Requirement (ET) for Irrigation Siebert and Döll (2009, 2008) Blue Water Requirement for Irrigation Based on Irrigated Areas of FAO/UF V4.0 km³/y A28	Water Green: Requirement (ET) for Irrigation Siebert and Döll (2009, 2008) Green Water Availability over Irrigated Areas Based on FAO/UF V4.0 km³/y A29	Water Blue: Requirement (ET) for Irrigation Siebert and Döll (2009, 2008) Total (Blue & Green) Water Requirement for Irrigation Based on FAO/UF V4.0 km³/y A30
65	Ecuador	432.0	2.1	9.6	2.7	2.1	4.8
63	Portugal	73.6	1.3	2.4	3.2	1.9	5.1
96	Ghana	53.2	0.5	0.1	0.1	0.1	0.1
47	Kyrgyzstan	46.5	2.6	4.1	3.1	2.6	5.7
110	Lithuania	24.5	0.1	0.0	0.0	0.0	0.0
51	Cuba	38.1	5.9	11.5	1.6	6.2	7.8
106	Cameroon	285.5	0.4	0.4	0.2	0.2	0.4
29	Turkmenistan	60.9	16.8	17.0	10.6	1.3	11.9
59	Mongolia	34.8	0.6	0.1	0.3	0.1	0.4
62	Guinea	226.0	2.8	1.5	0.1	0.1	0.2
158	Central African Republic	144.4	0.0	0.0	0.0	0.0	0.0
179	Congo	832.0	0.0	0.0	0.0	0.0	0.0
64	Senegal	39.4	3.2	0.6	0.6	0.2	0.8
45	Sri Lanka	50.0	19.4	11.2	2.2	3.0	5.2
44	Azerbaijan	30.3	7.5	10.9	3.8	1.2	5.0
99	Mali	100.0	0.9	5.1	1.3	0.4	1.7
137	Latvia	49.9	0.0	0.0	0.0	0.0	0.0
128	Burkina Faso	17.5	0.2	0.5	0.1	0.1	0.2

81	Laos	333.6	1.3	2.6	1.3	1.4	2.7
149	Malawi	17.3	0.0	0.1	0.3	0.4	0.7
87	Austria	84.0	0.3	0.1	0.0	0.2	0.2
50	Taiwan, Province of China	67.0	0.0	0.0	0.7	2.6	3.3
93	Slovakia	80.3/50.1	0.0	0.0	0.2	0.4	0.6
68	Moldova	11.7	0.9	1.1	0.7	0.6	1.3
57	Tajikistan	99.7	2.7	4.7	3.3	1.0	4.3
55	Belgium	20.8	0.8	0.3	0.0	0.0	0.0
192	Papua New Guinea	801.0	0.0	0.0	0.0	0.0	0.0
77	New Zealand	397.0	0.1	0.9	0.7	2.1	2.8
168	Liberia	232.0	0.1	0.0	0.0	0.0	0.0
109	Croatia	105.5	0.0	0.0	0.0	0.0	0.0
58	Somalia	15.7	3.0	1.9	1.0	0.7	1.6
39	Netherlands	89.7	1.3	1.0	0.1	0.8	0.9
89	Guatemala	111.3	0.0	0.0	0.7	1.2	1.9
40	Denmark	6.1	2.0	1.2	0.0	0.5	0.6
52	Syria	46.1	6.9	7.6	9.5	2.2	11.7
92	Honduras	95.9	0.3	0.4	0.4	0.6	0.9
86	Tunisia	4.6	0.7	2.3	2.6	0.7	3.3
115	Sierra Leone	160.0	0.2	0.2	0.1	0.1	0.2
130	Bosnia and Herzegovina	37.5	0.0	0.0	0.0	0.0	0.0
122	Nicaragua	196.7	0.2	0.5	0.3	0.4	0.7
97	Sweden	179.0	0.1	0.3	0.0	0.2	0.3
69	Albania	41.7	0.9	0.8	0.7	0.6	1.4
181	Gabon	164.0	0.0	0.0	0.0	0.0	0.0
108	Panama	148.0	0.0	0.0	0.1	0.3	0.4
129	Estonia	21.1	0.0	0.0	0.0	0.0	0.0
116	Chad	43.0	0.3	0.2	0.1	0.1	0.2
76	Georgia	63.3	0.8	1.6	0.5	0.6	1.1

(Continued)

TABLE 16.5 (*Continued*)

Global Water Withdrawal and Water Use by Irrigated Croplands: A Country-by-Country Assessment

Rank Based on Total IWMI GIAM Cropland Areas #	Country Name	Water: Renewable, Glieck et al. (2009), Annual Renewable Water Resource km³/y	Water: Withdrawal for Irrigation, Wisser et al. (2008), Total Water Withdrawal for Irrigation Based on IWMI GIAM Irrigated Areas km³/y	Water: Withdrawal for Irrigation, Wisser et al. (2008), Total Water Withdrawal for Irrigation Based on FAO/UF V4.0 Irrigated Areas km³/y	Water: Requirement (ET) for Irrigation, Siebert and Döll (2009, 2008), Blue Water Requirement for Irrigation Based on FAO/UF V4.0 of FAO/UF V4.0 km³/y	Water Green: Requirement (ET) for Irrigation, Siebert and Döll (2009, 2008), Green Water Availability over Irrigated Areas Based on FAO/UF V4.0 km³/y	Water Blue: Requirement (ET) for Irrigation, Siebert and Döll (2009, 2008), Total (Blue & Green) Water Requirement for Irrigation Based on FAO/UF V4.0 km³/y
A1	A2	A22	A26	A27	A28	A29	A30
79	Macedonia	6.4	0.0	0.0	0.2	0.1	0.3
136	Burundi	3.6	0.1	0.1	0.1	0.1	0.1
95	Finland	110.0	0.2	0.2	0.0	0.1	0.1
100	Rwanda	5.2	0.3	0.0	0.0	0.0	0.0
121	Togo	14.7	0.1	0.0	0.0	0.0	0.0
111	Switzerland	53.3	0.0	0.0	0.0	0.0	0.0
141	Niger	33.7	0.0	0.6	0.6	0.2	0.8
127	Benin	25.8	0.4	0.1	0.0	0.0	0.0
135	Namibia	45.5	0.0	0.1	0.1	0.0	0.1
185	Ireland	46.8	0.0	0.0	0.0	0.0	0.0
90	Dominican Republic	21.0	0.8	3.1	0.9	1.5	2.4
71	Libya	0.6	1.8	4.3	2.7	0.3	3.0
54	Saudi Arabia	2.4	16.9	25.7	12.4	0.7	13.1
145	Lesotho	5.2	0.0	0.0	0.0	0.0	0.0
88	Swaziland	4.5	0.5	0.3	0.3	0.3	0.6
165	Slovenia	32.1	0.0	0.0	0.0	0.0	0.1
104	Haiti	14.0	0.5	0.6	0.2	0.5	0.7
80	Armenia	10.5	1.1	2.3	0.6	0.3	0.9

155	Norway	381.4	0.0	0.1	0.0	0.1	0.2
131	Montenegro	0.0	0.0	0.0	0.0	0.0	0.0
75	Guinea Bissau	31.0	1.7	0.2	0.1	0.0	0.1
134	El Salvador	25.2	0.1	0.2	0.2	0.3	0.5
84	Guyana	241.0	0.5	1.1	0.3	1.2	1.6
91	Yemen	4.1	0.9	2.6	2.9	0.5	3.4
101	Gambia	0.0	0.0	0.0	0.0	0.0	0.1
132	Eritrea	6.3	0.1	0.5	0.0	0.0	0.1
147	Belize	18.6	0.0	0.0	0.0	0.0	0.0
143	East Timor	0.0	0.0	0.0	0.0	0.0	0.0
82	Israel	1.7	1.1	0.9	1.5	0.2	1.7
157	Bhutan	95.0	0.1	0.2	0.1	0.2	0.3
139	Cyprus	0.4	0.1	0.3	0.3	0.1	0.4
119	Lebanon	4.8	0.2	0.6	1.0	0.2	1.2
133	Puerto Rico	0.0	0.0	0.0	0.0	0.2	0.2
105	Jordan	0.9	1.0	0.8	0.8	0.1	0.9
125	Mauritania	0.0	0.0	0.0	0.2	0.0	0.2
187	Luxembourg	1.6	0.0	0.0	0.0	0.0	0.0
123	Suriname	122.0	0.2	0.3	0.1	0.3	0.4
114	Oman	1.0	0.5	2.3	1.1	0.1	1.1
160	Brunei	8.5	0.0	0.0	0.0	0.0	0.0
98	United Arab Emirates	0.2	1.8	4.7	3.3	0.1	3.4
140	Jamaica	9.4	0.0	0.0	0.1	0.3	0.3
154	West Bank	0.0	0.0	0.0	0.0	0.0	0.0
150	Equatorial Guinea	26.0	0.0	0.0	0.0	0.0	0.0
117	Qatar	0.1	0.8	0.2	0.1	0.0	0.1
118	Kuwait	0.0	0.8	0.1	0.1	0.0	0.1
148	French Guyana	0.0	0.0	0.0	0.0	0.0	0.0
153	Trinidad and Tobago	3.8	0.0	0.0	0.0	0.0	0.0

(Continued)

TABLE 16.5 (Continued)

Global Water Withdrawal and Water Use by Irrigated Croplands: A Country-by-Country Assessment

Rank Based on Total IWMI GIAM Cropland Areas #	Country Name	Water: Renewable Glieck et al. (2009) Annual Renewable Water Resource km³/y	Water: Withdrawal for Irrigation Wisser et al. (2008) Total Water Withdrawal for Irrigation Based on IWMI GIAM Irrigated Areas km³/y	Water: Withdrawal for Irrigation Wisser et al. (2008) Total Water Withdrawal for Irrigation Based on FAO/UF V4.0 Irrigated Areas km³/y	Water: Requirement (ET) for Irrigation Siebert and Döll (2009, 2008) Blue Water Requirement for Irrigation Based on Irrigated Areas of FAO/UF V4.0 km³/y	Water Green: Requirement (ET) for Irrigation Siebert and Döll (2009, 2008) Green Water Availability over Irrigated Areas Based on FAO/UF V4.0 km³/y	Water Blue: Requirement (ET) for Irrigation Siebert and Döll (2009, 2008) Total (Blue & Green) Water Requirement for Irrigation Based on FAO/UF V4.0 km³/y
A1	A2	A22	A26	A27	A28	A29	A30
195	Singapore	0.6	0.0	0.0	0.0	0.0	0.0
138	Gaza Strip	0.0	0.0	0.0	0.0	0.0	0.0
175	Andorra	0.0	0.0	0.0	0.0	0.0	0.0
186	Liechtenstein	0.0	0.0	0.0	0.0	0.0	0.0
162	San Marino	0.0	0.0	0.0	0.0	0.0	0.0
144	Mauritius	2.2	0.0	0.0	0.1	0.2	0.3
152	Guadeloupe	0.0	0.9	0.8	0.0	0.1	0.1
156	St. Kitts and Nevis	0.0	0.0	0.0	0.0	0.0	0.0
159	Virgin Islands	0.0	0.0	0.0	0.0	0.0	0.0
161	Reunion	5.0	0.0	0.0	0.0	0.1	0.1
163	Djibouti	0.3	0.0	0.0	0.0	0.0	0.0
166	Comoros	1.2	0.0	0.0	0.0	0.0	0.0
167	Anguilla	0.0	0.0	0.0	0.0	0.0	0.0
173	Monaco	0.0	0.0	0.0	0.0	0.0	0.0
169	Turks and Caicos Islands	0.0	0.0	0.0	0.0	0.0	0.0
170	Montserrat	0.0	0.0	0.0	0.0	0.0	0.0
171	St. Pierre and Miquelon	0.0	0.0	0.0	0.0	0.0	0.0

172	Cayman Islands	0.0	0.0	0.0	0.0	0.0	0.0
174	Seychelles	0.0	0.0	0.0	0.0	0.0	0.0
176	Bahrain	0.1	0.0	0.0	0.0	0.0	0.0
177	Barbados	0.1	0.0	0.0	0.0	0.0	0.0
178	Cape Verde	0.3	0.0	0.0	0.0	0.0	0.0
180	Fiji	28.6	0.0	0.0	0.0	0.0	0.0
183	Grenada	nd	0.0	0.0	0.0	0.0	0.0
184	Guam	0.0	0.0	0.0	0.0	0.0	0.0
188	Malta	0.1	0.0	0.0	0.0	0.0	0.0
189	Martinique	11.4	0.3	0.9	0.0	0.1	0.1
190	Northern Marianna Islands	0.0	0.0	0.0	0.0	0.0	0.0
193	Pitcairn Islands	0.0	0.0	0.0	0.0	0.0	0.0
194	Sao Tome and Principe	0.0	0.0	0.0	0.0	0.1	0.1
196	St. Lucia	0.0	0.0	0.0	0.0	0.0	0.0
197	St. Vincent and the Grenadines	0.0	0.0	0.0	0.0	0.0	0.0
198	Vatican City	0.0	0.0	0.0	0.0	0.0	0.0
	TOTAL	**54,695**	**3,798**	**3,091**	**1,180**	**919**	**2,099**

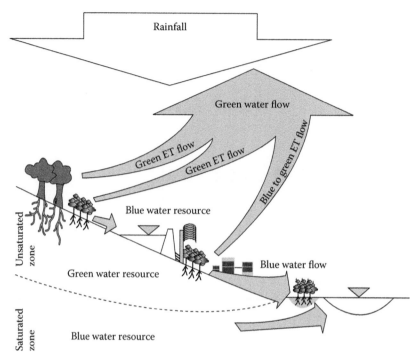

FIGURE 16.5
Concept of blue water versus green water. (Adapted from Falkenmark, M., and J. Rockström, *J Water Resour Plann Manage*, 132, 2006.)

variations are large; for instance, China is facing severe water shortages in the north, including the Yellow River Basin, while the south still has abundant water resources. Intracountry water transfers have been proposed as a solution to address such water scarcity issues (Xu, Ye, and Li 2008). However, it is not clear how to bring the water back to its origin to avoid potential environmental and socioeconomic consequences. In countries with large spatial variability in water availability, intracountry virtual water trade may help address the issue (Wichelns 2005), with a lower footprint on the environment albeit at a slightly higher economic cost (food transport may be more expensive than water transport; Sarkar et al. 2007). Shifting people instead of transferring water may be yet another alternative (Wichelns 2003), apart from global virtual water trade (Hanasaki et al. 2010). In any case, precise estimates of water use and withdrawals can help evaluate the impacts of these alternative strategies as well as optimize the productivity of croplands and their water use.

Water used by crops is different from water withdrawal. Also, there are no clear assessments of water used by rain-fed croplands. This calls for water-use assessment based on blue water use (water used by crops from rivers, reservoirs, lakes, and groundwater in the saturated zone) and green water use (water used by crops in saturated zone of the ground; Figure 16.5; Falkenmark 2007). Siebert and Döll (2008, 2009) made such an assessment over the irrigated areas of the world. They determined that global irrigated areas consume 2099 km^3/y, of which 919 km^3/y is green water use (precipitation that falls directly on irrigated croplands) and the 1180 km^3/y is blue water use (Table 16.4). A country-by-country

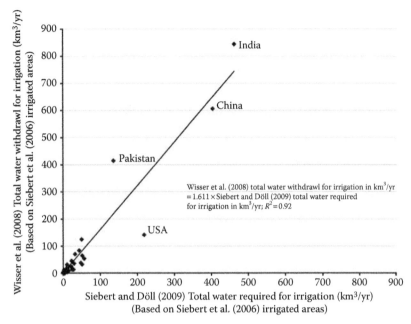

FIGURE 16.6
Water withdrawn versus water required—a country-by-country assessment.

water withdrawal versus water required (water approximately used) shows a consistent pattern of high water withdrawal to actual water required or used by irrigated croplands (Figure 16.6). Water withdrawals are even higher when irrigated areas computed by Thenkabail et al. (2009a) are used (Table 16.5). Typically, 1.6–2.5 times the water required (equivalent to the water used for optimal growing conditions) is actually withdrawn, thus achieving only 40–60% irrigation efficiency (Table 16.5). Clear assessments of green water use can improve the productivity of rain-fed agriculture and harness the potential of green water for increasing resilience to global environmental change (Hoff et al. 2010; Rockström et al. 2008).

16.6 Estimates from Regional Studies

Regional cropland area maps and statistics are the primary inputs for water use and food security modeling, environmental and natural resource management, assessment, planning, development, and better targeting of related policies and investments. A variety of georeferenced maps and data sets are needed to address global environmental change and water security issues, from regional to global scales to characterize current cropland patterns, document major changes in land and water use, and place renewed emphasis on land and water use in thematic classification schemes. Over the past decade, cropland mapping science has made considerable strides using remote sensing and nonsensing data. Examples include mapping irrigated areas over the continental United States, with a strong correlation and small bias, little over 2% of the total irrigated area in the United States, compared to existing large area irrigation database (Ozdogan and Gutman 2008); four LULC-related maps for the state of Kansas that progressively classified crop or non-crop, general crops, summer crops, and irrigated or rain-fed crops, with cropland areas generally having accuracies usually within 1–5% of USDA reported crop areas for most classes (Wardlow and Egbert 2008); use of field data and Landsat to map land cover at four sites in North America containing agricultural cropland, with overall accuracy of 81–95% (Cohen et al. 2003); modeling of water and carbon cycle for analyzing the seasonality of hydrologic processes, water management, and carbon cycles in California (Ichii et al. 2009); use of Advanced Very High Resolution Radiometer (AVHRR) imagery every 10 days at 1-km resolution and ground measurements to validate cropland map/leaf area index (LAI), with an accuracy of 50–75% for individual AVHRR and vegetation pixels, with mixed crop types found to cause some bias in a wide application in Canada (Chen et al. 2002), and land-cover classification that successfully compares with the Coordination of Information on the Environment database that serves as a reference for the reliability of the study over France (Han, Champeaux, and Roujean 2004); use of QuickBird satellite data for surface reflectance and water content retrieval for corn and potato croplands in Minnesota, with mean relative differences of 0.03% and 0.06% respectively (Wu, Wang, and Bauer 2005), and over Europe for the years 2000–2003, with ground measurement differences lower than 0.5 LAI units in most cases (Verger et al. 2009); comparison of nine sites varying widely in biome type and land use, but including croplands, using 1-km spatial resolution MODIS sensor to compare annual net primary production in multiple biomes (Turner et al. 2006); modeling of agricultural water and urban surface runoff based on remote sensing in Denmark, from field to macro scales in catchments dominated by agricultural, forest, and urban land use (Boegh et al. 2009); and identification of rapid transformation from natural vegetation to row-crop agriculture and conversion from single crops to double crops from 2000 to 2005 in southwestern Brazil (Galford et al. 2008).

Examples from Asia include the urbanization of Shenzhen City, China, between 1999 and 2005, and analysis of its impact on net primary productivity and annual reduction of carbon in croplands using MODIS, Landsat, meteorological and other field data (Deyong et al. 2009); spatial and temporal patterns of croplands in Chin during 1990–2000, showing that most of the lost cropland had good quality, whereas the most gained cropland had poor quality (Liu et al. 2005) and landscape changes for the Three Gorges Reservoir area of the Yangtze River from 1977 to 2005 (Zhang, Zhengjun, and Xiaoxia 2009); mapping of paddy rice agriculture in south and southeast Asia, with area estimates of paddy rice highly correlated at the national level and positively correlated with the subnational level statistical data sets (Xiao et al. 2006) as well as mapping of cropland, seasonal water use, and crop water productivity in the Indo-Gangetic river basin, including Bangladesh, India, Nepal, and Pakistan, by combining remotely sensed imagery, national statistics, and meteorological data (Cai and Sharma 2010); and monitoring 25 years of land-cover change dynamics in Africa through a sample-based remote sensing approach (Brink and Eva 2009).

Some other applications include land-cover generalizations and wetland detection (Cook et al. 2009), timing and extent of fires (Roy et al. 2008), agricultural burning and emissions from biomass (McCarty, Justice, and Korontzi 2007), and large-scale global disturbances such as hurricanes and ice storms (Mildrexler, Zhao, and Running 2009). Recent applications include the use of land cover, land use, and land functions and their relations (Verburg et al. 2009). Multifunctional productivity of croplands and water resources has not been assessed yet. Spatially explicit insights are needed in order to better assess and evaluate the potential welfare impacts. Multisource data fusion (Huang et al. 2010; You, Wood, and Wood-Sichra 2009) is a promising approach to improve global and regional cropland maps and their water use and can contribute to the predictive capabilities of global and regional water and carbon cycles and climate and ecosystem models (Sun et al. 2008). Spatial, temporal, spectral, multitype, or dimensional data fusion is quite new and poses significant challenges. Currently, a number of global and regional land-cover products exist, and they all have been produced from optical, moderate-resolution remote sensing and are thematically focused on characterizing the different cropland and vegetation types; some include temporal signature compliance to classification processes. However, they are not designed to be comparable and exist as independent data sets, and the lack of interoperability is a significant challenge (Herold et al. 2008) such that the uncertainties in estimates of cropland area across products are an issue (Heistermann, Müller, and Ronneberger 2006; Khan et al. 2010; Xiao et al. 2003).

16.7 Conclusions and Policy Implications

There are four major cropland area maps and statistics at the global level. One study primarily used multisensor remote sensing (Thenkabail et al. 2009a; Thenkabail et al. 2009c). The other three studies used a combination of national statistics and geospatial techniques (Goldewijk et al. 2009; Portmann, Siebert, and Döll 2009; Ramankutty et al. 2008; Siebert and Döll 2008, 2009). The total global cropland areas estimated by these four studies ranged between 1.3 and 1.54 billion hectares for the nominal year 2000. Only two studies, reported in four peer-reviewed papers (Portmann, Siebert, and Döll 2009; Siebert and Döll 2008, 2009; Thenkabail et al. 2009a; Thenkabail et al. 2009c), separated the irrigated areas from rain-fed areas. The global irrigated area estimates, without considering cropping intensity, ranged between 312 Mha (Portmann, Siebert, and Döll 2009) and 399 Mha (Thenkabail et al. 2009a). Only Thenkabail et al. (2009a) estimated the irrigated areas by considering cropping intensity as well, at 467 Mha. However, these studies varied significantly in providing precise spatial location of cropland areas and separating irrigated areas from rain-fed areas. Furthermore, none of the studies provide a proper assessment of crop type and/or crop dominance or irrigation by its source. A proper assessment and precise estimates of these aspects of global croplands are crucial given that 60–90% of all human water use is by croplands.

The global crop water use varied between 6685 and 7500 km^3 yr^{-1}; of this about 70% is by rain-fed croplands (green water use) and 30% by irrigated croplands (blue water use). However, irrigated croplands use blue water (water in rivers, reservoirs, lakes, and pumped groundwater from the saturated zone). About 80% of all blue water currently used by humans goes to irrigated areas. This highlights the need for continued focus on irrigated croplands and their water use for enhancing global food security.

Uncertainties in global and regional cropland areas, their water use, and the precise geographic location of these parameters are quite high at present. The need for high-resolution remote sensing products that can provide a greater geographic precision, crop types, and cropping intensities (by using high spatial resolution with high temporal resolution data) remains crucial for ensuring water and food security in the future.

References

Boegh, E., R. N. Poulsen, M. Butts, P. Abrahamsen, E. Dellwik, S. Hansen, C. B. Hasager et al. 2009. Remote sensing based evapotranspiration and runoff modeling of agricultural, forest and urban flux sites in Denmark: From field to macro-scale. *J Hydrol* 377:300–16.

Brink, A. B., and H. D. Eva. 2009. Monitoring 25 years of land cover change dynamics in Africa: A sample based remote sensing approach. *Appl Geogr* 29:501–12.

Cai, X. L., and B. R. Sharma. 2010. Integrating remote sensing, census and weather data for an assessment of rice yield, water consumption and water productivity in the Indo-Gangetic river basin. *Agric Water Manag* 97(2):309–16, doi:10.1016/j.agwat.2009.09.021.

Chen, J. M., G. Pavlic, L. Brown, J. Cihlar, S. G. Leblanc, H. P. White, R. J. Hall et al. 2002. Derivation and validation of Canada-wide coarse-resolution leaf area index maps using high-resolution satellite imagery and ground measurements. *Rem Sens Environ* 80:165–84.

Cohen, W. B., T. K. Maiersperger, Z. Yang, S. T. Gower, D. P. Turner, W. D. Ritts, M. Berterretche, and S. W. Running. 2003. Comparisons of land cover and LAI estimates derived from ETM+ and MODIS for four sites in North America: A quality assessment of 2000/2001 provisional MODIS products. *Rem Sens Environ* 88:233–55.

Cook, B. D., P. V. Bolstad, E. Næsset, R. S. Anderson, S. Garrigues, J. T. Morisette, J. Nickeson, and K. J. Davis. 2009. Using LiDAR and quickbird data to model plant production and quantify uncertainties associated with wetland detection and land cover generalizations. *Rem Sens Environ* 113:2366–79.

Deyong, Y., S. Hongbo, S. Peijun, Z. Wenquan, and P. Yaozhong. 2009. How does the conversion of land cover to urban use affect net primary productivity? A case study in Shenzhen city, China. *Agric For Meteorol* (Special Section on Water and Carbon Dynamics in Selected Ecosystems in China) 149:2054–60.

Dheeravath, V., P. S. Thenkabail, G. Chandrakantha, P. Noojipady, C. B. Biradar, H. Turral, M. Gumma, G. P. O. Reddy, and M. Velpuri. 2010. Irrigated areas of India derived using MODIS 500 m data for years 2001–2003. *ISPRS J Photogramm Rem Sens* 65(1):42–59.

Falkenmark, M. 2007. Shift in thinking to address the 21st century hunger gap: Moving focus from blue to green water management. *Water Res Manag* 21:3–18.

Falkenmark, M., and J. Rockström. 2006. The new blue and green water paradigm: Breaking new ground for water resources planning and management. *J Water Resour Plann Manage* 132:1–15.

Galford, G. L., J. F. Mustard, J. Melillo, A. Gendrin, C. C. Cerri, and C. E. P. Cerri. 2008. Wavelet analysis of MODIS time series to detect expansion and intensification of row-crop agriculture in Brazil. *Rem Sens Environ* (Soil Moisture Experiments 2004 (SMEX04) Special Issue) 112:576–87.

Goldewijk, K., A. Beusen, M. de Vos, and G. van Drecht. 2010. Holocene = Anthropocene? The HYDE database for integrated global change research over the past 12,000 years. *Global Ecol and Biogeogr*, doi:10.111/j.1466-8238.2010.00587.x.

Gordon, L. J., C. M. Finlayson, and M. Falkenmark. 2009. Managing water in agriculture for food production and other ecosystem services. *Agric Water Manag* 97(4):512–19, doi:10.1016/j.agwat.2009.03.017.

Gordon, L. J., G. D. Peterson, and E. Bennett. 2008. Agricultural modifications of hydrological flows create ecological surprises. *Trends Ecol Evol* 23:211–9.

Han, K. S., J. L. Champeaux, and J. L. Roujean. 2004. A land cover classification product over France at 1 km resolution using SPOT4/VEGETATION data. *Rem Sens Environ* 92:52–66.

Hanasaki, N., T. Inuzuka, S. Kanae, and T. Oki. 2010. An estimation of global virtual water flow and sources of water withdrawal for major crops and livestock products using a global hydrological model. *J Hydrol* 384(3–4):232–44.

Heistermann, M., C. Müller, and K. Ronneberger. 2006. Land in sight? Achievements, deficits and potentials of continental to global scale land-use modeling. *Agric Ecosyst Environ* 114:141–58.

Herold, M., P. Mayaux, C. E. Woodcock, A. Baccini, and C. Schmullius. 2008. Some challenges in global land cover mapping: An assessment of agreement and accuracy in existing 1 km datasets. *Rem Sens Environ* (Earth Observations for Terrestrial Biodiversity and Ecosystems Special Issue) 112:2538–56.

Hoff, H., M. Falkenmark, D. Gerten, L. Gordon, L. Karlberg, and J. Rockström. 2010. Greening the global water system. *J Hydrol* 384(3–4):177–86.

Huang, S., C. Potter, R. L. Crabtree, S. Hager, and P. Gross. 2010. Fusing optical and radar data to estimate sagebrush, herbaceous, and bare ground cover in Yellowstone. *Rem Sens Environ* 114(2):251–64, doi:10.1016/j.rse.2009.09.013.

Ichii, K., W. Wang, H. Hashimoto, F. Yang, P. Votava, A. R. Michaelis, and R. R. Nemani. 2009. Refinement of rooting depths using satellite-based evapotranspiration seasonality for ecosystem modeling in California. *Agric For Meteorol* (Special Section on Water and Carbon Dynamics in Selected Ecosystems in China) 149:1907–18.

IPCC. 2007. *The Fourth Assessment Report (The AR4 Synthesis Report)*. Geneva, Switzerland: Intergovernmental Panel on Climate Change.

Khan, M. R., C. A. J. M. de Bie, H. van Keulen, E. M. A. Smaling, and R. Real. 2010. Disaggregating and mapping crop statistics using hypertemporal remote sensing. *Int J Appl Earth Obs Geoinf* 12(1):36–46, doi:10.1016/j.jag.2009.09.010.

Khan, S., and M. A. Hanjra. 2008. Sustainable land and water management policies and practices: A pathway to environmental sustainability in large irrigation systems. *Land Degrad Dev* 19:469–87.

Khan, S., and M. A. Hanjra. 2009. Footprints of water and energy inputs in food production: Global perspectives. *Food Policy* 34:130–40.

Liu, J., M. Liu, H. Tian, D. Zhuang, Z. Zhang, W. Zhang, X. Tang, and X. Deng. 2005. Spatial and temporal patterns of China's cropland during 1990–2000: An analysis based on Landsat TM data. *Rem Sens Environ* 98:442–56.

Maraseni, T. M., S. M. Mushtaq, and J. Maroulis. 2009. Greenhouse gas emissions from rice farming inputs: A cross-country assessment. *J Agric Sci* 147(2):117–26, doi:10.1017/S0021859608008411.

McCarty, J. L., C. O. Justice, and S. Korontzi. 2007. Agricultural burning in the southeastern United States detected by MODIS. *Rem Sens Environ* (The Application of Remote Sensing to Fire Research in the Eastern United States) 108:151–62.

Melesse, A. M., Q. Weng, P. Thenkabail, and G. Senay. 2007. Remote sensing sensors and applications in environmental resources mapping and modelling. (Special Issue of Remote Sensing of Natural Resources and the Environment). *Sens J* 7:3209–41.

Mildrexler, D. J., M. Zhao, and S. W. Running. 2009. Testing a MODIS Global Disturbance Index across North America. *Rem Sens Environ* 113:2103–17.

Ozdogan, M., and G. Gutman. 2008. A new methodology to map irrigated areas using multi-temporal MODIS and ancillary data: An application example in the continental US. *Rem Sens Environ* 112:3520–37.

Portmann, F., S. Siebert, and P. Döll. 2009. MIRCA2000—Global monthly irrigated and rainfed crop areas around the year 2000: A new high-resolution data set for agricultural and hydrological modelling. *Global Biogeochem Cycles* 24, doi:2008GB0003435.

Ramankutty, N., A. T. Evan, C. Monfreda, and J. A. Foley. 2008. Farming the planet: 1. Geographic distribution of global agricultural lands in the year 2000. *Global Biogeochem Cycles* 22, doi:10.1029/2007GB002952.

Rockström, J., M. Falkenmark, L. Karlberg, H. Hoff, S. Rost, and D. Gerten. 2009. Future water availability for global food production: The potential of green water for increasing resilience to global change. *Water Resour Res* 45, doi:10.1029/2007WR006767.

Roy, D. P., L. Boschetti, C. O. Justice, and J. Ju. 2008. The collection 5 MODIS burned area product—Global evaluation by comparison with the MODIS active fire product. *Rem Sens Environ* 112:3690–707.

Sarkar, P., V. Mathur, V. Maitri, and K. Kalra. 2007. Potential for economic gains from Inland water transport in India. *Transp Res Rec* 2033:45–52.

Siebert, S., and P. Döll. 2008. The Global Crop Water Model (GCWM): Documentation and first results for irrigated crops. In *Frankfurt Hydrology Paper 07*. Frankfurt am Main, Germany: Institute of Physical Geography, University of Frankfurt.

Siebert, S., and P. Döll. 2009. Quantifying blue and green virtual water contents in global crop production as well as potential production losses without irrigation. *J Hydrol* 384(3–4):198–217, doi:10.1016/ j.j hydrol.2009.07.031.

Siebert, S., J. Hoogeveen, and K. Frenken. 2006. Irrigation in Africa, Europe and Latin America—Update of the Digital Global Map of Irrigation Areas to Version 4. *Frankfurt Hydrology Paper 05*, 134. Frankfurt am Main, Germany and Rome, Italy: Institute of Physical Geography, University of Frankfurt.

Sun, W., S. Liang, G. Xu, H. Fang, and R. Dickinson. 2008. Mapping plant functional types from MODIS data using multisource evidential reasoning. *Rem Sens Environ* 112:1010–24.

Tan, Z., S. Liu, L. L. Tieszen, and E. Tachie-Obeng. 2009. Simulated dynamics of carbon stocks driven by changes in land use, management and climate in a tropical moist ecosystem of Ghana. *Agric Ecosyst Environ* 130:171–6.

Thenkabail, P. S., C. M. Biradar, P. Noojipady, X. L. Cai, V. Dheeravath, Y. J. Li, M. Velpuri, M. Gumma, and S. Pandey. 2007. Sub-pixel irrigated area calculation methods. *Sens J* 7:2519–38.

Thenkabail, P. S., C. M. Biradar, P. Noojipady, V. Dheeravath, Y. J. Li, M. Velpuri, M. Gumma, Reddy, Get al. 2009a. Globl irrigated area map (GIAM), derived from remote sensing, for the end of the last millennium. *Int J Remote Sens* 30:3679–733.

Thenkabail, P. S., V. Dheeravath, C. M. Biradar, O. P. Gangalakunta, P. Noojipady, C. Gurappa, M. Velpuri, M. Gumma, and Y. Li. 2009b. Irrigated area maps and statistics of India using remote sensing and national statistics. *Rem Sens* 1:50–67.

Thenkabail, P. S., G. J. Lyon, H. Turral, and C. M. Biradar. 2009a. *Remote Sensing of Global Croplands for Food Security*. Boca Raton, FL: CRC Press/Taylor & Francis Group.

Thenkabail, P. S., M. Schull, and H. Turral. 2005. Ganges and Indus river basin land use/land cover (LULC) and irrigated area mapping using continuous streams of MODIS data. *Rem Sens Environ* 95:317–41.

Turner, D. P., W. D. Ritts, W. B. Cohen, S. T. Gower, S. W. Running, M. Zhao, M. H. Costa et al. 2006. Evaluation of MODIS NPP and GPP products across multiple biomes. *Rem Sens Environ* 102:282–92.

UNDP. 2009. Population Division of the Department of Economic and Social Affairs of the United Nations Secretariat. World Population Prospects: The 2008 Revision. http://esa.un.org/unpp (accessed October 31, 2009).

USEPA. 2006. *Global Anthropogenic Non-CO2 Greenhouse Gas Emissions: 1990–2002*. Washington, DC: Office of Atmospheric Programs, USEPA.

Verburg, P. H., J. van de Steeg, A. Veldkamp, and L. Willemen. 2009. From land cover change to land function dynamics: A major challenge to improve land characterization. *J Environ Manage* 90:1327–35.

Verger, A., F. Camacho, F. J. García-Haro, and J. Meliá. 2009. Prototyping of Land-SAF leaf area index algorithm with VEGETATION and MODIS data over Europe. *Rem Sens Environ* 113:2285–97.

Wardlow, B. D., and S. L. Egbert. 2008. Large-area crop mapping using time-series MODIS 250 m NDVI data: An assessment for the U.S. Central Great Plains. *Rem Sens Environ* 112:1096–116.

Wichelns, D. 2003. Moving water to move people: evaluating success of the Toshka project in Egypt. *Water Int* 28:52–6.

Wichelns, D. 2005. The virtual water metaphor enhances policy discussions regarding scarce resources. *Water Int* 30:428–37.

Wu, J., D. Wang, and M. E. Bauer. 2005. Image-based atmospheric correction of QuickBird imagery of Minnesota cropland. *Rem Sens Environ* 99:315–25.

Xiao, X., S. Boles, S. Frolking, C. Li, J. Y. Babu, W. Salas, and B. Moore III. 2006. Mapping paddy rice agriculture in South and Southeast Asia using multi-temporal MODIS images. *Rem Sens Environ* 100:95–113.

Xiao, X., J. Liu, D. Zhuang, S. Frolking, S. Boles, B. Xu, M. Liu, W. Salas, B. Moore, and C. Li. 2003. Uncertainties in estimates of cropland area in China: A comparison between an AVHRR-derived dataset and a Landsat TM-derived dataset. *Glob Planet Change* (Special Issue Asia Monsoon Environment System and Global Change) 37:297–306.

Xu, H., M. Ye, and J. Li. 2008. The water transfer effects on agricultural development in the lower Tarim River, Xinjiang of China. *Agric Water Manag* 95:59–68.

You, L., S. Wood, and U. Wood-Sichra. 2009. Generating plausible crop distribution maps for Sub-Saharan Africa using a spatially disaggregated data fusion and optimization approach. *Agric Syst* 99:126–40.

Zhang, J., L. Zhengjun, and S. Xiaoxia. 2009. Changing landscape in the Three Gorges Reservoir area of Yangtze River from 1977 to 2005: Land use/land cover, vegetation cover changes estimated using multi-source satellite data. *Int J Appl Earth Obs Geoinf* 11:403–12.

Zou, J., S. Liu, Y. Qin, G. Pan, and D. Zhu. 2009. Sewage irrigation increased methane and nitrous oxide emissions from rice paddies in southeast China. 2009. *Agric Ecosyst Environ* 129:516–22.

Section IV

Environmental Applications: Air, Water, and Land

17

Remote Sensing of Aerosols from Space: A Review of Aerosol Retrieval Using the Moderate-Resolution Imaging Spectroradiometer

Man Sing Wong and Janet Nichol

CONTENTS

17.1 Introduction

Aerosols are solid or liquid airborne particulates of variable composition, which reside in stratified layers of the atmosphere. Generally, they are defined as atmospheric particles of sizes between about 0.1 and 10 μm, though the sizes of condensation nuclei are typically about 0.01 μm. Under normal conditions, most atmospheric aerosol exists in the troposphere. Natural sources such as dust storms, desert and soil erosion, biogenic emissions, forest and grassland fires, and sea spray account for about 90% of aerosols, and the rest result from anthropogenic activity. The background (natural) tropospheric aerosols are temporally and spatially variable. The study of aerosols is important because of their effects on the Earth radiation budget, climate change, atmospheric conditions, and human health. Recent research has focused on fine aerosols due to their long-term damage to the respiratory system (Davidson, Phalen, and Solomon 2005; Dominici et al. 2006).

The most convenient unit of measurement for aerosols by remote sensing is aerosol optical thickness (AOT) because it represents the total attenuation due to scattering and absorption along a path measured vertically through atmosphere (Equation 17.1):

$$I = I_0 e^{-\tau_\lambda m} \tag{17.1}$$

where I is the observed intensity for a given path length, I_0 is the intensity of radiation at the source, τ_λ is the optical thickness as a function of wavelength, and m is air mass.

Due to the inadequacy of fixed air quality stations, which are unable to capture spatial variability, there is increasing interest in satellite sensors for synoptic measurement of atmospheric turbidity based on AOT. Retrieval of aerosols from satellite remote sensing is not straightforward because no single algorithm can function over all land surface types. The main state-of-the art aerosol retrieval algorithms have been devised for land (vegetation), land (bright surfaces), or ocean. Only a few sensors have been designed specifically for aerosol retrieval, but others that are not explicitly designed for this application have been used for aerosol retrieval. These include the Total Ozone Mapping Spectrometer (TOMS) whose primary purpose is monitoring the atmospheric ozone content (Herman et al. 1997; Hsu et al. 1999), the Advanced Very High Resolution Radiometer (AVHRR), which was designed for measuring sea surface temperature and vegetation, and the Sea-Viewing Wide Field-of-View Sensor (SeaWiFS), which was developed to study ocean color and marine biogeochemical processes. The Earth-Observing System (EOS) TERRA with the Moderate Resolution Imaging Spectroradiometer (MODIS) and Multiangle Imaging Spectroradiometer (MISR) also provides capabilities for atmosphere, as well as land and ocean, studies (Tanré et al. 1997; Wanner et al. 1997).

However, the detection of atmospheric parameters from remote sensing platforms is usually hindered by the signal from the ground surface, being stronger than that from the atmosphere, contrary to other applications, where the atmospheric effect is considered as noise and should be removed. Most algorithms for quantification of aerosol amounts are thus dependent on obtaining the surface reflectance at the bottom of atmosphere (BOA) and then subtracting that from the total remotely sensed signal at the top of atmosphere (TOA). In addition, aerosol remote sensing algorithms can be classified into three major types according to the type of sensing system: (1) multiwavelength retrieval, (2) polarization retrieval, and (3) active measurement using lidar. The most frequently used method is the multiwavelength retrieval, which is applied to sensors such as MODIS, AVHRR, TOMS, and SeaWiFS with considerable success; uncertainties are reported to be between 20% and 30%. The percentage of uncertainties is similar to that of polarization retrieval, with the Polarization and Directionality of the Earth's Reflectances (POLDER) instrument said to give an error of 30% (Herman et al. 1997). The newly launched Cloud-Aerosol Lidar and Infrared Pathfinder Satellite Observations (CALIPSO) conducts active measurement of vertical aerosol distribution at 5-km horizontal resolution, but the uncertainty of AOT retrieval from CALIPSO is approximately 40% (Vaughan et al. 2004). This chapter focuses on the use of multiwavelength algorithms with MODIS images.

The MODIS is a sensor aboard the TERRA and AQUA satellites. TERRA was launched in 1999, passing from north to south in the morning (approximately 10:30 A.M. local time), and AQUA was launched in 2002, passing from south to north in the afternoon (approximately 1:30 P.M. local time). With 36 wavebands at 250-m, 500-m and 1-km resolution, MODIS can be used for atmospheric, oceanic, and land studies at both global and local scales. MODIS also provides specific products such as atmospheric aerosols, ocean color, land-cover maps, and fire products. Calibration of the MODIS satellite observations for aerosols is currently achieved by a network of Aerosol Robotic Network (AERONET) stations distributed around the world. AERONET (Holben et al. 1998) is a federated network of ground sun photometers with more than 400 sites. An AERONET station consists of a Cimel sun photometer, which measures the aerosol extinction every 15 minutes using multiple wavelengths, a solar panel, and a controller. It provides real-time AOT, precipitable water, inversion products, including size distribution, single-scattering albedo (SSA), and refractive index, based on the solutions of radiative transfer equations. There are three levels of data: 1, 1.5, and 2, which represent the raw data, cloud-screened data, and cloud-screened and quality-assured data, respectively.

17.2 Aerosol Observations Using MODIS (Operational Products)

Before and following the launch of MODIS in 1999, a number of algorithms for aerosol retrieval were devised. These state-of-the art methodologies include (1) the dense dark vegetation (DDV) algorithm (known as collection-4 algorithm; Kaufman and Tanré 1998), (2) the second-generation MODIS operation algorithm (known as collection-5 algorithm; Levy et al. 2007), and (3) the deep blue algorithm (Hsu et al. 2004, 2006). These multiwavelength algorithms take advantage of different aerosol scattering properties at different wavelengths (e.g., longer wavelengths have smaller aerosol loadings). Thus, by virtue of their spectral differences, the amount and, to some extent, the type and size of aerosols can be inferred from a combination of longer and shorter wavelengths. This section will briefly describe the rationales and operational methodologies of these methods.

Kaufman and Tanré (1998) proposed the DDV method, using a shortwave infrared (SWIR; 2.1 μm) wavelength to estimate surface reflectances for shorter wavelength bands (red region: 0.66 μm and blue region: 0.47 μm) over dense forests. This methodology only works over dense forests and only with selected kernels containing >60% vegetated dark surfaces. Kaufman and Tanré (1998) and Kaufman and Sendra (1987) give equations for estimating the surface reflectances for red and blue wavelengths from their correlations with a SWIR band representing surface reflectances (Equations 17.2 and 17.3):

$$\rho_{0.47\mu m} = 0.25 \cdot \rho_{2\cdot12\mu m} \tag{17.2}$$

$$\rho_{0.66\mu m} = 0.5 \cdot \rho_{2\cdot12\mu m} \tag{17.3}$$

It is assumed that due to aerosol there is difference between the original TOA reflectances from the blue and red bands and the surface reflectances derived from the SWIR band. This difference is then fitted to a best-fit aerosol model, with the knowledge of the expected aerosol types in the study area, for example, continental (Lenoble and Brogniez 1984), industrial or urban (Remer et al. 1996), biomass burning (Hao and Liu 1994), and marine (Husar, Prospero, and Stowe 1997), to obtain an AOT value for each image waveband. Dubovik et al. (1998) suggested that a window with a size of 10 km gives the best signal-to-noise ratio for global aerosol retrieval using MODIS. However, this method has several limitations, including (1) only coarse resolution that is suitable for global monitoring can be achieved, (2) its operation is limited to vegetated areas and cannot operate over bright urban surfaces, and (3) it has low accuracy in southeast China (Kaufman and Tanré 1998). In addition, Chu et al. (2002) showed that collection-4 algorithm (DDV's algorithm, Equations 17.2 and 17.3) had a positive bias in comparison to the AERONET sun photometer data. Remer et al. (2005) and Levy et al. (2004) also reported certain inherent problems in determining surface reflectance using the MODIS collection-4 algorithm. Their results imply that inaccurate surface properties can lead to errors ($\pm0.05 \pm 0.2\tau$) in aerosol retrieval.

Due to these perceived errors, Levy et al. (2007) then modified the algorithm by considering band correlation based on the normalized difference vegetation index $NDVI_{SWIR}$, and the scattering angle, since Gatebe et al. (2001) and Remer, Wald, and Kaufman (2001) suggested the VISIBLE/SWIR ratio is angle dependent. The rationale of Levy et al.'s collection-5 method is to first identify the dark pixels. A kernel of 10-km size is used for scanning, and the dark pixels are identified as those with surface reflectance less than 0.25 at 2.12 μm wavelength. The darkest 20% and the brightest 50% of pixels inside the box are discarded, and the remaining 30%, or at least 12 pixels, inside the box are used for $NDVI_{SWIR}$ calculation (Equation 17.4). Following this, the pixels are classified into three categories based on the $NDVI_{SWIR}$ (Equation 17.5), and the *f* linear equations (Equation 17.6) are applied to those three categories with three sets of slope and intercepts. The values of linear equations are determined by band correlation analysis using atmospherically corrected MODIS images.

$$NDVI_{SWIR} = \frac{(\rho_{1.24\mu m} - \rho_{2.12\mu m})}{(\rho_{1.24\mu m} + \rho_{2.12\mu m})} \quad (17.4)$$

$$NDVI_{SWIR} < 0.25$$
$$0.25 < NDVI_{SWIR} < 0.75 \quad (17.5)$$
$$NDVI_{SWIR} > 0.75$$

$$\rho_{0.66\mu m} = f_1(\rho_{2.12\mu m})$$
$$\rho_{0.47\mu m} = f_2(\rho_{2.12\mu m}) \quad (17.6)$$

The collection-5 algorithm also operates over bright surfaces if the surface reflectance at 2.12 μm is less than 0.4 and the number of pixels inside the 10-km kernel is greater than 12. Then, a 0.47-μm channel is used for aerosol retrieval, and the continental model is assigned during the lookup table (LUT) calculation. This capability for bright surface aerosol retrieval combined with the increased threshold of surface reflectance for dark pixel selection (0.15 in DDV and 0.25 in collection-5) allows the collection-5 algorithm to work over semiurban and suburban areas, although the method still does not work well over large and very bright surfaces, such as deserts or complex land surfaces. Also, since only one band at the 0.47 μm wavelength is used and only one aerosol model is assigned for aerosol retrieval, the quality of AOT over bright surfaces is deemed poor, with greater uncertainties. Nevertheless, when the new collection-5 algorithm was evaluated (Li et al. 2007; Mi et al. 2007), a significant improvement was found, with an increase of 27% in accuracy over the original DDV algorithm, when correlated with ground measurements.

Aerosol retrieval over bright surfaces is challenging because the land surface and atmospheric aerosol contents are not easy to differentiate because both have high reflectance values. The operational DDV and collection-5 algorithms retrieve aerosols over land when the surface reflectances are less than 0.15 and 0.25 at a 2.12 μm wavelength, respectively. They are unable to retrieve aerosols over large bright surface areas like the Mongolian and Saharan deserts, which are the most important sources of dust in China and Africa. Hsu et al. (2004, 2006) recently developed the deep blue algorithm for aerosol retrieval over bright surfaces such as desert, arid, and semiarid areas using MODIS images. This algorithm makes use of ratio between two blue wavelengths (412 and 490 nm) since the surfaces are bright in the red region and darker in the blue region. The deep blue algorithm has been demonstrated successfully only for large homogeneous surfaces such as deserts and not for areas of complex land cover, like urban areas.

17.3 High-Resolution Aerosol Observations of Densely Urbanized Region (Case Study of Hong Kong and the Pearl River Delta Region)

17.3.1 Study Area

Hong Kong (Figure 17.1), a city with a service-based economy located in southeast China, has suffered serious air pollution over the last decade. The Hong Kong PolyU AERONET station shows aerosol levels to be high, compared with other urban stations worldwide, for example a mean AOT of 0.69 for 440-nm band, compared with 0.57 for Beijing, 0.55 for Singapore, 0.22 for Rome, and 0.24 for the Goddard Space Flight Center. The Pearl River Delta (PRD) region is often covered with haze and gray smoke, which is observed on daily MODIS satellite images. Wu et al. (2005) showed that the AOT in this region is often higher than 0.6 at 550 nm. Previous studies in the PRD region have measured a range of particle concentrations for PM10 (particulate matter with aerodynamic diameter less than or equal to 10 μm) of 70–234 μg/m^3, with high average PM10 concentrations of above 200 μg/m^3 in winter, and around 100 μg/m^3 for PM2.5 (particulate matter with aerodynamic diameter less than or equal to 2.5 μm) in autumn (Wei et al. 1999; Cao et al. 2003, 2004). These high concentrations of suspended particulates create low visibility and greatly affect the regional radiative budget (Wu et al. 2005). During the long dry season in winter, northeasterly air masses mainly bring continental pollution into the PRD region and Hong Kong (Gnauk et al. 2008). The consequent effects on visibility and health due to continuous bad air have appeared gradually since 2000. The Hong Kong Environmental Protection Department (EPD 2004) reported that an increase of 10 μg/m^3 in the concentration of NO$_x$, SO$_2$, respiratory suspended particulate (RSP), and ozone causes associated diseases such as respiratory, chronic pulmonary, and cardiovascular heart diseases to increase by 0.2–3.9%, respectively. Ko et al. (2007) demonstrated that air pollution is accompanied by increased hospital admissions for chronic obstructive pulmonary disease in Hong Kong, especially during winter. Wong et al. (1999) also found significant relationships between hospital admissions in Hong Kong for all respiratory diseases, all cardiovascular diseases, chronic obstructive pulmonary diseases, and heart failure, and concentrations of the following

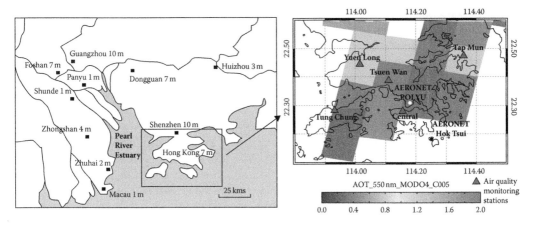

FIGURE 17.1
Left: Hong Kong and cities of the Pearl River Delta region, with population size (millions). Right: Study area of Hong Kong overlaid with five air quality monitoring stations, two AERONET stations, and a MODIS collection-5 aerosol optical thickness image on January 28, 2007.

four pollutants: nitrogen dioxide, sulfur dioxide, ozone, and PM10. The low visibility due to air pollutants in Hong Kong also affects marine and air navigation and affects the attractiveness of Hong Kong as a tourist destination.

The gathering of data over large regions such as the Hong Kong and Guangdong provinces of China (area of approximately 179,000 km²) is a major challenge to air pollution monitoring. The 16 air monitoring stations set up over this region are obviously insufficient for detailed observation. It is now realized that the only way to obtain measurements of aerosols over uninstrumented areas is from space (NASA 2009), and that this is the only way by which long-term global monitoring of aerosols can be done.

Images showing synoptic coverage over the PRD region as well as large parts of China often covered with gray haze are now available from the MODIS satellite sensor of the National Aeronautic and Space Administration (NASA), which provides images of the globe on a twice-daily basis, at spatial resolutions of 250 m, 500 m, and 1 km. From these, NASA has developed the MODIS level-2 AOT product (MOD04). However, the 10-km spatial resolution of this product only provides meaningful depictions on a regional scale, and aerosol monitoring over complex regions, such as urban and rural areas in Hong Kong territories (1095 km²), requires more spatial detail.

17.3.2 Methodology

17.3.2.1 Contrast Reduction Method and Li et al. Method

The heterogeneous land algorithm, also known as the contrast reduction method (Tanré et al. 1988), is based on the principle of measuring the blurring effect between highly contrasting adjacent pixels (Tanré et al. 1979; Mekler and Kaufman 1980). This has been used by Sifakis and Deschamps (1992), Sifakis, Soulakellis, and Paronis (1998), and Retalis, Cartalis, and Athanassiou (1999) for very "high"-resolution aerosol estimation over complex urban areas such as Athens using SPOT and Landsat images. A fairly high correlation of 0.76 was obtained between Landsat-derived AOT and SO_2 over Athens (Sifakis, Soulakellis, and Paronis 1998). The major drawback of this method for deriving high-resolution aerosol images is that it measures path radiance (aerosol scattering between adjacent image pixels) within a kernel of 15 × 15 pixels. Thus, for SPOT, a 20-m pixel produces an aerosol image of 300 × 300 m. Using MODIS, the resolution of the resulting aerosol product is 7.5 × 7.5 km. This resolution is too low for the spatial detail required over densely urbanized regions. In addition, accuracy is said to be sensitive to the selected aerosol model because particle shape and size distribution are crucial to the specular scattering and reflectance properties.

Li et al. (2005) developed a 1-km AOT algorithm for Hong Kong, using the same principles as that of the DDV algorithm of Kaufman and Tanré (1998), but under more stringent conditions for the cloud mask and vegetation screening. Error was within 15–20% compared with handheld sun photometer measurements, although too few validation sources were available in the region to obtain robust AOT measurements from the satellite data. Furthermore, the DDV algorithm used would not give accurate results over bright urban areas. Li et al. (2005) further suggested that based on 44 measurements from MICROTOPSII sun photometers, the sulfate and biomass burning models found in the second simulation of a satellite signal in the solar spectrum (6S) radiative transfer model were unsuitable for Hong Kong. Indeed, the diversity of aerosol sources in the region coupled with often high humidity poses challenges for finding a suitable model. Because the MICROTOPSII measurements used for validation in the study lack inversion data such as size distribution and

SSA, which are available from AERONET, more rigorous studies are needed to provide an operational aerosol retrieval method for complex regions.

17.3.2.2 Our Method

In order to estimate aerosols over variable cover types, including bright and dark surfaces, a newly developed methodology is described here for aerosol retrieval from MODIS 500-m data (Wong et al. 2009, 2010). For this study, five MODIS 500-m channels and two MODIS 250-m channels were acquired for aerosol retrieval, as well as for cloud and water masking. Figure 17.2 illustrates this AOT retrieval method.

The AERONET data (2005–2007) from the Hong Kong PolyU station were acquired and clustered to give four different aerosol models. The aerosol models coupled with relative humidity data and different viewing geometries were input into the Santa Barbara DISORT radiative transfer (SBDART; Ricchiazzi et al. 1998) code to build LUTs. The LUT construction was based on the four local aerosol models, namely, (1) mixed urban aerosol (which is a mixture of urban and marine pollutants), (2) polluted urban aerosol (which is dominated by local urban aerosol), (3) dust (which is long-distance Asian dust), and (4) heavy pollution

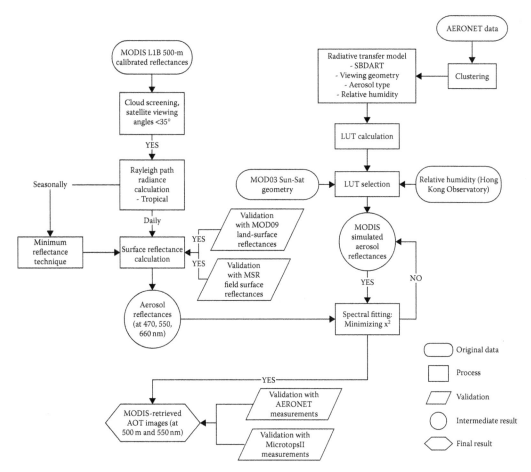

FIGURE 17.2
Schematic diagram for aerosol retrieval in the study.

(which has a high AOT value and represents a mixture of particle types such as very small carbonaceous and sulfate particles), 9 solar zenith angles ($0°–80°$, $\Delta = 10°$), 17 view zenith angles ($0°–80°$, $\Delta = 5°$), 18 relative sun/satellite azimuth angles ($0°–170°$, $\Delta = 10°$), and 8 RH values (RH = 0%, 50%, 70%, 80%, 90%, 95%, 98%, and 99%). The SBDART code uses the aerosol properties associated with a given model, with the combinations of values for the four parameters listed earlier (amounting to 264,384 combinations for three bands [470, 550, 660 nm]), to compute the hypothetical AOT. To minimize the computer memory, the specific relative humidity was first retrieved and LUT geometry was interpolated to the specific satellite geometry. Finally, the simulated aerosol reflectances and TOA reflectances were created as a function of AOT.

The MODIS 500-m calibrated reflectance images in 2007 were acquired from NASA's Distributed Active Archive Centers (DAAC). Geometric correction, reprojection, cloud and water screening, view angle screening, and Rayleigh correction were first applied to the images. The minimum reflectance technique (MRT) was then applied to the MODIS 500-m images in order to obtain the surface reflectance. The MRT was first developed for TOMS (Herman and Celarier 1997) and applied to the Global Ozone Monitoring Experiment (GOME; Koelemeijer, De haan, and Stammes 2003) data at coarse resolution (>1°). The MRT obtained the surface reflectance by extracting the minimum reflectance (or darkest) pixels for a land surface from many Rayleigh corrected images over a period. For validation purposes, the derived surface reflectance images were compared with field measurements and MODIS surface reflectance products (MOD09).

The aerosol reflectances can then be derived by decomposing the TOA reflectances from surface reflectance and Rayleigh path reflectances (Equation 17.7). The derived aerosol reflectances are then compared with simulated aerosol reflectances from LUTs using the spectral fitting technique. The aerosol model with minimum residual is selected, and the corresponding aerosol reflectance and AOT values are obtained. Finally, the AOT images at 550 nm are derived:

$$\rho_{TOA} = \rho_{Aer} + \rho_{Ray} + \frac{\Gamma_{Tot}(\theta_0) \cdot \Gamma_{Tot}(\theta_s) \cdot \rho_{Surf}}{1 - \rho_{Surf} \cdot r_{Hem}} \tag{17.7}$$

where θ_0 is the solar zenith angle, θ_s is the satellite zenith angle, ρ_{Aer} is the aerosol reflectance, ρ_{Ray} is the Rayleigh reflectance, $\Gamma_{Tot}(\theta_0)$ is the transmittance along the path from the sun to the ground, $\Gamma_{Tot}(\theta_s)$ is the transmittance along the path from the sensor to the ground, ρ_{Surf} is the surface reflectance, and r_{Hem} is the hemispheric reflectance.

When the derived AOT images are compared with a whole year's measurements with AERONET and MICROTOPSII, a strong correlation is observed between MODIS collection-5 AOT and AERONET data (Figure 17.3a). This is surprising because Kaufman and Tanré (1998) predicted AOT retrieval problems for the southern region of China, due to high humidity combined with a diversity of aerosol types. Similar correlations ($r^2 = 0.79$ for AERONET and $r^2 = 0.76$ for MICROTOPSII) are obtained for 500-m AOT data (Figures 17.3b and c), compared with $r^2 = 0.83$ for the MODIS collection-5. In addition, similar RMS errors (MODIS 500 m = 0.176, collection-5 = 0.167), similar MAD errors (MODIS 500 m = 0.142, collection 5 = 0.129), and similar bias estimators (MODIS 500 m = 0.011, collection-5 = 0.011) are also obtained from MODIS 500-m data compared with the MODIS operational collection-5 products at 10-km resolution. It is significant that 500-m AOT data not only retrieves AOT images at a much higher spatial resolution but also retrieves AOT over bright urban surfaces as well as dark vegetated areas. Both of these improvements are important for a topographically complex area with heterogeneous land cover like Hong Kong.

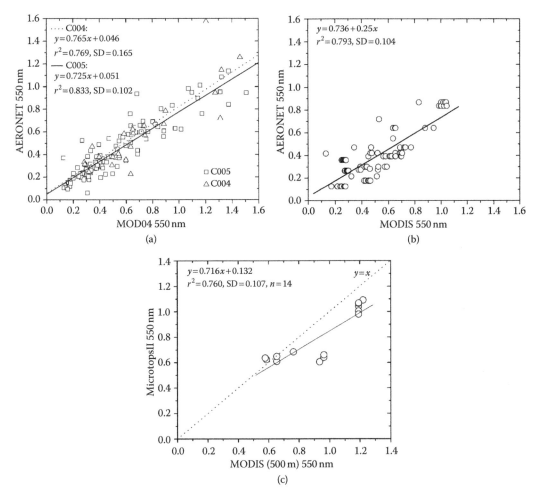

FIGURE 17.3
Scatter plots of (a) MODIS collection-4 and collection-5 aerosol optical thickness (AOT) data versus AERONET AOT measurements, (b) MODIS 500-m AOT data versus AERONET AOT measurements, and (c) MODIS 500-m AOT data versus MICROTOPSII AOT measurements.

Although the signal-to-noise ratio of the 10-km resolution data is 20 times higher than the 500-m resolution data (Kaufman and Tanré 1998), Henderson and Chylek (2005) showed that there are only small changes in the accuracy of aerosol retrieval with increasing pixel sizes from 40×80 m^2 to 2040×4080 m^2. Therefore, any loss of accuracy due to a decreased signal-to-noise ratio of 500-m AOT data is believed to be small enough to be compensated by an increased accuracy from higher spatial resolution.

The AOT distribution over Hong Kong and the PRD region on October 20, 2007 from MODIS 500-m data is shown in Figure 17.4b. Only approximately 15 pixels cover the entire territories of Hong Kong, with an AOT image at 10-km resolution (Figure 17.4a), while there are 400 times more using 500-m resolution (Figure 17.4b). The spatial pattern of aerosols, especially in Shenzhen (the Chinese city near Hong Kong), is much more precisely defined using the 500-m AOT image compared with the 10-km pixels of MOD04. Figure 17.4c shows the 500-m AOT images over Kowloon peninsula and part of Hong Kong island overlaid

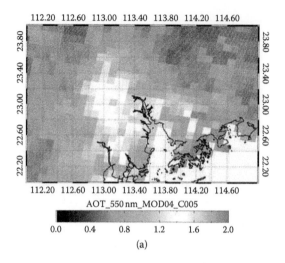

AOT_550 nm_MOD04_C005

0.0   0.4   0.8   1.2   1.6   2.0

(a)

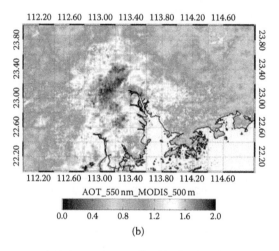

AOT_550 nm_MODIS_500 m

0.0   0.4   0.8   1.2   1.6   2.0

(b)

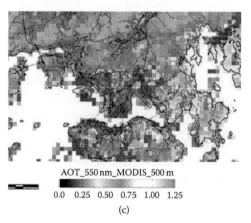

AOT_550 nm_MODIS_500 m

0.0   0.25   0.50   0.75   1.00   1.25

(c)

FIGURE 17.4
(See color insert following page 426.) Aerosol optical thickness image at 550 nm, (a) derived from MODIS collection-5 algorithm, (b) derived from MODIS 500-m data, and (c) derived from MODIS 500-m data overlaid with road layer.

with road networks. Urban districts like Hung Hom, Sham Shui Po, Kowloon Bay, and Ap Lei Chau observe high AOT values (~1.0), whereas the rural areas have relatively low AOT values (~0.3).

17.3.2.3 Applications of High-Resolution Aerosol Products

17.3.2.3.1 Monitoring Anthropogenic Emissions in the PRD Region

An example of rapid changes in AOT over the PRD region occurred on January 28, 2007 and January 30, 2007. Two MODIS 500-m AOT images are shown in Figures 17.5a and b. The AOT at 550 nm on January 28, 2007 is relatively low with a range of ~0.4 in rural areas to ~1.4 in urban areas. It is particularly notable that in the industrialized areas of the PRD, for example, in Guangzhou city and Shunde district, high AOT values are observed, but in other areas, low AOT values are observed due to strong wind speeds (~4 m/s). However, a marked increase in AOT occurred 2 days later on January 30, 2007. This extremely high

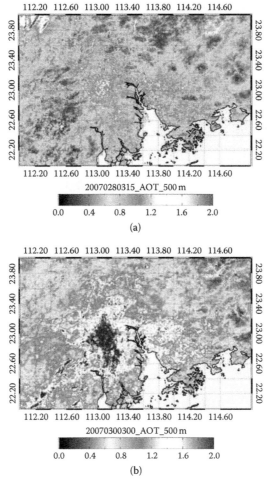

FIGURE 17.5
(See color insert following page 426.) Aerosol optical thickness image at 550-nm and 500-m resolution over Hong Kong and the Pearl River Delta region on (a) January 28, 2007 and (b) January 30, 2007.

AOT (~1.8), which is observed over most industrialized areas in the PRD, is shown in red in Figure 17.5b. Many industries and power plants are located there, and due to very low wind speeds (~1 m/s) on that day, pollutants were trapped in the PRD region. The pollutants would progressively accumulate as wind speeds decreased from 4 m/s on January 28, 2007 to 2 m/s on January 29, 2007 and 1 m/s on January 30, 2007.

17.3.2.3.2 Mapping Aerosols from Biomass Burning

China is still an agricultural country, and had a yield of 690 million tons of straw in 2000 (Wang et al. 2007), of which 36% was used for domestic fuel and 7% was disposed of by open fires (Gao et al. 2002). In the PRD region with extensive areas of dense forest, biomass burning (intentional or accidental) occurs frequently. This section demonstrates the application of MODIS 500-m AOT images to locate and pinpoint local sources of biomass burning.

Figure 17.6a shows the Rayleigh-corrected RGB image on November 30, 2007. Biomass burning is clearly evident on the left of the image (marked on the image), which is located

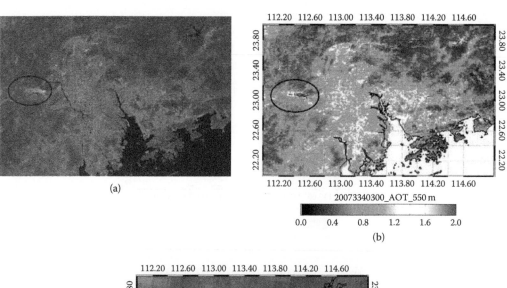

(a)

20073340300_AOT_550 m

0.0   0.4   0.8   1.2   1.6   2.0

(b)

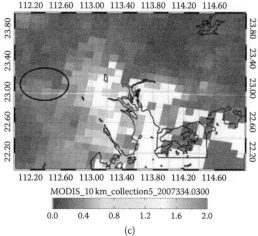

MODIS_10 km_collection5_2007334.0300

0.0   0.4   0.8   1.2   1.6   2.0

(c)

FIGURE 17.6
(See color insert following page 426.) (a) Rayleigh-corrected RGB image on November 30, 2007, (b) aerosol optical thickness (AOT) image at 500-m resolution, and (c) AOT collection-5 image at 10-km resolution.

in the dense forest area of Zhaoqing county. MODIS 500-m AOT images (Figure 17.6b) are also observed to have high AOT values (>1.8). The smoke plumes and the source of burning can be easily identified on the 500-m AOT image, whereas they cannot be identified on the MODIS 10-km AOT image. In addition, large patches over urban areas are masked out with no AOT data values on MODIS collection-5 algorithm (Figure 17.6c) due to their high surface reflectances not meeting the surface reflectance criteria in the collection-5 AOT algorithm. The same fire spots are also depicted on the Web Fire Mapper developed by the Geography Department of the University of Maryland. An easterly wind from a wind direction map confirms the direction of fire smoke. The images derived from our method not only can retrieve aerosols over bright urban surfaces but also can pinpoint small pollution sources such as biomass burning.

17.4 Summary

Satellite remote sensing for aerosol retrieval has been developed over 3 decades, and the techniques now permit aerosol mapping at global, regional, and local scales. Operational satellite aerosol products are now available from space agencies such as the National Oceanic and Atmospheric Administration (NOAA), NASA, and the European Space Agency (ESA). This chapter reviews different algorithms operating on MODIS images, including the DDV algorithm (known as collection-4 algorithm), the second-generation MODIS operation algorithm (known as collection-5 algorithm), the deep blue algorithm at 10-km resolution, Li et al.'s (2005) method at 1-km resolution, and Wong et al.'s (2009, 2010) method at 500-m resolution.

We described Wong et al.'s (2009, 2010) algorithm using the MODIS 500-m resolution images for the retrieval of aerosol properties over complex urbanized regions such as Hong Kong and the PRD region. Strong correlations with AERONET ($r^2 = 0.79$) and MICROTOPSII ($r^2 = 0.76$) sun photometer measurements, as well as low RMS error (0.176), low MAD error (0.142), and low bias estimator (0.011) were obtained for MODIS 500-m AOT data. The aerosol retrieval methodology presented here can be transferred to other mega cities. The MODIS 500-m AOT images are able to locate local-scale anthropogenic emissions such as traffic "black spots" and industrial emissions and to map rapid changes in AOT at regional scales. Moreover, aerosols from biomass burning can be identified using the MODIS 500-m AOT images, as the smoke plume and the source of burning can be easily identified. As such, with the high temporal resolution of MODIS, the 500-m AOT images can be used to monitor cross-boundary aerosols and the development of pollutant sources in the PRD region surrounding Hong Kong.

Two major impediments to the use of remote sensing for routine air quality monitoring over complex regions are the need for several images per day for accurate forecasting and the unknown relationship between AOT, which represents the whole atmospheric column, and air pollution levels near the ground. The first impediment may be overcome by the provision of geostationary satellites with multispectral sensors in the visible region or a constellation of satellites to supplement MODIS TERRA and AQUA, such as the forthcoming National Polar-Orbiting Operational Environmental Satellite System (NPOESS) program. The second one may be minimized as AOT retrievals become more accurate with improved future sensors and algorithms. Then, the inclusion of local meteorological data, including wind speed, humidity, and inversion height, should permit the retrieval of fractional column AOT concentrations, including the near-ground layers. The establishment of robust relationships between fractional column AOT concentrations and pollutants of concern, such as fine particulates (PM2.5), aerosol precursor gases, ozone, and oxides of nitrogen, in the future will make aerosol remote sensing an essential tool for city and regional environmental authorities.

Acknowledgments

The authors wish to thank the NASA Goddard Earth Science Distributed Active Archive Center for the MODIS level-1 and -2 data, Professor Zhanqing Li and Dr. Kwon Ho Lee for their valuable advice, Hong Kong CERG Grant PolyU 5253/07E, and PolyU PDF Research Grant G-YX1W.

References

Cao, J. J., S. C. Lee, K. F. Ho, X. Y. Zhang, S. C. Zou, K. Fung, J. C. Chow, and J. G. Watson. 2003. Characteristics of carbonaceous aerosol in PRD region, China during 2001 winter period. *Atmos Environ* 37(11):1451–60.

Cao, J. J., S. C. Lee, K. F. Ho, S. C. Zou, K. Fung, Y. Li, J. G. Watson, and J. C. Chow. 2004. Spatial and seasonal variations of atmospheric organic carbon and elemental carbon in Pearl River Delta region, China. *Atmos Environ* 38(27):4447–56.

Chu, A., Y. J. Kaufman, C. Ichoku, L. A. Remer, D. Tanré, and B. N. Holben. 2002. Validation of MODIS aerosol optical depth retrieval over land. *Geophys Res Lett* 29(12):1–4.

Davidson, C. I., R. F. Phalen, and P. A. Solomon. 2005. Airborne particulate matter and human health: A review. *Aerosol Sci Technol* 39(8):737–49.

Dominici, F., R. D. Peng, M. Bell, L. Pham, D. McDermott, J. Zeger, and J. Samet. 2006. Fine particulate air pollution and hospital admission for cardiovascular and respiratory diseases. *J Am Med Assoc* 295(10):1127–34.

Dubovik, O., B. N. Holben, Y. Kaufman, M. A. Yamasoe, A. Smirnov, D. Tanré, and I. Slutsker. 1998. Single-scattering albedo of smoke retrieved from the sky-radiance and solar transmittance measured from the ground. *J Geophys Res* 103(24):31903–24.

EPD. 2004. Study of short term health impact and costs due to road traffic-related air pollution. Environmental Protection Department, The Government of Hong Kong Special Administrative Region.

Gao, X. Z., W. Q. Ma, C. B. Ma, F. S. Zhang, and Y. H. Wang. 2002. Analysis on the current status of utilization of crop straw in China. *J Huazhong Agric Univ* 21:242–7.

Gatebe, C. K., M. D. King, S. C. Tsay, Q. Ji, G. T. Arnold, and J. T. Li. 2001. Sensitivity of off-nadir zenith angles to correlation between visible and near-infrared reflectance for use in remote sensing of aerosol over land. *IEEE Trans Geosci Rem Sens* 39(4):805–19.

Gnauk, T., K. Mueller, D. Van Pinxteren, L. Y. He, Y. Niu, M. Hu, and H. Herrmann. 2008. Size-segregated particulate chemical composition in Xinken, Pearl River Delta, China: OC/EC and organic compounds. *Atmos Environ* 42(25):6296–309.

Hao, W. M., and M. H. Liu. 1994. Spatial and temporal distribution of tropical biomass burning. *Global Biogeochem Cycles* 8:495–503.

Henderson, B. G., and P. Chylek. 2005. The effect of spatial resolution on satellite aerosol optical depth retrieval. *IEEE Trans Geosci Rem Sens* 43(9):1984–90.

Herman, J. R., and E. A. Celarier. 1997. Earth surface reflectivity climatology at 340–380 nm from TOMS data. *J Geophys Res* 102:28003–11.

Herman, M., J. L. Deuzé, C. Devaux, P. Goloub, F. M. Bréon, and D. Tanré. 1997. Remote sensing of aerosols over land surfaces, including polarization measurements; application to some airborne POLDER measurements. *J Geophys Res* 102:17039–49.

Holben, B. N., T. F. Eck, I. Slutsker, D. Tanré, J. P. Buis, A. Setzer, E. F. Vermote et al. 1998. AERONET-A federated instrument network and data archive for aerosol characterization. *Rem Sens Environ* 66(1):1–16.

Hsu, N. C., J. R. Herman, O. Torres, B. N. Holben, D. Tanré, T. F. Eck, A. Smirnov, B. Chatenet, and F. Lavenu. 1999. Comparison of the TOMS aerosol index with sun-photometer aerosol optical thickness: Results and application. *J Geophys Res* 23:745–8.

Hsu, N. C., S. C. Tsay, M. D. King, and J. R. Herman. 2004. Aerosol properties over bright reflecting source regions. *IEEE Trans Geosci Rem Sens* 42(3):557–69.

Hsu, N. C., S. C. Tsay, M. D. King, and J. R. Herman. 2006. Deep blue retrievals of asian aerosol properties during ACE-Asia. *IEEE Trans Geosci Rem Sens* 44(11):3180–95.

Husar, R. B., M. J. Prospero, and L. L. Stowe. 1997. Characterization of tropospheric aerosols over the oceans with the NOAA advanced very high resolution radiometer optical thickness operational product. *J Geophys Res* 102(D14):16889–909.

Kaufman, Y. J., and C. Sendra. 1987. Algorithm for automatic atmospheric corrections to visible and near-IR satellite imagery. *Int J Rem Sens* 9(8):1357–81.

Kaufman, Y. J., and D. Tanré. 1998. Algorithm for remote sensing of tropospheric aerosol from MODIS. NASA MOD04 Product Report.

Ko, F. W. S., W. Tam, T. W. Wong, D. P. S. Chan, A. H. Tung, C. K. W. Lai, and D. S. C. Hui. 2007. Temporal relationship between air pollutants and hospital admissions for chronic obstructive pulmonary disease in Hong Kong. *Thorax* 62:780–5.

Koelemeijer, R. B. A., J. F. De Haan, and P. Stammes. 2003. A database of spectral surface reflectivity in the range 335–772 nm derived from 5.5 years of GOME observations. *J Geophys Res* 108:4070.

Lenoble, J., and C. Brogniez. 1984. A comparative review of radiation aerosol models. *Beiträge Zur Atmosphärenphysik* 57:1–20.

Levy, R. C., L. A. Remer, J. V. Martins, Y. J. Kaufman, A. Plana-Fattori, J. Redemann, and B. Wenny. 2004. Evaluation of the MODIS aerosol retrievals over ocean and land during CLAMS. *J Atmos Sci* 62(4):974–92.

Levy, R. C., L. A. Remer, S. Mattoo, E. F. Vermote, and Y. J. Kaufman. 2007. Second-generation operational algorithm: Retrieval of aerosol properties over land from inversion of moderate resolution imaging spectroradiometer spectral reflectance. *J Geophys Res* 112:D13211.

Li, C. C., A. K. H. Lau, J. T. Mao, and D. A. Chu. 2005. Retrieval, validation, and application of the 1-km aerosol optical depth from MODIS measurements over Hong Kong. *IEEE Trans Geosci Rem Sens* 43(11):2650–8.

Li, Z., F. Niu, K. H. Lee, J. Xin, W. M. Hao, B. Nordgren, Y. Wang, and P. Wang. 2007. Validation and understanding of moderate resolution imaging spectroradiometer aerosol products (C5) using ground-based measurements from the handheld Sun photometer network in China. *J Geophys Res* 112:D22S07.

Mi, W., Z. Li, X. G. Xia, B. Holben, R. Levy, F. S. Zhao, H. B. Chen, and M. Cribb. 2007. Evaluation of the MODIS aerosol products at two AERONET stations in China. *J Geophys Res* 112:D22S08.

Mekler, Y., and Y. J. Kaufman. 1980. The effect of the earth atmosphere on contrast reduction for a nonuniform surface albedo and two-halves field. *J Geophys Res* 85:4067–83.

NASA. 2009. NASA TERRA (EOS AM-1) Fact Sheet of Aerosols. http://terra.nasa.gov/FactSheets/Aerosols/#top (accessed September 15, 2010).

Remer, L. A., S. Gasso, D. Hegg, Y. J. Kaufman, and B. Holben. 1996. Urban/industrial aerosols: Ground based sun/sky radiometer and airborne in situ measurements. *J Geophys Res* 2(D14):16849–60.

Remer, L. A., Y. J. Kaufman, D. Tanré, S. Mattoo, D. A. Chu, J. V. Martins, R. R. Li et al. 2005. The MODIS aerosol algorithm, products, and validation. *J Atmos Sci* 62(4):947–73.

Remer, L. A., A. E. Wald, and Y. J. Kaufman. 2001. Angular and seasonal variation of spectral surface reflectance ratios: Implications for the remote sensing of aerosol over land. *IEEE Trans Geosci Rem Sens* 39(2):275–83.

Retalis, A., C. Cartalis, and E. Athanassiou. 1997. Assessment of the distribution of aerosols in the area of Athens with the use of Landsat Thematic Mapper data. *Int J Rem Sens* 20(5):939–45.

Ricchiazzi, P., S. R. Yang, C. Gautier, and D. Sowle. 1998. SBDART: A research and teaching software tool for plane-parallel radiative transfer in the Earth's atmosphere. *Bull Am Meteorol Soc* 79(10):2101–14.

Sifakis, N., and P. Y. Deschamps. 1992. Mapping of air pollution using SPOT satellite data. *Photogramm Eng Rem Sens* 58:1433–7.

Sifakis, N., N. A. Soulakellis, and D. K. Paronis. 1998. Quantitative mapping of air pollution density using earth observations: A new processing method and applications to an urban area. *Int J Rem Sens* 19:3289–300.

Tanré, D., P. Y. Deschamps, C. Devaux, and M. Herman. 1988. Estimation of Saharan aerosol optical thickness from blurring effects in Landsat Thematic Mapper data. *J Geophys Res* 93:15955–64.

Tanré, D., M. Herman, P. Y. Deschamps, and A. de Leffe. 1979. Atmospheric modelling for space measurements of ground reflectances, including bidirectional properties. *Appl Optic* 18:3587–94.

Tanré, D., Y. J. Kaufman, M. Herman, and S. Mattoo. 1997. Remote sensing of aerosol properties over oceans using the MODIS/EOS spectral radiances. *J Geophys Res* 102:16971–88.

Vaughan, M., S. Young, D. Winker, K. Powell, A. Omar, Z. Liu, Y. Hu, and C. Hostetler. 2004. Fully automated analysis of space-based lidar data: An overview of the CALIPSO retrieval algorithms and data products. *Proc SPIE* 5575:16–30.

Wang, Q., M. Shao, Y. Liu, K. William, G. Paul, X. Li, Y. Liu, and S. Lu. 2007. Impact of biomass burning on urban air quality estimated organic tracers: Guangzhou and Beijing as cases. *Atmos Environ* 41:8380–90.

Wanner, W., A. H. Strahler, B. Hu, P. Lewis, J. P. Muller, X. Li, C. L. Barker Schaaf, and M. J. Barnsley. 1997. Global retrieval of bidirectional reflectance and albedo over land from EOS MODIS and MISR data: Theory and algorithm. *J Geophys Res* 102(17):143–62.

Wei, F., E. Teng, G. Wu, W. Hu, W. E. Wilson, R. S. Chapman, J. C. Pau, and J. Zhang. 1999. Ambient concentrations and elemental compositions of PM10 and PM2.5 in four Chinese cities. *Environ Sci Technol* 33(23):4188–93.

Wong, T. W., T. S. Lau, T. S. Yu, A. Neller, S. L. Wong, W. Tam, and S. W. Pang. 1999. Air pollution and hospital admissions for respiratory and cardiovascular diseases in Hong Kong. *Occup Environ Med* 56:679–83.

Wong, M. S., J. E. Nichol, K. H. Lee, and Z. Q. Li. 2009. High resolution aerosol optical thickness retrieval over the Pearl River Delta region with improved aerosol modeling. *Sci China Earth Sci* 52(10):1641–9.

Wong, M. S., J. E. Nichol, K. H. Lee, and Z. Q. Li. 2010. Retrieval of aerosol optical thickness using MODIS 500 × 500 m², a study in Hong Kong and Pearl River Delta region. *IEEE Trans Geosci Rem Sens* 46(8):3318–27.

Wu, D., X. Tie, C. Li, Z. Ying, A. K. H. Lau, J. Huang, X. Deng, and X. Bi. 2005. An extremely low visibility event over the Guangzhou region: A case study. *Atmos Environ* 39:6568–77.

18

Remote Estimation of Chlorophyll-a Concentration in Inland, Estuarine, and Coastal Waters

Anatoly A. Gitelson, Daniela Gurlin, Wesley J. Moses, and Yosef Z. Yacobi

CONTENTS

18.1 Introduction

Inland, estuarine, and coastal waters comprise only a small fraction of the Earth's aquatic component, but are extensively exploited by human activities. The water quality in these ecosystems is, therefore, of high ecological and economic importance, and in this respect, quantitative evaluation of phytoplankton biomass is a crucial endeavor. Despite the high variability of its composition, size, and forms (Reynolds 2006), phytoplankton may be relatively easily monitored by the estimation of the concentration of chlorophyll-a (chl-a), a pigment universally found in all phytoplankton species and routinely used as a substitute for biomass in all types of aquatic environments.

Remote sensing is an effective method for synoptic monitoring of chl-a concentration over potentially heterogeneous areas of phytoplankton distribution. Even a few remotely sensed images are useful in the design or improvement of in situ sampling strategies by identifying representative locations and optimizing the timing for sampling. Remote sensing studies typically involve mapping of constituent concentrations in water bodies using water-leaving radiance or reflectance collected by a sensor held above or below the water surface. The estimation of constituent concentrations usually requires the development of a model, which is a mathematical combination of reflectances at different wavelengths, in such a way that the model is maximally sensitive to changes in the concentration of the constituent of interest (e.g., chl-a) and is minimally sensitive to changes in concentrations of other constituents present in the water.

In this chapter, we will present algorithms for the remote estimation of chl-a concentration in turbid protective waters and show how they work at close range and at satellite altitude. This chapter also contains

- Brief background information on the commonly used remote sensing models for estimating chl-a concentration
- A description of a semianalytical model that uses reflectances in the red and near-infrared (NIR) wavelengths for estimating chl-a concentration
- The data and methods used
- The results of calibrating and validating the NIR–red models using reflectance data measured in situ and from satellites
- The results supporting the potential for a universal NIR–red algorithm
- A discussion on the challenges and limitations in developing a universal NIR–red algorithm for accurately estimating chl-a concentration from the satellite data routinely acquired over turbid productive waters from around the globe

18.2 Background

Historically, remote sensing of chl-a concentration has been commonly used for open ocean, case I waters (Morel and Prieur 1977) using reflectances in the blue and green spectral regions (Gordon and Morel 1983; Kirk 1994; Mobley 1994). In turbid productive waters, however, the reflectances in these spectral regions cannot be used for estimating chl-a concentration due to overlapping and uncorrelated absorption by colored dissolved organic matter (CDOM) as well as scattering and absorption by detritus and tripton, which are higher in turbid waters than in open oceans (Figure 18.1; GKSS 1986; Gitelson 1992; Dekker 1993; Gons 1999; Gons et al. 2000; Dall'Olmo and Gitelson 2005; Schalles 2006; Gitelson, Schalles, and Hladik 2007).

Many investigations have been directed toward the development of remote sensing techniques for the estimation of the concentration of chl-a and other water constituents in turbid productive waters (Bukata et al. 1979; Vasilkov and Kopelevich 1982; Gitelson, Keydan, and Shishkin 1985; Vos, Donze, and Bueteveld 1986; Gitelson, Kondratyev, and Garbusov 1987; Gitelson and Kondratyev 1991; Gitelson 1992; Dekker 1993; Gitelson et al. 1993a; Gitelson, Szilagyi, and Mittenzwey 1993b; Jupp, Kirk, and Harris 1994; Bukata et al. 1995; Gege 1995; Gons 1999; Gons et al. 2000; Pierson and Strömbäck 2000; Kallio et al. 2001; Kutser et al. 2001;

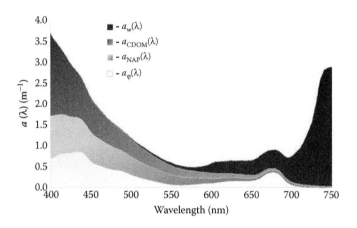

FIGURE 18.1

Spectra of the absorption coefficients of phytoplankton, $a_\varphi(\lambda)$, nonalgal particles, $a_{NAP}(\lambda)$, CDOM, $a_{CDOM}(\lambda)$, and water, $a_w(\lambda)$ for a moderately turbid lake with a chl-a concentration of 27.8 mg $\cdot$ m^{-3} and total suspended solids (TSS) concentration of 6.5 mg $\cdot$ L^{-1}. (Values of $a_w(\lambda)$ taken from Mueller, J. L. 2003. Inherent optical properties: Instrument characterizations, field measurements and data analysis protocols. In *Ocean Optics Protocols for Satellite Ocean Color Sensor Validation*, Revision 4, Volume IV, Erratum 1 to Pegau, S., J. R. V. Zaneveld, and J. L. Mueller. 2003. Inherent optical property measurement concepts: Physical principles and instruments. In *Inherent Optical Properties: Instruments, Characterizations, Field Measurements, and Data Analysis Protocols. Ocean Optics Protocols for Satellite Ocean Color Sensor Validation*, Revision 4, Volume IV, ed. J. L. Mueller, G. S. Fargion, and C. R. McClain. Goddard Space Flight Center Technical Memorandum 2003-211621.)

Strömbäck and Pierson 2001; Ruddick et al. 2001; Thiemann and Kaufmann 2002; Kallio, Koponen, and Pulliainen 2003; Albert 2004). The two main approaches used in the construction of algorithms for the remote estimation of chl-a concentration in turbid productive waters were analytical and empirical/semianalytical. The analytical approach is based on specific inherent optical properties (IOP), such as absorption and scattering coefficients per unit concentration of constituents, which are used to simulate reflectance spectra by using a radiative transfer technique. Then, by a process of optimization based on minimizing the difference between the simulated and measured reflectance spectra, the concentrations of different constituents are adjusted and subsequently determined. If the values of the specific IOPs included in the model are fairly close to those of the water body where the reflectance data were measured, this approach may yield accurate results. However, specific IOPs vary widely in space and time even within a water body. Thus, the assumption of a priori IOP determination is not often valid for turbid productive waters. The empirical or semianalytical approach involves algorithms that are based on relationships between physically based models and experimental results.

The spectral features of turbid productive waters have been studied for a wide range of chl-a concentrations from 3 to more than 200 mg $\cdot$ m^{-3} (Gitelson, Keydan, and Shishkin 1985; Gitelson et al. 1986; Gitelson, Kondratyev, and Garbusov 1987; Gitelson and Kondratyev 1991; Gitelson 1992; Quibell 1992; Dekker 1993; Gitelson 1993; Gitelson et al. 1993a; Gitelson, Szilagyi, and Mittenzwey 1993b; Gitelson et al. 1994; Han et al. 1994; Han and Rundquist 1994; Matthews and Boxall 1994; Millie et al. 1995; Rundquist, Schalles, and Peake 1995; Yacobi, Gitelson, and Mayo 1995; Han and Rundquist 1996; Vos, Donze, and Bueteveld 1986; Han and Rundquist 1997; Gons 1999; Gitelson et al. 2000; Gons et al. 2000; Dall'Olmo, Gitelson, and Rundquist 2003; Dall'Olmo and Gitelson 2005; Dall'Olmo 2006; Dall'Olmo and Gitelson 2006; Schalles 2006; Gitelson, Schalles, and Hladik 2007; Gitelson et al. 2008;

Gitelson et al. 2009; Moses 2009; Moses et al. 2009a,b). Three main features of chl-a are potentially important in the context of concentration estimation using spectral reflectance. First, chl-a has a strong absorption band around 670 nm, forming a trough in the reflectance spectrum. The magnitude of reflectance around 670 nm (ρ_{670}) is related to chl-a concentration. However, chl-a absorption is often not the sole factor controlling ρ_{670}, as it depends also on the concentration of inorganic and organic suspended solids (ISS and OSS). Thus, ρ_{670} alone cannot be used for a reliable estimation of chl-a concentration (Dekker 1993; Gitelson et al. 1993a; Gitelson, Szilagyi, and Mittenzwey 1993b; Gitelson et al. 1994; Yacobi, Gitelson, and Mayo 1995).

A peak due to solar-induced chl-a fluorescence near 685 nm is the second significant spectral feature in the red region (Neville and Gower 1977; Gower 1980; Doerffer 1981). With an increase in chl-a concentration, the fluorescence peak near 685 nm increases; therefore, it was used as an indicator of chl-a concentration, which was calculated as the height above the baseline positioned from 650 through 730 nm (Neville and Gower 1977; Gower 1980; Doerffer 1981; GKSS 1986; Fischer and Kronfeld 1990). However, the quantitative accuracy of this approach is limited by the varying fluorescence efficiencies of different phytoplankton populations and changes in water absorption, which reduce the available light. Another limitation is the reabsorption of the fluoresced light by chl-a, resulting in a decrease in the emitted signal; this happens when the chl-a concentration increases above 10–15 mg·m^{-3} (Kishino, Sugihara, and Okami 1986; Gitelson 1992; Gitelson 1993). Thus, although the use of chl-a fluorescence at 685 nm seems to be useful and effective for the estimation of low chl-a concentrations, it is not expedient for the development of algorithms that yield consistent and accurate results for a wide range of chl-a concentrations in turbid productive waters with highly variable optical properties.

The third reflectance feature specific to chl-a is a peak in the NIR region around 700 nm. The magnitude of the peak, as well as its position, depends on the chl-a concentration (Vasilkov and Kopelevich 1982; Gitelson et al. 1986; Vos, Donze, and Bueteveld 1986; Gitelson 1992; Gitelson 1993; Gitelson et al. 1994; Yacobi, Gitelson, and Mayo 1995, Schalles et al. 1998) but is also affected by absorption and scattering by other constituents.

Most of the algorithms developed to quantify chl-a concentration are based on the properties of the peak near 700 nm. These include the ratio of the reflectance peak (ρ_{max}) to ρ_{670} (ρ_{max}/ρ_{670}), the ratio ρ_{705}/ρ_{670} (Gitelson et al. 1986; Gitelson and Kondratyev 1991; Dekker 1993; Gitelson et al. 1993a; Gitelson, Szilagyi, and Mittenzwey 1993b), and the position of this peak (Gitelson 1992). Gons (1999) used the ratio of reflectances at 704 and 672 nm, the absorption coefficients of water at these wavelengths, and the backscattering coefficient at 776 nm to estimate chl-a concentrations ranging from 3 to 185 mg·m^{-3}. In many studies, close relationships have been found between chl-a concentrations and NIR-to-red reflectance ratios, with the red wavelength around 675 nm and the NIR wavelength varying between 700 and 725 nm (Hoge, Wright, and Swift 1987; Yacobi, Gitelson, and Mayo 1995; Pierson and Strömbäck 2000; Pulliainen et al. 2001; Ruddick et al. 2001; Oki and Yasuoka 2002; Dall'Olmo and Gitelson 2005). Using vector analysis, Stumpf and Tyler (1988) showed that the ratio of reflectances in the NIR and red bands of the Advanced Very High Resolution Radiometer (AVHRR) and Coastal Zone Color Scanner (CZCS) can be used to identify phytoplankton blooms and to estimate chl-a concentrations above 10 mg·m^{-3} in turbid estuaries.

These methods are based on the assumption that optical parameters such as the specific absorption coefficient of phytoplankton, $a^*_\phi(\lambda)$, and the chl-a fluorescence quantum yield, η, remain constant. In reality, these parameters depend on the physiological state and structure of the phytoplankton community and can vary widely. It was shown that $a^*_\phi(675)$ can vary up to fourfold for chl-a concentrations ranging from 0.02 to 25 mg·m^3 (Bricaud et al. 1995). Fluorescence quantum yield is affected by the taxonomic composition of phytoplankton, illumination conditions, light adaptation, nutritional status, and temperature and can vary by eightfold for chl-a concentrations ranging from 2 to 30 mg·m^{-3}, which is typical for inland and coastal waters (GKSS 1986). Therefore, the assumptions of constant $a^*_\phi(\lambda)$ and η are a significant source of uncertainty in models for the remote estimation of chl-a concentrations.

18.3 Semianalytical NIR–Red Model

A fundamental relationship between the remote sensing reflectance (ρ_{rs}) and IOPs was formulated as follows (Gordon, Brown, and Jacobs 1975):

$$\rho_{rs}(\lambda) \propto \frac{b_b(\lambda)}{a(\lambda) + b_b(\lambda)} \tag{18.1}$$

where $a(\lambda)$ is the absorption coefficient and $b_b(\lambda)$ is the backscattering coefficient.

Recently, a conceptual model based on Equation 18.1 was developed and used to estimate pigment concentration in terrestrial vegetation at leaf and canopy levels (Gitelson, Gritz, and Merzlyak 2003; Gitelson et al. 2005):

$$\text{Pigment content} \propto [\rho^{-1}(\lambda_1) - \rho^{-1}(\lambda_2)] \times \rho(\lambda_3) \tag{18.2}$$

where $\rho(\lambda_1)$, $\rho(\lambda_2)$, and $\rho(\lambda_3)$ are the reflectance values at wavelengths λ_1, λ_2, and λ_3 respectively. λ_1 is in a spectral region such that $\rho(\lambda_1)$ is maximally sensitive to absorption by the pigment of interest, although it is still affected by absorption by other pigments and scattering by all particulates. λ_2 is in a spectral region such that $\rho(\lambda_2)$ is minimally sensitive to absorption by the pigment of interest and its sensitivity to absorption by other constituents is comparable to that of $\rho(\lambda_1)$. Thus, the difference $[\rho^{-1}(\lambda_1) - \rho^{-1}(\lambda_2)]$ is related to the concentration of the pigment of interest. However, the difference is still potentially affected by variations in scattering by particles. Consequently, information on λ_3 is required. Wavelength λ_3 is located in a spectral region where the reflectance $\rho(\lambda_3)$ is minimally affected by absorption due to any constituent and is therefore used to account for the variability in scattering between samples.

Dall'Olmo, Gitelson, and Rundquist (2003) suggested the use of this conceptual model (Equation 18.2) for estimating chl-a concentration in turbid productive waters. In Equation 18.1, the absorption coefficient, $a(\lambda)$, is the sum of the absorption coefficients of water, $a_w(\lambda)$, phytoplankton, $a_\phi(\lambda)$, nonalgal particles, $a_{NAP}(\lambda)$, and CDOM, $a_{CDOM}(\lambda)$. Following Gordon's concept, the model presented in Equation 18.2 (called henceforth the three-band NIR–red model) was designed by choosing three optimal wavelengths, such that the contributions due to absorption by constituents other than chl-a and backscattering by particles are kept to a negligible minimum, and the model output is maximally sensitive to chl-a concentration. The red region around 670 nm, where chl-a absorption is maximal (but the reflectance may be affected also by other constituents), was chosen as λ_1. λ_2 is longer than λ_1, where absorption by chl-a, $a_\phi(\lambda)$, is minimal and the absorption by other constituents, $a_{NAP}(\lambda)$ and $a_{CDOM}(\lambda)$, is about the same as at λ_1. Thus, $\rho^{-1}(\lambda_1)$ is a measure of absorption by chl-a and other constituents, and $\rho^{-1}(\lambda_2)$ is a measure of absorption by constituents other than chl-a. λ_3 is at a wavelength beyond λ_2 in the NIR region, where the absorption by all particles and dissolved constituents is null. The backscattering coefficient is considered spectrally uniform across the range of wavelengths considered (from λ_1 through λ_3; Dall'Olmo and Gitelson 2005), which is a fundamental assumption in the model.

The subtraction of $\rho^{-1}(\lambda_2)$ from $\rho^{-1}(\lambda_1)$ isolates the absorption by chl-a as follows:

$$\rho^{-1}(\lambda_1)-\rho^{-1}(\lambda_2) \propto \frac{a_{w\lambda_1}+a_{\phi\lambda_1}+a_{NAP\lambda_1}+a_{CDOM\lambda_1}+b_b(\lambda)}{b_b(\lambda)} - \frac{a_{w\lambda_2}+a_{\phi\lambda_2}+a_{NAP\lambda_2}+a_{CDOM\lambda_2}+b_b(\lambda)}{b_b(\lambda)}$$

$$\rho^{-1}(\lambda_1)-\rho^{-1}(\lambda_2) \propto \frac{a_\phi+a_{w\lambda_1}-a_{w\lambda_2}}{b_b(\lambda)} \tag{18.3}$$

Another assumption is that the absorption by water at λ_3 is much greater than the total backscattering, such that $a_w(\lambda_3) \gg b_b(\lambda)$ and $a(\lambda) \cong a_w(\lambda_3)$.

$$\rho(\lambda_3) \propto \frac{b_b(\lambda)}{a_{w\lambda_3}} \tag{18.4}$$

Considering the fact that the absorption by water, $a_w(\lambda)$, is independent of the constituent concentrations and ignoring its dependence on temperature, the model has the following form:

$$[\rho^{-1}(\lambda_1)-\rho^{-1}(\lambda_2)] \times \rho(\lambda_3) \propto a_\phi(\lambda) \tag{18.5}$$

Absorption by phytoplankton is related to chl-a concentration as follows:

$$a_\phi(\lambda) = a_\phi^*(\lambda) \times C_{chl-a} \tag{18.6}$$

where $a_\phi^*(\lambda)$ is the chl-a specific absorption coefficient and C_{chl-a} is the concentration of chl-a.

Thus, the three-band NIR–red model was finally formulated as

$$[\rho^{-1}(\lambda_1)-\rho^{-1}(\lambda_2)] \times \rho(\lambda_3) \propto chl-a \tag{18.7}$$

Dall'Olmo and Gitelson (2005) have found that the optimal wavelengths for the accurate estimation of chl-a concentrations in the range of 2 to 180 mg·m^{-3} were as follows: $\lambda_1 = 670$ nm, $\lambda_2 = 710$ nm, and $\lambda_3 = 740$ nm. Testing the model for several data sets collected in inland and estuarine waters, Gitelson, Schalles, and Hladik (2007) and Gitelson et al. (2008) found relatively wide optimal spectral bands of wavelengths of $\lambda_1 = 660$–670 nm, $\lambda_2 = 700$–720 nm, and $\lambda_3 = 730$–760 nm, which provided accurate estimations of chl-a concentration with the three-band NIR–red model.

For waters that do not have significant concentrations of nonalgal particles and colored dissolved organic matter, the subtraction of $\rho^{-1}(\lambda_2)$ in the model may be omitted (Dall'Olmo and Gitelson 2005), which leads to the special case of a two-band NIR–red model (Stumpf and Tyler 1988):

$$\rho^{-1}(\lambda_1) \times \rho(\lambda_3) \propto chl-a \tag{18.8}$$

where λ_1 is in the red region and λ_3 in the NIR region beyond 730 nm.

Another two-band model, which is different in its formulation from the previously mentioned two-band model (Equation 18.8), is (Gitelson 1992; Gitelson, Szilagyi, and Mittenzwey 1993)

$$\rho^{-1}(\lambda_1) \times \rho(\lambda_2) \propto \text{chl-a} \tag{18.9}$$

where λ_1 is in the red region and λ_2 is in the region of the reflectance peak, around 700–710 nm. Recently, a four-band model was suggested (Le et al. 2009) for the estimation of chl-a concentration in productive waters with very high concentrations of inorganic suspended matter:

$$[\rho^{-1}(662) - \rho^{-1}(693)] \times [\rho^{-1}(740) - \rho^{-1}(705)]^{-1}$$

The three-band NIR–red model was modified by including one more spectral band, ρ_{705}, in order to reduce the effect of variations in scattering by suspended matter. The Medium Resolution Imaging Spectrometer (MERIS), the Moderate Resolution Imaging Spectroradiometer (MODIS), and the Sea-Viewing Wide Field-of-View Sensor (SeaWiFS) are three commonly used spaceborne optical sensors, whose data may be used for the estimation of chl-a concentration using NIR–red models. The spectral bands in the red and NIR regions for these sensors are as follows:

- MERIS—Spectral bands centered at 665 nm (band 7), 681 nm (band 8), 708 nm (band 9), and 753 nm (band 10)
- MODIS—Spectral bands centered at 667 nm (band 13), 678 nm (band 14), and 748 nm (band 15)
- SeaWiFS—Spectral bands centered at 670 nm (band 6) and 765 nm (band 7)

The proximity of the 681-nm MERIS band and the 678-nm MODIS band to the chl-a fluorescence wavelength at 685 nm means that the variable quantum yield of fluorescence (Dall'Olmo and Gitelson 2006) might affect the accuracy of chl-a concentration estimated using these bands. Therefore, these bands were eliminated as candidates for inclusion in NIR–red models for estimating chl-a concentration. Thus, the NIR–red models for estimating chl-a concentration using satellite data are as follows:

Three-band MERIS NIR–red model based on Equation 18.7:

$$\text{chl-a} \propto [(\rho_{\text{band7}})^{-1} - (\rho_{\text{band9}})^{-1}] \times (\rho_{\text{band10}}) \tag{18.10}$$

Two-band MERIS NIR–red model based on Equation 18.9:

$$\text{chl-a} \propto (\rho_{\text{band7}})^{-1} \times (\rho_{\text{band9}}) \tag{18.11}$$

Two-band MODIS NIR–red model based on Equation 18.8:

$$\text{chl-a} \propto (\rho_{\text{band13}})^{-1} \times (\rho_{\text{band15}}) \tag{18.12}$$

An appropriate equivalent for Equation 18.12 for SeaWiFS should be based on bands 6 and 7 of that sensor. Note that the two-band MERIS NIR–red model (Equation 18.11) is fundamentally different from the two-band MODIS NIR–red model (Equation 18.12), yielding significantly different results.

18.4 Data and Methods

18.4.1 Field Measurements

The field data were collected at 89 stations from July through November 2008 and at 63 stations from April through July 2009 at the Fremont Lakes State Recreation Area in eastern Nebraska, and included a standard set of optical water quality parameters. In the field, water transparency was measured with a standard Secchi disk, and turbidity was measured with a HACH 2100 portable turbidimeter. Surface water samples were collected at a depth of 0.5 m and stored on ice in a dark container. Water samples for wet laboratory analysis of chl-a concentrations, total particulate absorption coefficients, $a_p(\lambda)$, absorption coefficients of non-algal particles, $a_{NAP}(\lambda)$, and absorption coefficients of phytoplankton, $a_\varphi(\lambda)$, were filtered through 25-mm Whatman GF/F filters within 24 hours after collection. The filters for the extraction of chl-a were stored in a freezer at a temperature of $-18°C$ for a maximum of 4 weeks. Surface water samples for laboratory analysis of TSS, ISS, and OSS were filtered through 47-mm Whatman GF/F filters within 42 hours. The filtrates were filtered through 47-mm Whatman 0.2-μm nylon membranes for laboratory analysis of the absorption coefficients of CDOM, $a_{CDOM}(\lambda)$.

Backscattering coefficients, $b_b(\lambda)$, were measured in the field in 2009 with a customized ECO Triplet sensor. The $b_b(\lambda)$ measurements were corrected for salinity, and absorption by water (Mueller 2003), particulates, and CDOM (absorption coefficients were taken from laboratory measurements).

Hyperspectral reflectance measurements were taken by means of two intercalibrated Ocean Optics USB2000 miniature fiber optic spectrometers. Data were collected over optically deep water in the range of 400–900 nm in intervals of ~0.3 nm with a spectral resolution of ~1.5 nm. Radiometer 1, equipped with a 25° field-of-view optical fiber, was pointed downward to measure the below-surface upwelling radiance, $L_u(\lambda)$, at nadir. The tip of the optical fiber was kept just below the water surface by means of a 2-m long, hand-held dark pole on the sunlit side of the boat. To simultaneously measure the incident irradiance $E_d(\lambda)$, radiometer 2, connected to an optical fiber fitted with a cosine collector, was pointed upward and mounted on a 2.5-m mast. The integration time of radiometer 2 was up to 10 times shorter than the integration time of radiometer 1. Hyperspectral reflectance measurements were collected from 10 A.M. to 2 P.M. The solar zenith angles ranged from a maximum of 66.44° in November 2008 to a minimum of 18.16° in June 2009.

The critical issue with regard to the dual-fiber approach is that the transfer functions of the radiometers used for measuring upwelling and downwelling fluxes, which describe the relationship of the incident flux measured by the sensor to the data numbers produced by the radiometers, should be identical. We studied the identity of the two radiometers used in this study and found that the difference between their transfer functions did not exceed 0.4% (Dall'Olmo and Gitelson 2005).

To match their transfer functions, the radiometers were intercalibrated by measuring simultaneously the upwelling radiance, $L_{ref}(\lambda)$, from a white Spectralon® reflectance standard (Labsphere, Inc., North Sutton, NH), and correspondingly, the irradiance incident on the reflectance standard, $E_{ref}(\lambda)$. The remote sensing reflectance at nadir was computed by

$$\rho_{rs}(\lambda) = \left[\frac{L_u(\lambda)}{E_d(\lambda)} \times \frac{E_{ref}(\lambda)}{L_{ref}(\lambda)} \times 100 \times \frac{\rho_{ref}(\lambda)}{\pi} \times \frac{t}{n^2} \times F(\lambda) \right]$$

where $L_u(\lambda)$ is the below-surface upwelling radiance at nadir, $E_d(\lambda)$ is the incident irradiance, t is the water-to-air transmittance (taken as equal to 0.98), $\rho_{ref}(\lambda)$ is the irradiance reflectance of the Spectralon® reflectance standard linearly interpolated to match the wavelength of each radiometer, π is used to transform the irradiance reflectance into remote sensing reflectance, n is the refractive index of water relative to air (taken as equal to 1.33), and $F(\lambda)$ is the spectral immersion factor computed after Ohde and Siegel (2003). The reflectance spectra were collected and processed using the CALMIT Data Acquisition Program (CDAP) developed at the Center for Advanced Land Management Information Technologies (CALMIT) at the University of Nebraska—Lincoln.

18.4.2 Laboratory Measurements

Pigments were extracted in subdued light conditions in a laboratory at the University of Nebraska—Lincoln. The samples were extracted for 5 minutes in 10 mL of 99.5% ethanol at 78°C and cooled in the dark for 4 hours (modified from Nusch 1980). The samples were then centrifuged for 5 minutes in a Cole-Parmer EW-17250-10 fixed-speed centrifuge, and chl-a concentrations were quantified fluorometrically with a Turner 10-AU-005 CE fluorometer on the same day (Welschmeyer 1994). The instrument was calibrated every 3 months with a 100 $\mu g \cdot L^{-1}$ chl-a standard prepared from C6144-1MG *Anacystis nidulans* chl-a (Sigma-Aldrich). The *Anacystis nidulans* chl-a was dissolved in 1 L of 99.5% ethanol, and the concentration was determined spectrophotometrically (Ritchie 2008). Standard curves with a series of 10 dilutions were made at the time of the calibration to study the linearity of the single point calibration of the instrument for chl-a concentrations from 0 to 200 $\mu g \cdot L^{-1}$. Concentrations of TSS, ISS, and OSS were determined gravimetrically (Eaton et al. 2005).

Particulate absorption coefficients were measured by the quantitative filter technique (Mitchell et al. 2003). The suspended particles were concentrated on 25-mm Whatman GF/F filters, and spectral measurements of the optical density (OD) were made within 1.5 hours after the filtration of the water samples with a Cary 100 Varian spectrophotometer. The filters were scanned in the range 400–800 nm at intervals of 1 nm and the signal from a MilliQ water-saturated reference filter was subtracted automatically from the measurements of the OD. Total particulate absorption coefficients, $a_p(\lambda)$, were calculated as follows:

$$a_p(\lambda) = \frac{\ln(10)}{V/A} [0.3893 \times [OD_{fp}(\lambda) - OD_{null}] + 0.4340 \times [OD_{fp}(\lambda) - OD_{null}]^2]$$

where $OD_{fp}(\lambda)$ is the OD of the sample on the filter, OD_{null} is the average OD of the sample in the range 780–800 nm, which was applied for the null point correction of the OD measurement, V is the volume of water filtered in cubic meters, and A is the area of the filter in square meters. The equation includes a quadratic function for the pathlength amplification correction of the measurements (Cleveland and Weidemann 1993) derived by Dall'Olmo (2006) for water samples from lakes and reservoirs in Nebraska and laboratory cultures of *Microcystis* and *Synechococcus*. Removal of the light absorption by pigments for the measurement of the absorption coefficient of nonalgal particles, $a_{\text{NAP}}(\lambda)$, followed the modified approach presented by Ferrari and Tassan (1999). The samples were treated with 120 µL sodium hypochlorite solution in MilliQ water (0.1%–0.2% active Cl) and rinsed with 50 mL MilliQ water after a 20-minute reaction time. The absorption coefficients of nonalgal particles were calculated similarly to the total particulate absorption coefficients. The absorption coefficients of phytoplankton, $a_{\phi}(\lambda)$, were finally calculated by the subtraction of the absorption coefficients of nonalgal particles from the particulate absorption coefficients.

The absorption coefficients of CDOM were measured spectrophotmetrically immediately after the filtration of the water samples. The optical densities of the filtrates were measured in a 0.1-m cuvette in the range 200–800 nm at intervals of 1 nm with the Cary 100 Varian spectrophotometer and the signal from a MilliQ water reference sample was subtracted automatically from the measurements. Filtrates and MilliQ water reference samples were kept at 24°C (the temperature in the sample compartment of the instrument) to minimize the effects of a temperature-dependent water absorption feature at a wavelength of 750 nm. The absorption coefficients of CDOM, $a_{\text{CDOM}}(\lambda)$, were calculated by

$$a_{\text{CDOM}}(\lambda) = \frac{\ln(10)}{l}[\text{OD}_s(\lambda) - \text{OD}_{\text{null}}]$$

where $\text{OD}_s(\lambda)$ is the OD of the sample, OD_{null} is the average OD of the sample in the range 780–800 nm, which was applied for the null point correction, and l is the pathlength of the cuvette in meters.

18.4.3 Descriptive Statistics of Water Quality Parameters

The descriptive statistics of Fremont Lakes water quality parameters indicate the typical range of variations for turbid productive inland, estuarine, and coastal waters (Tables 18.1 and 18.2) with chl-a concentrations that ranged from 2.3 to 200.8 mg·m^{-3} in 2008 and from 4.0 to 196.4 mg·m^{-3} in 2009. The distributions of the chl-a concentrations in 2008 and 2009 were significantly different from normal distributions ($p < .001$ for 2008 and 2009, Shapiro-Wilk test) and skewed toward lower chl-a concentrations. TSS concentrations ranged from 1.2 to 15.0 mg·L^{-1} in 2008 and from 1.3 to 22.9 mg·L^{-1} in 2009 and were uncorrelated with chl-a concentrations.

18.4.4 Model Calibration and Validation Using Satellite Data

To test the performance of NIR–red models for the estimation of chl-a concentration using the data acquired by spaceborne sensors, seven data collection campaigns were undertaken (in April, July, September, and October of 2008 and March, April, and June of 2009) on the Taganrog Bay and the Azov Sea by the crew at the Southern Scientific Centre of the Russian Academy of Sciences in Rostov-on-Don, Russia. Water samples were collected at each station, filtered through Whatman GF/F filters, and analyzed for chl-a and TSS. Chl-a was extracted in hot ethanol and its concentration was quantified spectrophotometrically. TSS concentrations were determined gravimetrically.

TABLE 18.1

Descriptive Statistics of the Optical Water Quality Parameters Measured in the Fremont Lakes, Nebraska in 2008

	N	Min	Max	Median	Mean	Standard Deviation	Coefficient of Variation
Chl-a (mg·m⁻³)	89	2.27	200.81	27.44	33.540	29.439	0.878
Secchi disk depth (m)	89	0.51	4.20	0.98	1.205	0.697	0.579
Turbidity (NTU)	89	1.51	19.20	6.79	7.585	4.386	0.578
TSS (mg·L⁻¹)	89	1.19	15.00	6.80	7.298	3.258	0.446
ISS (mg·L⁻¹)	87	0.15	5.85	0.80	1.109	0.951	0.857
OSS (mg·L⁻¹)	87	0.81	12.80	6.00	6.259	2.972	0.475
a_p(670) (m⁻¹)	80	0.045	4.333	0.471	0.5583	0.5436	0.9737
a_{NAP}(670) (m⁻¹)	80	0.003	0.311	0.063	0.0759	0.0583	0.7670
a_ϕ(670) (m⁻¹)	80	0.042	4.245	0.369	0.4823	0.5370	1.1133
a^*_ϕ(670) (m²·mg⁻¹)	80	0.005	0.026	0.013	0.0144	0.0048	0.3301
a_{CDOM}(440) (m⁻¹)	80	0.455	1.453	0.895	0.8806	0.2468	0.2803
a_{CDOM}(670) (m⁻¹)	80	0.012	0.050	0.024	0.0249	0.0084	0.5561

TABLE 18.2

Descriptive Statistics of the Optical Water Quality Parameters Measured in the Fremont Lakes, Nebraska in 2009

	N	Min	Max	Median	Mean	Standard Deviation	Coefficient of Variation
Chl-a (mg·m⁻³)	63	3.97	196.39	16.07	30.027	35.187	1.172
Secchi disk depth (m)	63	0.39	3.32	1.12	1.374	0.770	0.560
Turbidity (NTU)	63	1.08	23.40	4.73	6.077	5.311	0.874
TSS (mg·L⁻¹)	63	1.32	22.89	5.83	6.668	4.672	0.701
ISS (mg·L⁻¹)	63	0.10	6.22	1.00	1.392	1.395	1.002
OSS (mg·L⁻¹)	63	0.85	17.00	4.50	5.276	3.988	0.756
a_p(670) (m⁻¹)	63	0.063	2.659	0.282	0.4458	0.4990	1.1192
a_{NAP}(670) (m⁻¹)	63	0.008	0.120	0.047	0.0533	0.0286	0.5368
a_ϕ(670) (m⁻¹)	63	0.038	2.551	0.214	0.3925	0.4829	1.2302
a^*_ϕ(670) (m²·g⁻¹)	63	0.008	0.019	0.013	0.0126	0.0020	0.1566
a_{CDOM}(440) (m⁻¹)	63	0.353	1.349	0.648	0.6852	0.2203	0.3215
a_{CDOM}(670) (m⁻¹)	63	0.003	0.042	0.015	0.0172	0.0088	0.5121
b_b(660) (m⁻¹)	44	0.016	0.095	0.056	0.0529	0.0248	0.4697

MERIS images acquired up to 2 days before or after the date of in situ data acquisition were used in cases where the same-day images were not available. For the whole data set, the average temporal difference between the times of in situ and satellite data acquisitions was less than 1 day. The remote sensing reflectance was retrieved through the bright pixel atmospheric correction procedure (Moore, Aiken, and Lavender 1999). Surface reflectance at the appropriate wavelengths was subsequently used in the three-band (Equation 18.10) and two-band (Equation 18.11) MERIS models (for details, see Moses et al. 2009a).

18.5 Results

18.5.1 Optical Properties of the Constituents

The spectra of $a_p(\lambda)$, $a_{NAP}(\lambda)$, and $a_\phi(\lambda)$ indicate different sources of nonalgal particles (inorganic particles, detritus, and organisms other than phytoplankton) and a diverse phytoplankton species composition (the presence of *Cyanophyta* is indicated by absorption around 625 nm in several spectra) in the Fremont Lakes (Figure 18.2a through c). The relationship of the absorption coefficients of phytoplankton at a wavelength of 676 nm, $a_\phi(676)$, and chl-a concentration in the Fremont Lakes was

$$\text{chl-a} = 72.39 \times a_\phi(676) + 0.195 \quad \text{with} \quad R^2 = 0.99$$

It was only slightly different from the relationship published by Oubelkheir et al. (2005) for the Mediterranean Sea (Figure 18.2d):

$$\text{chl-a} = 67.8 \times a_\phi(676)^{1.16} \quad \text{with} \quad R^2 = 0.94$$

The spectra of the absorption coefficients of CDOM, $a_{CDOM}(\lambda)$, displayed a uniform pattern, but a wide range of values (Figure 18.3). We found a conspicuous intra-annual difference as the absorption coefficients of CDOM at a wavelength of 440 nm, $a_{CDOM}(440)$ ranged from 0.45 to 1.45 m^{-1} in 2008 and from 0.35 to 1.35 m^{-1} in 2009.

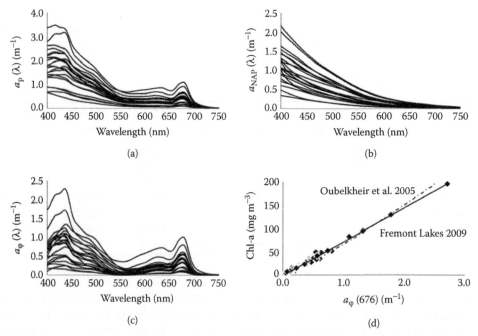

FIGURE 18.2
Representative spectra of the absorption coefficients of (a) total particulates, (b) nonalgal particles, and (c) phytoplankton. (d) Scatterplot of $a_\phi(676)$ versus chl-a concentration for the Fremont Lakes 2009 data set (solid line). Dashed line is the relationship $a_\phi(676)$ versus chl-a for the Mediterranean Sea. (From Obelkheir, K. et al. *Limnol Oceanogr*, 50, 6, 2005.)

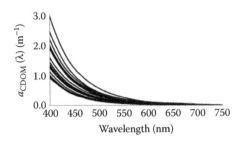

FIGURE 18.3
Representative spectra of the absorption coefficients of CDOM.

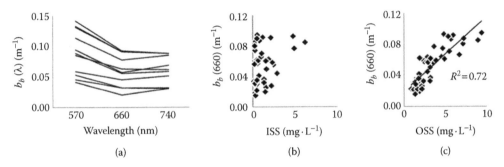

FIGURE 18.4
Representative spectra of the backscattering coefficient, $b_b(\lambda)$, (a) taken in 2009 and relationships of (b) $b_b(660)$ and ISS and (c) b_b and OSS.

The spectra of the backscattering coefficients, $b_b(\lambda)$ (Figure 18.4a), were similar to the spectra for different moderately turbid inland and coastal waters (Kutser et al. 2009). Ten of 44 spectra had a higher b_b at 660 nm than at 740 nm. This indicated that the assumption of the spectrally uniform backscattering coefficient across the range of the wavelength from λ_1 to λ_3 (Dall'Olmo and Gitelson 2005) might not hold for turbid productive waters. The backscattering coefficients at 660 nm, $b_b(660)$, did not relate to the ISS concentration (Figure 18.4b), but had a close relationship with the concentration of OSS (Figure 18.4c).

18.5.2 Spectral Reflectance Properties

The reflectance spectra of the waters studied were highly variable over the visible and NIR spectral regions (Figure 18.5). The spectra were quite similar in magnitude and shape to the reflectance spectra collected in turbid productive waters (Lee et al. 1998; Gitelson et al. 2000; Dall'Olmo and Gitelson 2005; Schalles 2006). In the blue spectral range between 400 and 500 nm, chlorophylls and carotenoids strongly absorb light (Bidigare et al. 1990). The minimum, near 440 nm, used in algorithms for the estimation of chl-a in oligotrophic waters (Gordon and Morel 1983), was almost indistinct, and the reflectance in this range was low, with no pronounced spectral features over the broad range of turbidity and phytoplankton densities. In addition to the strong absorption by all thylakoid-bound

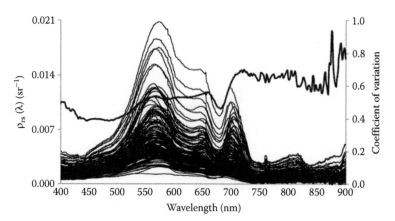

FIGURE 18.5
Remote sensing reflectance spectra for the waters sampled in 2008 and 2009. The thick line represents the coefficient of variation of reflectance.

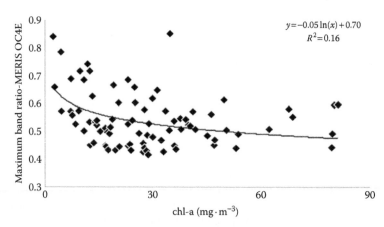

FIGURE 18.6
Maximum band ratio calculated for simulated MERIS bands versus chl-a concentration for the Fremont Lakes 2008 data set.

pigments, the absorption by dissolved organic matter and tripton influences the reflectance in the range of 400–500 nm. As a result, the maximum band ratios, based on the wavelengths of 442.5, 490, 510, and 560 nm, that is, ($\rho_{rs}[442.5]/\rho_{rs}[560]$, $\rho_{rs}[490]/\rho_{rs}[560]$, and $\rho_{rs}[510]/\rho_{rs}[560]$), used for the estimation of chl-a concentrations in case I ocean waters (O'Reilly et al. 1998), were poorly related to chl-a concentrations in the case of the productive system of Fremont Lakes (Figure 18.6). Reciprocal of reflectance, which directly relates to the absorption coefficient,

$$\rho_{rs}^{-1}(\lambda) \propto \frac{a(\lambda) + b_b(\lambda)}{b_b(\lambda)} \qquad (18.13)$$

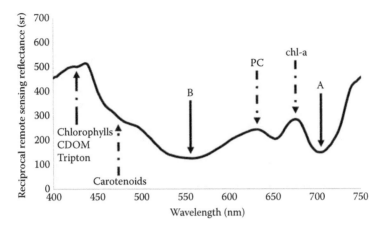

FIGURE 18.7
Reciprocal reflectance spectrum typical for productive waters (chl-a = 80.2 mg · m⁻³) taken at the Fremont Lakes, Nebraska. Chlorophylls: position of the maximal absorption by all chlorophylls present; carotenoids: position of the maximal absorption by carotenoids; PC: position of the maximal absorption by phycocyanin; chl-a: maximal absorption by chlorophyll-a in the red spectral region; A: minimal combined absorption by chlorophyll-a, inorganic suspended solids, and water; B: minimal absorption by phytoplankton pigments.

showed five distinct features (Figure 18.7): (1) high ρ_{rs}^{-1} values in the blue range due to absorption by chlorophylls, carotenoids, CDOM, and tripton; (2) a trough in the green region near 550–570 nm (B in Figure 18.7) due to minimal absorption by all algal pigments and scattering by inorganic suspended matter and detritus, as well as by phytoplankton cells that control the magnitude of reflectance in this range; (3) a peak near 625 nm due to phycocyanin absorption (PC in Figure 18.7) that typically covaries with cyanobacterial abundance and seasonality (Schalles et al. 1998); (4) a peak at 670–680 nm corresponding to the in situ red chl-a absorption maximum (chl-a in Figure 18.7), and (5) a minimum in the NIR near 700 nm (A in Figure 18.7).

The prominent trough in the reciprocal of the reflectance around 700 nm (A in Figure 18.7) corresponds to a peak in the reflectance spectrum (Figure 18.5). The nature of this reflectance peak is elucidated by the examination of the absorption spectra of phytoplankton and pure water ($a_\varphi(\lambda) + a_w(\lambda)$) for variable chl-a concentrations (Figure 18.8). Chl-a absorbs up to approximately 730 nm, although at that point the absorption is only 0.3% and 0.5% of the absorption at 440 and 675 nm, respectively (Bidigare et al. 1990). Nevertheless, in productive waters, chl-a concentration is sufficient to exert a detectable impact on the optical characteristics of water, as chl-a absorption interacts with the absorption of pure water. With the increase of chl-a concentration, the curve of the absorption coefficient becomes wider, and when the absorptions by chl-a and pure water become equal, the combined absorption by all constituents is minimal (Vasilkov and Kopelevich 1982; Gitelson et al. 1986; Vos, Donze, and Bueteveld 1986; Gitelson 1992; Gitelson et al. 1993a; Gitelson, Szilagyi, and Mittenzwey 1993b). Consequently, with the increase in chl-a, this intersection point $a_\varphi(\lambda) = a_w(\lambda)$ takes place at progressively longer wavelengths (Figures 18.8 and 18.9).

The magnitude of the reflectance peak near 700 nm (Figure 18.5) is related to chl-a concentration: an increase in chl-a concentration means an upsurge of phytoplankton biomass, which results in an increase in the cell surface area, leading to enhanced scattering (Gitelson 1992; Gitelson et al. 1993a; Gitelson et al. 1994; Yacobi, Gitelson, and

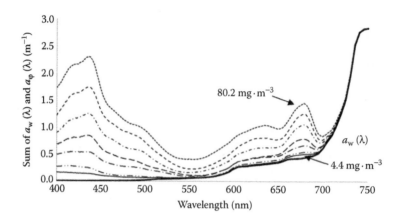

FIGURE 18.8
Spectra of the sum of the absorption coefficients of water and phytoplankton for chl-a concentrations of 4.4 (bottom), 7.9, 15.2, 27.8, 39.2, 53.7, and 80.2 mg·m^{-3} (top). The thick line is the absorption coefficient of water. (From Mueller, J. L. 2003. Inherent optical properties: Instrument characterizations, field measurements and data analysis protocols. In *Ocean Optics Protocols for Satellite Ocean Color Sensor Validation*, Revision 4, Volume IV, Erratum 1 to Pegau, S., J. R. V. Zaneveld, and J. L. Mueller. 2003. Inherent optical property measurement concepts: Physical principles and instruments. In *Inherent Optical Properties: Instruments, Characterizations, Field Measurements, and Data Analysis Protocols*. Ocean Optics Protocols for Satellite Ocean Color Sensor Validation, Revision 4, Volume IV, ed. J. L. Mueller, G. S. Fargion, and C. R. McClain. Goddard Space Flight Center Technical Memorandum 2003-211621.)

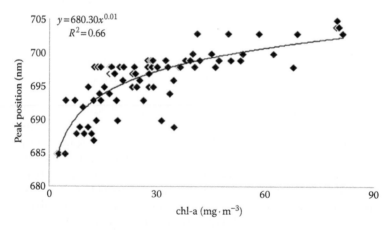

FIGURE 18.9
Reflectance peak position versus chl-a concentration for the Fremont Lakes 2008 data set.

Mayo 1995; Schalles et al. 1998). However, in the waters studied, the correlation coefficient of the relationship "peak magnitude versus chl-a concentration" was less than 0.1 (not shown), indicating the crucial (if not dominant) role of scattering by inorganic suspended matter and detritus in that spectral region.

As absorption by all constituents decreases beyond 715–720 nm, the reflectance in the NIR region (ρ_{NIR}) is controlled by scattering by all particulate matter and relates closely to the total suspended matter concentration (Figure 18.10).

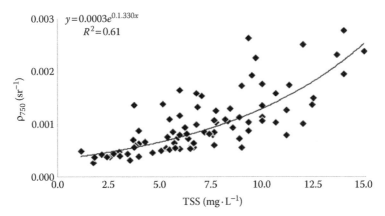

FIGURE 18.10
Remote sensing reflectance at 750 nm versus the concentration of total suspended solids for the Fremont Lakes 2008 data set.

18.5.3 Calibration and Validation of NIR–Red Models Using Proximal Sensing

Using an optimization procedure for the three-band model (Equation 18.7), for the 2008 data set, the optimal spectral region for λ_1 was found to be around 670 nm, which is in accordance with previous studies (Dall'Olmo and Gitelson 2005: Gitelson, Schalles, and Hladik 2007; Gitelson et al. 2008). The magnitude of the reciprocal reflectance at 670 nm was poorly correlated with chl-a concentration (Figure 18.11a). Since $(\rho_{670})^{-1}$ is directly related to chl-a absorption (Equation 18.13), an increase in chl-a concentration should lead to an increase in $(\rho_{670})^{-1}$, but Figure 18.11a shows just the opposite for stations with chl-a concentrations < 20 mg $\cdot$ m^{-3}; that is, $(\rho_{670})^{-1}$ decreases as chl-a concentration increases. This is due to the role of suspended solids. The relationship between $(\rho_{670})^{-1}$ and the concentration of TSS (Figure 18.11b) shows a steep decrease in $(\rho_{670})^{-1}$ as TSS concentration increases, confirming that scattering by suspended solids was a main factor controlling reflectance at 670 nm.

To reduce the effects of scattering by suspended particles and absorption by nonalgal particles and CDOM on the reflectance at 670 nm, $\rho^{-1}(\lambda_2)$ at 710 nm was used. $\rho^{-1}_{710} \propto (a_{CDOM}(\lambda) + a_{NAP}(\lambda) + a_{water}(\lambda) + b_b(\lambda))/b_b(\lambda)$ was strongly related to the TSS concentration (not shown) as well as to absorption by all constituents except chl-a. The difference $\rho^{-1}_{670} - \rho^{-1}_{710} \propto a_{chl-a}/b_b(\lambda)$ was, therefore, more closely related to chl-a (Figure 18.12). However, it was still dependent on backscattering, $b_b(\lambda)$, and thus was strongly affected (especially for chl-a < 10 mg $\cdot$ m^{-3}) by the scattering of all suspended particles. The reflectance in the NIR region of the spectrum is clearly influenced by the concentration of TSS (Figure 18.10) and is closely related to $b_b(\lambda)$; so $\rho_{750} \propto b_b(\lambda)$ was used to remove the effect of the differences between samples in scattering by suspended particles.

Taking into account all the aforementioned considerations and using them in the algorithm construction, we have found a very close relationship between chl-a concentration and the three-band NIR–red model (Equation 18.7) with simulated MERIS bands: determination coefficient $R^2 \approx 0.94$ (Figure 18.13a and b). We also established relationships between the chl-a concentration and the two-band models with simulated MODIS and MERIS bands. The two-band NIR–red model with simulated MERIS bands (Equation 18.9) had a close relationship ($R^2 \approx 0.94$) with chl-a concentration (Figure 18.13c) in the range typical for

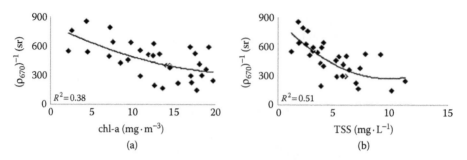

FIGURE 18.11
Reciprocal remote sensing reflectance at 670 nm, $(\rho_{670})^{-1}$, versus chl-a (a), and $(\rho_{670})^{-1}$ versus total suspended solids (b) concentrations for stations with chl-a concentrations, <20 mg · m^{-3}.

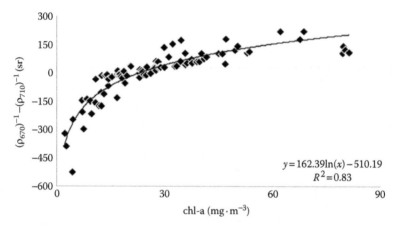

FIGURE 18.12
Relationship of chl-a concentration and the difference of the reciprocal reflectances at 670 and 710 nm.

productive waters and slightly lower when chl-a concentrations were limited to 30 mg · m^{-3}, the range typical for coastal and estuarine waters (Figure 18.13d). The two-band NIR–red model with simulated MODIS bands (Equation 18.8) also had a close linear relationship, with a chl-a concentrations ranging from 2 to 90 mg · m^{-3} and $R^2 = 0.75$ (Figure 18.13e). However, the model was almost not sensitive to chl-a concentrations below 30 mg · m^{-3} (Figure 18.13f). The R^2 was below 0.18, which shows that the two-band MODIS NIR–red model is not reliable for estimating low to moderate chl-a concentrations. Due to the low accuracy and unreliability of the two-band MODIS NIR–red model at low to moderate chl-a concentrations, no attempt was made to calibrate this model for potential use with the satellite data.

Thus, we calibrated the NIR–red models with simulated MERIS bands and established the algorithms for estimating the chl-a concentration for the range of chl-a concentrations from 2 to 120 mg · m^{-3} measured in 2008:

$$\text{chl-a} = 243.862 \times (3\text{-Band MERIS NIR-red model}) + 27.219 \qquad (18.14)$$

$$\text{chl-a} = 72.66 \times (2\text{-Band MERIS NIR-red model}) - 46.535 \qquad (18.15)$$

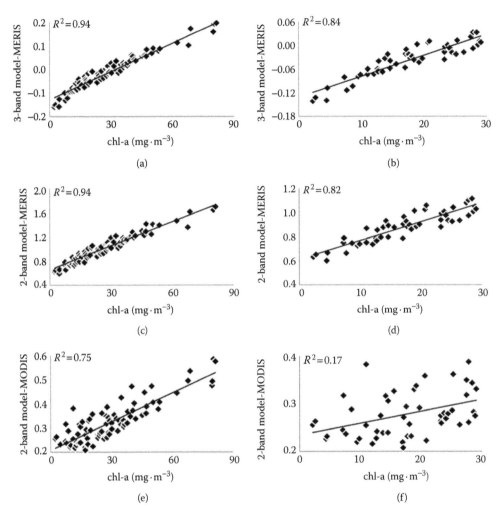

FIGURE 18.13
Performance of (a and b) three-band and (c and d) two-band models with simulated MERIS bands, and (e and f) two-band model with simulated MODIS bands for the estimation of chl-a in the Fremont Lakes 2008 in 2008. a, c, and e show chl-a concentrations from 0 to 90 mg · m^{-3}, while b, d, and f present chl-a concentrations below 30 mg · m^{-3} typical for coastal and estuarine waters.

The next step was to validate the algorithms (Equations 18.14 and 18.15) using the data set obtained in 2009. The relationships between the three-band and two-band MERIS NIR–red models and chl-a concentrations measured in 2009 were very close ($R^2 > 0.94$), and the best fit functions of these relationships were very close to those obtained for the 2008 data. Using simulated MERIS band reflectances measured in 2009, we estimated chl-a concentrations, chl-a$_{est}$ (Equations 18.14 and 18.15), and compared them with the concentrations measured analytically, chl-a$_{meas}$. The relationships between the estimated and the measured chl-a concentrations were very close for both models, thereby allowing a highly accurate estimation of chl-a concentration (Figure 18.14).

The three-band MERIS NIR–red model is

$$\text{chl-a}_{est} = 0.9475 \times \text{chl-a}_{meas} + 1.3693$$

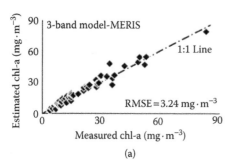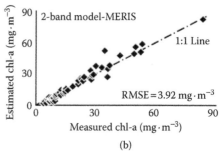

FIGURE 18.14

Estimation of chl-a concentrations in Fremont Lakes in 2009 by the (a) three-band and (b) two-band NIR–red models with simulated MERIS bands, calibrated using the Fremont Lakes 2008 data set (Equations 18.14 and 18.15).

The root mean square error (RMSE) of chl-a estimation was below 3.89 mg · m^{-3} for chl-a concentrations ranging from 4 to 104 mg · m^{-3}.

The two-band MERIS NIR–red model is

$$\text{chl-a}_{est} = 0.9823 \times \text{chl-a}_{meas} - 0.6848$$

The RMSE of chl-a estimation was below 4.72 mg · m^{-3}.

18.5.4 Calibration and Validation of NIR–Red Models Using Satellite Data

Data obtained in four campaigns in 2008 and three campaigns in 2009 in Azov Sea were used to test the performance of the two-band and three-band MERIS NIR–red models for the estimation of chl-a concentration using MERIS data. Of all the stations where in situ data were collected, the stations that satisfied the following criteria were considered for comparisons:

- The station is at least at a two-pixel length from the shoreline.
- The station is in a cloud-free/haze-free pixel in an image acquired within 2 days before or after the date of in situ data collection.
- The atmospheric correction procedure did not produce reflectance spectra with negative values beyond 443 nm.

Altogether from the 7 in situ data collection campaigns, there were 18 stations from the 2008 data set and 19 stations from the 2009 data set that satisfied the above criteria. The stations from the 2008 data set were used to establish and calibrate the relationship between chl-a concentrations and the model values, and the stations from the 2009 data set were used to test the validity of the algorithms. The minimum, maximum, median, and mean in situ chl-a concentrations of the 18 stations used for calibration were 0.63 mg · m^{-3}, 65.51 mg · m^{-3}, 24.35 mg · m^{-3}, and 26.97 mg·m^{-3}, respectively. The corresponding chl-a concentrations for the 19 stations used for validation were 0.2 mg · m^{-3}, 79.67 mg · m^{-3}, 13.97 mg · m^{-3}, and 19.76 mg·m^{-3}, respectively. The TSS concentration ranged from 0.4 g · m^{-3} to 27.4 g · m^{-3} for the entire data set.

For the stations chosen for calibration, the three-band and two-band MERIS NIR–red model values had very close linear relationships with in situ chl-a concentrations, with R^2 higher than 0.96 (Figure 18.15). The calibrated MERIS NIR–red algorithms were as follows:

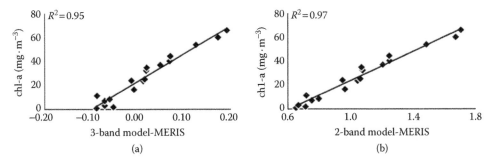

FIGURE 18.15
Calibration of (a) three-band and (b) two-band MERIS models using MERIS data collected over Azov Sea in 2008 (Equations 18.16 and 18.17).

The three-band MERIS NIR–red algorithm was

$$\text{chl-a} = 232.29 \times \left[\left(\rho_{665}^{-1} - \rho_{708}^{-1} \right) \times \rho_{753} \right] + 23.174 \tag{18.16}$$

The two-band MERIS NIR–red algorithm was

$$\text{chl-a} = 61.324 \times \left(\rho_{665}^{-1} \times \rho_{708} \right) - 37.94 \tag{18.17}$$

It is important to note that the slope and intercept of both MERIS NIR–red algorithms were close to the slope and intercept of the algorithms calibrated using in situ reflectances collected in 2008 from the Fremont Lakes in Nebraska (Equations 18.14 and 18.15).

The algorithms thus calibrated were used to estimate the chl-a concentration at the 19 stations from the 2009 data set, which was marked for validation. The validation procedure included (1) the estimation of chl-a concentrations by applying the calibrated algorithms (Equations 18.16 and 18.17) to the remote sensing reflectances, retrieved for the stations in the validation data set; and (2) the comparison between the estimated chl-a concentrations and the in situ chl-a concentrations. The comparison showed that the chl-a concentrations estimated using the calibrated algorithms were remarkably accurate.

The three-band MERIS NIR–red model was

$$\text{chl-a}_{\text{est}} = 0.97 \times \text{chl-a}_{\text{meas}} - 0.65$$

The two-band MERIS NIR–red model was

$$\text{chl-a}_{\text{est}} = 0.99 \times \text{chl-a}_{\text{meas}} - 1.17$$

The three-band MERIS NIR–red algorithm yielded an RMSE of 7.97 mg·m⁻³, whereas the two-band MERIS NIR–red algorithm had an even smaller RMSE of 6.93 mg·m⁻³.

The two-band MERIS NIR–red model was more accurate than the three-band MERIS NIR–red model for the estimation of chl-a concentration. We assume that the reason for the difference originates in the reflectance at λ_3 (753 nm in the three-band MERIS NIR–red model), which does not depend on chl-a concentration, but is susceptible to variations due to scattering by ISS and OSS. Significantly, the reflectance at 753 nm bears a considerably different relationship with the concentration of suspended matter than do the reflectances at 665 nm and 708 nm (Moses 2009), thereby invalidating the assumption of spectral independence of scattering by suspended particles in the wavelength range from λ_1 through λ_3 in the three-band model (Dall'Olmo, Gitelson, and Rundquist 2003). Therefore, the effects of scattering by ISS and OSS might not be fully removed using the three-band MERIS NIR–red model. Therefore, scattering potentially introduces uncertainties in chl-a estimation by the three-band MERIS NIR-red model, especially at low to moderate chl-a concentrations, where ρ_{665} is highly affected by the scattering of suspended solids, and ρ_{750} is very small and minor differences in its magnitude cause significant changes in the output of the three-band model.

In addition, the two-band MERIS NIR–red model takes full advantage of both the reflectance trough around 665 nm (due to absorption by chl-a) and the reflectance peak near 700 nm, which is related to both chl-a and suspended solids concentrations. The effect of scattering by suspended matter on the reflectance at 665 nm is similar to that on the reflectance at 708 nm, and therefore, the ratio ρ_{708}/ρ_{665} virtually cancels out the effect of scattering by suspended particles. Consequently, the two-band MERIS NIR-red model is sensitive to chl-a concentration and stable, reliable, and accurate over a wide range of chl-a concentrations, and perhaps is the best-suited model for application to data acquired by satellite-carried sensors.

Moreover, λ_3 in the three-band MERIS model (Equation 18.10) is at a longer wavelength (753 nm) than λ_2 in the two-band MERIS NIR–red model (Equation 18.12). Hence, the three-band MERIS NIR–red model was more sensitive than the two-band MERIS NIR–red model to uncertainties in the atmospheric correction procedure, due to the low signal-to-noise ratio, especially for stations with low chl-a concentrations and low magnitudes of reflectance in the NIR region. This, in addition to the reasons described previously, may explain the looser fit of points with chl-a concentrations below 10 mg · m^{-3} and the slightly higher RMSE for the three-band MERIS NIR–red model. Hence, even though both the algorithms yield high accuracies, the two-band MERIS NIR–red algorithm is preferred over the three-band MERIS NIR–red algorithm.

18.6 Toward a Universal NIR–Red Algorithm

The presented analysis indicates that there was no need to reparameterize the MERIS NIR–red algorithms for different water bodies. The slopes and intercepts of the two-band and three-band MERIS NIR–red algorithms derived from MERIS satellite data (Equations 18.16 and 18.17) were similar to corresponding figures for the two-band and three-band MERIS NIR–red algorithms derived from the data collected from the Fremont Lakes in 2008 (Equations 18.14 and 18.15). When the two-band MERIS NIR–red algorithm developed using the Fremont Lakes 2008 data set (reflectance spectra measured using field spectrometers) was applied to the MERIS data acquired over the Azov Sea in 2008 and 2009, the estimated chl-a concentrations closely matched with the chl-a concentrations measured in situ, with a very low RMSE of 3.64 mg · m^{-3} (Figure 18.16). This remarkable result illustrates the insensitivity of the algorithm to differences in the kind of remote sensor and the type of data processing, and strongly presents a case for the universal applicability of the two-band MERIS NIR–red algorithm. This is further illustrated in Figure 18.17, which shows plots of the in situ–measured chl-a concentration versus chl-a concentrations estimated by the two-band MERIS NIR–red algorithm developed using the 2008 Fremont Lakes data, for Lake Kinneret (Israel), Chesapeake Bay, Azov Sea, and Fremont Lakes (see also Table 18.3).

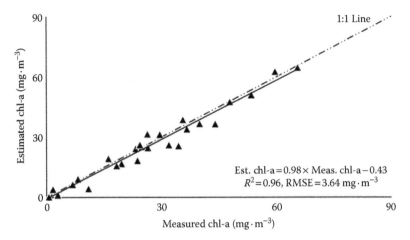

FIGURE 18.16
Estimation of chl-a concentrations in the Azov Sea in 2009 by the two-band MERIS model calibrated using the Fremont Lakes 2008 data set (Equation 18.15).

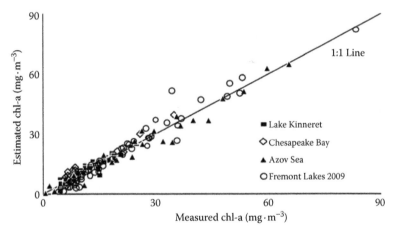

FIGURE 18.17
Estimation of chl-a concentrations in the Azov Sea, Lake Kinneret (Israel), Chesapeake Bay, and the Fremont Lakes in 2009 by the two-band MERIS model, calibrated using the Fremont Lakes 2008 data set.

TABLE 18.3

Accuracy of the Estimation of Chl-a by the Two-Band MERIS NIR–Red Model Calibrated by Data Taken at the Fremont Lakes in 2008

Water Body	Number of Stations for Calibration	Chl-a ($mg \cdot m^{-3}$)		Coefficient of Variation	Coefficient of Variation
		Min	Max		
Lake Kinneret	58	4.60	20.75	1.46	0.13
Chesapeake Bay	11	6.20	34.89	3.42	0.24
Azov Sea	26	0.63	65.51	3.64	0.13
Fremont Lakes 2009, Nebraska	84	3.97	83.18	3.92	0.19

The algorithm is remarkably consistent and highly accurate for data from different waters and different remote sensors (field spectrometers and satellite-carried sensors).

18.7 Limitations and Challenges in Developing Satellite Algorithms

The results presented here illustrate the high potential of the three-band and two-band NIR–red models to accurately estimate chl-a concentration in turbid productive waters, using the reflectance measured by field spectrometers as well as MERIS satellite data. Nevertheless, challenges still remain in calibrating the models for universal application with satellite data (Moses et al. 2009a,b). The MERIS NIR–red algorithms were developed and validated with a rather small data set (18 stations for calibration and 19 stations for validation). The algorithms should be tested using a larger set of data from water bodies with a wider variability of constituent composition and from different geographic locations. Some of the limitations and challenges involved in developing such a universal algorithm are discussed next.

18.7.1 Atmospheric Correction

A successful correction for atmospheric effects on satellite data and an accurate retrieval of surface reflectance are crucial to the success of the NIR–red model. The slope and offset of the relationship between chl-a concentration and the NIR–red model values are affected by atmospheric effects on the satellite images. This is pronouncedly seen in multitemporal data sets in which the atmospheric effects are not uniform on all the images. A reliable atmospheric correction procedure that is able to uniformly correct the nonuniform atmospheric effects across multitemporal data from multiple geographic locations is mandatory before an attempt to apply NIR–red algorithms on a universal scale. Significant differences have been observed in the shape and magnitude of surface reflectances (especially, the chl-a absorption in the red and the reflectance peak in the NIR region) retrieved through different atmospheric correction procedures for the same station (Moses 2009; Moses et al. 2009b). This means that the NIR–red model will produce very different estimates for the same chl-a concentration, depending on the particular atmospheric correction procedure used.

For the data analyzed in this research, the bright pixel atmospheric correction procedure, implemented in the standard processing of MERIS data, has given the most consistent and reliable results. However, inconsistencies still exist, and the procedure sometimes yields negative reflectances, especially for very turbid waters. The procedure needs to be tested for data from other geographic locations with variations in the type and quantity of aerosol loading. In situ reflectances measured at the time of satellite image acquisition will help to analyze the consistency of atmospheric correction procedures and their effect on the performance of the NIR–red models.

18.7.2 Temporal Variation of Water Quality

A satellite captures its entire swath within a few seconds, whereas it takes several hours to collect in situ data. As natural waters are normally highly dynamic, it is conceivable that the investigated water body might have undergone considerable changes in its biophysical and optical characteristics during these few hours. In our studies, differences in chl-a concentrations of up to a factor of two have been observed within a few hours (Moses 2009; Moses et al. 2009b). Thus, it is important that the temporal variations in the concentrations of optically active constituents, such as chl-a, TSS, ISS, and CDOM, have to be accounted for. This problem is magnified when there is no cloud-free satellite image available for the day of *in situ* data collection, and one has to use the image acquired a day or two before or after.

With the in situ stations spread quite far from each other and considering the satellite pixel dimension and the necessity to have stations separated by at least two pixel lengths, it has been rather difficult to collect in situ data using a single vessel at more than 10–12 stations within a time frame of a few hours surrounding the satellite overpass. The biophysical and optical characteristics at some of these stations might be different at the time of measurement from what they were at the time of satellite overpass. Furthermore, some of these stations might fall under cloud cover or haze. Thus, the number of stations available for comparison with the same-day images is very few, thereby making it difficult to develop reliable calibration equations for the model.

The effect of temporal variability is not uniform for all water bodies, but is rather case-specific. As indicated in some of our results, there have been cases where a temporal difference of up to 2 days did not adversely affect the estimation of chl-a concentration due to the stable biophysical condition of the water body. Nevertheless, it is still essential to account for the temporal variations in the water quality between the time of in situ data collection and the time of satellite image acquisition when attempting to calibrate or validate chl-a algorithms.

18.7.3 Within-Pixel Spatial Heterogeneity

Often, the spatial heterogeneity in the water body might be such that the point in situ station may not be truly representative of the satellite pixel area (260 m × 290 m for MERIS and 1 km × 1 km for MODIS) surrounding the station. In analyzing fluorescence measurements taken continuously along a transect in the Azov Sea in June 2005, significant variations were found in fluorescence values within every 300-m and 1-km lengths along the transect (Moses et al. 2009b; Moses 2009). When the water within each satellite pixel is not truly homogeneous, it becomes difficult to confidently and reliably compare the satellite-derived values to point in situ observations.

18.7.4 Need for a Modified In Situ Data Collection Strategy

The significance of the effects of the factors mentioned earlier and the difficulty in isolating them necessitate the development of in situ data collection techniques that may help to understand and account for these factors. In order to reliably assess the accuracy of atmospheric correction procedures and its effect on the performance of the NIR–red models, actual measurements of the water-leaving radiance should be collected in situ at the time of satellite overpass. Within-pixel spatial heterogeneity and temporal variation have to be accounted for by taking multiple measurements around each station so as to characterize the spatial variation within the satellite pixel area around the station and taking repeated measurements (at least twice, covering the length of time elapsed between the satellite overpass and the in situ data collection) at each station to characterize the temporal variation. If these factors are not accounted for, they present inherent hurdles to the development of reliable regression equations for the calibration of models. Of course, the rigor and the extent to which the in situ data collection procedures need to be adapted depend on the particular conditions at the water body.

18.8 Conclusions

NIR–red models are consistently highly accurate over a wide range, including low to moderate chl-a concentrations. They need not be reparameterized for different water bodies. The two-band MERIS NIR–red model is the most suitable model for estimating chl-a concentration in turbid productive waters using satellite data. The two-band MODIS NIR–red model can be used for estimating chl-a concentrations exceeding 20 mg · m^{-3} (e.g., for detecting phytoplankton blooms). However, it is not accurate in estimating low to moderate chl-a concentrations.

Our work suggests that universal NIR–red algorithms are possible and may be applied to decipher information acquired by spaceborne sensors for the estimation of chl-a concentration in inland, estuarine, and coastal waters around the globe.

References

Albert, A. 2004. Inversion technique for optical remote sensing in shallow water. PhD thesis, Hamburg University, Hamburg, 188.

Bidigare, R. R., M. E. Ondrusek, J. H. Morrow, and D. A. Kiefer. 1990. In vivo absorption properties of algal pigments. *SPIE, Ocean Opt X* 1302:290–302.

Bricaud, A., M. Babin, A. Morel, and H. Claustre. 1995. Variability in the chlorophyll-specific absorption-coefficients of natural phytoplankton: Analysis and parameterization. *J Geophys Res Oceans* 100(C7):13321–32.

Bukata, R. P., J. H. Jerome, J. E. Bruton, and S. C. Jain. 1979. Determination of inherent optical properties of Lake Ontario coastal waters. *Appl Opt* 18(23):3926–32.

Bukata, R. P., J. H. Jerome, K. Y. Kondratyev, and D. V. Pozdnyakov. 1995. *Optical Properties and Remote Sensing of Inland and Coastal Waters*. New York: CRC Press.

Cleveland, J. S., and A. D. Weidemann. 1993. Quantifying absorption by aquatic particles: A multiple scattering correction for glass-fiber filters. *Limnol Oceanogr* 38(6):1321–7.

Dall'Olmo, G. 2006. Isolation of optical signatures of phytoplankton pigments in turbid productive waters: Remote assessment of chlorophyll-*a*. PhD thesis, University of Nebraska—Lincoln, Lincoln, NE.

Dall'Olmo, G., and A. A. Gitelson. 2005. Effect of bio-optical parameter variability on the remote estimation of chlorophyll-*a* concentration in turbid productive waters: Experimental results. *Appl Opt* 44:412–22, see also erratum *Applied Optics* 44(16):3342.

Dall'Olmo, G., and A. A. Gitelson. 2006. Effect of bio-optical parameter variability and uncertainties in reflectance measurements on the remote estimation of chlorophyll-*a* concentration in turbid productive waters: Modeling results. *Appl Opt* 45(15):3577–92.

Dall'Olmo, G., A. A. Gitelson, and D. C. Rundquist. 2003. Towards a unified approach for remote estimation of chlorophyll-*a* in both terrestrial vegetation and turbid productive waters. *Geophys Res Lett* 30:1038.

Dekker, A. 1993. Detection of the optical water quality parameters for eutrophic waters by high resolution remote sensing. PhD thesis, Free University, Amsterdam, Netherlands.

Doerffer, R. 1981. Factor analysis in ocean color interpretation. In *Oceanography from Space*, ed. J. F. R. Gower, 339–45. New York: Plenum Press.

Eaton, A. D., L. S. Clesceri, E. W. Rice, A. E. Greenberg, and M. A. H. Franson, eds. 2005. *Standard Methods for the Examination of Water and Wastewater: Centennial Edition*. Baltimore, MD: United Book Press, Inc.

Ferrari, G. M., and S. Tassan. 1999. A method using chemical oxidation to remove light absorption by phytoplankton pigments. *J Phycol* 35(5):1090–8.

Fischer, J., and V. Kronfeld. 1990. Sun-stimulated chlorophyll fluorescence. 1: Influence of oceanic properties. *Int J Rem Sens* 11:2125–47.

Gege, P. 1995. Water analysis by remote sensing: A model for the interpretation of optical spectral measurements. European Space Agency Technical translation ESA-TT-1324. PhD thesis, *Institut für Optoelektronik Oberpfaffenhofen, Deutsche Forschungsanstalt für Luft- und Raumfahrt, e.V.*

Gitelson, A. A. 1992. The peak near 700 nm on reflectance spectra of algae and water: Relationships of its magnitude and position with chlorophyll concentration. *Int J Rem Sens* 13:3367–73.

Gitelson, A. 1993. The nature of the peak near 700 nm on the radiance spectra and its application for remote estimation of phytoplankton pigments in inland waters. *Opt Eng Rem Sens, SPIE* 1971:170–9.

Gitelson, A. A., and K. Y. Kondratyev. 1991. Optical models of mesotrophic and eutrophic water bodies. *Int J Rem Sens* 12:373–85.

Gitelson, A. A., U. Gritz, and M. N. Merzlyak. 2003. Relationships between leaf chlorophyll content and spectral reflectance and algorithms for non-destructive chlorophyll assessment in higher plant leaves. *J Plant Physiol* 160:271–82.

Gitelson, A., G. Keydan, and V. Shishkin. 1985. Inland waters quality assessment from satellite data in visible range of the spectrum. *Sov J Rem Sens* 6:28–36.

Gitelson, A., K. Kondratyev, and G. Garbusov. 1987. New approach to monitoring aquatic ecosystem quality. *Proc USSR Acad Sci* 295:825–7.

Gitelson, A. A., J. F. Schalles, and C. M. Hladik. 2007. Remote chlorophyll-*a* retrieval in turbid, productive estuaries: Chesapeake Bay case study. *Rem Sens Environ* 109:464–72.

Gitelson, A., F. Szilagyi, and K. Mittenzwey. 1993. Improving quantitative remote sensing for monitoring of inland water quality. *Water Res* 7:1185–94.

Gitelson, A., A. M. Nikanorov, G. Sabo, and F. Szilagyi. 1986. Study of water surface quality using remote sensing. *IAHS Publ* 157:111–21.

Gitelson, A., M. Mayo, Y. Z. Yacobi, A. Parparov, and T. Berman. 1994. The use of high spectral radiometer data for detection of low chlorophyll concentrations in Lake Kinneret. *J Plankton Res* 16:993–1002.

Gitelson, A., G. Garbuzov, F. Szilagyi, K. H. Mittenzwey, A. Karnieli, and A. Kaiser. 1993. Quantitative remote sensing methods for real-time monitoring inland water quality. *Int J Rem Sens* 14:1269–95.

Gitelson, A. A., Y. Z. Yacobi, J. F. Schalles, D. C. Rundquist, L. Han, R. Stark, and D. Etzion. 2000. Remote estimation of phytoplankton density in productive waters. *Arch Hydrobiol Spec Issues Adv Limnol* 55:121–36.

Gitelson, A. A., A. Viña, V. Ciganda, D. C. Rundquist, and T. J. Arkebauer. 2005. Remote estimation of canopy chlorophyll content in crops. *Geophys Res Lett* 32:L08403, doi:10.1029/2005GL022688.

Gitelson A. A., G. Dall'Olmo, W. Moses, D. C. Rundquist, T. Barrow, T. R. Fisher, D. Gurlin, and J. Holz. 2008. A simple semi-analytical model for remote estimation of chlorophyll-*a* in turbid waters: Validation. *Rem Sens Environ* 112:3582–93.

Gitelson, A. A., D. Gurlin, W. J. Moses, and T. Barrow. 2009. A bio-optical algorithm for the remote estimation of the chlorophyll-*a* concentration in case 2 waters. *Environ Res Lett* 4(4), doi:10.1088/1748-9326/4/4/045003.

GKSS. 1986. The use of chlorophyll fluorescence measurements from space for separating constituents of sea water. ESA Contract No. RFQ3-5059/84/NL/MD Vol II, GKSS Research Centre, Germany.

Gons, H. J. 1999. Optical teledetection of chlorophyll a in turbid inland waters. *Environ Sci Technol* 33:1127–32.

Gons, H., J. M. Rijkeboer, S. Bagheri, and K. G. Ruddick. 2000. Optical teledetection of chlorophyll a in estuarine and coastal waters. *Environ Sci Technol* 34:5189–92.

Gordon, H. R., O. B. Brown, and M. M. Jacobs. 1975. Computed relationships between the inherent and apparent optical properties of a flat homogeneous ocean. *Appl Opt* 14(2):417–27.

Gordon, H., and A. Morel. 1983. Remote assessment of ocean color for interpretation of satellite visible imagery. A review. In *Lecture Notes on Coastal and Estuarine Studies 4*, ed. R. T. Barber, C. N. K. Mooers, M. J. Bowman, and B. Zeitzschel, 7. New York: Springer-Verlag.

Gower, J. F. R. 1980. Observations of *in-situ* fluorescence of chlorophyll-*a* in Saanich Inlet. *Boundary-Layer Meteorol* 18:235–45.

Han, L. H., and D. C. Rundquist. 1994. The response of both surface reflectance and the underwater light-field to various levels of suspended sediments: Preliminary results. *Photogramm Eng Rem Sens* 60(12):1463–71.

Han, L. H., and D. C. Rundquist. 1996. Spectral characterization of suspended sediments generated from two texture classes of clay soil. *Int J Rem Sens* 17(3):643–9.

Han, L. H., and D. C. Rundquist. 1997. Comparison of NIR/RED ratio and first derivative of reflectance in estimating algal-chlorophyll concentration: A case study in a turbid reservoir. *Rem Sens Environ* 62(3):253–61.

Han, L., D. C. Rundquist, L. L. Liu, R. N. Fraser, and J. F. Schalles. 1994. The spectral responses of algal chlorophyll in water with varying levels of suspended sediment. *Int J Rem Sens* 15(18):3707–18.

Hoge, E. F., C. W. Wright, and R. N. Swift. 1987. Radiance ratio algorithm wavelengths for remote oceanic chlorophyll determination. *Appl Opt* 26(11):2082–94.

Jupp, D. L. B., J. T. O. Kirk, and G. P. Harris. 1994. Detection, identification and mapping of cyanobacteria: Using remote-sensing to measure the optical-quality of turbid inland waters. *Aust J Mar Freshwater Res* 45(5):801–28.

Kallio, K., S. Koponen, and J. Pulliainen. 2003. Feasibility of airborne imaging spectrometry for lake monitoring: A case study of spatial chlorophyll alpha distribution in two meso-eutrophic lakes. *Int J Rem Sens* 24(19):3771–90.

Kallio, K., T. Kutser, T. Hannonen, S. Koponen, J. Pulliainen, J. Vepsäläinen, and T. Pyhälahti. 2001. Retrieval of water quality from airborne imaging spectrometry of various lake types in different seasons. *Sci Total Environ* 268(1–3):59–77.

Kirk, J. T. O. 1994. *Light and Photosynthesis in Aquatic Ecosystems*. Cambridge, UK: Cambridge University Press.

Kishino, M., S. Sugihara, and N. Okami. 1986. Theoretical analysis of the in-situ fluorescence of chlorophyll-*a* on the underwater spectral irradiance. *Bulletein de la Societe Franco-Japanaise d'Oceanographie* 24:130–8.

Kutser, T., A. Herlevi, K. Kallio, and H. Arst. 2001. A hyperspectral model for interpretation of passive optical remote sensing data from turbid lakes. *Sci Total Environ* 268(1–3):47–58.

Kutser, T., M. Hiire, L. Metsamaa, E. Vahtmae, B. Paavel, and R. Aps. 2009. Field measurements of spectral backscattering coefficient of the Baltic Sea and boreal lakes. *Boreal Environ Res* 14:305–12.

Le, C., Y. Li, Y. Zha, D. Sun, C. Huang, and H. Lu. 2009. A four-band semi-analytical model for estimating chlorophyll a in highly turbid lakes: The case of Taihu Lake, China. *Rem Sens Environ* 113:1175–82.

Lee, Z. P., K. L. Carder, C. D. Mobley, R. G. Steward, and J. S. Patch. 1998. Hyperspectral remote sensing for shallow waters. I: A semianalytical model. *Appl Opt* 37(27):6329–38.

Matthews, A. M., and S. R. Boxall. 1994. Novel algorithms for the determination of phytoplankton concentration and maturity. *Proceedings of the Second Thematic Conference on Remote Sensing for Marine and Coastal Environments* 1:173–80.

Millie, D. F., B. T. Vinyard, M. C. Baker, and C. S. Tucker. 1995. Testing the temporal and spatial validity of site-specific models derived from airborne remote sensing. *Can J Fish Aquat Sci* 52:1094–107.

Mitchell, B. G., M. Kahru, J. Wieland, and M. Stramska. 2003. Determination of spectral absorption coefficients of particles, dissolved material and phytoplankton for discrete water samples. In *Inherent Optical Properties: Instruments, Characterizations, Field Measurements and Data Analysis Protocols*. Ocean Optics Protocols for Satellite Ocean Color Sensor Validation, Revision 4, Volume IV, ed. J. L. Mueller, G. S. Fargion, and C. R. McClain. Goddard Space Flight Center Technical Memorandum 2003-211621.

Mobley, C. D. 1994. *Light and Water: Radiative Transfer in Natural Waters*. San Diego, CA: Academic Press.

Moore, G. F., J. Aiken, and S. J. Lavender. 1999. The atmospheric correction of water colour and the quantitative retrieval of suspended particulate matter in case II waters: Application to MERIS. *Int J Rem Sens* 20(9):1713–33.

Morel, A., and L. Prieur. 1977. Analysis of variations in ocean color. *Limnol Oceanogr* 22:709–22.

Moses, W. J. 2009. Satellite-based estimation of chlorophyll-*a* concentration in turbid productive waters. PhD thesis, University of Nebraska—Lincoln, Lincoln, NE.

Moses, W. J., A. A. Gitelson, S. Berdnikov, and V. Povazhnyy. 2009a. Satellite estimation of chlorophyll-*a* concentration using the red and NIR bands of MERIS—the Azov Sea case study. *IEEE Geosci Rem Sens Lett* 6(4):845–9.

Moses, W. J., A. A. Gitelson, S. Berdnikov, and V. Povazhnyy. 2009b. Estimation of chlorophyll-*a* concentration in case II waters using MODIS and MERIS data-successes and challenges. *Environ Res Lett* 4(4):045005(8), doi:10.1088/1748-9326/4/4/045005.

Mueller, J. L. 2003. Inherent optical properties: Instrument characterizations, field measurements and data analysis protocols. In *Ocean Optics Protocols for Satellite Ocean Color Sensor Validation*, Revision 4, Volume IV, Erratum 1 to Pegau, S., J. R. V. Zaneveld, and J. L. Mueller. 2003. Inherent optical property measurement concepts: Physical principles and instruments. In *Inherent Optical Properties: Instruments, Characterizations, Field Measurements, and Data Analysis Protocols*. Ocean Optics Protocols for Satellite Ocean Color Sensor Validation, Revision 4, Volume IV, ed. J. L. Mueller, G. S. Fargion, and C. R. McClain. Goddard Space Flight Center Technical Memorandum 2003-211621.

Neville, R. A., and J. F. R. Gower. 1977. Passive remote sensing of phytoplankton via chlorophyll *a* fluorescence. *J Geophys Res* 82:3487–93.

Nusch, E. A. 1980. Comparison of different methods for chlorophyll and phaeopigment determination. *Adv Limnol–Fundam Appl Limnol Special Issues* 14:14–36.

Ohde, T., and H. Siegel. 2003. Derivation of immersion factors for the hyperspectral Trios radiance sensor. *J Opt A: Pure Appl Opt* 5(3):12–4.

Oki, K., and Y. Yasuoka. 2002. Estimation of chlorophyll concentration in lakes and inland seas with a field spectroradiometer above the water surface. *Appl Opt* 41(30):6463–9.

O'Reilly, J. E., S. Maritorena, B. G. Mitchell, D. A. Siegel, K. L. Carder, S. A. Garver, M. Kahru, and C. McClain. 1998. Ocean color chlorophyll algorithms for SeaWiFS. *J Geophys Res Oceans* 103(C11):24937–53.

Oubelkheir, K., H. Claustre, A. Sciandra, and M. Babin. 2005. Bio-optical and biogeochemical properties of different trophic regimes in oceanic water. *Limnol Oceanogr* 50(6):1795–809.

Pierson, D., and N. Strömbäck. 2000. A modelling approach to evaluate preliminary remote sensing algorithms: Use of water quality data from Swedish great lakes. *Geophys* 36(1–2):177–202.

Pulliainen, J., K. Kallio, K. Eloheimo, S. Koponen, H. Servomaa, T. Hannonen, S. Tauriainen, and M. Hallikainen. 2001. A semi-operative approach to lake water quality retrieval from remote sensing data. *Sci Total Environ* 268(1–3):79–93.

Quibell, G. 1992. Estimation chlorophyll concentrations using upwelling radiance from different freshwater algal genera. *Int J Rem Sens* 13(14):2611–21.

Reynolds, C. S. 2006. *Ecology of Phytoplankton*. Cambridge, UK: Cambridge University Press.

Ritchie, R. J. 2008. Universal chlorophyll equations for estimating chlorophylls *a*, *b*, *c*, and *d* and total chlorophylls in natural assemblages of photosynthetic organisms using acetone, methanol, or ethanol solvents. *Photosynthetica* 46(1):115–26.

Ruddick, K. G., H. J. Gons, M. Rijkeboer, and G. Tilstone. 2001. Optical remote sensing of chlorophyll a in case 2 waters by use of an adaptive two-band algorithm with optimal error properties. *Appl Opt* 40(21):3575–85.

Rundquist, D. C., J. F. Schalles, and J. S. Peake. 1995. The response of volume reflectance to manipulated algal concentrations above bright and dark bottoms at various depths in an experimental pool. *Geocarto Int* 10:5–14.

Schalles, J. F. 2006. Optical remote sensing techniques to estimate phytoplankton chlorophyll a concentrations in coastal waters with varying suspended matter and CDOM concentrations. In *Remote Sensing of Aquatic Coastal Ecosystem Processes: Science and Management Applications*, ed. L. Richardson and E. Ledrew, 27–79. New York: Springer-Verlag.

Schalles, J. F., A. Gitelson, Y. Z. Yacobi, and A. E. Kroenke. 1998. Chlorophyll estimation using whole seasonal, remotely sensed high spectral-resolution data for an eutrophic lake. *J Phycol* 34(2):383–90.

Strömbäck, N., and D. C. Pierson. 2001. The effects of variability in the inherent optical properties on estimations of chlorophyll a by remote sensing in Swedish freshwaters. *Sci Total Environ* 26(1–3):123–37.

Stumpf, R. P., and M. A. Tyler. 1988. Satellite detection of bloom and pigment distributions in estuaries. *Rem Sens Environ* 24(3):385–404.

Thiemann, S., and H. Kaufmann. 2002. Lake water quality monitoring using hyperspectral airborne data: A semiempirical multisensor and multitemporal approach for the Mecklenburg Lake District, Germany. *Rem Sens Environ* 81(2–3):228–37.

Vasilkov, A., and O. Kopelevich. 1982. Reasons for the appearance of the maximum near 700 nm in the radiance spectrum emitted by the ocean layer. *Oceanology* 22:697–701.

Vos, W. L., M. Donze, and H. Bueteveld. 1986. On the reflectance spectrum of algae in water: The nature of the peak at 700 nm and its shift with varying concentration. Technical Report, 86–22, Communication on Sanitary Engineering and Water Management, Delft, Netherlands.

Welschmeyer, N. A. 1994. Fluorometric analysis of chlorophyll a in the presence of chlorophyll b and paeopigments. *Limnol Oceanogr* 39(8):1985–92.

Yacobi, Y. Z., A. A. Gitelson, and M. Mayo. 1995. Remote sensing of chlorophyll in Lake Kinneret using high spectral resolution radiometer and Landsat TM: Spectral features of reflectance and algorithm development. *J Plankton Res* 17(11):2155–73.

19

Retrievals of Turbulent Heat Fluxes and Surface Soil Water Content by Remote Sensing

George P. Petropoulos and Toby N. Carlson

CONTENTS

19.1 Introduction

Earth's land surface and atmosphere are under a constant exchange of energy, momentum, and water via the flux of sensible heat (H; the heat energy transferred between the surface and air when there is a difference in temperature between them) and latent heat (LE; the flux of heat from the Earth's surface to the atmosphere that is associated with evaporation of water at the surface and plant transpiration). Soil water content is defined as the water content available in the soil profile of a specific depth. Accurate knowledge of both LE and H as well as soil water content are of great importance in a large number of regional- and global-scale applications, such as monitoring of plant water requirements, plant growth and productivity, and management of irrigation and cultivation procedures (Dodds, Meyer, and Barton 2005; Consoli, Urso, and Toscano 2006). Such data are also of key significance in the numerical modeling and prediction of atmospheric and hydrologic cycles and in improving the accuracy of weather forecasting models (Jacob et al. 2002).

Furthermore, quantitative information on these parameters is important for monitoring the degradation and desertification of land (Xu and Chen 2005; McCabe and Wood 2006), for understanding the processes that control ecosystem carbon dioxide (CO_2) exchange (Yepez et al. 2003), and for understanding the interactions between parameters in different ecosystem processes (Wever, Flanagan, and Carlson 2002).

The advent of satellite-based remote sensing over the last nearly 4 decades has led to a considerable amount of work in determining whether such systems can provide spatially explicit maps of these parameters from space. The general attributes that make such satellite remote sensing techniques attractive for the retrieval of land-surface fluxes and surface moisture content are summarized, for instance, by De Troch et al. (1996) and Engman and Schultz (2000). These traits include their ability to provide synoptic views in a spatially contiguous fashion and in a repetitive manner, without a disturbing influence on the area to be surveyed and without accessibility issues to the site. Optical remote sensing–based methods can provide information on vegetation health and biomass amount, whereas thermal methods detail the temperature structure of the land surface, which has a direct relationship to heat flux parameters. Microwave-based methods offer all-weather capability and daytime and nighttime observations which, combined with their strong dependence on the dielectric properties of the target (for soils, which is largely a function of the amount of soil water present), make them potentially very powerful for estimating various hydrometerological parameters, mainly soil moisture (Schmugge et al. 2002). Particularly, the combined use of satellite data from optical and thermal infrared radiometers has shown a promising avenue in the retrieval of both LE and H fluxes and soil surface moisture content (the latter is defined as the water contained at the first 5 cm of the soil depth; Moran et al. 2004; Stisen et al. 2008).

The aim of this chapter is twofold: First, to provide a comprehensive overview of the development of remote sensing–based methods currently used in the estimation of land-surface atmosphere fluxes and surface soil water content. Second, to present in more detail how these methods work, showing a paradigm from the use of one such methodology. A variant of the method we describe has been proposed in the operational retrieval of soil water content by the National Polar-Orbiting Operational Environmental Satellite System (NPOESS), in a series of platforms planned to be launched in the next 12 years starting in 2013 (Chauhan, Miller, and Ardanuy 2003). To facilitate an effective understanding of the topic with which this chapter is concerned, we first provide a brief discussion that addresses what type of information a remote sensing radiometer measures and how these measures can be interpreted. In the largest part, this chapter is aimed at describing the use of satellite remote sensing data, although it is equally applicable to airborne remote sensing observations, provided that appropriate data is available from such sensor systems.

19.2 Remote Sensing of Surface Energy Fluxes and Soil Water Content: An Overview

In the following sections is made available an overview of the remote sensing techniques which have been employed for the estimation of surface energy fluxes and of soil surface moisture content. However, before that is discussed what a remote sensing radiometer actually measures is rather important in being able to understand the basis of the different approaches.

19.2.1 What Does a Remote Sensing Radiometer Measure?

Remote sensing radiometers do not directly measure either soil water content or LE and H fluxes. The spectral radiance measures they provide should be combined in some form of retrieval algorithm or model in order to estimate these parameters. Therefore, let us first consider what type of information one can measure from a remote sensing sensor using information from optical, thermal, or microwave sensors. A frequently used sensor measures the upwelling long-wave flux, from which one can calculate a blackbody temperature (T_{BB}), or the temperature of an object that has not been corrected for emissivity and atmospheric effects. An often misused term in thermal remote sensing is a quantity called the "skin temperature." This is actually the blackbody radiometric surface temperature (T_s), called kinetic temperature, (T_{kin}), and sometimes referred to as the "dynamic" temperature, and is not necessarily the air temperature (T_{air}). The term "skin temperature" on the other hand may erroneously be assigned to the temperature of a uniform surface, whereas it is actually derived from the radiant flux emitted from a mixture of vegetation canopy and bare soil. While such skin temperatures may reflect the turbulent heat flux from that surface pixel, the soil water content may be highly misrepresented in the measurement. A detailed description of the terminology used in thermal infrared remote sensing for natural surfaces, detailing the differences between the terms used, has been provided by Norman and Becker (1995).

In order to understand the ambiguity of such measurements, we refer to Figure 19.1, which is a schematic interpretation of what one actually "sees" in making an optical, thermal, or microwave remote measurement from a satellite. The dotted line on the right represents the vertical profile of substrate temperature below a sunlit bare soil. The greatest temperature variation belowground (and aboveground) is, of course, nearest the surface. At the surface itself, the radiant surface temperature is given by the value at point g. Point d denotes the surface air temperature T_{air} (say at 10 m elevation) for this case, in which the

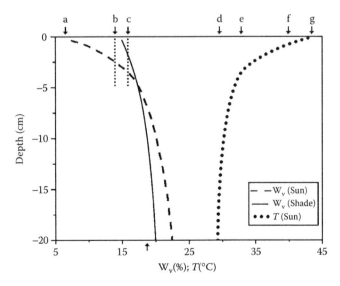

FIGURE 19.1
Schematic of vertical profiles of soil water content (W_v) and soil temperature (T) at different conditions. The thin pair of dotted lines near the top of figure represents soil water content averages over the top 5 cm. The letters at the top of the diagram are explained in the text (see also Carlson, T.N. et al. 2004).

surface may be either bare or shaded by vegetation; at that level, the horizontal variation of T_{air} will not vary greatly between points above vegetation and above nearby bare soil. Points e and f, respectively, represent the radiometric surface temperature of sunlit vegetation and the radiometric surface temperature of a mixture of sunlit vegetation and sunlit bare soil.

The dot-dashed and solid lines represent the soil water content, respectively, below the sunlit bare soil surface and a surface consisting of bare soil and vegetation and therefore partly shaded. Herein, we will discuss in Sections 19.2.2 and 19.2.3 how soil water content might be inferred from optical and thermal measurements, though the subject of microwave soil water content will be largely avoided in this chapter. We do know that microwave measurements can be inverted to yield a vertically averaged soil water content estimate over a depth, usually no more than several centimeters below the land surface. An inversion of the radiometric surface temperature over the bare soil would yield a soil water content very close to that of the surface as given by the point labeled a, a rather low amount. Points b and c, however, might represent the soil water content obtained from microwave measurements on the same day, respectively, for the bare soil and surfaces partially shaded by vegetation, essentially average values for the top 5 cm. Of course, the depth over which the microwave measurements apply depends on the wavelength of the sensor and is therefore not necessarily equal to 5 cm. Here, we have assumed a depth of 5-cm measurement simply to illustrate that the soil water content is some sort of average, which is not equal to the surface value that is supposedly obtained from the temperature measurement (g). The point here is that values b and c obtained from microwaves neither represent the full soil column average nor do they resemble that obtained from the temperature for the sunlit bare soil. Indeed, taken as an ensemble of bare soil and soil shaded by an arbitrary amount of vegetation, almost any soil water content value can be obtained. Even if each type of measurement, optical, thermal, or microwave, were to yield a "correct" value, one could not be sure over what depth that supposedly accurate soil water content would apply.

Clearly, comparing one type of measurement with another or with an in situ soil sample is fraught with ambiguities. One way to reduce ambiguity in estimating soil water content is to separate the bare soil fraction of the satellite pixel from the vegetated fraction and to assume that the derived surface soil water content is represented by that obtained from the bare soil temperature. This approach will constitute an underpinning for deriving all the subsequent parameters from satellite measurements. Let us therefore now concentrate on reviewing briefly how one obtains a few useful land-surface parameters such as the LE and H fluxes and soil water content with as little ambiguity as possible using mainly optical and thermal remote sensing techniques for a mix of bare soil and vegetation in clear-sky conditions.

19.2.2 Overview of Estimation of Turbulent Fluxes by Remote Sensing

Several algorithms have been developed in the last 4 decades for estimating the exchange of moisture and heat between the surface and the atmosphere from space- or airborne systems; these are often used in combination with ancillary surface and atmospheric observations. Table 19.1 summarizes some of the available satellite sensors providing data suitable for the retrieval of surface energy fluxes and soil water content. Overviews of the available methodologies can be found in studies by Diak et al. (2003), Courault, Seguin, and Olioso (2005), and Verstraeten, Veroustraete, and Feyen (2008).

The vast majority of remote sensing–based methods employed today for estimating energy fluxes are fundamentally residual-based approaches, working based on the principle

TABLE 19.1

Examples of Spaceborne Sensors Currently in Orbit Providing Observations Appropriate to Derive Surface Heat Fluxes and Soil Surface Moisture

Sensor Name	Manufacturer	Platform	Spatial Resolution	Spectral Resolution	Revisit Period
ASTER	NASA/ERSDAC	Terra	VNIR: 15 m SWIR: 30 m TIR: 90 m	VNIR: 4 SWIR: 6 TIR: 5	16 days
Landsat TM/ETM+	NASA/U.S. Department of Defense	Landsat	VNIR: 30 m SWIR: 30 m TIR: 120 m (TM)/60 m (ETM+)	VNIR: 4 SWIR: 2 TIR: 1	16 days
MODIS	NASA	Terra and Aqua	VNIR: 250 m/500 m SWIR: 500 m TIR: 1 km	VNIR: 18 SWIR: 2 TIR: 16	2 daytime/ 2 nighttime
AVHRR	NASA	NOAA	VNIR: 1.1 km SWIR: 1.1 km TIR: 1.1 km	VNIR: 2 SWIR: 1 TIR: 2/3	2 daytime/ 2 nighttime
AATSR	ESA	ENVISAT	VNIR: 1 km SWIR: 1 km TIR: 1 km	VNIR: 3 SWIR: 1 TIR: 3	2 daytime/ 2 nighttime
SEVIRI	EUMETSAT/ESA	Meteosat-2	VNIR: 1.1 and 3.0 km, SWIR: 3 km	VNIR: 4 TIR: 8	96 scenes per day (every 15')

of energy conservation. In such methods, net radiation (R_n) and soil heat flux (G) are computed using well-established approaches (see the review of Diak et al. 2003), whereas H flux is estimated based on the evaluation of difference between the surface radiometric temperature (T_s) and the T_{air} gradient at a single time, using only T_s (or a derivative quantity) as the surface boundary condition. Subsequently, LE flux is computed from the difference between R_n and G and H fluxes, based on the principle of energy conservation.

The simplest scheme was originally proposed by Seguin and Itier (1983) and elaborated on by Carlson, Capehart, and Gillies (1995). This scheme uses the midday T_s and the net radiation to estimate the mean daily evapotranspiration. The two empirical coefficients for the simple equation are themselves variables, depending on the wind speed and fractional vegetation cover (Fr). More physically based (but equally simple) estimation schemes propose the so-called one-layer models (Hall et al. 1992; Inoue and Moran 1997). In these models, energy balance, temperature, and vapor pressure regimes of the vegetation canopy and the soil are not distinguished. Those models typically use T_s in place of the aerodynamic temperature and link H flux to the difference between T_s and T_{air} through a single aerodynamic resistance. However, with these models major problems exist in the interpretation of the derived results when the soil surface is partially covered by vegetation (e.g., discussions by Kustas et al. 1989; Moran et al. 2005). Because of partial plant cover, the surface temperature measured by a thermal infrared sensor will be a composite temperature between that of the vegetation and the soil substrate. Relevant studies have shown that these models tend to overestimate the H flux term, especially over sparse canopies, because the resistance to heat transport from the soil component within the sensor's field of view is often significantly larger than the resistance above the canopy.

Another drawback in the use of such methods is that no distinction is made between soil and vegetation components, which in turn makes impossible the identification of vegetation stress conditions (see the review by Schmugge et al. 2002).

As an improvement to these simple "one-layer" models, "two-layer (two-source)" models have been developed. These models have included treatment for the temperature and energy balance regimes separately for the vegetation canopy and soil surface components, accounting for the variation in surface resistance due to the variation in vegetation cover and surface roughness (Norman, Kustas, and Humes 1995; Anderson et al. 1997; Brasa et al. 1998; Norman et al. 2000; Chehbouni et al. 2000; French, Schmugge, and Kustas 2002). Validation of the LE and *H* estimates from such models has demonstrated a varying degree of accuracy, despite the complexity with which the soil and vegetation components are treated in these types of models (Norman et al. 2003). As noted by Kustas, Zhan, and Schmugge (1998), an important advantage of two-layer models is that they can be useful in interpreting aggregated flux estimates using bulk atmospheric boundary layer approaches over heterogeneous surfaces (Hipps, Swiatek, and Kustas 1994; Kustas and Norman 1996). Furthermore, another important advantage of some of these two-layer models (e.g., that of Mecikalski et al. 1999) is that they accommodate a view-angle dependence of surface brightness temperature. Consideration of the latter has shown that it can have a pronounced effect on the accuracy of the LE and *H* retrievals, especially over sparse vegetation where changes in the view angle cause large differences in the fractions of vegetation and bare soil visible within the radiometer footprint. Nonetheless, one of the main drawbacks of two-layer models in comparison with one-layer models is their increased architectural complexity, which results in difficulties in their implementation, requiring a larger number of shelter-level meteorological inputs (primarily wind speed, aerodynamic resistance, friction velocity, air and aerodynamic temperature) and introducing errors in their representativeness (Beven and Fisher 1996; Jacob et al. 2002).

A different approach for deriving spatial maps of land-surface heat fluxes from remote sensing observations—and in some cases also soil water content—is to place theoretical boundary lines on the observed inverse relationship between an estimate of the land radiometric temperature and a spectral vegetation index (VI). VI is an index related to the amount of vegetation present, often taken as F_r (Jiang and Islam 1999; Sandhold, Rasmussen, and Andersen 2002). In such a scatter plot, the boundary lines may resemble a triangle (or trapezoid). Even though it has been demonstrated that the derivation of spatially distributed estimates of turbulent heat fluxes using the T_s/VI triangular scatter plot is feasible without the use of a boundary layer model (Moran et al. 1994, 1996), more sophisticated approaches (Gillies et al. 1997; Carlson 2007a) have proposed the retrieval of the above parameters from the combined use of the contextual interpretation of the T_s/VI domain with thermodynamic principles embodied in a two-layer surface/boundary layer energy balance (in particular, soil–vegetation–atmosphere transfer [SVAT]) models. This type of approach has certain advantages over one- and two-layer models, including a potentially improved ability to deal with surface heterogeneity (because this can be encapsulated in the VI measure), their potential to provide easier transformation between instantaneous and daytime average fluxes (which is often based on the conservation of a flux ratio, the Bowen ratio, during the day), ability to avoid dependence on external surface and meteorological parameters, and that the key input data are relatively easy to obtain from space- or airborne data over large areas (i.e., VI and surface radiometric temperature and nominal T_{air}), and that they allow for the correlation between the input (i.e., F_r, T_s) and output variables (soil water content and surface heat fluxes) to be nonlinear, contrary to the majority of all other analogous methods that are based on the T_s/VI pixel envelope or one- or two-layer models that assume linear interpretations of the T_s/VI domain. Nevertheless, an important limitation of these approaches is the assumption that the temperature of live leaf surfaces is close to that for potential transpiration. This restriction, however, does not significantly affect the results as long as the vegetation is not seriously stressed. Another minor difficulty that conceivably restricts the use of these methods relates to the difficulty in choosing the appropriate parameters for the SVAT model, as their use requires some familiarity by the users; the latter will be shown to be relatively unimportant, however. Finally, the T_s/VI methods require a large number of pixels to be sampled over a varied terrain.

Despite these potential impediments, the method of Gillies and Carlson (1995) and Gillies et al. (1997), the so-called triangle method, has been applied in a number of studies that have highlighted its potential for mapping surface heat fluxes and that have shown its ability to provide distributed estimates of LE and H with an accuracy of around 25–55 W · m⁻², or about 10–30% (Gillies et al. 1997, Brunsell and Gillies 2003). This accuracy is generally comparable to other T_s/VI scatter plot–based methods (Jiang and Islam 2001) and/or also some two-layer models (Norman et al. 2003). These numbers are to be compared with the accuracy in the measurement of these fluxes using ground instrumentation, which is around 10–15% (as referred to in Jiang, Islam, and Carlson 2004; and also in Kustas and Norman 1996; Wilson et al. 2002). Thus, current methods are pressing the limit of accuracy in the use of remote measurements to estimate LE, H, and soil surface water content.

19.2.3 Overview of Estimation of Soil Water Content by Remote Sensing

Substantial research has been carried out indicating that soil water content, or at least "surface wetness," can be estimated by several methods using mainly visible (VIS) channels (Whalley, Leeds-Harrison, and Bowman 1991; Leone and Sommer 2000; Schlesinger et al. 1996), thermal infrared (TIR; Cracknell and Xue 1996; Gillies et al. 1997; Sobrino and Raissouni 2000) or microwave data (Quesney et al. 2000; Wang et al. 1997; Biftu and Gan 1999; Griffiths and Wooding 1996; Oldak et al. 2003; Njoku and Entekhabi 1996). Comprehensive reviews of the basic approaches can be found in Kostov and Jackson (1993) and Moran et al. (2004).

The relationship between spectral reflectance particularly and soil water content has been discussed in many studies (Weidong et al. 2002), and several research workers have demonstrated that observations from optical sensors could be used for the retrieval of soil water content, particularly over bare soil surfaces (Idso et al. 1975; Sommer, Hill, and Meiger 1998; Whalley, Leeds-Harrison, and Bowman 1991). However, the use of visible and near-infrared data is limited by the fact that such data offer a measurement of target reflectance of only the top several millimeters of the surface, as already shown in Figure 19.1. Another problem with the use of this type of data is related to the inference of both clouds and vegetation canopy with the optical signal, of which the latter has a limited capability to penetrate. Furthermore, soil reflectance measurements are not only a function of the soil water content but also of the soil composition, physical structure, and observation conditions. Optical methods require that the satellite pixel contain at least some fraction of bare soil. Finally, while such data have a high spatial resolution, they usually have quite low temporal coverage if made at high spatial resolution (30 m), which makes them less suitable for watershed management applications (Muller and Decamps 2000).

Soil water content estimates can also be derived using TIR data, utilizing the physically based negative association between surface temperature (T_s) and soil water content (Friedl and Davis 1994; Schmugge 1978). A number of studies have also explored the added value of multiview T_s observations for this purpose (Chehbouni et al. 2001; Francois 2002).

Other studies have been based on the correlations between T_s, spectral vegetation indices, and soil water content, resulting in many variations in this approach (Lambin and Ehrich 1996; Carlson, Capehart, and Gillies 1995; Prihodko and Goward 1997; Sandhold, Rasmussen, and Andersen 2002). Although such approaches have a high spatial resolution and have shown potential, they are confronted by several caveats in their implementation. First, they are characterized by the same problems as the optical-based methods (i.e., low depth of signal penetration degrading the influence of vegetation canopy and cloud cover and infrequent coverage). The most serious problem in using such models is that they are empirical in nature and their performance is a function of local meteorological conditions (e.g., wind speed, T_{air}, and humidity; Nemani et al. 1993) and/or local relief (Gillies and Carlson 1995). They have demonstrated a varying degree of accuracy in retrieving soil water content across different land-cover types (Smith and Choudhury 1991; Czajkowski et al. 2002).

Several authors have suggested that the microwave domain is currently the optimum spectral region for deriving soil water content (Quesney et al. 2000; Wang et al. 1997), with accuracies that can be within 5% (vol/vol) for bare soils (Mancini, Hoeben, and Troch 1999; Hoeben and Troch 2000). Microwave sensors have several advantages over both the optical and thermal domains, including their capability for cloud penetration, their all-weather and day/night coverage capability, and their signal independence from solar illumination variations. Both active and passive microwave sensors are used in these approaches (Zribi et al. 2005). The use of passive microwave sensors is limited by their coarse resolution (on the order of tens of kilometers), making such data inappropriate for watershed-scale applications and appropriate only for meteorological and climate models on a more global scale (Schmugge et al. 2002). Active microwave sensors can provide high-resolution data (on the order of tens of meters), but they are rather sensitive to soil surface roughness and vegetation-cover variations and are often also dependent on the terrain structure as a function of instrument look angle (Dobson and Ulaby 1998; Walker and Houser 2004). According to a review by Moran et al. (2004), the only satellite systems that currently meet the spatial resolution (10–100 km over a swath width of 50–500 km) and coverage (repeat time/revisit every 2–3 days) required for monitoring surface moisture content at watershed scales are the active microwave sensors. Active microwave remote sensing via Synthetic Aperture Radar (SAR) offers the potential to map soil water content at high resolutions over large areas (Verhoest et al. 1998; Troch et al. 1997). A significant limitation of SAR for watershed-scale applications is that the sun-synchronous satellites can provide at best only weekly or even longer temporal coverage for the same orbital path. Lastly, note that the accuracy of current remote sensing technology in microwave sensing instruments may be much lower over vegetated systems as a result of the interaction of the microwave signal with the vegetation canopy, whereas the obtained results also relate to the thin near-surface layer rather than the entire soil profile (Moran et al. 2004). Moreover, as pointed out in regard to Figure 19.1, the interpretation of microwave soil water content measurements and their comparison with other methodologies is not necessarily unambiguous.

In view of the practical and theoretical difficulties associated with the direct use of observations from VIS, TIR, and microwave remote sensing for the retrieval of soil water content, recent investigations have focused on the combined use of remote sensing data and hydrologic models, including two-layer SVAT models. These techniques aim to improve estimates of the soil water content profile by combining the horizontal coverage and spatial resolution of remote sensing data with the vertical coverage and temporal continuity of hydrologic models. Moradkhani (2008) reviews the mathematical and data assimilation approaches adopted in such methods, whereas Olioso et al. (1999) provide a comprehensive discussion on the variety of methods used for incorporating remote sensing data specifically into SVAT models. A sizeable body of literature also advocates the improved ability of methods assimilating information from remote sensing observations into a land-surface model for deriving the soil water content at much greater depths (Crosson et al. 2002; Heathman et al. 2003). However, data assimilation methods have several difficulties in their practical use. Nevertheless, many support the idea that a combined approach which uses remotely sensed data with a SVAT model scheme is the most promising research direction for satellite-estimation of surface soil water content at watershed scales (e.g. see review by Moran et al. 2004). The method of Gillies et al. (1997) and Carlson (2007a), which was briefly mentioned in the previous paragraph, belongs in this category.

19.3 The "Triangle" Method

In the following sections is made available a more detailed description of the principles and working of this so-called "triangle" method as well as examples from its implementation to date. This discussion also includes a brief reference to the biophysical properties encapsulated within the T_s/VI domain.

19.3.1 Theoretical Basis of the T_s/VI Methods

The emergence of the triangular (or trapezoid) shape in T_s/VI feature space is the result of the low variability of T_s and its relative insensitivity to soil water content variations over areas covered by dense vegetation, but its increased sensitivity (and thus larger spatial variation) over areas of bare soil. The right-hand side border of the triangle (or trapezoid; the so-called dry edge or warm edge) shown in Figure 19.2 is defined by the locus of points of highest temperatures that contain differing amounts of bare soil and vegetation and are assumed to represent conditions of limited surface soil water content and zero evaporative flux from the soil. Likewise, the left-hand border (the so-called wet edge or cold edge) corresponds to a set of cooler pixels that have varying amounts of vegetation, which represent those pixels at the limit of maximum surface soil water content. Variation along the lower edge (i.e., the "base") of the triangle (or trapezoid) represents pixels of bare soil and is assumed to reflect the combined effects of soil water content variations and topography, while the triangle's (or trapezoid's) apex equates to full vegetation cover (as expressed by the highest VI value). Points within the triangular space correspond to pixels with varying VI (i.e., F_r) and surface soil water content between those with bare soil and those with dense vegetation. For data points having the same VI, T_s can range markedly. As vegetation transpires, the vegetation surface is cooled, but as vegetation undergoes water stress, the plant closes its stomata and the resulting transpiration decrease causes leaf temperature to increase. However, this effect is difficult to determine from a satellite image of a vegetation ensemble, as the leaf temperature of such an ensemble tends to remain close to that of potential transpiration until severe wilting has occurred. Thus, for pixels with the same VI, those with minimum T_s represent the strongest evaporative cooling, while those with maximum T_s represent the weakest evaporative cooling. In this way, the triangle's (or trapezoid's) "dry edge" is considered to represent the lower limit of evapotranspiration for the different vegetation conditions found at that value of F_r within the scene, whereas the reverse is implied for the "wet edge." That the high vegetation end of the triangle (or trapezoid) exhibits either a small or a vanishing variation in a scaled

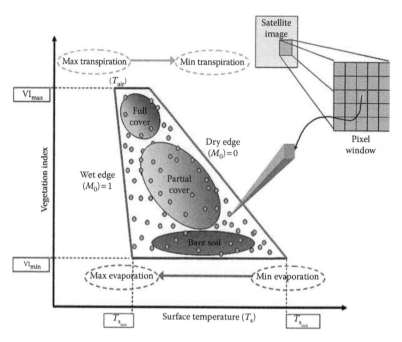

FIGURE 19.2

(See color insert following page 426.) Summary of the main physical properties and interpretations of the satellite-derived (or airborne) T_s/VI feature space. Dots represent the measurements at pixels observed by a VIS/IR radiometer at various fractional vegetation covers (F_r) and surface temperatures (T_s). In this illustration, pixels classified as water or clouds are assumed to have been masked out. (Adapted from Petropoulos, G. et al. *Adv Phys Geogr*, 33, 2, 2009a.)

radiometric surface temperature is an indication that the spatial variation in transpiration over dense vegetation is unimportant, the implication being that surface soil water content exerts little effect on the transpiration. Accordingly, all the T_s/VI models assume potential transpiration over the vegetated part of the pixel, regardless of the value of F_r, while soil evaporation is allowed to vary. These assumptions are consistent with the fact that M_0 lines tend to merge near the top of the triangle (Figure 19.2), thus making it impossible to resolve the spatial variation of soil water content in the presence of dense vegetation or to anticipate the effect of severe water stress on the vegetation. The presence of a trapezoidal shape, rather than a perfectly triangular shape in the T_s/VI plot, is the result of the soil thermal inertia variation, which changes with changing soil water content, thus affecting the soil heat storage and therefore the soil temperature.

Several studies have also been concerned with the examination of the main factors driving the shape of the T_s/VI scatter plot, an overview of which was made available recently by Petropoulos, Carlson et al. (2009). According to their overview, the main factors affecting the shape of the T_s/VI pixel envelope include the F_r, soil surface moisture content, synoptic state of the atmosphere (T_{air}, vapor pressure deficit), atmospheric forcing, and the characteristics of the specific location of the study area (e.g., soil type, landform, local climate, spatial heterogeneity in surface attributes, geographical location). A way to circumvent having to account for these environmental factors is to scale the surface temperature between the maximum and minimum values for pixels forming the triangular pattern. This transformation results in values assigned to pixels within the triangle tending to remain approximately in the same relative location with respect to the triangle boundaries for differing values of the environmental parameters, for example, surface albedo, surface emissivity, bulk stomatal resistance, atmospheric temperature, and net radiation. A possible weakness in the triangle method is that it requires a large number of pixels, some of which should contain bare, dry soil (as in an urban center) and some of which should contain dense vegetation. Given a sufficiently large sampling area, these criteria are likely to be met.

Section 19.3.2 provides a more in-depth description of one such methodology—the so-called triangle method of Gillies et al. (1997)—which is employed today in the retrieval of spatially explicit maps of the turbulent energy fluxes and surface soil water content via the interpretation of the remotely sensed T_s/VI scatter plot.

19.3.2 Triangle Method Implementation

The steps followed for the implementation of the so-called triangle method for the retrieval of both M_0 and surface heat fluxes are summarized in Figure 19.3, whereas details concerning the working of this approach can be found in the studies by Gillies and Temesgen (2000) and Carlson (2007a). As seen from Figure 19.3, a preliminary data preprocessing includes data resampling to a common spatial resolution, masking clouds and water. The first step in the method implementation is the computation of the normalized difference vegetation index (NDVI), which was originally proposed by Deering et al (1975):

$$\text{NDVI} = \frac{\rho_{\text{NIR}} - \rho_{\text{RED}}}{\rho_{\text{NIR}} + \rho_{\text{RED}}} \tag{19.1}$$

where ρ_{NIR} and ρ_{RED} denote the near-infrared and the red surface spectral reflectances, respectively. NDVI values can range between −1 and +1; more typically, NDVI values over a varying mixture of bare soil and vegetation vary between 0 and 0.8. Those for water are below 0, whereas those for bare soils typically range between 0 and 0.1 (Jensen 2000).

The NDVI is then scaled to an N^* value:

$$N^* = \frac{\text{NDVI} - \text{NDVI}_0}{\text{NDVI}_s - \text{NDVI}_0} \tag{19.2}$$

where NDVI_0 and NDVI_s are the minimum and maximum values of NDVI at minimum (0%) and maximum (100%) vegetation cover, respectively. These values are generally computed from the scatter plot of the T_s versus the NDVI maps, as shown in Figure 19.4.

Then, N^* is related to the F_r, following Gillies and Carlson (1995) and Choudhury et al. (1994):

$$F_r = N^{*2} \tag{19.3}$$

where F_r is the vegetation fraction and N^* is the linearly scaled NDVI. Transformation of N^* to F_r allows us to plot both the SVAT-simulated and the measured surface radiant temperatures from the satellite sensor on the same scale.

The next step of the implementation of the method includes the T_s normalization using the following equation:

$$T_{\text{scaled}} = \frac{T_0 - T_{\text{min}}}{T_{\text{max}} - T_{\text{min}}} \tag{19.4}$$

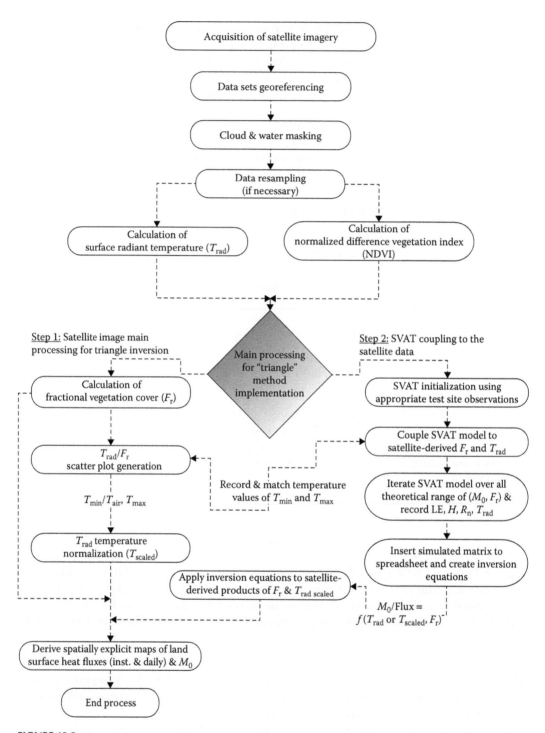

FIGURE 19.3
Overview of the workings of the triangle method implementation. Full details of the method implementation were recently summarized by Carlson (2007a).

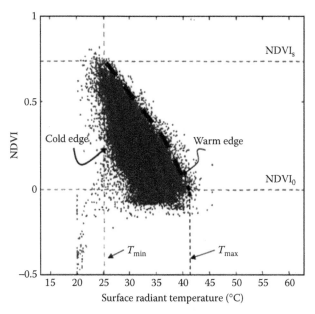

FIGURE 19.4
Scatter plot of normalized difference vegetation index (NDVI) versus surface radiant temperature for an NS001 airborne image over Walnut Gulch, Arizona, during summertime. Salient features of the triangle are the maximum and minimum temperatures, as vertical, dashed lines (T_{max} and T_{min}); the warm edge (heavy dashed lines); the cold edge; and the limits for dense vegetation ($NDVI_s$) and bare soil ($NDVI_0$). (Courtesy of Carlson, T.N., *Sens*, 7, 2007a.)

where T_{min} and T_{max} are the expected minimum and the maximum T_s for wet, vegetated pixels and for the dry, bare soil, respectively, interpolated from the scatter-plot bounds, and T_s corresponds to the radiometric temperature value of any pixel in the scene (Figure 19.4).

As stated by Carlson (2007a), T_{min} is the temperature of a dense clump of vegetation in well-watered soil and T_{max} is the temperature of dry, bare soil represented by the highest temperatures in the image. T_{scaled} is a generalized measure of the degree to which the surface temperature resembles that of either a wet, vegetated surface or a dry, bare soil surface. As such, T_{scaled} is believed to provide a more fundamental variable than does T_s, which varies with atmospheric conditions and time of day. The temperature normalization tends to impose a uniformity or universality on triangles that might otherwise have different shapes, allowing all unmasked pixel data to be inserted in the simulated triangle. It also offers the possibility to compare composite data from different observation periods, which in turn permits the monitoring of key parameters implicated in land-surface processes and land-use change with time.

Subsequently, in step 2 of Figure 19.3, the satellite observations of T_s (or equally T_{scaled}) and F_r are coupled with a SVAT model in order to derive the inversion equations that will provide the spatially explicit maps of land-surface fluxes and M_0. This process is described in detail by Gillis and Temesgen (2000). Briefly, the process is composed of the following steps:

1. First, the SVAT model is parameterized using the time and geographic location as well as the general soil and vegetation characteristics of the study site, together with the appropriate atmospheric profile data. Parameters are adjusted principally based on field observations.

2. Then, the SVAT model's input parameters are further adjusted, and the SVAT is iterated repetitively until the extreme values of F_r and T_s in the T_s/F_r scatter plot (see Figure 19.4) between the simulated (modeled with the SVAT) and observed (obtained from the satellite-derived data) are matched. In other words, initial model simulations endeavor to align observed T_s with two end points (NDVI$_0$, NDVI$_s$) where they intersect the "dry" edge. This extrapolation to NDVI$_0$ and NDVI$_s$ guarantees that the implied temperatures along the "dry" edge for bare soil and full vegetation cover are consistent with simulations for M_0 of zero. Note that the first model is run for dry, bare soil conditions. Alternately, scaling the temperatures as described above requires only that the model temperatures be scaled in the same manner, thereby eliminating the necessity of iterating the model in order to match its output with the measured T_{max}.

3. Once the model tuning is completed, the SVAT model is repetitively run keeping the time (corresponding to the satellite overpass) constant but varying F_r and M_0 over all possible values (0%–100% and 0–1 respectively), here in increments of 10 and 0.1, respectively, for all possible theoretical combinations of M_0, F_r. The result is a matrix of model outputs, including the columns of the following simulated parameters: M_0, F_r, T_{scaled} (or equally T_{kin}), LE, and H, all for the time of satellite overpass and calculated for each combination of F_r and M_0. An example of such matrix is shown in Figure 19.5.

F_r	M_0	$M_{0actual}$	R_n	LE	H	T_{kin}
0	0.00	0.04	312.56	17.39	184.09	24.79
0	0.10	0.13	333.29	49.97	165.06	23.07
0	0.20	0.22	351.66	75.28	151.62	21.77
0	0.30	0.31	368.26	95.81	140.70	20.81
0	0.40	0.40	383.94	113.47	132.30	20.03
0	0.50	0.50	398.83	128.94	125.45	19.39
0	0.60	0.59	413.24	142.95	119.74	18.86
0	0.70	0.68	427.02	155.42	114.75	18.44
0	0.80	0.77	440.71	167.55	110.94	18.04
0	0.90	0.87	453.95	178.48	107.43	17.72
0	1.00	0.96	467.02	188.83	104.36	17.43
10	0.00	0.04	327.89	20.08	196.70	23.76
.	.	.	.	.	.	.
.	.	.	.	.	.	.
.	.	.	.	.	.	.
.	.	.	.	.	.	.
.	.	.	.	.	.	.
.	.	.	.	.	.	.
90	1.00	0.97	428.85	74.27	212.11	15.52
100	0.00	0.04	408.45	42.61	270.60	18.41
100	0.10	0.13	411.52	47.36	262.42	17.94
100	0.20	0.23	414.15	50.87	255.67	17.54
100	0.30	0.32	416.53	53.62	250.22	17.18
100	0.40	0.41	418.75	55.98	245.44	16.87
100	0.50	0.51	420.62	57.96	241.19	16.62
100	0.60	0.60	422.34	59.67	237.44	16.40
100	0.70	0.69	423.93	61.28	234.12	16.20
100	0.80	0.79	425.30	62.53	230.93	16.04
100	0.90	0.88	426.70	63.75	228.16	15.87
100	1.00	0.97	428.15	64.97	225.52	15.71

FIGURE 19.5

Example of the matrix of model outputs created from the iteration of the M_0 and F_r after the soil–vegetation–atmosphere fransfer (SVAT) initialization.

4. Next, this output matrix is used to derive a series of nonlinear (quadratic) equations empirically, relating F_r and scaled surface temperature (T_{scaled}) to each of the other variables of interest: H, LE, latent heat flux ratio (LE/R_n), and sensible heat flux ratio (H/R_n). By this method, the set of physically based relationships between various surface-atmosphere parameters, as described by the detailed biophysical descriptions included in the SVAT model, are inherent in the matrix outputs, and are used to derive a series of simple, empirical relations relating each of these parameters to just the locations F_r and T_{scaled} recorded at that location. Because these parameters of F_r and T_{scaled} are derivable from the satellite data, these empirical equations could then be used to derive the required spatially explicit maps of the land-surfaces fluxes LE and H as well as of M_0 from the satellite products of F_r and T_{scaled}. The quadratic polynomial equations derived from the SVAT model matrix model outputs have the general form as follows (here shown for the version relating M_0 to F_r and T_s and/or T_{scaled}):

$$M_0 = \sum_{p=0}^{3}\sum_{q=0}^{3} a_{pq}(T_{scaled}{}^*)^p (F_r)^q \qquad (19.5)$$

where the coefficients $a_{p,q}$ are derived from nonlinear regression between the matrix values of F_r, T_{scaled}, and M_0 and p and q vary from 0 to 3.

Equation (19.5) is expanded functionally as (here for example for M_0)

$$
\begin{aligned}
M_0 = 1.0 &\times [\alpha_{00} + \alpha_{10}\times F_r^1 + \alpha_{20}\times F_r^2 + \alpha_{30}\times F_r^3] + \\
T_{scaled} &\times [\alpha_{01} + \alpha_{11}\times F_r^1 + \alpha_{12}\times F_r^2 + \alpha_{13}\times F_r^3] + \\
(T_{scaled})^2 &\times [\alpha_{02} + \alpha_{21}\times F_r^1 + \alpha_{22}\times F_r^2 + \alpha_{23}\times F_r^3] + \\
(T_{scaled})^3 &\times [\alpha_{03} + \alpha_{31}\times F_r^1 + \alpha_{32}\times F_r^2 + \alpha_{33}\times F_r^3]
\end{aligned}
\qquad (19.6)
$$

Thus, by this method, the set of physically based relationships between the different surface-atmosphere parameters (as described by the detailed biophysical descriptions included in the SVAT model and inherent in the matrix outputs) are used to derive a series of simple, empirical relations whereby F_r and T_s (or equally T_{scaled}) recorded at each location are used to generate output values for H and LE for a range of measured values of F_r and T_s. Because these variables of F_r and T_{scaled} are derivable from the satellite data, empirical equations such as this can then be used to obtain the required spatially explicit maps of the LE and H fluxes as well as of M_0 from satellite observations. Also, although these coefficients pertain to a set of specific environmental conditions, they may be used with some caution for a variety of initial atmospheric conditions without much loss of accuracy, provided that the radiometric surface temperatures are scaled as in Equation 19.4 and a scaled NDVI (Equation 19.2) or F_r (Equation 19.3) is used; a table of the coefficients for polynomials relating M_0 and H to T_{scaled} and F_r is provided by Carlson (2007a). As such they can be applied by the user without the need to run a SVAT model. Figure 19.6 shows the interior of the triangle mapped as a function of scaled radiometric surface temperatures (Equation 19.4) versus F_r (Equation 19.3). Note that although the isopleths of M_0 are nearly straight lines, both M_0 and EF vary in a highly nonlinear fashion between the cold (wet) and warm (dry) edges.

19.3.3 Overview of the SVAT Model Architecture Used in the Triangle Method

This section briefly overviews the SVAT model architecture that has been used to generate the polynomials referred to in the triangle method implementation, although, in general, any similar SVAT model can be used. The different facets of the SVAT model's overall structure, namely, the physical, vertical, and horizontal, are illustrated in Figure 19.7 (left). This model is essentially a one-dimensional boundary layer model with a plant component. It

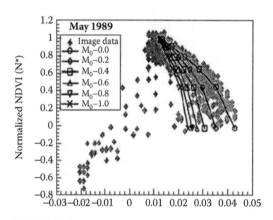
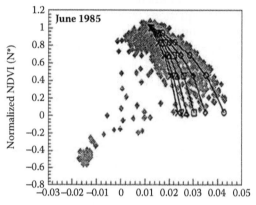

FIGURE 19.6
Scatter plots of scaled normalized difference vegetation index (NDVI; produced from a simple normalization of NDVI) versus normalized surface radiant temperature (T_s) derived in the study of Gillies and Carlson (1995) for different days on which the method was implemented using NOAA-AVHRR imagery. Isopleths of constant M_0 derived from a series of soil–vegetation–atmosphere transfer (SVAT) model runs are superimposed on the pixel-derived data. Slanting, nearly straight, lines represent surface soil moisture availability, M_0 labeled at intervals of 0.1, increasing from 0 on the right side (the warm edge). Curved lines labeled as fractions represent the soil water content (M_0). Note that pixels lying outside the triangular array of M_0 contours are likely to be contaminated by clouds or water.

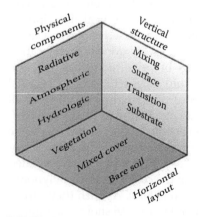
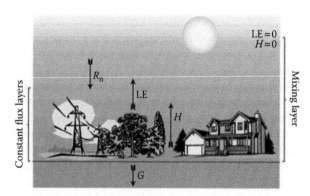

FIGURE 19.7
The figure on the left shows the different layers of the soil–vegetation–atmosphere transfer (SVAT) model in the vertical domain, whereas the figure on the right provides a schematic representation of the surface energy balance components computation in the SVAT model. (Adapted from SimSphere User's manual available at https://courseware.e-education.psu.edu/simsphere/workbook/)

has been developed to simulate the various physical processes that take place as a function of time in a column that extends from the root zone below the soil surface up to a level higher than the surface vegetation canopy. It performs simulations over a 24-hour cycle, starting from a set of initial conditions given in the early morning (at 5:30 A.M. local time).

The underlying constraint in the model is that the energy fluxes at the Earth's surface and within the plant canopy must balance appropriately (Figure 19.7, right). Initial forcing of the model begins with the calculation of solar radiation, determined from a one-dimensional solar radiation model. For a mixture of soil vegetation and bare soil patches, the shortwave incoming radiation and downward long-wave radiation are calculated in an identical way for the bare soil and vegetation regimes, and the radiation partitioning is computed as a function of the foliage density. In a similar way, the LE and H fluxes, the upward flux of long-wave radiation above the plant canopy, and the G flux are taken as weighted averages of the bare soil and vegetation components according to the F_r value set within the model. The partitioning of the surface turbulent fluxes of momentum, heat, and mass is parameterized as a function of the dynamic stability of air, canopy structure, and water evaporative capability of the soil and vegetation layers. Simulated fluxes are expressed in the units of watts per square meter (W · m^{-2}) of leaf area in order that they can be related to the surface energy balance via a shelter factor, which is a function of the leaf area index (LAI). The soil and vegetation temperatures are obtained by solving the energy budget equations simultaneously at the ground and canopy levels. T_s and the fluxes above the canopy are then computed from a weighted average of bare soil and vegetation components of upward long-wave radiation fluxes. Flux per unit leaf area is converted into flux per unit surface area by scaling the fluxes by the LAI divided by a "shelter factor" (see Mascart et al. 1991). The shelter factor accounts for the fact that not all leaves transpire at the sunlit amount because available solar radiation decreases with height beneath the top of the canopy.

An extensive mathematical account of the model basis has been provided by Carlson and Boland (1978) and Carlson et al. (1981), bare soil component of the model is described by Carlson et al. (1981), its vegetation component by Taconet et al. (1986) and Mehrez et al. (1992), and its plant hydraulics by Lynn and Carlson (1990) and Olioso, Carlson, and Brisson (1996). An overview of the model use to date can also be found in a study by Petropoulos, Carlson, and Wooster (2009). The most recent version of the SVAT model, called SimSphere, is freely available from Web site of the Department of Meteorology of Pennsylvania State University (http://www.agry.purdue.edu/climate/dev//simsphere.asp).

19.3.4 Example Applications of the Triangle Method

19.3.4.1 Testing the Setting

In the remaining part of the chapter we summarize some results of a study conducted as a validation exercise concerning the examination of the ability of the triangle method to provide estimates of LE and H fluxes as well as M_0 in a variety of ecosystem, environmental, and topographical conditions in Europe. In the framework of this research study, the triangle method was applied using satellite data from the Advanced Spaceborne Thermal Emission and Reflection Radiometer (ASTER) satellite radiometer and the most recent version of the Penn State two-layer SVAT model, called SimSphere. In order to provide an all-inclusive analysis of the triangle method performance, this study aimed at evaluating individually the performance of the triangle method in deriving the energy fluxes and M_0 and subsequently the ability of the SVAT model itself to produce the diurnal turbulent heat fluxes and other key land-surface parameters. Also, as an integral part of the SVAT model verification that was executed, a cutting-edge sensitivity analysis (SA) methodology was implemented, allowing derivation of quantitative measures of the sensitivities of key model outputs simulated by the model with respect to the model inputs. SA can be defined broadly as the process of determining the effect of changing the value of one or more input variables and observing the effect of that process on the model output. SA is generally regarded by many researchers as a necessary part of any model building, and its use should be considered in any field in which models are used (Saltelli et al. 2004). SA can be used for several purposes in modeling practice, including the understanding of a model's behavior and its coherence with the real world. SA is also necessary to establish the dependency of a model's outputs on its input parameters and illuminate the internal relationships of the different parameters within the model.

Validation of the results produced from the triangle method and the simulations from the SVAT model alone were performed using in situ validated observations taken from the selected observations collected from selected CarboEurope flux tower network (Aubinet et al. 2000), the largest ground-based measurement network operating at present in Europe. Agreement between the observed and the predicted parameters for each case was initially examined by directly comparing the predictions and the simulations for all days of the experiment, but subsequently by comparing predictions and observations in relation to land use and terrain type. In addition, another aspect of analysis concerned the assessment of the effect of clouds between the model predictions and observations in agreement. Judgment on which days (or time periods) were cloud-free was based on analysis of the observations of R_g, where cloud-free days were flagged as those having smoothly varying R_g curves, a property signifying clear-sky conditions. Finally, additional comparisons were also performed with the data stratified by the degree of energy balance closure (EBC), the assessment of which was accomplished by widely recognized methods (Wilson et al. 2002; Liu, Hiyama, and Yamaguchi 2006). To quantify the level of agreement between the parameters compared each time, a series of appropriate statistical measures were computed, such as root-mean-square difference (RMSD) and the mean absolute difference (MAD); a detailed description of these can be found in the studies of Silk (1979), Burt and Barber (1996), and Wilmott (1982).

In terms of SA, this was conducted to SimSphere using a software platform called the Gaussian Emulation Machine for Sensitivity Analysis (GEM SA), which performs global sensitivity analysis (GSA) based on Bayesian analysis of computer code outputs (BACCO; Kennedy and O'Hagan 2001) and is a freely available software tool (http://www.tonyohagan.co.uk/academic/GEM/index.html). Details concerning the statistical emulation process can be found in a study by Kennedy and O'Hagan (2000), while a tutorial introducing the method workings is available in a study by O'Hagan (2006).

For convenience and efficiency, selected results from this verification exercise study are presented separately for the verification of the triangle method and the SVAT model alone, whereas this section closes by presenting studies in which the concept of the triangle method has been extended toward the retrieval of other parameters associated with the already discussed biophysical parameters computed from the T_s/VI feature space.

19.3.4.2 Validation of the Triangle Method Predictions

For the evaluation of the triangle-inverted maps of M_0 and LE and H fluxes at the selected CarboEurope sites, point-by-point comparisons were performed. Such point-based comparisons have been the most common approach followed in analogous validation experiments of satellite-derived maps of surface energy fluxes and M_0, including even past verification exercises of the triangle method (Gillies et al. 1997; Brunsell and Gillies 2003;

Chauhan, Miller, and Ardanuy 2003). In addition, mean retrievals of the inverted parameters from a 3×3 pixel area surrounding each tower location were also compared to the in situ data in order to minimize spatial registration errors and reduce random "noise" in the satellite-derived retrievals. Figures 19.8 and 19.9 illustrate the results from the

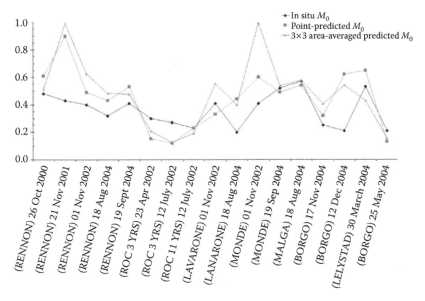

FIGURE 19.8
Agreement between in situ, point, and 3×3-area average M_0 (vol/vol) at various sites for all the case days on which LE fluxes were available at the ASTER satellite overpass times.

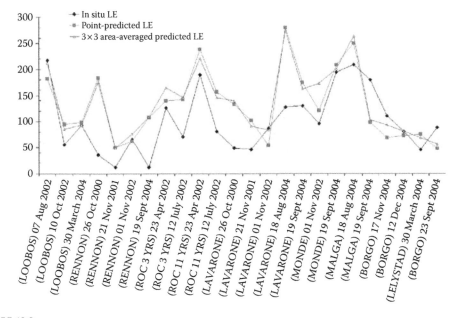

FIGURE 19.9
Same as Figure 19.8 but for area average instantaneous latent heat (LE) fluxes (W · m⁻²) for the case days on which LE observations at the time of the ASTER satellite overpass times.

comparisons of both M_0 and LE fluxes derived from triangle method implementation versus the corresponding measurements from the selected validation sites and all days (including those marked as "cloud-passing") taken from the CarboEurope sites.

Validation results from the triangle method indicated that M_0 and LE fluxes were in general overestimated by the triangle method, whereas the opposite was found for H fluxes. It is also worth noting that in the same data set, 56% of predicted M_0 and 42% of predicted LE fell within ~35% of the corresponding in situ values. Analogous results were also reported in 3 × 3 area-averaged comparisons for all the triangle-inverted parameters that were compared. Closer agreements with the ground observations were generally found when comparisons were limited to cloud-free days at flat terrain sites. Indeed, for such conditions, the triangle method was found to estimate the instantaneous LE fluxes with a mean RMSD of 35 W · m^{-2} (or equally mean MAE of 27 W · m^{-2}), whereas M_0 comparisons showed a mean RMSD of 0.22 and a mean MAD of 0.17 and a general overestimation of observed M_0 by a mean bias of 0.13 vol/vol. Analogous results were also found in general in the comparisons concerning H fluxes. Overall, agreement found for both M_0 and the instantaneous LE fluxes to a large extent was also comparable to accuracy levels reported in previous verification studies of the triangle method using satellite data from different sensors and at different implementation conditions (Carlson, Capehart, and Gillies 1995; Gilles et al. 1997; Capehart and Carlson 1997; Brunsell and Gillies 2003).

19.3.4.3 Validation of the SVAT Model Predictions

The Penn State SVAT model (SimSphere) has been developed over a period of more than 2 decades. There exist a number of implementations that have been applied in many studies investigating interactions between the land surface and the atmosphere, with results evaluated based on comparison of both field measurements and outputs from other models or via the study of scientific scenarios. A review of SVAT model use to date was provided by Petropoulos, Carlson, and Wooster (2009). Currently, the focus has been on expanding the range of model or in situ data intercomparisons to other regions, time periods, and comparison of data sets.

In this framework, SimSphere was applied to simulate the diurnal evolutions of various parameters (including LE and H fluxes), and the results were compared against measurements from flux towers at the selected CarboEurope test sites that covered different environmental and ecosystem conditions. In addition, the SA assisted in identifying and characterizing quantitatively the sensitivity of key model outputs to the input parameters, allowing one to study their interactions and derive absolute sensitivity measures appropriate to the structure of the SimSphere model.

Results for the comparisons that were performed during this study for a 30-minute mean average of LE and H fluxes for all days of comparison between SimSphere and the CarboEurope observations are shown in Figures 19.10 and 19.11. Table 19.2 summarizes the SimSphere model inputs, which were included in the sensitivity analysis conducted. Figures 19.12 and 19.13 present results also for SA that was conducted in SimSphere key model outputs, namely, the daily average latent heat flux ($\overline{LE_{daily}}$) and the daily average sensible heat flux ($\overline{H_{daily}}$).

In terms of the diurnal comparisons performed (Figures 19.10 and 19.11), despite the nonideal conditions of the SimSphere evaluation in some of the days of comparison (i.e., cloud-passing days, rugged terrain), SimSphere was able to produce simulations that agreed reasonably with the observations in terms of the diurnal changes and seasonal patterns. Generally, SimSphere was found to frequently overestimate the relatively small LE fluxes observed during the dry days for which comparisons were made. These days were

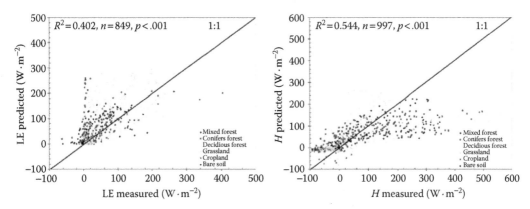

FIGURE 19.10
Comparisons of in situ measured and predicted values of latent heat (LE) fluxes for all the simulations days (left) and for the days flagged as cloud-free only (right) separated by land-use type. Each point in the figure is a single 30-minute flux measurement.

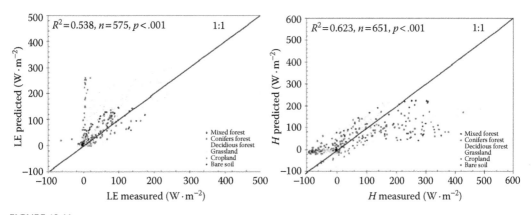

FIGURE 19.11
Comparisons of in situ measured and predicted values of sensible heat flux (H) for all the simulations days (left) and for the days flagged as cloud-free only (right) separated by land-use type. Each point in the figure is a single 30-minute flux measurement.

characterized by low in situ soil volumetric moisture and high T_{air} and H fluxes. The values of LE were underestimated comparatively on moist days. Comparisons of all cases of days (including those days marked previously "cloud-passing") showed that specifically LE and H fluxes were almost consistently underestimated by approximately 35%. Observed LE flux was frequently overestimated by SimSphere, with a mean bias of +14 W $\cdot$ m^{-2} and a mean RMSD of 29.5 W $\cdot$ m^{-2} (d-index of 0.889), whereas observed H flux was underestimated, with a mean bias of −11 W $\cdot$ m^{-2} and a mean RMSD of 52.2 W $\cdot$ m^{-2} (d-index of 0.826). H fluxes were found to be almost consistently underestimated by the model by approximately 25–33%. However, comparisons performed for just the subset of data of cloud-free days of flat terrain sites where the degree of EBC of the flux tower data had been confirmed indicated a closer agreement between the parameters. Under such conditions, the RMSD for LE and H were 27.11 and 68.75 W $\cdot$ m^{-2}, respectively (d-index of 0.881 and 0.762, respectively). Other studies (e.g., Gillies et al. 1997) show much smaller values of RMSD for H fluxes.

In summary, the agreement reported here, especially for the cloud-free days at flat terrain sites, was comparable to, if not sometimes better than, those reported in analogous

TABLE 19.2

Summary of the SimSphere Inputs Included in the Sensitivity Analysis Study

Short Name of Model Input	Actual Name of the Model Input	Process in Which Each Parameter Is Involved	Min	Max	Units
X1	Slope	Time and location	0	45	Degrees
X2	Aspect	Time and location	0	360	Degrees
X3	Station height	Time and location	0	4.92	m
X4	Fractional vegetation cover	Vegetation	0	100	0/0
X5	Leaf area index	Vegetation	0	10	$m^2 \cdot m^{-2}$
X6	Foliage emissivity	Vegetation	0.951	0.99	–
X7	[Ca] (external [CO_2] in the leaf)	Vegetation	250	710	ppmv
X8	[Ci] (internal [CO_2] in the leaf)	Vegetation	110	400	ppmv
X9	[o³] (ozone concentration in the air)	Vegetation	0	0.25	ppmv
X10	Vegetation height	Vegetation	0.021	20	m
X11	Leaf width	Vegetation	0.012	1	m
X12	Minimum stomatal resistance	Plant	10	500	$s \cdot m^{-1}$
X13	Cuticle resistance	Plant	200	2000	sm^{-1}
X14	Critical leaf water potential	Plant	–5	–30	bar
X15	Critical solar parameter	Plant	25	300	$W \cdot m^{-2}$
X16	Stern resistance	Plant	0.011	0.15	$s \cdot m^{-1}$
X17	Surface moisture availability	Hydrologic	0	1	vol/vol
X18	Root zone moisture availability	Hydrologic	0	1	vol/vol
X19	Substrate maximum volume water content	Hydrologic	0.01	1	vol/vol
X20	Substrate climatological mean temperature	Surface	3.5	30	°C
X21	Thermal inertia	Surface	0.951	0.98	$W \cdot m^{-2} \cdot K^{-1}$
X22	Ground emissivity	Surface	0.05	5	–
X23	Atmospheric precipitable water	Meteorological	0.02	2	cm
X24	Surface roughness	Meteorological	1	10	m
X25	Obstacle height	Meteorological	0.02	2	m
X26	Fractional cloud cover	Meteorological	1	10	0/0
X27	RKS (saturated thermal conductivity)	Soil	0	10	See Cosby et al. 1984 for units.
X28	Cosby B (Cosby et al. 1984 "b" parameter)	Soil	2	12	
X29	THM (saturated volume water content)	Soil	0.3	0.5	
X30	PSI (saturated water potential)	Soil	1	7	

past verifications of this specific SVAT model (e.g., Todhunter and Terjung 1986; Carlson and Boland 1978; Gillies et al. 1997). Moreover, when considering interpreting these results, instrumental uncertainty related to the flux tower measures should be borne in mind (for R_n–10% and for turbulent heat fluxes ~10–20%), which can explain partially the disagreement reported here between model predictions and ground measurements. Overall, despite the occasionally inferior performance of the model in simulating the examined parameters (mainly the underestimation of H fluxes), SimSphere was able to identify the

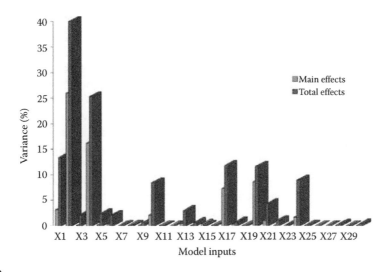

FIGURE 19.12
Results of SA done for the $\overline{LE_{daily}}$ using the BACCO GEM GSA method. Here are illustrated the results from the computed main effects and total effects for each of the SimSphere inputs. The large total effect in comparison to the corresponding main effect of each model input indicates the presence of high interaction effects. Parameters Xi shown in the x-axis are explained in Table 19.2.

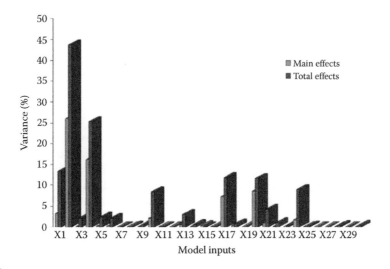

FIGURE 19.13
Results sensitivity analysis done for $\overline{H_{daily}}$ using the BACCO GEM GSA method. Here are illustrated the results from the computed main effects and total effects for each of the SimSphere inputs. The large total effect in comparison to the corresponding main effect of each model input indicates the presence of high-interaction effects. Parameters Xi shown in the x-axis are explained in Table 19.2.

patterns of the change expected, if not always the magnitudes. SimSphere evaluation results evidenced the use of the model as a tool that identifies the patterns of the change expected, if not always the magnitudes, thus indicating the usefulness of the model in practical applications either as a stand-alone tool or in combination with remote sensing via the implementation of the triangle inversion method of Carlson, Capehart, and Gillies (1995) and Gillies and Carlson (1995).

In terms of the SA, as seen from Figures 19.12 and 19.13, the BACCO GEM GSA method was employed in identifying the most responsive inputs of SimSphere and also capturing their key interactions for each of the simulated target quantities on which the GSA was conducted. Figures 19.11 and 19.12 illustrate the computed main effects and the total effects for two of the simulated parameters in the model in which their sensitivity was examined. The main effect represents the expected amount of variance that would be removed from the total output variance if we were able to learn the true value of each model input parameter (within its uncertainty range). Thus, the main effect provides a quantitative indication of the relative importance of an individual input variable Xi in driving the total output uncertainty. The total effect collects in one single term all the interactions involving one input parameter and represents the expected amount of output variance that would remain unexplained if the considered model input and only that was left free to vary over its range, the value of all other input parameters variables having been learned.

Results of the SA indicate that only a small fraction of the model input parameters exert appreciable influence on the target quantities. Both the simulated H and LE fluxes were found to be most sensitive to aspect ratio, F_r, M_0 terrain slope, surface roughness, and vegetation height as well as to the substrate climatological mean temperature and thermal inertia. Notable was the sharing of the interaction effects between the model parameters, reflecting the complexity of the model architecture. Results, however, did suggest the presence of highly complex interaction structures within SimSphere, which drove a considerable fraction of the variance of the studied model outputs. Other influential input parameters for all model outputs of which their sensitivity was examined were reported to be surface roughness, vegetation height, substrate climatological mean temperature (the latter governing the deep soil temperature), and thermal inertia. The detailed results of the SA, which was conducted herein, are presented in a study by Petropoulos, Wooster et al. (2009).

19.3.4.4 Other Applications

A number of studies were focused on demonstrating the use of remote sensing, especially utilizing the triangle method, such as for the study of urbanization. More specifically, Owen, Carlson, and Gillies (1998) implemented the triangle method proposed by Carlson, Capehart, and Gillies (1995), Gillies and Carlson (1995), and Gillies et al. (1997) using NOAA-AVHRR and Landsat TM imagery. Carlson and Sanchez-Azofeifa (1999), studying urban growth in San Jose, Costa Rica, showed the effect of development in specific residential areas during a 9-year period. They were able to relate the temporal movement of particular image pixels within the T_s/F_r triangular envelope to changes in urbanization, expressed as increases in T_s as a result of the decline in both F_r and M_0; we suggested that a scaled temperature along with F_r would permit us to view the time history of pixels in F_r/T_{scaled} space. They suggested the use of vectors, that is, trajectories, for visualization of the temporal movement of such pixels. Carlson and Arthur (2000) extended the above concept by relating the area of impervious surfaces, specifically the impervious surface area (ISA) fraction, referred to earlier, and the surface runoff to the triangle domain. Estimating ISA requires an additional step beyond the aforementioned image analyses. Although several techniques currently exist for estimating ISA, the method we favor requires one to perform a classification of the image, separating the pixels into different land-use categories, such as urban, forest, and water. We then assume that ISA exists only within the pixels classified as urban and that within the urban pixels ISA is equal to the fractional part of the pixel not covered by vegetation $(1 - F_r)$.

This method of calculating ISA has been demonstrated using the triangle approach with Landsat TM data on a case-study site over the seriously degraded Conestoga Watershed in Lancaster County, PA (Carlson 2007b), a region that has undergone rapid urban development over the past decade. In Figure 19.14, the light-shaded areas pertain to ISA values generally between 30% and 70%, which is characteristic of residential development. The darker areas within the light shading denote ISA in excess of 70%, characteristic of commercial and industrial development; the latter typically resides in or near the center of a city, in this case the city of Lancaster. Besides being a sensitive monitor of urbanization and the human use of the land, ISA is highly useful for estimating urban storm water runoff and for assessing the possible effects of urban development on stream water quality.

Owen, Carlson, and Gillies (1998) suggested a scheme for formulating equations to predict the effects of various urban development scenarios on the microscale surface temperature and moisture. Arthur-Hartanft, Carlson, and Clarke (2003) extended the work of Carlson and Arthur (2000) by showing how another parameter, storm water runoff, can be linked to the triangular domain. They demonstrated the potential of coupling the triangle method outputs with the SLEUTH cellular automata urban growth model (Clarke, Hoppen, and Gaydos 1996) for potential use in urban planning and policy decision making. Another application particularly suited to urban hydrology makes use of land classification by satellite to estimate surface runoff in streams from rainfall events. This approach uses the concept of ISA to calculate the runoff potential (Carlson 2007b).

From a very different perspective, Crombie et al. (1999) used NOAA-AVHRR data to implement the triangle method in the area of Egypt's Nile Delta and used the derived spatially explicit maps of M_0 as an index to assess the prevalence of diseases such as malaria and filariasis, because the mosquitoes that carry the parasite require standing water for breeding. Their results unfortunately failed to show a very significant correlation between filariasis infection rate and retrieved M_0 ($R^2 = 0.37$). In another study, Ray et al. (2002) employed the triangle approach of Gilles et al. (1997) to investigate the differences in cumulus cloud formation, M_0, surface energy fluxes, and top of the atmosphere shortwave irradiance between areas of agricultural development and native vegetation in southwestern Australia. Using three Terra MODIS scenes obtained in December 2000, they found that local effects due to land use had a significant influence on the formation

FIGURE 19.14
Satellite-derived (Landsat TM) analysis of impervious surface area in percent for the Conestoga Watershed in Lancaster County, eastern Pennsylvania, for the year 2000. The arrow denotes the town of Lancaster.

of cumulus clouds. The authors used the monthly maps of cumulus frequency for the year 2000 and showed the preferential development of cumulus clouds over areas of native vegetation in summer and over agricultural areas in winter, but for the dry season, the inverted M_0 and surface energy fluxes showed areas of native vegetation to be moister. Their study demonstrated that enhanced LE over native vegetation implied a moister boundary layer that could be responsible for preferential cloud formation over native vegetation during the dry season.

19.4 Summary and Future Outlook

This chapter for the most part has been confined to providing an overview of the retrieval of H and LE heat fluxes as well as soil water content using information derived from thermal and optical remote sensing sensors, with an emphasis placed on the so-called triangle method of Gillies et al. (1997) and Carlson (2007a). As was clearly evidenced from the review of the T_s/VI methods, which was also undertaken here, the state-of-the-art in the retrieval of surface heat fluxes from optical and thermal remote sensing yields the retrieval of LE and H with a 10–30% accuracy, a range probably approaching the limit of accuracy currently possible using satellite measurements, as was also indicated by Jiang, Islam, and Carlson (2004). This has been generally considered to be reasonable, given that the accuracy in the measurement of these fluxes using ground instrumentation is generally around 10–15% (as referred to in Jiang, Islam, and Carlson 2004). However, in terms of the surface soil water content estimation, unlike radar-derived estimates, it appears that optical and thermal remote sensing are not able to achieve the accuracy of approximately ±4% vol/vol in the retrieval of surface soil water content recommended for a large range of applications, as reported, for instance, by Engman (1992), Calvet and Noilhan (2000), and Walker and Houser (2004).

From all the methods employed today in the retrieval of the LE and H fluxes as well as M_0 using optical and thermal remote sensing data, those based on the triangular (or trapezoid) space that emerges from a satellite-derived surface temperature (T_s) and VI appear to possess certain advantages. The ability of these methods to relate the patterns encapsulated by the T_s/VI pixel envelope to key biophysical properties explains the large number of studies concerned with their implementation for retrieving spatially explicit maps of H, LE fluxes, and M_0. Of course, care must be taken in interpreting the surface moisture availability, as the precise meaning of any soil water content derived from satellite is fundamentally uncertain, as pointed out in reference to Figure 19.1.

The most efficient and possibly the most consistently accurate one of these T_s/VI methods for estimating surface variables is the so-called triangle method of Gillies and Carlson (1995) and Gillies et al. (1997), the workings of which are presented in this chapter, along with some examples of validation exercises. Results from various validation exercises that have been conducted have shown that this method is able to provide results of at least similar, or in some occasions, better accuracy compared to other methods available. However, this method has several advantages in its use compared to other methods discussed in this chapter, which make it ideal for use by in the operational estimation of soil water content from the Visible/Infrared Imager/Radiometer Suite (VIIRS) and the Microwave Imager Sounder (MIS) under NPOESS starting in the year 2016 (Chauchan et al. 2003).

It is understandable that arguments for a further comprehensive validation of this method are now of a high priority and scientific interest. It also appears that to best promote the use of T_s/VI approaches, a number of technological/theoretical/practical hurdles must be overcome. The spatial and temporal resolution of satellite instruments with appropriate specifications for T_s/VI methods should be improved to allow the study of the surface heat fluxes and soil surface moisture at the spatiotemporal frequencies required. Also, the development of techniques for the implementation of these methods over cloudy conditions and conditions representative of a non-full-range of surface conditions, which is at present holding these methods from operational applications, should be further investigated. Further work toward the development of methods for the operational retrieval of surface heat fluxes and surface soil water content utilizing remotely sensed data becomes more indispensable considering the forthcoming launch of new satellites. The VIIRS instrument planned to be placed in orbit by the NPOESS/NASA Preparatory Project in 2011, as well as the Sentinel-3 mission of the European Space Agency (ESA), which is planned to be launched in 2012, providing thermal infrared observations from space at 750 and 1 km, respectively, are expected to be highly valuable in estimating land atmosphere energy fluxes and surface soil water content from remote sensing in the coming years. The importance of the use of data from the above sensors is even strengthened by the fact that no future mission has been yet formalized regarding the continuation of spaceborne thermal infrared data acquisitions at very high spatial resolutions and at a global scale, as a succession of the Landsat and ASTER missions.

Acknowledgments

The authors wish to thank the Greek Scholarships Foundation (IKY) of the Ministry of Education of Greece for providing PhD scholarships that assisted in completing the inclusive verification exercises of some of the results presented here. In addition, the authors are grateful to the site managers of the CarboEurope IP sites for providing the validated ground truth measurements and to the ASTER Remote Sensing Data Analysis Center (ERSDAC) team of Japan for the free provision of the ASTER images over the CarboEurope IP sites used in this verification study. The authors also wish to express their gratitude to Dr. Marc Kennedy from DEFRA, United Kingdom, for providing the BACCO GEM SA software, which allowed the SA study implementation to the SimSphere model. Closing, Dr. Petropoulos wishes to thank Professor Martin Wooster and Dr. Nick Drake from King's College, London, for their contributions to the content of this manuscript.

References

Anderson, M. C., J. M. Norman, G. R. Diak, and W. P. Kustas. 1997. A two-source time Integrated model for estimating surface fluxes for thermal infrared satellite observations. *Remote Sens Environ* 60:195–216.

Arthur-Hartanft, T., S. T. N. Carlson, and K. C., Clarke. 2003. Satellite and ground-based microclimate and hydrologic analyses coupled with a regional urban growth model. *Remote Sens Environ* 86:385–400.

Aubinet, M., A. Grelle, A. Ibrom, Ü. Rannik, J. Moncrieff, T. Foken, A. S. Kowalski, P. et al. 2000. Estimates of the annual net carbon and water exchange of forests: The EUROFLUX methodology. *Adv Ecol Res* 30:113–75.

Beven, K. J., and J. Fisher. 1996. Remote sensing and scaling in hydrology. In *Scaling in Hydrology Using Remote Sensing*, ed. J. B. Stewart, E. T. Engman, R. A. Feddes, and Y. Kerr, 270. New York: John Wiley & Sons.

Biftu, G. F., and T. Y. Gan. 1999. Retrieving near-surface soil moisture from Radarsat SAR data. *Water Resour Res* 35:1569–79.

Brasa, A., F. Martın de Santa Olalla, V. Caselles, and M. Jochum.1998. Comparison of evapotranspiration estimates by NOAA-AVHRR images and aircraft flux measurement in a semiarid region of Spain. *J Agric Eng Resour* 70:285–94.

Brunsell, N. A., and R. R., Gillies. 2003. Scale issues in land-atmosphere interactions: Implications for remote sensing of the surface energy balance. *Agric For Meteorol* 117:203–221.

Burt, J. E., and G. M. Barber. 1996. *Elementary Statistics for Geographers*. London: Longman.

Calvet, J. C., and Noilhan J. 2000. From near-surface to root-zone soil moisture using year-round data. *J Hydrol* 1(5):393–411.

Capehart, W. J., and T. N. Carlson. 1997. Decoupling of surface and near-surface soil water content: A remote sensing perspective. *Water Resour Res* 33(6):1383–95.

Carlson, T. N. 2007a. An overview of the "triangle method" for estimating surface evapotranspiration and soil moisture from satellite imagery. *Sens* 7:1612–29.

Carlson, T. N. 2007b. Impervious surface area and its effect on water abundance and water quality. In *Remote Sensing of Impervious Surfaces*, ed. Q. Weng, 353–67. Boca Raton: CRC Press.

Carlson, T. N., and S. T., Arthur. 2000. The impact of land use- land cover changes due to urbanization on surface microclimate and hydrology: A satellite perspective. *Glob Planet Change* 25:49–65.

Carlson, T. N., and F. E. Boland. 1978. Analysis of urban-rural canopy using a surface heat flux/temperature model. *J Appl Meteorol* 17:998–1014.

Carlson, T. N., W. J. Capehart, and R. R. Gillies. 1995. A new look at the simplified method for remote sensing of daily evapotranspiration. *Remote Sens Environ* 54:161–7.

Carlson, T. N., J. K. Dodd, S. G. Benjamin, and J. N. Cooper. 1981. Satellite estimation of the surface energy balance, moisture availability and thermal inertia. *J Appl Meteorol* 20:6–87.

Carlson, T. N. and G. A. Sanchez-Azofeifa. 1999. Satellite remote sensing of land use changes in and around San Jose, Costa Rica. *Remote Sens Environ* 70:247–56.

Chauhan, N. S., S. Miller, and P. Ardanuy. 2003. Spaceborne soil moisture estimation at high resolution: A microwave-optical/IR synergistic approach. *Int J Remote Sens* 22:4599–46.

Chehbouni, A., Y. Nouvellon, J.-P. Lhomme, C. Watts, G. Boulet, Y. H. Kerr, M. S. Moran, and D. C. Goodrich. 2001. Estimation of surface sensible heat flux using dual angle observations of radiative surface temperature. *Agric For Meteorol* 108:55–65.

Chehbouni, A., C. Watts, Y. H. Kerr, G. Dedieu, J.-C. Rodriguez, F. Santiago, P. Cayrol, G. Boulet, and D. C. Goodrich. 2000. Mehtods to aggregate turbulent fluxes over heterogeneous surfaces: Applications to SALSA data set in Mexico. *Agric For Meteorol* 105:133–44.

Choudhury, B. J, N. U. Ahmed, S. B. Idso, R. J. Reginato, and C. S. T., Daughtry. 1994. Relations between evaporation coefficients and vegetation indices studied by model simulations. *Remote Sens Environ* 50(1):1–17.

Clarke, K. C., S. Hoppen, and L. J. Gaydos. 1996. Methods and techniques for rigorous calibration of a cellular automaton model of urban growth. In *Proceedings of the Third International Conference/Workshop on Integrating GIS and Environmental Modelling CD-Rom*, Santa Fe, NM, January 21–26, 1996.

Consoli, S., G. Urso, and A. Toscano. 2006. Remote sensing to estimate ET-fluxes and the performance of an irrigation district in southern Italy. *Agric Water Manag* 81:295–314.

Courault, D., B. Seguin, and A. Olioso. 2005. Review on estimation of evapotranspiration from remote sensing data: From empirical to numerical modeling approaches. *Irrigation Drainage Syst* 19:223–24.

Cracknell, A. P., and Y. Xue. 1996. Thermal determination from space—A tutorial review. *Int J Remote Sens* 17:431–61.

Crombie, M. K., R. R. Gillies, R. E. Arvidson, P. Brookmeyer, G. J. Weil, M. Sultan, and M. Harb. 1999. An application of remotely-derived climatological fields for risk assessment of vector-borne diseases—A spatial study of filariasis prevalence in the Nile Delta, *Egypt Photogramm Eng Remote Sens* 65(12):1401–9.

Crosson, W. L., C. A. Laymon, R. Inguva, and M. P. Schamschula. 2002. Assimilating remote sensing data in a surface flux-soil moisture model. *Hydrol Process* 16:1645–62.

Czajkowski, K., S. Goward, D. Shirey, and A. Walz. 2002. Thermal remote sensing of near-surface water vapor. *Remote Sens Environ* 79(2–3):253–65.

De Troch, F. P., P. A. Troch, Z. Su, and D. S. Lin. 1996. Application of remote sensing for hydrological modelling. In *Distributed Hydrological Modelling*, ed. M. B. Abbott, and J. C. Refsgaard, 165–91. Dordecht: Kluwer Academic Publishers.

Deering, D. J., J. W. Rouse, R. H. Haas, and J. A. Schell. 1975. Measuring production of grazing units from Landsat MSS data. In *Proceedings of the 10th International Symposium of remote Sensing of Environment*, ERIM, Ann Arbor, MI, August 23–25, 1975, 1169–78.

Demšar, U. 2005. A strategy for observing soil moisture by remote sensing in the Murray-Darling basin. In *Proceedings of the 8th AGILE Conference on Geographic Information Science*, Estoril, Portugal, May 2005. http://www.infra.kth.se/~demsaru/publications/50_Urska%20Demsar_AGILE2005.pdf (accessed August 4, 2009).

Diak, G., J. R. Mecikalski, M. C. Anderson, J. M. Norman, W. P. Kustas, R. D. Torn, and R. L. DeWolf. 2003. Estimating land surface energy budgets from space: Review and current efforts at the University of Wisconsin-Madison and USDA-ARS. Document Prepared for the Bulletin of American Meteorological Society. http://ams.allenpress.com/amsonline/?request=get-abstract&doi=10.1175%2FBAMS-85-1-65 (accessed August 14, 2009).

Dobson, M. C., and F. T. Ulaby. 1998. Mapping soil moisture distribution with imaging radar. In *Principles and Applications of Imaging Radar. Manual of Remote Sensing, American Society for Photogrammetry and Remote Sensing*, ed. F. M. Henderson and A. J. Lewis, 407–433. New York: John Wiley & Sons.

Dodds, P. E., W. S. Meyer, and A. Barton. 2005. A review of methods to estimate irrigated reference crop evapotranspiration across Australia. Document prepared for Irrigation Features Technical Report, Australia. http://www.clw.csiro.au/publications/consultancy/2005/CRCIFtr04-05-CropEvapotranspiration.pdf#search=%22A%20review%20of%20methods%20to%20esti-mate%20irrigated%20reference%20crop%20evapotranspiration%20across%20Australia.%20%22 (accessed August 24, 2009).

Engman, E. T. 1992. Soil moisture needs in Earth sciences. *Proceedings of International Geoscience and Remote Sensing Symposium (IGARSS)*, 477–9.

Engman, E. T., and G. A. Schultz. 2000. Future perspectives. In *Remote Sensing in Hydrology and Water Management*, chap. 20, ed. G. A. Schultz and E. T. Engman. Berlin: Springer Verlag.

François, C. 2002. The potential of directional radiometric temperatures for monitoring soil and leaf temperature and soil moisture status. *Remote Sens Environ* 80:122–33

French, A., T. Schmugge, and W. Kustas. 2002. Estimating evapotranspiration over El-Reno, Oklahoma with ASTER imagery. *Agronomie* 22:105–6.

Friedl, M. A., and F. W. Davis. 1994. Sources of variation in radiometric surface temperature over a tall grass prairie. *Remote Sens Environ* 48:1–17.

Gillies, R. R., and T. N. Carlson. 1995. Thermal remote sensing of surface soil water content with partial vegetation cover for incorporation into climate models. *J Appl Meteorol* 34:745–56.

Gillies, R. R., T. N. Carlson, J. Cui, W. P. Kustas, and K. S. Humes. 1997. Verification of the "triangle" method for obtaining surface soil water content and energy fluxes from remote measurements of the NDVI and surface radiant temperature. *Int J Remote Sens* 18:3145–66.

Gillies, R. R. and B. Temesgen. 2000. Coupling thermal infrared and visible satellite measurements to infer biophysical variables at the land surface. In *Thermal Remote Sensing in Land Surface Processes*, Chapter 5, 160–183, New York: CRC Press.

Goetz, S. J. 1997. Multi-sensor analysis of NDVI, surface temperature and biophysical variables at a mixed grassland site. *Int J Remote Sens* 18(1):71–94.

Goward, S. N., G. D. Cruickhanks, and A. S. Hope. 1985. Observed relation between thermal emission and reflected spectral radiance of a complex vegetated landscape. *Remote Sens Environ* 18:137–46.

Griffiths, G. H., and M. G. Wooding. 1996. Temporal monitoring of soil moisture using ESAR-1 SAR data. *Hydrol Process* 10:1127–38.

Hall, F. G., K. F. Huemmrich, S. J. Goetz, P. J. Sellers, and J. E. Nickeson. 1992. Satellite remote sensing of the surface energy balance: Success, failures and unresolved issues in FIFE. *J Geophys Res* 97(D17):19061–89.

Heathman, G. C., P. J. Starks, L. R. Ahura, and T. H. Jackson. 2003. Assimilation of surface soil moisture to estimate profile soil water content. *J Hydrol* 279:1–17.

Hipps, L. E., E. Swiatek, and W. P., Kustas. 1994. Interactions between regional surface fluxes and the atmospheric boundary layer over a heterogeneous watershed. *Water Resour Res* 30:1387–92.

Hoeben, R., and P. A. Troch. 2000. Assimilation of active microwave observation data for soil moisture profile estimation. *Water Resour Res* 36 2805–19.

Idso, S. B., R. D. Jackson, R. J. Reginato, B. A. Kinball, and F. S. Nakayama. 1975. The utility of surface temperature measurements for the remote sensing of soil water status. *J Geophys Res* 80:3044–9.

Inoue, Y., and M. S. Moran. 1997. A simplified method for remote sensing of daily canopy transpiration—a case study with direct measurements of canopy transpiration in soybean canopies. *Int J Remote Sens* 18:139–52.

Jacob, F., A. Olioso, Z. Su, and B. Sequin. 2002. Mapping surface fluxes using airborne visible, near infrared, thermal infrared remote sensing data and a specialized surface energy balance model. *Agronomie* 22:669–80.

Jensen, J. R. 2000. *Remote Sensing of the Environment.* Upper Saddle River, NJ: Prentice Hall.

Jiang, L., and S. Islam. 1999. A methodology for estimation of surface evapotranspiration over large areas using remote sensing observations. *Geophys Res Lett* 26:2773–6.

Jiang, L., and S. Islam. 2001. Estimation of surface evaporation map over southern Great Plains using remote sensing data. *Water Resour Res* 37:329–40.

Jiang, L., S. Islam, and T. N. Carlson. 2004. Uncertainties in latent heat flux measurement and estimation: Implications for using a simplified approach with remote sensing data. *Can J Remote Sens* 30(5):769–87.

Kennedy, M. C., and A. O'Hagan. 2000. Predicting the output from a complex computer code when fast approximations are available. *Biometrika* 87:1–13.

Kennedy, M. C and A. O'Hagan. 2001. Bayesian calibration of computer models. *J R Stat Soc Series B Stat Methodol* 63(3):425–64

Kostov, K. G., and T. J. Jackson. 1993. Estimating profile soil moisture from surface layer measurements–a review. In *Proceedings of the International Society of Optical Engineering*, 125–36. Orlando, FL: International Society of Optical Engineering.

Kustas, W. P., B. J. Choudhury, M. S. Moran, R. J. Reginato, R. D. Jackson, L. W. Gay, and H. L. Weaver. 1989. Determination of sensible heat flux over sparse canopy using thermal infrared data. *Agric For Meteorol* 44:197–216.

Kustas W. P., and J. M. Norman. 1996. Use of remote sensing for evapotranspiration monitoring over land surfaces. *Hydrolo Sci* 41(4):495–516.

Kustas, W. P., X. Zhan, and T. J. Schmugge. 1998. Combining optical and microwave remote sensing for mapping energy fluxes in a semiarid watershed. *Remote Sens Environ* 64:116–31.

Lambin E. F., and D. Ehrlich. 1996. The surface temperature-vegetation index space for land cover and land-cover change analysis. *Int J Remote Sens* 17(3):463–87.

Leone, A. P., and S. Sommer. 2000. Multivariate analysis of laboratory spectra for the assessment of soil development and soil degradation in the southern Apennines (Italy). *Remote Sens Environ* 72(3):346–359

Li, F., W. P. Kustas, M. C. Anderson, J. H. Prueger, and R. Scott. 2008. Effect of remote sensing spatial resolution on interpreting tower-based flux observations. *Remote Sens Environ* 112:337–49.

Liu, Y., T. Hiyama, and Y. Yamaguchi. 2006. Scaling of land surface temperature using satellite data: A case examination on ASTER and MODIS products over a heterogeneous terrain area. *Remote Sens Environ* 105(2):115–128.

Lynn, B., and T.N. Carlson. 1990. A stomatal resistance model illustrating plant versus external control of transpiration. *Agric For Meteorol* 52:5–43.

Mancini, M., R. Hoeben, and P. Troch. 1999. Multi-frequency radar observations of bare surface soil moisture content: A laboratory experiment. *Water Resour Res* 35(6):1827–38.

Mascart, P. O. Taconet; J. P. Pinty, and M. B. Mehrez. 1991. Canopy resistance formulation and its effect in mesoscale models: A HAPEX perspective. *Agric For Meteorol* 54:319–351.

McCabe, M. F., and E. F. Wood. 2006. Scale influences on the remote estimation of evapotranspiration using multiple satellite sensors. *Remote Sens Environ* 105:271–85.

Mecikalski, J. R., G. R. Diak, M. C. Anderson, and J. M. Norman. 1999. Estimating fluxes on continental scales using remotely-sensed data in an atmospheric-land exchange model. *J Appl Meteorol* 38:1352–69.

Mehrez, M. B., O. Taconet, D. Vidal-Madjar, and C. Valencogne. 1992. Estimation of stomatal resistance and canopy evaporation during the HAPEX-MOBILHY experiment. *Agric For Meteorol* 58:285–313.

Moradkhani, H. 2008. Hydrologic remote sensing and land surface data assimilation. *Sensors* 8: 2986–3004.

Moran, M. S., T. R. Clarke, Y. Inoue, and A. Vidal. 1994. Estimating crop water deficit using the relation between surface-air temperature and spectral vegetation index. *Remote Sens Environ* 49:246–63.

Moran, S. M., S. McElroy, J. M. Watts, and C. D. Peters-Lidard. 2005. Radar remote sensing for estimation of surface soil moisture at the watershed scale. In *Modelling and Remote Sensing Applied in Agriculture (US and Mexico)*, Chapter 7, ed. C. W. Richardson, A. S. Baez-Gonzalez and M. Tiscareno, 91–106. Aquascalientes, Mexico: INIFAP Publ. Available at http://www.tucson.ars.ag.gov/unit/Publications/PDFfiles/1566.pdf (accessed September 14, 2009).

Moran, M. S., C. D. Peters-Lidard, J. M. Watts, and S. McElroy. 2004. Estimating soil moisture at the watershed scale with satellite-based radar and land surface models. *Can J Remote Sens* 30(5):805–24.

Moran, M. S., A. F. Rahman, J. C. Washburne, D. C. Goodrich, M. A. Weltz, and W. P. Kustas. 1996. Combining the Penman-Monteith equation with measurements of surface temperature and reflectance to estimate evaporation rates of semi-arid grassland. *Agric For Meteorol* 80:87–109.

Muller, E., and H. Decamps. 2000. Impact of riparian vegetation on hydrological processes. *Hydrol Process* 14:2959–76.

Nemani, R. R., L. Pierce, S. Running, and S. Goward. 1993. Estimation of regional surface resistance to evapotranspiration from NDVI and thermal–IR AVHRR data. *J Appl Meteorol* 32:548–57.

Njoku, E. G., and N.-A. Kong. 1977. Theory for passive microwave remote sensing of near-surface soil moisture. *J Geophys Res* 82:3108–18.

Njoku, E. G., and D. Entekhabi. 1996. Passive microwave remote sensing of soil moisture. *J Hydrol* 184(1–2):101–29.

Norman, J. M., M. C. Anderson, W. P. Kustas, A. N. French, J. R. Mecikalski, R. Torn, G. R. Diak, T. Schmugge, and B. C. W. Tanner. 2003. Remote sensing of surface energy fluxes at 10-m pixel resolutions. *Water Resour Res* 39:1221.

Norman, J. M., and F. Becker. 1995. Terminology in thermal infrared remote sensing of natural surfaces. *Agric For Meteorol* 77(3–4):153–66.

Norman J. M., L. C. Daniel, G. R. Diak, T. E. Twine, W. P. Kustas, and A. French. 2000. Satellite estimates of evapotranspiration on the 100-m pixel scale. *IEEE IGARRS 2000* IV:1483–5.

Norman, J. M., W. P. Kustas, and K. S. Humes. 1995. Two source approach for estimating soil and vegetation energy fluxes in observations of directional radiometric surface temperature. *Agric For Meteorol* 77:263–93.

O'Hagan, A. 2006. Bayesian analysis of computer code outputs: A tutorial. *Reliab Eng Syst Saf* 91:1290–300.

Oldak, A., T. J. Jackson, P. Starks, and R. Elliott. 2003. Mapping near surface soil moisture on a regional scale using ERS-2 SAR data. *Int J Remote Sens* 24:1887–905.

Olioso, A., T. N. Carlson, and N. Brisson. 1996. Simulation of diurnal transpiration and photosynthesis of a water stressed soybean crop. *Agric For Meteorol* 81:41–59.

Olioso, A., H. Chauki, D. Courault, and J.-P. Wigneron. 1999. Estimation of evapotranspiration and photosynthesis by assimilation of remote sensing data into SVAT models. *Remote Sens Environ* 68:341–356.

Owen, T. W., T. N. Carlson, and R. R. Gillies. 1998. Remotely sensed surface parameters governing urban climate change. *Int J Remote Sens* 19:1663–81.

Petropoulos, G., T. N. Carlson, M. J. Wooster, and S. Islam. 2009. A review of Ts/VI remote sensing based methods for the retrieval of land surface fluxes and soil surface moisture content. *Adv Phys Geogr* 33(2):1–27.

Petropoulos, G., T. N. Carlson, and M. J. Wooster. 2009. A review of a 1D SVAT model for the estimation of hydro-meteorological and related land surface parameters. *Sensors* 9(6):4286–308.

Petropoulos, G., M. J. Wooster, M. Kennedy, T. N. Carlson, and M. Scholze. 2009. A global sensitivity analysis study of the 1d SimSphere SVAT model using the GEM SA software. *Ecol Modell* 220(19):2427–40.

Price, J. C. 1990. Using spatial context in satellite data to infer regional scale evapotranspiration. *IEEE Trans Geosci Remote Sens* 28:940–48.

Prihodko, L., and S. N. Goward. 1997. Estimation of air temperature from remotely sensed surface observations. *Remote Sens Environ* 60(3):335–46.

Quesney, A., S. Le Hégarat-Mascle, O. Taconet, D. Vidal-Madjar, J. P. Wigneron, and C. Loumagne. 2000. Estimation of watershed soil moisture index from ERS/SAR data. *Remote Sens Environ* 72(3):290–303.

Ray, D. K., U. S. Nair, R. M. Welch, W. Su, and T. Kikutchi. 2002. Influence of land use on the regional climate of southwest Australia. *13th Symposium on Global Change and Climate Variations and 16th Conference on Hydrology*, Australia, January 17, 2002. http://ams.confex.com/ams/annual2002/techprogram/paper_29880.htm (accessed October 16, 2009).

Saltelli, A., S. Tarantola, F. Campologno, and M. Ratto. 2004. *Sensitivity Analysis in Practice: A Guide to Assessing Scientific Models*. London: John Wiley & Sons.

Sandhold, I., K. Rasmussen, and J. Andersen. 2002. A simple interpretation of the surface temperature/vegetation index space for assessment of surface moisture status. *Remote Sens Environ* 79:213–24.

Schlesinger, W. H., J. A. Raikes, A. E. Hartley, and A. F. Cross. 1996. On the spatial pattern of soil nutrients in desert ecosystems. *Ecology* 77:364–74.

Schmugge, T. J. 1978. Remote sensing of soil moisture. *J Appl Meteorol* 17:1549–57.

Schmugge, T. J., W. P. Kustas, J. C. Ritchie, T. J. Jackson, and A. Rango. 2002. Remote sensing in hydrology. *Adv Water Resour* 25:1367–85.

Seguin, B., and B. Itier. 1983. Using midday surface temperature to estimate daily evaporation from satellite thermal IR data. *Int J Remote Sens* 4:371–83.

Silk, J. 1979. *Statistical Concepts in Geography*. New York: HarperCollins.

Smith, R. C. G. and B. J. Choudhury, 1991. Analysis of normalized difference and surface temperature observations over southeastern Australia. *Int J Remote Sens* 12(10):2021–44.

Sobrino, J. A. and N. Raissouni. 2000. Toward remote sensing methods for land cover dynamic monitoring: Application to Morocco. *Int J Remote Sens* 21(2):353–66.

Sommer, S., J. Hill and J. Megier. 1998. The potential of remote sensing for monitoring rural land use changes and their effects on soil conditions. *Agric Ecosyst Environ* 67(2):197–209(13).

Stisen, S., I. Sandholt, A. Norgaard, R. Fensholt, and K. H. Jensen. 2008. Combining the triangle method with thermal inertia to estimate regional evapotranspiration—applied to MSG SEVIRI data in the Senegal River basin. *Remote Sens Environ* 112:1242–55.

Sun, Y.-J., J.-F.Wang, R. R. Gillies, Y. Xue, and Y.-C. Bo. 2005. Air temperature retrieval from remote sensing data based on thermodynamics. *Theor Appl Climatol* 80(1):37–48.

Taconet, O., T. Carlson, R. Bernard, and D. Vidal-Madjar. 1986. Evaluation of a surface/vegetation parameterisation using satellite measurements of surface temperature. *J Clim Appl Meteorol* 25:1752–67.

Todhunter, P., and W. Terjung. 1988. Intercomparison of three urban climate models. *Bound Layer Meteorol* 42:181–205.

Troch, P. A., F. F. De Troch, R. Grayson, A. Western, A. Derauw, and C. Barbier. 1997. Spatial organisation of hydrological processes in small catchments derived from advanced SAR image processing: Field work and preliminary results. *Proceedings of Third ERS Symposium on Space at the Service of Our Environment*, European Space Agency, SP-414, Florence, Italy, 93–97.

Verhoest, N., P. A. Troch, C. Paniconi, and F. P. De Troch. 1998. Mapping basin scale variable source areas from multitemporal remotely sensed observations of soil moisture behavior. *Water Resour Res* 34(12):3235–44.

Verstraeten, W. W., F. Veroustraete, and J. Feyen. 2008. Assessment of evapotranspiration and soil moisture content across different scales of observation. *Sensors* 8:70–117.

Walker, J. P. and P. R. Houser. 2004. Requirements of a global near-surface soil moisture satellite mission: Accuracy, repeat time and spatial resolution. *Adv Water Resour* 27:785–801.

Wang, J., A. Hsu, J. C. Shi, P. O'Neil, and T. Engman. 1997. Estimating surface soil moisture from SIR-C measurements over the Little Washita River watershed. *Remote Sens Environ* 59:308–20.

Weidong, L., G. Xingfa, T. Qingxi, Z. Lanfen, and Z. Bing. 2002. Relating soil moisture to reflectance. *Remote Sens Environ* 81:238–46.

Wever, L., L. Flanagan, and P. J. Carlson. 2002. Seasonal and inter-annual variation in evapotranspiration, energy balance and surface conductance in a northern temperate grassland. *Agric For Meteorol* 112:31–49.

Whalley, W. R., P. B. Leeds-Harrison, and G. E. Bowman. 1991. Estimation of soil moisture status using near infrared reflectance. *Hydrol Process* 5:321–7.

Wilmott, C. J. 1982. Some comments on the evaluation of model performance. *Bul Am Meteorol Soc* 63(11):1309–13.

Wilson, K. B., A. H. Goldstein, E. Falge, M. Aubinet, D. Baldocchi, P. Berbigier, C. Bernhofer et al. 2002. Energy balance closure at FLUXNET sites. *Agric For Meteorol* 113:223–34.

Xu, C. Y., and D. Chen. 2005. Comparison of seven models for estimation of evapotranspiration and groundwater recharge using lysimeter measurement data in Germany. *Hydrol Process* 19:3717–34.

Yepez, E. A., D. G. Williams, R. L. Scott, and G. Lin. 2003. Partitioning overstorey and understorey evapotranspiration in a semiarid savanna woodland from the isotopic composition of water vapour. *Agric For Meteorol* 119:53–68.

Zribi, M., N. Baghdadi, N. Holah, and O. Fafin. 2005. New methodology for soil surface moisture estimation and its application to ENVISAT-ASAR multi-incidence data inversion. *Remote Sens Environ* 96:485–96.

20

Remote Sensing of Urban Biophysical Environments

Qihao Weng

CONTENTS

20.1 Introduction

Urban environmental problems have become unprecedentedly significant in the twenty-first century. This is not a simple consequence of ever-increasing urban population and land, but also because urbanization is one of the most profound examples of human modification of the Earth. Urbanization may have an impact on local energy, water and carbon exchanges, climate, habitat, and biodiversity. Depending on the size of the area affected, the impacts may be on a local, regional, or global scale. Continued urbanization and the associated environmental impacts are receiving great attention in the remote sensing community and beyond. It has been suggested that urban environment should be defined as a "new science" to be focused on U.S. satellite missions such as Hyperion in the near future. Driven by societal needs and improved spatial, spectral, and geometric (e.g., light detection and ranging [LiDAR]) resolutions in sensor technology and image processing algorithms, in recent years we have witnessed a great increase in the number of publications, special issues, and books on urban remote sensing. The Decadal Survey (National Research Council 2007) further suggests improving the temporal resolution of satellites capable of urban imaging. Therefore, we may well be entering an era of "high-definition" urban remote sensing. This chapter reviews recent research progresses in several aspects of urban remote sensing, including urban landscape, impervious surface, urban air quality, and vegetation. The chapter ends with the author's prospects on future developments and emerging trends in urban remote sensing.

20.2 Remote Sensing of Urban Landscapes

Urban areas are composed of a variety of materials, including different types of artificial materials (concrete, asphalt, metal, plastic, glasses, etc.), soils, rocks and minerals, and green and nonphotosynthetic vegetation. These materials comprise land cover and are used in different manners for various purposes by humans. Land cover can be defined as the biophysical state of the Earth's surface and immediate subsurface, including biota, soil, topography, surface and groundwater, and man-made structures (Turner et al. 1995). In other words, it describes both natural and man-made coverings of the Earth's surface. Land use can be defined as human use of the land. Land use involves both the manner in which the biophysical attributes of the land are manipulated and the purpose for which the land is used (Turner et al. 1995). Remote sensing technology has often been applied to map land use or land cover, instead of materials. Each type of land cover may possess unique surface properties (material); however, mapping land covers and materials have different requirements. Land-cover mapping needs to consider surface characteristics in addition to those from the material (Herold et al. 2006). The surface structure (roughness) may influence the spectral response as much as the intraclass variability (Gong and Howarth 1990; Myint 2001; Shaban and Dikshit 2001; Herold et al. 2006). Two different land covers, for example asphalt roads and composite shingle or tar roofs, may have very similar materials (hydrocarbons) and thus may be difficult to discern, although from a material perspective, these surfaces can be mapped accurately with hyperspectral remote sensing techniques (Herold et al. 2006). Therefore, land-cover mapping needs to take into account the intraclass variability and spectral separability. On the other hand, analysis of land-use classes would nearly be impossible with spectral information alone. Additional information, such as spatial, textural, and contextual information, is usually required in order to have a successful land-use classification in urban areas (Gong and Howarth 1992; Stuckens, Coppin, and Bauer 2000; Herold, Liu, and Clark 2003).

Traditional classification methods of land use and land cover (LULC) based on detailed fieldwork suffered two major common drawbacks: confusion between LULC and the lack of uniformity or comparability in classification schemes, leaving behind a sheer difficulty for comparing land-use patterns over time or between areas (Mather 1986). The use of aerial photographs and satellite images after the late 1960s does not solve these problems, because these techniques are based on the formal expression of land use rather than on the actual activity itself (Mather 1986). In fact, many land-use types cannot be identified from the air. As a result, mapping of the Earth's surface tends to present a mixture of LULC data with an emphasis on the latter (Lo 1986). This problem is reflected in the title of the classification developed in the United States for the mapping of the country at a scale of 1:100,000 or 1:250,000, commencing in 1974 (Anderson et al. 1976). Moreover, the U.S. Geological Survey (USGS) LULC Classification System has been designed as a resource-oriented one. Therefore, eight out of nine of the first-level categories relate to nonurban areas. The 2001 National

Land-Cover Database developed by the USGS reflects both problems (Homer et al. 2004). Alternative to the USGS scheme, the Land-Based Classification Standard developed by the American Planning Association emphasizes extracting urban or suburban land-use information. The parcel-level land-use information is obtained from in situ survey, aerial photography, and high-resolution satellite imagery (HRSI), based on the characteristics of activity, function, site development, structure, and ownership (American Planning Association 2004). Generally speaking, the success of most land-use or land-cover mapping is typically measured by the ability to match remote sensing spectral signatures to the Anderson classification scheme, which, in urban areas, is mainly land use (Ridd 1995). The confusion between LULC contributes to the low classification accuracy (Foody 2002), while less emphasis on land cover in urban areas weakens the ability of digital remote sensing as a research tool for characterizing and quantifying the urban ecological structure and process (Ridd 1995).

Another major problem in urban LULC classification is related to so-called mixed pixels. It is rare that urban land classification can yield an accuracy of greater than 80% by using per-pixel classification (i.e., "hard classification") algorithms (Mather 1999). The low accuracy of LULC classification in urban areas is largely attributed to the mixed-pixel problem, in which several types of LULC are contained in one pixel. The mixed-pixel problem results from the fact that the scale of observation (i.e., pixel resolution) fails to correspond to the spatial characteristics of the target (Mather 1999). Therefore, the "soft"/fuzzy approach of LULC classifications has been applied, in which each pixel is assigned a class membership of each LULC type rather than a single label (Wang 1990). Nevertheless, as Mather (1999) suggested, neither "hard" nor "soft" classifications were an appropriate tool for analyzing heterogeneous landscapes. Both Ridd (1995) and Mather (1999) maintained that characterization, rather than classification, should be applied in order to provide a better understanding of the compositions and processes of heterogeneous landscapes such as urban areas. To do so, one must be able to quantify accurately the spatial pattern of the landscape and its temporal changes (Wu et al. 2000). Therefore, it is necessary to have a standardized method to define theses component surfaces and to detect and map them in repetitive and consistent ways so that a global model of urban morphology may be developed, and monitoring and modeling of their changes over time may be possible (Ridd 1995).

Ridd (1995) proposed a conceptual model for remote sensing analysis of urban landscapes, that is, the vegetation–impervious surface–soil (V-I-S) model. This model assumes that land cover in urban environments is a linear combination of three components, namely, vegetation, impervious surface, and soil. Ridd believed that this model can be applied to spatial–temporal analyses of urban morphology, biophysical, and human systems. While urban land-use information may be more useful in socioeconomic and planning applications, biophysical information that can be directly derived from satellite data is more suitable for describing and quantifying urban structures and processes (Ridd 1995). The V-I-S model was actually developed for Salt Lake City, Utah, but has been tested in other cities (Ward, Phinn, and Murray 2000; Madhavan et al. 2001; Setiawan, Mathieu, and Thompson-Fawcett 2006). All of these studies employed the V-I-S model as the conceptual framework to relate urban morphology to medium-resolution satellite imagery, but "hard classification" algorithms were applied. Weng and Lu (2009) applied the V-I-S model for characterizing urban landscapes and analyzing their dynamics in Indianapolis, Indiana between 1991 and 2000. The technique of linear spectral mixture analysis (LSMA) was employed to extract landscape (V-I-S) components from Landsat images, which were further classified into urban thematic classes. Their results indicated that the reconciliation between the V-I-S model with LSMA for Landsat imagery was an effective approach for characterizing and quantifying the spatial and temporal changes of urban landscape compositions.

20.3 Remote Sensing of Impervious Surfaces

Impervious surfaces are anthropogenic features through which water cannot infiltrate into the soil, such as roads, driveways, sidewalks, parking lots, and rooftops. In recent years, impervious surface area (ISA) has emerged not only as an indicator of the degree of urbanization but also as a major indicator of environmental quality (Arnold and Gibbons 1996). ISA is found to be inversely related to vegetation cover in urban areas. In other words, as impervious cover increases within a watershed or an administrative unit, vegetation cover decreases. The percentage of land covered by impervious surfaces varies significantly with land-use categories and subcategories (Soil Conservation Service 1975). Detecting, monitoring, and mapping impervious surfaces is valuable not only for environmental management, for example, water quality assessment and storm water taxation, but also for urban planning, for example, building infrastructure and sustainable urban growth.

Many techniques have been applied to characterize and quantify impervious surfaces using either ground measurements or remotely sensed data. Field surveys with global positioning systems, although expensive and time-consuming, can provide reliable information on impervious surfaces. Manual digitizing from hard-copy maps and remote sensing imagery (especially aerial photographs) have also been used for mapping imperviousness. This technique has become more heavily involved with automation methods such as scanning and the use of feature extraction algorithms in recent years. Various digital remote sensing approaches have been developed to measure impervious surfaces, including mainly multiple regression, subpixel classification, artificial neural network, and the classification and regression tree (CART) algorithm. The multiple regression approach relates percentage of ISA to remote sensing and/or geographic information system (GIS) variables (Chabaeva, Civco, and Prisloe 2004; Bauer et al. 2004). Subpixel classification divides an image pixel into fractional components, assuming that the spectrum measured by a remote sensor is a linear or nonlinear combination of the spectra of all components within the pixel (Ji and Jensen 1999; Wu and Murray 2003; Lu and Weng 2004, 2006; Weng, Hu, and Lu 2008; Weng, Hu, and Liu 2009). The artificial neural network approach applies advanced machine learning algorithms to derive impervious surface coverage (Flanagan and Civco 2001; Weng and Hu 2008; Hu and Weng 2009). The CART approach produces a rule-based model for prediction of continuous variables based on training data and yields the spatial estimates of subpixel percent imperviousness (Yang et al. 2003). Spectral mixture analysis (SMA) as a subpixel classifier has been gaining great interest in the remote sensing community in recent years. As a physically based image analysis procedure, it supports repeatable and accurate extraction of quantitative subpixel information (Roberts et al. 1998). Because of its effectiveness in handling spectral mixture problems, LSMA has been widely used in estimating impervious surfaces (Ward, Phinn, and Murray 2000; Madhavan et al. 2001; Phinn et al. 2002; Wu and Murray 2003; Lu and Weng 2006; Weng, Hu, and Lu 2008; Weng, Hu, and Liu 2009). Different methods of impervious surface extraction based on the SMA model have been developed. For example, impervious surface may be extracted as one of the endmembers in the standard SMA model (Phinn et al. 2002). Impervious surface estimation can also be done by adding high-albedo and low-albedo fraction images, with both as the SMA endmembers (Weng, Hu, and Lu 2008). Moreover, a multiple endmember SMA method has been developed (Rashed et al. 2003), in which several impervious surface endmembers can be extracted and combined. However, these SMA-based methods have a common problem, that is, the impervious surface is often overestimated in areas with a small amount of impervious surface, but is underestimated in areas with a large amount of impervious surface. The similarity in spectral properties among nonphotosynthetic vegetation, soil, and different kinds of impervious surface materials makes it difficult to distinguish impervious from pervious materials. In addition, shadows caused by tall buildings and large tree crowns in urban areas may lead to underestimation of ISA. Lu and Weng (2006) employed Landsat thermal infrared data to remove pervious cover from impervious cover based on their distinct thermal response. Weng, Hu, and Liu (2009) found that using LULC and land-surface temperature (LST) maps as image masks, the accuracy of impervious surface estimation can be significantly improved.

Previous research has largely used medium spatial resolution images such as Landsat Thematic Mapper (TM)/Enhanced Thematic Mapper+ (ETM+) and Terra's Advanced Spaceborne Thermal Emission and Reflectance Radiometer (ASTER) images for extraction of impervious surfaces (Wu and Murray 2003; Yang et al. 2003; Lu and Weng 2006; Weng, Hu, and Liu 2009). However, both spatial and spectral resolution are regarded as too coarse for use in urban environments because of the heterogeneity and complexity of urban impervious surface materials. Urban areas may have substantially distinct types and amounts of impervious surfaces. Identifying one suitable endmember to represent all types of impervious surfaces is often found to be problematic. Lu and Weng (2004) suggested that three possible approaches may be taken to overcome these problems: (1) stratification, (2) use of multiple endmembers, and (3) use of hyperspectral imagery. In the SMA model, the maximum number of endmembers is directly proportional to the number of spectral bands used. The vastly increased dimensionality of a hyperspectral sensor may remove the sensor-related limit on the number of endmembers available. More significantly, the fact that the number of hyperspectral image channels far exceeds the likely number of endmembers for most applications readily permits the exclusion of any bands with low signal-to-noise ratios or with significant atmospheric absorption effects from the analysis (Lillesand, Kiefer, and Chipman 2004). Weng, Hu, and Lu (2008) found that the Hyperion image was more powerful in discerning low-albedo surface materials and that the improvement mainly came from the additional bands in the mid-infrared spectrum.

High spatial resolution satellite images, such as IKONOS and QuickBird images, offer great potential for accurate urban mapping, but proper algorithms need to be developed and applied (Lu and Weng 2009; Tong, Liu, and Weng 2009). As the spatial resolution increases, the proportion of pure pixels most likely increases and mixed pixels decreases (Hsieh, Lee, and Chen 2001). Therefore, subpixel classifiers may not be appropriate. Moreover, traditional image classification methods are mostly based on the color and tone of the pixels. Other important information rich in high-resolution imagery, such as texture, shape, and context, are completely neglected (Sharma and Sarkar 1998). As a result, it is not suitable to employ traditional classifiers for feature extraction from high spatial resolution imagery. An important step with these high-resolution images is to separate dark ISAs and shadowed impervious surfaces from water and shadows cast by tree crowns. Lu and Weng (2009) demonstrated that a hybrid approach based on a decision tree classifier and an unsupervised ISODATA classifier can effectively extract impervious surfaces from IKONOS images, which provided significantly better results than the maximum likelihood classifier.

In earlier studies, various image segmentation techniques were developed and applied for feature extraction with a fair amount of success (Mayer et al. 1997; Wei, Zhao, and Song 2004; Karimi and Liu 2004; Guo, Weeks, and Klee 2007; Cao and Jin 2007; Yun and Uchimura 2007). However, most of the segmentation techniques are not robust enough for a spectrally complex environment (Pal and Pal 1993), which makes them less suitable for urban classification. Therefore, it is necessary to develop new techniques to tackle these problems. The object-based image analysis (GEOBIA) approach uses not only the spectral properties but also characteristics such as shape, texture, context, relationship with neighbors, and super- and subpixels. Successful results have been obtained with this approach (Voorde et al. 2004). GEOBIA operates on objects that are composed of many pixels grouped by image segmentation (Shackelford and Davis 2003). GEOBIA has been developed and applied with varying degrees of success (Blaschke 2010). For example, Van Coillie, Verbeke, and De wulf (2007) developed a three-step object-based classification that included image segmentation, feature selection by generic algorithms, and joint neural network–based object classification. Zhou and Wang (2008) developed an algorithm of multiple agent segmentation and classification that included four steps: (1) image segmentation, (2) shadow-effect, (3) multivariate analysis of variance (MANOVA)-based classification, and (4) postclassification. This algorithm was applied for impervious surface extraction in Rhode Island. In addition, rule-based classification is another method to classify image objects. However, traditional rule-based classification is based on strict binary rules. Objects are assigned to a class if the objects meet the rules of that class. These rules may not be suitable for classifying objects because the attributes of different features may overlap (Jin and Paswaters 2007). Fuzzy logic can better cope with the uncertainties inherent in the data and vague in human knowledge (Jin and Paswaters 2007).

20.4 Remote Sensing of Urban Climate and Air Quality

Chapter 6 reviewed recent remote sensing literatures on urban LST and urban heat island (UHI) using thermal infrared data or a combined use of visible and infrared data. This section briefly discusses some recent studies of urban air quality using remotely sensed and other geospatial data. Urban areas are associated with sources of a variety of air pollutants and regional pollution problems, such as acid rain and photochemical smog. Cities are also major contributors to global air pollution related to ozone depletion and carbon dioxide (CO_2) warming. Within an urban area, the pollution level varies with the distance to pollution sources, including both stationary and mobile sources (e.g., vehicles). Local pollution patterns in cities are mainly related to the distribution of different LULC categories, occurrence of water bodies and parks, building and population densities, division of functional districts, layout of transportation network, and air flushing rates. It is well known that pollution levels rise with land-use density, which tends to increase toward a city center (Marsh and Grossa 2002). Therefore, there is generally an urban–rural gradient in the concentrations of air pollutants. For example, the concentrations of particulates, CO_2, and nitrate ion (an oxide as in acid rain) in the inner city are typically two to three times higher than in suburban areas and five times higher than in rural areas (Marsh and Grossa 2002).

Furthermore, urban areas experience another type of pollution—heat pollution. Because of the construction of tall and closely spaced buildings, the flushing capability of the air at the ground level is largely reduced. Thermal variations within an urban area mainly relate to different LULC classes, surface materials, and air flushing rates (Marsh and Grossa 2002). However, the relationship between air pollution and urban heat (and thus UHIs) is not fully understood, although both relate to the pattern of urban LULC. UHIs favor the development of air pollution problems, but are not an indicator of air pollution (Ward and Baleynaud 1999). Higher urban temperatures generally result in higher ozone levels due to increased ground-level ozone production (DeWitt and Brennan 2001). Moreover, higher urban temperatures mean increased energy use, mostly due to a greater demand for air conditioning. As power plants burn more fossil fuels, the pollution level is driven up. A few studies have so far examined the correlation between LST and air pollution measurements. Poli et al. (1994) investigated the relationship between satellite-derived apparent temperatures and daily sums of total suspended particulates (TSP) and sulfur dioxide (SO_2) in the winter in five locations of Rome, Italy. They found that apparent temperatures had a strong negative correlation with TSP, but a weak correlation with SO_2. Brivio et al. (1995) used three Advanced Very High Resolution Radiometer (AVHRR) images of Milan, Italy, acquired on February 12–14, 1993, to study the correlation of apparent temperatures with air quality parameters, including TSP and SO_2. A weak correlation was found with both TSP and SO_2, which could be explained by the large pixel size of the image. Ward and Baleynaud (1999) explored the correlation between Landsat TM band-6 digital counts and the concentrations of pollutants, including black particulates (BP), SO_2, nitrogen dioxide (NO_2), nitrogen monoxide (NO), and strong acidity (AF), in Nantes, France, based on the measurements of daily sums, individual measurements every 15 minutes, and daily mean values taken on May 22, 1992. Apparent temperatures were highly positively correlated with BP and moderately correlated with SO_2 and daily means of NO_2, NO, and AF, but weakly correlated with instantaneous measurements of NO_2 and NO. Weng and Yang (2006) investigated the relationship of patterns of air pollution with patterns of urban land use and thermal landscape in Guangzhou, China, by using GIS analysis. Ambient air quality measurements for SO_2, nitrogen oxide, carbon monoxide, TSP, and dust level were obtained between 1981 and 2000. They found that the spatial patterns of air pollutants probed were positively correlated with urban built-up density, and with satellite-derived LST values, particularly with measurements taken during summer. These cited studies contribute to the literature by providing more evidence on the correlation between air pollution and urban thermal patterns.

20.5 Remote Sensing of Urban Vegetation

Various remote sensing techniques and methods have been developed and applied to monitoring, mapping, and modeling vegetation. Aerial photos are used to identify individual plants by crown shape and size, while photogrammetry techniques are used to measure tree height, crown diameter, and closure. At the advent of satellite multispectral remote sensing imagery, many efforts are being made to understand the spectral behavior of leaves and the spectral response pattern of canopies. To measure the biomass and vegetative vigor, various vegetation indexes have been developed that are often grouped into slope-based (e.g., normalized difference vegetation index [NDVI]), distance-based (e.g., the perpendicular vegetation index), and those through orthogonal transformation techniques (e.g, the Kauth-Thomas tasseled cap transformation). However, most of the studies that utilize satellite multispectral imagery have been focused on nonurban vegetation at the regional or global scales due to their coarse spatial and spectral resolutions. The emergence of hyperspectral sensors (imaging spectrometers) alters this situation and provides an excellent opportunity to study urban vegetation. These instruments can acquire images in many very narrow, contiguous spectral bands throughout the visible, near-infrared, mid-infrared, and thermal-infrared portions of the spectrum and thus can give information on the diagnostic absorption and reflection characteristics of tree species that are "lost" within the bands of conventional multispectral scanners (Goetz et al. 1985). Hyperspectral sensors have been applied to analyze the spectral properties of numerous plant biophysical parameters (e.g., leaf area index, crown closure, biomass, and net primary productivity) and biochemical parameters (e.g., plant pigments and nutrients) by statistical analysis or the physically based modeling approach (Pu et al. 2008). LSMA, a physically based modeling method, has recently been applied to estimate urban vegetation abundance in several studies (Small 2001; Weng et al. 2004; Song 2005). High spatial resolution images (less than 5 m) from IKONOS and QuickBird were used to validate the results of estimation based on medium-resolution multispectral images (Small and Lu 2006; Nichol and Wong 2007). More recently, LiDAR data has been regarded as a critical data source in urban vegetation studies (Popescu, Wynne, and Nelson 2003; Secord and Zakhor 2007); it provides land-surface elevation information by emitting a laser pulse and providing high vertical and horizontal resolutions of less than 1 m. When combined with an object-oriented method and hyperspectral imagery, LiDAR data shows great potential for identification of tree species (Voss and Sugumaran 2008).

In urban environmental studies, the relationship between LST and vegetation indices (e.g, NDVI) has been extensively documented (Weng 2009). The LST-vegetation index relationship has been used by Carlson, Gillies, and Perry (1994) to retrieve surface biophysical parameters, by Kustas et al. (2003) to extract subpixel thermal variations, and by Lambin and Ehrlich (1996) and Sobrino and Raissouni (2000) to analyze land-cover dynamics. Many studies observed a negative relationship between LST and vegetation indices. This finding has stimulated research in two major directions, namely, statistical analysis of the LST-vegetation abundance relationship (e.g., Weng et al. 2004) and the thermal vegetation index approach, which is a multispectral method of combining LST and a vegetation index in a scatter plot to observe their associations (Quattrochi and Ridd 1994). Another interesting relationship among urban biophysical variables is found between vegetation vigor and LST. Weng, Hu, and Liu (2009) found that vegetation phenology had a fundamental impact on impervious surface estimation when linear spectral unmixing technique was employed. Plant phenology can cause changes in the variance partitioning and thus affect the mixing space characterization, leading to a less accurate estimation of impervious surfaces. Yuan and Bauer (2007) made a correlation analysis between ISA and NDVI and suggested that ISA showed a higher stability and a lower seasonal variability; they recommended it as a complementary measure to NDVI in UHI studies. Xian Carne (2006) suggested that the combined use of ISA, NDVI, and LST can explain temporal thermal dynamics across the cities.

20.6 Improved Sensors and Algorithms Integral to Urban Remote Sensing

20.6.1 Ultra High-Resolution Satellite Imagery and LiDAR Data

With the recent advent of very HRSI, such as IKONOS (launched in 1999), QuickBird (2001), and OrbView (2003) images, great efforts have been made in the applications of these remote sensing images in urban and environmental studies. HRSIs have been widely applied in urban land-cover mapping (Thomas, Hendrix, and Congalton 2003; Warner and Steinmaus 2005; Im, Jensen, and Hodgson 2008; Wulder et al. 2008; Lu and Weng 2009), 3D shoreline extraction and coastal mapping (Di, Ma, and Li 2003; Ma, Di, and Li 2003), earthquake damage assessment (Al-Khudhairy, Caravaggi, and Glada 2005; Miura and Midorikawa 2006), digital terrain modeling and digital elevation model (DEM) generation (Toutin 2004a, b), 3D object reconstruction (Tao and Hu 2002), and topographic mapping and change detection (Birk et al. 2003; Holland, Boyd, and Marshall 2006). Many GIS databases also use high-accuracy geopositioning HRSI images as the base maps, providing both metric and thematic information.

These fine spatial resolution images contain rich spatial information, providing a greater potential to extract much more detailed thematic information (e.g., LULC, impervious surface, and vegetation), cartographic features (e.g., buildings and roads), and metric information with stereoimages (e.g., height and area), which are ready to be used in GIS. However, some new problems come with the HRSI image data, notably shades caused by topography, tall buildings, or trees, and high spectral variation within the same land-cover class. The shade problem increases the difficulty to extract both thematic and cartographic information. The shade problem and high spectral variation are common with the high degree of spectral heterogeneity in complex landscapes, such as in urban areas (Lu and Weng 2009). These disadvantages may lower image classification accuracy if the classifiers used cannot effectively handle them (Irons et al. 1985; Cushnie 1987). In addition, the huge amount of data storage and the computer display of HRSI images can also affect image processing in general and the selection of classification algorithms in particular.

With respect to the extraction of cartographic features, HRSI image data provide a great possibility to achieve the effectiveness and efficiency of extraction through automated extraction methods. But the issues of shade and image distortion can affect the resultant accuracy to a certain degree. LiDAR data, which provides land-surface elevation information by emitting a laser pulse and providing high vertical and horizontal resolutions of less than 1 m, has been increasingly used in many geospatial applications due to its high data resolution, short time consumption, and low cost. Unlike other remotely sensed data, LiDAR data focus solely on geometry rather than radiometry. Some typical products derived include DEM, the surface elevation model, triangulated irregular network, and intermediate return information. As a result, feature classification and extraction based on LiDAR data are widely performed (Filin 2004; Forlani et al. 2006; Clode et al. 2007; Lee, Lee, and Lee 2008).

LiDAR data show great potential for building and road extraction because elevation data can be derived quickly and at a high resolution in comparison to photogrammetric techniques (Miliaresis and Kokkas 2007). The DEM and associated products and LiDAR-derived cartographic information have become an important GIS data source in recent years. It should also be noted that many researchers have used LiDAR in conjunction with optical remote sensing and GIS data in urban, environment, and resource studies. These new data sources also require a processing functionality that is not currently standard in many systems (Poulter 1995). Techniques for information or pattern extraction from such remote sensing data sets and data analysis need to be developed, for example, self-organizing neural networks, integrated spatial and temporal representation and analysis, and data mining (Wilkinson 1996).

20.6.2 Enhanced and New Image Analysis Algorithms

20.6.2.1 Knowledge-Based Expert Systems

This approach is now increasingly becoming attractive due to its ability to accommodate multiple sources of data, such as satellite imagery and GIS data. GIS plays an important role in developing knowledge-based classification approaches, because of its ability to manage different sources of data and spatial modeling. As different kinds of ancillary data, such as DEM, soil map, housing and population density, road network, temperature, and precipitation, become readily available, they may be incorporated into a classification procedure in different ways. One approach is to develop knowledge-based classifications based on the experts' knowledge of the spatial distribution pattern of land-cover classes and selected ancillary data. For example, elevation, slope, and aspect are related to vegetation distribution in mountainous regions. Data on terrain features are thus useful for separation of vegetation classes. Population, housing, and road densities are related to urban land-use distribution and may be very helpful in the distinctions between commercial or industrial lands and high-intensity residential lands, between recreational grassland and pasture or crops, or between residential areas and forest land (Lu and Weng 2006). Similarly, temperature, precipitation, and soil-type data are related to land-cover distribution at a large scale. Effective use of these relationships in a classification procedure has proven to be very helpful in improving classification accuracy. Expert systems are considered to have great potential for providing a general approach to the routine use of image ancillary data in image classification (Hinton 1996). A critical step is to develop rules that can be used in an expert system or a knowledge-based classification approach. Three methods have been employed to build rules for image classification: (1) explicitly eliciting knowledge and rules from experts and then refining the rules (Stefanov, Ramsey, and Christensen 2001; Hung and Ridd 2002; Stow et al. 2003), (2) implicitly extracting variables and rules using cognitive methods (Hodgson 1998; Lloyd, Hodgson, and Stokes 2002), and (3) empirically generating rules from observed data with automatic induction methods (Huang and Jensen 1997; Hodgson et al. 2003; Tullis and Jensen 2003).

20.6.2.2 Artificial Neural Network

Artificial neural network (ANN) has been increasingly applied in recent years. The neural network has several advantages, including a nonparametric nature, arbitrary decision boundary capability, ability to adapt to different types of data and input structures, fuzzy output values, and generalization for use with multiple images (Paola and Schowengerdt 1995). The fact that ANN behaves like general pattern recognition systems and assumes no prior statistical model for the input data makes it an excellent technique for integrating remote sensing and GIS data. Although many neural network models have been developed, the multilayer perceptron (MLP) feed-forward neural network is most frequently used (Kavzoglu and Mather 2003). MLP has been applied in LULC classifications (Foody et al. 1997; Zhang and Foody 2001; Kavzoglu and Mather 2003), impervious surface estimation (Weng and Hu 2008), and change detection (Li and Yeh 2002). Other applications include water properties estimation (Schiller and Doerffer 1999; Zhang et al. 2002; Corsini et al. 2003), forest structure mapping (Ingram, Dawson, and Whittaker 2005), understory bamboo mapping (Linderman et al. 2004), cloud detection (Jae-Dong et al. 2006), and mean monthly ozone prediction (Chattopadhyay and Bandyopadhyay 2007). Although MLP has been widely applied, some drawbacks have been raised by previous research. For instance, how to design the number of hidden layers and the number of hidden layer nodes in the model are still challenges. Another issue is that MLP requires training sites to include both presence and absence data. The desired output must contain both true and false information, so that the network can learn all kinds of patterns for a study area in order to classify accordingly (Li and Eastman 2006). However, in some cases, absence data is not always readily available. Finally, MLP has the local minima problem in the training process, which may significantly affect the accuracy of the result.

Another neural network approach, Kohonen's self-organizing map (SOM), has not been applied as widely as MLP (Pal, Laha, and Das 2005). SOM can be used for both supervised and unsupervised classifications and has the properties for both vector quantization and projection (Li and Eastman 2006). SOM has been used for both "hard" classification and "soft" classification in previous studies. Ji (2000) compared a Kohonen self-organizing feature map and MLP for image classification at the per-pixel level. Seven classes were identified, and the results showed that SOM provided an excellent alternative to the MLP neural network in "hard" classification. Lee and Lathrop (2006) conducted a SOM-LVQ-GMM to extract urban land cover from Landsat ETM+ imagery at the subpixel level. They found that SOM can generate promising results in "soft" classification and that SOM had several advantages over MLP. Hu and Weng (2009) compared the two neural networks to three ASTER images of Marion County, Indiana, and found that SOM outperformed MLP slightly for every season of image data, especially in the residential areas.

20.6.2.3 Object-Based Image Analysis

The object-oriented concept was first introduced to the GIS community in the late 1980s (Egenhofer and Frank 1992), and since then, especially after the 1990s, a great deal of research has been conducted with the object-oriented approach (Bian 2007). In remote sensing, image segmentation, which is usually applied before image classification, has a longer history and has its roots in industrial image processing, but had not been used extensively in the geospatial community in the 1980s and 1990s (Blaschke 2010). Object-oriented image analysis has been increasingly used in remote sensing applications due to the advent of HRSI data and the emergence of commercial software such as eCognition (Benz et al. 2004; Wang et al. 2004). Image segmentation merges pixels into objects, and a classification is then implemented based on objects, instead of individual pixels. In the process of creating objects, a scale determines the occurrence or absence of an object class, and the size of an object affects the classification result. This approach has proven to be able to provide better classification results than per-pixel-based classification approaches, especially for fine spatial resolution data. Object-oriented image analysis holds great potential in the development of a fully integrated GIS (Ehlers 2007). Two key difficulties lie in (1) algorithms for smoothing and thinning to produce acceptable vectors, and (2) the automated assignment of attributes to vectors, points, nodes, and so on (Faust, Anderson, and Star 1991).

20.6.2.4 Data Mining

Data mining is a field that has been developed by encompassing principles and techniques from statistics, machine learning, pattern recognition, numeric search, and scientific visualization to accommodate new data types and data volumes being generated (Miller and Han 2001). The tasks of data mining vary, but the premise, to discover unknown information from large databases, remains the same. In short, data mining can be defined as analysis of (often large) observational data sets to find unsuspected relationships and to summarize the data in novel ways that are both understandable and useful to the data owners (Hand, Mannila, and Smyth 2001). Over the last few years, the techniques of data mining have been pushed by three major technological factors that have advanced in parallel. First, the growth in the amount of data has led to the development of mass storage devices. Second, the problem of accessing this information has led to the development of advanced and improved processors. Third, the need for automating the tasks involved in data retrieval and processing led to advancements in statistic and machine learning algorithms. Data mining tasks can be broadly classified into five categories based on their tasks, that is, (1) segmentation, (2) dependency analysis, (3) outlier analysis, (4) trend detection, and (5) characterization. In order to do these tasks, various techniques, such as cluster analysis, neural networks, genetic algorithms, Bayesian networks, decision trees, and so on are applied. Some of these techniques are also good at executing more than one task and have their own advantages and disadvantages (Rajasekar, Bijker, and Stein 2007). Data mining techniques have been successfully applied to the combined data sets of GIS and remote sensing (Mennis and Liu 2005; Mennis 2006; Rajasekar and Weng 2009a,b) and may be a good approach for the processing and analysis of HRSI or other image data of high volume.

20.6.2.5 Data Fusion

Images from different sensors contain distinctive features. Data fusion or integration of multisensor or multiresolution data takes advantage of the strengths of distinct image data for improving visual interpretation and quantitative analysis. In general, three levels of data fusion can be identified (Gong 1994): (1) pixel (Luo and Kay 1989), (2) feature (Jimenez, Morales-Morell, and Creus 1999), and (3) decision (Benediktsson and Kanellopoulos 1999). Data fusion involves two merging procedures: (1) geometric coregistration of two data sets and (2) mixture of spectral and spatial information contents to generate a new data set that contains enhanced information from both data sets. Accurate registration between the two data sets is extremely important for precisely extracting the information contents from both data sets, especially for line features, such as roads and rivers. Radiometric and atmospheric calibrations are also needed before multisensor data are merged.

Many methods have been developed to fuse spectral and spatial information in previous studies (Gong 1994; Pohl and Van Genderen 1998; Chen and Stow 2003; Lu and Weng 2005). Solberg, Taxt, and Jain (1996) broadly divided data fusion methods into four categories: (1) statistical, (2) fuzzy logic, (3) evidential reasoning, and (4) neural network. Dai and Khorram (1998) presented a hierarchical data fusion system for vegetation classification. Pohl and Van Genderen (1998) provided a literature review on the methods of multisensor data fusion. The methods, including color-related techniques (e.g., color composite, intensity–hue–saturation [IHS], and luminance-chrominance), statistical/numerical methods (e.g., arithmetic combination, principal component analysis, high-pass filtering, regression variable substitution, canonical variable substitution, component substitution, and wavelets), and various combinations of the above methods were examined. IHS transformation was identified as the most frequently used method for improving the visual display of multisensor data (Welch and Ehlers 1987), but the IHS approach can only employ three image bands, and the resultant image may not be suitable for further quantitative analysis such as image classification. Principal component analysis is often used for data fusion because it can produce an output that can better preserve the spectral integrity of the input data set. In recent years, wavelet-merging techniques have also shown to be an effective approach to enhance spectral and spatial information contents (Li, Kwok, and Wang 2002; Simone et al. 2002; Ulfarsson, Benediktsson, and Sveinsson 2003). Previous research works indicated that integration of Landsat TM and radar (Ban 2003; Haack et al. 2002), SPOT HRV and Landsat TM (Welch and Ehlers 1987; Munechika et al. 1993; Yocky 1996), and SPOT multispectral and panchromatic bands (Garguet-Duport et al. 1996; Shaban and Dikshit 2001) can improve classification results. An alternative integrated use of multiresolution images, such as Landsat TM or SPOT and Moderate Resolution Imaging

Spectroradiometer (MODIS) or AVHRR, is to refine the estimation of LULC types from coarse spatial resolution data (Moody 1998; Price 2003).

20.6.2.6 Hyperspectral Imaging

The spectral characteristics of land surfaces are the fundamental principles for land-cover classification using remotely sensed data. The spectral features include the number of spectral bands, spectral coverage, and spectral resolution (or bandwidth). The number of spectral bands used for image classification can range from a limited number of multispectral bands (e.g., four bands in SPOT data and seven for Landsat TM) to a medium number of multispectral bands (e.g., ASTER with 14 bands and MODIS with 36 bands) and to hyperspectral data (e.g., Airborne Visible/Infrared Imaging Spectrometer and EO-1 Hyperion images with 224 bands). The large number of spectral bands provides the potential to derive detailed information on the nature and properties of different surface materials on the ground, but it also means difficulty in image processing and a large data redundancy due to high correlation among the adjacent bands. High-dimension data also require a larger number of training samples for image classification. Increase of spectral bands may improve classification accuracy, but only when those bands are useful in discriminating the classes (Thenkabail, Enclona et al. 2004a). In previous research, hyperspectral data have been successfully used for LULC classification (Benediktsson, Sveinsson, and Arnason 1995; Hoffbeck and Landgrebe 1996; Platt and Goetz 2004; Thenkabail, Enclona, et al. 2000a,b) and vegetation mapping (McGwire, Minor, and Fenstermaker 2000; Schmidt et al. 2004). As spaceborne hyperspectral data such as EO-1 Hyperion become available, research and applications with hyperspectral data will increase. Weng, Hu, and Lu (2008) found that a Hyperion image was more powerful in discerning low-albedo surface materials, which has been a major obstacle for impervious surface estimation with medium-resolution multispectral images. A sensitivity analysis of the mapping of impervious surfaces using different scenarios of the Hyperion band combinations suggested that the improvement of mapping accuracy in general and the better ability in discriminating low-albedo surfaces mainly came from additional bands in the mid-infrared region (Weng, Hu, and Lu 2008).

20.7 Conclusions

The majority of previous remote sensing studies of urban biophysical environment have used medium-spatial resolution (10–100 m) images. The concepts and image analysis algorithms in urban remote sensing have been developed based on the images of such resolutions. The advent of high spatial resolution satellite images (especially those with less than 5-m resolution), spaceborne hyperspectral images, and LiDAR data has provided an unprecedented opportunity for urban remote sensing, and in the meantime, is challenging the traditional remote sensing concepts, models, and image processing algorithms. The desire to take advantage of the opportunity to combine ever-increasing computational power, modern telecommunications technologies, more plentiful and capable digital data, and more advanced algorithms has driven the field of urban remote sensing into a new frontier of scientific inquiry. Various emerging trends in image processing and data analysis will advance urban remote sensing, including attribute analysis or classification of remote sensing data by more powerful ANN models and knowledge-based expert systems, object-based image analysis, object search by data mining techniques, and enhanced environmental mapping via data fusion of different sensors and hyperspectral imaging.

References

Al-Khudhairy, D. H. A., I. Caravaggi, and S. Glada. 2005. Structural damage assessments from IKONOS data using change detection, object-oriented segmentation, and classification techniques. *Photogramm Eng Rem Sens* 71(8):825–37.

American Planning Association. 2004. *Land-Based Classification System*. Washington , D.C.: American Planning Association. http://www.planning.org/lbcs (accessed January 20, 2007).

Anderson, J. R., E. E. Hardy, J. T. Roach, and R. E. Witmer. 1976. *Land-Use and Land-Cover Classification Systems for Use with Remote Sensing Data*. USGS Professional Paper 964. Washington DC: United States Geological Survey.

Arnold Jr., C. L., and C. J. Gibbons. 1996. Impervious surface coverage: The emergence of a key environmental indicator. *J Am Plann Assoc* 62:243–58.

Ban, Y. 2003. Synergy of multitemporal ERS-1 SAR and Landsat TM data for classification of agricultural crops. *Can J Rem Sens* 29:518–26.

Bauer, M. E., N. J. Heiner, J. K. Doyle, and F. Yuan. 2004. Impervious surface mapping and change monitoring using Landsat remote sensing. In *ASPRS Annual Conference Proceedings*, Denver, Colorado. (Unpaginated CD-ROM.)

Benediktsson, J. A., and I. Kanellopoulos. 1999. Classification of multisource and hyperspectral data based on decision fusion. *IEEE Trans Geosci Rem Sens* 37:1367–77.

Benediktsson, J. A., J. R. Sveinsson, and K. Arnason. 1995. Classification and feature extraction of AVIRIS data. *IEEE Trans Geosci Rem Sens* 33:1194–205.

Benz, U. C., P. Hofmann, G. Willhauck, I. Lingenfelder, and M. Heynen. 2004. Multi-resolution, object-oriented fuzzy analysis of remote sensing data for GIS-ready information. *ISPRS J Photogramm Rem Sens* 58:239–58.

Bian, L. 2007. Object-oriented representation of environmental phenomena: Is everything best represented as an object? *Ann Assoc Am Geogr* 97(2):266–80.

Birk, R. J., T. Stanley, G. I. Snyder, T. A. Hennig, M. M. Fladeland, and F. Policelli. 2003. Government programs for research and operational uses of commercial remote sensing data. *Rem Sens Environ* 88(1–2):3–16.

Blaschke, T. 2010. Object based image analysis for remote sensing. *ISPRS J Photogramm Rem Sens* 65:2–16.

Brivio, P. A., G. Gempvese, S. Massari, N. Mileo, G. Saura, and E. Zilloli. 1995. Atmospheric pollution and satellite remotely sensed surface temperature in metropolitan areas. In *EARSeL Advances in Remote Sensing Pollution Monitoring and Geographical Information Systems*, ed. EARSeL, 40–6. Paris: EARLSeL.

Cao, G., and Y. Q. Jin. 2007. Automatic detection of main road network in dense urban area using microwave SAR images. *Imaging Sci J* 55(4):215–22.

Carlson, T. N., R. R. Gillies, and E. M. Perry. 1994. A method to make use of thermal infrared temperature and NDVI measurements to infer surface water content and fractional vegetation cover. *Rem Sensing Rev* 9(1–2):161–73.

Chabaeva, A. A., D. L. Civco, and S. Prisloe. 2004. Development of a population density and land use based regression model to calculate the amount of imperviousness. In *ASPRS Annual Conference Proceedings*, Denver, Colorado. American Society for Photogrammetry and Remote Sensing, Bethesda, Maryland. (Unpaginated CD ROM.)

Chattopadhyay, S., and G. Bandyopadhyay. 2007. Artificial neural network with backpropagation learning to predict mean monthly total ozone in Arosa, Switzerland. *Int J Rem Sens* 28(20):4471–82.

Chen, D., and D. A. Stow. 2003. Strategies for integrating information from multiple spatial resolutions into land-use/land-cover classification routines. *Photogramm Eng Rem Sens* 69:1279–87.

Clode, Simon, Rottensteinerb, F., P. Kootsookosc, and E. Zelniker. 2007. Detection and vectorization of roads from LiDAR data. *Photogramm Eng Rem Sens* 73(5):517–35.

Corsini, G., M. Diani, R. Grasso, M. De Martino, P. Mantero, and S. B. Serpico. 2003. Radial basis function and multilayer perceptron neural networks for sea water optically active parameter estimation in case II waters: A comparison. *Int J Rem Sens* 24(20):3917.

Cushnie, J. L. 1987. The interactive effect of spatial resolution and degree of internal variability within land-cover types on classification accuracies. *Int J Rem Sens* 8:15–29.

Dai, X., and S. Khorram. 1998. A hierarchical methodology framework for multisource data fusion in vegetation classification. *Int J Rem Sens* 19:3697–701.

DeWitt, J., and M. Brennan. 2001. Taking the heat. *Imaging Notes* 16(6):20–3.

Di, K., R. Ma, and R. Li. 2003. Geometric processing of IKONOS geo stereo imagery for coastal mapping applications. *Photogramm Eng Rem Sens* 69(8):873–9.

Egenhofer, M. F., and A. Frank. 1992. Object-oriented modeling for GIS. *J Urban Reg Inf Syst Assoc* 4:3–19.

Ehlers, M. 2007. Integration taxonomy and uncertainty. In *Integration of GIS and Remote Sensing*, ed. V. Mesev, 17–42. England: West Sussex.

Faust, N. L., W. H. Anderson, and J. L. Star. 1991. Geographic information systems and remote sensing future computing environment. *Photogramm Eng Rem Sens* 57(6):655–68.

Filin, S. 2004. Surface classification from airborne laser scanning data. *Comput Geosci* 30(9–10):1033–41.

Flanagan, M., and D. L. Civco. 2001. Subpixel impervious surface mapping. In *ASPRS Annual Conference Proceedings*, St. Louis, Missouri, American Society for Photogrammetry and Remote Sensing, Bethesda, Maryland. (Unpaginated CD-ROM.)

Foody, G. M. 2002. Status of land cover classification accuracy assessment. *Remote Sens Environ* 80:185–201.

Foody, G. M., R. M. Lucas, P. J. Curran, and M. Honzak. 1997. Non-linear mixture modelling without end-members using an artificial neural network. *Int J Rem Sens* 18:937–53.

Forlani, G., C. Nardinocchi, M. Scaioni, and P. Zingaretti. 2006. Complete classification of raw LiDAR data and 3D reconstruction of buildings. *Pattern Anal Appl* 8(4):357–74.

Garguet-Duport, B., J. Girel, J. Chassery, and G. Pautou. 1996. The use of multiresolution analysis and wavelet transform for merging SPOT panchromatic and multispectral image data. *Photogramm Eng Rem Sens* 62:1057–66.

Goetz, A. F. H., G. Vane, J. E. Solomon, and B. N. Rock. 1985. Imaging spectrometry for earth remote sensing. *Science* 228(4704):1147–53.

Gong, P. 1994. Integrated analysis of spatial data from multiple sources: An overview. *Can J Rem Sens* 20:349–59.

Gong, P., and P. J. Howarth. 1990. The use of structure information for improving land-cover classification accuracies at the rural-urban fringe. *Photogramm Eng Rem Sens* 56(1):67–73.

Gong, P., and P. J. Howarth. 1992. Frequency-based contextual classification and gray-level vector reduction for land-use identification. *Photogramm Eng Rem Sens* 58(4):423–37.

Guo, D., A. Weeks, and H. Klee. 2007. Robust approach for suburban road segmentation in high-resolution aerial images. *Int J Rem Sens* 28(1–2):307–18.

Haack, B. N., E. K. Solomon, M. A. Bechdol, and N. D. Herold. 2002. Radar and optical data comparison/integration for urban delineation: A case study. *Photogramm Eng Rem Sens* 68:1289–96.

Hand, D., H. Mannila, and P. Smyth. 2001. *Principles of Data Mining, Vol. 1, A Bradford Book.* Cambridge, MA: The MIT Press.

Herold, M., X. Liu, and K. C. Clark. 2003. Spatial metrics and image texture for mapping urban land use. *Photogramm Eng Rem Sens* 69(9):991–1001.

Herold, M., S. Schiefer, P. Hostert, and D. A. Roberts. 2006. Applying imaging spectrometry in urban areas. In *Urban Remote Sensing*, ed. Q. Weng and D. Quattrochi, 137–61. Boca Raton, FL: CRC Press/Taylor & Francis.

Hinton, J. C. 1996. GIS and remote sensing integration for environmental applications. *Int J Geogr Inf Syst* 10(7):877–90.

Hodgson, M. E. 1998. What size window for image classification? A cognitive perspective. *Photogramm Eng Rem Sens* 64:797–808.

Hodgson, M. E., J. R. Jensen, J. A. Tullis, K. D. Riordan, and C. M. Archer. 2003. Synergistic use LiDAR and color aerial photography for mapping urban parcel imperviousness. *Photogramm Eng Rem Sens* 69:973–80.

Hoffbeck, J. P., and D. A. Landgrebe. 1996. Classification of remote sensing having high spectral resolution images. *Rem Sens Environ* 57:119–26.

Holland, D. A., D. S. Boyd, and P. Marshall. 2006. Updating topographic mapping in Great Britain using imagery from high-resolution satellite sensors. *ISPRS J Photogramm Rem Sens* 60:212–23.

Homer, C., C. Huang, L. Yang, B. Wylie, and M. Coan. 2004. Development of a 2001 national land-cover database for the United States. *Photogramm Eng Rem Sens* 70(7):829–40.

Hsieh, P. F., L. C. Lee, and N. Y. Chen. 2001. Effect of spatial resolution on classification errors of pure and mixed pixels in remote sensing. *IEEE Trans Geosci Rem Sens* 39:2657–63.

Hu, X., and Q. Weng. 2009. Estimating impervious surfaces from medium spatial resolution imagery using the self-organizing map and multi-layer perceptron neural networks. *Rem Sens Environ* 113(10):2089–102.

Huang, X., and J. R. Jensen. 1997. A machine-learning approach to automated knowledge-base building for remote sensing image analysis with GIS data. *Photogramm Eng Rem Sens* 63:1185–94.

Hung, M, and M. K. Ridd. 2002. A subpixel classifier for urban land-cover mapping based on a maximum-likelihood approach and expert system rules. *Photogramm Eng Rem Sens* 68:1173–80.

Im, J., J. R. Jensen, and M. E. Hodgson. 2008. Optimizing the binary discriminant function in change detection applications. *Rem Sens Environ* 112(6):2761–76.

Ingram, J. C., T. P. Dawson, and R. J. Whittaker. 2005. Mapping tropical forest structure in southeastern Madagascar using remote sensing and artificial neural networks. *Rem Sens Environ* 94(4):491–507.

Irons, J. R., B. L. Markham, R. F. Nelson, D. L. Toll, D. L. Williams, R. S. Latty, and M. L. Stauffer. 1985. The effects of spatial resolution on the classification of Thematic Mapper data. *Int J Rem Sens* 6:1385–403.

Jae-Dong, J., A. A. Viau, A. Francois, and E. Bartholome. 2006. Neural network application for cloud detection in SPOT VEGETATION images. *Int J Rem Sens* 27(3/4):719–36.

Ji, C. Y. 2000. Land-use classification of remotely sensed data using Kohonen self-organizing feature map neural networks. *Photogramm Eng Rem Sens* 66:1451–60.

Ji, M., and J. R. Jensen. 1999. Effectiveness of subpixel analysis in detecting and quantifying urban imperviousness from Landsat Thematic Mapper imagery. *Geocarto Int* 14:33–41.

Jimenez, L. O., A. Morales-Morell, and A. Creus. 1999. Classification of hyperdimensional data based on feature and decision fusion approaches using projection pursuit, majority voting, and neural networks. *IEEE Trans Geosci Rem Sens* 37:1360–6.

Jin, X., and S. Paswaters. 2007. A fuzzy rule base system for object-based feature extraction and classification. In *Signal Processing, Sensor Fusion, and Target Recognition XVI*, ed. K. Ivan, 6567: 65671H.

Karimi, H. A., and S. Liu. 2004. Developing an automated procedure for extraction of road data from high-resolution satellite images for geospatial information systems. *J Transp Eng-Asce* 130(5):621–31.

Kavzoglu, T., and P. M. Mather. 2003. The use of backpropagating artificial neural networks in land cover classification. *Int J Rem Sens* 24:4907–38.

Kustas, W. P., J. M. Norman, M. C. Anderson, and A. N. French. 2003. Estimating subpixel surface temperatures and energy fluxes from the vegetation index-radiometric temperature relationship. *Rem Sens Environ* 85(4):429–40.

Lambin, F. F., and D. Ehrlich. 1996. The surface temperature-vegetation index space for land use and land cover change analysis. *Int J Rem Sens* 17(3):463–87.

Lee, S., and R. G. Lathrop. 2006. Subpixel analysis of Landsat ETM+ using Self-Organizing Map (SOM) neural networks for urban land cover characterization. *IEEE Trans Geosci Rem Sens* 44(6):1642–54.

Lee, D. H., K. M. Lee, and S. U. Lee. 2008. Fusion of LiDAR and imagery for reliable building extraction. *Photogramm Eng Rem Sens* 74(2):215–25.

Li, Z., and J. R. Eastman. 2006. Commitment and typicality measurements for the self-organizing map. In *Proceedings of SPIE*. The International Society for Optical Engineering, 64201I-1–64201I-4. Bellingham, WA.

Li, S., J. T. Kwok, and Y. Wang. 2002. Using the discrete wavelet frame transform to merge Landsat TM and SPOT panchromatic images. *Inf Fusion* 3:17–23.

Li, X., and A. G. O. Yeh. 2002. Neural-network-based cellular automata for simulating multiple land use changes using GIS. *Int J Geogr Inf Sci* 16(4):323–43.

Lillesand, T. M., R. W. Kiefer, and J. W. Chipman. 2004. *Remote Sesning and Image Interpretation*, 614. New York: John Wiley & Sons.

Linderman, M., Liu, J., Qi, J., An, L., Ouyang, Z., Yang, J., and Tan, Y. 2004. Using artificial neural networks to map the spatial distribution of understorey bamboo from remote sensing data. *Int J Rem Sens* 25(9):1685–1700.

Lloyd, R. E., M. E. Hodgson, and A. Stokes. 2002. Visual categorization with aerial photography. *Ann Assoc Am Geogr* 92:241–66.

Lo, C. P. 1986. *Applied Remote Sensing*. New York: Longman Inc.

Lu, D., and Q. Weng. 2004. Spectral mixture analysis of the urban landscape in Indianapolis with Landsat ETM+ imagery. *Photogramm Eng Rem Sens* 70:1053–62.

Lu, D., and Q. Weng. 2005. Urban land-use and land-cover mapping using the full spectral information of Landsat ETM+ data in Indianapolis, Indiana. *Photogramm Eng Rem Sens* 71(11):1275–84.

Lu, D., and Q. Weng. 2006. Use of impervious surface in urban land use classification. *Rem Sens Environ* 102(1–2):146–60.

Lu, D., and Q. Weng. 2009. Extraction of urban impervious surfaces from IKONOS imagery. *Int J Rem Sens* 30(5):1297–311.

Luo, R. C., and M. G. Kay. 1989. Multisensor integration and fusion for intelligent systems. *IEEE Trans Syst Man Cybern* 19:901–31.

Ma, R., K. Di, and R. Li. 2003. 3-D shoreline extraction from IKONOS satellite. *J Mar Geod* 26(1–2):107–15.

Madhavan, B. B., S. Kubo, N. Kurisaki, and T. V. L. N. Sivakumar. 2001. Appraising the anatomy and spatial growth of the Bangkok Metropolitan area using a vegetation-impervious soil model through remote sensing. *Int J Rem Sens* 22:789–806.

Marsh, W. M., and J. M. Grossa Jr. 2002. *Environmental Geography: Science, Land Use, and Earth Systems*. 2nd ed. New York: John Wiley & Sons.

Mather, A. S. 1986. *Land Use*. London: Longman.

Mather, P. M. 1999. Land cover classification revisited. In *Advances in Remote Sensing and GIS*, ed. P. M. Atkinson and N. J. Tate, 7–16. New York: John Wiley & Sons.

Mayer, H., I. Laptev, A. Baumgartner, and C. Steger. 1997. Automatic road extraction based on multi-scale modeling, context and snakes. *Int Arch Photogramm Rem Sens* Band XXXII, part 3-2W32:106–13.

McGwire, K., T. Minor, and L. Fenstermaker. 2000. Hyperspectral mixture modeling for quantifying sparse vegetation cover in arid environments. *Rem Sens Environ* 72:360–74.

Mennis, J. 2006. Socioeconomic-vegetation relationships in urban, residential land: The case of Denver, Colorado. *Photogramm Eng Rem Sens* 72:911–21.

Mennis, J., and J. W. Liu. 2005. Mining association rules in spatio-temporal data: An analysis of urban socioeconomic and land cover change. *Trans GIS* 9:5–17.

Miliaresis, G., and N. Kokkas. 2007. Segmentation and object-based classification for the extraction of the building class from LiDAR DEMs. *Comput Geosci* 33(8):1076–87.

Miller, H. J., and J. Han. 2001. *Geographic Data Mining and Knowledge Discovery*, 372. New York: Taylor & Francis.

Miura, H., and S. Midorikawa. 2006. Updating GIS building inventory data using high-resolution satellite images for earthquake damage assessment: Application to metro Manila, Philippines. *Earthquake Spectra* 22:151–68.

Moody, A. 1998. Using Landsat spatial relationships to improve estimates of land-cover area from coarse resolution remote sensing. *Rem Sens Environ* 64:202–20.

Munechika, C. K., J. S. Warnick, C. Salvaggio, and J. R. Schott. 1993. Resolution enhancement of multispectral image data to improve classification accuracy. *Photogramm Eng Rem Sens* 59:67–72.

Myint, S. W. 2001. A robust texture analysis and classification approach for urban land-use and land-cover feature discrimination. *Geocarto Int* 16:27–38.

National Research Council, 2007. *Earth Science and Applications from Space: National Imperatives for the Next Decade and Beyond*. Washington, DC: The National Academy Press.

Nichol, J. E., and M. S. Wong. 2007. Remote sensing of urban vegetation life form by spectral mixture analysis of high-resolution IKONOS satellite images. *Int J Rem Sens* 28(5):985–1000.

Pal, N. R., A. Laha, and J. Das. 2005. Designing fuzzy rule based classifier using self-organizing feature map for analysis of multispectral satellite images. *Int J Rem Sens* 26(10):2219–40.

Pal, N. R., and S. K. Pal. 1993. A review on image segmentation techniques. *Pattern Recognit* 26(9):1277–94.

Paola, J. D., and R. A. Schowengerdt. 1995. A review and analysis of back propagation neural networks for classification of remotely sensed multispectral imagery. *Int J Rem Sens* 16:3033–58.

Phinn, S., M. Stanford, P. Scarth, A. T. Murray, and P. T. Shyy. 2002. Monitoring the composition of urban environments based on the vegetation-impervious surface-soil (VIS) model by subpixel analysis techniques. *Int J Rem Sens* 23:4131–53.

Platt, R. V., and A. F. H. Goetz. 2004. A comparison of AVIRIS and Landsat for land use classification at the urban fringe. *Photogramm Eng Rem Sens* 70:813–9.

Pohl, C., and J. L. van Genderen. 1998. Multisensor image fusion in remote sensing: Concepts, methods, and applications. *Int J Rem Sens* 19:823–54.

Poli, U., F. Pignatoro, V. Rocchi, and L. Bracco. 1994. Study of the heat island over the city of Rome from Landsat-TM satellite in relation with urban air pollution, In *Remote Sensing-From Research to Operational Applications in the New Europe*, ed. R. Vaughan, 413–22. Berlin: Springer Hungarica.

Popescu, S. C., R. H. Wynne, and R. F. Nelson. 2003. Measuring individual tree crown diameter with LiDAR and assessing its influence on estimating forest volume and biomass. *Can J Rem Sens* 29(5):564–77.

Poulter, M. 1995. Integrating remote sensing and GIS: Designs for an operational future. In *Proceedings of a Seminar on Integrated GIS and High Resolution Satellite Data*, ed. M. Palmer. Farnborough UK: Defense Research Agency.

Price, J. C. 2003. Comparing MODIS and ETM+ data for regional and global land classification. *Rem Sens Environ* 86:491–99.

Pu, R., M. Kelly, G. L. Anderson, and P. Gong. 2008. Using CASI hyperspectral imagery to detect mortality and vegetation stress associated with a new hardwood forest disease. *Photogramm Eng Rem Sens* 74(1):65–75.

Quattrochi, D. A., and M. K. Ridd. 1994. Measurement and analysis of thermal energy responses from discrete urban surface using remote sensing data. *Int J Rem Sens* 15(10):1999–2024.

Rajasekar, U., W. Bijker, and A. Stein. 2007. Image mining for modeling of forest fires from meteosat images. *IEEE Trans Rem Sens* 45:246–53.

Rajasekar, U., and Q. Weng. 2009a. Urban heat island monitoring and analysis by data mining of MODIS imageries. *ISPRS J Photogramm Rem Sens* 64(1):86–96.

Rajasekar, U., and Q. Weng. 2009b. Application of association rule mining for exploring the relationship between urban land surface temperature and biophysical/social parameters. *Photogramm Eng Rem Sens* 75(4):385–96.

Rashed, T., J. R. Weeks, D. Roberts, J. Rogan, and R. Powell. 2003. Measuring the physical composition of urban morphology using multiple endmember spectral mixture models. *Photogramm Eng Rem Sens* 69:1011–20.

Ridd, M. K. 1995. Exploring a V-I-S (vegetation-impervious surface-soil) model for urban ecosystem analysis through remote sensing: comparative anatomy for cities. *Int J Rem Sens* 16(12):2165–85.

Roberts, D. A., G. T. Batista, J. L. G. Pereira, E. K. Waller, and B. W. Nelson. 1998. Change identification using multitemporal spectral mixture analysis: Applications in eastern Amazônia. In *Remote Sensing Change Detection: Environmental Monitoring Methods and Applications*, ed. R. S. Lunetta and C. D. Elvidge, 137–61. Ann Arbor, MI: Ann Arbor Press.

Schiller, H., and R. Doerffer. 1999. Neural network for emulation of an inverse model operational derivation of Case II water properties from MERIS data. *Int J Rem Sens* 20(9):1735–46.

Schmidt, K. S., A. K. Skidmore, E. H. Kloosterman, H. van Oosten, L. Kumar, and J. A. M. Janssen. 2004. Mapping coastal vegetation using an expert system and hyperspectral imagery. *Photogramm Eng Rem Sens* 70:703–15.

Secord, J., and A. Zakhor. 2007. Tree detection in urban regions using aerial LiDAR and image data. *IEEE Geosci Rem Sens Lett* 4(2):196–200.

Setiawan, H., R. Mathieu, and M. Thompson-Fawcett. 2006. Assessing the applicability of the V-I-S model to map urban land use in the developing world: Case study of Yogyakarta, Indonesia. *Comput Environ Urban Syst* 30(4):503–22.

Shaban, M. A., and O. Dikshit. 2001. Improvement of classification in urban areas by the use of textural features: The case study of Lucknow city, Uttar Pradesh. *Int J Rem Sens* 22:565–93.

Shackelford, A. K., and C. H. Davis. 2003. A combined fuzzy pixel-based and object-based approach for classification of high-resolution multispectral data over urban areas. *IEEE Trans Geosci Rem Sens* 41(10):2354–63.

Sharma, K. M. S., and A. Sarkar. 1998. A modified contextual classification technique for remote sensing data. *Photogramm Eng Rem Sens* 64(4):273–80.

Simone, G., A. Farina, F. C. Morabito, S. B. Serpico, and L. Bruzzone. 2002. Image fusion techniques for remote sensing applications. *Inf Fusion* 3:3–15.

Small, C. 2001. Estimation of urban vegetation abundance by spectral mixture analysis. *Int J Rem Sens* 22(7):1305–34.

Small, C., and J. W. T. Lu. 2006. Estimation and vicarious validation of urban vegetation abundance by spectral mixture analysis. *Rem Sens Environ* 100(4):441–56.

Sobrino, J. A., and N. Raissouni. 2000. Toward remote sensing methods for land cover dynamic monitoring: Application to Morocco. *Int J Rem Sens* 21:353–66.

Soil Conservation Service. 1975. Urban hydrology for small watersheds. Washington, DC: USDA Soil Conservation Service Technical Release No. 55.

Solberg, A. H. S., T. Taxt, and A. K. Jain. 1996. A Markov random field model for classification of multisource satellite imagery. *IEEE Trans Geosci Rem Sens* 34:100–12.

Song, C. 2005. Spectral mixture analysis for subpixel vegetation fractions in the urban environment: How to incorporate endmember variability? *Rem Sens Environ* 95(2):248–63.

Stefanov, W. L., M. S. Ramsey, and P. R. Christensen. 2001. Monitoring urban land cover change: An expert system approach to land cover classification of semiarid to arid urban centers. *Rem Sens Environ* 77:173–85.

Stow, D., L. Coulter, J. Kaiser, A. Hope, D. Service, K. Schutte, and A. Walters. 2003. Irrigated vegetation assessment for urban environments. *Photogramm Eng Rem Sens* 69:381–90.

Stuckens, J., P. R. Coppin, and M. E. Bauer. 2000. Integrating contextual information with per-pixel classification for improved land cover classification. *Rem Sens Environ* 71:282–96.

Tao, C. V., and Y. Hu. 2002. 3D reconstruction methods based on the rational function model. *Photogramm Eng Rem Sens* 68(7):705–14.

Thenkabail, P. S., E. A. Enclona, M. S. Ashton, C. Legg, and M. J. de Dieu. 2004a. Hyperion, IKONOS, ALI, and ETM+ sensors in the study of African rainforests. *Rem Sens Environ* 90:23–43.

Thenkabail, P. S., E. A. Enclona, M. S. Ashton, and B. van der Meer. 2004b. Accuracy assessments of hyperspectral waveband performance for vegetation analysis applications. *Rem Sens Environ* 91:354–76.

Thomas, N., C. Hendrix, and R. G. Congalton. 2003. A comparison of urban mapping methods using high-resolution digital imagery. *Photogramm Eng Rem Sens* 69:963–72.

Tong, X., S. Liu, and Q. Weng. 2009. Geometric processing of Quickbird stereo imagery for urban land use mapping: A case study in Shanghai, China. *IEEE J Sel Top Appl Earth Obs Rem Sens* 2(2):61–6.

Toutin, T. 2004a. DTM generation from IKONOS in-track stereo images using a 3D physical model. *Photogramm Eng Rem Sens* 70(6):695–702.

Toutin, T. 2004b. Comparison of stereo-extracted DTM from different high-resolution sensors: SPOT-5, EROS-A, IKONOS-II and Quickbird. *IEEE Trans Geosci Rem Sens* 42(10):2121–9.

Tullis, J. A., and J. R. Jensen. 2003. Export system house detection in high spatial resolution imagery using size, shape, and context. *Geocarto Int* 18:5–15.

Turner, B. L. II., D. Skole, S. Sanderson, G. Fisher, L. Fresco, and R. Leemans. 1995. Land-use and land-cover change: Science and research plan. International Geosphere-Biosphere Program and the Human Dimensions of Global Environmental Change Programme (IGBP Report No. 35 and HDP Report No. 7). Stockholm and Geneva: Royal Swedish Academy of Sciences.

Ulfarsson, M. O., J. A. Benediktsson, and J. R. Sveinsson. 2003. Data fusion and feature extraction in the wavelet domain. *Int J Rem Sens* 24:3933–45.

Van Coillie, F. M. B., L. P. C. Verbeke, and R. R. De Wulf. 2007. Feature selection by genetic algorithms in object-based classification of IKONOS imagery for forest mapping in Flanders, Belgium. *Rem Sens Environ* 110(4):476–87.

Van de Voorde, T., W. De Genst, F. Canters, N. Stephenne, E. Wolff, and M. Binard. 2004. Extraction of land use/land-cover related information from very high resolution data in urban and suburban areas, *Remote Sensing in Transition* (R. Goossens, editor), *Proceedings of the 23rd Symposium of the European Association of Remote Sensing Laboratories*, 02–05 June 2003, Ghent, Belgium, Millpress Rotterdam, Netherlands, pp. 237–244.

Voss, M., and R. Sugumaran. 2008. Seasonal effect on tree species classification in an urban environment using hyperspectral data, LiDAR, and an object-oriented approach. *Sensors* 8(5):3020–36.

Wang, F. 1990. Fuzzy supervised classification of remote sensing images. *IEEE Trans Geosci Rem Sens* 28(2):194–201.

Wang, L., W. P. Sousa, P. Gong, and G. S. Biging. 2004. Comparison of IKONOS and QuickBird images for mapping mangrove species on the Caribbean coast of panama. *Rem Sens Environ* 91:432–40.

Ward, L., and J. M. Baleynaud. 1999. Observing air quality over the city of Nates by means of Landsat thermal infrared data. *Int J Rem Sens* 20(5):947–59.

Ward, D., S. R. Phinn, and A. T. Murray. 2000. Monitoring growth in rapidly urbanizing areas using remotely sensed data. *Prof Geogr* 53:371–86.

Warner, T. A., and K. Steinmaus. 2005. Spatial classification of orchards and vineyards with high spatial resolution panchromatic imagery. *Photogramm Eng Rem Sens* 71(1):179–87.

Wei, Y., Z. Zhao, and J. Song. 2004. Urban building extraction from high-resolution satellite panchromatic image using clustering and edge detection. *Proceedings of 2004 IEEE International Geoscience and Remote Sensing Symposium* 3:20–24.

Welch, R., and M. Ehlers. 1987. Merging multi-resolution SPOT HRV and Landsat TM data. *Photogramm Eng Rem Sens* 53:301–3.

Weng, Q. 2009. Thermal infrared remote sensing for urban climate and environmental studies: Methods, applications, and trends. *ISPRS J Photogramm Rem Sens* 64(4):335–44.

Weng, Q., and X. Hu. 2008. Medium spatial resolution satellite imagery for estimating and mapping urban impervious surfaces using LSMA and ANN. *IEEE Trans Geosci Rem Sens* 46(8):2397–406.

Weng, Q., X. Hu, and H. Liu. 2009. Estimating impervious surfaces using linear spectral mixture analysis with multi-temporal ASTER images. *Int J Rem Sens* 30(18):4807–30.

Weng, Q., X. Hu, and D. Lu. 2008. Extracting impervious surface from medium spatial resolution multispectral and hyperspectral imagery: A comparison. *Int J Rem Sens* 29(11):3209–32.

Weng, Q., and D. Lu. 2009. Landscape as a continuum: An examination of the urban landscape structures and dynamics of Indianapolis city, 1991–2000. *Int J Rem Sens* 30(10):2547–77.

Weng, Q., Lu, D. and J. Schubring. 2004. Estimation of land surface temperature-vegetation abundance relationship for urban heat island studies. *Remote Sensing of Environment*, 89(4):467–483.

Weng, Q., and S. Yang. 2006. Urban air pollution patterns, land use, and thermal landscape: An examination of the linkage using GIS. *Environ Monit Assess* 117(1–3):463–89.

Wilkinson, G. G. 1996. A review of current issues in the integration of GIS and remote sensing data. *Int J Geogr Inf Syst* 10:85–101.

Wu, J. G., D. E. Jelinski, M. Luck, and P. T. Tueller. 2000. Multiscale analysis of landscape heterogeneity: Scale variance and pattern metrics. *Geogr Inf Sci* 6(1):6–19.

Wu, C., and A. T. Murray. 2003. Estimating impervious surface distribution by spectral mixture analysis. *Rem Sens Environ* 84:493–505.

Wulder, M. A., J. C. White, N. C. Coops, and C. R. Butson. 2008. Multi-temporal analysis of high spatial resolution imagery for disturbance monitoring. *Rem Sens Environ* 112(6):2729–2740.

Xian, G., and M. Crane. 2006. An analysis of urban thermal characteristics and associated land cover in Tampa Bay and Las Vegas using Landsat satellite data. *Rem Sens Environ* 104(2):147–56.

Yang, L., C. Huang, C. G. Homer, B. K. Wylie, and M. J. Coan. 2003. An approach for mapping large-scale impervious surfaces: Synergistic use of Landsat-7 ETM+ and high spatial resolution imagery. *Can J Rem Sens* 29:230–40.

Yocky, D. A. 1996. Multiresolution wavelet decomposition image merger of Landsat Thematic Mapper and SPOT panchromatic data. *Photogramm Eng Rem Sens* 62:1067–74.

Yuan, F., and M. E. Bauer. 2007. Comparison of impervious surface area and normalized difference vegetation index as indicators of surface urban heat island effects in Landsat imagery. *Rem Sens Environ* 106(3):375–86.

Yun, L., and K. Uchimura. 2007. Using self-organizing map for road network extraction from Ikonos imagery. *Int J Innovative Comput Inf Control* 3(3):641–56.

Zhang, J., and G. M. Foody. 2001. Fully-fuzzy supervised classification of sub-urban land cover from remotely sensed imagery: Statistical neural network approaches. *Int J Rem Sens* 22:615–28.

Zhang, Q., J. Wang, X. Peng, P. Gong, and P. Shi. 2002. Urban built-up land change detection with road density and spectral information from multitemporal Landsat TM data. *Int J Rem Sens* 23:3057–78.

Zhou, Y. Y. and Wang, Y. Q. 2008. Extraction of impervious, surface areas from high spatial resolution imagery by multiple agent segmentation and classification. *Photogramm Eng Rem Sens* 74(7):857–68.

21

Development of the USGS National Land-Cover Database over Two Decades

George Xian, Collin Homer, and Limin Yang

CONTENTS

21.1 Introduction

Land-cover composition and change have profound impacts on terrestrial ecosystems. Land-cover and land-use (LCLU) conditions and their changes can affect social and physical environments by altering ecosystem conditions and services. Information about LCLU change is often used to produce landscape-based metrics and evaluate landscape conditions to monitor LCLU status and trends over a specific time interval (Loveland et al. 2002; Coppin et al. 2004; Lunetta et al. 2006). Continuous, accurate, and up-to-date land-cover data are important for natural resource and ecosystem management and are needed to support consistent monitoring of landscape attributes over time. Large-area land-cover information at regional, national, and global scales is critical for monitoring landscape variations over large areas.

Previous studies have demonstrated that LCLU activities have significant impacts on biophysical conditions such as soil moisture availability (Li et al. 2000), length of growing season (White et al. 2002), and local and regional climate conditions including temperature, precipitation, and severe storm frequency (Pielke 2001; Stone and Weaver 2003; Marshall, Pielke, and Steyaert 2003; Changnon 2003; Rozoff, Cotton, and Adegoke 2003; Georgescu et al. 2009). Understanding the impacts of landscape change on ecosystems requires both better observations and more comprehensive modeling studies (Xian and Crane 2005; Findell et al. 2007; Ge et al. 2007). Contemporary, accurate, and consistently repeatable land-cover characterizations such as land-cover classifications and continuous field predictions (e.g., impervious surface and tree canopy [TC] cover) are important for global change studies (DeFries, Field, and Fung 1995; Seneviratne et al. 2006; Seto and Shepherd 2009). However, at regional or national scales, such efforts face a number of challenges, including the timely acquisition of data, the high cost of creating national products, and the development of appropriate analytical techniques to successfully evaluate the current conditions and associated changes.

The National Land-Cover Database (NLCD) encompasses three major successful data releases to date. These include a 1992 conterminous U.S. land-cover data set with one thematic layer of land cover (NLCD 1992), an updated U.S. 50-state and Puerto Rico land-cover database with three thematic layers including land cover, percent imperviousness, and percent TC (NLCD 2001), and an NLCD 1992–2001 retrofit land cover change product that was specially designed to identify land-cover changes between the two data sets (i.e., NLCD 1992 and 2001). All NLCD products have been developed under the auspices of the Multiresolution Land Characteristics Consortium (MRLC; Loveland and Shaw 1996; Homer et al. 2004). MRLC was originally formed in early 1990 and consisted of several U.S. federal agencies, including the U.S. Geological Survey (USGS), U.S. Environmental Protection Agency (EPA), National Oceanic and Atmospheric Administration (NOAA), and U. S. Forest Service (USFS; Loveland and Shaw 1996). Under the coordination of MRLC 1990, National Land Cover Data 1992 (NLCD 1992) was developed (Vogelmann et al. 2001).

In 1999, the MRLC initiated an expansion that resulted in six new agencies joining the consortium: National Aeronautics and Space Administration (NASA), National Park Service, Natural Resources Conservation Service, Bureau of Land Management, U.S. Fish and Wildlife Service, and the Office of Surface Mining. This expansion enabled the creation of a more comprehensive land-cover database for the nation, encompassing all 50 states and Puerto Rico (Homer et al. 2004). NLCD 2001 was produced as a three-product suite including products that identified 1 of 16 classes of land cover, the percent urban impervious surface, and the percent TC for every 30-m cell for all 50 states (Homer et al. 2007). The primary data used to create NLCD 2001 included data from both Landsat Thematic Mapper (TM) and Enhanced Thematic Mapper (ETM+) in circa 2001. An NLCD 1992 and 2001 land-cover change product was also produced to offer users direct land-cover change analysis for the 9 years between NLCD 2001 (nominal year 2001) and NLCD 1992 (nominal year 1992). This product was developed to provide more accurate and useful land-cover change data than would be possible by direct comparison of NLCD 1992 and NLCD 2001 (Fry et al. 2009). The successful completion of these large-area land-cover characterization activities is attributable primarily to improved algorithms available for land-cover characterization and the availability of high-quality remote sensing and geospatial data sets. Currently, a new update of NLCD 2001 to circa 2006 is under production to provide 5-year land-cover change information for the nation.

The primary goal of this chapter is to briefly summarize the methods used and experiences learned from the activities of NLCD development. We also illustrate the merits of the new framework for developing NLCD 2011, which is planned to be accomplished by 2016 for the nation.

21.2 Development of NLCD 1992 and NLCD 2001

21.2.1 Development of NLCD 1992

In early 1990, several federal agencies recognized the need for national land-cover data of medium spatial resolution to support their environmental and natural resource management programs. MRLC was convened to pool resources for the purchase and processing of Landsat data for the conterminous United States to U.S. federal agencies in a standard processed format. These data included leaf-on and leaf-off seasonal Landsat 5 imagery mosaics for the nation and were used to develop a seamless national land-cover product for the conterminous United States. In addition to Landsat 5 imagery, other ancillary geospatial data were compiled to support land-cover mapping, including USGS Digital Terrain Elevation Data and its derivatives (slope, aspect, and shaded relief), U.S. Census Bureau population density data, USGS land use data analysis of 1970s, the National Wetlands Inventory (NWI) data, and the USDA State Soil Geographic (STATSGO) Database.

The land-cover classification scheme designed for NLCD 1992 was a modified Anderson LCLU classification system with 21 classes (Anderson et al. 1976). Land cover was classified by each of the 10 federal regions using either an unsupervised or supervised classification method (Vogelmann et al. 2001). For the unsupervised method, a *k*-mean clustering algorithm was applied to either a leaf-on or leaf-off Landsat image to generate spectral clusters, with clusters labeled subsequently by interpreters. In many cases, ancillary data (e.g., census, slope, aspect, and elevation) were used to resolve spectral confusion, so that each Landsat TM pixel could be labeled as one of the 21 land-cover classes. A supervised approach using a classification tree algorithm was used at the later stage of the project. Details of the mapping procedures used for NLCD 1992 can be found in Vogelmann, Sohl, and Howard (1998) and Vogelmann et al. (2001). Figure 21.1 presents land-cover maps from NLCD 1992 of the southeastern part of the Seattle, Washington, and Sioux Falls, South Dakota areas, displaying mostly forest in the Seattle area and urban and agricultural classes in the Sioux Falls area.

Following the completion of the land-cover mapping, an accuracy assessment was conducted using a procedure based on a statistical sampling design. The sampling incorporated three layers of stratification and a two-stage cluster sampling protocol (Stehman et al. 2003). Each mapping region constituted a stratum and was sampled independently. The reference data used for assessing land-cover data quality were the aerial photographs acquired by the National Aerial Photography Program in 1990 (Zhu et al. 2001). The overall agreement between the reference data and the mapped land-cover type for the regions ranges from 38% to 70% for full legend (21 classes) and from 70% to 85% for aggregated legend (7 classes), which resembles the Anderson level I classification scheme (Yang et al. 2001; Stehman et al. 2003; Wickham et al. 2004).

The production of NLCD 1992 represented the first ever medium-resolution consistent land cover produced for the conterminous United States. This effort identified many technological and data constraints and revealed areas that still needed improvement for a new effort. First, better stratification of the spatial mapping units was needed to further optimize land-cover mapping. Second, a more complete data stream was required that encompassed three dates of Landsat imagery and comprehensive ancillary data sets. Third, new classification algorithms were needed to successfully mine this expanded data stream. Fourth, the products themselves needed to be represented as multiple database layers rather than a single product.

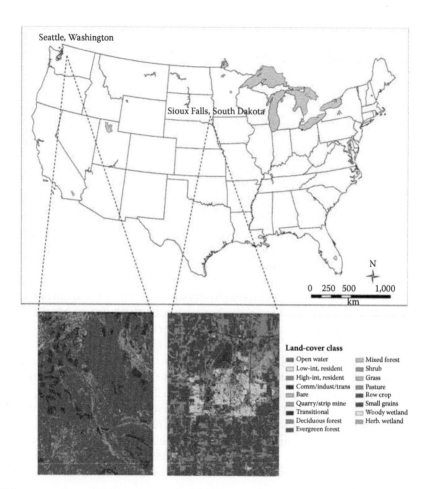

FIGURE 21.1
(See color insert following page 426.) NLCD1992 land cover for the Seattle, Washington and Sioux Falls, South Dakota areas.

21.2.2 Development of NLCD 2001

NLCD 2001 was designed to incorporate many of the lessons learned from NLCD 1992 production, as well as target new MRLC 2001 member requirements. NLCD 2001 is a multiattribute, multisource database that includes a suite of data layers (30-m resolution) intended for many applications at both national and local scales. NLCD 2001 followed a database approach that moved beyond the traditional remote sensing classification of land cover, which focuses on a single-legend classification system and a single land-cover layer that meet only specific requirements. Traditional land-cover products, while useful, had often been developed according to the specific project needs and did not meet general application requirements or allow crosswalk to other land-cover schemes for broader user applications. The local product focus had restricted the broad development of remote sensing data sets, especially for nationwide implementations. The NLCD 2001 was designed as a database approach and consisted of a suite of three products: land cover based on the Anderson level II classification scheme, percent impervious surface, and percent TC characterized for each 30-m cell for the nation.

The NLCD 2001 adopted several innovative approaches to develop the database. Ecologically based mapping zone delineation was used to define NLCD 2001 mapping regions that are relatively homogeneous with distinct ecological features. Six factors were considered in defining these mapping zones: physiography, LCLU characteristics, spectral feature uniformity, edge-matching feasibility among mapping zones, the size of each mapping zone, and the number of Landsat images required to make a mosaic (Homer et al. 2004). A total of 66 mapping zones were identified in the conterminous United States.

A vegetation phenology–based Landsat scene selection strategy was implemented in order to establish the image base that was best suitable for land-cover characterization. Landsat image selection was based on knowledge of the LCLU features and the phenology of vegetation within each mapping zone. For each mapping zone, a preferred image acquisition time window was identified based on the maximum separation of the major land-cover types using Advanced Very High Resolution Radiometer (AVHRR) normalized differential vegetation index (NDVI) time-series data. Following the identification of the target dates, three Landsat scenes for each path and row were selected as consistent remote sensing data sources for land-cover mapping in each mapping zone.

In order to ensure that the combination of three dates of imagery for each path and row was effective, a consistent satellite image preprocessing strategy was used to produce good-quality data for multiscene mosaics, spectral transformations, and information extraction. The image preprocessing follows a standard procedure to ensure the consistency of image data used in land-cover mapping. Specifically, preprocessing steps include radiometric and geometric calibration, orthorectification of the images through terrain correction with improved accuracy of ground control points and digital elevation model (DEM) data, conversion of digital number (DN) to at-sensor reflectance and radiant temperature, and a tasseled cap transformation of the original spectral bands into greenness, brightness, and wetness components using a new set of transformation coefficients based on Landsat ETM+ at-satellite reflectance (Huang et al. 2002).

Consistent and efficient classification methods were also implemented for the NLCD 2001. To improve the consistency and efficiency of the classification process, a data mining technique of using a classification and regression tree (CART) was adopted to develop NLCD 2001. For land-cover mapping, a classification tree algorithm was chosen for the land-cover classification. A classification tree is a nonparametric statistical approach that does not require the normality of input variables and can handle both categorical and continuous data. The algorithm provides a robust and efficient way to extract information from a large quantity of satellite and ancillary data by creating classification rules that can be readily interpreted and applied spatially. Additional algorithms chosen for NLCD 2001 also included regression models to calculate percent impervious surface area (ISA) and TC estimates. The decision tree model develops a sequence of classification trees, in which each subsequent tree attempts to reduce the misclassification errors resulting from the previous one. The algorithm also provided a cross-validation procedure for conducting training and rule-based modeling to ensure better quality assurance while products were being produced. To quantify percentages of urban impervious surfaces and TC at Landsat subpixel level, a regression tree algorithm was utilized. This algorithm generalizes a set of rules based on training data to predict percent imperviousness or TC density for each pixel of an image. Essentially, the algorithm establishes a statistical relationship between known imperviousness (or TC density) at a location and the corresponding Landsat spectral reflectance. The regression tree model then uses a sequence of multivariate linear equations to calculate the percent impervious surface in urban areas and the percent TC in forest lands.

Extensive and high-quality training data were collected for the use of CART in NLCD 2001. The success of land-cover characterization using CART relies greatly on the availability of abundant high-quality training data. A major effort was devoted to the collection of training data for land-cover classification and imperviousness and TC density estimates. For land-cover classification, the training data were compiled from several data sources through collaborations with MRLC members and state and local organizations. Sources of training data included NLCD field collections, MRLC partner programs, such as forest plot data collected by the USDA Forest Service's Forest Inventory and Analysis (FIA) program, information from state and local organizations, and other miscellaneous sources. For modeling imperviousness and TC density, a set of high spatial resolution training data was developed on imagery from the USGS Digital Ortho-Quads and satellite-based IKONOS imagery provided by NASA's Science Data Purchase Program (Yang et al. 2003). Typically six to eight small representative urban and forest areas were classified on these high-resolution sources for subsequent training of the regression tree models.

NLCD 2001 products for all 50 states and Puerto Rico were delivered in 2008. Figure 21.2 shows NLCD 2001 land cover (Figure 21.2a) and TC (Figure 21.2b) in the Seattle area, and land cover (Figure 21.2c) and impervious surface (Figure 21.2d) in the Sioux Falls area. The displayed extents in Figure 21.2 are the same as in Figure 21.1. In the NLCD 1992, the vegetated land in the Sioux Falls airport was incorrectly classified as small grains. In the NLCD 2001, this part of land was corrected as developed, open space. The roads missed in the NLCD 1992 for the Seattle area were added in the 2001 product. The NLCD 2001 products provide more detail and close-to-true land-cover data in these regions.

Among the 16 classes of land cover over the conterminous United States, shrub and scrub is the most common class, occupying 21.03% of the total area (Homer et al. 2007). The second largest class is cultivated crops, at 15.67% of the total area. For TC, cells representing canopy density from 91% to 100% represent the largest proportion (16.36% of the total area), and canopy groupings from 1% to 10% represent the smallest proportion (1.96% of the total area). The total area of TC is approximately 2,691,988 square kilometers. For impervious cells, groupings from 1% to 10% represent the largest proportion (47.13% of the total area; likely due to the large number of tertiary roads outside the urban areas), and groupings from 91% to 100% represent the smallest proportion (1.49% of the total area). The total spatial extent of impervious surface is approximately 457,059 square kilometers.

An accuracy assessment of NLCD 2001 (conterminous United States) was conducted using a similar sampling procedure, reference data collection, and analysis protocol that were implemented for the NLCD 1992 accuracy assessment. The sampling design was a two-stage cluster sample with three levels of stratification. Ten mapping regions within the conterminous United States were sampled and evaluated independently. Nationwide overall accuracies of NLCD 2001 were 78.5% (Anderson level I) and 85% (modified Anderson level II), respectively, as compared with the overall accuracies of the NLCD 1992 of 58% and 80% (level II). The overall accuracy for 10 geographic regions of the United States ranges from 68% to 86% at level II and from 79% to 91% at the Anderson level I classification scheme (Wickham et al. 2010). The assessment for percent impervious surface and tree cover estimates was also conducted for the conterminous United States (Greenfield, Nowak, and Walton 2009). The NLCD mapping zones were aggregated into five large regions. Four mapping zones from each of five regions were randomly selected. Within each of the mapping zones, 200 random points were placed in each selected geographic zone for photo interpretation. Overall, NLCD 2001 underestimated the tree cover when compared with the photo interpretation method by a mean of 9.7%, and it underestimated the impervious surface cover by a mean of 5.1%.

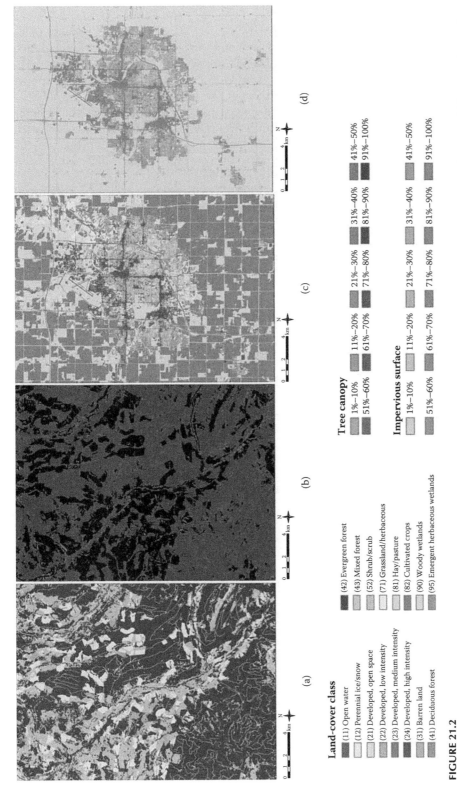

Land-cover class

- (11) Open water
- (12) Perennial ice/snow
- (21) Developed, open space
- (22) Developed, low intensity
- (23) Developed, medium intensity
- (24) Developed, high intensity
- (31) Barren land
- (41) Deciduous forest

- (42) Evergreen forest
- (43) Mixed forest
- (52) Shrub/scrub
- (71) Grassland/herbaceous
- (81) Hay/pasture
- (82) Cultivated crops
- (90) Woody wetlands
- (95) Emergent herbaceous wetlands

Tree canopy

1%–10%	11%–20%	21%–30%	31%–40%	41%–50%
51%–60%	61%–70%	71%–80%	81%–90%	91%–100%

Impervious surface

1%–10%	11%–20%	21%–30%	31%–40%	41%–50%
51%–60%	61%–70%	71%–80%	81%–90%	91%–100%

FIGURE 21.2

(See color insert following page 426.) NLCD 2001 (a) land cover and (b) percent tree canopy cover for the Seattle, Washington; (c) land cover and (d) percent impervious surface for the Sioux Falls, South Dakota area.

21.2.3 Development of the NLCD 1992–2001 Land-Cover Change Retrofit Product

The NLCD 1992–2001 land-cover change retrofit product was developed to provide more accurate and useful land-cover change data than would be possible by the direct comparison of NLCD 1992 and NLCD 2001. Substantial differences in imagery, legends, and methods between these two land-cover products needed to be overcome in order to support comparison. For the change analysis method to meet the requirements of both the national spatial scale and the nearly decadal temporal range, implementation of the NLCD 1992–2001 land-cover products required production across many Landsat TM and ETM+ path and rows simultaneously. To meet these requirements, a hybrid change analysis process was developed to incorporate both postclassification comparison and specialized ratio-differencing change analysis techniques (Fry et al. 2009).

At a resolution of 30 m, the completed NLCD 1992–2001 land-cover change retrofit product contains unchanged pixels from the NLCD 2001 land-cover data set that have been crosswalked to a modified Anderson level I class code and changed pixels labeled with a "from-to" class code. Figure 21.3 presents the NLCD 1992–2001 land-cover change retrofit product for the Seattle and Sioux Falls areas. The displayed extents in Figure 21.3 are the same as in Figure 21.1. The change class represents the changes of land-cover type from 1992 to 2001 within the Anderson level I class scheme because of the inconsistency of the two products in the level II scheme.

Analysis of the results for the conterminous United States indicated that about 3% of the land-cover data set changed between 1992 and 2001. Variations among the forest, grass and shrub, and agriculture classes accounted for the majority of mapped land-cover change (Fry et al. 2009).

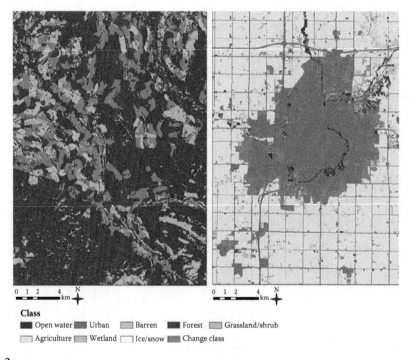

FIGURE 21.3
(See color insert following page 426.) The NLCD 1992–2001 retrofit land cover changes for the Seattle and Sioux Falls areas. The change class represents changes in land-cover type from 1992 to 2001.

21.3 NLCD 2006 and Beyond

21.3.1 NLCD 2006

The NLCD 1992 and 2001 products have been released based on a 10-year cycle. NLCD has been widely recognized as an important data source to quantitatively describe the terrestrial ecosystem conditions and support landscape change research. With these national data layers, there has been nearly a 5-year time lag between the image capture date and product release. However, in some areas, terrestrial ecosystems often experience significant natural and anthropogenic disturbances during production time, resulting in products that may be out of date. To keep the NLCD as temporally relevant as possible, a new frequency of 5 years was introduced for a more timely capture of land-cover change. NLCD 2006 represents the first database conceived to address the needs of the user community for more frequent land-cover monitoring and provide a reduction of production time between image capture and product release. NLCD 2006 was designed to provide updated land-cover, ISA, and TC data and additional information that can be used to identify the pattern, nature, and magnitude of changes that occurred between 2001 and 2006.

To achieve this goal in a cost-effective manner, the selected approach seeks to identify the areas of landscape change occurring after 2001 and to update land-cover data only for those changed areas. For the areas that have not changed, the NLCD 2006 products would remain the same. One advantage in monitoring land-cover change with remotely sensed data is that temporal sequences of images can accurately indicate spectral changes based on variations in the surface physical condition, assuming that digital values are radiometrically consistent for all scenes (Cakir, Khorram, and Nelson 2006). Multitemporal remotely sensed data can be used as primary sources to identify such spectral changes and extrapolate land-cover types for updating land-cover classification and the continuous variables related to LCLU change over a large geographic region.

A prototype method was developed by the USGS for updating NLCD 2006 (Xian, Homer, and Fry 2009). Here, we illustrate the prototype method used to update NLCD 2001 products, including land cover, ISA, and TC, to circa 2006. The method involves the use of change vector (CV) analysis to identify changed pixels, coupled with decision tree classification (DTC) and regression tree modeling (RTM) trained by using NLCD 2001 data designated in unchanged pixels, to label changed pixels in the updated 2006 land cover and to calculate updated continuous variables in the ISA and TC. This prototype method has been implemented for NLCD 2006 production.

The NLCD 2006 method depends upon two dates of Landsat imagery in 2001 and 2006 acquired in the same season by satellite paths and rows. This method focuses on radiometric spectral change detection from satellite images and change identification for all potential land-cover changes. Although the method is designed for using two-date image pairs for change detection, multiple-date image pairs can also be used to improve the change detection performance if images are available for different seasons. Figure 21.4 illustrates the overall procedures for obtaining NLCD 2006 land cover, ISA, and TC products. In this method, the land-cover product must be produced prior to the imperviousness and TC products.

21.3.1.1 *Landsat Imagery Selection and Processing*

To satisfy the preprocessing requirement for change detection, multitemporal image geometric precision, radiometric, and atmospheric corrections are accomplished first.

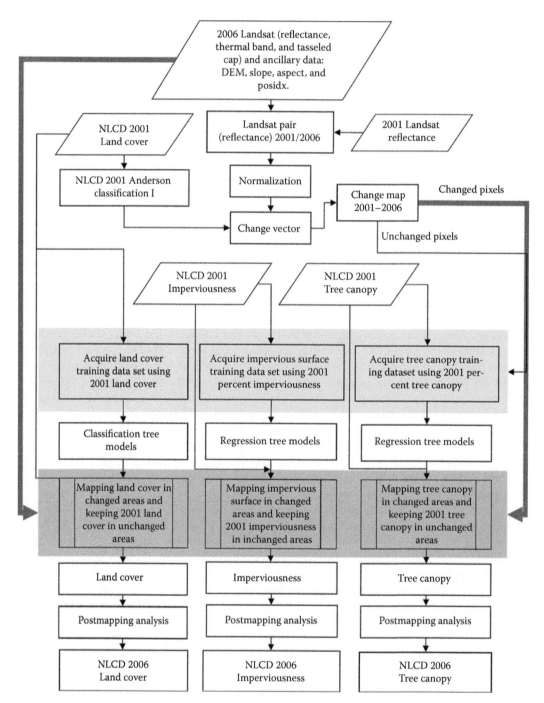

FIGURE 21.4
Flow chart of NLCD 2006 prototyping method for updating land cover, impervious surface, and tree canopy products.

The atmospheric correction converts a DN to the top-of-atmosphere reflectance using the same procedures as for NLCD 2001 (Homer et al. 2004). Images are then normalized using a linear regression normalization algorithm to reveal changes in the surface reflectance from multidate satellite images. The normalized image is used to reduce false changes caused by seasonal phenology and atmospheric effects from two-date images. Accordingly, scene pairs are normalized using the following linear regression formula:

$$s_i = a_i x_i + b_i \tag{21.1}$$

where x_i is the DN of band i in the image that is to be normalized (subject image), s_i is the normalized DN of band i in the subject image, a_i is the slope or gain, and b_i is the intercept or offset accounting for the difference in the mean and variance between radiance values in different dates. The transformation coefficients a_i and b_i are computed from a linear regression carried out on reference and subject images, in which clouds and shadows are excluded for the whole scene.

21.3.1.2 Change Vector Analysis and Land-Cover Change Detection

The normalized and reference images were used to calculate a CV image. A CV represents the spectral feature differences that may represent changes in LCLU types between two dates. If a pixel's values in two images on dates t_1 and t_2 are represented by a reference vector **(R)** and a subject vector **(S)**, which are expressed as

$$\mathbf{R} = (r_1, r_2, \ldots r_n)^T, \quad \mathbf{S} = (s_1, s_2, \ldots s_n)^T \tag{21.2}$$

where n is the number of bands, then the change magnitude is calculated with the following equation:

$$\|\Delta \mathbf{V}\| = [(r_1 - s_1)^2 + (r_2 - s_2)^2 + \cdots + (r_n - s_n)^2]^{1/2} \tag{21.3}$$

Generally, a greater value of a CV indicates a higher possibility of real change in the land-cover type, and a specific threshold is usually used to determine pixels of change or no-change. The normalized process reduces the variance range for pixels in unchanged areas.

A multithreshold approach was developed to determine the validity of pixels of change or no-change by calculating a specific threshold for each land-cover type. To simplify the class legend during threshold calculations, the NLCD 2001 product was recategorized to eight land-cover classes according to the Anderson level I classification. After all pixels were labeled as change or no-change, they were combined to create a binary mask file for the construction of training data sets from NLCD 2001 baseline products (Xian, Homer, and Fry 2009). One advantage of using CV to identify the changed or unchanged areas between two dates is that both potential changes among the categories and within a category are incorporated in changed areas.

21.3.1.3 Land-Cover Classification

The land cover is classified using similar procedures as the NLCD 2001 land-cover classification. The overall approach of the strategy includes three key procedures: training data construction, decision tree model creation, and land-cover classification for each Landsat path and row. The training samples are selected from the portion of the NLCD 2001 land-cover product that is unchanged. Samples of each land-cover class are held proportional to the total number of pixels in that class to ensure that adequate pixels of each land-cover class are included in the training data set for a path and row. These samples are then used to create an environment file that serves as the dependent variable for the decision tree model. Independent variables used as input to the model are the 2006 Landsat reflectance (bands 1–5 and 7), the thermal band, the tasseled cap derivative, and digital elevation data. Separate and unique classification models are developed for each path and row. A set of decision rules is produced to determine the class of a target variable (land-cover type) based on this training data set. Each rule set defines the conditions under which decisions are established for labeling land-cover types. Land-cover classifications are processed only for identified changed pixels, with baseline land-cover type remaining static for unchanged pixels. The NLCD 2001 full-legend land-cover product is used as a baseline to construct training data sets for each area. Modeling results (usually having cross-validation errors retained within 4–10%) are used to update the land-cover status for changed pixels on NLCD 2006.

21.3.1.4 Impervious Surface Estimation

The NLCD 2006 percent impervious surface is estimated using similar procedures as the NLCD 2001 imperviousness calculation, which quantifies the spatial distribution of ISA as a continuous variable from 1% to 100%. The overall approach consists of three key procedures: training data collection, RTM, and imperviousness estimation. The training data sets are selected from the unchanged impervious surfaces of the NLCD 2001 baseline. The regression tree model is constructed using the training data sets as the dependent variable and the Landsat image and ancillary data sets as the independent variables. A set of rules is produced to predict a target variable (percent imperviousness) based on the training data set. Each rule set defines the condition under which a multivariate linear regression model is established for prediction. The linear model is a simplified equation to fit the training data covered by the rule. Models based on the regression tree algorithm provide a propositional logic representation of these conditions in the form of numbers of tree rules.

Imperviousness estimations are conducted only for identified changed pixels, with baseline ISA values remaining static for unchanged pixels. The NLCD 2006 land-cover product is used to create a mask file to remove false estimations in nonurban areas defined by land-cover classes that would not logically include urban development.

21.3.1.5 Tree Canopy Estimation

The overall approach to update the NLCD 2006 TC product is similar to that for the NLCD 2006 ISA product updating. The spatial distribution of TC is quantified as a continuous variable from 1% to 100%. The procedures include training data collection, RTM, and TC estimation. The training data sets are determined from unchanged tree coverage areas, and the regression tree model is constructed using the training data set as the dependent variable (percent TC) and the Landsat image and ancillary data sets as the independent variables. Multiple rule-based regression equations are produced for predicting the target variables (percent TC) based on the training data set. After TC is extrapolated to the entire satellite scene, the change mask file can then be implemented to ensure that updating is accomplished only for targeted changed areas, with baseline TC values remaining static for unchanged pixels. The NLCD 2006 land-cover product is used as a reference to remove false estimations in nontree areas determined by land-cover classes that would not logically include significant trees.

The TC updating strategy was also explored using coarser spatial resolution remotely sensed data to allow updating to be conducted over an even larger geographic region for further efficiency and lower cost. The 56-m resolution advanced wide field sensor (AWiFS) imagery, which has four reflectance bands and a larger spatial coverage, was prototyped for TC change estimations. Assessment of the TC training data sets obtained from Landsat versus TC training data sets obtained from AWiFS revealed that products produced from AWiFS were reasonably comparable to those obtained from Landsat imagery. Although a TC procedure for NLCD 2006 has been prototyped, there is currently no TC production underway for NLCD 2006.

21.3.1.6 Examples of the NLCD 2006 Updates

To demonstrate the updated results, land cover, ISA, and TC changes in same areas presented previously in the Seattle and Sioux Falls regions are presented in Figure 21.5. Both 2001 and 2006 Landsat images (Figure 21.5a and b), newly updated 2006 land cover (Figure 21.5c), and TC (Figure 21.5d) are displayed for the Seattle area. Figure 21.5e through h include 2001 and 2006 Landsat images, land cover, and ISA in 2006 for the Sioux Falls area. The land-cover changes observed in the Seattle region are due to forest harvest patterns of cutting and regrowth in various stages. Changes associated with forest disturbances including both regrowth and new forest-cutting were observed and classified. Many new urban developments occurred on agricultural land in the Sioux Falls area. The land-cover map shows the new urban land use emerging on the edge of the city. The new ISA map also captures the new growth in the region. Figure 21.5h shows that the ISA variations in the southeastern and northwestern parts of Sioux Falls are identified on the map. New growths of ISA are associated with residential housing developments in suburban areas and emerged at the edge of existing urban areas.

TC was first estimated using a 2006 Landsat image and the 2001 baseline TC data set. Figure 21.6a and c show TC in 2006 and changes that occurred between 2001 and 2006 in an area on the southeastern side of Seattle. Both TC and its changes are in 30-m resolution. Areas having current and previous forest-cutting activities showed relatively lower and higher TC coverage in 2006, respectively. Meanwhile, in the 2001 and 2006 change graphics (Figure 21.6b), areas that have TC decreases and increases are colored as red and green, respectively. The TC amounts were also estimated using a 2006 AWiFS image that has a larger spatial extent than a Landsat image (Figure 21.6c). TC was estimated using the AWiFS image in the overlap area covered by both images. The TC training and ancillary data sets were subsets and were rescaled to 56 m. The TC was estimated following the same procedures as used for Landsat except the input Landsat image was replaced by an AWiFS image. Figure 21.6d displays the TC change that occurred between the 2001 baseline and the new 2006 image in 56-m resolution. Our preliminary results show that the comparison agreement between the Landsat and AWiFS predictions (both portrayed in 56 m) is 94% for TC decreasing areas and 92% for increasing areas, respectively. Our preliminary conclusions are that the AWiFS imagery provides a larger, more efficient spatial coverage and can be used for updating percent TC without losing the primary change features of the forest land.

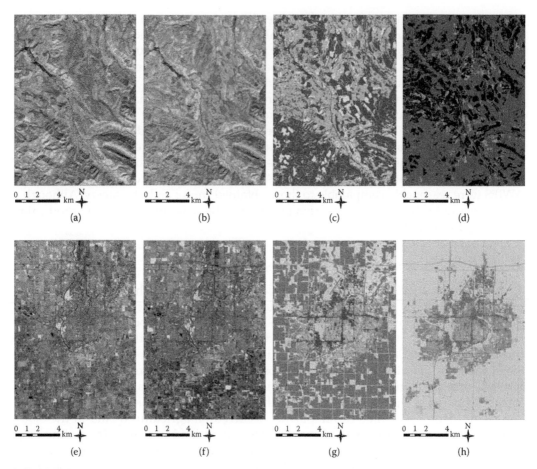

FIGURE 21.5
(See color insert following page 426.) Landsat imagery in (a) 2001 (b) 2006, (c) 2006 land cover, (d) 2006 percent tree canopy in the Seattle area. Landsat imagery in (e) 2001 (f) 2006, (g) 2006 land cover, and (h) 2006 impervious surface in the Sioux Falls area. The color legends for land cover, tree canopy, and impervious surface are the same as in Figure 21.2.

21.3.2 NLCD 2011

With just two eras of delivered land cover, and a third in production, the NLCD data provide fundamental support for a wide variety of federal and state requirements and management programs. However, as most federal and state environmental programs are periodic, they must produce updated information at specified intervals and so are dependent on an accurate portrayal of current land cover. Furthermore, because land-cover change is always occurring, long time lags between land-cover updates may result in unacceptable accuracies for their applications. Hence, the NLCD has, by necessity, evolved toward an operational land-cover monitoring program that provides land cover and land-cover change products on 5-year intervals. The NLCD 2011 is the next-generation proposed design to ensure future NLCD products, which support national needs. Key to the design will be better integration of multiple scales (250 m, 30 m, and potentially 1–2 m), additional thematic richness, and full integration of a land-cover change paradigm.

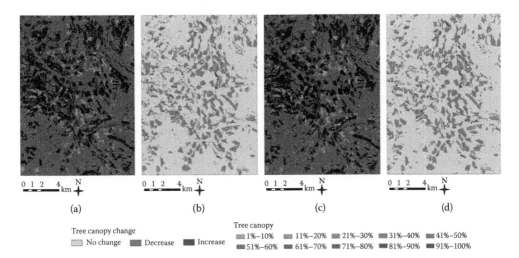

(a)   (b)   (c)   (d)

Tree canopy change   Tree canopy
☐ No change ■ Decrease ■ Increase   ■ 1%–10% ■ 11%–20% ■ 21%–30% ■ 31%–40% ■ 41%–50%
                                       ■ 51%–60% ■ 61%–70% ■ 71%–80% ■ 81%–90% ■ 91%–100%

FIGURE 21.6
(See color insert following page 426.) (a) Tree canopy in 2006 and (b) the tree canopy change from 2001 to 2006 in the Seattle area estimated by using Landsat image. (c) Tree canopy estimated using 2006 AWiFS image. (d) The tree canopy change from 2001 to 2006 in 56-m resolution for the same area. In both (b) and (d), red and green represent tree canopy decrease and increase from 2001 to 2006, respectively.

21.3.2.1 Principles of NLCD 2011

According to the objectives of the NLCD 2011 planned initiative, a set of guiding principles have been established to focus the development of NLCD 2011 products, including the following: (1) the NLCD 2011 should be a multiscale and multitemporal database capable of monitoring the nation's land cover and land-cover change; (2) the products should be flexible enough with acceptable accuracy to support a variety of MRLC requirements; (3) the products must be well defined, not currently under production by other entities, and accommodate maximum leveraging of existing relevant data or products that are already in production; (4) the NLCD 2011 will be centered around medium spatial resolution (30-m) data as the backbone of the database and will be complemented by those of a coarser (250-m) and finer (1–5-m) resolutions; (5) methods and algorithms used for modeling the attributes of the NLCD 2011 should be based on scientifically sound concepts and should be quantifiable, scalable, and repeatable; (6) the products will require identical protocols across the country, spanning spectral sources, preprocessing, classification, models, algorithms, synthesis, and quality assessment; (7) a team-oriented research and development process will be used to prototype an initial design and to ensure that the best science is used in product generation; and (8) the products will be developed to ensure full MRLC support for subsequent operational production.

21.3.2.2 Overall Design Considerations of NLCD 2011

The development of the NLCD 2011 is planned to be based on an integrated framework that encompasses LCLU change products at various thematic, spatial, and temporal resolutions. The main deliverables will include land-cover products of 30 m every 5 years, 250 m annually, and 1–5 m as needed in selected locations. The temporal and spatial resolutions of these products will be designed to be seamlessly integrated to enable the investigation, confirmation, calibration, and assessment of a wide variety of land-cover change issues across a broad spectrum of user-defined scales. The overall design should yield products that enable users to identify a variety of land-cover changes, including those caused by intense, local events (e.g., fire, urban development, drought, and flooding), and others caused by gradual, broad-based events (e.g., climate change, disease, succession, urban intensification, and invasive species). Since the NLCD 2011 is envisioned as a multiresolution, multiattribute, and multisource database that is intended for many applications at both national and local scales, the following six key components are proposed and illustrated in Figure 21.7: (1) a comprehensive reference database for NLCD mapping, modeling, and product evaluation, including data sets available from national and regional land and vegetation monitoring and mapping programs or projects, reference data derived from the high spatial resolution remote sensing data, and ancillary geospatial data sets; (2) a well-calibrated remote sensing database including Moderate Resolution Imaging Spectroradiometer (MODIS) 250-m imagery, Landsat 30-m (or Landsat-like) imagery and 15-m imagery at selected locations; (3) land-cover classes, estimates of TC height, percent ISA, bare ground, TC, and shrub canopy at 30 m; (4) land-cover classes, estimates of percent ISA, bare ground, TC, and shrub canopy at 250 m annually; (5) changes of spectral, land cover, ISA, bare ground, and canopy density at 250 m every 2 years and those at 30 m

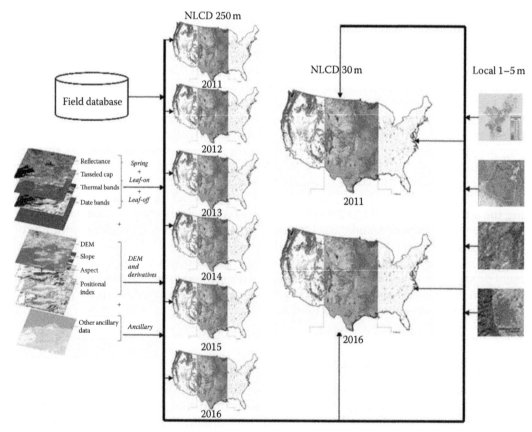

FIGURE 21.7
A potential product framework proposed for NLCD 2011.

every 5 years; and (6) land cover, imperviousness, bare ground, tree canopy, and shrub canopy estimates at 1–5 m resolution for selected sample sites.

21.4 Conclusions

The U.S. NLCD products have been developed over two decades and have allowed unprecedented land-cover analysis to occur in thousands of applications across federal, state, regional, and private entities. Continuation of the MRLC partnership, evolution of more abundant data streams, new implementation strategies, and improved algorithms have all been important contributions to NLCD development. The NLCD is now evolving beyond simple land-cover characterization, with new emphasis on capturing land-cover change between eras. The NLCD 1992–2001 retrofit land-cover change product represents the first effort to harmonize two eras of NLCD to discover the amount of land-cover change, but technically could only be accomplished on Anderson level I classes. The NLCD 2006 represents the first suite of NLCD products to integrate the land-cover change assessment as part of the characterization update process on all NLCD classes. The NLCD 2006 provides a cost-effective and faster way to enable more frequent land-cover product updates by using semiautomated data-processing protocols to reduce the cost, improve production efficiencies, and increase objectivity. The NLCD 2006 is now in operational production for the conterminous United States, with a completion date of early 2011.

With the advent of the NLCD 2011, now in the planning stage, and free of the Landsat data costs that hindered our 1992 and 2001 efforts, we expect that the NLCD product suite will have fully matured to better serve the nation's land-cover monitoring needs. The NLCD 2011 design, shown in Figure 21.7, will enable a better integration of multiple scales, add additional thematic richness, and represent full implementation of a land-cover change paradigm. Continuing the evolution of the NLCD within the MRLC umbrella will also ensure that future products are not only actually produced, but remain application relevant.

Acknowledgments

The authors thank Drs. Darrell Napton from the South Dakota State University and James Vogelmann from the USGS EROS Center for their constructive suggestions and comments. George Xian's work is performed under USGS contract 08HQCN0007.

References

Anderson, J. R., E. E. Hardy, J. T. Roach, and R. E. Witmer. 1976. A land use and land cover classification system for use with remote sensor data. U.S. Geological Survey Professional Paper 964, 28.
Cakir, H. I., S. Khorram, and S. A. C. Nelson. 2006. Correspondence analysis for detecting land cover change. *Remote Sens Environ* 102:306–17.

Changnon, S. A. 2003. Urban modification of freezing-rain events. *J Appl Meteorol* 42:863–70.

Coppin, P., I. Jonckheere, K. Nackerts, B. Muys. 2004. Digital change detection methods in ecosystem monitoring: A review. *Int J Remote Sens* 10:1565–96.

DeFries, R., C. B. Field, and I. Fung. 1995. Mapping the land surface for global atmosphere–biosphere models: Towards continuous distributions of vegetation's functional properties. *J Geophys Res* 100:867–920.

Findell, K. L., E. Shevliakova, P. C. D. Milly, and R. J. Stouffer. 2007. Modeled impact of anthropogenic land cover change on climate. *J Climate* 20:3621–34.

Fry, J. A., M. J. Coan, C. G. Homer, D. K. Meyer, and J. D. Wickham. 2009. Completion of the national land cover database (NLCD) 1992–2001 land cover change retrofit product. U.S. Geological Survey Open-File Report 2008-1379, 18.

Ge, J., Qi, J., B. M. Lofgren, N. Moore, N. Torbick, and J. Olson. 2007. Impacts of land use/cover classification accuracy on regional simulation. *J Geophys Res* 112: D05107, doi: 10.1029/2006JD007404.

Georgescu, M., L. T. Steyaert, C. P. Weaver, and G. Miguez-Macho. 2009. Climatic effects of 30 years of landscape change over the Greater Phoenix, Arizona, region: 1. Surface energy budget changes. *J Geophys Res D: Atmospheres* 114: D05110, doi:10.1029/2008JD010762.

Greenfield, E. J., D. J. Nowak, and J. T. Walton, 2009. Assessment of 2001 NLCD percent tree and impervious cover estimates. *Photogramm Eng Remote Sens* 75:1279–86.

Homer, C., J. Dewitz, J. Fry, M. Coan, N. Hossain, C. Larson, N. Herold, A. McKerrow, J. N. VanDriel, and J. Wickham. 2007. Completion of the 2001 National Land Cover Database for the conterminous United States. *Photogramm Eng Remote Sens* 73:337–41.

Homer, C., C. Huang, L. Yang, B. Wylie, and M. Coan. 2004. Development of a 2001 national land cover database for the United States. *Photogramm Eng Remote Sens* 70:829–40.

Huang, C., B. Wylie, C. Homer, L. Yang, and G. Zylstra. 2002. Derivation of a tasseled cap transformation based on Landsat 7 at-satellite reflectance. *Int J Remote Sens* 23:1741–8.

Li, S. G., Y. Harazono, H. L. Zhao, Z. Y. He, and X. Y. Chang. 2000. Grassland desertification by grazing and the resulting micrometeorological changes in Inner Monogolia. *Agric For Meteorol* 102:125–37.

Loveland, T. R., and D. M. Shaw. 1996. Multi-resolution land characterization—building collaborative partnerships. In *GAP Analysis—A Landscape Approach to Biodiversity Planning*, ed. J. M. Scott, T. H. Tear, and F. W. Davis, 79–85. Bethesda, MD: American Society for Photogrammetry and Remote Sensing.

Loveland, T. R., T. L. Sohl, S. L. V. Stehman, A. L. Gallant, K. L. Sayler, and D. E. Napton. 2002. A strategy for estimating the rates of recent United States land-cover changes. *Photogramm Eng Remote Sens* 68:1091–9.

Lunetta, R. S., J. F. Knight, J. Ediriwickrema, J. G. Lyon, and L. D. Worthy. 2006. Land-cover change detection using multi-temporal MODIS NDVI data. *Remote Sens Environ* 105:142–154.

Marshall, C. H., R. A. Pielke, and L. T. Steyaert. 2003. Wetlands: Crop freezes and land-use change in Florida. *Nature* 426:29–30.

Pielke, R. A. Sr. 2001. Influence of the spatial distribution of the spatial distribution of vegetation and soils on the prediction of cumulus convective rainfall. *Rev Geophys* 39:151–77.

Rozoff, C. M., W. R. Cotton, and J. O. Adegoke. 2003. Simulation of St. Louis, Missouri, land use impacts on thunderstorms. *J Appl Meteorol* 42:716–38.

Seneviratne, S., D. Lüthi, M. Litschi, and C. Schär. 2006. Land-atmosphere coupling and climate change in Europe. *Nature* 443:205–9.

Seto, K. C., and J. M. Shepherd. 2009. Global urban land-use trends and climate impacts. *Curr Opin Environ Sustain* 1:89–95.

Stehman, S., J. Wickham, J. Smith, and L. Yang. 2003. Thematic accuracy of the 1992 National Land-Cover Data for the eastern United States: Statistical methodology and regional results. *Remote Sens Environ* 86:500–516.

Stone, D. A., and A. J. Weaver. 2003. Factors contributing to diurnal temperature range trends in twentieth and twenty-first century simulations of the CCCma coupled model. *Climate Dynamics* 20:435–45.

Vogelmann, J. E., S. M. Howard, L. Yang, C. R. Larson, B. K. Wylie, and N. Van Driel. 2001. Completion of the 1990's National Land Cover Data Set for the conterminous United States from Landsat Thematic Mapper Data and ancillary data sources. *Photogramm Eng Remote Sens* 67:650–62.

Vogelmann, J. E., T. Sohl, and S. M. Howard. 1998. Regional characterization of land cover using multiple sources of data. *Photogramm Eng Remote Sens* 64:45–57.

White, M. A., R. R. Nemani, P. E. Thornton, and S. W. Running. 2002. Satellite evidence of phonological differences between urbanized and rural areas of the eastern United States deciduous broadleaf forest. *Ecosystems* 5:260–77.

Wickham, J. D., S. V. Stehman, J. A. Fry, J. H. Smith, and C. G. Homer. 2010. Thematic accuracy of the NLCD 2001 land cover for the conterminous United States. *Remote Sens Environ* 113:1236–49.

Wickham, J. D., S. V. Stehman, J. H. Smith, and L. Yang. 2004. Thematic accuracy of the 1992 national land-cover data for the western united states. *Remote Sens Environ* 91:452–68.

Xian, G., and M. Crane. 2005. Assessments of urban growth in the Tampa Bay watershed using remote sensing data. *Remote Sens Environ* 97:203–15.

Xian, G., C. Homer, and J. Fry. 2009. Updating the 2001 National Land Cover Database land cover classification to 2006 by using Landsat imagery and change detection methods. *Remote Sens Environ* 113:1133–47.

Yang, L., C. Huang, C. Homer, B. Wylie, and M. Coan. 2003. An approach for mapping large-area impervious surfaces: Synergistic use of Landsat 7 ETM+ and high spatial resolution imagery. *Can J Remote Sens* 29:230–40.

Yang, L., S. Stehman, J. Smith, and J. Wickham. 2001. Thematic accuracy of MRLC land cover for the eastern United States. *Remote Sens Environ* 76:418–22.

Zhu, Z., L. Yang, S. V. Stehman, and R. L. Czaplewski. 2001. Accuracy assessment for the U.S. geological survey regional land-cover mapping program: new york and new jersey region. *Photogramm Eng Remote Sens* 66:1425–35.

Index

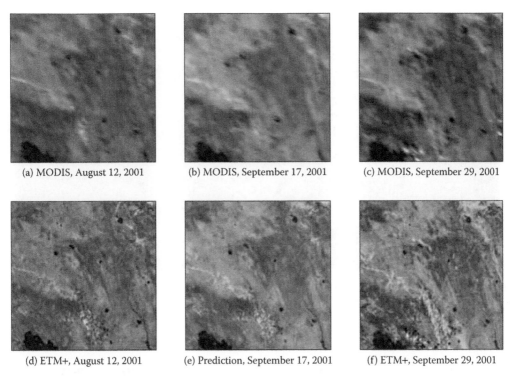

(a) MODIS, August 12, 2001 (b) MODIS, September 17, 2001 (c) MODIS, September 29, 2001

(d) ETM+, August 12, 2001 (e) Prediction, September 17, 2001 (f) ETM+, September 29, 2001

FIGURE 1.8
Predicted Landsat surface reflectance (e) using STARFM from daily Moderate Resolution Imaging Spectrad iometer (MODIS) reflectance imagery (b) and Landsat/MODIS image pairs (a and d, c and f). (Reprinted from Gao, F., Masek, J., Schwaller, M., and Forrest, H., On the blending of the Landsat and MODIS surface reflectance: Predicting daily Landsat surface reflectance, *IEEE Trans Geosci Remote Sens* 44(8):2207–18. © (2006) IEEE.)

FIGURE 3.10
Lidar point cloud over coniferous forests in the western United States.

FIGURE 3.11
Ground-based lidar data collected over Mesquite trees in central Texas.

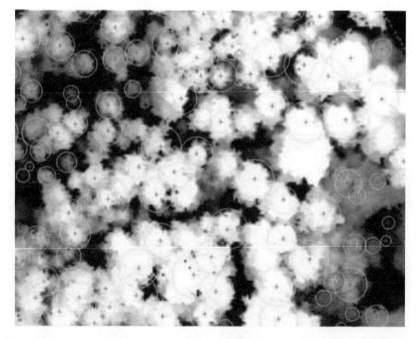

FIGURE 3.12
Automatically measuring individual trees on a lidar-derived canopy height model. Circles represent computer-measured crown diameters, whereas each cross sign indicates identified individual trees.

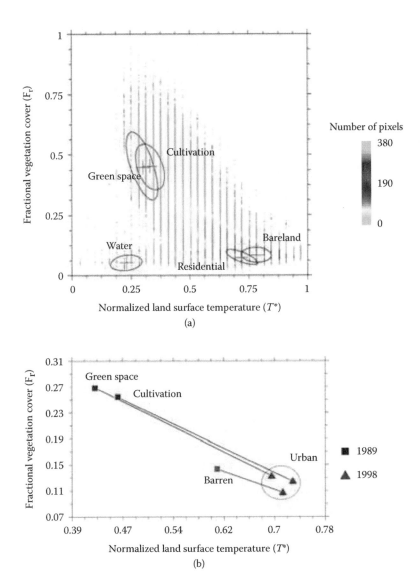

FIGURE 6.1
Fractional vegetation cover (Fr)/T^* scatter plot (thermal-vegetation index [TVX] space) with sample land use and land-cover (LULC) classes from a Landsat thematic mapper (TM) image of the city of Tabriz and change trajectory in the TVX space for a specific period: (a) The scatter plot with sample LULC classes from a Landsat TM image of Tabriz (38°05′, 46°17′) in northwestern Iran, which was acquired on August 18, 1998; (b) change trajectory in the TVX space for a long (1989–1998) period (June 30, 1989–August 18, 1998). The vectors show the magnitude of change associated with LULC change from green space, cultivation, and barren pixels to urbanized pixels. (From Amiri, R., Q. Weng, A. Alimohammadi, and S. K. Alavipanah, *Remote Sens Environ*, 113, 12, 2009. With permission.)

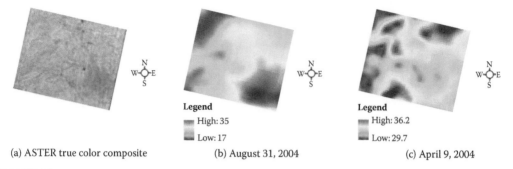

| (a) ASTER true color composite | (b) August 31, 2004 | (c) April 9, 2004 |

Legend (b): High: 35, Low: 17

Legend (c): High: 36.2, Low: 29.7

FIGURE 6.2
The results of kernel convolution for two advanced spaceborne thermal emission and reflection radiometer (ASTER) images of Beijing: (a) A true color composite of Beijing using ASTER acquired on August 31, 2004; (b) and (c) the results of convoluted images (with a smoothing parameter of 0.6) showing thermal landscape pattern of Beijing on August 31, 2004 and April 9, 2004, respectively. The temperature is given in degrees Celsius.

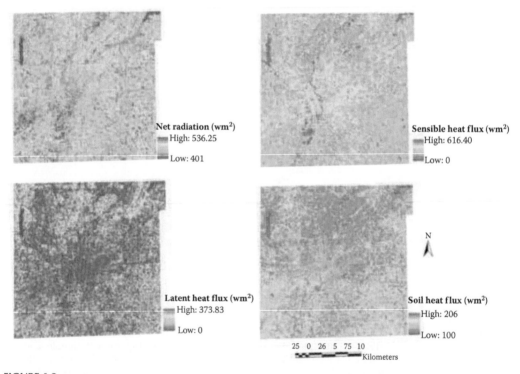

Net radiation (wm^2): High: 536.25, Low: 401

Sensible heat flux (wm^2): High: 616.40, Low: 0

Latent heat flux (wm^2): High: 373.83, Low: 0

Soil heat flux (wm^2): High: 206, Low: 100

25 0 26 5 75 10 Kilometers

FIGURE 6.3
Net radiation, sensible heat flux, latent heat flux, and soil heat flux on October 13, 2006 in Indianapolis estimated by the combined use of advanced spaceborne thermal emission and reflection radiometer image and ground meteorological data.

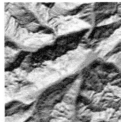

FIGURE 7.5
Example of a combined atmospheric and topographic correction of a SPOT-5 scene from a part of the Swiss Alps. Left to right: Original SPOT-5 scene (color coding for red, green, and blue bands is 1650, 840, and 660 nm, respectively), illumination map, and combined atmospheric and topographic correction. (From Richter, R. et al., *Rem Sens*, 1, 2009. With permission.)

FIGURE 7.7
Deshadowing of SPOT-5 imagery (dated May 22, 2005; color coding of red, green, and blue bands is 830, 660, and 555 nm, respectively). Left and right: Original and deshadowed scene, respectively.

FIGURE 8.1
Geometry of viewing of a satellite scanner in orbit around the Earth. (Courtesy and copyright Serge Riazanoff, VisioTerra, 2009.)

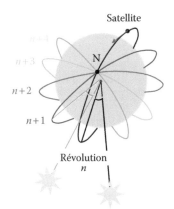

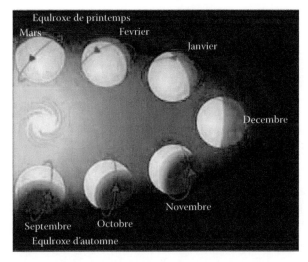

FIGURE 8.2
Near-Earth, quasi-circular, quasi-polar, sun-synchronous orbit for EO satellites. The different revolutions around the poles with a constant illumination angle (top) showing the same illumination condition all the year (bottom). (Courtesy and copyright Serge Riazanoff, VisioTerra, 2009.)

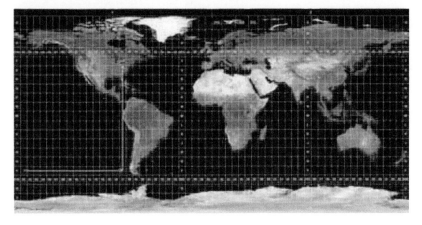

FIGURE 8.7
Example of a cylindrical conformal projection: the universal transverse mercator projection with its 3° longitudinal (1–60) zones and latitudinal (A–Z) zones.

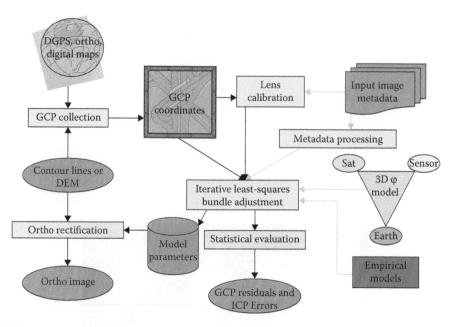

FIGURE 8.9
Description of the geometric correction method and its processing steps. The ellipse symbols denote input/ output data and the box symbols denote processes.

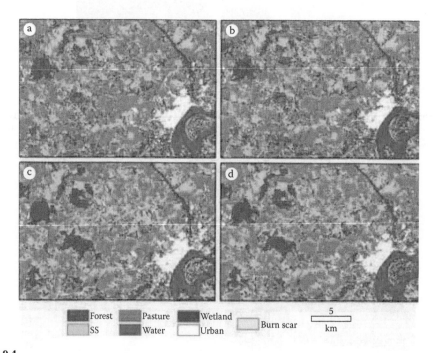

	Forest		Pasture		Wetland		Burn scar
	SS		Water		Urban		

FIGURE 9.4
Comparison of classification results among different scenarios with the maximum likelihood classifier: (a) six Thematic Mapper spectral bands, (b) combination of spectral bands and two vegetation indices, (c) combination of spectral bands and two textural images, and (d) combination of spectral bands, two vegetation indices, and two textural images.

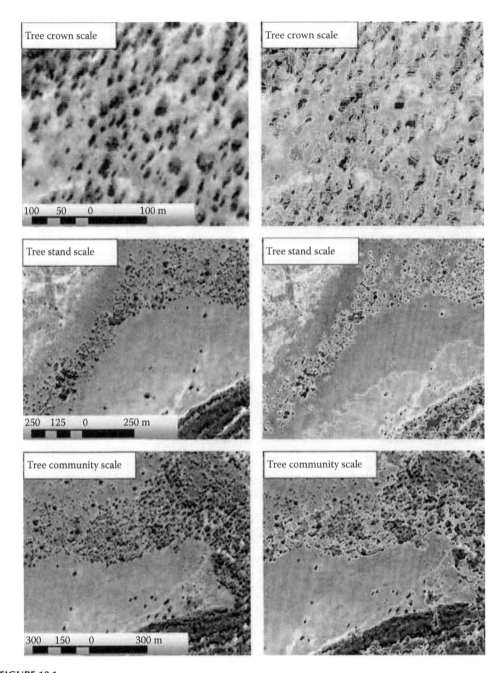

FIGURE 10.1
Different spatial scales of observations in high spatial resolution QuickBird image data shown as a false-color composite. The three spatial scales show individual tree crowns and stands and associated feature segmentation within an Australian tropical savanna landscape as well as a tree community segmentation level showing riparian vegetation, savanna woodlands, and rangelands. This figure illustrates the multiscale concept by creating multiple scales of segmentation through successive grouping of image pixels into homogeneous image objects, providing a more intuitive and hierarchical partitioning of the image results, which cannot effectively be achieved in per-pixel approaches.

FIGURE 10.2
Location and field photos of the Mimosa Creek savanna woodland riparian study site in central Queensland.

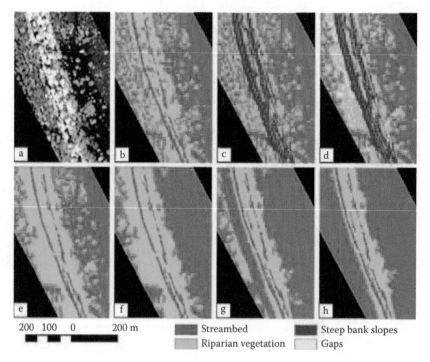

FIGURE 10.4
Object-based image analysis steps for mapping the extent of the riparian zone from light detection and ranging data ((a) through (h)).

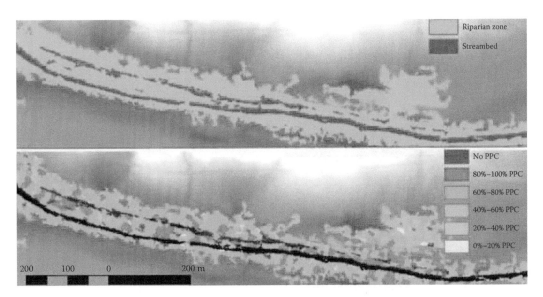

	Riparian zone
	Streambed

	No PPC
	80%–100% PPC
	60%–80% PPC
	40%–60% PPC
	20%–40% PPC
	0%–20% PPC

200 100 0 200 m

FIGURE 10.5
Mapped riparian zone extent and plant project cover overlaying the digital terrain model.

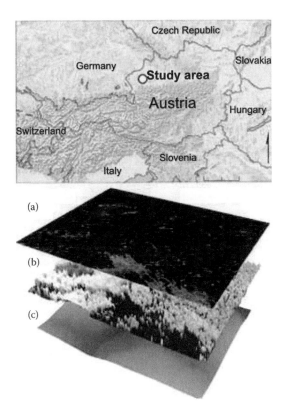

(a)

(b)

(c)

FIGURE 10.7
Location of the study area in the Austrian state of Upper Austria. Data sets used in the case study are
(a) UltraCamX digital infrared orthophotos, (b) normalized digital surface model derived from the UltraCamX
stereo imagery, and (c) an existing light detection and ranging–based digital elevation model.

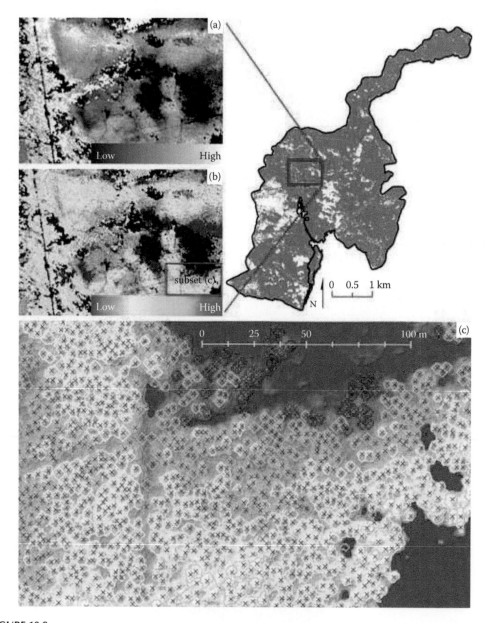

FIGURE 10.9
Results of individual tree crown extraction and delineation for the whole study area (right) and subsets showing (a) the normalized digital surface model, (b) the overlaid tree crowns, and (c) the tree crowns with the extracted local maxima (color coded according to the extracted height values).

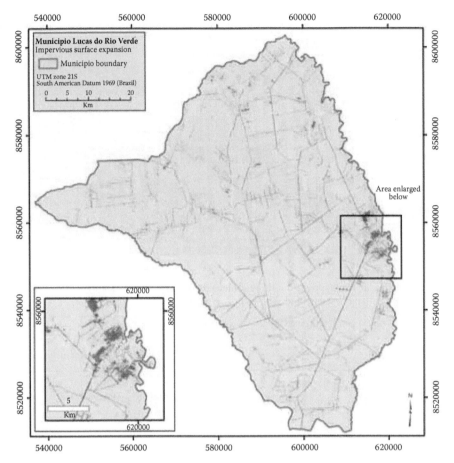

FIGURE 11.3
Color composite of three dates of impervious surface distribution in 2008, 1996, and 1984 by assigning them as red, green, and blue, illustrating the spatial distribution and patterns of impervious surface changes.

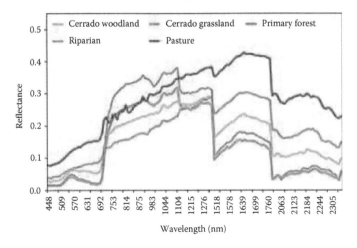

FIGURE 12.3
Canopy level spectral reflectance signatures measured by an EO-1 Hyperion sensor over Araguaia National Park, Brazil.

FIGURE 12.8
Example of airborne Lidar for mapping the three-dimensional properties of canopy surfaces (Adapted with permission from Macmillan Publishers Ltd., Tollefson, J., *Nature* 461:1048, copyright 2009.)

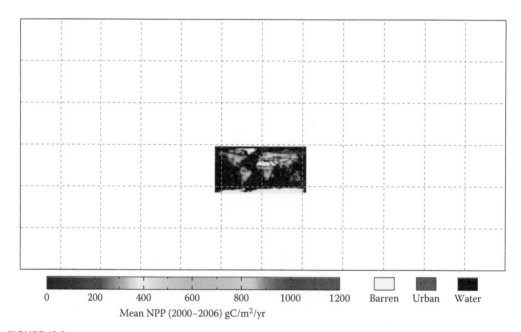

0   200   400   600   800   1000   1200   Barren   Urban   Water

Mean NPP (2000–2006) gC/m^2/yr

FIGURE 12.9
MODIS annual NPP global product averaged for the years 2000–2006 (From Running, S., Numerical Terradynamic Simulation Group. http://www.ntsg.edu. Accessed September 2010. With permission.)

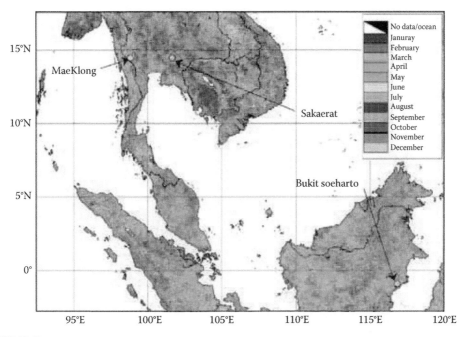

FIGURE 12.12
Maximum greenness date phenology image derived from MODIS over Southeast Asia. (From *Agric and Forest Meteorol*, 148, Huete, A.R. et al., Multiple site tower flux and remote sensing comparisons of tropical forest dynamics in Monsoon Asia, 748. Copyright (2008), with permission from Elsevier.)

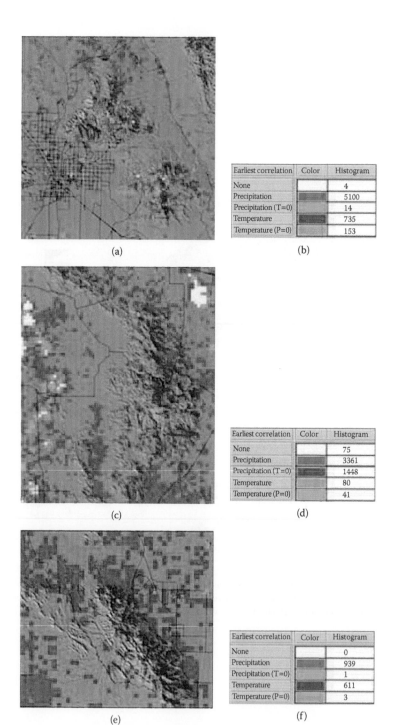

Earliest correlation	Color	Histogram
None		4
Precipitation		5100
Precipitation (T=0)		14
Temperature		735
Temperature (P=0)		153

(a)   (b)

Earliest correlation	Color	Histogram
None		75
Precipitation		3361
Precipitation (T=0)		1448
Temperature		80
Temperature (P=0)		41

(c)   (d)

Earliest correlation	Color	Histogram
None		0
Precipitation		939
Precipitation (T=0)		1
Temperature		611
Temperature (P=0)		3

(e)   (f)

FIGURE 13.3
(a;c;e) Earliest correlating climate drivers across the three sites. Precipitation is most often the earliest correlating climate driver in low elevations. (b;d;f) Color-coding in this figure indicates (1) pixels where FMSI did not correlate significantly to either driver, (2) pixels where FMSI correlated first to preceding winter precipitation, (3) pixels where FMSI correlated only to precipitation, (4) pixels where FMSI correlated first to temperatures and (5) pixels that correlated only to temperature.

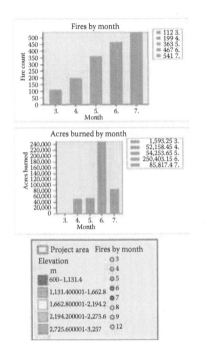

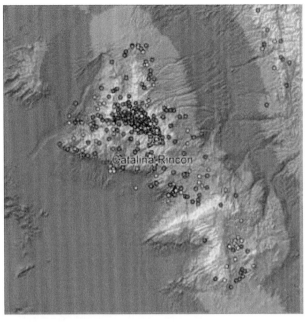

(a)

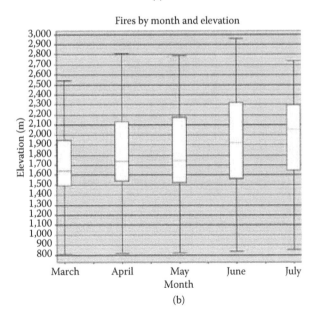

(b)

FIGURE 13.4

(a) Actual fires for the period of record covered by the FMSI. Earliest fires (e.g., March), occur at lower elevations in the Catalina-Rincons. The FMSI shows lower elevations remain vulnerable as the fire season "progresses" in elevation. By July, all elevations are vulnerable; this is evident in Figure 13.4b, which shows July fires span high and low elevations; and in Figure 13.4c, where the FMSI shows the early, sustained vulnerability of low-elevation fuels. (b) Box plot showing fire counts by elevation per month. The medians are the dotted lines within the box. Each box contains 50% of the values. The "whiskers" denote minimum and maximum values. Median fire counts tend to increase in elevation by month. (c) The length of fire season (LOFS) as determined by the FMSI, which shows live fuels cure later at higher elevations; fire season thus is later at higher elevations, and this is consistent with Figures 13.4a and 13.4b.

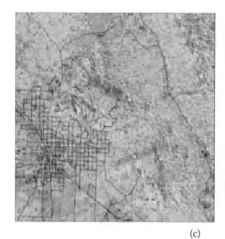

Color	LOFS(wks)	Histogram
	No corr	4
	2	189
	4	461
	6	513
	8	610
	10	566
	12	728
	14	988
	16	880
	18	683
	20	274
	22	110

(c)

FIGURE 13.4   (*Continued*)

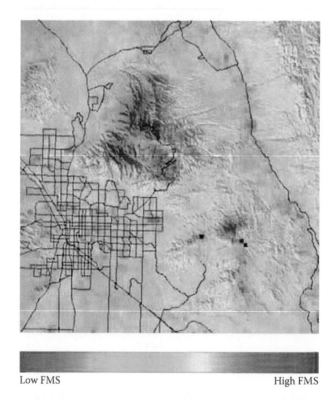

Low FMS   High FMS

FIGURE 13.6
Fire season summary map for the Catalina-Rincon site for the year 2004, clearly showing fire scars from the previous year's Aspen fire in the Catalinas and the smaller Helens II fire in the Rincons.

Fire season summaries in chronological sequence for the three southeastern Arizona study sites

The figure below shows 15–18 years of fire season fuel moisture stress in three sky island regions of southeastern Arizona. Each column represents a year and each row represents a site.

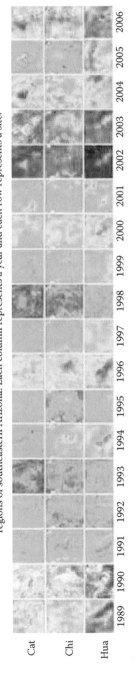

Notes:
– Because satellite imagery was not available for all periods spanning March through July, the following fire season summaries are approximations:
  – Catalina: 1990, 1999, and 2003
  – Chiricahua: 1990, 1991, and 1999
  – Huachuca: 1990
– White pixels indicate nondata.

FIGURE 13.7
Fire season summaries in chronological sequence.

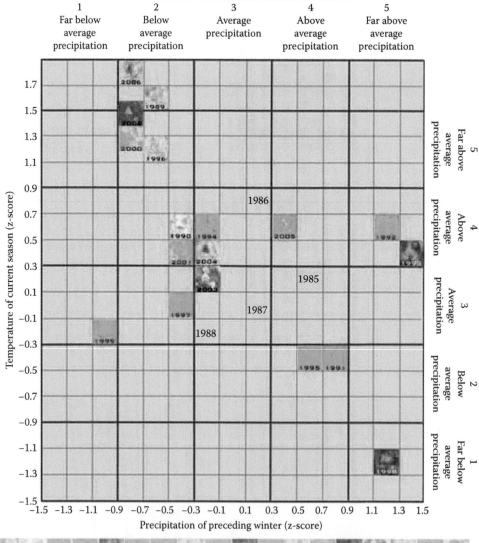

Catalina FMS/climate grid
updated 1/3/07

Fuel moisture stress (FMS): 1989 to 2006
Low FMS ■■■■■■■■■■■■■■■■■■■■ High FMS
Temperature climatology: 1959–2002
Precipitation climatology: 1985–2002

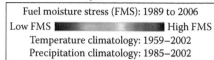

1989 1990 1991 1992 1993 1994 1995 1996 1997 1998 1999 2000 2001 2002 2003 2004 2005 2006

Note: – White pixels indicate nondata (large regions of 1990 and 2003 are nondata).
        – 1990, 1999, and 2003 show approximated fire seasons truncated out to gaps in satellite imagery.
        – 2000 and 2002 share the same grid cell.

FIGURE 13.8
The FMSI/climate grid for the Catalina-Rincon study site. Fire season summaries for 18 years are plotted using precipitation and temperature z-scores for each year.

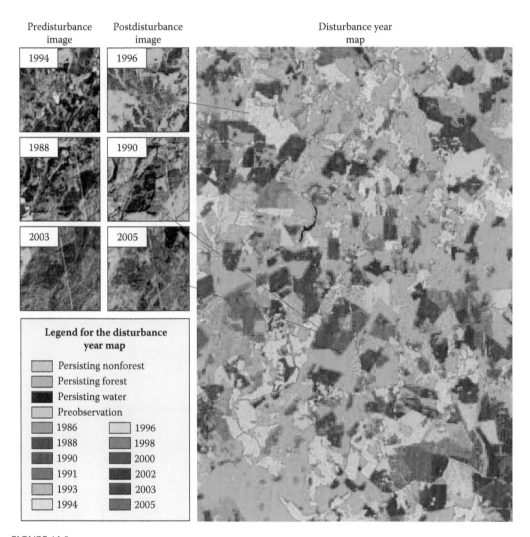

Predisturbance image   Postdisturbance image   Disturbance year map

1994   1996

1988   1990

2003   2005

Legend for the disturbance year map

- Persisting nonforest
- Persisting forest
- Persisting water
- Preobservation

	1986		1996
	1988		1998
	1990		2000
	1991		2002
	1993		2003
	1994		2005

FIGURE 14.9

Visual validation of three mapped disturbances using pre- and post-disturbance Landsat images. The disturbance year map was selected from a 17.1 × 11.4 km area in the Uwharrie national forest located in North Carolina
(WRS path 16/row 36). The size of each Landsat image chip shown to the left is 2.85 × 2.85 km. (From Huang, C.
et al. *Remote Sens Environ*, 113, 7, 2009. With permission.)

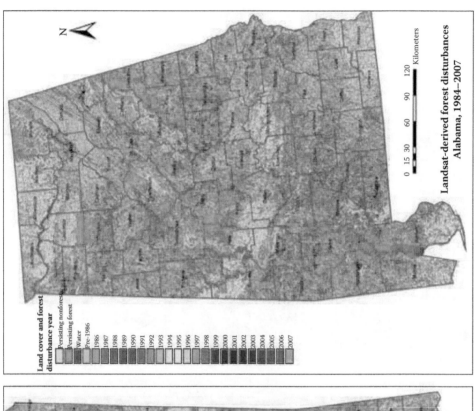

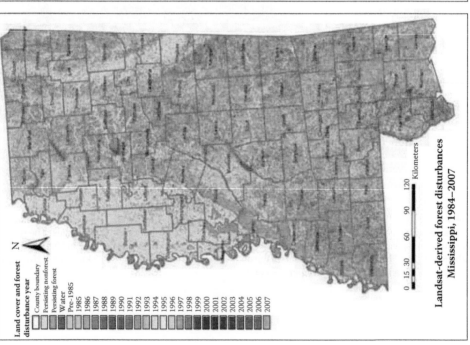

FIGURE 14.10

Disturbance year map derived using the LTSS-VCT approach for Mississippi (left) and Alabama (right).

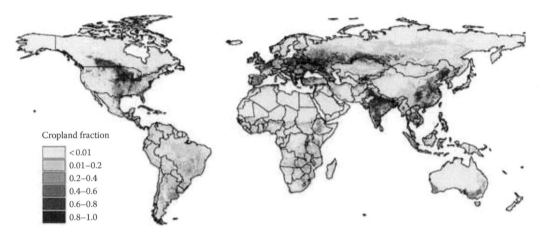

FIGURE 16.1
Global cropland map at nominal 5-minutes (0.083333 decimal degrees) resolution using national statistics and geospatial techniques for the nominal year 2000. Total area of croplands is 1.47 billion hectares. (Adopted from Ramankutty, N. et al. *Global Biogeochem Cycles*, 22, 2008.)

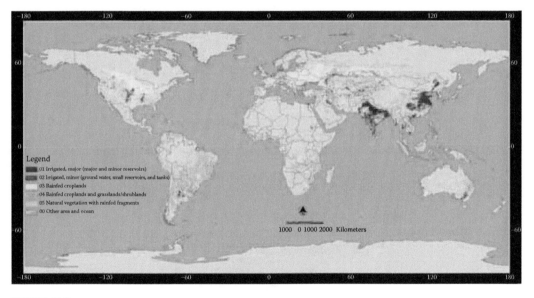

FIGURE 16.2
Global cropland map at nominal 1-km resolution using remote sensing for the nominal year 2000. Total cropland area was determined to be 1.53 billion hectares, of which 399 Mha was irrigated area. Because irrigated areas often had more than one crop per year, the total annualized irrigated area was 467 Mha. (Adapted from Thenkabail, P.S. et al. *Rem Sens*, 1, 2009b. http://www.mdpi.com/2072-4292/1/2/50; Thenkabail, P.S. et al. *Remote Sensing of Global Croplands for Food Security*, CRC Press/Taylor & Francs, Boca Raton, FL, 2009c.)

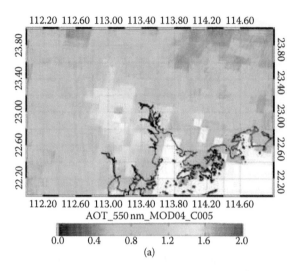

(a)

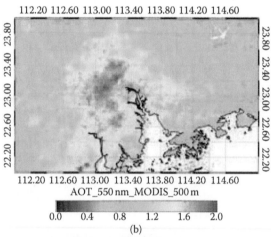

(b)

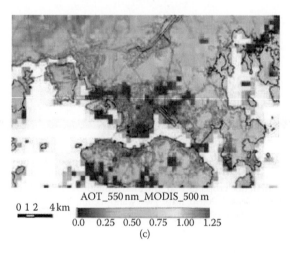

(c)

FIGURE 17.4
Aerosol optical thickness image at 550 nm, (a) derived from MODIS collection-5 algorithm, (b) derived from MODIS 500-m data, and (c) derived from MODIS 500-m data overlaid with road layer.

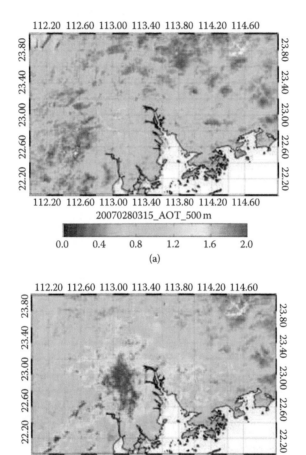

FIGURE 17.5
Aerosol optical thickness image at 550-nm and 500-m resolution over Hong Kong and the Pearl River Delta region on (a) January 28, 2007 and (b) January 30, 2007.

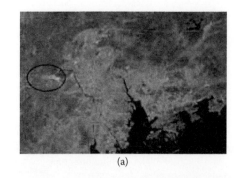

(a)

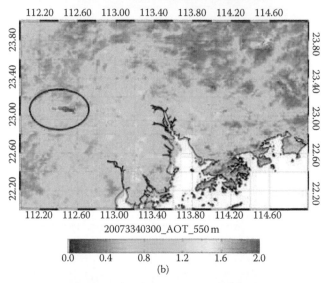

20073340300_AOT_550 m

(b)

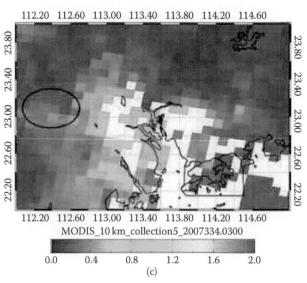

MODIS_10 km_collection5_2007334.0300

(c)

FIGURE 17.6
(a) Rayleigh-corrected RGB image on November 30, 2007, (b) aerosol optical thickness (AOT) image at 500-m resolution, and (c) AOT collection-5 image at 10-km resolution.

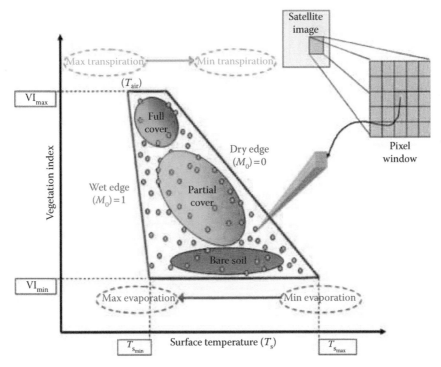

FIGURE 19.2
Summary of the main physical properties and interpretations of the satellite-derived (or airborne) T_s/VI feature space. Dots represent the measurements at pixels observed by a VIS/IR radiometer at various fractional vegetation covers (F_r) and surface temperatures (T_s). In this illustration, pixels classified as water or clouds are assumed to have been masked out. (Adapted from Petropoulos, G. et al. *Adv Phys Geogr,* 33, 2, 2009a.)

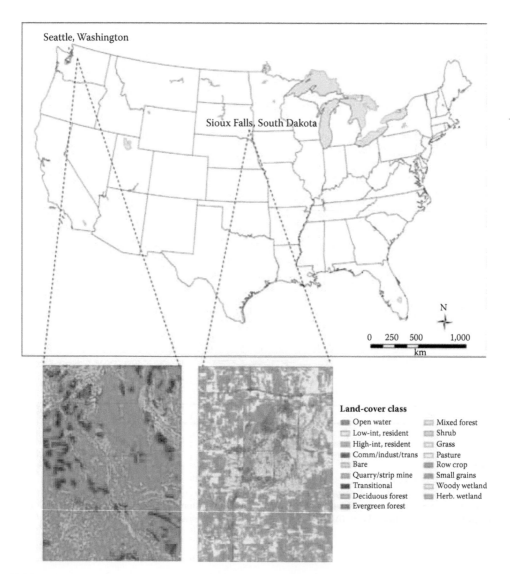

FIGURE 21.1
NLCD1992 land cover for the Seattle, Washington and Sioux Falls, South Dakota areas.

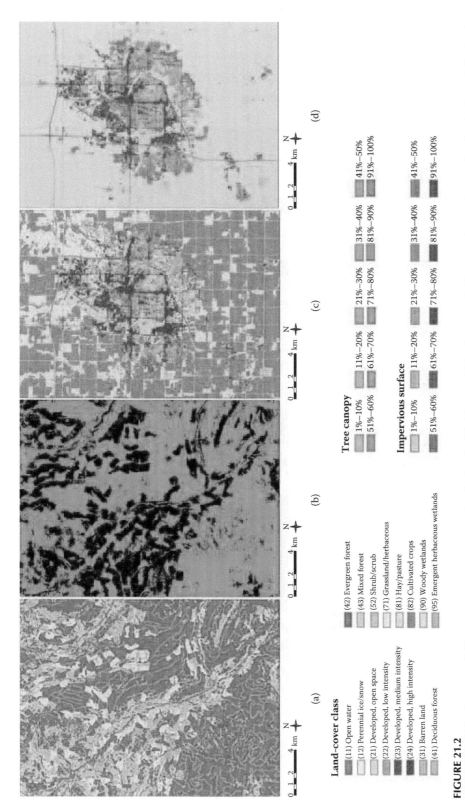

Land-cover class

(11) Open water
(12) Perennial ice/snow
(21) Developed, open space
(22) Developed, low intensity
(23) Developed, medium intensity
(24) Developed, high intensity
(31) Barren land
(41) Deciduous forest

(42) Evergreen forest
(43) Mixed forest
(52) Shrub/scrub
(71) Grassland/herbaceous
(81) Hay/pasture
(82) Cultivated crops
(90) Woody wetlands
(95) Emergent herbaceous wetlands

Tree canopy

1%–10%   11%–20%   21%–30%   31%–40%   41%–50%
51%–60%   61%–70%   71%–80%   81%–90%   91%–100%

Impervious surface

1%–10%   11%–20%   21%–30%   31%–40%   41%–50%
51%–60%   61%–70%   71%–80%   81%–90%   91%–100%

FIGURE 21.2

NLCD 2001 (a) land cover and (b) percent tree canopy cover for the Seattle, Washington; (c) land cover and (d) percent impervious surface for the Sioux Falls, South Dakota area.

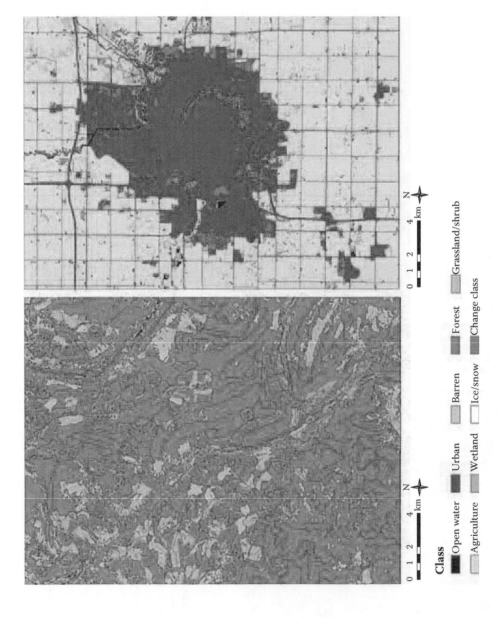

FIGURE 21.3

The NLCD 1992–2001 retrofit land cover changes for the Seattle and Sioux Falls areas. The change class represents changes in land-cover type from 1992 to 2001.

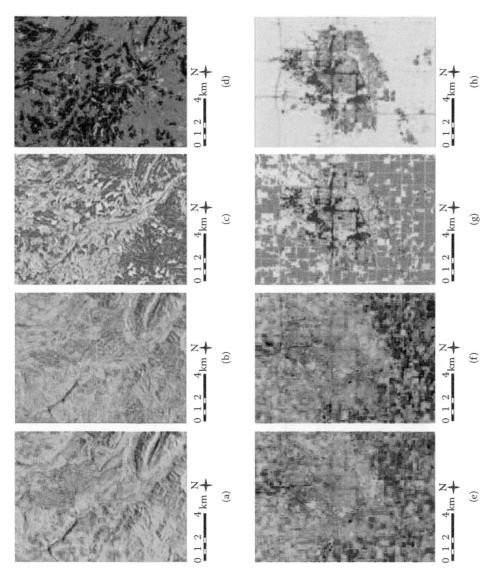

FIGURE 21.5

Landsat imagery in (a) 2001 (b) 2006, (c) 2006 land cover, (d) 2006 percent tree canopy in the Seattle area. Landsat imagery in (e) 2001 (f) 2006, (g) 2006 land cover, and (h) 2006 impervious surface in the Sioux Falls area. The color legends for land cover, tree canopy, and impervious surface are the same as in Figure 21.2.

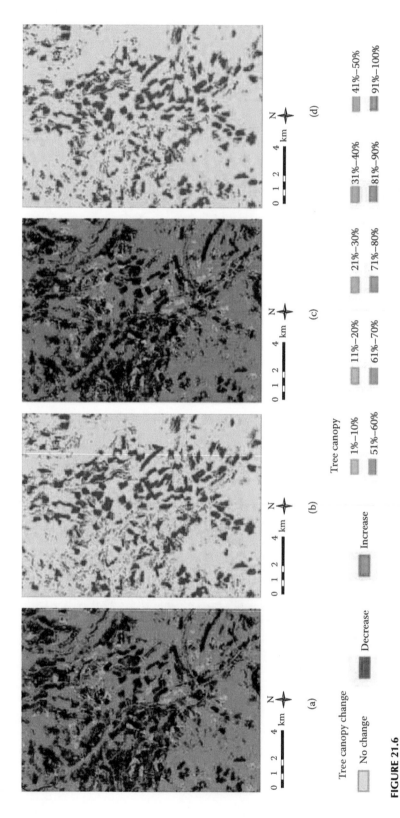

FIGURE 21.6

(a) Tree canopy in 2006 and (b) the tree canopy change from 2001 to 2006 in the Seattle area estimated by using Landsat image. (c) Tree canopy estimated using 2006 AWiFS image. (d) The tree canopy change from 2001 to 2006 in 56-m resolution for the same area. In both (b) and (d), red and green represent tree canopy decrease and increase from 2001 to 2006, respectively.

Printed and bound by CPI Group (UK) Ltd, Croydon, CR0 4YY

18/10/2024

01776210-0011